Energy Conversion

Energy Conversion

Systems, Flow Physics and Engineering

REINER DECHER

New York Oxford
OXFORD UNIVERSITY PRESS
1994

Oxford University Press

Oxford New York Toronto
Delhi Bombay Calcutta Madras Karachi
Kuala Lumpur Singapore Hong Kong Tokyo
Nairobi Dar es Salaam Cape Town
Melbourne Auckland Madrid

and associated companies in
Berlin Ibadan

Published by Oxford University Press, Inc.,
200 Madison Avenue, New York, New York 10016

Library of Congress Cataloging-in-Publication Data
Decher, Reiner.
Energy conversion: systems, flow physics and engineering/Reiner Decher.
p. cm. Includes bibliographical references and index.
ISBN 0-19-507959-0
1. Thermodynamics. 2. Heat—Transmission. 3. Power (Mechanics)
I. Title. TJ265.D43 1994 621.402′1—dc20 93-31900

9 8 7 6 5 4 3 2 1

Printed in the United States of America
on acid-free paper

PREFACE

With the understanding of thermodynamics over the past two centuries or so, the resultant technology has affected much of modern man's activity and accomplishments. The engines that propel society's vehicles on fields, roads, rails, sea lanes, through air, and into space attest to the ingenuity of humankind to use understanding to his or her advantage. The recent revolution in human activity has made heavy demands on heat resources and has exacted environmental costs. This book celebrates the accomplishments, expresses a concern for the environment, and is motivated by the reality that these resources will become scarcer in the future; technology will be asked to alleviate the effects of this scarcity. If one overriding point is to be made by this work, it is that demands for increased efficiency require costlier energy conversion processes. If a smooth transition toward more efficient use of thermal resources is to be made, it will require investment in the capital plant before the economic consequences of resource shortages are felt by the society using diminishing resources. This investment may well determine whether the desired changes can indeed be accomplished. A look around in the modern world clearly shows that access to power generated by machines and resources assists people with the burden of physical and economic survival, allowing time and resources for the needs of society, and enrichment, recreation, and art for its members. Planning for a day with reduced access to these resources is important, but investing for that day even more so. The transition may well lead to a reordering of the list of who is wealthy and who is less so.

This work is intended to convey an understanding of the physical processes involved in the transformation of one energy form to another. The emphasis is on a description of models of the elementary processes to allow assessment of performance potential and to allow a determination of the sensitivity to design choices. Since many energy conversion processes involve the manipulation of gaseous substances, there is a heavy emphasis on the description of fluids and of gases in particular. Energy conversion processes involve heat and work interactions between a system and its environment, friction, as well as state and property descriptions. In order to arrive at simple, understandable relations. simplifications are made that allow description at the expense of some numerical accuracy. More accurate descriptions can be made with more sophisticated computational tools. Nearly all numerical calculations presented are made with the equations developed, so that the reader can implement them on a computer and reproduce them with his or her own choice of parameters. The specific reduction to practice is generally not covered here. The reader is encouraged to consult the bibliography as well as the cited and current literature for coverage in greater depth.

The reader is assumed to be proficient in the calculus and the physical sciences, including an introduction to thermodynamics and fluid mechanics.

At the conclusion of most of the chapters and some appendices in this book, there are problems provided for the serious student. The nature of these is rather unusual in that the problems may be open-ended or may ask for functional relationships that force the development of assumptions, and seek techniques for solution. It is hoped that an inquiry will follow as to whether the conclusions drawn are physically meaningful in the context of the assumptions. This approach makes the problems more difficult for the student. Thus, before using these problems, an instructor should work out the problems to a satisfactory degree to ensure that they meet teaching goals.

CHAPTER OUTLINE

In the study of engines for the production of power, the student must begin with classical thermodynamics. That introduction is usually insufficient to bridge the gap between the analysis introduced there and the practical implementation. This book is an attempt to describe an approach to determining the performance of real engines for electromechanical power or propulsive thrust. Thus the approach is to describe the following elements of the broad and universally important subject of energy and power. The chapters, their principal goals, and additional background requirements for the reader are listed below.

1. Energy forms, energy utilization and resources available: physics, history, and earth's resource limitations.
2. Conversion process characterization: How good is the process and how good can it be? Review of elementary thermodynamics.
3. Characteristics of power systems: fuel consumption rate, compactness, cost, weight, vehicular versus stationary power sources. How is energy used?
4. Description of the gas used as the working medium in thermodynamic cycles. How complex does its decription have to be for satisfactory description of processes?
5. Combustion: How much heat is available from a fuel and at what temperature?
6. Heat exchangers: What are the limitations of devices designed to transfer heat from one fluid to another?
7. Heat engine modeling: Irreversible thermodynamics plays an important

role in determining overall engine design and performance characteristics: components and processes.

8. Otto and Diesel cycle engines are examples of fuel energy limited engine cycles which can be economically operated at part power: cycle variations can bring performance improvement.
9. Brayton cycle engines are examples of a temperature-limited cycle for power on earth and in space: design point operation and closed cycles.
10. Stirling cycle engines are examples of engines whose performance is strongly influenced and limited by the manner in which the environment of a thermodynamic system interacts with the system: thermodynamic description and performance.
11. The performance of some cycles is limited by the working fluid properties: The Rankine cycle and its modifications can be made to operate with other cycles. The MHD work interaction as a high performance turbine makes unusual demands on the cycle into which it is incorporated.

To this point the material is primarily applied thermodynamics, an extension of that normally covered in a first semester course. In Chapters 12–15, it is assumed that the student has had an exposure to one-dimensional gas dynamics and an introduction to airfoil and wing theory.

12. Fluid kinetic energy is an energy form of interest in many internal and external flows: Convertibility of thermal energy to kinetic energy.
13. Propellers and wind turbines are mechanical means employed to add power to an unbounded fluid or extract power from it: Performance limits of such aerodynamic devices.
14. Compressors: Steady flow cycles require efficient means of raising pressure of an internal flow; description of these aerodynamic devices, their performance and limitations.
15. Turbines: power extraction from internal flows with an emphasis on the difference between compressors and turbines.
16. Part power operation of the Brayton cycle: What role do the cycle configuration and the choice and characteristics of the components (described in Chapters 14 and 15) play in determining the part-power performance.
17. Energy storage for matching to user needs: Physics and limitations of practical devices.
18. Environmental impact of energy conversion: Chemical emissions and thermal and radioactive waste. Maximizing performance of power systems has consequences for the resource, system economics and the environment. Solar power is discussed here as a power resource alternative with its own set of important environmental consequences.

Not covered is the topic of nuclear power generation. This topic requires coverage of physics and engineering which is extensive and is well covered in a number of books. Omission of nuclear power from this text is not intended to imply that it is unimportant. Future recognition of the environmental and resources questions associated with fossil fuels may well serve to reestablish nuclear power to its needed place, in spite of valid concerns regarding the storage of nuclear waste and the safety of plant operations. Also not included is a discussion of so-called *direct energy conversion*, from heat to electricity. A companion book covering this topic, including electrochemistry (batteries and fuel cells), thermoelectric and thermionic conversion, and magnetohydrodynamics, is currently in preparation

Seattle R. D.

ACKNOWLEDGMENTS

The author is indebted to countless individuals who have contributed to the understanding of the heat engines we have today; we build on the shoulders of those who have gone before us. Many passed on their understanding through the written word, others through their deeds, yet others by having touched our lives. My father Siegfried H. Decher, who was part of the engineering team which built the first jet engine to be used in service, was largely responsible for my interest in engineering. The influence of two teachers and colleagues, Jack L. Kerrebrock and Gordon C. Oates, is an important part of this work: their ideas and views have found their way onto these pages, in one form or another. In particular, I wish to credit part of the material in Chapter 4 to Gordon Oates and that in Chapters 14 and 15 to Jack Kerrebrock.

With thanks and love, this work is dedicated to my understanding and supporting wife Mary (who prior to this time did not appreciate the acknowledgments made in the front of books), and our daughters Laura and Meika.

Thanks are due a number of people and organizations who helped make this writing project possible. Colleagues at University of Washington, Department of Aeronautics and Astronautics deserve thanks for their understanding and assistance. Although much of the writing took place over recent years, concentrated energies were spent in residence at the Gas Turbine Laboratory at the Massachusetts Institute of Technology and at the Laboratoire de Thermique Appliquée at the Ecole Polytechnique Fédérale de Lausanne (EPFL-LTT). A semester in each of these institutions proved productive and personally very satisfying. Special thanks to Mary I. Bunting-Smith for the use of her peaceful cottage in New Hampshire which, for an autumn, was a delightful environment for writing.

Special thanks to Diana Park for her assistance and expertise with a portion of the figures, and Anita Lekhwani for making this interaction with Oxford University Press a most pleasant experience.

Lastly, a heartfelt thanks to the individuals who agreed to provide critical comments: N. A. Cumpsty, M. J. Dunn, K. L. Garlid, J. R. Senft, as well as the students who struggled with this text in the form of class notes.

CONTENTS

SYMBOLS

The symbols used in this text are as uniform and conventional as practical. However, traditions have evolved in various fields which make use of differing symbols for the same quantities across differing disciplines. Further, a particular symbol may have two or more different meanings. To that extent, the meaning of symbols may vary from one chapter to another and within a chapter. It is hoped that the context of the use of a quantity makes its meaning clear. Special symbols are defined in the context of their use. The following is a list of symbols that appear in the text together with their definitions and references. The reference is absent if the symbol usage is very common. Other references may be by chapter, for example (18), or section (18.1.4), or by the equation (eq. 18–75) where first used. Neither chemical symbols nor general-purpose constants of local utility are listed here. Some symbols, uniformly accepted to stand for various physical constants, are given in Appendix B, together with numerical values. In the interest of space, some quantities of interest in thermodynamics that are proportional to mass are not given twice: For example, while the enthalpy per unit mass h appears, the true enthalpy H does not because of its direct relation to h.

Units for all quantities will generally be in the SI system, although there are exceptions when tradition calls for it. A number of useful conversion factors are given in Appendix A.

LATIN SYMBOLS

A	area inlet guide vane orientation constant (eq. 14–39)
AR	aspect ratio (13.1.8)
a	speed of sound acceleration (3) Van der Waals gas constant (eq. 4–36) constant to describe turbine cooling effectiveness (eq. 9–33) parameter characterizing airfoil drag (eq. 13–35)
b	Van der Waals gas constant (eq. 4–36) propeller blade chord (eq. 13–19)

	turbine nozzle area ratio (eq. 15–2) ρv constant (eq. 12–44)
B	number blades (eq. 13–22) rotor blading turning angle parameter (eqs. 14–36, 14–46) magnetic field strength (eqs. 18–30, 18–35)
B_0	Beale number (10.2.1)
c	speed of light
c_c	polytropic compression process group of terms (eq. 9–29)
c_t	polytropic expansion (turbine) process group of terms (eq. 9–29)
C	capacitance (1, 17) energy cost (3.2)
C_p	specific heat at constant pressure power coefficient (eq. 13–45)
C_d	drag coefficient (3) nozzle discharge coefficient (eq. 12–3)
C_F	propeller thrust coefficient (eq. 12–35)
C_g	nozzle thrust coefficient (eq. 12–35)
C_Q	propeller torque coefficient (eq. 13–24)
C_R	work recovery factor (eq. 15–16)
$C_{S/M}$	system/mission parameter (eq. 3–16)
C_v	specific heat at constant volume nozzle velocity coefficient (eq. 12–34)
d	length parameter characterizing type of compressor (16.6)
d.r.	degree of reaction (eq. 15–15)
D	diameter drag force (3, 13)
e	electronic charge internal energy (u is also used) polytropic efficiency (eq. 7–42) mechanism effectiveness (eq. 10–50) exponent to characterize compressor operating line (eq. 16–31)
e_{ct}	polytropic compression and expansion parameter (eq. 9–29)
E	energy elastic modulus (1, 17) short-hand for an exponential term (eq. 10–26)
F	force, thrust
f	ratio of actual to maximum power (eq. 3–20) Helmholtz free energy (eq. 4–14) friction factor (6, 10) fuel/air mass flow rate ratio (7, 9, 16)

g	acceleration of gravity Gibbs free energy
g^0	temperature-dependent portion of g (eq. 5–28)
h	enthalpy Planck constant length or height heat transfer coefficient (6, 9, 10) cumulative drilling footage (1.9)
H	heating value (8, 9, 16)
I	influence coefficient (3.5.1)
I_N	cost factor (eq. 3–7)
j	modified advance ratio (eq. 13–30)
J	electric current advance ratio (eq. 13–15)
k	thermal conductivity Boltzmann constant chemical rate constant (eq. 5–54) spring constant (14.7)
k_r	radiator mass parameter (eq. 9–87)
k_s	space power system mass parameter (eq. 9–86)
K_p	equilibrium constant (eq. 5–33)
K	cruise power parameter (eq. 3–45)
L	length inductance (1, 17) lift (3, 13) loss constant (eq. 8–40)
LF	load factor (eq. 3–5)
m	mass lift curve slope (13.1.8)
$\dot{m}$	mass flow rate
M	Mach number
MW	molecular weight (mass)
mep	mean effective pressure (8)
n	number density or moles number of tubes in a heat exchanger (eq. 6–54) polytropic exponent (eqs. 7–41, 7–46, 7–48)
N	revolutions per unit time number of intercooling or reheat stages (eq. 9–60)
N_0	Avogadro's number
N_c	corrected speed (eq. 14–25)
N_{ST}	Stanton number (6)

NPR	nozzle pressure ratio (eq. 12–18)
p	pressure (absolute scale)
P	power (only in 3.4.3, otherwise $\dot{W}$)
q	heat per unit mass charge dynamic pressure (13) torque (9, 13, 16) heat (1, 2, 5, 7, 11)
R	gas constant range (eq. 3–30) rotor radius (13, 14, 15)
r	radial coordinate
R_e	Reynolds number
R_v	volume ratio (8, 10)
s	entropy pressure parameter (eq. 12–38)
S	area
s^0	$= \int (C_p/T)\, dT$
SFC	specific fuel consumption (2, 3, 9, 16)
T	temperature (absolute scale)
t	time tube wall thickness (eqs. 6–57, 9–125, 17–12)
$t_{1/2}$	half-life (eq. 17–26)
t_{max}	energy production period (3, 17)
u	internal energy (e is also used) uncertainty (3.5) overall thermal conductance (eq. 6–10)
u, v, w	velocity components
v	specific volume velocity ratio (eq. 3–43)
V	velocity, speed volume voltage
x, y, z	coordinate directions
x	mole fractions (5) two-phase mixture quality C_p ratio in combustion process (eq. 7–19) throttle pressure ratio (eq. 8–18)

	mass fraction (11)
	nondimensional radius (13.1.8)
W	weight
W_F	fuel
$\dot{W}$	weight flow rate
	power
W	wing or blade width (13.1.8)
w	nondimensional specific work (eq. 3–10)
w_R	nonuniform flow parameter for total mass flow rate (eq. 12–52)
$\dot{w}_f$	fuel weight flow rate
z	power multiplier (3)
Z	partition functions (5.10)
	pressure multiplier (eq. 9–118)

GREEK SYMBOLS

α	power density parameter (eq. 3–24)
	degree of dissociation or ionization (5)
	capacity rate ratio (eqs. 6–7, 10–28)
	group of polytropic exponents (eq. 8–10)
	mass flow rate ratio (eq. 12–37)
	angle of attack (eq. 13–15)
	turbine nozzle exit angle (15.1)
β	heat transfer rate ratio (6)
	group of polytropic exponents (eq. 8–10)
	blade orientation angle (13)
	blading angle (13, 14, 15)
	radial flow lean angle (eq. 14–5)
	control variable parameter (eq. 16–8)
γ	specific heat ratio ($=C_p/C_v$)
Γ	algebraic collection of γ's (eq. 6–52)
	system mass parameter (eq. 9–94)
δ	general-purpose small parameter
	pressure loss parameter (eqs. 9–51, 9–65, 9–74)
	boundary layer displacement thickness (eq. 12–32)
	reduced pressure (eqs. 14–28, 15–38)
	degree of reaction parameter (eq. 15–22)
ΔH_f	heat of formation (5.2)
ΔG_{RP}	reaction Gibbs free energy change (5)
ε	general purpose small parameter
	black body emissivity

ε_i ionization energy (eq. 5–40)

ζ fractional area (12)

η efficiency, with subscripts
c, E, t adiabatic efficiency for compressor, expander, turbine
f fuel utilization
m mechanical
th thermal
x regenerator, heat exchanger, or coolant temperature parameter (eq. 9–40) others listed under subscripts

η^* transport efficiency (3)

θ cylindrical or spherical coordinate angle
ratio of T_t/T_0 (total to ambient static)
boundary layer momentum thickness (eq. 12–32)
azimuth angle (13.2.4)
blading angle (14, 15)
reduced temperature (eqs. 14–28, 15–38)

λ mean free path
decay time constant (17)
heat exchanger aspect ratio (eq. 6–58)

μ viscosity
friction coefficient
turbine coolant mass fraction (eq. 9–33)
Mach number function (eqs. 12–25, 15–8)

μ_0 permeability

ν kinematic viscosity

ν_i stochiometric coefficients (eq. 5–32)

ξ nondimensional x
nondimensional radius (eq. 14–38)

π stagnation pressure ratio across a component $= p_t(\text{out})/p_t(\text{in})$
stagnation pressure nonuniformity parameter (eq. 12–37)

Π nonuniformity parameter for total pressure (eq. 12–85)

ρ mass density

σ stress level (1, 9, 17)
Stefan-Boltzmann constant (9, 17)
nonuniformity parameter for total pressure (eq. 12–41)
solidity (eq. 14–72)

τ stagnation temperature ratio across a component $= T_t(\text{out})/T_t(\text{in})$
mission time (3, 10)
shear stress (6, 12)
stagnation temperature nonuniformity parameter (eq. 12–37)

τ_a acceleration time (3)

ϕ	cylindrical or spherical coordinate angle equivalence ratio (5.1) $= fL/D$ (6, 9, 10) relative thermal efficiency of Rankine cycle (9, 11) effective pitch angle (eq. 13.14) radial flow compressor turning angle (14.5)
Φ	nonuniformity parameter for temperature (eq. 12–65)
χ	gas/metal heat capacity ratio (eq. 10–29)
ψ	stream function (eq. 14–60) short hand for an exponential term (eq. 10–30)
ψ_d	loading-induced disc velocity increment (eq. 13–39)
Ψ_m	mass flow function (eq. 12–58)
ω	angular speed (14, 15, 16, 17)
Ω	nondimensional angular speed (eq. 13.51) wheel speed parameter (eq. 15–40)

SUBSCRIPTS

a	available
b	burner or combustion chamber
B	Brayton (9)
c	compression based on chord length (14) corrected (14, 15, 16)
C	cold
D	disk (13) design value (16)
e	exit, exhaust expansion, turbine
E	expander
f	fuel liquid phase property subscript (11) turbocharger turbine adiabatic efficiency subscript (eq. 8–45)
fg	liquid-vapor phase change (11)
g	neutral gas atoms (5, 18) vapor phase property subscript (11) gross thrust (12)
G	gyration
h, hub	hub or inner annular location (14)
H	hot

HX	heat exchanger
i	associated with species "i"
i, id	general number, ideal
i, in	in, inlet
j	jet, general number
KE	kinetic energy (12)
L	load
l	lift
max	maximum
m	material mission mean mixer (12.6) mechanical mhd (11)
min	minimum
ne	nozzle expansion (12)
o	out
p	pressure propeller
PH	preheater
r	reduced (eq. 4–52) radiator (9)
R	Rankine (9) rotor (eq. 15–32)
R, Rel	relative (13)
rev	reversible
s	constant entropy supercharger (eq. 8–42) heat source (9) state (14)
t	total or stagnation expansion, turbine
t, tip	outer annulus (14)
th	thermal
tot	total (9)
T	tip (17), outer element
v	volume
w	wetted, wall referring to maximum w

WT	wind turbine
x	heat exchanger
x, y, z	relating to direction indicated
∞	far field
η	referring to maximum efficiency
0	ambient
1, e	primary stream at exit (12.6)
1, m	primary stream at mixer entry (12.6)
numbers	state of physical location defined as needed

SUPERSCRIPTS

*	sonic referring to H − H(298) in Chapter 5
tt	total to total
ts	total to static
~	(tilde) is used to distinguish among properties which are per mole and per mass (12.8)

Energy Conversion

1

ENERGY AND POWER: TECHNOLOGY AND RESOURCES

Access to energy is critical to the establishment of an advanced standard of living and to the wealth of people with access to it. This chapter is an examination of the sources of energy and the power that it creates with an eye toward historical evolution in its use and toward future availability.

1.1. POWER, ENERGY, AND ENERGY FORMS

Energy conversion is the engineering technology that deals with the production of mechanical and/or thermal power for industrial processes and for comfort heat from other available power sources or from energy resources. The First Law of Thermodynamics states that there is a quantity called the "energy" of a system which is altered when it experiences a heat or work interaction with its environment. The equivalence of heat and work is implicit in the first law statement, although convertibility between these two energy forms is limited. This limitation is embodied in the Second Law of Thermodynamics.

The concepts of energy and power are carefully described in the sections that follow. The ability to generate heat and produce mechanical power is fundamental to animal life forms. The result of power production is active motion for food acquisition and processing, for locomotion, and (in some creates) for the generation of heat to control the individual's environment. Thus biological systems have always been engaged in the generation of power. The utilization of resources with greater energy content can allow adaptation to changes in life processes.

The energy resource for the biological system is the chemical energy content of food. All animals operate as energy converters, similar to the heat engines and processes discussed here. By contrast, plants are the primary storage means for solar energy, the aboriginal energy source for nearly all life on Earth.

The power output capability of conversion devices to be discussed here are significantly larger (by many orders of magnitude) than that associated with biological processes, as measured by indicators such as energy or power density. However, the abilities of biological systems, humans in particular, form the basis for understanding and relevance of man-made systems. It is in this context that one may ask how powerful and how efficient is a given creature, such as man? Researchers in the field of human powered vehicles have examined this question: Peak power output of circa one-half horsepower (ca. 400 watts) has been recorded, while steady metabolic power consumption rate of man is about

100 watts, ref. 1–1. Sport performance records in running, swimming, rowing, etc. give a clue as to what the human body is capable of in a given time interval. Efficiencies of converting food energy to work are of the order of 25% (ref. 1–2).

It is no coincidence that the power measurement unit prior to standardization on the SI units was the "horsepower." The ability to provide power was and is the value of working animals. Reference 1–1 notes that metabolic power production scales roughly with the 0.75 power of an animal's body mass (the mouse to elephant curve), making larger animals more useful to man as power sources. The large muscle mass and breathing capacity of the horse provides more power than a human can, for tasks that are useful to man, and for carrying the man, faster at that. Best of all, from a human viewpoint, the animal provides the work with a requirement for a lower grade of fuel than that usable by man: The food it eats is more readily available.

The performance of biological systems as energy converters is a fascinating field from the viewpoint of sport and survival in difficult environments. Its understanding allows valid challenges of nonordinary tasks. Our concern here is to put man's capability in the context of the utilization of energy in a modern industrial society. The difference between actual energy consumption per capita and minimum human need is a measure of the dependence on energy resources man has developed. The depletion of these resources and the consequent rise in cost will require examination of more efficient ways of maintaining an economic "lifestyle" that resembles the current one. Some of these means are described here.

1.1.1. *Heat and Work*

Heat is that which flows from a body at a given temperature to another at a lower temperature. This definition will be revisited on numerous occasions. Work is often a desirable end product and is the scalar product of force and distance (vectors in bold type face):

$$W = \int \mathbf{F} \cdot d\mathbf{s} \tag{1–1}$$

In words, work is done on a body, or system, when a force acting on the body is moved through a distance. Figure 1–1 shows a force applied through a

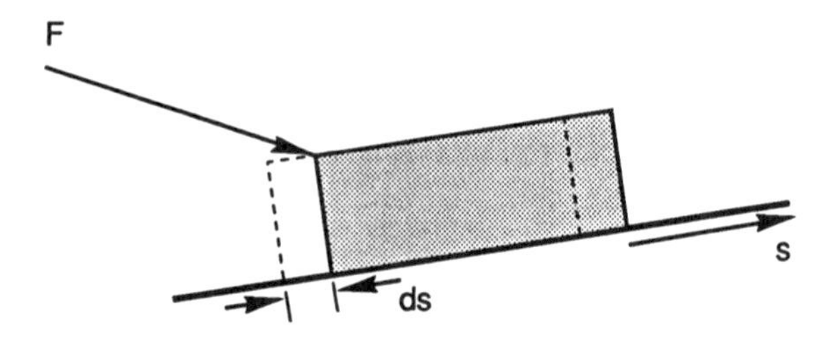

FIG. 1–1. Body sliding along an inclined plane. The applied force does work on the body.

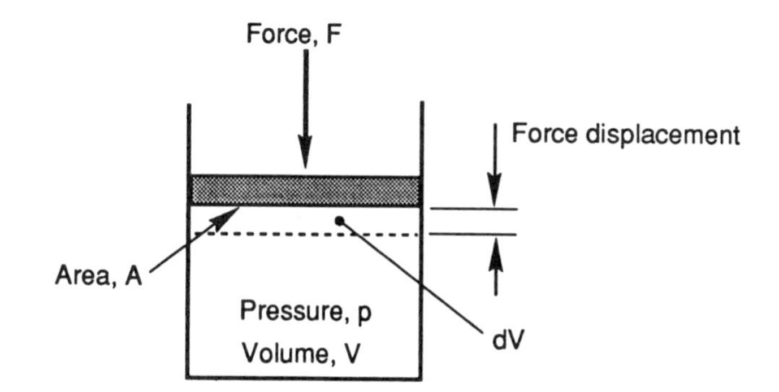

Fig. 1–2. Work associated with volume changes (displacement of boundary) of a closed system and its pressure (resisting force).

distance as the body slides. Here only the force component along the surface does work. The other component normal to the surface sets up a reaction force. The work done here results in (1) the raising of the object and (2) the overcoming of sliding friction. The raising of a weight in a geopotential field is the simplest example of doing work in that the force (weight) and the vertical displacement are colinear. This idea can be applied to a system which is defined as the (homogeneous) substance in a piston cylinder arrangement shown in Fig. 1–2. Evidently, the force is the pressure times the piston area and the distance moved is the swept volume divided by the piston area,

$$dW = F\,ds = (pA)\left(\frac{dV}{A}\right) = p\,dV \qquad (1\text{–}2)$$

A process involving a reduction of volume is compression and requires work to be provided. The production of work *by* a system with an increase of system volume is an expansion process. Work done *by* a system is considered positive.

The thermodynamic notion of work and the force that generates it, is somewhat in conflict with the human experience. To clarify this idea, the following from ref. 1–2 may be helpful.

> A large muscle is composed of a large number of individual fibers. Each fiber, like the muscle itself, cannot contract; a muscle cannot "push." Neither can an individual muscle exert a continuous force. It is caused to contract by the nervous system's release of adenosine triphosphate (ATP) After contraction, a fiber will again relax. If a muscle is required to exert a continuous force, for instance in holding up a weight, muscle fibers will "fire" sequentially. Even if the weight is not lifted, which means that in the thermodynamic sense no external work is being done, the muscle will require energy either from its stores or from the blood stream. We call this "isometric" exercise, because there is no change in the measurement of the muscle or of the body.
>
> The ATP, which is the muscle fiber's immediate fuel, can be supplied in two ways. For almost immediate short-term use, the muscle can draw on its

> own stored phosphoryl creatine and glycogen. It can use these without the need for blood [oxygen]; hence, we call this muscle action *anaerobic*. The muscle fibers that work anaerobically are termed type II fibers. They are developed by sprinters and by animals who rely on a sudden spurt of activity to escape from their predators. These fibers are found in the white meat of turkey. Anaerobic-muscle use in humans can last for up to five minutes. Because there is a restricted amount of energy available (proportional to the mass of the muscle), the duration of its use depends on the power output demanded. For longer term use, in the so-called steady state, the ATP needed by the muscle must be supplied from glucose and fatty acids that are supplied by the blood and oxidized. The muscle fibers that can work for long periods, which use the blood and work with oxygen and therefore work *aerobically* are termed type I fibers and are dark brown, like the dark meat of a turkey's legs.

Before one can discuss energy conversion, a number of the energy forms of interest must be discussed. These forms are the basis of means of storing energy between production and use, Chapter 17.

1.1.2. *Kinetic Energy*

Kinetic energy is associated with motion. An object of mass m, initially at rest, has kinetic energy (KE) after it has suffered a work input, if the result of the work input is *only* motion of the body.

Consider a body of fixed mass on a frictionless surface and an applied force along the surface, Fig. 1–3. From Newton's second law, the force exerted on a body of fixed mass m results in acceleration,

$$\mathbf{F} = m\mathbf{a} = m\frac{d\mathbf{V}}{dt} \qquad (1\text{–}3)$$

If the mass is not constant, the additional term $\mathbf{V}\, dm/dt$ arises. This term is important in propulsion of vehicles that do not have contact with the "ground" (i.e., flight vehicles, see Chapter 12). If this force is applied while the body moves through a distance $\mathbf{s}$, then the work input is

$$W = \int_1^2 \mathbf{F}\cdot d\mathbf{s} = \int_{V_1}^{V_2} m\frac{d\mathbf{s}}{dt}\cdot d\mathbf{V} = \tfrac{1}{2}m(V_2^2 - V_1^2) = \Delta KE \qquad (1\text{–}4)$$

so that work input to a body of mass m can result in a change in its kinetic

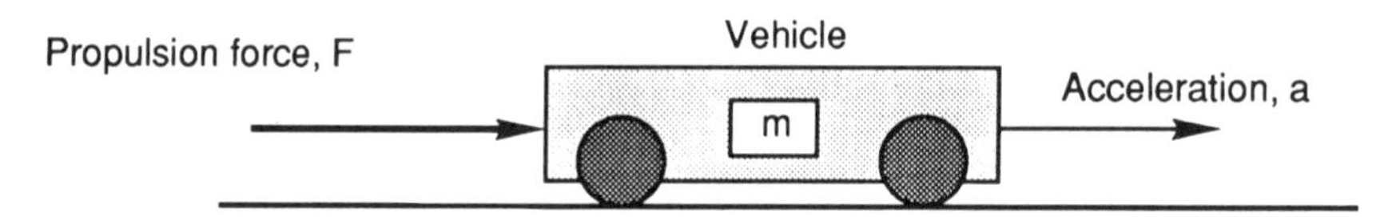

FIG. 1–3. Propulsion force on a vehicle to overcome resistive forces.

energy. Here the rate of change of position, ds/dt is the velocity, V. With friction present, the force would have to be greater and that implies that the work input would be larger. In general, the change in kinetic energy is less than the work input,

$$\Delta KE \leq W \tag{1-5}$$

Since velocity can be angular about an axis, kinetic energy is also associated with rotation about the axis, and the equivalent to "mass" is the "moment of inertia."

1.1.3. *Potential Energy*

Potential energy is that energy associated with position in a force field, such as gravity or electrostatic forces. An object is said to possess this energy form when the result of work done on the object results only in a change in position. In nature there are a number of forces experienced by bodies which depend on their positional relation. Examples are the gravitational forces acting on bodies with mass, and the electrostatic forces acting on electrically charged bodies (Fig. 1–4). The forces involved are observed to be dependent on the distance r between the bodies which varies as

$$\mathbf{F} = \frac{\text{constant}}{r^n} \mathbf{e}_r \tag{1-6}$$

where the exponent n is most commonly 2, although other force law exponents are found to be descriptive in nature, such as the force interaction between colliding molecules of a gas. The force $\mathbf{F}$ is radial under many circumstances so that the work done in moving the bodies through a distance $r_2 - r_1$ relative

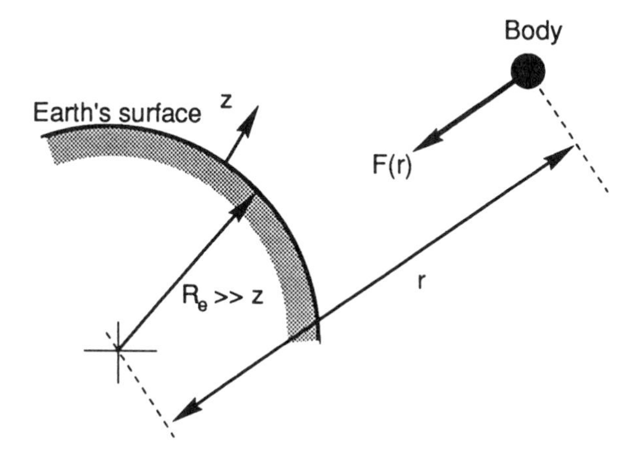

FIG. 1–4. Central force field (gravity) on an object near a planetary body such as the Earth.

to each other is the scalar product

$$W = \int_1^2 \mathbf{F} \cdot d\mathbf{r} = \text{constant} * \left[\frac{1}{r_2^{(n-1)}} - \frac{1}{r_1^{(n-1)}} \right] \tag{1–7}$$

When considering the special case of object displacement on a scale that is small compared to the dimensions of the attracting body (the Earth, for example), the r-dependence of the force vanishes because the radius of the Earth is the dominant part of the distance r ($r = r_e + z$, $z \ll r_e$). In that special situation, the work done is

$$W = \int_1^2 F\, dr = \text{constant} * [r_2 - r_1] \tag{1–8}$$

which can be obtained from eq. 1–7 by linearizing this equation by $r_2 = r_1 + z$, where $z \ll r_1$ is a small elevation coordinate in a cartesian system with the origin on the Earth's surface. Equation 1–8 can then be written

$$W = \text{constant}\ [z_2 - z_1] \tag{1–9}$$

where the constant is identified as the weight of the body.

$$\text{constant} = \text{weight} = \text{mg} \tag{1–10}$$

Thus from eqs. 1–8 and 1–10, the work input to the body can result in a change in its (gravitational) potential energy. Equation 1–7 gives the change in potential energy for the general force law. Note that potential energy is always stated relative to a reference position.

1.1.4. *Strain Energy*

Strain energy is an energy form that can be altered in a body by application of a force that deforms the body. This energy form is relevant primarily to solid objects. The work input is calculated in a manner equivalent to the compression work discussed above. An example is the compression of a cube of matter by a force which reduces its length in the direction of the force. The resisting force per unit area, or stress, is related to the percentage of the length reduction, or strain, that has been measured for many materials. The stress-strain relation may be linear, much like the spring constant k of a coiled spring. Such a material is termed *elastic*.

$$F = kx \qquad \text{or} \qquad ks \tag{1–11}$$

In that case, the relation between work done and strain energy is of the form

$$W = \int_1^2 ks\, ds = \tfrac{1}{2}k[x_2^2 - x_1^2] \tag{1–12}$$

For a volumetric element of material, the relation between stress (σ, force/area) and strain (e, elongation/length) is written as

$$\sigma = Ee \tag{1–13}$$

For an elastic material, E, the modulus of the material, is a constant. E has the units of pressure. For steel for example, E is of the order of 2×10^6 atm. For an elastic material, the strain energy per unit volume is

$$w = \tfrac{1}{2}Ee^2 \tag{1–14}$$

(relative to an unstrained state) which is equivalent to eq. 1–12. Energy input into a structural element (beam) experiencing bending may be similarly described. In such cases, the stress field is nonuniform and therefore the energy is not uniformly distributed in the material volume. Figure 1–5 shows the stress field in a bending beam and the relations for strain energy. An expression for

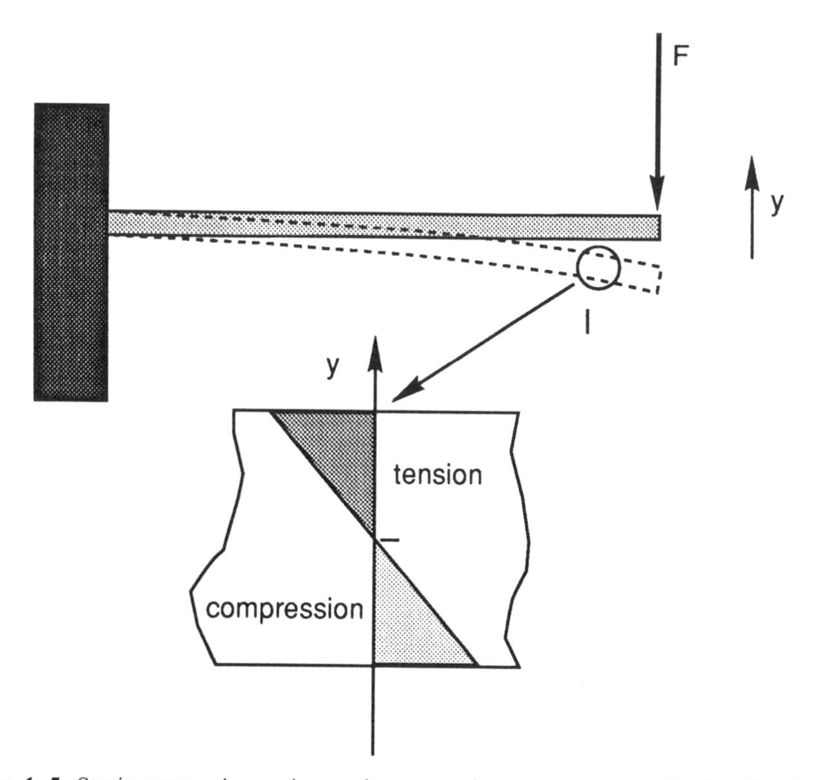

Fig. 1–5. Strain energy in tension and compression on a beam under bending load. Extreme fibers are in greatest tension and compression.

the energy per unit deflection may be found by an appropriate integration of the stress field (including normal and shearing stresses) over the specified geometry of the beam.

The energy associated with *surface tension* on an elastic sheet is a distributed force over a length. This is analogous to the volumetric strain energy, except that the "thinness" of the material allows mathematical simplifications in the description of the stress field.

1.1.5. *Thermal Energy*

Thermal energy is associated with the random motion of atoms and molecules. This energy is the kinetic and potential energy associated with the atoms and molecules themselves. The potential energy is associated with the relative proximity of the molecules to one another, see Section 4.2 and eq. 4–43 in particular. One distinguishes between kinetic energy of finite size objects and that of molecules because one cannot *freely* convert the random kinetic energy of atoms and molecules to other energy forms (such as potential energy).

The motion of the molecules may be free, as in a gas, except for collisions with other molecules. In a liquid molecules can move, but never far from the influence of other molecules and thus have limited freedom. Even more restricted is the motion of molecules in a solid where motion takes place within a "sphere" at a fixed location due to the strong interaction forces between molecules. Such configurations are termed *phases.* Changes in phase often involve significant amounts of energy. Such changes include evaporation/condensation and fusion/solidification as a substance is made to pass from one phase to another by heating or cooling.

Thermal energy, like kinetic and potential energies, is extensive (i.e. proportional to the mass). In general, the temperature (T) and specific volume (v) of the material describe the thermal energy, or, more precisely, the internal energy. The temperature is a measure of kinetic energy while the specific volume is a measure of intermolecular separation which influences the potential energies of the molecules relative to each other. Thus one has

$$U = \mathrm{mu}; \; u = u(T, v) \tag{1–15}$$

U is the internal energy and u is the specific internal energy (i.e., per unit mass).

1.1.6. *Chemical Energy*

Atoms can have a varying affinity for each other in a combined state as molecules. Transformations from one such state to another are termed *chemical reactions.* These involve primarily the outer, or least tightly bound, electrons or shells of the atoms. Such reactions can result in a release of energy which may be thermal, electrical, or photonic. The *heat of formation* of a molecule is energy required to form the molecule from more elementary building blocks such as atoms. The notion that chemical energy is relative to a reference state is detailed in Chapter 5.

1.1.7. *Electromagnetic Field Energy*

The work required for the creation of electric or magnetic fields is termed field energy. Examples may be the electrostatic field of a capacitor and the magnetic field associated with a coil (Fig. 1–6). This is discussed in greater detail in Chapter 17 in connection with the possibilities for practical energy storage.

One may think of the electric field energy as the aggregate of the potential energies associated with the separation of unlike charges. Similarly one may visualize the magnetic field as the work that was required to establish the steady flow of current through the wire (i.e., accelerate the charges comprising the current).

Electromagnetic energy exists in very small entities called photons. Photons are "bundles" of energy that may be emitted when atoms change configuration or when charged elementary particles are decelerated. For example, when an atomic nucleus changes from a configuration with an energy level 1 to a level 2: the photon emitted is a gamma (γ) ray. When the level change occurs in the inner electron structure of an atom, the photon is called an x-ray, while it is a visible light photon when the levels affected are the outer electrons (Fig. 1–7). At the lowest photon energy extreme of this scale is the thermal photon of infrared heat radiation. Such photons are emitted by the accelerated charges that make up atoms or molecules in a solid. The motion is intimately associated with the finite temperature of the solid.

The photon as a fundamental packet of energy has characteristics that enable one to describe it either as a (light) wave or as a particle (wave/particle duality), depending on the nature of the experimental observation.

The magnitude of the energy carried by a photon is measured solely by the frequency, ν, (or wave length, $\lambda = c/\nu$) of the light wave:

$$E = h\nu = \frac{hc}{\lambda} \tag{1–16}$$

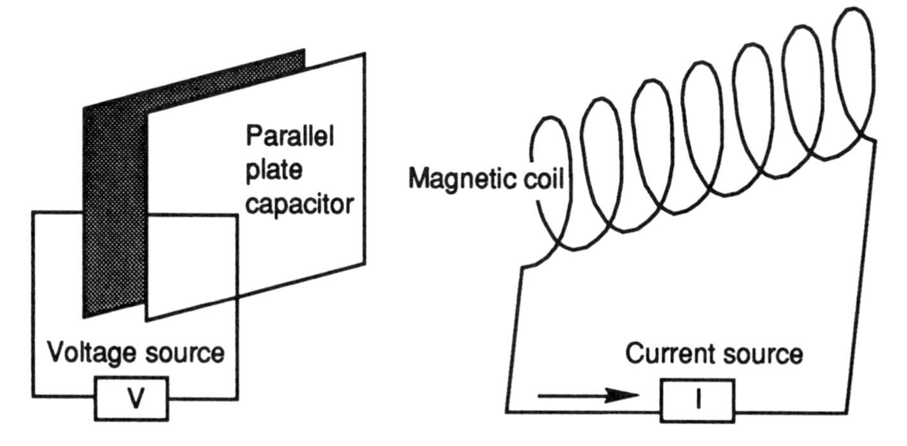

Fig. 1–6. Devices for the storage of electromagnetic field energy: coil and capacitor.

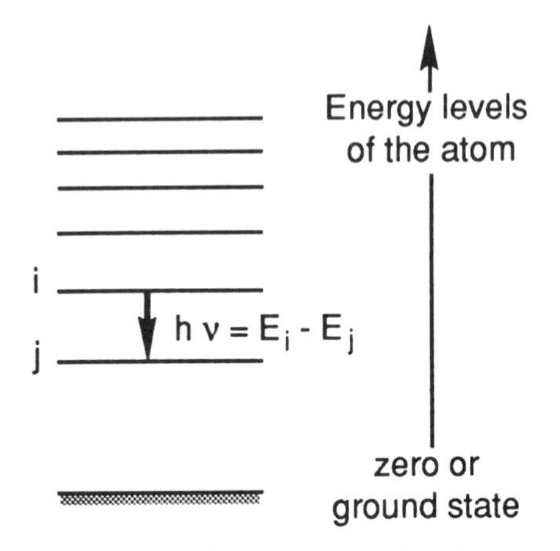

FIG. 1–7. Photon emission from an atom changing energy states.

Here h is Planck's constant and c is the speed of light. Values for these constants are given in Appendix A.

1.1.8. *Nuclear Energy*

Kinetic energy and photons are liberated when heavy atomic nuclei disintegrate to more stable daughter nuclei through the fission process. The mass of the products of a nuclear disintegration is less than the mass of the original nucleus. The change in mass is converted to energy. The equivalence between mass and energy was established by Einstein with his famous relation:

$$E = [\Delta m]c^2 \tag{1–17}$$

where Δm is the change in mass involved in the reaction. In a nuclear reaction, this energy is shared with the environment as kinetic energy of the fission fragments which ultimately degrades to thermal energy by collisions with the molecules making up the environment.

The energy forms listed above all play basic roles in the production of useful power as resources or as means to store energy.

1.1. UNITS OF MEASUREMENT

For most of the discussion in this text, the system of units to be used is the "Système International" or SI. In this system of units, the force is measured in Newtons (N) and the distance in meters (m). The work is obtained in Joules (J) since 1J = 1 Newton * meter (Nm). The unit of power is the watt (W) = 1 joule/second or, for greater convenience, the kilowatt (1kW = 1.36 horsepower). For the thermochemistry section, Chapters 4, 5, and Appendix C, the data are given in calories or kilocalories. English units are

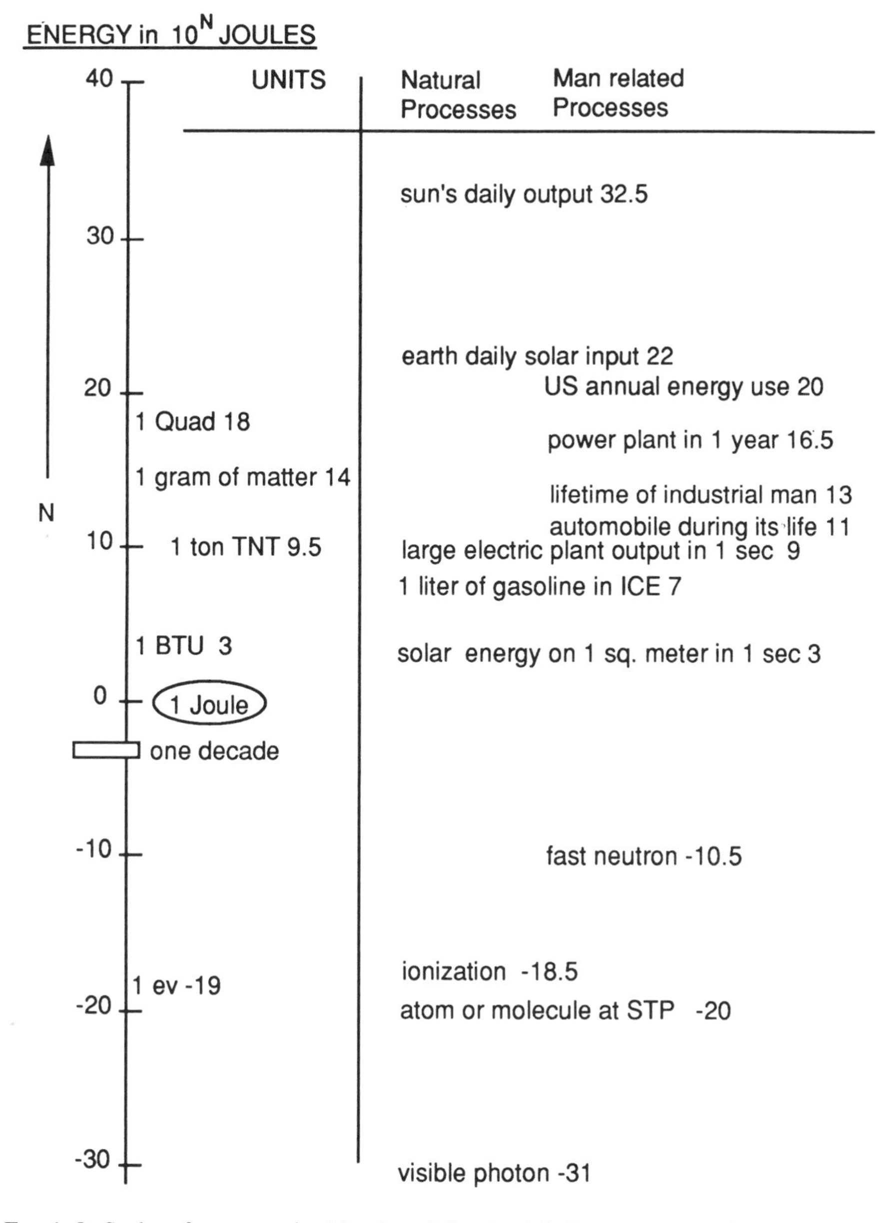

FIG. 1–8. Scales of energy units. Numbers following labels are an approximate value of N in the decade exponent of the energy in Joules.

quoted for some quantities where such use is common in the United States. In some fields, use of units is made which arose from traditional units of measurements or which are defined to avoid use of numbers involving large powers of 10. Figure 1–8 shows the magnitude relationship between a number of common units of energy measurement. Appendix A gives the conversion factors among common units of energy and power measures.

1.3. POWER

Mechanical power is the *rate* of doing work. When a constant force moves steadily, the power is given by

$$\dot{W} = \frac{dW}{dt} = \lim_{\Delta t \to 0} \frac{\Delta W}{\Delta t} = \mathbf{F} \cdot \frac{\Delta s}{\Delta t} = \mathbf{F} \cdot \mathbf{v} \tag{1–18}$$

Thermal power is the time rate of heat delivery. Since the common unit of power measurement is the kilowatt and thermal and mechanical power can both be described by the same unit, one distinguishes between these power forms by the abbrevations kW_e (electrical or mechanical) and kW_t (thermal).

In moving many kinds of objects, an unavoidable resistance is experienced. This resistance could be the force of a saw blade moving along a wood surface or an airplane through air. This force must be provided at the chosen speed so that it is mechanical power which must be supplied.

Such mechanical forms of power are available in nature and have been exploited over many centuries of human activity. Examples are wind, moving water, and human as well as animal labor. Since such forms are not often directly useful, they may be made to undergo a transformation by what is termed a machine or device. The machinery is the physical manifestation of technological innovation. The evolution of man's abilities and achievements over the past millennia and more dramatically over the past few centuries bears witness to the importance of having power at one's disposal.

The economic and physical well-being of mankind is intimately associated with wealth and power. While it is difficult to quantify wealth in overall terms, the economic gross domestic product (GDP) is a measure of economic activity, and thus indirectly of wealth. Figure 1–9 shows the annual energy consumption rate and GDP for a number of countries in the world and thus the relationship between power (sic!) and wealth (ref. 1–3).

1.4. TECHNOLOGY

During the eighteenth and the nineteenth centuries, formulation of the physical laws led to an understanding of the convertibility of heat to work and the limitations inherent in that process. This understanding is expressed in statements termed the "First and Second Laws of Thermodynamics." The contributions made by the many individuals to this profound understanding is reflected in the names of processes, mathematical functions, and relations that permeate the field of energy conversion. The motivation for understanding the relationship between heat and power is clear: to make power available where there was previously little or none available. The understanding of the Second Law raised the value of heat resources and promoted their exploitation. Today, energy conversion is concerned primarily, but not exclusively, with transformation of thermal energy sources to useful mechanical power. Figure 1–10

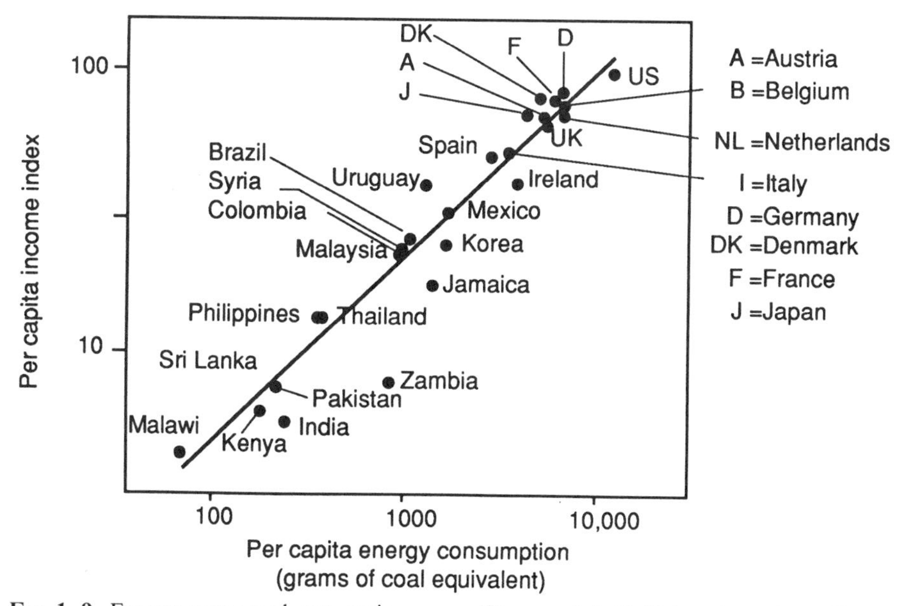

FIG. 1–9. Energy usage and economic power. (From ref. 1–3) Original caption: *Close connection between energy consumption and standard of living emerges from comparisons of energy use and purchasing power (vertical axis) in 28 countries. The figures were compiled from data for 1979; the calculation of purchasing power was developed by Irving B. Kravis, Alan Heston, and Robert Summers of the University of Pennsylvania, who relied on estimates of the actual costs (in local currency) of a long list of goods and services.*

Table 1–1. Applications of Various Types of Heat Engines

Application	Otto (Gasoline)	Diesel	Brayton (Gas Turbine)	Rankine
Automobiles	X	x		
Trucks		X		
General Utility	X			
Large Aircraft			X	
Small Aircraft	X			
Ships		X	x	X
Peak Load Power			X	
Commercial Electricity				X (base load)

shows a number of ways of converting between energy forms. The heat engines that carry out this process are everywhere around us: automotive engines, steam power plants, gas turbines, to mention a few. Table 1–1 lists a number of heat engine types and their principal applications. Reference 1–4 gives an excellent historical review of the development of heat engines, both successful and unsuccessful.

Final energy form \ Initial energy form	Mechanical	Electrical	Kinetic	Potential	Thermal	Chemical
Mechanical	Gears Pulleys Torque converters	Electric motor	Impulse turbine	Spring	Turbine Wind Heat engine	Explosive
Electrical	Electric generator	Rectifier Inverter Transformer	MHD		Thermoelec. Thermionic Photovoltaic	Fuel cell Battery
Kinetic	Propeller	MHD	Molecular collisions	Pendulum	Nozzle	Gun
Potential	Lifts		Pendulum			
Thermal	Friction	Heater Light bulb	Diffuser		Absorption	Combustion
Chemical		Electrolysis				

FIG. 1–10. A matrix of energy conversion schemes. Cited are some examples of conversion between energy forms.

1.5. ENERGY RESOURCES

The thermal power input to heat engines can be obtained from a number of sources. These can be classified into (1) depletable energy resources, which can be transformed into thermal power sources, and (2) thermal power sources from which the heat is "continuously" available. The key to differentiating between these kinds of sources is the time scale of availability and its relation to human time scales.

1.5.1. *Depletable Energy Resources*

These sources of energy are primarily remainders from much earlier events on Earth. The formation of the solar system and the Earth left a number of heavy elements that are in the process of decaying to stable forms by means of nuclear transformations. Humanity is fortunate to be able to take advantage of these unstable products before they disappear. The process whereby heat is generated from the fissioning of heavy nuclei such as U^{235} must take place in a nuclear reactor. For our purposes, the reactor is a heat source using a depletable fuel. The possible utilization of much more abundant U^{238} or thorium (see Chapter 18) which must take place in reactors that have yet to be accepted as viable from a number of points of view could hardly be described as utilizing a depletable resource: The energy supply as converted from these isotopes of uranium is enough to last into the very distant future. Similarly, the fusion of deuterium with itself or tritium may also become an engineering and economic reality in the future. When this day dawns, the conversion machine will run strictly speaking, on a depletable resource, but the depletion time scale is also so long that it may as well be considered a steady resource (ref. 1–5).

Fossil fuels are the most important energy resources in use today, although not without problems (ref. 1–6). The reasons are simple: They are abundant, low in cost, and relatively easy to convert to power. Originally these chemical combinations of carbon and hydrogen were vegetable matter that grew in prehistoric times. Geologic events preserved them for current and future use. As resources, man utilizes them by means of chemical oxidation or combustion. This process is incorporated into most heat engines for the production of mechanical power. Fossil fuels range from coal to natural gas (methane) depending on the details of their formation history. The origins and characteristics of fossil fuels as well as techniques for estimating the magnitude of the resources are discussed in Section 1.8.

Hydrogen and alcohol are not energy resources since they must be made using other energy forms or from less useful forms of hydrocarbon resources (refs. 1–7 and 1–8). Their usefulness lies in their convenience for use as a transport fuel or as a transportable energy form.

1.5.2. *Steady Power Sources*

A number of natural energy flow processes can be intercepted by the interposition of devices. These power streams include beamed radiative power (photons) from the sun and the heat which is gradually leaving the Earth's interior (Fig. 1–11). The time scale for these resources is not infinite, but large compared to time scales of interest to present-day humans.

The solar energy stream can generate heat, chemically stored energy (refs. 1–9, 1–10, and 1–11), electric power, or mechanical power indirectly by means of its interaction with the atmosphere (wind, ocean waves) and with the water cycle (rivers). The ability for humans and the required equipment to travel to space makes for interesting future power generation possibilities using the solar resource (ref. 1–12).

The process of the Earth's interior heat radiating to space is felt as a weak gradient in temperature in the Earth's crust. The global mean gradient is sufficiently weak that exploitation of this geothermal resource is economically possible only in regions where the interior material is conveniently accessible by

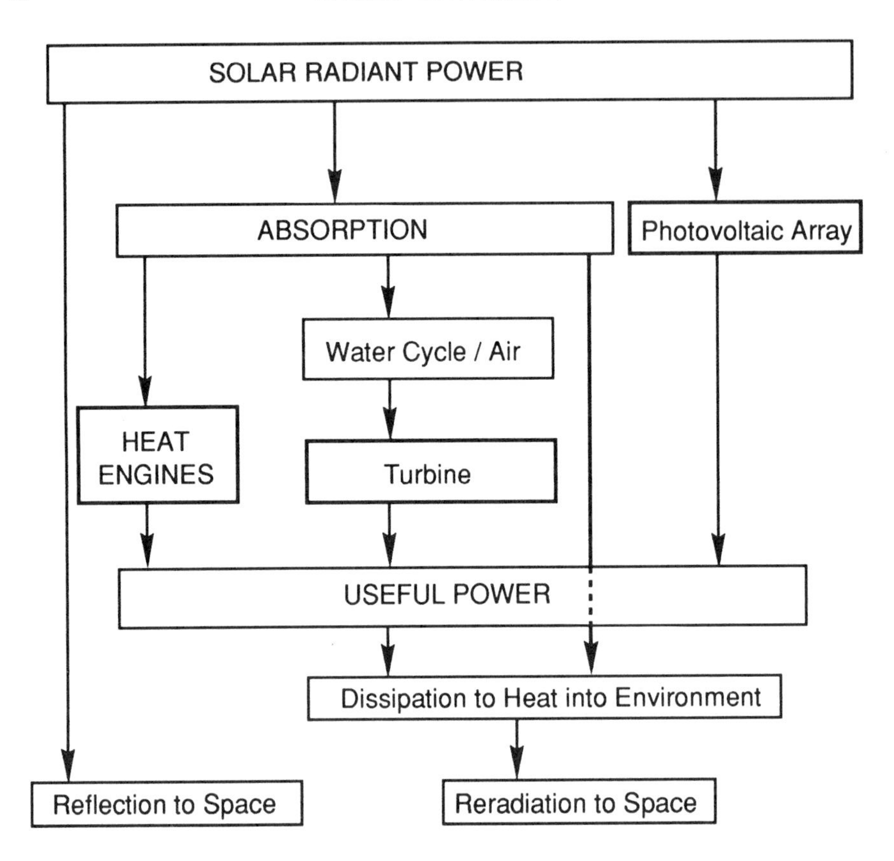

FIG. 1–11. Natural solar energy flow and man-made interception possibilities.

surface water penetrating this region and being brought to the surface as steam or hot water (ref. 1–13).

The moon-Earth dynamic system causes the Earth's oceans to experience tides. The periodic rise and fall of the water relative to the shore presents an opportunity for generating power if a dam is built to trap high tide water which is released when the tide is low. Such facilities have been built (ref. 1–14 and 1–15), although many suitable sites do not exist.

Figure 1–12 summarizes the available energy resources and their availability time scales. Manufactured fuels such as methane, alcohol, and hydrogen may be made from any of these energy forms and the raw materials. Table 1–2 gives a quantitative summary of resources, by type, for North America.

1.6. USER REQUIREMENTS

The energy consumption of human beings has varied through the ages. Figure 1–13 shows typical energy use rates in the major epochs of human development. Note that a distinction is made between the end uses of the

RESOURCE AVAILABILITY TIME SCALE	NATURAL RESOURCE
SHORT (decades)	oil coal
↓	U 235, thorium U 238 deuterium
LONG (millenia)	geothermal solar

FIG. 1–12. Time scales for resource availability.

Table 1–2. Proved recoverable resources in North America

	Energy reserves (10^{18} J)	Annual production (10^{18} J)	Production projection (years)
Crude oil	287	22	18
Natural gas	383	27	14
Shale oil	7345		
Bituminous sands	2266	0.14	
Coal, total:	5226	16.6	315
anthracite	193	0.265	
bituminous	3573	16.1	
subbituminous	990	0.1	
lignite	470	0.14	
Uranium	444 or 26,700*		
Thorium	0.1 or 6800*		

* higher number quoted is with breeding, otherwise without breeding.
Source: from Hill, P. G.

energy. Over the past 200 years, the consumption rate can be specified more accurately and is shown in Fig. 1–14 for a number of cultures as well as for the world as a whole.

The earliest energy conversion machines were crude. They had nonetheless an enormous impact on life in the nineteenth century. They eased many tasks such as forging and textile work and made a variety of new processes possible such as forming, milling, and so on. The industrial revolution changed forever the economic structure of the world.

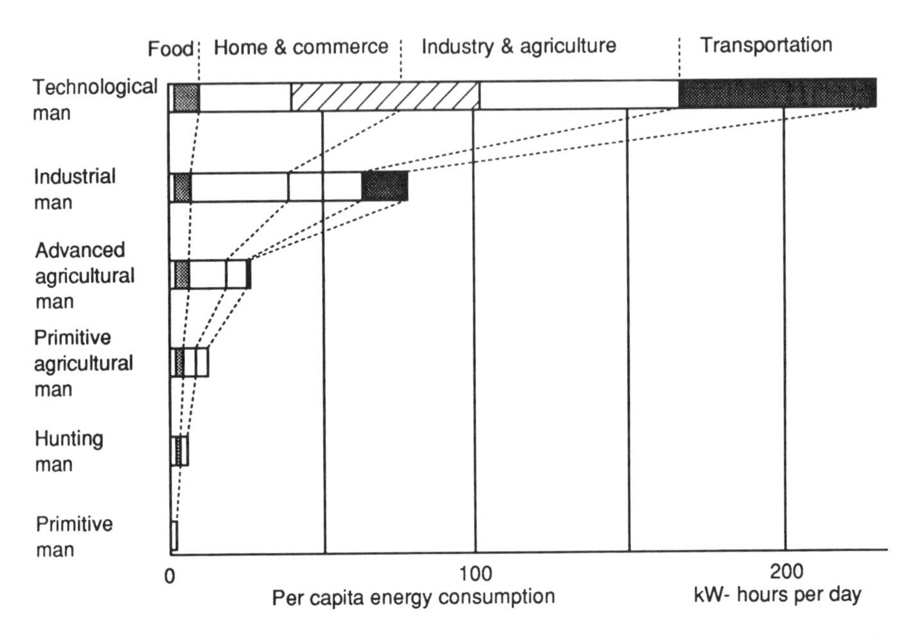

FIG. **1–13.** Daily consumption of energy of man through the ages. (From ref. 1–19.) Original caption: *Daily consumption of energy per capita was calculated by the author for six stages in human development (and with an accuracy that decreases with antiquity). Primitive man (East Africa about 1,000,000 years ago) without the use of fire had only the energy of the food he ate. Hunting man (Europe about 100,000 years ago) had more food and also burned wood for heat and cooking. Primitive agricultural man (Fertile Crescent in 5000 B.C.) was growing crops and had gained animal energy. Advanced agricultural man (northwestern Europe in A.D. 1400) had some coal for heating, some water power, and wind power and animal transport. Industrial man (in England in 1975) had the steam engine. In 1970 technological man (in the U.S.) consumed 230,000 kilocalories per day, much of it in the form of electricity (hatched area). Food is divided into plant foods (far left) and animal foods (or foods fed to animals).*

1.6.1. *Requirement for Energy Storage*

Efficient use of energy resources requires the energy conversion process to have the characteristic of adaptability to the user: Varying amounts of power must be deliverable at various times. A successful energy conversion system must be able to deliver a maximum as well as lesser amounts of power as needed. This requirement is generally fulfilled by an ability to store energy during the utilization process. The fuel used in a typical energy conversion process is a storable material which can be utilized as needed. The heat produced could also be stored for later use, although it could be partially lost if the time interval between storage and use is sufficiently long. The work produced by the heat engine could be stored in any one of a number of devices which may store various forms of energy, such as kinetic energy in a flywheel, potential energy in a storage reservoir, and so on. The best means of carrying

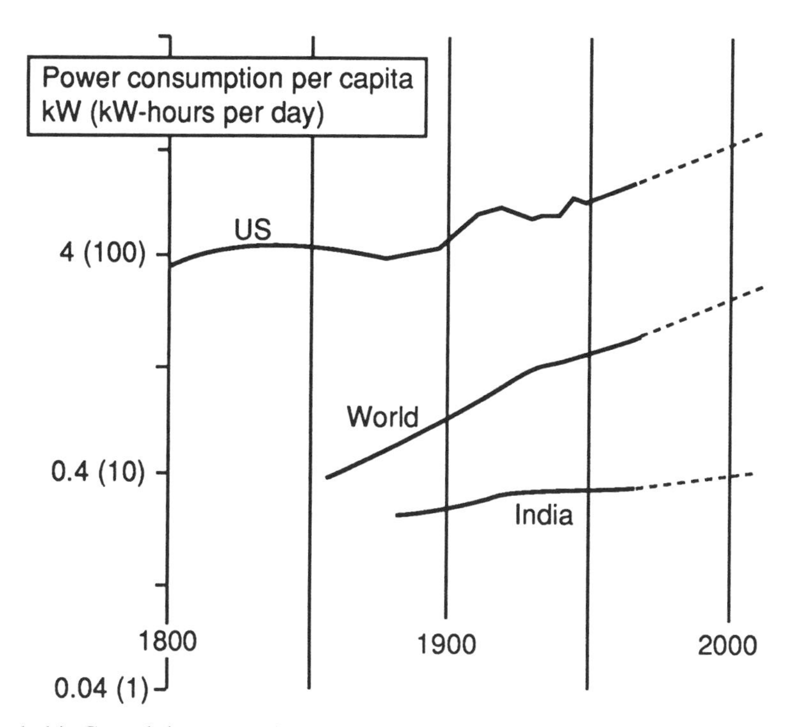

Fig. 1–14. Growth in energy demand in a number of world sectors. (From ref. 1–19.) Original caption: *Growth in energy demand in the United States is at the annual rate of about 1 percent per capita. For the world as a whole per capita consumption is growing about a third faster. Even so, for the world supply of energy per capita in A.D. 2000 will be less than a fourth of the projected U.S. figure. In India the rate of increase is only about a third of the U.S. rate.*

out this process is usually determined by the costs involved. The most common and easiest storage method is to store the *fuel* and use it when and in the amount required.

The development of heat engines that can accommodate a varying fuel input rate and thus provide a power level consistent with demand, and simultaneously achieve high power-to-weight ratios, led to the applications of heat engines to vehicles previously drawn by animals or pushed by wind. The resulting inventions were made rapidly in the years between 1850 and 1910. These, in turn, allowed the development of the ships, trains, automobiles, and airplanes that exist today.

1.7. HEAT GENERATION FROM COMBUSTION

The chemical energy stored in hydrocarbons is derived from the solar energy captured through photosynthesis eons ago. In combustion, this energy is released by the process of combining the hydrocarbon with oxygen from the atmosphere. The chemical reaction is ideally:

$$C_nH_m + \left(n + \frac{m}{4}\right)O_2 \rightarrow n\,CO_2 + \frac{m}{2}\,H_2O \qquad (1\text{–}19)$$

It turns out that the formation of CO_2 and H_2O rather than other combustion products releases the greatest amount of heat. In the combustion process, the important reactions occur in the gas phase. Chapter 4 addresses the properties of gases as chemical reactants and as working fluids for heat engines, whereas the energy release and the temperature of the resulting combustion gas as well as the formation of "pollutants" and other products are examined in Chapter 5.

1.8. FOSSIL FUEL ORIGIN AND COMPOSITION

For purposes of classification, one distinguishes between coal, petroleum, and natural gas. The substances are produced by the partial oxidation of vegetable matter in a process that is typically terminated by a sedimentary cover. Coal refers to solid materials. Petroleum includes tar sands and oil shales, but consists primarily of substances that are low enough in average molecular weight to be a liquid that can be pumped from wells. An intermediate product with properties between oil (petroleum) and natural gas (which is a mixture of methane, CH_4 and ethane C_2H_6) is the so-called natural gas liquid, NGL. It has a chemical formula C_nH_{2n+2}, with n ranging between 3 and 5.

Oil and gas are moved by environmental forces (water pressure, geostatic forces, etc.) through porous medium to positions of equilibrium. The rock porosity is nominally 15–20%. The equilibrium position is one where a cap of impervious material prevents escape to the atmosphere. The spaces are referred to as oil or gas entrapment domes (Fig. 1–15).

1.8.1. *Coal Formation*

Coal is geologically aged plant material that consisted originally of woody plants, composed essentially of cellulose and lignin, whose original chemical identity has been lost through the effects of bacterial action, time, temperature, and pressure. Certain physical characteristics remain, however, and definite evidence of the origin of coal is found in macroscopic and microscopic residues of plant structures occurring in the coal samples. The progressive metamorphic changes involved in the formation of coals are indicated in terms of rank or age as follows:

Plant material (wood)	→ Peat	→ Lignite	→
Sub-bituminous	→ Bituminous	→ Semi-anthracite	→
Anthracite	→ Graphite		

Coal and its associated materials are found widely distributed in areas

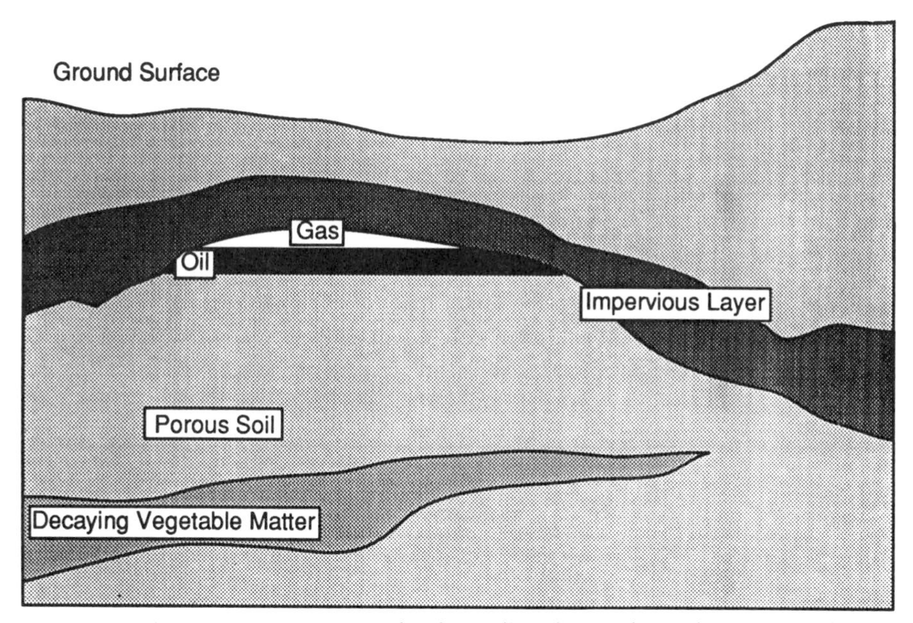

FIG. 1–15. Oil and gas geology showing how oil and gas migrate in porous soils and are trapped by impervious layers.

Table 1–3. Chemical composition of coal

Substance	Carbon	Hydrogen	Oxygen
Wood fiber	52.6	5.3	42.1
Peat	59.6	6.0	34.5
Lignite	66.0	5.3	28.7
Earthy brown coal	73.2	5.7	21.1
Bituminous coal	75.1	5.8	19.1
Semi-anthracite coal	89.3	5.0	5.7
Anthracite coal	91.6	4.0	4.5

throughout the world. If coal contained only constituents that were present in the original plant material, it would be uniformly low in both sulfur and ash, but during the early stages of its formation underneath the surface of swamps or lakes, streams or rivulets carried silt and sediment over the decomposing bed of vegetation. This sediment settled down and became an integral and varying constituent of coal. Sulfur in solution in the water, coming into contact with salts of iron and reducing organic compounds, resulted in the formation and precipitation of pyrite, whereas other reactions, not clearly understood, produced variable quantities of organic compounds of sulfur as a constituent of the coal.

Coal has no fixed chemical formula. The gradation in the formation of coal from its original state is shown in Table 1–3.

Table 1–4. Key properties of fossil fuels

Substance	H/C	HHV 60°F, 1 atm 10^3 BTU/lb[a]	Specific gravity[b]	Energy density[c] 10^3 Joules/m^3
Hydrogen	—	52	0.071 (20 K)	8.6
Methane	4.0	23.8	0.425 (111 K)	23.5
Ethane	3.0	22.3		
Propane	2.67	21.6	0.582 (231 K)	29.2
n-Butane	2.5	21.3		
Gasoline	2.0–2.2	20.2–20.6	0.72	34
No. 1 fuel oil (kerosene)	1.9–2.0	19.8–20.0	0.8	37
No. 2 fuel oil	1.7–1.9	19.3–19.8		
Crude shale oil	1.6	18.6–18.8		
No. 6 fuel oil	1.3–1.6	18.0–18.9		
Lignite	0.7–1.0	9.7–11.3		
Sub-bituminous	0.7–1.0	10.3–12.3		
Bituminous	0.6–0.9	12.0–15.0	1.2–1.5	41
Anthracite	0.2–0.5	11.9–14.0	1.4–1.8	48
Wood (dry)		8.4–8.6	0.4–0.7	11
Wood (10–15% H_2O)		7.7–8.0		

Note: H/C is the molar hydrogen to carbon ratio. HHV is the co-called higher heating value obtained *with* the condensation of the water vapor.

[a] Multiply by 2.324 to convert to kJ/kg or by 0.555 to convert to kcal/kg.

[b] For gases, liquid phase density is given as well as the boiling temperature.

[c] Solid packing efficiency of the bulk material reduces the approximate values given..

In North America;,three periods existed that were favorable to the accumulation of plant material for coal formation.

1. Toward the close of the Paleozoic Era, Eastern U.S. (most extensive accumulations)
2. During Cretaceous times, Rocky Mountain coals
3. During the Tertiary Era, lignite coals west of Rocky Mountains

For energy conversion purposes such fuels are burned with air to produce heat. The heating value of typical fuels is given in Table 1–4.

1.9. RESOURCE QUANTITY ESTIMATION

The following approach to estimating resource quantity is due to Hubbert of the U.S. Geological survey (1967, refs. 1–17 and 1–18). Let Q = resource quantity and P = production rate so that

$$P = \frac{dQ}{dt} \tag{1–20}$$

The production rate, P, must follow a curve of the type shown in Fig. 1–16 [the rate plotted is given by $\exp - (t - t_0)^2$] with zero production before

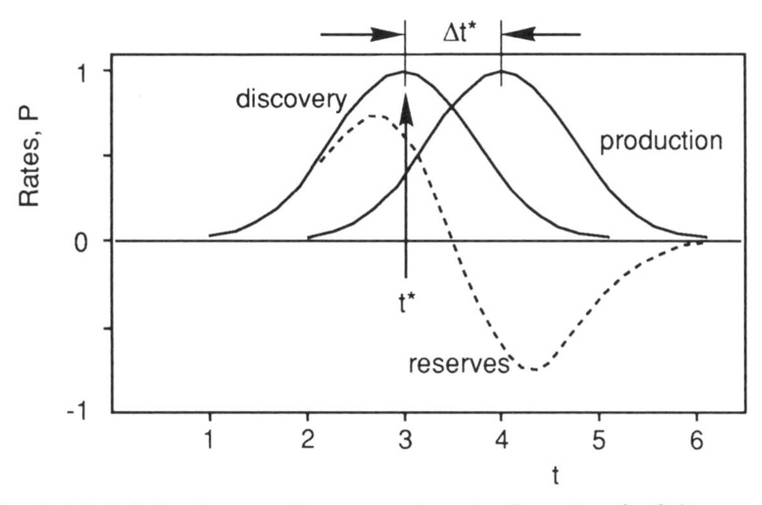

Fig. 1–16. Relation between discovery and production rates of a finite resource.

discovery and after depletion. For the arbitrary time scales, t^* (peak discovery) is at 3 and Δt^* is between 3 and 4 (peak production). The total quantity of a resource can be divided between the amount produced and the amount yet to be exploited, or reserve.
Thus

$$Q_{\text{discovered}} = Q_{\text{produced}} + Q_{\text{reserve}} \tag{1–21}$$

Differentiating with respect to time, one obtains,

$$\frac{dQ_{\text{d}}}{dt} = \frac{dQ_{\text{p}}}{dt} + \frac{dQ_{\text{r}}}{dt} \quad \text{or} \quad P_{\text{d}} = P_{\text{p}} + P_{\text{r}} \tag{1–22}$$

By plotting Q's or their derivatives one may identify several characteristic times. At the time $(t^* + \Delta t^*)$, the peak production rate will occur. Half of the resource will have been used at that time. The actual discovery rate data for North American oil is shown in Fig. 1–17 (ref. 1–17) for comparison. The actual data are much more erratic, although the trend follows projection. The time, t^*, was approximately 1960 for U.S. oil production and Δt^* in that time period was 10–12 years. Such an estimate of the total resource gives $Q_\infty(\text{US}) \sim 170\text{–}173 \times 10^9$ barrels.

Another method of estimating Q_∞ is to look at exploratory drilling. Each new well that brings in oil will do so after an increasing amount of exploratory drilling. Thus one may say

$$Q_\infty = \int_0^\infty \frac{dQ}{dh}\, dh \tag{1–23}$$

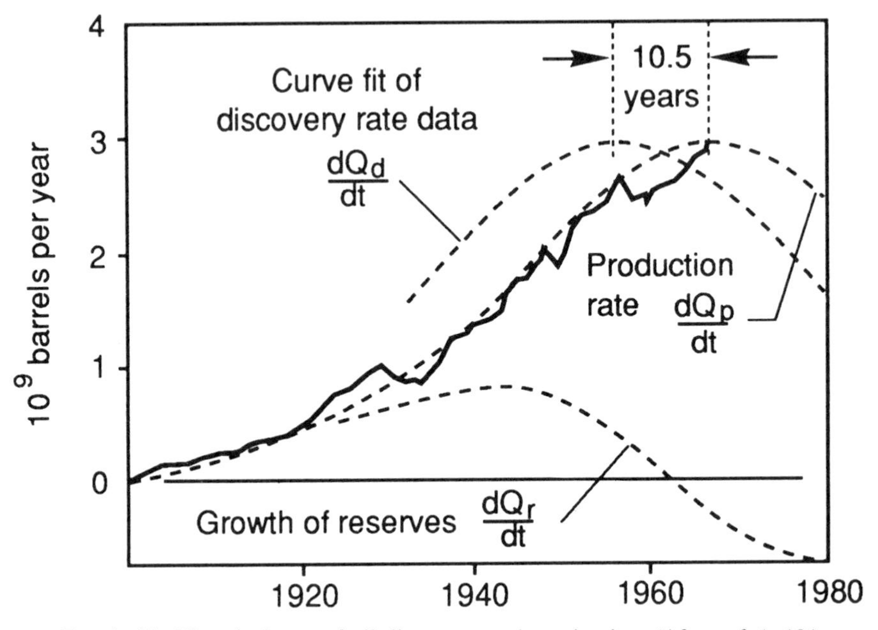

FIG. 1–17. Historical rate of oil discovery and production. (After ref. 1–18.)

Table 1–5. Resource summary: Q_∞ (US) and energy total

Substance and measure	Q_∞	Total heat value 10^{18} J	Δt (80%)
Oil (barrels)	190×10^9	1300	1935–2000
NGL (barrels)	41×10^9	1300	1935–2000
Gas (cubic feet)	1290×10^{12}	1290	1950–2015
Coal (metric tons)	1486×10^{19}	38,600	2040–2440

Note: There are 1054 J/BTU. The heating values of various cited measures of quantity are given in Appendix A.

Source: From ref.1–19.

where dQ/dh is the oil quantity produced per exploratory foot drilled and h is cumulative footage drilled.

This method incorporates technological factors that are insensitive to economic disturbances hence this method might be somewhat better than Hubbert's. Note that a realistic h_{max} exists. The actual variation dQ/dh is shown in Fig. 1–18 (ref. 1–17). Such a plot, terminated with an exponential decay gives, when integrated, Q_∞ (US) ~ 153–164×10^9 barrels.

Such estimates of Q_∞ permit estimating the possible variation of the resource with time. Table 1–5 is a summary of such estimates. Figure 1–19 shows plausible variations for oil and coal production rates. The two curves are for various estimates of Q_∞. Note the relatively short oil availability time span and

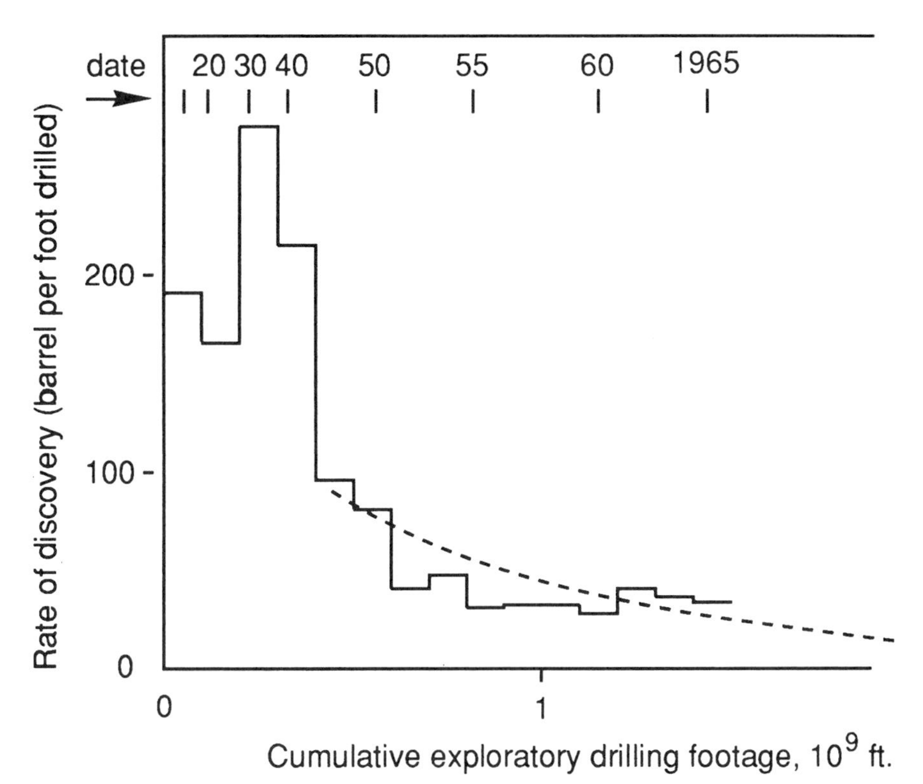

FIG. 1–18. Rate of oil discovery with cumulative exploratory drilling rate. After ref. 1–17, Hubbard, M. K., "Degree of Advancement of Petroleum Exploration in the United States," *Bulletin of the American Association of Petroleum Geologists*," 1967, reprinted by permission of the American Association of Petroleum Geologists.) Original caption: *Crude oil discoveries per foot of exploratory drilling versus cumulative exploratory footage in the United States, exclusive of Alaska, 1860–1967.*

the lack of sensitivity of the time span where 80% of the resource is used [$\Delta t(80\%)$] to Q_∞, ref. 1–19].

1.10. OTHER MEANS OF HEAT PRODUCTION

This text emphasizes the heat production processes from fossil resources because these play a dominant role in the production of useful power in the world today and will for some time to come. However, fossil reserves are finite. As these resources become more scarce, the price will rise, leading to oil-like fuels and possibly hydrogen produced from other primary sources, coal in particular, and greater utilization of electric power generated from coal wherever possible. Other commercially important heat generation methods include nuclear power which is discussed briefly in Sections 1.10.2 to 1.10.4.

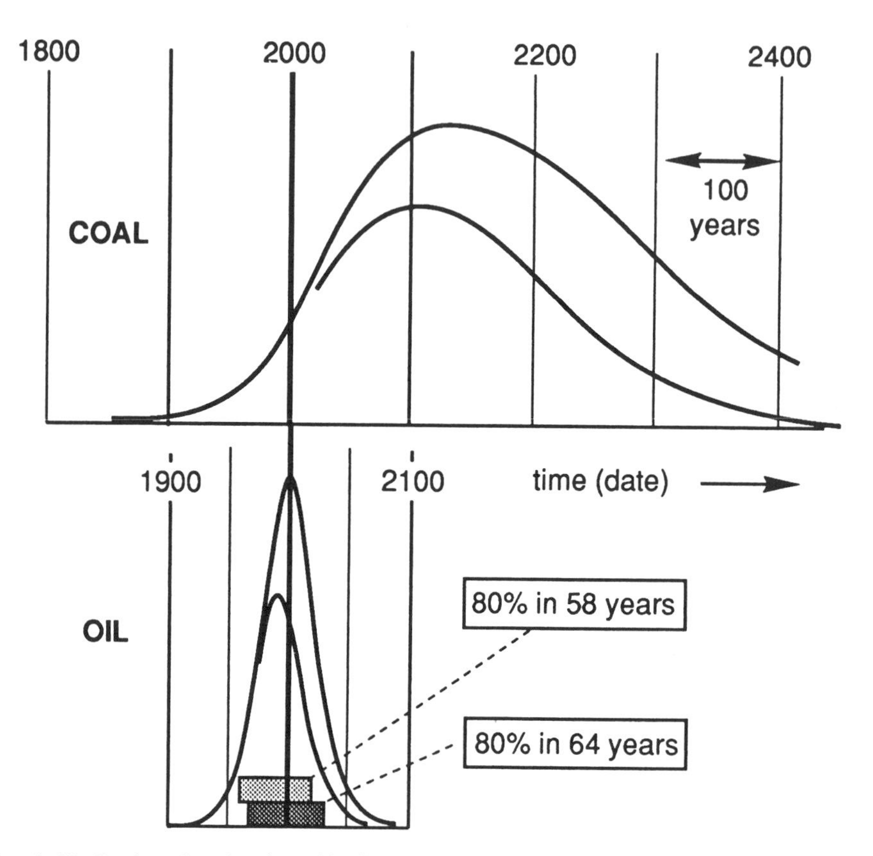

FIG. 1–19. Projected cycle of world oil and coal production. (Adapted from ref. 1–19). Original caption: *Cycle of world coal production is plotted on the basis of estimated supplies and rates of production. The top curve reflects Averitt's estimate of* 7.6×10^{12} *metric tons as the initial supply of minable coal; the bottom curve reflects an estimate of* 4.3×10^{12} *metric tons. The upper curve for oil reflects Ryman's estimate of* 2100×10^9 *barrels and the lower curve* 1350×10^9 *barrels.*

The heat of the Earth's interior may be exploited for the production of useful thermal power, see Section 1.10.1.

1.10.1. *Geothermal Power*

The Earth's interior temperature is elevated above that of the surface. In locations where the interior heat is close to the surface, the natural heat flux may be greater locally and therefore utilized to humanity's benefit. Appropriate regions are generally located in seismically active regions and are sometimes identified by natural seepage of water from the surface that is heated there and then vented to the atmosphere as steam. Such a natural flow or man-made injection can lead to steam generation that can be captured at relatively high pressure for expansion through a turbine. The temperature and pressure of the steam are typically modest ($\sim$400–500 K, $\sim$2–10 atm) relative to conditions in steam power plants, so that larger mass flow rates of steam are required per unit of mechanical power.

An operational difficulty with the use of steam from the Earth's interior is associated with the carry over of soluble minerals which can and will precipitate in the process machinery, that is, turbines or heat exchangers. The latter component is the important component when a heat engine such as the supercritical Rankine cycles described in Chapter 11 are used to capture the energy. These heat exchangers (Chapter 6) are typically constructed of long and narrow tubes which are compromised by the buildup of scale on these tubes from the viewpoint of (1) area available for steam flow and (2) heat transfer across the tube wall.

The fraction of presently used thermal power from all sources that can potentially be filled by geothermal heat is quite small because cost rises as the need for drilling to reach hot rock regions increases and because of the need to deal with mineral deposits in the machinery. On the positive side, the heat resource can be locally important, the cost of the heat per se is zero, and the environmental impact is limited to disposal of the minerals.

1.10.2. *Nuclear Reactions*

Nuclear power is possible as a result of understanding the physical structure of the atomic nucleus. In particular, that the nucleus is made up of nucleons, that is, protons (charge +1) and neutrons (charge 0) which are all held together by short-range nuclear forces. The atomic mass is the sum of the number of protons and neutrons. To a small but significant degree the mass of a nucleus differs from that of the separate nucleons, and this small mass difference is proportional to the binding energy through the Einstein formula (eq. 1–17). The binding energy is the work associated with the nuclear forces holding the nucleus together. The binding energy per nucleon varies with atomic mass and is highest for the average midweight nuclei and lowest for the very lightest and the very heaviest. This implies that nuclei near the middle of the mass range are more stable and that state can be reached by combining two light nuclei (fusion) or making a heavy nucleus heavy enough by forcing absorption of a neutron (or other particle), thus inducing spontaneous fission which results in two (more stable) fission fragments. In either case the decrease in mass of the products of the reaction is transformed to energy.

1.10.3. *Nuclear Fission*

The discussion of fission here is limited to the important issues related to the physics underlying practical utilization. For a thorough discussion of reactor physics and engineering, the student is referred to the technical literature. The fission reaction may be stated as

$$n + \text{fissile nucleus} \rightarrow \text{fission fragments} + \sim 2.5\, n + Q \qquad (1\text{–}24)$$

where n is the neutron and Q in the heat liberated. The neutron is the easiest to use to penetrate a heavy nucleus because its charge is zero and thus suffers no electrical repulsion forces as a proton would. A chain reaction may be maintained because there is a net production of neutrons. Fissile atoms are U^{233}, U^{235} and Pu^{239}. U^{235} is naturally occurring whereas the Pu and U^{233} are man-made using the fertile-to-fissile conversion of U^{238} and Th^{232} respectively as raw materials. Uranium is the primary source of commercial reactor fuel and is found on Earth with an isotopic abundance of 0.7% U^{235}, 99.28% U^{238}, and a small remainder of U^{234}. Uranium is left over from the Earth's creation process and will decay away in the future due to its slight radioactivity. The decay half-lives (see eq. 17–26 and Tables 18–2 and 18–3) and 4.5×10^9 years for U^{238} and $.71 \times 10^9$ years for U^{235}. These time scales are of a similar order as the age of the Earth. An important aspect of fission energy production is that the more abundant U^{238} is not directly useful because it is not fissile.

Neutron conservation in the typical power reactor (i.e., achieving criticality) dictates that the fraction of U^{235} in the uranium must be higher than the natural abundance: U must be enriched to have an isotopic abundance of 2.5 to 3.5 percent of U^{235} for the reactor to maintain a sustaining chain reaction. Some reactor configurations, however, can operate with natural uranium.

The fission reaction may be carried out with low-energy (thermal) or with (relatively) high-energy neutrons. Low-energy neutrons are obtained by allowing the high-energy neutrons resulting from the fission process (initially created with $\sim 2 \times 10^6$ ev) to collide with a moderator material which shares the energy and thus cools the neutron until it achieves thermal equilibrium with the reactor materials at their temperature (~ 0.023 ev, see Fig. 1–8). Reactors using thermal neutrons are classified according to the moderator because it plays an important role in determining the probability that a neutron survives the slowing-down process and is thus available to participate in the next fission generation. The best moderators are light mass atoms such as hydrogen and its isotopes (as water), carbon (graphite), etc., because the energy transfer from neutron to a low mass atom is higher than for a heavy collision partner. Thus H_2O (light water) and D_2O (deuterium: heavy water) are suitable moderators. They also serve well as a fluid to carry the heat generated to an energy conversion cycle. Solid moderators must be cooled by convection of another fluid, often gases such as helium, which can be directly used in the energy conversion cycle (see Chapters 9 and 16). The majority of nuclear reactors in commercial use are of the light water moderated thermal type.

The moderator and the fuel together are assembled in a reactor whose function is to maintain and control the chain reaction by minimizing neutron losses by absorption through the boundaries and in the materials which make up the reactor. Typically, the engineering design of the reactor dictates the performance which the conversion system can achieve by limiting the output temperature and pressure of the coolant.

Fast neutrons are used in reactors where fertile materials (such as U^{238})

may be transformed into fissile materials such as Pu^{239}. This process is primarily

$$\begin{aligned} &U^{238} + n \rightarrow U^{239} + \gamma \\ &U^{239}\,(23\text{ min}) \rightarrow Np^{239} + \beta^- \\ &Np^{239}\,(2.3\text{ days}) \rightarrow Pu^{239} + \beta^- \end{aligned} \tag{1–25}$$

Here γ and β^- are energetic photons and electrons (β particles) involved in the reaction. The use of U^{238} in this way makes a very large energy resource available for power reactors (see Table 1–2).

The so-called fast breeder reactor utilizing the U^{238}-plutonium cycle uses heavier moderators (typically sodium metal) to avoid slowing the neutrons as rapidly as a thermal reactor. Such reactors have been built and may play a significant role in the energy future of modern nations.

The environmental consequences of nuclear power are discussed in Chapter 18.

1.10.4. *Nuclear Fusion*

The fusion of light nuclei to release thermal power as from the combination of hydrogen and its isotopes to form helium occurs naturally in stars as in our sun and can be made to happen in a bomb. The reactants must be heated to very high energies and held in a compressed state for a sufficiently long time to allow the reaction to take place. These conditions are met in the fission reaction-initiated bomb which is not of interest for power production. The controlled nuclear fusion community has long been at work to develop a reactor configuration that allows very small amounts of the reactants to generate manageable thermal power levels. It is beyond the scope of this book to describe the physics and engineering of these reactors, and the student is encouraged to consult the technical literature. It may well be that before the year 2000 net energy production will be realized and within a few decades it may be possible to build power plants. The motivation is simply that the material resources (the isotopes of H) are abundant on Earth and may be utilizable for a very long time into the future. The environmental aspects of this energy source are also described briefly in Chapter 18.

Since solar power is a recurring topic of discussion in connection with alternative power generation, some aspects of this method of power production are discussed in Chapter 17 (since storage of solar energy is an important issue) and in Chapter 18 since an environmental impact is unavoidable when intercepting solar power.

1.11. SOLAR ENERGY COLLECTION

Solar power is the primary energy source for life on Earth. It is absorbed by plants through the photosynthesis reaction. These are eaten by herbivores which, in turn, are eaten by carnivorous creatures, including omnivorous humans. Solar power thus plays a critical role in driving the carbon cycle wherein carbon is removed from the atmosphere by plants and then serves as a fuel for power production by living creatures.

From an energy conversion viewpoint, solar power has several characteristics that are important in its utilization as a power source. These are its power density and its temporal variability.

1.11.1. *Characteristics of Solar Power*

For the discussion to follow, the solar power is viewed as thermal power. The source temperature is so high that the thermodynamic value of the thermal power is very high: It has a potentially high efficiency in the conversion to work. This is also true of the power of radiation since physical devices (solar cells) can be used to produce electrical power directly. Such devices are described briefly in Section 18.6.4. The reader is referred to the specialist literature for coverage of this topic.

Power Density

The insolation is the power incident on a unit area of surface. Near equatorial regions the radiant power density is about 1.4 kW/m^2 on a horizontal surface at noon. This quantity varies with time of day and with latitude in two time cycles: the day (8.64×10^4 sec) and the year (3.15×10^7 sec). At the higher latitudes, the lower thermal input results in a colder climate. At any time, solar power is attenuated by the atmosphere and its components—molecular gases, clouds, and dust. The fraction absorbed or reflected by the atmosphere depends on meteorological, geological (e.g., volcanoes), and geometric aspects. The geometric factors result in variations in the solar power over time and location as shown in Fig. 1–20.

An important aspect of solar power is that a large surface area is required to collect amounts of it which compare to that currently provided by fossil fuels. Consider each person's continuous need for 1–10 kW the range of which depends on economic status from Third World inhabitant to industrial man (see Section 1.6). The atmosphere absorbs on average 30% of the solar power (this fraction depends strongly on the local climate), which is available only about 35% of the time (the rest of the time is twilight and night), and can be collected for present or later use with an efficiency of say 10%. These considerations alone dictate that, on a per capita basis, the surface area requirement is of the order of 30–300 m^2 (300–3000 ft^2). This area requirement is significantly larger than that required for food production alone. Setting aside the question of cost for this area, it is doubtful that it would even be available where it is needed: in or near cities and towns. If one takes the view that solar energy is to be used for supplying all of the needs of US inhabitants (200–300 million), then the land area required is about one entire large western state. The environmental impact is certainly severe. Even if such power-gathering systems were distributed throughout the land, it would be difficult to have a place where it would not have at least a visual impact. The ocean surface has been

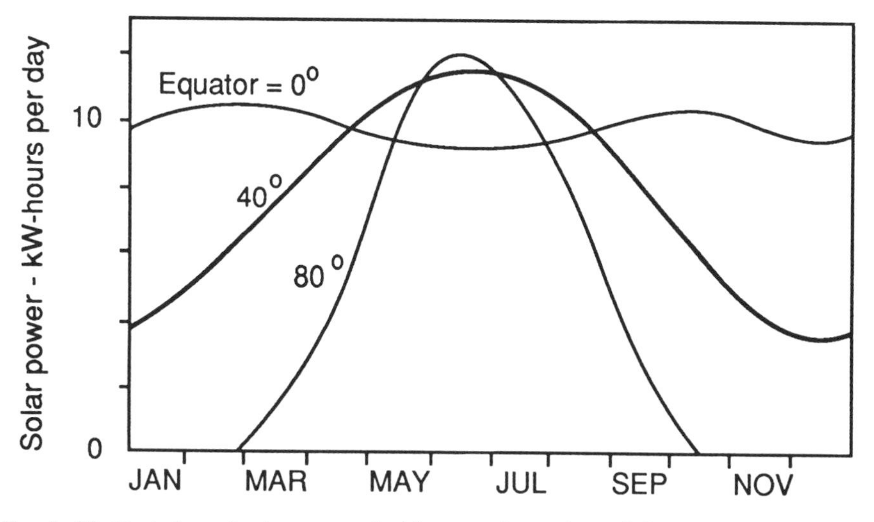

Fig. 1–20. Variation of solar energy incident on the surface of the Earth as a function of time of year for three latitudes.

suggested for such a purpose. If such numbers are applied to western Europe or some regions of Asia, where the population density is large, the difficulty of replacing present fossil fuel exploitation with solar power becomes apparent.

Temporal Variability

For the purpose of generating electromechanical power or comfort heat, the need for solar energy is typically much smaller when it is available than when it is absent. Effective use of solar energy requires storage of energy on the daily time cycle and also on the yearly cycle. Daily cycle storage can be realized to some extent for the storage of comfort heat in the building structure or in special (low cost) materials from which the heat can be recovered (e.g., water tanks or hot rock beds). For the yearly cycle, the sole economically effective means for storing the summer heat is presently as firewood or, in the recent past, ice storage from winter.

In view of these characteristics, low power density and need for storage, the conclusion must be that solar power cannot support our economy as it presently operates. In some specific applications, the limitations of solar power may be acceptable and a partial reliance on solar power is very practical. For a heavy reliance, per capita energy consumption will have to decrease, the number of consumers will have to decrease, and/or technology will have to play a role in increasing the utilization efficiencies. The latter is the only element that can be sensibly discussed here.

1.11.2. *Collection Methods*

The methods available for collection depend on the end use of the energy. Broadly, solar energy collection is classified in terms of the temperature of the material in which the energy is captured, stored, or used. Low temperature collection to about 100°C is used primarily in sensible heat or latent storage for use at a later time. Some drying processes use the solar power directly: hay, food stuffs, wood products, salt production, etc. Intermediate-level temperatures reaching about 400°C make it possible to store the heat for conversion to electric power by use of heat engines with Carnot efficiencies of the order of 50%, whereas real engines realize only a fraction of this value. Higher temperature heat makes some thermal processes possible or reduces the cost of storage by reducing the mass involved. High-temperature collection also makes possible consideration of thermal power conversion to work by means of heat engines which are limited by the mechanisms cited in Chapters 7–11.

The temperature of the material in which the power is collected is determined by the design of the collection system. Flat plate collectors (usually fixed orientation) are used for low-temperature heat in the form of hot water. For higher temperature heat collection, concentrating systems (usually sun following with sensors and drive mechanism) are required. At the highest performance end of the spectrum of collection technologies are the very sophisticated collectors using steered optics, such as the solar power tower where a central receiver is illuminated from a field array of controlled mirrors (heliostats). These systems are described in the literature (e.g., ref. 1–20) but no matter what collection method is used, the surface area per unit power required is large. Figure 1–21 shows a typical flat plate collector cross section and a concentrating system.

When radiant power is absorbed by a collector, for example (Fig. 1–22), it results in heating of the absorber to a temperature (T_R) determined by a balance between heat input rate, removal for useful purposes, and losses.

The heat balance equation that describes a solar collector is a variant of the following

$$Q_{solar} = Q_{useful} + Q_{loss} \tag{1–26}$$

where

$$Q_{useful} = \dot{m} C_p \, \Delta T$$

which is delivered at a temperature T for the purposes intended, one of which may be conversion to work at some fraction, ϕ, of the Carnot efficiency:

$$W = Q_{useful} \, \phi \left(1 - \frac{T_0}{T}\right) \tag{1–27}$$

where T_0 is the ambient temperature of the environment where the heat engine operates. The losses may consist of conduction and convection losses as well as radiation losses:

$$Q_{loss} = aT_R^4 + b(T_R - T_0) \tag{1–28}$$

The constants a and b are determined by the geometry. The temperature T_R

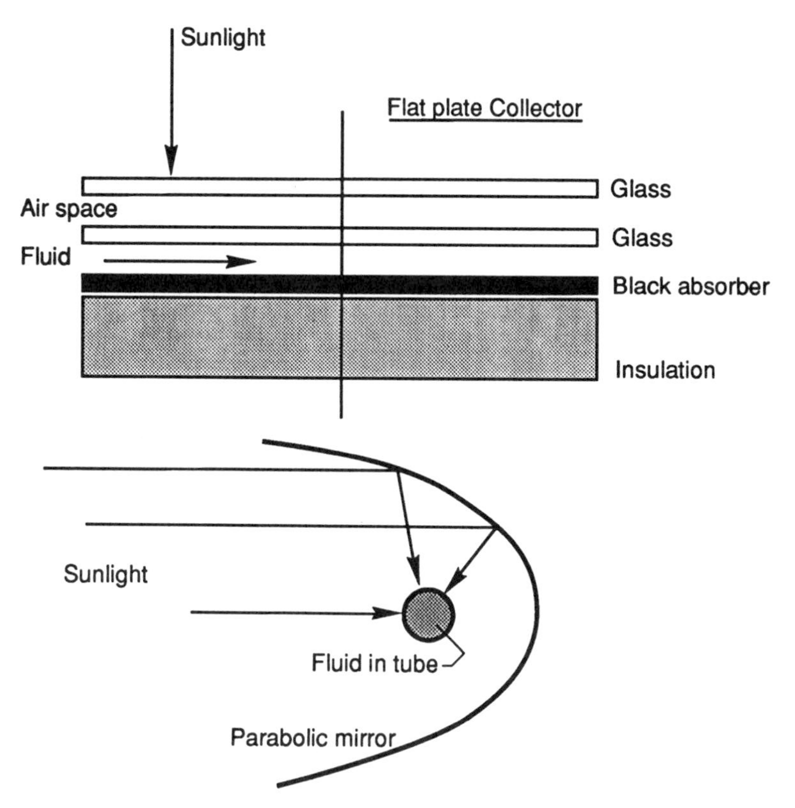

FIG. 1–21. Flat plate and concentrating solar power collectors.

must be larger than the working fluid collector outlet temperature, $T + \Delta T$. From these elements of the work calculation procedure, it should be evident that the production of work from solar radiant power is affected by the following irreversibilities:

1. Reflection losses from first surfaces, if present
2. Reradiation loss at T_R and losses to environment
3. Temperature drop from T_R to $T + \Delta T$
4. Temperature drop to the heat engine, ΔT
5. Carnot loss
6. Heat engine irreversibilities

and storage and transmission losses, if and wherever they occur.

Figure 1–23 is a sketch of the collection efficiency (ratio of collected to incident energy) as a function of collection temperature. The lowest collection efficiency systems are typically designed to supply high performance heat engines whose Carnot efficiency rises with increasing temperature.

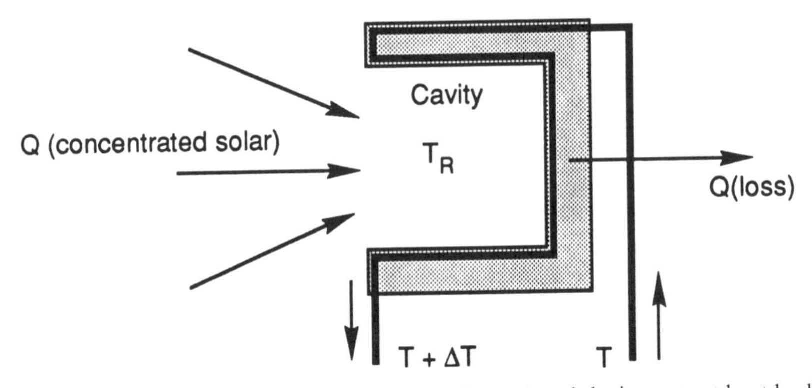

FIG. 1–22. Cavity solar collector (for a field of heliostats) and the important heat budget elements.

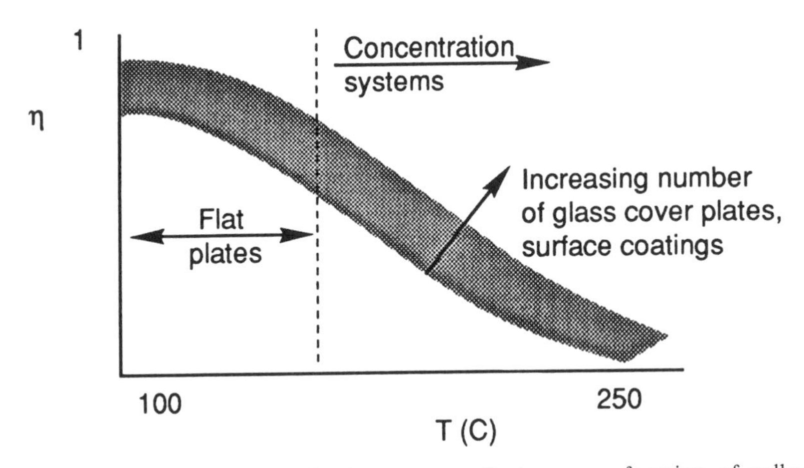

FIG. 1–23. Performance trend of solar power collectors as a function of collection temperature.

To date, the most economically successful method for solar energy collection is the flat plate hot water heater in common use where the climate favors it and because sufficient surface area is available, such as on the roofs of buildings. This approach is also describable as local collection and use in contrast to the high-temperature systems that distribute their electricity to remote users. Other successful methods for integrating solar power use into buildings for comfort heat is through the architectural design of windows and walls. Passive design features such as building and window orientation, high thermal inertia walls, and greenhouselike features have been widely used and accepted as effective.

A promising approach to the utilization of solar energy for commercial power (or hydrogen) production is through the exploitation of the difference in temperature between the ocean's surface and that at greater depths

(400–800 m). The region where this difference is greatest is near the equatorial deep ocean. This so-called ocean thermal energy conversion (OTEC) takes advantage of the low heat collection cost but is limited to cycles with low specific work and efficiency. The high power plant cost, the remote location of the plant, and the not-well-understood environmental impact have precluded this option from being exercised to the present time. Solar ponds may be designed for the same purpose on land. Losses from the surface, particularly those resulting from a static instability associated with a temperature-determined density variation with depth, are a critical performance issue.

The performance analysis of solar power collection systems is typically a demanding application of heat transfer analysis with conduction, convection, and radiation playing a role, all of which vary with the time over the sun's daily course through the sky and as cloud cover may vary. Often economic factors play an important part in the design of systems. For example, the degree to which a collector tracks the sun, the heat losses and so on, all affect cost. The design details and the environmental conditions tend to play a role in governing the efficiency of the collection system. Typically, one determines overall performance efficiency by averaging over some sensible time period so that temporal vagaries associated with weather are averaged out. Generally, the efficiency of collection decreases with increasing temperature because the losses are more severe at high temperatures (Fig. 1–23).

For the high-temperature collection systems, the steady flow cycles described in Chapters 7–11 may be used and are most often considered. The fact that the energy source is solar is relatively unimportant, except that the day/night cycle may demand particular design consideration.

1.11.3. *Solar Power from Space*

The relatively small insolation fraction of daylight may be increased to near 100% when the sunlight is intercepted in space and reflected to a collection site on the ground. This energy may be in the form of visible light, or it may be transformed to microwaves by a conversion system in space. Visible light suffers the disadvantage that it is subject to occlusion by clouds. The microwaves can be generated in space by conversion of the electric power collected there. On the ground the microwave power may be reconverted to electricity. Any method of space-based power collection adds to the total energy input of the Earth. To the extent that this is a very small fraction of the natural power flow, it may have a negligible global effect. Locally, however, the impact is likely to be much more severe. Sunlight for a 24 hr day will impact the ecology of the area, more so if the sunlight is concentrated so that a small receiver can be used. When microwaves are used, the efficiency of collection is improved by the removal of cloud influence and by the reconversion process itself. However, microwaves (in the domain where they avoid cloud occlusion) are hazardous to life forms by the heating experienced. Thus people and animals will have to be excluded from the vicinity of the collection site.

The reader should consult ref. 1–21 as source material for a more thorough examination of space-based power technologies.

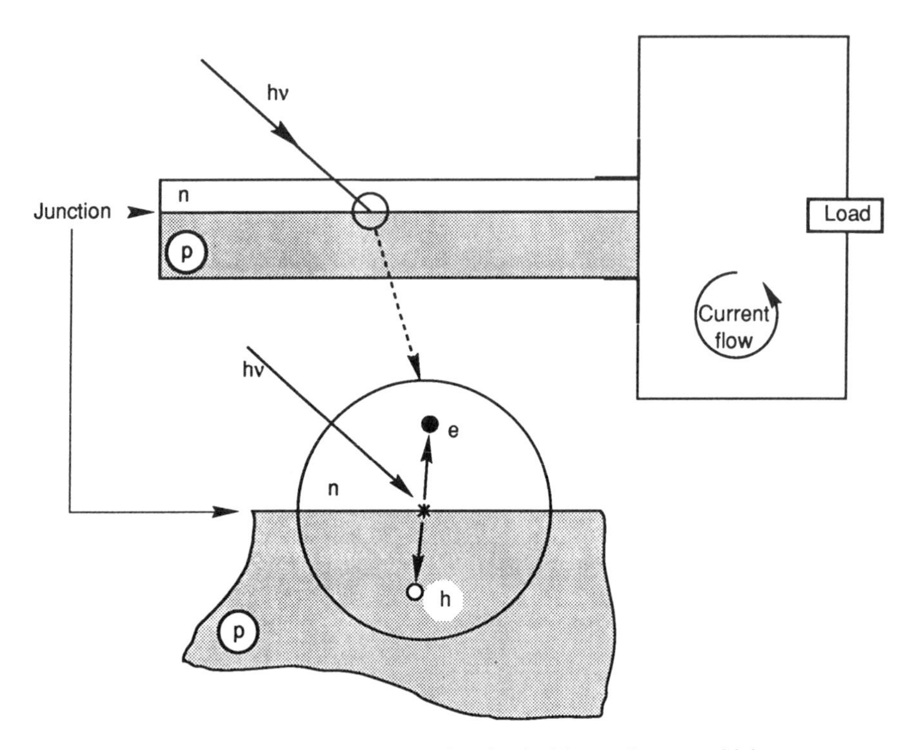

FIG. 1–24. Solar cell showing external circuit, incident photon which creates an electron-hole pair. Electron current direction is indicated. Only the useful photon interaction is shown.

1.11.4. *Solar Cells*

Solar cells are a means of converting the photon energy of sunlight to electrical energy directly. A cell and its external circuit are shown in Fig. 1–24. The cell is a semiconductor junction with dissimilar materials. An energy band gap for electrons must exist that is small enough for a photoexcitation collision to move the electron across the gap. The photons with the correct energy produce an electron-hole pair. The electron and the hole diffuse to the electrodes through the semiconductor. The result is a current flow across the cell at, or close to, the potential of the gap. The energy barrier "height" varies between 1.1 ev for silicon and 2.4 ev for cadmium sulfide (CdS). This characteristic makes it clear that the lower energy photons (longer wavelength) in sunlight cannot be effective in transferring their energy to semiconductor electrons or holes. A portion of the electromagnetic sunlight spectrum is therefore not available for solar cell power production. At the high-energy end of the spectrum, photon energies that are effective are also limited (by reflection and transmission characteristics of the cell material) so that the range of effective energies is limited and varies with choice of the semiconductor material(s). The result is that a theoretically maximum efficiency is attainable for a material at a given temperature (ref. 1–22). At ambient temperatures such maximum efficiencies may exceed 25% for materials like Si, InP, GeAs, and CdTe which have a band gap voltage near 1–1.4 ev. The cell temperature has an influence on the performance of these devices because higher temperature allows random motion across the gap by thermal excitation which degrades the photocurrent.

The typical cell therefore operates with low voltage so that arrays must be constructed to match the load current and voltage requirements. Most cell arrays are designed to operate for maximum power since that point is relatively close to an open circuit condition where efficiency is high. The magnitude of the efficiency is of the same order or somewhat less as a heat engine (see, in particular, Section 9.10 where the related space power generation problem is discussed). Note that the solar cell, like the heat engine, requires a radiator to maintain the cell temperature. The singular advantage of the solar cell is in the absence of moving parts so that wear and rotation dynamics are not cell-lifetime-determining issues. The efficiency is determined by losses suffered by the cell as an electric power producer which include

1. First surface reflection
2. Rejection of long-wavelength energy
3. Absorption of upper cell layer which is converted to heat
4. Junction loss associated with internal resistance
5. Hole-electron recombination

and others.

The net cell efficiency turns out to be of the order of 0.15 for Si, and somewhat higher for other materials. The fact that suitable materials are semiconductors and require a complex manufacturing process has kept the cost of solar cells high. Cost and efficiency have to be considered together for any practical application. Significant performance improvements have been made in the recent past, and more are possible. The direct production of electric power by solar cells is presently limited to special-purpose applications where alternatives are lacking or impractical. Examples are near earth space, remote sensing on Earth, communications, and so on. The physics and technology associated with solar cells are actively evolving and specialized to the point where the reader is encouraged to review background and progress by consulting the literature. An example is ref. 1–22 where performance analysis is described.

1.12. MECHANICAL POWER SOURCES

While thermal power is available from a number of sources and the technology exists for converting it to work, there are also a number of energy forms that are already in mechanical form and that are directly convertible to other high-grade energy such as electricity (see Section 2.2). These include the wind driven by air currents in the atmosphere, ocean waves driven by the wind, and the tides. On a global scale, these resources are limited in that siting is not widely practical and estimates of the energy that is associated with such resources may be found in the literature. Locally, however, these resources may be very important.

1.12.1. *Wind*

The utilization of this energy form requires use of surfaces on which reaction forces may be developed. These forces are of an aerodynamic nature, and there are limitations of flow energy convertibility. This topic is covered in Chapter 13.

1.12.2. *Hydropower*

The river as part of the hydrological cycle transports water from positions of high potential energy to the lowest available, the ocean. The construction of a dam holds water at an elevated potential energy level above the dam base where a turbine may be located. The potential and kinetic energies of a fluid are readily convertible one to the other by such a turbine so that the efficiency (η_T) of generation of mechanical power by these means is fundamentally very high, limited only by the losses that must be accepted in order to make costs manageable. The power from a river dam is given by the mass flow rate of water and the potential energy drop per unit mass:

$$\dot{W} = \eta_T \rho \dot{V} g \, \Delta h \approx 10^4 \, \dot{V} \, \Delta h \text{ (SI)} \tag{1-29}$$

in watts with river volume flow rate in cubic meters per second and dam height in meters. Because this performance is cost dependent and involves no thermodynamic limitations, this subject is not covered here.

The run of river turbines such as those one might use if a dam is to be avoided is described by mechanics similar to that of the wind turbine. An important difference between water and wind is associated with the densities that result in much higher dynamic pressure in water. This leads to significant mechanical design challenges in water hydrodynamics. Further, since water is a liquid, it can change phase if the pressure is made sufficiently low. The resulting cavitation (vapor bubble formation) imposes limits on machinery performance, and on operating lifetime, and presents significant design challenges.

1.12.3. *Tidal Power*

The Earth-moon dynamic system forces water levels in large bodies of water on Earth, such as oceans, to fluctuate in time. When water at high tide is retained and discharged through a turbine in the dam, power may be obtained. Thus a system may be visualized as a dam along a section of ocean front with appropriate turbines and gates to control the flow. Such systems may generate power on both inlet and outlet of water, or only on outlet if the turbine flow is restricted to one direction. From eq. 1–29 it is evident that the power from a tidal power dam is related to the volume of the tidal basin [i.e. its (vertical) range, which is site dependent, and the length of the dam]. The relevant time scale is the tidal cycle or the half cycle with reversible turbine. The power is evidently proportional to the tidal range squared.

The environmental impact of such systems is primarily of an ecological nature.

1.12.4. *Ocean Wave Power*

The wind-driven waves on the oceans may be used to drive a floating object upward so that its energy is transferred to an electrical generator during the return to lower level. The power is evidently related to wave height (times the fraction of the height that the float is raised during a wave passage), the float or wave segment length, and the wave frequency. Acceptability of this form of energy conversion is limited by the economics and possibly aesthetics. It is not in significant use due in part the difficulties of mechanical design for high reliability.

1.13. CONCLUDING REMARKS

A serious dimension in the use of oil for running a country's economy is the dependence on external sources. This is an issue in the United States and in Europe which import increasing fractions of their oil. This makes these countries vulnerable to supply disruptions, both economic and those brought on by conflict. Safe approaches include source diversity, oil, coal, nuclear, and renewable resources. Oil will likely be the fuel of choice for aircraft and trucks because alternatives are not very attractive. Coal and nuclear power, however, can supply utility systems and large transport systems such as ships, railroads, and to some extent cars, the latter two via electricity distribution. National planning policy regarding energy use and transportation should recognize these dimensions of the energy supply picture.

It should be noted that the idea that methane is intimately associated with oil and coal is not universally accepted. Gold (ref. 1–23) argues that the origin of much of the methane in the Earth's crust may have originated and continues to originate from more primeval processes deep in the Earth's interior. The implication is that, if true, the world supply of this resource may be significantly larger than the estimates discussed in this chapter.

PROBLEMS

1. Carry out the linearization of eq. 1–7 to obtain eq. 1–8. Relate the constant in the equation to the gravitational constant G, given in Appendix B.
2. Estimate the global energy consumption and the lifetimes of petroleum

and coal, if everyone on our planet were using that consumed by modern industrial man.

3. Compare the volumes of the fuel requirements for a commercial airliner like a Boeing 747 on a transatlantic mission if one were to substitute for kerosene, liquid methane or liquid hydrogen. Assume that the energy requirements involved would be identical. Assume that the 747 burns about 8000 kg per hour in flight and that the flight takes 10 hours.
4. Determine the wavelength cutoff beyond which solar radiation cannot elevate an electron in a solar cell junction above 1.1 and 2.4 volts.

BIBLIOGRAPHY

Energy and Power, Freeman Press, New York, 1971.

Energy Technology to the Year 2000, Technology Review, Massachusetts Institute of Technology, October 1971–January 1972.

Pedley, T. J., Ed., *Scale Effects in Animal Locomotion*, Academic Press, London, 1977.

Hill, P. G., *Power Generation, Resources, Hazards, Technology and Costs*, MIT Press, Cambridge, Mass., 1977.

REFERENCES

1–1. Schmidt-Nelson, K., *Scaling: Why Is Animal Size So Important?* Cambridge University Press, Cambridge, 1984.

1–2. Whitt, F. R., Wilson, D. G., *Bicycling Science*, MIT Press, Cambridge, Mass., 1989.

1–3. Claassen, R. S., Girifalco, L. A., "Materials for Energy Utilization," *Scientific American*, **244**(4): 51–8, October 1986.

1–4. Kolin, I., *Evolution of the Heat Engine*, Longman Group Ltd., London, 1972.

1–5. Kulcinski, G. L., Kessler, G., Holdren, J., Häfele, W., "Energy for the Long Run: Fission or Fusion?," *American Scientist*, **67**, January–February 1979.

1–6. Morris, S. C. *et al*, "Coal Conversion Technologies: Some Health and Environmental Effects," *Science*, **206**, November 9, 1979.

1–7. Wentorf, R. H., "Hydrogen Generation," General Electric Company Report No. 75-CRD119, May 1975.

1–8. Wentorf, R. H., Hanneman, R. E., "Thermochemical Hydrogen Generation," *Science*, **185**, July 26, 1974.

1–9. Bolton, J. R., "Solar Fuels" *Science*, **202**, November 17, 1978.

1–10. Wrighton, M. S., "The Chemical Conversion of Sunlight," *Technology Review*, May 1977.

1–11. Bard, A. J., "Photoelectrochemistry" *Science*, **207**, January 11, 1980.

1–12. *Space: A Resource for Earth, An AIAA Review*, American Institute of Aeronautics & Astronautics, New York, 1977.

1–13. Milora, S. L., Tester, J. W., *Geothermal Energy as a Resource for Electric Power*, MIT Press, Cambridge, Mass., 1977.

1–14. Cotillon, J., "La Rance: Six Years of Operating a Tidal Power Plant in France," *Water Power and Dam Construction*, **26**(9): 314–22, 1974.

1–15. Bernshtein, L. B., "Tidal Energy for Electric Power Production," Israel Program for Scientific Translation Limited, Jerusalem, 1965.
1–16. World Energy Conference, US National Committee, Survey of Energy Resources, 1974.
1–17. Hubbard, M. K., "Degree of Advancement of Petroleum Exploration in the United States," *Bulletin of American Association of Petroleum Geologists*, **51**, 2207–27, 1967.
1–18. Hubbard, M. K., Chapter 8, "Energy Resources," *Resources and Man*, Freeman Press, 1969.
1–19. Hubbard, M. K., "The Energy Resources of the Earth," *Scientific American*, September 1971.
1–20. Kreith, F., Kreider, J. F., *Principles of Solar Engineering*, Hemisphere Publishing, 1978.
1–21. *Solar Energy for Earth, An AIAA Assessment*, American Institute of Aeronautics & Astronautics, New York, 1975.
1–22. Merrigan, J. A., *Sunglight to Electricity: Prospects for Solar Energy Conversion by Photovoltaics*, MIT Press, Cambridge, Mass., 1975.
1–23. Gold, T., "The Origin of Methane in the Crust of the Earth," U.S. Geological Survey, May 1993.

2

EFFICIENCY: PROCESS PERFORMANCE

In the process of converting power from one available form to another desirable form, one necessarily deals with the possibility that part of the power is not used to achieve the desired changes. This chapter addresses the performance measures used to quantify the quality of the process: efficiency. It is shown that the concept of efficiency incorporates the notion of energy conservation (a consequence of the First Law of Thermodynamics), and the idea (based on the Second Law) that no process can be better than the reversible process. The review of thermodynamics described here is necessarily brief.

2.1. THE LAWS OF THERMODYNAMICS

The first and second laws of thermodynamics are the foundation of energy system analysis.

2.1.1. *The First Law of Thermodynamics Applied to a Closed System*

The First Law of Thermodynamics results from observation of a closed system. A closed system is an aggregate of matter that can be enclosed by a "control surface" (see Fig. 2–1). Such a system can be made to undergo changes in state by work interactions or by heat interactions. Work is "added" to the system if its volume is reduced, while heat is said to be added to the system when the environment causes the energy of the system to increase (in the absence of work). The state of the system is that collection of measurable quantities which is sufficient to describe the system fully. For a simple system, specification of any two thermodynamic variables is sufficient to identify the state of the system. Uniform aggregates of a single substance, mixtures of chemically nonreacting substances or two-phase mixtures of a single chemical substance are examples of simple systems. Figure 2–2 shows the state representation in the pressure-(specific) volume plane of a unit mass in a variety of conditions.

A point on a plane of two independent state variables can be used to describe the state of a system. With heat and work interactions, the state may change resulting in a "thermodynamic path" being traced. The path may be thought of as a succession of equilibrium states through which the system passes. Figure 2–3 illustrates the paths and the end states for a work interaction process of compression and a heat interaction process (cooling) as depicted in Fig. 2–1. The pure work process involves motion of a force through a distance or, in thermodynamic terms, a pressure force displacing the boundaries of a volume

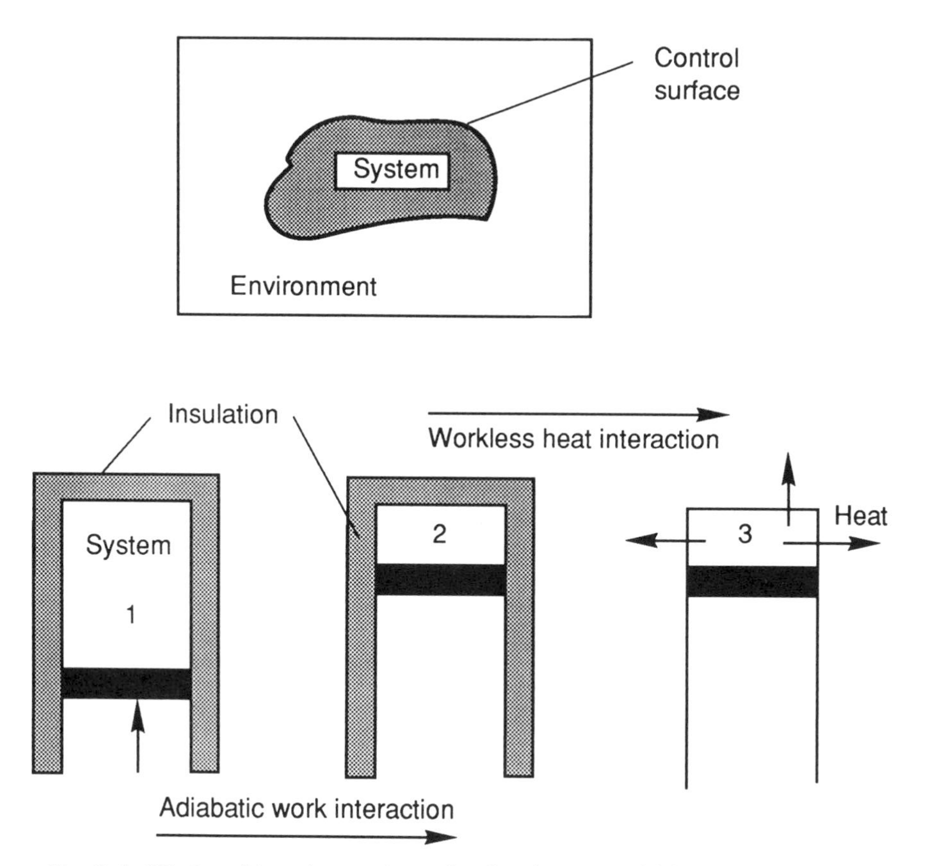

FIG. 2–1. Work and heat interactions of a closed system with its environment.

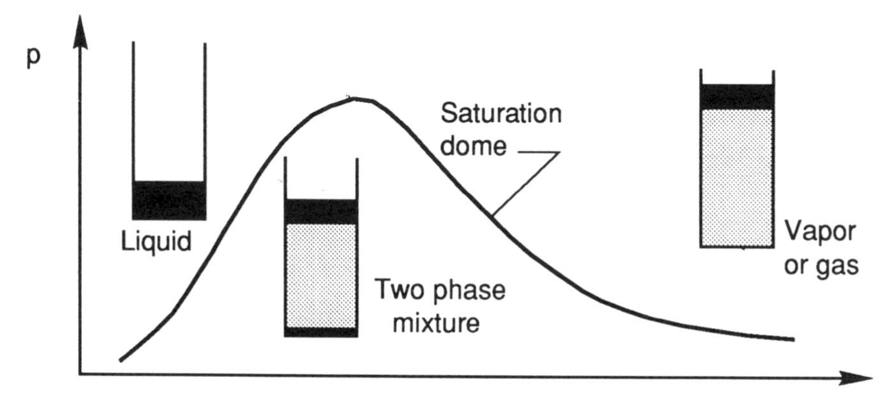

FIG. 2–2. System phase configurations on the *p-v* plane.

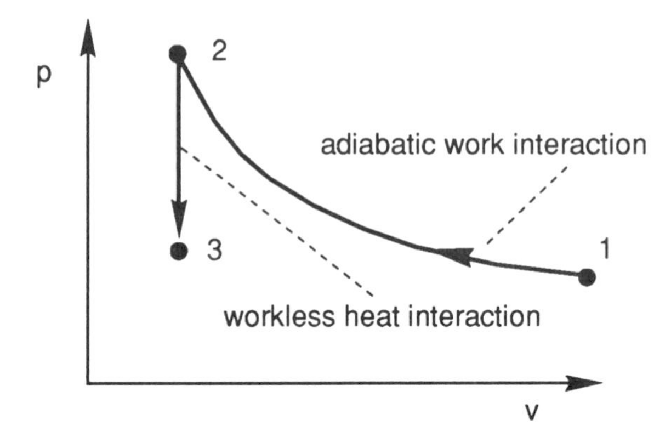

FIG. 2–3. State paths for process illustrated in Fig. 2–1.

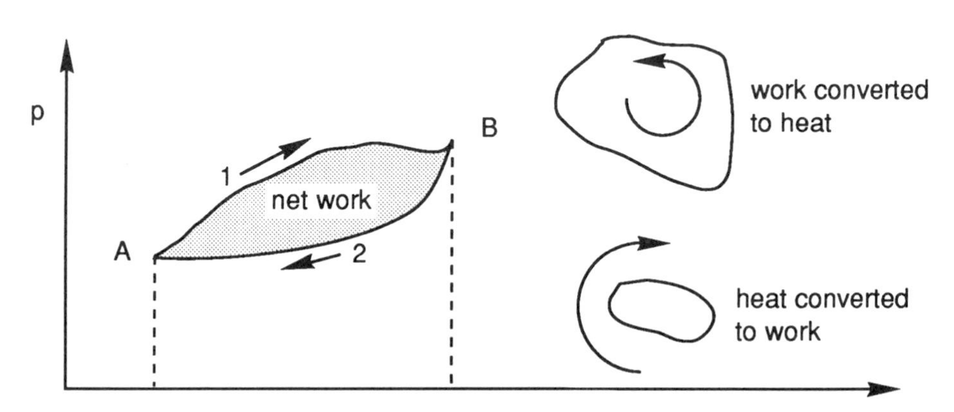

FIG. 2–4. Paths on the *p-v* plane with work produced or absorbed.

(see Section 1.1). Thus

$$dW_{\text{rev}} = p\,dV \tag{2–1}$$

In the absence of friction, the work invested in the system is recoverable and the path followed is the reverse of the input process. This work is termed reversible and is represented as the area under a path in the p–V plane. The real work done on a system always exceeds the reversible work by that dissipated through friction or equivalently converted to heat. Figure 2–4 shows a closed path in the p–V plane of a simple system. The work done is evidently not zero. The first law describes the observation that the line integral of *heat minus work* is zero. In effect, because work is nonzero, work has been converted to heat, or, depending on the sense of the path, heat is converted to work (Fig. 2–4). The quantity $d'Q$–$d'W$ must therefore be a property. This property is the "internal energy" of the system. In most texts in thermodynamics, a distinction is made between state changes denoted by the perfect differential (d) whereas differential amounts of heat and work input or output to a system are described with the imperfect differential, (d'). This first law of thermodynamics reads,

$$dU = d'Q - d'W \tag{2–2}$$

which states that heat and/or work interactions of a closed system with the environment result in changes in the "internal energy" of the system. This equation may be rewritten when a distinction is made in $d'W$ between reversible work ($p\,dV$) and the irreversible work done, which is heat. Thus

$$d'W = d'W_{\text{rev}} + d'W_{\text{irrev}} = p\,dV + d'Q \tag{2–3}$$

or from eq. 2–2:

$$d'Q = dU + p\,dV \tag{2–4}$$

The first law is a statement describing the conversion of energy forms. In some devices it may be desired to convert work to internal energy and not to heat. It is therefore appropriate to define the quality of the process by forming the ratio of (in this example) the increase of internal energy to the work supplied. Such a ratio is a "first law efficiency" relevant to the conversion process.

2.1.2. *First Law Applied to an Open System*

An open system is a "control volume" bounded by a control surface through which matter can enter and/or leave the volume. In many engineering applications, such as steady flow processes, there is no mass accumulation in the volume. The fluid flow rates entering and leaving are equal, although they may (and usually do) have different properties. This difference gives rise to the possibility that the work done in pushing the entering fluid in is different from the work realized by the fluid in the process of leaving. The proper accounting for this flow work leads to the logical definition of a new property, called the enthalpy. The steady flow mass conservation statement and first law for a control volume with power input in the form of shaft power and external thermal power input read as follows:

$$\dot{m}_{\text{out}} = \dot{m}_{\text{in}} = \dot{m} \tag{2–5}$$

$$\dot{m}[h_{t2} - h_{t1}] = -\dot{W}_{\text{shaft}} + \dot{Q}_{\text{external}} \tag{2–6}$$

The total enthaply h_t written here is defined as the sum of the static enthalpy defined by (eq. 4–13)

$$h = u + pv \tag{2–7}$$

and the other relevant energy modes which may be important in a specific situation, such as kinetic or potential energies. The specific volume, v, is the reciprocal of the density, $v = 1/\rho$. In the extensive form, that is, on a per unit mass basis (lower case symbols), the total enthalpy is

$$h_t = h + \tfrac{1}{2}V^2 + gz + \cdots \tag{2–8}$$

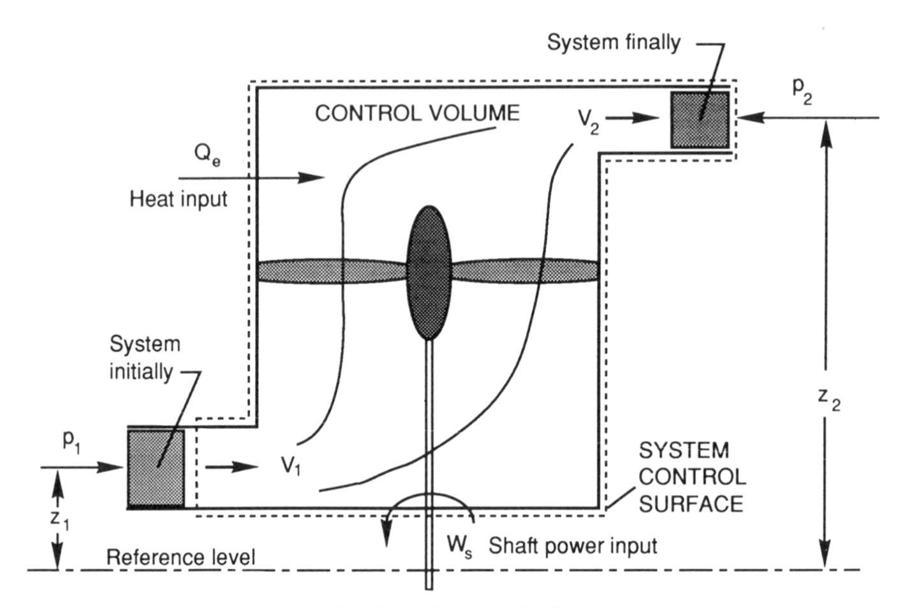

Fig. 2–5. First law control volume in a steady flow, open system analysis.

Here V is the fluid velocity, g is the gravitational acceleration, and z is an elevation coordinate referenced to an arbitrary datum. Figure 2–5 illustrates the control volume idea applied to a steady flow process. Shown are shaft power input, thermal power input, and the resultant state changes as measured by the state property changes in steadily flowing gas or liquid.

2.1.3. *The Second Law of Thermodynamics for a Closed System*

An observation is that heat travels only from a body of higher "temperature" to one with a lower temperature. It is possible to define the temperature as a quantity that is proportional to the heat that flows to and from a reversible heat engine. The reversible heat engine is nothing more than a system operated in a cyclic manner with (1) $dW = p\,dV$ and (2) infinitesimally small temperature differences between the system and the heat reservoirs. Thus, for a reversible cyclic process, an amount of heat is accepted by the system dQ_{in} and an amount is rejected, dQ_{out}. Since the temperature of the heat reservoirs from which and to which these heat quantities flow are defined to be proportional to the heat transferred, it follows that

$$\frac{dQ_{\text{rev, in}}}{T_{\text{H}}} + \frac{dQ_{\text{rev, out}}}{T_{\text{C}}} = 0 \tag{2–9}$$

since both quantities are just a proportionality constant. The heats involved have opposite signs by adoption of any logical convention regarding the meaning of positive Q (usually into the system). For cyclic, reversible processes, one may generalize by stating that the integral of dQ_{rev}/T around a closed thermodynamic path is zero or,

$$\oint \frac{dQ_{rev}}{T} = 0 \tag{2–10}$$

If a quantity undergoes *no* change when the state is altered from one point and returned to the same point through a (reversible) path, then the quantity must be a property. This property can be expressed as having a value that depends only on state variables. The property of the system defined through the reversible heat transfer process is called the entropy:

$$dS \equiv \frac{dQ_{rev}}{T} \tag{2–11}$$

For a process with end states, the change in entropy is given by

$$\int_1^2 dS = \int_1^2 \frac{dQ_{rev}}{T} = S_2 - S_1 \tag{2–12}$$

It can be said, therefore, that the entropy exists because it is found that heat cannot flow from a cooler to a warmer body since the two terms of eq. 2–9 must have unlike signs. The notion of warmer and cooler is quantified by an absolute temperature scale, whose existence is sometimes referred to as the "Zeroth Law of Thermodynamics."

For irreversible processes, the argument is made that the conversion efficiency of a real engine (operating between the same temperature limits) must be lower than that of a reversible engine, and the result is that more heat is required by a real engine for the same rejected heat. If 1 and 2 describe the beginning and end of the heat input process for a real cycle, one has, by eliminating Q/T for the reversible rejection process in both the real and reversible engine:

$$S_2 - S_1 = \frac{Q_{irrev,\,in}}{T_H} - \frac{Q_{rev,\,in}}{T_H} \quad \text{or} \quad dS > \frac{dQ_{irrev}}{T} \tag{2–13}$$

It follows that entropy is useful for quantifying the reversibility of a process. In Section 2.3, this idea is exploited for the purpose of defining the "Second Law efficiencies" of a number of processes.

2.2. CONVERSION OF MECHANICAL TO ELECTRICAL POWER

The mechanical power produced by a heat engine can be converted to electrical power by means of an electrical generator. The inverse process is carried out by an electric motor. The quality of either process is measured by a (first law) efficiency which is defined as

$$\eta = \frac{\text{power output in desired form}}{\text{power input in original form}} \tag{2–14}$$

The fraction (1-η) of the input is converted to heat. Thus a generator efficiency is

$$\eta_G = \frac{\text{electrical power output}}{\text{mechanical power input}} \tag{2–15}$$

and a motor efficiency is

$$\eta_M = \frac{\text{mechanical power output}}{\text{electrical power input}} \tag{2–16}$$

In practice, it is found that these efficiencies are very high (i.e., close to unity). This is true because electrical work, which involves the motion of charges in an electrical field, is analogous to work involving physical objects of larger dimensions in a gravitational field. In both cases, work is done by moving an object in a field that exerts a force on the object. The motion of the object(s) involves changes in energy which, by appropriate means, could be transferred from the physical object to the electron or vice versa. The motor and the generator are just such means. The movement of such charges along the conductors involves little generation of heat, and, with good mechanical bearings, little of the mechanical power is wasted. The result is that the conversion efficiency of equivalent energy forms can be made quite high.

In practice, therefore, one can transform mechanical power to electrical in order to be able to transport the power easily and with little loss. The user can reconvert the electrical power to mechanical if this is what is needed. Figure 2–6 shows how these first law efficiencies are related to the notion of reversibility.

The mechanical energy forms such as wind kinetic energy, tidal, and hydro potential energy can be readily transformed to electrical power by means of the appropriate machinery. The performance of this machinery closely approaches the theoretical limits imposed by the laws of mechanics. The performance of the specific devices is considered in Chapter 13.

2.3. CONVERSION OF HEAT TO MECHANICAL POWER

Classical thermodynamics can be used to show that the efficiency of converting heat to work is strongly influenced by the temperature of the heats involved in the conversion process. This suggests that thermal energy has a "quality" in its ability to be converted to work which is related to the temperature.

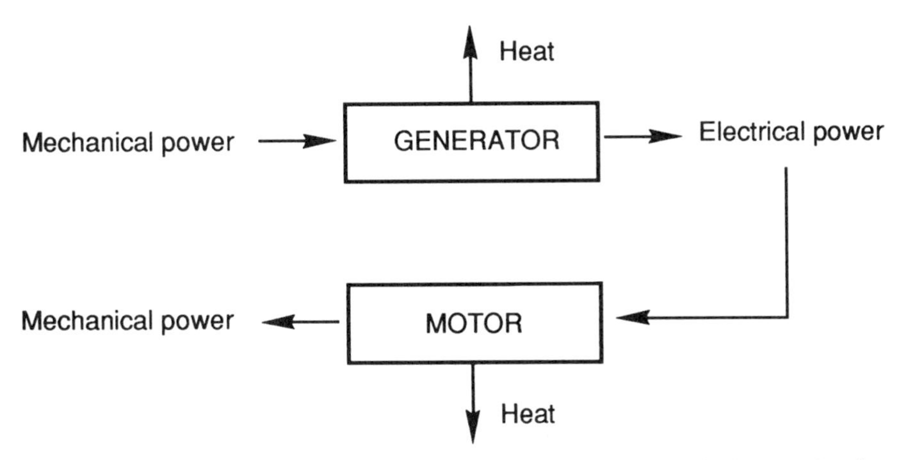

FIG. 2–6. High conversion efficiency between electrical and mechanical power implies near reversibility.

High-temperature heat is thus more useful than low-temperature heat and, as such, is more valuable. This notion can also be extended to the idea that work (i.e. mechanical or electrical energy forms) can be thought of as heat at an infinite temperature.

The efficiency of a heat engine is defined by the ratio:

$$\eta_{\text{th}} = \frac{\text{mechanical power output}}{\text{thermal power input}} \tag{2–17}$$

Such an efficiency is called a thermal efficiency, which is evidently similar to the process efficiency defined above. The First Law of Thermodynamics states that for a system undergoing a cyclic process, the *net* heat input must equal the net work produced. The second law states that one must distinguish between heat input and rejected heat. Thus the *net* heat input is the supplied heat less the rejected heat (waste heat):

$$Q_{\text{net}} = Q_{\text{in}} - Q_{\text{rej}} = W \tag{2–18}$$

The thermal efficiency can thus be written as:

$$\eta_{\text{th}} \equiv \frac{W}{Q_{\text{input}}} = 1 - \frac{Q_{\text{rej}}}{Q_{\text{in}}} \tag{2–19}$$

The input heat is derived from a "heat reservoir" at temperatures T_{H} and the rejected heat flows to a reservoir at T_{L} (Fig. 2–7). For a reversible engine, the linear relationship between the temperature and the heat (Kelvin's postulate) allows one to write the thermal efficiency of the reversible engine as:

$$\eta_{\text{th}}(rev) = \eta_{\text{th}}(\text{CARNOT}) = 1 - \frac{T_{\text{L}}}{T_{\text{H}}} \tag{2–20}$$

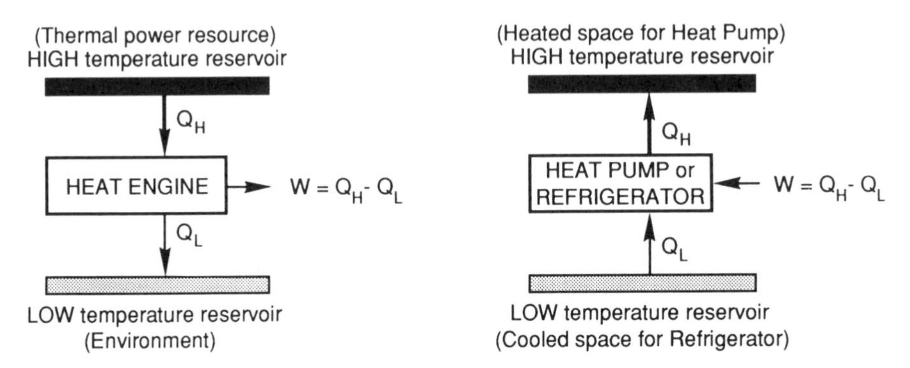

FIG. 2–7. Heat flow in a heat engine producing work and in a heat upgrade engine producing high-grade heat (heat pump) or withdrawing heat from a cooled space (refrigerator).

The reversible engine (to be described in Chapter 7) is referred to as a Carnot engine, named after the contributions of Sadi Carnot to the understanding of the convertibility of heat to work in the 1800s.

The thermal efficiency of real (i.e., irreversible) heat engines can also be shown to depend strongly on the ratio of maximum to minimum temperatures (see Chapters 7–11). This important relationship makes it crucial that the temperatures associated with the thermal power source and the heat sink be known.

In the Earth environment, the minimum temperature, T_L, is of the order of 300 K. The temperature associated with thermal power sources depends on the means used to utilize this reservoir. Typically the limiting temperature is imposed by (1) a material wall confining the working fluid required to have certain strength or elastic properties; or (2) the energy content of the thermal resource, or finally, (3) the properties of the working fluid. The variation in these limitations has important consequences for the design of heat engines, and these consequences are described in detail in Chapters 8–11.

Thermal power sources with temperatures in the range 700 to 2500 K are generally referred to as high temperature sources. On the other hand, thermalized solar energy, geothermal, and other heat resources with temperatures lower than 500 K are described as low-temperature resources. Low-temperature resources are primarily useful for comfort heat used locally and for some industrial processes. High-temperature sources are very valuable for their efficient convertibility to work.

Solar energy has a source temperature of around 5800 K so that, in principle, it has a high "quality," that is, convertibility to electric power. However, practical machines to take advantage of this high potential are difficult to build and justify economically because the power of collecting this dilute power form is subject to losses and involves a relatively large capital investment (Section 1.11).

A low temperature reservoir may be used to run a heat engine between the high-temperature environment and the heat sink. In such a case it is the low-temperature reservoir heat (q_L) that has value and an efficiency for conversion to work must recognize this value. Thus a "sink" thermal efficiency may be defined as

$$\eta_s \equiv \frac{w}{q_L} \tag{2–21}$$

This sink efficiency may and should be significantly larger than unity. This efficiency is applied to cryogenic heat in connection with energy storage concepts described in Section 17.3.4.

2.4. OTHER MEASURES OF HEAT ENGINE EFFICIENCY

The thermal efficiency of a heat engine was defined in the previous section. Since the Second Law of Thermodynamics prescribes how well a heat engine may convert heat to work, it may be appropriate to define a "relative" efficiency, sometimes called an "exergy" efficiency (ref. 2–1), as the ratio of work produced to work producible with a reversible or Carnot engine operating with the same heat input and between the same temperature extremes:

$$\eta_{REL} = \frac{\eta_{th}}{\eta_{th,rev}} = \frac{\eta_{th}}{1 - T_L/T_H} \tag{2–22}$$

such an efficiency gives a measure of how well a given heat engine produces work compared to the ideal (i.e., reversible engine). This efficiency is not in common usage, although the usefulness of "exergy analysis" of complex systems are often cited.

For many types of engines, the fuel used is uniform and invariant. It represents a certain amount of heat energy per unit mass when used in the combustion process of that engine. It is often more convenient to speak of fuel consumption rate rather than heat generation rate as a measure of efficiency because the physical quantity one buys is the fuel and the physical characteristics (such as density) have direct consequences on the mission which the conversion of this heat is designed to achieve. Thus a convenient measure of efficiency is "specific fuel consumption (rate)," SFC. This is defined as

$$\text{SFC} = \text{kg of fuel per sec/kilowatts of (mechanical) power} \tag{2–23}$$

which would normally carry units of kg/J if the SI system is used. The energy density of fuels is fortunately high so that this number is quite small. To avoid dealing with small numbers, it is often altered to read in g or mg/(kW_e hr) or similar units.

For applications where the heat engine is used to generate a propulsive force, SFC is more convenient if it is defined in terms of thrust produced rather than power produced. The implication is that the speed with which the vehicle travels is dictated by factors independent of the thrust, such as drag characteristics of an airplane, traffic considerations, etc. For such instances the units of SFC (or sometimes TSFC) are SFC = kg of fuel per sec/Newton thrust or equivalently, for greater convenience,

$$\text{SFC} = \frac{\text{kg}}{\text{N hr}} \quad \text{or} \quad \frac{\text{mg}}{\text{N sec}} \tag{2–24}$$

If the heating value of the fuel as measured in thermal kilowatt-seconds per kilogram, $(\text{kW}_t\text{-sec})/\text{kg}$, is H (noted as HHV in Table 1–4), then the relation between efficiency and SFC is

$$\eta_{\text{th}} = \frac{1}{\text{SFC H}} \tag{2–25}$$

Note that H for a typical hydrocarbon liquid fuel is of the order of 8,600 BTU/lb or 43.2×10^6 J/kg. This number is obtained from combustion of the fuel with air to a state where the H_2O is in either the gaseous or the liquid state (see Section 5.2). The heat of vaporization is included in the heating value when the water condenses. The higher heating value (HHV, liquid H_2O) and lower heating value (LHV, gaseous H_2O) differ slightly because the physical end states differ.

In the field of propulsion, particularly rocket propulsion, the commonly used measure of efficiency is the "specific impulse" defined by

$$I_{\text{sp}} = \frac{\text{thrust}}{\text{weight flow rate of propellant}}$$

Evidently the specific impulse is related to SFC by

$$I_{\text{sp}} = \frac{1}{g_0} \frac{1}{\text{SFC}} \tag{2–26}$$

Here the numerical value of the Earth's gravitational attraction at sea level is used. The units of I_{sp} are "seconds" in either English or SI units.

A third measure of efficiency is the so-called *heat rate*, which is appropriate when discussing the performance of heat engines with various kinds of fuels. This measure is common in the U.S. utility industry and is stated in the mixed units of BTU/kW-hr. Since 3413 BTU are equivalent to a kW-hr, a heat rate of 7000 corresponds to an efficiency of about 50%.

Finally, use of a combustion process in some cycles may involve generation of that which cannot be transferred to a working fluid and is carried away in the combustion gas without involvement in the heat engine. An efficiency associated with this process is discussed in Section 7.5.2.

2.5. HEAT UPGRADE CYCLES: CONVERSION OF WORK TO HEAT

A heat engine that utilizes work and accepts low-temperature heat to supply higher temperature heat to a user is called a heat pump. A refrigerator is a similar heat engine with the implication that the desired end product is the heat removal *from* some space (the low temperature reservoir) rather than the delivery of heat *to* a space (Fig. 2–7). The measure of performance of such a device is the "coefficient of performance" (COP), defined for a heat pump as

$$\text{COP (Heat pump)} = \frac{\text{high temperature side thermal power output}}{\text{electromechanical power input}} = \frac{Q_H}{W} \tag{2–27}$$

because the heat is the desired product and the supply of electromechanical power is the cost. For a reversible heat pump operating between a low temperature T_L and a high temperature T_H, this parameter is

$$\text{COP(HP, rev)} = \left(1 - \frac{T_L}{T_H}\right)^{-1}$$

which should be significantly larger than unity for an irreversible (real) heat pump to make economic sense.

For a refrigeration cycle, the desired end product is the removal of heat from the low temperature space (Fig. 2–7). Thus a coefficient of performance is

$$\text{COP (Refrigerator)} = \frac{\text{low temperature side thermal power input}}{\text{electromechanical power input}} = \frac{Q_L}{W} \tag{2–28}$$

Since $Q_L + W = Q_H$, the relation between these two kinds of coefficients of performance is

$$\text{COP (REFR)} = \text{COP (HP)} - 1 \tag{2–29}$$

2.6. THE ENERGY CASCADE

All power relevant to citizens of the Earth flows from the sun (and to a minor extent, the moon and the Earth's interior). Figure 2–8 shows the stages of energy utilization, possible penalties, and possible energy storage opportunities. *Direct energy conversion* describes processes that bypass the step involving production of electromechanical work by means of a heat engine. Examples are the fuel cells, thermoelectric, and thermionic generators.

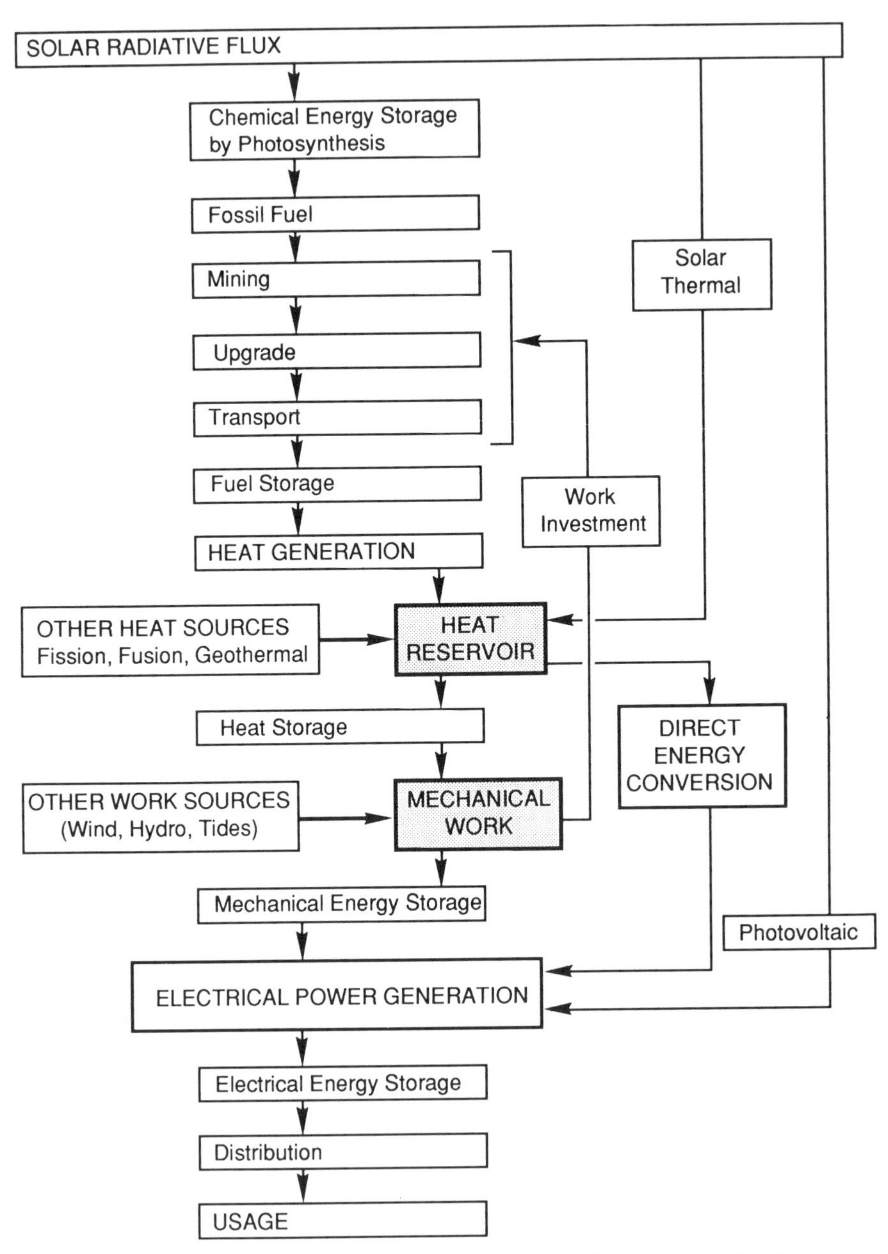

Fig. 2–8. The energy cascade from source to user showing opportunities for power production and need for investment.

2.6.1. *Concept of Net Energy*

The process of tapping and using an energy resource requires investment of work resources to start the process of allowing the flow of the resource down the above cascade. The end result must be that at least as much energy is produced as is used. In economic terms, this means that the product must be

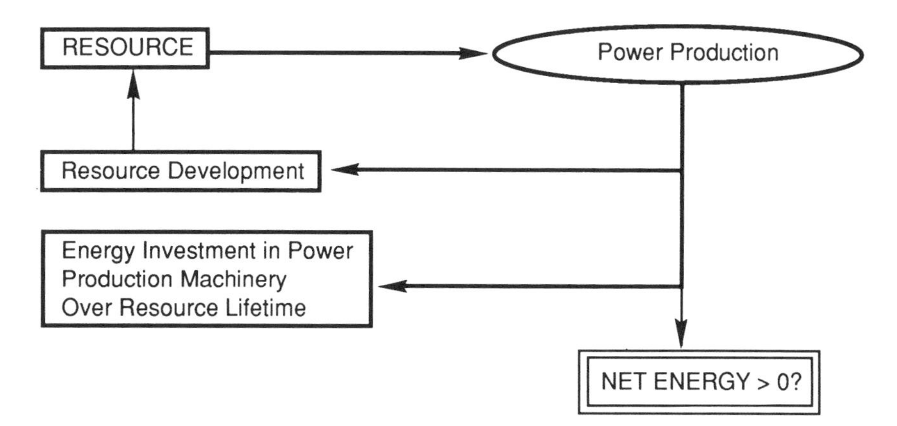

FIG. 2–9. The resource net energy must be positive for it be economically viable.

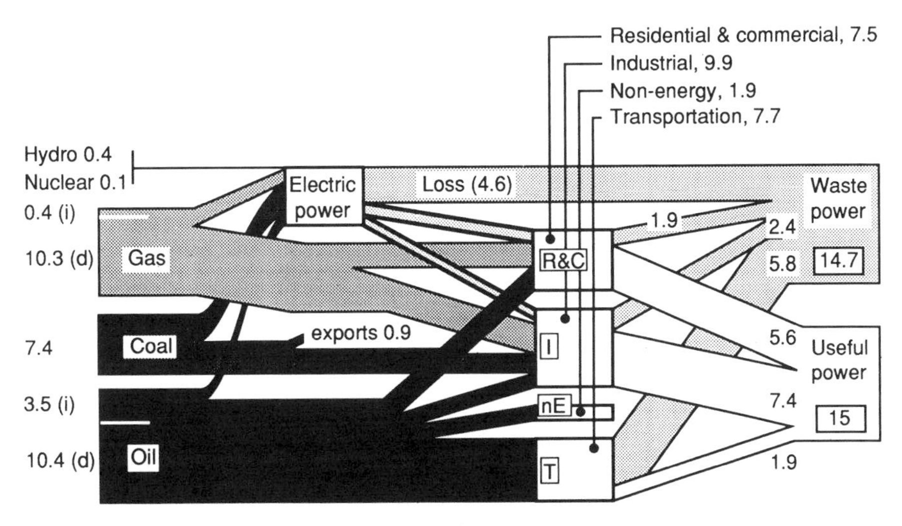

FIG. 2–10. Power flow from source to end use for the United States in 1970. Units are millions (10^6) barrels per day or its equivalent. Smaller flow paths are omitted. (i) = imports, (d) = domestic. (Adapted from ref. 2–4.)

valuable enough to warrant the investment. References 2–2, and 2–3 raise these questions in connection with specific energy technologies. Figure 2–9 shows the energy flow path which should show a positive outcome for a system under consideration.

2.6.2. *Power Flow in an Industrial Economy*

Figure 2–10 (adapted from ref. 2–4) shows the source and destination of power in its various forms as used in the United States. Worthy of note is specifically the distribution of energy in the broad end-use categories. The minor elements are omitted from the figure for clarity. The data are for 1970, which is a historical benchmark year since the oil embargo followed shortly thereafter in 1973, and the year lies in the period of the beginning of the nuclear power age decline. From an examination of this figure (and the original data), one may infer energy utilization efficiencies in the various use sectors. Comparison with earlier and later data for the United States, and with data from other countries may be of interest for energy policy development.

PROBLEMS

1. A gas turbine operating between 300 and 1200 K produces power at half the Carnot efficiency. Find the specific fuel consumption if CH_4 is used as a fuel.
2. A refrigerator operates between 0°C and 40°C. Determine the reversible COP.
3. Determine the heat rate in BTU per kW-hr, and the specific fuel consumption of an engine using Diesel fuel at an efficiency of 30%.
4. Determine the reversible sink efficiency of liquid nitrogen and liquid hydrogen operating in a 288 K environment.

REFERENCES

2–1. Bejan, A., "A Second Look at the Second Law," *Mechanical Engineering*, May 1988.

2–2. Chambers, R. S., Herendeen, R. A., Joyce, J. J., Penner, P. S., "Gasohol: Does It or Doesn't It Produce Net Energy?" *Science*, **206**, November 16, 1979.

2–3. Da Silva, J. G. *et al*, "Energy Balance for Ethyl Alcohol Production from Crops," letters in *Science*, **201**, September 8, 1978.

2–4. Joint Committee on Atomic Energy, "Certain Background Information for Consideration when Evaluating the National Energy Dilemma," U.S. Government Printing Office, Washington D.C., 1973.

3

POWER SYSTEMS

The power system intended for a particular use must meet a number of objectives in a satisfactory manner. These include the ability to meet the users' needs as they vary in time and being able to deliver the power in an economic manner. This chapter addresses consideration of these aspects of a system design for a utility type of power system and for a power plant designed to be used in a transportation application. This latter kind of power plant must make it possible for the transport vehicle to meet a number of additional performance goals.

3.1. TIME VARIATION OF DELIVERED POWER

Since the user of electrical power is likely to have a time varying demand, the supplier must optimally match this demand. Utility power systems are generally large enough to supply many users with different variations of demand. This fact allows the supplier to deliver an amount of power which is an average of the needs of many customers. An individual's need is very erratic in time, but the averaged need of a large number of users is fortunately much smoother. Figures 3–1 and 3–2 show the electrical power delivered by Seattle City Light (ref. 3–1), a representative electric utility company, during typical summer and winter weeks.

Data of this type may be presented in a way that shows the maximum power that is delivered and the fraction of time over which varying power levels are delivered (Fig. 3–3). The overall time period over which such a presentation is made may be days, weeks, or a year, often expressed in percentage of the time period. The length of time is chosen to suit the purpose of the discussion. The area under this load curve is the energy delivered by the system. Since the system must have a capacity to deliver the maximum power demanded, the deliverable energy is the area under the capacity line. The ratio of these two energies is defined as a load factor (see eq. 3–5).

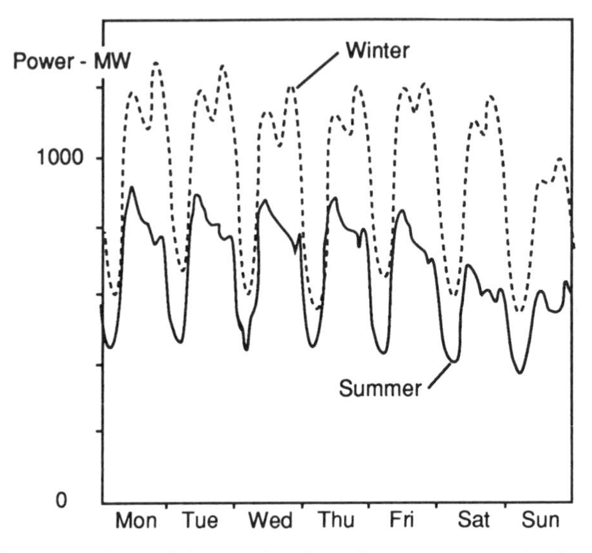

FIG. 3–1. Daily power demand for an electric utility, summer and winter. (Data from ref. 3–1.)

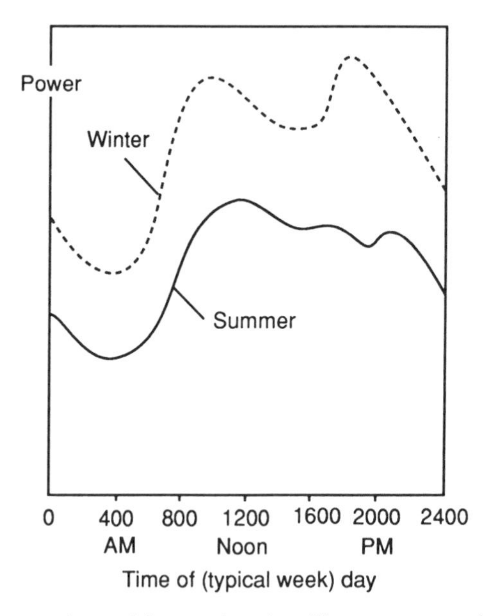

FIG. 3–2. Weekly power demand for an electric utility, summer and winter. (Data from ref. 3–1.)

3.2. ECONOMICS

For a plant that uses fuel to generate heat and delivers electrical power, the cost of the fuel and the physical plant are the major contributors to the cost of the power delivered. On the basis of these costs, the cost of the electricity is

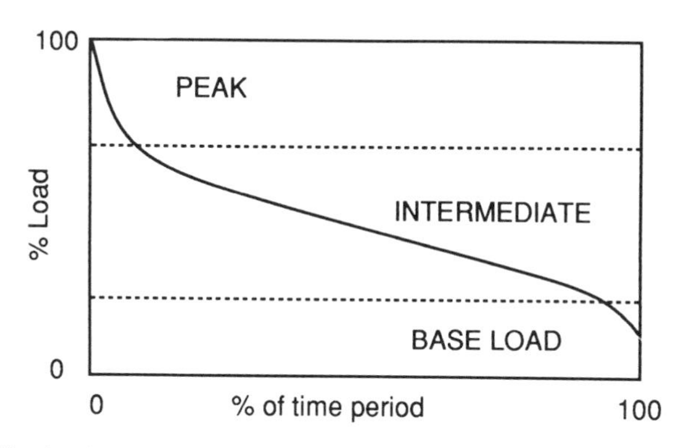

Fig. 3–3. Utility load curve and terms describing the power delivered over various time scales.

called "bus-bar" cost. The consumer also pays for the cost of distribution, taxes, management, and profit. The bus-bar cost is usually measured in $/kW$_e$-hr.

Bus-bar cost = fuel cost + plant cost + operating cost

$$C_{bb} = C_f + C_c + C_o \tag{3–1}$$

Neglecting the operating cost allows one to focus on the principal issues in understanding the relationship between the economics and the performance characteristics of the system for delivering an upgraded form of energy for sale. For accounting purposes one may let the fuel cost (in $/kW$_t$-hr) be:

$$C_{tf} = \text{cost of thermal fuel energy} \tag{3–2}$$

The plant thermal efficiency is the ratio of electrical power output to thermal power input (see eq. 2–19):

$$\eta_{th} = \frac{\dot{W}_e}{\dot{Q}_{in}} \tag{3–3}$$

The fuel cost component of the electric power cost ($/kW$_e$-hr) is therefore

$$\text{Fuel power cost} = C_f = \frac{C_{tf}}{\eta_{th}} \tag{3–4}$$

The plant load factor is the ratio of energy delivered to the energy deliverable (see Fig. 3–3).

$$\text{LF} = \frac{W_e}{\dot{W}_{e,\text{Cap}}\,\tau} \tag{3–5}$$

Here $\dot{W}_{e,\text{Cap}}$ is the rated maximum power output of which the plant is capable and τ is the time period of interest. In order to estimate the plant cost, one sums all components, assembled to enable start of operation, $C_{c,\,Total}$ (in \$). The capital cost of a plant includes:

Heat generation equipment: nuclear reactor or combustor
Boiler or heat exchangers
Condenser or cooling heat exchangers
Rotating fluid machinery: pumps, compressors and turbines
Electric power generator
Control system
Emissions control equipment
High voltage transformers and distribution equipment
Starting and standby power equipment
Buildings and site

It turns out in practice, that $C_{c,\,Total}$ is roughtly proportional to the capacity of the plant. The reason is that the capacity determines the size and therefore the cost of plant components. As a consequence of linear scaling, the ratio of

$$\frac{C_{c,\,Tot}}{\dot{W}_{e,\,Cap}} = C_{c,\,w} \tag{3-6}$$

(in \$/kW$_e$) is a convenient parameter to measure the capital cost of the plant. One would expect that this parameter would depend on the kind of power plant, its design and operating parameters and, to a small extent, its size. Typically, the benefits of large-scale plants arise from a smaller value of this cost parameter.

The *type* of system plays a very important role in determining the capital cost parameter: a typical automotive power plant might have $C_{c,\,w}$ near \$20/kW, a gas turbine might exceed \$200/kW, and a steam power plant \$2000/kW.

In addition to the simple cost of the equipment involved, there is also the economic cost of the capital required. This is particularly important in the construction of central station power plants where the economic lifetime is long, the capital needs are large, and the time to finished construction can be many years. These and other factors make the economics of power plants very complex and an integral part of system design. Hill (ref. 3–2) discusses some of the cost details which apply to the construction of central station power plants since 1960. Figure 3–4 (based on ref. 3–3) is not current data but aims to show the relative costs associated with plant complexity and the effect of scale.

A very simple incorporation of economic consideration is to calculate the capital cost escalation assuming the borrowed capital is paid back during the life of the plant. Thus one might borrow the amount $C_{c,\,Tot}$ from a source which would charge an interest rate i, per year. The amount that would

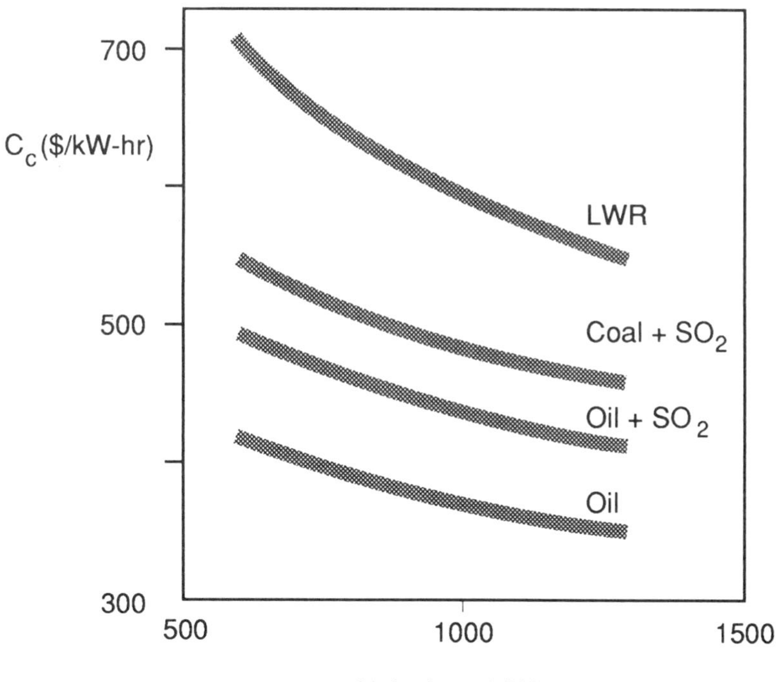

FIG. 3–4. Capital cost parameters for electric power plants. Cost basis: 1981 startup, natural draft cooling towers, and 2 identical units. Reference to SO_2 is an inclusion for its control.

have to be paid to amortize the loan per hour during each of the N years (8766 hours per year) of life is $I_N\ C_{c,Tot}/8766$ where I_N is a capital escalation factor associated with repayment of borrowed principal,

$$I_N = i\frac{(1+i)^N}{(1+i)^N - 1} \qquad (3\text{–}7)$$

Here N is the payback period in years, which one may take to be the "economically useful" life of the plant. In this formulation, this life includes also the construction period. Note that the factor NI_N is the number of times the borrowed capital has to be paid back, an amount I_N during each year. The factor I_N becomes proportional to the interest rate, i, for large N, so that the cost is almost proportional to the interest rate. For vanishingly small interest rate, I_N approaches $1/N$.

The capital cost of the plant expressed in terms of cost per unit energy delivered can therefore be written,

$$\text{Capital component cost} = \frac{\text{capital cost/hour}}{\text{electrical energy delivered/hour}}$$

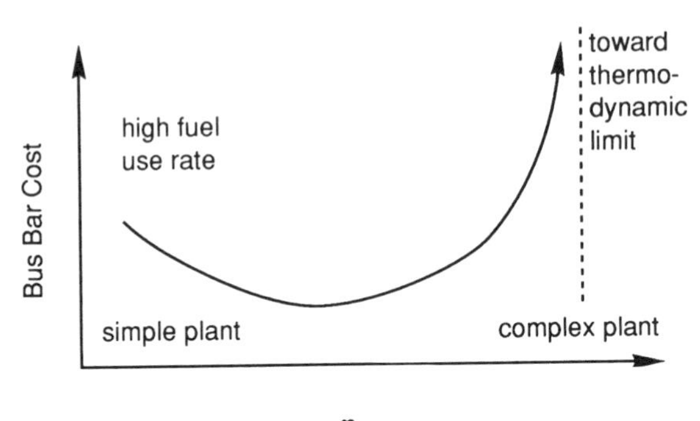

FIG. 3–5. Role of plant thermal efficiency on bus bar electric energy cost.

or

$$C_c = \frac{C_{c,\mathrm{Tot}}}{\dot{W}_{e,\mathrm{Cap}}} \frac{1}{LF} \frac{I_N}{8766} = \frac{C_{c,w}}{LF} \frac{(NI_N)}{8766\,N} \tag{3–8}$$

since the energy delivered per hour is $W_{e,\mathrm{Cap}} \times LF$. The bus-bar cost of electricity is therefore

$$C_{bb} = \frac{C_{f,t}}{\eta_{th}} + \frac{I_N\, C_{c,w}}{8766\, LF} \tag{3–9}$$

In economically sound designs, this quantity is near a minimum (see Fig. 3–5). The equation shows the relationship among efficiency, fuel type (reflected in $C_{f,t}$), and complexity of the plant machinery which is reflected in $C_{c,w}$. For a given fuel type, it is desirable to make the plant more efficient by making it more complex (and costly), but only to the point where the plant cost increase causes the bus-bar cost to rise above the minimum. The break-even value of efficiency is obtained by setting the derivative $dC_{bb}/d\eta_{th} = 0$.

3.2.1. *Waste Heat Utilization*

Thermodynamically speaking, it is worthwhile to take advantage of the waste heat generated by any heat engine for satisfying needs for low-grade heat. Selling this waste heat amounts to raising the ratio of useful power (formerly electric only) to thermal power from the fuel. Note that raising the minimum cycle temperature of the heat engine in order to make the "waste" heat useful lowers the classical thermal efficiency of the engine because the spread in temperature limits between which the engine operates is reduced.

If waste heat, at say 350 K, is useful, then perhaps as much as 70–80% of the fuel's heat may be sold. This reduces the fuel cost, but the cost of distribution of this low-grade heat is generally very high. The reason is that the mass of material (usually water) which must be transported to the user is large, which leads to friction losses and heat losses in the piping. Most important is the

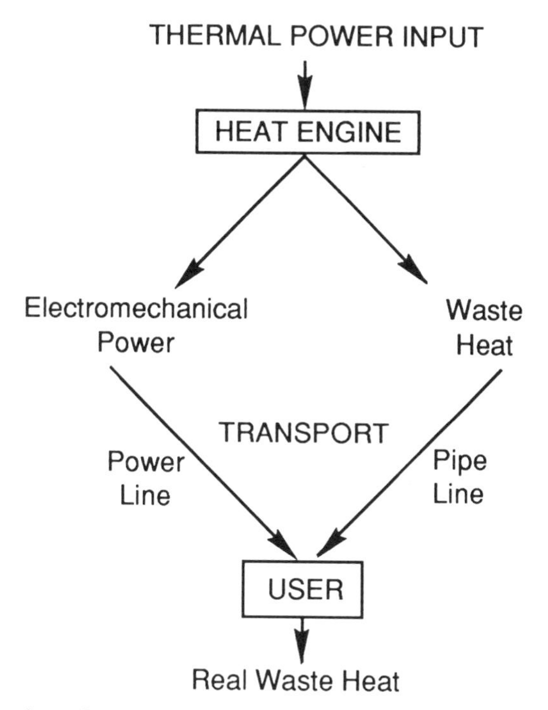

FIG. 3–6. Utilization of waste heat is difficult because of the material bulk involved.

requirement for an expensive distribution network which increases $C_{c,w}$. Currently such transport of low grade heat can be economic only where the users are located close to the thermal source. Figure 3–6 illustrates the penalty or benefit associated with the use of waste heat from a power plant. In a number of applications, so-called combined power systems that deliver both electric and thermal power are in use. Such systems are typically compact and may be used in ships, large buildings, and so on. Horlock (ref. 3–4) gives a detailed engineering economic discussion of this topic.

Consider the transport of 1000 MW_t of heat in 350 K water ($\Delta T = 50$ K) which has to be carried by 4800 kg/sec of water. This requires a pipe 1 m in diameter when the water flows with a velocity of 6 m/sec.

A competitive alternative has been the employment of heat pumps at the user's end, utilizing environmental heat (see Section 2.5). This is equivalent to the use of the environment as the physical transport mechanism to bring heat to the user.

An interesting dimension of power system economics arises in connection with user-generated power (such as from wind, for example), when a primary power system serves as a backup or as the recipient of excess power. The issue is that the owner of the wind generator should not expect to receive the same retail price for power he produces as that from the utility because part of the price of power delivered includes the delivery system and the utility provides insurance that power will be available at all times. The local generator must sell energy at a lower rate to the backup utility than the utility charges for energy it delivers. This must be resolved by a fair accounting for the various cost elements and is an issue that discourages interties between user and utility systems.

3.3. SYSTEM MASS: POWER DENSITY PARAMETERS OF HEAT ENGINE SYSTEMS

The mass of an energy conversion system consists of the mass of the heat engine and the mass of the fuel required to do the task assigned. This total mass is important in determining the cost of carrying out the task, particularly in transportation systems. In the following, the parameter describing the compactness of a power system will be developed, highlighting in particular the role of efficiency and specific work of the thermodynamic cycle used.

3.3.1. *Engine Mass*

Consider a steady flow heat engine whose power output is given by the specific work, w, and the mass flow rate of working fluid (considered to be ambient air, subscript 0) processed, $\dot{m}$,

$$\dot{W} = \dot{m} C_p T_0 w \tag{3–10}$$

Here the product of $C_p T_0$ is included to nondimensionalize this definition of the specific work, w. This nondimensional parameter is a measure of compactness of the machinery and may be obtained for any cycle from thermodynamic considerations, as is shown in Chapters 7, 8 and 9.

The mass flow rate, $\dot{m}$, can be written in terms of a characteristic velocity, u (which is a fraction of the sonic speed in aerodynamic machines or a piston velocity in a displacement machine) and a characteristic size (inlet flow area or piston area, see Fig. 3–7, ref. 3–5):

$$\dot{m} = \rho_0 u \frac{\pi}{4} D^2 \approx \rho_0 u D^2 \tag{3–11}$$

The power output may therefore be written as

$$\dot{W} = u D^2 \left[\frac{\gamma}{\gamma - 1} p_0 w \right] \tag{3–12}$$

Here p_0 is the environmental pressure that follows from the ideal gas law. In the internal combustion engine community the square brackets term is referred to as the "mean effective pressure." The specific heat ratio, γ, of the

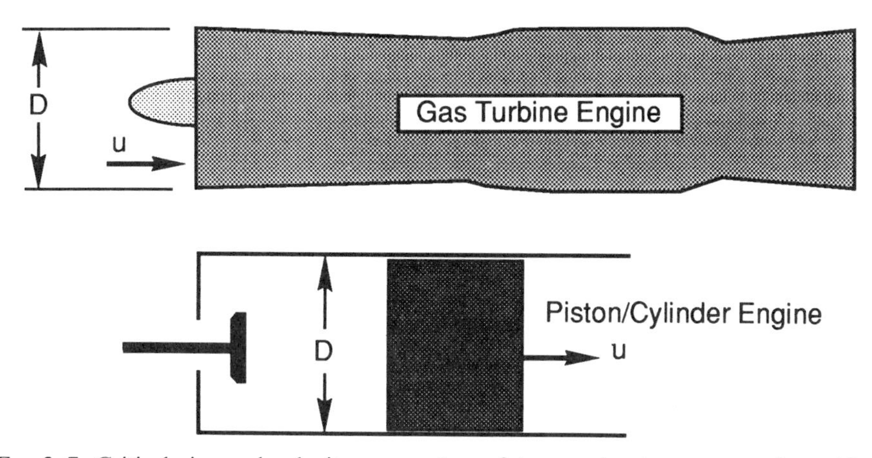

FIG. 3–7. Critical size and velocity parameters of two engine types: an engine with aerodynamic compressor (gas turbine) and a displacement engine.

working fluid arises from the relation between C_p and R which is described in Chapter 4.

Consequently the characteristic size, D, of an engine required to produce a required power output and a realistically limited flow speed, u, is

$$D = \sqrt{\frac{\gamma - 1}{\gamma} \frac{\dot{W}}{p_0} \frac{1}{w} \frac{1}{u}} \tag{3–13}$$

Within a class of machines of similar type, D may be used to determine the mass of the engine when the material density, ρ_m, of the engine is assumed known. Thus

$$m_E = \rho_M D^3 = \rho_m \left(\frac{\gamma - 1}{\gamma} \frac{\dot{W}}{p_0} \frac{1}{w} \frac{1}{u} \right)^{3/2} \tag{3–14}$$

This mass is seen to vary with the 3/2 power of a number of variables, the most important of which are the specific work, w, and the limiting speed, u.

3.3.2. *Fuel Mass*

The mass of fuel used depends on the heating value of the fuel, H (J/kg or BTU/lb), the efficiency of the cycle which is obtained from the thermodynamic analysis, and the duration of the power production process, τ_m. Thus

$$m_F = \frac{\dot{W} \tau_m}{H \eta_{th}} \tag{3–15}$$

3.3.3. *System Mass*

Combining the fuel and engine mass, $m_{Tot} = m_F + m_E$, and nondimensionalizing result in the following expression for the total system mass, or, with eqs. 3–14 and 3–15:

$$\frac{m_{TOT}}{\dot{W}} = \left(\frac{1}{\eta_{th}} + \frac{C_{S/M}}{w^{3/2}}\right)\frac{\tau_m}{H} \tag{3–16}$$

which illuminates the roles of efficiency and specific work in a total system design. Here $C_{S/M}$ is a system/mission parameter consisting of:

$$C_{S/M} = \rho_m \frac{\sqrt{\dot{W}H}}{\tau_m}\left(\frac{\gamma - 1}{\gamma}\frac{1}{p_0 u}\right)^{3/2} \tag{3–17}$$

In eq. 3–16, note in particular the nonlinear way in which the two performance indices combine to give minimum system mass. In general, one may conclude that a thermodynamic performance optimum between maximum w and maximum efficiency is desirable. The weighting is provided by a relation such as the one developed here. The parameter $C_{S/M}$ combines the mission objectives ($\dot{W}$ annd τ_m) and the particular engine characteristics (u and ρ_m). For a Diesel engine producing 100 kW for 4 hr and a gas turbine producing 10,000 kW for a similar period, the values of $C_{S/M}$ are of the order of 6 and 0.3, respectively, showing how both elements contribute to the total mass.

3.3.4. *Part Power Operation: Fuel Heat and System Cost*

The role of the load curve and the part-load efficiency may be examined from the viewpoint of minimizing the fuel heat requirement or minimizing system cost.

Energy conversion systems operation with a time-varying demand for power output. It is usually possible to identify a time scale over which the power demand variation is repeated. This may be a day, a week, or a typical journey of a transportation vehicle. For our purposes, it will be assumed that this time constitutes a "power production cycle time." In the discussion to follow, all fractional times will be of this cycle time. For purposes of modeling engine performance, it is assumed that this time scale is very long compared to heat engine fluid process times so that quasi-steady operating conditions may be taken to apply.

Typical temporal variations (load curve) of power production by heat engines are described in Fig. 3–3. The part load performance problem is to optimize the system design, considering that the system with a fixed configuration and size must be operated at reduced power output for a fraction of the cycle time. One design goal might be to minimize the fuel utilized.

The variation of efficiency with power level depends on the nature of the cycle. This is quantified for the simple Brayton cycle in Chapter 16. Typically, the efficiency versus percent load of a heat engine follows a relationship such

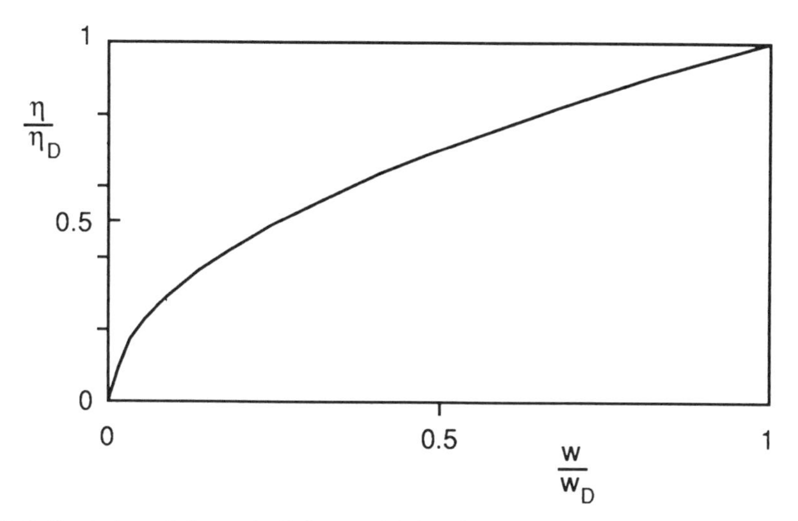

Fig. 3–8. Variation of thermal efficiency with fractional power output: $\eta/\eta_D = (w/w_D)^n$. Here $n = 0.5$.

as the examples shown in Fig. 16–1, although some cycles have an increasing efficiency with reduced power. The Diesel cycle has this characteristic as shown in Chapter 8. The application dictates that the power delivered versus (percent) time be as expressed in a load curve shown in Fig. 3–3 or in Fig. 3–8 which gives the data for the example to follow.

Consider the problem of minimizing the thermal resources to be used for supplying a time-varying load. The total thermal power used is:

$$\int_0^1 dQ = \int_0^1 \dot{Q}\, dt \tag{3–18}$$

where the time-span of interest is unity. This may be related to the power delivered

$$\dot{Q} = \frac{1}{\eta} \dot{W} \tag{3–19}$$

where η is the thermal efficiency of the heat engine. The time-varying fractional power is

$$f = \frac{\dot{W}}{\dot{W}_{max}} \tag{3–20}$$

where $\dot{W}_{max}$ is the maximum available power of the engine. The total thermal energy requirement during the unit time is therefore:

$$Q_{TOT} = \dot{W}_{max} \int_0^1 \frac{f(t)}{\eta(f)}\, dt \tag{3–21}$$

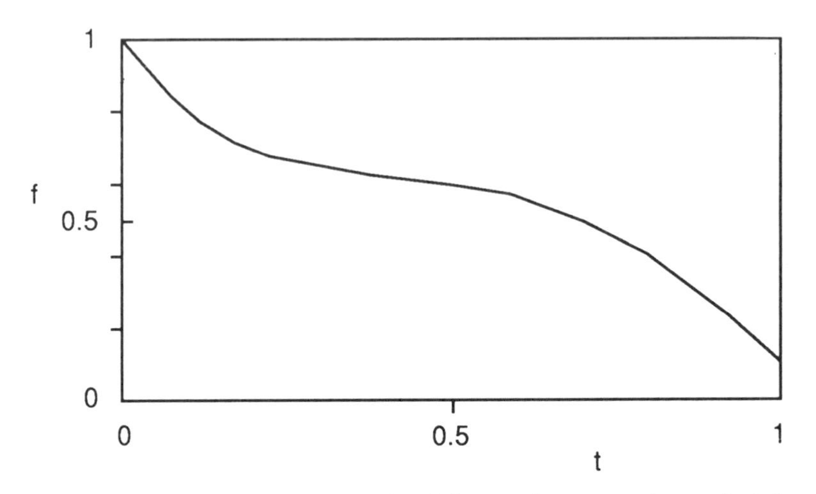

FIG. 3–9. A representative load curve. A minimum of y power is required for x% of the time.

Examples of $\eta(f)$ and $f(t)$ are illustrated in Figs. 3–8 and 3–9, respectively. The role of part power efficiency may be illustrated parametrically through the value of n in

$$\frac{\eta}{\eta_{\max} w} = \left(\frac{w}{w_{\max}}\right)^{n} = f^{n} \tag{3–22}$$

where $n = 0.5$ is plotted in Fig. 3–9.

For some purposes, it is appropriate to minimize Q_{TOT}. For example, if $f(t) = 1$, then the device operates at maximum power continuously and the minimum thermal requirement is achieved with η as large as possible.

For the general case, minimizing the total heat requirement amounts to minimizing the area under the curve formed by the ratio of the two curves as shown in Figs. 3–9 and 3–10. Figure 3–10 shows the ratio of $\eta(f)$ and $f(t)$ for 3 values of the exponent used for the efficiency. $n = 0.5$ is shown in Fig. 3–9, whereas $n = 0$ is the constant η case and $n = 1$ corresponds to a linear variation that is indeed quite poor. These curves are obtained by taking various time intervals and finding the (interpolated) value of η that corresponds to the power level during this interval. Under some circumstances an analytical approach may work. In this example, the fuel consumption rate is high for small fractional times (because of the high power required) and then falls to intermediate values. The area under the curves represents the heat resource used in the time span 0–1. Consequently, one may conclude that both the user requirements, $f(t)$, and the part-load characteristics, $\eta(f)$, are important for the design of economically sound energy systems.

The problem of minimizing the cost of an energy system over its lifetime or over a duty cycle requires at the very least a way of relating system cost to efficiency. To show how to arrive at a system configuration, one may take the

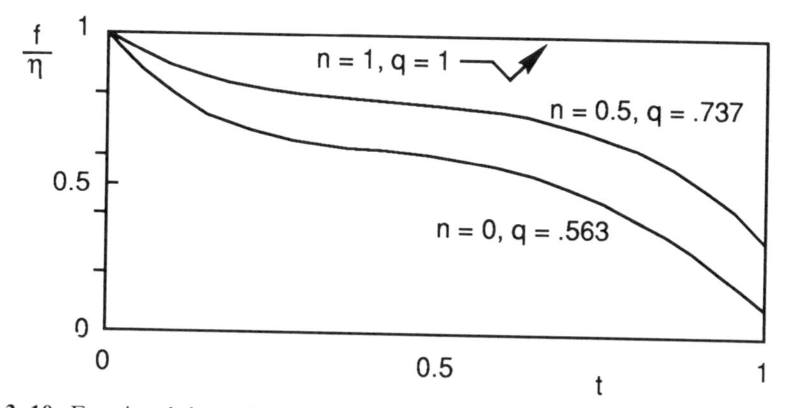

FIG. 3–10. Fractional thermal resource requirement. Plot is of the integrand of eq. 3–21 for data given in Figs. 3–9 and 3–10. The total heat required, Q_{TOT}, is the area under the curve and is indicated for the three cases.

view that, near some design point, the cost increases with increasing design point efficiency. The cost of the energy product is then minimized when the sum of system and thermal resource cost is minimized:

$$\$ \sim F_1(\eta_D) + \frac{1}{\eta_D} \int_0^1 f \left(\frac{\eta}{\eta_D} \right)^{-1} dt \tag{3–23}$$

The (increasing) function F_1 of η_D and the variation of η with load allows an optimum design to be identified when the user characteristics are specified through f.

3.4. VEHICULAR PERFORMANCE: ENERGY AND POWER

The performance of transport vehicles depends to a large extent on the characteristics of the energy supply carried on the vehicle and the power source used to develop the thrust necessary to overcome the resistive force experienced as a result of motion. The resistive force is often suitably described as the sum of one component that is independent of vehicle speed and another that depends on the square of the vehicle speed. For example, vehicles that are in contact with the ground, such as automobiles or trains, experience rolling friction, which is roughly independent of speed and aerodynamic drag. On the other hand, vehicles that are supported by a fluid, such as airplanes or ships, experience only the fluid-mechanical drag. In the following, the role of energy and power in determining range and acceleration is examined for two types of transportation vehicles.

3.4.1. *Power Density of Chemical Energy Conversion System*

The mission time plays an important role in determining the optimal type of power system. Its mass includes primarily the mass of the thermal or

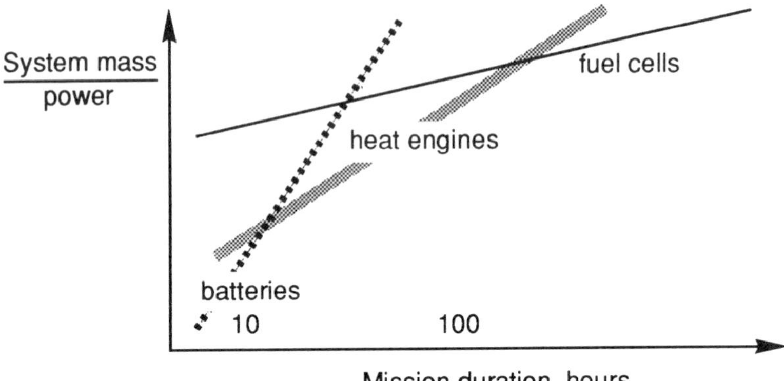

FIG. 3–11. Specific mass characteristics for a number of power systems.

chemical energy source and the mass of the conversion machinery. Thus one may write, as before in eq. 3–16,

$$\text{power system total mass} = \frac{\text{power}}{\text{converter power density}} + \frac{\text{energy}}{\text{energy density}}\,\frac{1}{\text{energy efficiency}}$$

The density terms are either power/mass or (thermal or chemical) energy/mass which are parameters that are, to some extent, scale independent, but vary among types of fuel or conversion machinery. Denoting the power density as α and the fuel energy density as H, the mass per unit power of the system is (eq. 3–16):

$$\frac{m_{\mathrm{TOT}}}{\dot{W}} = \frac{1}{\eta_{\mathrm{th}}}\frac{\tau_{\mathrm{m}}}{H} + \frac{1}{\alpha} \tag{3–24}$$

where τ_{m} is again the mission duration time. This reciprocal power density parameter can be plotted versus mission duration time. It is generally desirable to have as low a mass as possible in order to maximize the range of a vehicle as will be shown below. For a comparison of battery, heat engine, and fuel cell power systems, the general variation of "specific mass" is shown in Fig. 3–11. Evidently, batteries with their low energy density, but high power density, are suitable for short missions. Fuel cells, on the other hand, are best for long missions because of their high effective energy density (high efficiency), but relatively low power density which is the result of a limited electric current per unit area. A discussion of batteries and fuel cells as conversion devices may be found in the follow-on volume to this book. Heat engines fall between these extremes. The important ones of these are discussed in Chapters 7–10 and 16.

In the following, the performance of a transport system is examined from the viewpoint of highlighting the importance of the parameters identified. For a given system, the mass of the energy conversion machinery is invariant so that its mass may be included in the empty weight of the vehicle.

3.4.2. *Range and Energy Use*

For a vehicle in ground contact (Fig. 3–12) moving with a steady speed, the resistive force balanced by a propulsive force is

$$F = \text{rolling friction} + \text{drag} = \mu W + \tfrac{1}{2}\rho_0 V^2 C_d A \qquad (3\text{–}25)$$

where V is the speed of the vehicle, ρ_0 is the air density, μ the resistance coefficient, and W the vehicle weight ($=\text{m}\ \text{g}_0$). The terms C_d and A are the drag coefficient and frontal area respectively.

Most vehicles in common use have propulsion force requirements to overcome forces that vary with velocity to various powers (V^n), with n ranging from 0 for pure rolling friction to 2 for aerodynamic drag, and higher under some circumstances. The forces may also vary with the weight of the vehicle to various powers. The rolling friction is proportional to weight whereas aerodynamic form drag is independent of it. An intermediate case is found in the drag experienced by a ship in water. Skin friction and wave drag are important here, and the skin friction portion is related to the wetted area which is

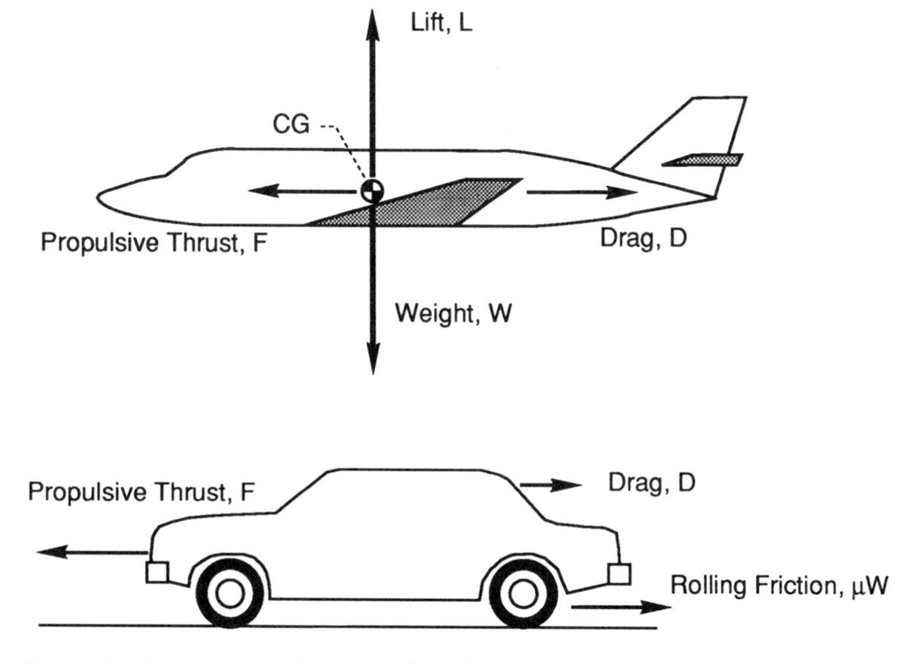

Fig. 3–12. Forces on vehicles propelled through a fluid (aircraft) and on land (automobile).

related to the displacement of the hull (and thus its weight) and the shape of the hull. The velocity and weight dependence of the force law differs among various transport vehicle types, although the approach that can be used to analyze the system performance is similar to that considered in the following.

As a first example consider an aircraft (Fig. 3–12) in steady, level, cruise flight, for which an equation similar to 3–25 applies to the force balance, with $\mu = 0$. For steady and level flight, a vertical force balance requires that

$$\text{Lift } (=L) = W = mg \tag{3–26}$$

For the airplane, the drag is related to the lift, L, through

$$D = \frac{W}{(L/D)} \tag{3–27}$$

where the lift/drag ratio, L/D, is carefully maximized for an airplane design which is configured to achieve a certain mission objective. The value of L/D thus typically varies near a maximum and thus can be assumed not to vary much, even though the airplane weight changes over the mission. The most common design objective is the attainment of a certain range, when flying at a prescribed speed. During the cruise segment of the flight, the airplane gross weight will decrease from the take-off value (Take-off gross weight, TOGW $= W_0$), reaching the weight of $W_0 - W_F$, where W_F is the weight of the fuel burned:

$$\begin{aligned}\text{initial weight} &= W_0 \\ \text{final weight} &= W_0 - W_F\end{aligned}$$

The fuel consumption rate of the engine(s) whose thrust (F) equals the drag (D) is well approximated by stating that the rate is proportional to the power generated. This allows writing the fuel consumption rate in terms of specific fuel consumption (SFC) defined by:

$$\text{SFC} = \frac{\dot{m}_f\, g}{FV} \tag{3–28}$$

The product FV is the power delivered to the vehicle. The fuel consumption rate equals the rate of airplane weight decrease:

$$\dot{m}_f\, g = -\frac{dW}{dt} \tag{3–29}$$

The fact that SFC and L/D are carefully maximized by design, allows one to say that these parameters will not change significantly during a mission; that is, they may be assumed constant near their minimum and maximum, respectively. Furthermore, the velocity that the vehicle uses to cruise is often dictated by other considerations (such as speed limits, traffic, severe drag increases, etc.) so that the velocity may also be assumed constant during the mission.

The range of the vehicle is given by

$$R = \int_0^{\tau_m} V\, dt \tag{3–30}$$

where the time increment is obtained from eq. 3–29, so that

$$R = -\int_0^{\tau_m} \frac{V}{\dot{m}_f g}\, dW \tag{3–31}$$

Solving eq. 3–28 for $\dot{m}_f$ gives,

$$R = -\frac{1}{\text{SFC}}\frac{L}{D}\ln\left(1 - \frac{W_F}{W_0}\right) \tag{3–32}$$

This is the Breguet range equation. Before interpreting this fundamental vehicle performance equation, it is useful to derive the same expression for a second example, the land vehicle, where the rolling friction term plays a more important role.

From eq. 3–25, the power produced by the propulsion system is

$$FV = (\mu W + D)V \tag{3–33}$$

This gives the fuel consumption rate and the range;

$$\dot{m}_f g = (\text{SFC})FV = (\text{SFC})V\mu\left(W + \frac{D}{\mu}\right) \tag{3–34}$$

$$R = -\frac{1}{\text{SFC}}\frac{1}{\mu}\ln\left(1 - \frac{W_F}{(D/\mu + W_0)}\right) \tag{3–35}$$

From a comparison of eqs. 3–32 and 3–35, it appears that the resistance coefficient, μ, and the ratio, D/L, play identical roles in determining the range of these two vehicle types. The automobile has the additional burden that the influence of drag is to increase the apparent empty weight of the vehicle. The energy expended by the vehicle during a time interval dt is

$$dE = F\, dR = FV\, dt = \frac{\dot{m}_f g}{\text{SFC}}\, dt \tag{3–36}$$

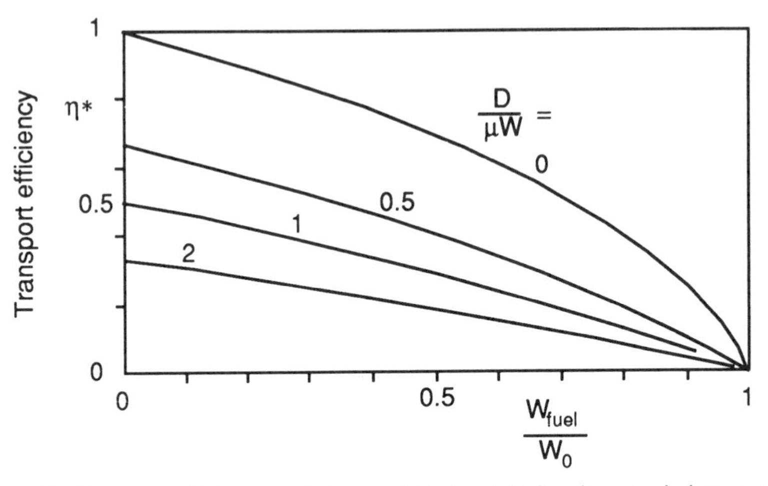

Fig. 3–13. Fransport efficiency variation with fuel weight fraction at mission start. Line for parameter = 0 applies to an airplane.

and the total energy used W_F/SFC. It is useful to compare the minimum work that could have been done to the energy actually expended.

One may define a transport efficiency (η^*) based on this ratio. The delivered weight is $(W_0 - W_F)$. The resistive force is therefore μ (or D/L here) times the delivered weight so that the minimum work done may be written as

$$\text{minimum work} = (W_0 - W_F)\left(\frac{L}{D}\right)^{-1} R$$

The ratio of the minimum work to energy is

$$\eta^* = \left(1 - \frac{W_0}{W_F}\right) \ln\left(1 - \frac{W_F}{W_0 + D/\mu}\right) \tag{3–37}$$

which is plotted in Fig. 3–13 for several values of $D/(\mu W_0)$. This calculation is generalized to include the automotive drag term if such a vehicle is considered. For the airplane, this term is absent and the efficiency depends only on the fuel/gross weight ratio. The figure shows the variation of this efficiency. Evidently, it reduces efficiency to be in a hurry since the drag, D, is proportional to V^2. Note that the transport efficiency thus defined does not distinguish between payload and empty weight, which requires that an additional parameter be specified if the measure of system quality is to be an efficiency based on the work required to deliver only the weight of the payload.

It should be noted that implicit in the ratio W_F/W_0 is the *energy density* of the fuel used during the mission.

$$\frac{W_F}{W_0} = \frac{(\text{energy} = \text{work done})}{W_0 * \text{energy/mass}} \tag{3–38}$$

In the limit of very large energy density, the mission is accomplished with little change in mass, so that the minimum energy is used. In other words, the inefficiency is in transporting the last drop of fuel over the entire mission. From this point of view, the ideal fuel is one with high energy content per unit mass. The electric train and battery vehicles have a high transport efficiency.

3.4.3. *Acceleration and Power*

The minimum power required in a vehicle is that necessary to provide the thrust force at cruising speed. If a vehicle has just that amount available, the time required to accelerate to cruising speed is usually unacceptably long. In practice, a power plant is designed to be able to produce z times the power required to cruise. This is done for a variety of reasons, including traffic safety for automotive vehicles, short runway lengths for aircraft, among others. To see how this power ratio z is related to the time for acceleration, the equations of motion are examined, with a constant weight vehicle. It is assumed that the time scale is sufficiently short for fuel used to be a small fraction of the vehicle gross weight. Thus the vehicle weight is constant during the acceleration process. The cruise power is P_c, and the thrust available to accelerate the vehicle is

$$F_c = \frac{P_c}{V_c} = D_c + \mu W \tag{3–39}$$

Here, contrary to convention, P is used in lieu of $\dot{W}$ to avoid confusion with the symbol for weight. The weight is W, and the cruise speed is V_c. In general, the thrust (at maximum power) is the ratio of the maximum power delivered by the varying speed,

$$F = \frac{P_{max}}{V} = z\frac{P_c}{V} \tag{3–40}$$

The acceleration is obtained from Newton's law,

$$\frac{W}{g}\frac{dV}{dt} = F - D - \mu W \tag{3–41}$$

The drag, D, is related to the cruise drag by:

$$D = D_c v^2 \tag{3–42}$$

where v is defined as the ratio

$$v = \frac{V}{V_c} \tag{3–43}$$

which varies between 0 and 1. Equation 3–41 can be rewritten (using eqs. 3–39 to 3–43) as

$$\frac{V_c}{g}\frac{dv}{dt} = K\left[\frac{z}{v} - v^2 - \frac{\mu}{K}(1 - v^2)\right] \tag{3–44}$$

which is a differential equation for v as a function of time. The nondimensional parameter K is constant and defined by

$$K = \frac{P_c}{WV_c} \tag{3–45}$$

The time required to reach cruising speed ($v = 1$) is obtained by the integration of this equation. Letting this time be τ_a, one finds that τ_a is scaled by V_c/gK and is a function of the nondimensional parameters z, μ, and K:

$$\tau_a \frac{gK}{V_c} = f(K, \mu, z) \tag{3–46}$$

The time scale parameter (V_c/g_0K) may be interpreted as (twice) the ratio of the kinetic energy at cruise to the power required at cruise since,

$$\frac{V_c}{gK} = 2\frac{\frac{1}{2}\frac{W}{g}V_c^2}{P_c} \tag{3–47}$$

Figure 3–14 shows the variation of the nondimensional acceleration time (eq. 3–46) to cruise speed (expressed as $\tau_a gK/V_c$) versus the power parameter, z, and the resistance factor, μ/K, which is zero for the aircraft. By way of realistic values, a unit value of τ_a is about 30 sec for an automobile, and about ten times that value for a commercial jet aircraft. Evidently z of the order 2–3 gives a significant reduction in acceleration time. For larger z, the benefit becomes marginal.

3.5. SENSITIVITY ANALYSIS AND UNCERTAINTY

The analysis of a complex system depends on a variety of inputs determining a result. It is often useful to know which inputs play an important role in determining that result and furthermore, given that any input is in some way estimated, one must know the "uncertainty" built into the result by the calculation process. This allows one to concentrate one's efforts and resources to reduce the uncertainties of the inputs that have the greatest influence on the result and narrow its uncertainty.

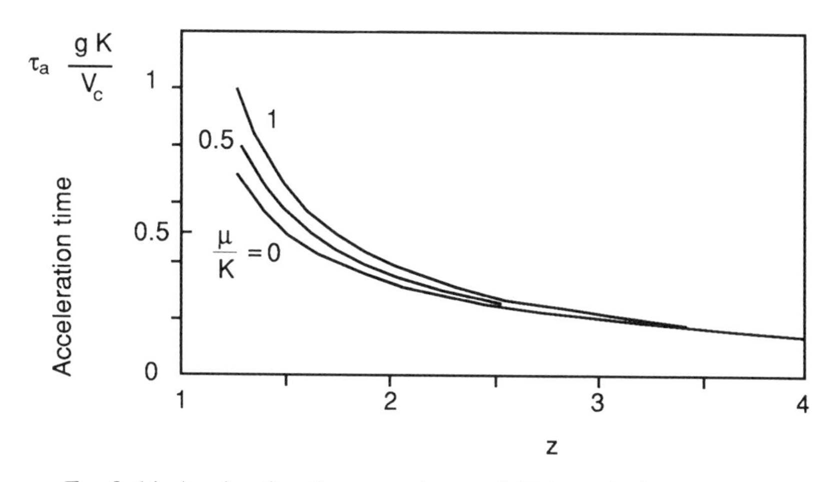

FIG. 3–14. Acceleration time to cruise speed. Values of μ/K are noted.

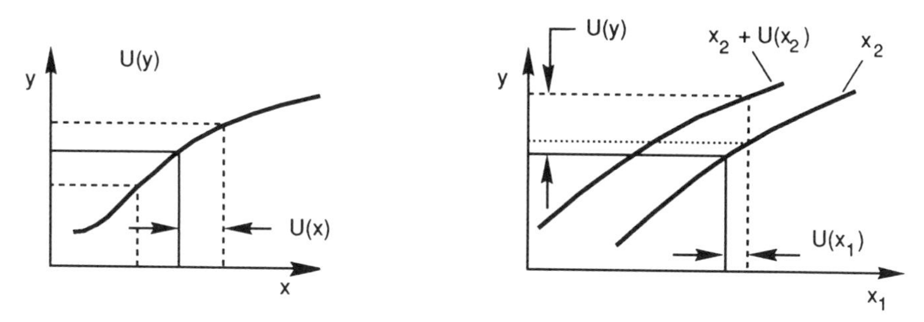

FIG. 3–15. Uncertainty in y due to uncertainty in x or in x_1 and x_2.

3.5.1. *Influence Coefficients*

Let the calculated result be y and the required inputs be x_i, all of which are independent. In mathematical terms the calculation process may be stated (see Fig. 3–15)

$$y = f(x_1, x_2, x_3, \ldots, x_N) \tag{3–48}$$

The values of the independent variables are almost always uncertain to some extent. This means that in a statistical sense, x_i is known to be

$$x_i = \text{best value of } x_i \pm \text{an uncertainty in } x_i$$

or

$$x_i = x_i \pm U[x_i]$$

The resulting dependent variable, y, is consequently uncertain by some amount, $U[y]$. When y depends on a single x, the sensitivity of y on x is given by the derivative, dy/dx, or the partial derivative, $\partial y/\partial x$, when there is more than one independent variable. It follows that the uncertainty in y is related to the uncertainty of a single parameter x_i by

$$U[y] = \frac{\partial y}{\partial x_i} U[x_i] \tag{3–49}$$

It is often useful to express the uncertainty in any variable as a fraction or, equivalently, a percentage, that is,

$$U\%[y] = \frac{U[y]}{y} \tag{3–50}$$

The uncertainty relation above can then be written in the following way:

$$\frac{U[y]}{y} = \frac{x_i}{y}\frac{\partial y}{\partial x}\frac{U[x_i]}{x_i}$$

or

$$U\%[y] = I[y, x_i]\, U\%[x_i] \tag{3–51}$$

where the influence coefficient $I[y, x]$ is the logarithmic derivative of y with respect to x:

$$I[y, x] = \frac{dy/y}{dx/x} = \frac{d\ln(y)}{d\ln(x)} \tag{3–52}$$

The influence coefficient is a measure of the sensitivity of y on x which is equal to the value of the exponent n in a relation between y and x of the form,

$$y = x^n \qquad \text{for which } I[y, x] = n$$

Obviously, $I[y, x]$ equal to $+1$ means a proportional relationship, $I[y, x]$ equal to -1 is an inverse relationship and $I[y, x] = 0$ implies no relationship. Note that this value of $I[y, x]$ is a "local" one, meaning that for a different set of parameters, the numerical value of $I[y, x]$ might be different. Further, if a minimum or maximum in y has been found for a value of $x = x_0$, then $I[y, x_0]$ is necessarily zero. The greatest care for determining uncertainties of independent variables should be employed for those variables that have the largest influence coefficients at the condition of interest because these particular variables can contribute the largest uncertainty in y.

3.5.2. *Uncertainty Analysis*

In most engineering analyses, the uncertainty is a small fraction of the value of the variable. One can therefore write for the expected value of y in terms of the estimated value y_0,

$$y - y_0 = (x_1 - x_{1,0})\frac{\partial y}{\partial x_1} + (x_2 - x_{2,0})\frac{\partial y}{\partial x_2} + \cdots N \text{ terms} \tag{3–53}$$

in terms of series expansion involving the possible departure of x_i from its estimated values. The difference, $x_i - x_{i,0}$, between the possible value and the estimated value is the uncertainty in x_i, which should be zero, but may not be. In a statistical sense then, the uncertainty will have a zero mean when averaged over a large number of measurements:

$$\overline{(x_i - x_{i,0})} = 0 \tag{3–54}$$

The square of the above expression can be written as

$$(y - y_0)^2 = (x_1 - x_{1,0})^2\left(\frac{\partial y}{\partial x_1}\right)^2 + \cdots + 2(x_1 - x_{1,0})(x_2 - x_{2,0})\frac{\partial y}{\partial x_1}\frac{\partial y}{\partial x_2} + \cdots \tag{3–55}$$

which, when averaged, becomes an expression written in terms of two groups of averaged terms. The first is squares of $(x_i - x_{i,0})$, and the second the product of terms like $(x_i - x_{i,0}) * (x_j - x_{j,0})$. Taking the mean of this expression leads to the disappearance of the second group because x_i and x_j are independent:

$$\overline{(x_i - x_{i,0})(x_j - x_{j,0})} = \overline{(x_i - x_{i,0})}\ \overline{(x_j - x_{j,0})} = 0 \tag{3–56}$$

The squared terms survive and the equation can be nondimensionalized by dividing by y (or equivalently, y_0). When higher-order terms are neglected:

$$\frac{\overline{(y - y_0)^2}}{y_0^2} = \left(\frac{x_{i,0}}{y_0}\frac{\partial y}{\partial x_i}\right)^2 \frac{\overline{(x_i - x_{i,0})^2}}{x_0^2} \tag{3–57}$$

Written this way, it appears that the equation is a relation between the uncertainties in x_i and those in the desired result y. Thus for more than one independent variable x_i, eq. 3–51 is modified to read:

$$U\%[y] = \sqrt{\sum I^2[y, x_i] * U\%^2[x_i]} \tag{3–58}$$

A very useful application for determining engine condition in service based on the idea of influence coefficients is described in refs. 3–6 and 3–7.

3.5.3. *Numerical Determination of Influence Coefficients*

The desired result, y, is usually calculated by means of an algebraically complex process for which it is difficult to obtain the derivative $\partial y/\partial x$ (or the logarithmic form) in an analytic way. Since computers are usually used to determine y, it is logical to determine the derivative numerically. A logical procedure would be to arrange the independent variables in such a way as to make it possible to vary each x by a small amount (of the order of 1 to 2%), say $x(1 \pm \varepsilon)$ and note the percentage change in y. An approximation to the influence coefficient is then obtained from

$$I[y, x] \cong \frac{x\left(1 + \frac{\varepsilon}{2}\right)}{y\left[x\left(1 + \frac{\varepsilon}{2}\right)\right]} \frac{y[x(1 + \varepsilon)] - y[x]}{\varepsilon} \tag{3-59}$$

Here the square brackets denote a functional relationship whereas the parentheses form an algebraic group.

3.5.4. *Graphical Presentation of Results*

Figure 3–16 shows the result of an influence coefficient or sensitivity analysis where the performance of a closed Brayton cycle engine is examined (ref. 3–8). Note that when the percentage deviations are not small, the curvature of the lines shown becomes apparent.

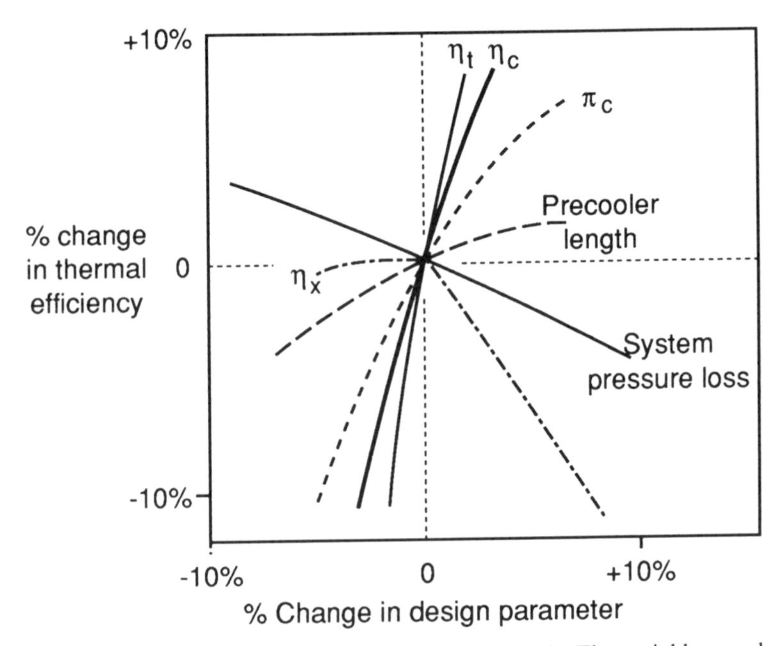

FIG. 3–16. Sensitivity analysis results of a gas turbine cycle. The variables are described in greater detail in Chapter 9.

PROBLEMS

1. A ship can be said to experience a drag force that is proportional to the wetted area of the hull. If this area varies as W^n where $n = 0.6$ (depends on hull shape), determine the equivalent to the Breguet range equation for this ship. Assume drag coefficient, speed, and engine SFC are constant.
2. Using the equations developed in Section 9.3 for the performance of a regenerated, nonideal Brayton cycle, determine the influence coefficients to follow at a pressure ratio for maximum specific work:

$$\frac{\partial \ln y}{\partial \ln x} \quad \text{where } y = w \text{ and } \eta_{th} \quad \text{and} \quad x = \pi_c, \theta_4, \eta_c, \eta_E, \eta_x$$

for values for the x variables $= \pi_c$ value for maximum w, $\theta_4 = 5$, $\eta_c = 0.85$, $\eta_E = 0.9$ and $\eta_x = 0.1$ and 0.9 as separate cases.
3. Estimate the transport efficiencies of real systems currently in use.
4. Show how, in the limit of a linear relation between power and part load efficiency, eq. 3–23 is related to eq. 3–9. What is particular about such a linear relationship?
5. Calculate the energy required to accelerate a 1.5 mt automobile from 90 to 120 km/hr, ignoring friction. If one assumes that power in excess of 40 kw is available and used, what is the time required to achieve this acceleration? By carrying out such a calculation for various speeds, determine $V(t)$ and estimate the work involved in overcoming aerodynamic drag during the time interval of the acceleration process. Assume a realistic value of $C_d A$ and compare the energy required for these two aspects of the power expenditure. 1 mt = 1 metric ton = 1000 kg.
6. Carry out a calculation similar to that in problem 5 for a semi-trailer truck with appropriate numerical values: 75–100 km/hr, 40 mt, $C_d A = 10\ \text{m}^2$, $\Delta \dot{W} = 200$ kW. Include rolling friction resistance.
7. The Breguet range equation, eq. 3–32, was derived for a constant L/D since this parameter remains near a maximum during the mission where the airplane weight changes from W_0 to W_f. Consider that L/D actually varies as

$$\frac{L}{D} = \left(\frac{L}{D}\right)_{\max}\left[1 - a\left(\frac{W - W_m}{W_m}\right)^2\right]$$

where a is a small parameter that was assumed to be zero in the formulation of eq. 3–32. Find an expression for the range in terms of parameters identified in eq. 3–32 and a. Determine the optimal value of W_m in relation to W_0 to W_f so that maximum range is achieved even though L/D at the very beginning and end of the mission are less than the maximum. For $a = 1$ and $W_f/W_0 = 0.7$, find the relative magnitudes of the range when $W_m/W_0 = 0.7$, 0.85 and 1, respectively.

8. Consider a train set consisting of a locomotive of mass M and coaches of mass m. The locomotive applies a propulsion force that is limited to the friction its wheels can exert on the rails. That force equals the coefficient of static friction times its weight ($\mu_L Mg$). The resistance of the coaches involves the coefficient of rolling friction in a similar way. The resistance force is therefore $\mu_T m g$. At low speed, eq. 3–40 states that the propulsion force is very large. This is obviously not possible so that adhesion-limited performance at low speed and power-limited performance at high speed are described by different governing equations. Derive these equations for accelerating motion from rest to a cruise condition (velocity = V_c) and show that the characteristic time scale and distance travelled scale are $V_c/\mu_L g$ and $(V_c)^2/\mu_L g$, respectively. Integrate these equations to find the time required and the distance traveled neglecting aerodynamic drag and assuming that the power available is z times the cruise power required. The integrals involved here and in part 2 are available in standard tables. For definite answers use the following data: $V_c = 200$ km/hr, $\mu_L = 0.25$, $\mu_T = 0.0025$, $m = 500$ mt, $M = 100$ mt (1 mt = 1000 kg). For $z = 2$, what is the ratio of energy required to reach cruise speed in relation to the kinetic energy in cruise?

   The train may be slowed by absorbing the energy of motion through the wheels of the locomotive. Considering the adhesion limitation again and that the engine power would be reduced to zero during deceleration, determine the time and distances traveled as well as the energy that would be absorbed by the locomotive. Compare to the acceleration results. Extend the analysis to include aerodynamic drag.

BIBLIOGRAPHY

Fraas, A. P., *Engineering Evaluation of Energy Systems*, McGraw-Hill, New York, 1982.

REFERENCES

3–1. *Power Generation Alternatives*, City of Seattle, Department of Lighting, 1972.

3–2. Hill, P. G., *Power Generation, Resources, Hazards, Technology and Costs*, MIT Press, Cambridge, Mass., 1977.

3–3. U.S. Atomic Energy Commission, Division of Reactor Research and Development, Power Plant Capital Costs: Current Trends and Sensitivity to Economic Parameters, WASH-1345, Washington D.C., October 1974.

3–4. Horlock, J. H., *Cogeneration: Combined Heat and Power-Thermodynamics and Economics*, Pergamon Press, Oxford, 1987.

3–5. Decher, R., "Power Scaling Characteristics of a Displacement Brayton Cycle Engine," *International Journal of Turbo and Jet Engines*, **2**(2): 141–8, 1985.

3–6. Urban, L. A., "Gas Turbine Engine Parameter Interrelationships," Technical Report Hamilton-Standard, United Aircraft Corp. (Now United Technologies), 1969

3–7. Urban, L. A., "Gas Path Analysis Applied To Turbine Engine Condition Monitoring," AIAA paper 72-1082, AIAA/SAE 8th Joint Propulsion Specialist Conference, 1972.

3–8. Schuster, J. R., Vrable, D. L., Huntsinger, J. P., "Binary Plant Cycle Studies for the Gas Turbine HTGR," ASME paper 76-GT-39, 1976.

4

PROPERTIES OF GASES AND GAS MIXTURES

The largest changes in volume for practical changes in pressure or temperature are experienced by substances in the gaseous phase. This means that the work obtainable from state changes for such a substance may be large enough to be of interest in energy conversion machines. Most heat engines devised, and certainly all that have a commercial importance, have significant portions of the state path in the gas or vapor phase. The motivation for dealing with other phases may be to take advantage of: (1) the reversibility of heat rejection at constant temperature when a substance changes phase from the gas to the liquid or vice versa, and/or (2) the small work required to raise the pressure of liquids. Such processes are considered in Section 7.6 and engines based on them in Chapter 11. This chapter is concerned with the quantitative description of gases in terms of thermodynamic variables and approximations that may be useful.

4.1. THE EQUATION OF STATE FOR A GAS

Two variables are needed to specify the state of a "simple" substance. A gas or vapor obeys the descriptive equation of state

$$v = v(p, T) \tag{4-1}$$

When the molecules or atoms are sufficiently far apart as measured by the ratio of mean free path to length scale characterizing the force field associated with collisional encounters (see Fig. 4–1), the gas is described as ideal. Under such circumstances, the equation of state reads,

$$pv = RT \tag{4-2}$$

where R is a constant that has a characteristic value for each gas or gas mixture. This equation of state is simple and fortunately applies to many situations where the description of gaseous systems is required. The regime of validity for this equation is discussed in Section 4.2.

4.1.1. *Forms of the Ideal Gas Equation of State*

An ideal gas consisting of N molecules of mass m is contained in a volume V. From statistical mechanics one can show that the pressure exerted on the

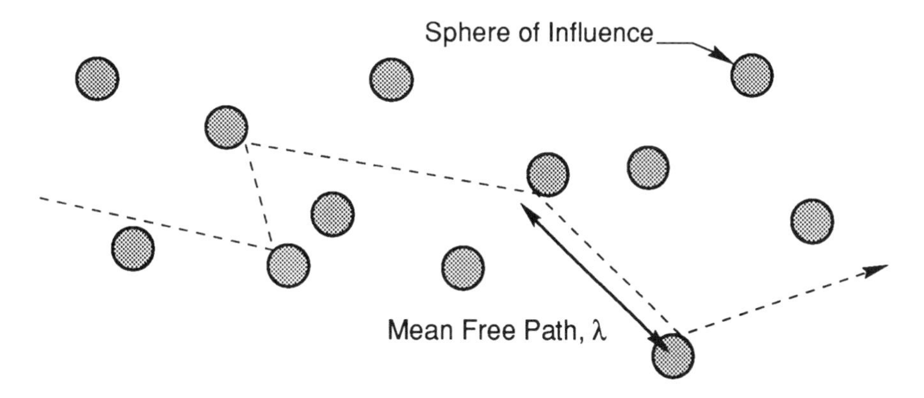

Fig. 4–1. Path of a gas atom or molecule in an ideal gas, $D \ll \lambda$. D is the diameter of the sphere of influence where appreciable intermolecular forces are experienced.

confining wall is:

$$p = \frac{NkT}{V} \quad \text{or} \quad p = \frac{Nm}{V}\frac{k}{m}T \tag{4–3}$$

where k is Boltzmann's constant. In SI units, $k = 1.38 \times 10^{-23}\ J/K$. The group Nm/V is the mass per unit volume or density, ρ. The ratio k/m is a combination that depends only on the mass of the individual gas molecule and is the specific gas constant, $R = k/m$. Hence it follows,

$$p = \rho RT \tag{4–4}$$

One may also multiply and divide the right side of eq. 4–3 by a constant as follows:

$$p = \frac{Nm}{V}\frac{m^*}{m}\frac{k}{m^*}T = \rho\frac{1}{MW}R_u T$$

where m^* is $\frac{1}{12}$ of the carbon atom mass = 1 atomic mass unit (or 1 a.m.u.) which is approximately equal to the mass of a hydrogen atom. Here $R_u = k/m$ is the ratio of the two constants and is called the *universal gas constant*. The value of R_u is given in Appendices B and C. MW is the molecular mass or, more commonly, the molecular weight of the gas; $MW = m/m^*$. It follows when this equation and eq. 4–4 are compared, one may conclude:

$$R = \frac{R_u}{MW}. \tag{4–5}$$

Taking N_0 to be the standard number of molecules in a mass mole, one may say:

$$pV = \frac{N}{N_0} N_0 m^* \frac{k}{m^*} T \quad \text{or} \quad pV = \tilde{n} R_u T \tag{4–6}$$

Since m^* is the mass per molecule of the substance with molecular weight unity, one may use

$$N_0 m^* = 1. \tag{4–7}$$

Thus $\tilde{n}$ is the number of moles defined as N/N_0. The molar density, defined by $n = \tilde{n}/V$, gives

$$p = nR_u T \tag{4–8}$$

A kg-mole contains $N_0 = 6.022 \times 10^{26}$ molecules (a gm-mole 6.022×10^{23} and a lb-mole 2.74×10^{26}). N_0 is Avogadro's number. One may readily show that a kg-mole of any ideal gas at 1 atm pressure and at 0°C occupies 22.414 m^3.

4.2. PROPERTIES

The First Law of Thermodynamics identifies the internal energy per unit mass, u, as an energy form of interest. A property of a simple substance is a function of two variables such as T and v:

$$u = u(T, v) \tag{4–9}$$

Changes in u may be determined for a substance of interest through heat and work interactions by means of the First Law of Thermodynamics (see eq. 2.2):

$$du = d'q - d'w \tag{4–10}$$

Another property of interest is the entropy which may be obtained from its definition,

$$ds = \frac{dq_{rev}}{T} \tag{4–11}$$

and the First Law of Thermodynamics for a reversible process (eq. 2.4) is:

$$dq_{rev} = du + p\,dv$$

which implies:

$$ds = \frac{du}{T} + \frac{p\,dv}{T} \tag{4–12}$$

This relation holds for all processes, reversible or irreversible. With $u(T, v)$ and the state equation $p(T, v)$ known, this equation gives the property $s(T, v)$ when integrated from an initial (reference) condition to any other.

4.2.1. *Enthalpy, Gibbs Function, and Helmholtz Function*

It is convenient to define another state variable, the enthalpy:

$$h = u + pv; \qquad dh = du + p\,dv + v\,dp = dq_{\text{rev}} + v\,dp \tag{4–13}$$

from which the definition of ds (eq. 4–12) reads, when rearranged,

$$T\,ds = dh - v\,dp \tag{4–14}$$

This combination of the First and Second Laws of Thermodynamics is the *Gibbs equation.* Similarly the Gibbs (g) and Helmholtz (f) functions and their differentials are defined as:

$$g = h - Ts; \qquad dg = dh - T\,ds - s\,dT \tag{4–15}$$

$$f = u - Ts; \qquad df = du - T\,ds - s\,dT \tag{4–16}$$

From eq. 4–14 and the function definitions, the differentials of f and g may be found. The differential forms of the four state functions are:

$$du = T\,ds - p\,dv \tag{4–17a}$$

$$dh = T\,ds + v\,dp \tag{4–17b}$$

$$dg = v\,dp - s\,dT \tag{4–17c}$$

$$df = -p\,dv - s\,dT \tag{4–17d}$$

These four functions are useful in describing processes where one or both of the independent variables [such as s and/or v and, for the internal energy, $u(s, v)$] may be held constant. These equations are of the form

$$dz = \left.\frac{\partial z}{\partial x}\right)_y dx + \left.\frac{\partial z}{\partial y}\right)_x dy \tag{4–18}$$

for a function

$$z = z(x, y) \tag{4–19}$$

Since $\partial^2 z/\partial x\, \partial y$ must be unique for the state function z to be single valued, the following must be true:

$$\frac{\partial}{\partial y}\left(\left(\frac{\partial z}{\partial x}\right)_y\right) = \frac{\partial}{\partial x}\left(\left(\frac{\partial z}{\partial y}\right)_x\right) \tag{4–20}$$

Applying this condition to the state functions yields the so-called Maxwell's relations: from eqs. 4–17 and 4–20

$$\left.\frac{\partial T}{\partial v}\right)_s = -\left.\frac{\partial p}{\partial s}\right)_v \tag{4–21a}$$

$$\left.\frac{\partial \mathrm{T}}{\partial \mathrm{p}}\right)_s = \left.\frac{\partial v}{\partial s}\right)_p \tag{4–21b}$$

$$\left.\frac{\partial v}{\partial T}\right)_p = -\left.\frac{\partial s}{\partial p}\right)_T \tag{4–21c}$$

$$\left.\frac{\partial p}{\partial T}\right)_v = \left.\frac{\partial s}{\partial v}\right)_T \tag{4–21d}$$

These relations will be shown to be useful for the development of the functional dependence of variables of interest. While not all of these are used here, they are cited for completeness. From eqs. 4–17 and 4–18, it follows

$$\left.\frac{\partial u}{\partial s}\right)_v = T = \left.\frac{\partial h}{\partial s}\right)_p \tag{4–22a}$$

$$\left.\frac{\partial g}{\partial p}\right)_T = v = \left.\frac{\partial h}{\partial p}\right)_s \tag{4–22b}$$

$$\left.\frac{\partial u}{\partial v}\right)_s = -p = \left.\frac{\partial f}{\partial v}\right)_T \tag{4–22c}$$

$$\left.\frac{\partial f}{\partial T}\right)_v = -s = \left.\frac{\partial g}{\partial T}\right)_p \tag{4–22d}$$

These equations give the relation between elementary thermodynamic variables, such as p, T, etc., and the higher level functions, h, g, etc. They are very useful for providing insight into the characteristics of more physical variables of interest. One important set of such variables are the specific heats.

4.2.2. *Specific Heats*

The specific heat at constant volume ($dv = 0$) is defined by

$$C_v = \lim_{\Delta T \to 0} \left.\frac{\Delta q}{\Delta T}\right)_v = \left.\frac{\partial u}{\partial T}\right)_v \text{ from eq. 4–10 with } d'w = 0 \tag{4–23}$$

and may be interpreted as the heat required to raise the temperature of the substance by one degree, under the restriction that the volume of the substance remains constant. Similarly the specific heat at constant pressure is

$$C_p = \lim_{\Delta T \to 0} \left.\frac{\Delta q}{\Delta T}\right)_p = \left.\frac{\partial h}{\partial T}\right)_p \text{ from eq. 4–13 with } dp = 0 \tag{4–24}$$

These specific heats differ numerically because the heat supplied at constant pressure also results in expansion work being done by the substance against the environment. No such work can be done by the substance when its volume is held fixed.

The ratio of these specific heats (always > 1), defined as γ, may be written using eqs. 4–23 and 4–24 and rewriting these to give the variable relations derived as eq. 4–22, (4–22a, in this case):

$$\gamma = \frac{C_p}{C_v} = \frac{\left.\frac{\partial h}{\partial s}\right)_p \left.\frac{\partial s}{\partial T}\right)_p}{\left.\frac{\partial u}{\partial s}\right)_v \left.\frac{\partial s}{\partial T}\right)_v} = \frac{T\left.\frac{\partial s}{\partial T}\right)_p}{T\left.\frac{\partial s}{\partial T}\right)_x} \text{ using eq. 4–22a} \tag{4–25}$$

The cyclic rule on partial derivatives for a single valued function $z(x, y)$ is

$$\left.\frac{\partial z}{\partial x}\right)_y \left.\frac{\partial x}{\partial y}\right)_z \left.\frac{\partial y}{\partial z}\right)_x = -1 \tag{4–26}$$

so that

$$\gamma = \frac{\left.\frac{\partial T}{\partial v}\right)_s \left.\frac{\partial v}{\partial s}\right)_T}{\left.\frac{\partial p}{\partial s}\right)_T \left.\frac{\partial T}{\partial p}\right)_s} = \frac{\left.\frac{\partial T}{\partial v}\right|}{\left.\frac{\partial T}{\partial p}\right|_s} \frac{\left.\frac{\partial v}{\partial s}\right|}{\left.\frac{\partial p}{\partial s}\right|_T} = \frac{\left.\frac{\partial p}{\partial v}\right)_s}{\left.\frac{\partial p}{\partial v}\right)_T} \tag{4–27}$$

Thus γ is seen to be the local value of the ratio of slopes of an isentrope (s = constant line) and an isotherm (T = constant line) in the p-v plane.

Further relations involving the specific heats may be obtained. Consider specifically the difference between the specific heats which may be written in terms of state variables in a form $C_p - C_v = f(p, T)$. From $s(T, v)$

$$ds = \left.\frac{\partial s}{\partial T}\right)_v dT + \left.\frac{\partial s}{\partial v}\right)_T dv = \left.\frac{\partial s}{\partial u}\right)_v \left.\frac{\partial u}{\partial T}\right)_v dT + \left.\frac{\partial s}{\partial v}\right)_T dv \tag{4–28}$$

From the definition of the specific heat (eq. 4–23) and eq. 4–22a, and the Maxwell relation (eq. 4–21d), eq. 4–28 becomes

$$ds = \frac{C_v\, dT}{T} + \left.\frac{\partial p}{\partial T}\right)_v dv \tag{4–29}$$

Similarly, from $s(T, p)$, one obtains:

$$ds = \frac{C_p\, dT}{T} - \left.\frac{\partial v}{\partial T}\right)_p dp \tag{4–30}$$

Multiplying eqs. 4–29 and 4–30 by T and subtracting the results gives

$$(C_p - C_v)\, dT = T \left.\frac{\partial p}{\partial T}\right)_v dv + T \left.\frac{\partial v}{\partial T}\right)_p dp$$

At this point, it is convenient to adopt a shorthand notation for the derivatives:

$$p' = \left.\frac{\partial p}{\partial T}\right)_v \quad \text{and} \quad v' = \left.\frac{\partial v}{\partial T}\right)_p$$

so that

$$\frac{(C_p - C_v)}{T}\, dT = p'\, dv + v'\, dp$$

Differential changes in temperature may also be expressed in terms of dv and dp using the state equation:

$$dT = \left.\frac{\partial T}{\partial v}\right)_p dv + \left.\frac{\partial T}{\partial p}\right)_v dp = \frac{1}{v'} dv + \frac{1}{p'} dp = \frac{p'\, dv + v'\, dp}{p'v'}$$

Eliminating dT between these last two equations yields:

$$C_p - C_v = Tp'v' = T \left.\frac{\partial p}{\partial T}\right)_v \left.\frac{\partial v}{\partial T}\right)_p \tag{4–31}$$

which shows how the specific heats are related to each other and the state equation.

Equating the cross derivatives of eqs. 4–29 and 4–30 yields

$$\frac{\partial}{\partial v}\left(\frac{1}{T}C_v\right) = \left.\frac{\partial^2 p}{\partial T^2}\right)_v \quad \text{or} \quad \left.\frac{\partial C_v}{\partial v}\right)_T = T\left.\frac{\partial^2 p}{\partial T^2}\right)_v \tag{4–32}$$

and

$$\left.\frac{\partial C_p}{\partial p}\right)_T = -T\left.\frac{\partial^2 v}{\partial T^2}\right)_p \tag{4–33}$$

A gas for which C_p is a function of T only is termed "thermally perfect" while one for which C_p may be assumed constant is "calorically perfect".

The relations 4–31, 4–32, and 4–33 provide insight into the consequences of dealing with a nonideal gas. If the equation of state is known, then the volume dependence of C_v and the pressure dependence of C_p may be readily assessed. An example is illustrated in the following section. Before doing so, however, the differential relation for the internal energy and enthalpy are stated. From eqs. 4–12 and 4–29 one has

$$du = C_v\,dT + \left(T\left.\frac{\partial p}{\partial T}\right)_v - p\right)dv \tag{4–34}$$

and similarly from eqs. 4–14 and 4–30

$$dh = C_p\,dT + \left(v - T\left.\frac{\partial v}{\partial T}\right)_p\right)dp \tag{4–35}$$

In these expressions, the dependence of u on v and of h on p can be clearly seen using the state equation by evaluation of the terms in parentheses. For the ideal gas, these terms are zero.

4.2.3. *The Van Der Waals Fluid*

A number of analytical equations of state have been developed to incorporate nonideal gas effects in the description of gases. The Van Der Waals formulation is discussed here because the terms introduced are readily interpreted physically. It has a number of shortcomings some of which are discussed here. The reader is encouraged to review a classical thermodynamics discussion of the p-v-T description of the Van Der Waals gas, the shape of isotherms on the p-v plane and the relation between the triple valued isotherms and the saturation dome.

The Van Der Waals (VDW) equation of state is given by

$$\left(p + \frac{a}{v^2}\right)(v - b) = RT \tag{4–36}$$

where a, b, and R are constants that may be identified for a particular gas.

Values for the a and b terms are tabulated and may be found in, for example, ref. 4–1. The equation is written in this form in an attempt to include the effect of the finite size of the molecules through the "covolume" b term which reduces the volume available to the molecules. The mutual interaction of the molecules is accounted for through the a term. Presumably, when the mean distance between molecules is small, the forces attracting the molecules to each other reduce the pressure experienced by a boundary. In the limit of large v, the ideal gas equation of state (eq. 4–4) is recovered:

$$pv = RT \tag{4–37}$$

The VDW equation of state is not a particularly accurate model, but it does give a good qualitative description in the state plane region (p, v or T, s, for examples) between the saturation line and the ideal gas. The reader is referred to classical texts on thermodynamics for a discussion of other state equation models. The VDW equation is not a valid description of a real substance in the region where an isotherm is triple valued. In this region, real substances exist in two phases. The value of the VDW equation lies in the understanding it provides for a gas *near* the saturation dome and the role played there by finite volume effects and intermolecular force fields. By writing

$$p = \frac{RT}{v-b} - \frac{a}{v^2}$$

the derivatives of interest for the determination of u (see eq. 4–34), and h, as well as specific heats, may be obtained:

$$\left.\frac{\partial p}{\partial T}\right)_v = \frac{R}{v-b} \tag{4–38a}$$

$$\left.\frac{\partial p}{\partial v}\right)_T = -\frac{RT}{(v-b)^2} + \frac{2a}{v^3} \tag{4–38b}$$

$$\left.\frac{\partial^2 p}{\partial T^2}\right)_v = 0 \tag{4–38c}$$

Equation 4–38b will be useful in the association of the constants a and b with the critical constants of the substance under consideration. Equation 4–38c is needed for the evaluation of eq. 4–32. Further, the form

$$T = \frac{1}{R}\left[p(v-b) + a\left(\frac{1}{v} - \frac{b}{v^2}\right)\right]$$

allows evaluation of the derivatives required for eqs. 4–31, 4–33, and 4–35 which are

$$\left.\frac{\partial v}{\partial T}\right)_p = \frac{R}{p + a\left(-\frac{1}{v^2} + \frac{2b}{v^3}\right)} \tag{4–39a}$$

$$\left.\frac{\partial 2v}{\partial T^2}\right)_p = \frac{-R^2 \frac{2a}{v^3}\left(1 - \frac{3b}{v}\right)}{\left[p + a\left(-\frac{1}{v^2} + \frac{2b}{v^3}\right)\right]^3} \tag{4–39b}$$

The dependence of C_v on v is obtained from eqs. 4–32 and 4–38c:

$$\left.\frac{\partial C_v}{\partial v}\right)_T = T \left.\frac{\partial^2 p}{\partial T^2}\right)_v = 0 \tag{4–40}$$

Thus one may conclude that for a VDW gas $C_v = C_v(T)$ only. On the other hand,

$$\left.\frac{\partial C_p}{\partial p}\right)_T = -T \left.\frac{\partial^2 v}{\partial T^2}\right)_p = \frac{R^2 T \frac{2a}{v^3}\left(1 - \frac{3b}{v}\right)}{\left(p - \frac{a}{v^2} + \frac{2ab}{v^3}\right)^3} \tag{4–41}$$

shows that C_p is a function of both T and p (or v). Note that it is the force interaction term a that causes nonperfect gas behavior. The difference between C_p and C_v (eq. 4–31) for a VDW gas is

$$C_p - C_v = \frac{R}{1 - \frac{2a}{RTv^3}(v - b)^2} \tag{4–42}$$

which evidently approaches R as v becomes large. Equations 4–41 and 4–42 are restated below as eq. 4–54, to show what nondimensional parameters are relevant to determine whether the gas is or is not perfect.

4.2.4. *Equations for Internal Energy and Enthalpy for a VDW Fluid*

From eq. 4–34 giving du, from the state equation (eq. 4–36), and from $\left.\frac{\partial p}{\partial T}\right)_v$ (eq. 4–38a), it follows that

$$du = C_v\, dT + \frac{a}{v^2}\, dv \tag{4–43}$$

Note here that the a/v^2 term represents the energy stored in the intermolecular force field. The differential enthalpy from eq. 4–35 is

$$dh = C_p\, dT + v\, dp - T \left.\frac{\partial v}{\partial T}\right)_p dp$$

or, in terms of C_v which is a function of T only,

$$dh = du + d(pv) = C_v\, dT + \left(T \left.\frac{\partial p}{\partial T}\right)_v - p \right) dv + v\, dp + p\, dv$$

which becomes, when the differential of $p(T, v)$ is used,

$$dh = \left(v \left.\frac{\partial p}{\partial T}\right)_v + C_v \right) dT + \left(v \left.\frac{\partial p}{\partial v}\right)_T + T \left.\frac{\partial p}{\partial T}\right)_v \right) dv \tag{4–44}$$

From eq. 4–44, the dependence of h on v may be deduced:

$$\left.\frac{\partial h}{\partial v}\right)_T = v \left.\frac{\partial p}{\partial v}\right)_T + T \left.\frac{\partial p}{\partial T}\right)_v = -\frac{RTb}{(v - b)^2} + \frac{2a}{v^2}$$

where the last portion is for the VDW fluid (eqs. 4–38a and 4–38b). This equation can be integrated to give

$$h = f(T, v) + \frac{RTb}{(v - b)} - \frac{2a}{v} \tag{4–45}$$

which shows that *both* the a and b terms play a role in the pressure or specific volume dependence of h in a VDW gas. Here $f(T, v)$ is the integral of

$$f(T, v) = \int \left(C_v + v \left.\frac{\partial p}{\partial T}\right)_v \right) dT = \int \left(C_v + R \frac{v}{v - b} \right) dT \tag{4–46}$$

which becomes $\int C_p\, dT$ in the limit of $v \gg b$. Table 4–1 summarizes these results for a VDW gas and contrasts them to the ideal gas. For other state relations, analogous results may be expected.

4.2.5. *The Perfect Gas Assumption*

Differential changes in the enthalpy can be written from eqs. 4–45 and 4–46 with the derivatives evaluated for a VDW fluid

$$dh = \left(C_v + R \frac{v}{v - b} \right) dT + \left(-\frac{bRT}{(v - b)^2} + \frac{2a}{v^2} \right) dv \tag{4–47}$$

Table 4–1. Functional dependence of ideal and Van Der Waals state variables

	Van Der Waals	Ideal Gas
C_v	$f(T)$ only	$f(T)$ only
C_p	$f(T, p)$ or $f(T, v)$	$f(T)$ only
$C_p - C_v$	not constant	constant
u, h	$f(T, p)$ or $f(T, v)$	$f(T)$ only
du	$\neq C_v\,dT$ if $dv \neq 0$	$= C_v\,dT$
dh	$\neq C_p\,dT$ if $dp \neq 0$	$= C_p\,dT$

which reduces to the following, when v becomes large:

$$\begin{aligned} dh_i &= (C_v + R)\,dT \quad \text{for an ideal gas (where } v \text{ is large)} \\ &= C_p\,dT \end{aligned} \tag{4–48}$$

The error in the enthalpy made when assuming an ideal gas undergoes a process described by a path of specified dv/dT is

$$\text{err}(h) \equiv \frac{dh}{dh_i} - 1 = \frac{R}{C_p}\left[\frac{b}{v-b} + \frac{dv}{dT}\left(-\frac{bT}{(v-b)^2} + \frac{2a}{Rv^2}\right)\right] \tag{4–49}$$

The error is seen to be associated with nonzero a and/or b. An expression such as this can be more easily interpreted when the constants a and b are evaluated in terms of the values at the critical point. At this point the isotherm has both a zero slope and an inflection point. Thus, at the critical point

$$p = p_c \qquad v = v_c \qquad T = T_c$$

and

$$\left.\frac{\partial p}{\partial v}\right)_T = 0; \qquad \left.\frac{\partial^2 p}{\partial v^2}\right)_T = 0 \tag{4–50}$$

Numerical values for the critical values for nitrogen and for water are given in Appendix B and may be readily found elsewhere, including ref. 4–1. For a VDW fluid, the first and second conditions give, respectively,

$$-\frac{RT_c}{(v_c - b)^2} + \frac{2a}{v_c^3} = 0 \qquad \text{and} \qquad \frac{2RT_c}{(v_c - b)^3} - \frac{6a}{v_c^4} = 0$$

The critical variables may thus be expressed in terms of a, b and R:

$$p_c = \frac{a}{27b^2}; \qquad v_c = 3b; \qquad T_c = \frac{8a}{27Rb} \tag{4–51}$$

The so-called "reduced" quantities are defined as

$$p_r = \frac{p}{p_c}, \qquad T_r = \frac{T}{T_c} \text{ etc.} \tag{4–52}$$

The equation of state written in terms of reduced variables then reads

$$\left(p_r + \frac{3}{v_r^2}\right)(3v_r - 1) = 8T_r \tag{4–53}$$

This formulation allows one to write the eqns. 4–41 and 4–42 to reflect the simple nondimensional variables which cause a gas to be imperfect. Thus these equations become in nondimensional form:

$$\frac{p}{R}\frac{\partial C_p}{\partial p}\bigg)_T = \frac{2x(1+x)\left(1-\dfrac{1}{v_r}\right)\left(1-\dfrac{1}{3v_r}\right)}{\left(1-x\left(1-\dfrac{2}{3v_r}\right)\right)^3} \qquad \text{where } x \equiv \frac{3}{p_r v_r^2} \tag{4–54}$$

and

$$C_p - C_v = \frac{R}{1 - \dfrac{2x}{1+x}\left(1-\dfrac{1}{3v_r}\right)}$$

By examining the mathematical limit of a small value of the shorthand parameter, x, it should be evident that both quantities given above, depart from perfect gas behavior with increasing x, that is, with $3/(p_r v_r^2) \approx 1/(v_r T_r)$. Thus as v_r increases (i.e., x decreases), the gas becomes more ideal.

The error in dh obtained in eq. 4–49 is

$$\text{err}(h) = \frac{R}{C_p}\frac{1}{3v_r - 1}\left\{1 + \frac{d\ln v_r}{d\ln T_r}\left[-\left(1-\frac{1}{3v_r}\right)^{-1} + \frac{3}{4}\frac{1}{T_r}\left(1-\frac{1}{3v_r}\right)\right]\right\} \tag{4–55}$$

This error depends on the thermodynamic path as described by $d\ln v_r/d\ln T_r$. As an example, consider an isentropic path (where $\gamma \cong$ constant) described by

$$Tv^{\gamma-1} = \text{constant}$$

then

$$\frac{d\ln v_r}{d\ln T_r} = -\frac{1}{\gamma - 1} = -\frac{R}{C_v}$$

Since $d\ln v_r/d\ln T_r$ is of order unity but could be zero, the error is proportional to $1/(3v_r - 1)$ and is the largest of the two terms in the square bracket of eq. 4–55.

For large v_r, the two terms in parentheses approach unity. For the vapor region, the reduced temperature, T_r, may be quite low, so that the last term is likely to be dominant. In that case,

$$\operatorname{err}(h) \sim \frac{2}{T_r v_r} \frac{R}{C_p} \left.\frac{d \ln v_r}{d \ln T_r}\right)_{\text{PATH}} \tag{4–56}$$

This error vanishes when the reduced volume is large.

4.2.6. *The Ideal Gas Assumption*

The VDW equation of state may also be used to estimate the validity of the use of the ideal gas equation of state. Suppose one knew p and T and wished to use the ideal gas equation of state to find the specific volume, v. In reduced form, this gives

$$\text{approximate } v_r = v_{r,a} = \frac{8}{3} \frac{T_r}{p_r} \tag{4–57}$$

when expressed in reduced form. The actual specific volume is obtained from eq. 4–53. By setting $v_r = v_{r,a}(1 + \varepsilon)$, one defines an error increment which one may limit to be small. The VDW equation then becomes

$$\left(p_r + \frac{3}{[v_{r,a}(1+\varepsilon)]^2}\right)([3v_{r,a}(1+\varepsilon)] - 1) = 8T_r = 3p_r v_{r,a}$$

By linearizing for small ε,

$$\operatorname{err}(v) = \varepsilon = \frac{1}{3v_{r,a}} \left(\frac{1 - \dfrac{9}{p_r v_{r,a}} + \dfrac{3}{p_r v_{r,a}^2}}{1 - \dfrac{3}{p_r v_{r,a}^2} + \dfrac{2}{p_r v_{r,a}^3}} \right) = \frac{1}{3v_{r,a}} \left(\frac{1 - x(3v_{r,a} - 1)}{1 - x\left(1 - \dfrac{2}{3v_{r,a}}\right)} \right) \approx \frac{1}{3v_{r,a}} \tag{4–58}$$

Equation 4–58 shows that for p_r of order 1 and $v_{r,a}$ large, the error in assuming the fluid to be an ideal gas is of the order of $1/(3v_{r,a})$.

In summary, it may be said that gases behave as ideal and perfect [$pv = RT$ and C_p is a function of T only] when the density is sufficiently low (or, equivalently, if specific volume is sufficiently large, $v_r \gg 3$). For specific circumstances, this conclusion may have to be examined with a more accurate equation of state.

4.3. PROPERTIES OF IDEAL GAS MIXTURES

In many energy conversion devices, the hot gaseous products from combustion are used in flow devices to transfer heat or extract work. For these purposes it is necessary to describe the properties of the gas mixture. Of interest are enthalpy, internal energy, specific heats, and molecular weight.

For combustion problems, gas component conditions are usually such that the temperature and pressure are far from the critical point, and the description of these components as *ideal* gases ($pv = RT$) is an excellent one. If one assumes further that the gas is always in thermodynamic equilibrium so that a temperature can be defined, the temperatures of each component of the mixture will be equal to each other and therefore to that of the mixture.

For a gas mixture of volume V, the state equation for each ideal component gives the number of moles of each component in terms of the partial pressure, p_i, which is the pressure which the i-component would exert on the walls if it were alone in volume V. The mass-mole is defined as the ratio of the number of molecules to the number of molecules which have a mass of MW units, where MW = the molecular weight. Hence the number of moles per unit volume is

$$n_i = \frac{p_i}{R_u T} \tag{4-59}$$

where $T_i = T$ is used as justified above. The total number of moles is N or

$$n = \sum n_i = \frac{1}{R_u T} \sum p_i$$

Since each component of the mixture is ideal, so must be the mixture or

$$p = nR_u T$$

These two expressions may be combined to read

$$p = \sum p_i \tag{4-60}$$

which is the Gibbs-Dalton Law of Partial Pressures. This law is not valid if the gas components are not ideal which is when the molecules experience significant intermolecular forces.

For ideal gas components, the internal energy of a mixture is the sum of the internal energies of the components, that is

$$U = \sum U_i \tag{4-61}$$

Similarly the enthalpy of a mixture of ideal gases equals the sum of the enthalpies of the components. This follows directly from:

$$H = U + pV = \sum U_i + V \sum p_i = \sum H_i \tag{4–62}$$

The entropy of a gas mixture of ideal gases is also the sum of entropies of the components. This can be shown by considering the following hypothetical experiment which could be carried out in reality, if truly semipermeable membranes existed and one could wait indefinitely for the process to be completed. Figure 4–2 shows a container of volume V containing two ideal gases. The containers are expanded very slowly as shown.

The semipermeable membranes allow only molecules of type A or type B to cross. The other molecule type is deflected back into the mixture as if the membranes were walls. The membrane experiences, therefore, a pressure p_A or p_B as shown, which is equal and opposite to the pressure felt by the wall on the opposite side. The *net* force (left minus right) required to double the total volume while maintaining the volume available to either species constant is zero. Hence no work is done. By considering the solid walls to be perfectly

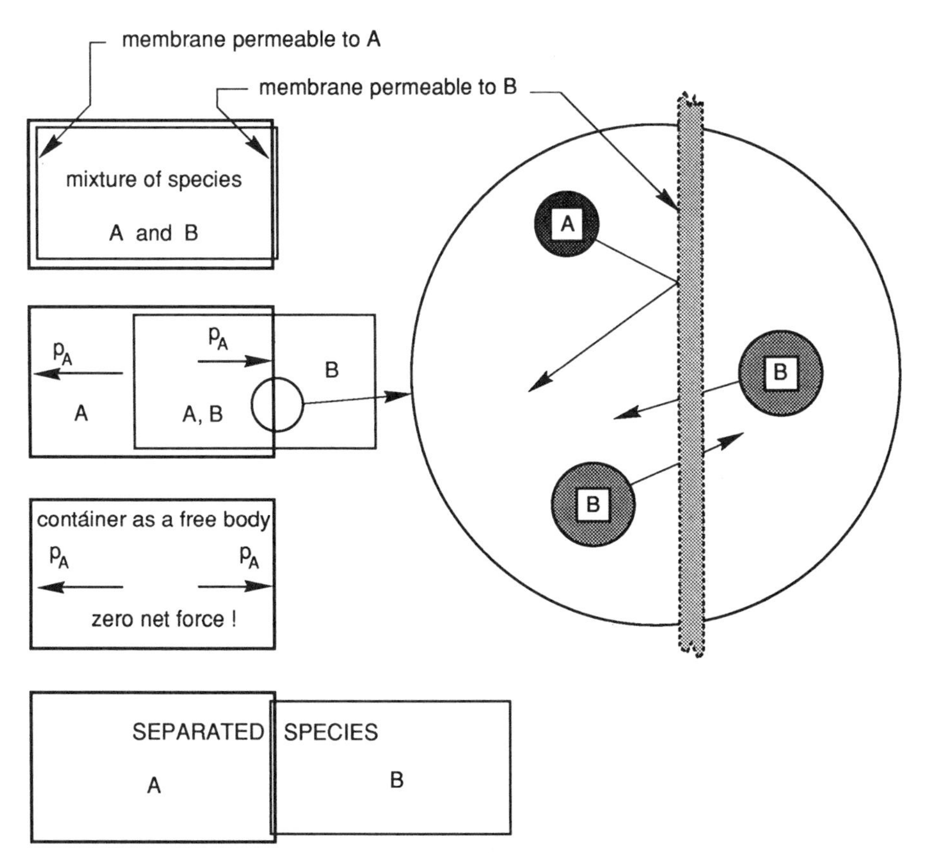

Fig. 4–2. Work done in separating two ideal gases is zero: Gibbs rule. Container walls other than membranes are impermeable and do not allow heat transfer.

insulated, the process may be made adiabatic as well. Hence there are no temperature changes for either A or B from the temperature of the mixture. Since an ideal gas is a simple system describable in terms of two thermodynamic state variables (p and T), the entropy of A when separated is equal to the entropy of A when mixed. Thus

$$\left.\begin{aligned} &S_{\text{A, mixed}} = S_{\text{A, separated}} \quad \text{and} \quad S_{\text{B, mixed}} = S_{\text{B, separated}} \\ \text{or} \\ &S_{\text{mixture}} = \sum S_{\text{i}} \end{aligned}\right\} \tag{4–63}$$

This is known as the Gibbs rule.

4.3.1. *Extensive Properties*

The internal energy, enthalpy, and entropy of a substance is proportional to the amount of the substance involved; these quantities are termed extensive.

The amount of a substance may be measured by the number of molecules or by the mass. The number of molecules is an awkward quantity because their numbers are typically very large. By introducing the mass-mole one deals with the specific number of molecules which comprise the mass equal to the molecular weight. It is customary to denote u, h, s to mean per mole or per unit mass (the context usually clarifies this ambiguity) (the capital symbols U, H, S per kg-mole are used in the JANAF tables, see Section 4.4). If the number of moles of component i is n_{i}, then the mixture properties are:

$$U = \sum n_{\text{i}} u_{\text{i}}; \qquad H = \sum n_{\text{i}} h_{\text{i}}; \qquad S = \sum n_{\text{i}} s_{\text{i}} \tag{4–64}$$

where the units of u, h, and s must be per mole. With the total number of moles, $n = \sum n_{\text{i}}$ these extensive quantities may be written

$$nu = \sum n_{\text{i}} u_{\text{i}}; \qquad nh = \sum n_{\text{i}} h_{\text{i}}; \qquad ns = \sum n_{\text{i}} s_{\text{i}}$$

or

$$u = \sum x_{\text{i}} u_{\text{i}}; \qquad h = \sum x_{\text{i}} h_{\text{i}}; \qquad s = \sum x_{\text{i}} s_{\text{i}}$$

where x_{i} is the mole fraction ($= n_{\text{i}}/n$) and it follows that

$$\sum x_{\text{i}} = 1 \tag{4–65}$$

In a similar way the ratio $m_{\text{i}}/\sum m_{\text{i}}$ is referred to as a mass fraction. Extensive quantities may be also measured per unit mass of the component. Hence if m_{i} is the mass of component i

$$U = \sum m_{\text{i}} u_{\text{i}}; \qquad H = \sum m_{\text{i}} h_{\text{i}}; \qquad S = \sum m_{\text{i}} s_{\text{i}} \tag{4–66}$$

where the u, h, and s are per unit mass.

The number of moles of a substance is related to the mass by

$$n_{\mathrm{i}} = \frac{m_{\mathrm{i}}}{\mathrm{MW}_i} \tag{4–67}$$

where MW_i is the mass per mole, that is, the molecular weight. Although one could assign different symbols for extensive quantity per mole from those assigned to quantity per mass, no standard is in use. The reason is that conversion from one to the other is readily made and the units quoted usually indicate the basis of the description. For example,

$$h_i \text{ per mass} = \frac{h_i \text{ per mole}}{\text{mass per mole}} = \frac{h_i \text{ per mole}}{M_i} \tag{4–68}$$

4.3.2. *Average Molecular Weight*

Since the number of moles of a mixture is the sum of the number of moles of each of the constituents (recall that a mole is a specific number of molecules),

$$n = \sum n_{\mathrm{i}} \qquad \text{and} \qquad m = \sum m_{\mathrm{i}}$$

One may define the average molecular weight as the mass per mole of the mixture or

$$\mathrm{MW} \equiv \frac{m}{n} = \frac{\sum m_i}{\sum n_i} = \frac{\sum n_i \, \mathrm{MW}_i}{\sum n_i} = \sum x_i \, \mathrm{MW}_i \tag{4–69}$$

4.3.3. *Specific Heats*

The specific heats are useful quantities for gases because u and h can often be approximated as being a function of temperature only. This is the perfect gas approximation which means specifically that for the general functional dependence,

$$h = h(T, p)$$

one has

$$\left.\frac{\partial h}{\partial p}\right)_T = 0 \qquad \text{and} \qquad \left.\frac{\partial h}{\partial T}\right)_p \equiv C_{\mathrm{p}}$$

If the gas is ideal, the internal energy, $u = h - RT$, is also a function of T only. Writing $u(T, v)$, it follows that

$$\left.\frac{\partial u}{\partial v}\right)_T = 0 \qquad \text{and} \qquad \left.\frac{\partial u}{\partial T}\right)_v = \frac{du}{dT} \equiv C_{\mathrm{v}}$$

which is substantiated in the classical Joule-Thompson experiment.

Note that the units on C_p and C_v must be similar to those of h and u, that is, either per mole or per mass. Using a "per mole" formulation, one has for component i

$$h_i = \int_0^T C_{pi}\, dT, \qquad u_i = \int_0^T C_{vi}\, dT$$

with the arbitrary zero reference state at $T = 0$.

With H equal to the enthalpy of the ideal gas mixture

$$\left.\begin{aligned} &H = nh = \sum h_i \qquad n\frac{\partial h}{\partial T} = \sum n_i \frac{\partial h_i}{\partial T} \qquad nC_p = \sum n_i C_{pi} \\ &\text{or} \\ &C_p = \sum x_i C_{pi} \end{aligned}\right\} \tag{4–70}$$

(If C_{pi} and C_{vi} are per unit mass, then the x_i's in the above expressions must be mass fractions.) The specific heat ratio of the mixture

$$\gamma = \frac{C_p}{C_v} = \frac{\sum x_i C_{pi}}{\sum x_i C_{vi}} \tag{4–71}$$

Since $h = u + R_u T$ for the ideal and perfect gas with constant specific heats, it follows that

$$\frac{\partial h}{\partial T} = \frac{\partial u}{\partial T} + R_u$$

$$C_p = C_v + R_u$$

This becomes, on division by molecular weight,

$$C_p = C_v + R \tag{4–72}$$

These two relations are seen to be consistent with eqs. 4–31 (ideal gas) and 4–42 (VDW with large reduced specific volume). In the last equation, C_p and C_v are specific heats per mass and $R = R_u/\text{MW}$ is the *specific* gas constant.

Thermochemical information for various substances is available from a number of sources. Here attention is restricted to the widely available JANAF (Joint Army-Navy-Air Force) tables which were developed in part by Dow Chemical under contract to the U.S. government. From the JANAF tables described below, it is clear that, over the temperature range tabulated, C_p for gases of interest in energy conversion and hence C_v ($= C_p - R_u$) are functions of T and not constants. γ therefore varies with temperature, and *its usefulness is limited to processes* where temperature changes are sufficiently small so that γ is approximately constant.

4.4. JANAF THERMOCHEMICAL DATA TABLES

Appendix C is a compilation of the JANAF thermochemical tables of interest to problems related to combustion of gases. The species for which data are included consists of the following molecules (in alphabetical order):

$$C, CH_4, CO, CO_2, H, H_2, H_2O, N, N_2, NO, NO_2, O, O_2, OH$$

The atoms involved include only carbon, hydrogen, oxygen, and nitrogen, the principal actors in the question of energy release, use of air, and pollutant formation in combustion.

4.4.1. *Tabular Data*

Table 4–2 shows a portion of the table for H_2O. The data are given for each constituent, assuming the conditions to be such that the ideal gas assumption is valid. This means that the temperature must be above the critical value and/or the pressure is sufficiently low. Under these conditions, C_p, the enthalpy h, and the integral of $(C_p/T)\,dT$ are functions of T only. The tables are given with absolute temperature in K as the independent variable. The complete tables show a dashed line near the critical temperature to advise caution below that value. Table 4–2 shows in particular the variation of C_p with temperature in the range where combustion processes are likely to occur. The constant C_p approximation (say at 298 K) leads to the values noted parenthetically ($=C_p\Delta T$ and $=s^0(298) + C_p \ln T/298$, respectively). The values lead to errors that may be permitted for the purpose of modeling physics but require recognition in precise performance calculations.

4.4.2. *Thermodynamic Standard State*

The thermodynamic standard state is taken to be the phase of the homogeneous substance which exists at 1 atm pressure at the temperature in question. The circular superscript on property values refers to the thermodynamic standard state. The reference state is denoted by a value of the temperature as a subscript.

Table 4–2. Excerpt from the JANAF thermochemical data table for H_2O

T(K)	C_p	$H(T) - H(298)$	s^0
298	8.025	0.0	45.106
500	8.415	1.654	49.334
1000	9.851	6.209	55.592
1500	11.233	11.495 (9.65)	59.859 (58.07)
2000	12.214	17.373	63.234
2500	12.863	23.653	66.034
3000	13.304	30.201	68.421

In the JANAF tables, the value is commonly 298.15 K (with 298 used as a subscript). The symbols and units used in the tables are described in Appendix C.

4.5. THE ADIABATIC REVERSIBLE PROCESS

One of the important and elementary processes of interest is that wherein entropy changes are absent due to the absence of mechanisms that produce it: friction and heat transfer. Real processes designed to be close to adiabatic and reversible, such as work interactions with a thermodynamic system's surroundings almost always involve some entropy production. To the extent that the isentropic process is the goal of good engineering design, it is of interest.

Consider the general relationship (Gibbs equation) between entropy and other state variables such as the pressure and temperature,

$$T\,ds = dh(T, p) - v(T, p)\,dp \tag{4–73}$$

The following is a discussion of the relation between T and p for an isentropic process given the various descriptive models that may apply to the working fluid: ideal and calorically perfect gas. For such a process one has

$$0 = \frac{1}{T}\left[\left(\frac{\partial h}{\partial p}\right)_T - v\right]dp + \frac{C_p\,dT}{T} \tag{4–74}$$

which describes the relation between p and T for the general case. Here this relationship is made more concrete for special cases.

The ideal gas obeys an equation of state (eq. 4–37) which directly allows the conclusion that h and C_p are functions of T only (eq. 4–41). Thus eq. 4–74 can be written

$$\left.\begin{aligned} &\frac{dp}{p} = \frac{C_p(T)\,dT}{RT} \quad \text{for an ideal gas} \\ &\text{or} \\ &\ln\frac{p}{p_1} = \frac{1}{R}\int_{T_1}^{T}\frac{C_p(T)\,dT}{T} = \frac{1}{R}\left[s^0(T) - s^0(T_1)\right] \end{aligned}\right\} \tag{4–75}$$

Here s^0 is the integral indicated from a reference temperature ($=0$ K) to T as given in the JANAF tables, Appendix C. Since s^0 and C_p are per mole, R is the universal gas constant.

EXAMPLE

Suppose H_2 is expanded from 2000 K through a pressure ratio of 10. The problem is to determine the final temperature.

$$T_1 = 2000 \text{ K and } s^0(2000) = 45.004$$

Thus according to eq. 4–75,

$$s^0(T_2) = s^0(2000) - 1.987 \ln 10$$

or

$$T_2 = 1105 \text{ K}$$

from the table for H_2.

Now if the change from 2000 K to 1105 K were small enough for the variation of C_p (as seen in the JANAF tables) to be sufficiently small that it may be considered constant, then the s^0 integral may be obtained analytically. Thus with C_p constant, eq. 4–75 may be integrated:

$$\ln \frac{p}{p_1} = \frac{C_p}{R} \ln \frac{T}{T_1} = \frac{\gamma}{\gamma - 1} \ln \frac{T}{T_1} \tag{4–76}$$

or

$$\frac{p}{p_1} = \left(\frac{T}{T_1}\right)^{\gamma/(\gamma-1)} \tag{4–77}$$

The fraction involving γ in eq. 4–76 results from the relation between specific heats (eq. 4–72) and the definition of γ as given by eq. 4–25.

$$\gamma \equiv \frac{C_p}{C_v} \tag{4–78}$$

The usefulness of γ as a descriptive parameter for the gas and the process it is undergoing is therefore limited to situations where the specific heats may be considered constant over the changes involved in the process. The very simple and convenient eq. 4–77 is similarly limited.

EXAMPLE

The 2000 K hydrogen gas may be seen to have a (local!) value of $C_p = 8.195$ so that $\gamma = 8.195/(8.195 - 1.987) = 1.32$. The final temperature for an expansion process over a pressure ratio of 10 is 1145 K according to eq. 4–77 which compares rather well with the previous, more exact calculation. It should be evident that *if* one were to determine the right value of average C_p

for the process, the calculations could yield identical answers. In changing from 2000 to 1100 K the variation of C_p for hydrogen is about 15% as seen from the tables.

This discussion applies to single-component gases and to gas mixtures. When chemical reaction may take place, the adiabatic reversible process may be handled analytically under two circumstances. One is where the process is so rapid that reaction rates are sufficiently slow that reactions do not take place, in the second they are so fast that chemical equilibrium is maintained throughout the process. In the first case, the methodology is to treat the composition as an unreacting mixture of unvarying or "frozen" composition. In the second, the composition adjusts to the changing conditions. The methodology for describing "equilibrium" process is discussed in Chapters 5 and 12.

An important application where nonperfect and chemically reacting gases undergo an isentropic process is the expansion in a nozzle for production of a high-speed jet. Section 12.7 is a discussion of the process and the associated calculation procedures.

PROBLEMS

1. For the isotherm of water $T_r = 0.7$, find v_r on the vapor saturation line. Find the reduced pressure at this state. Find two values of v_r assuming that the gas (1) is ideal and (2) obeys the VDW equation of state. Note in particular that the VDW description is close to the ideal gas result and thus not a very good approximation of the real value. Repeat for $T_r = 0.9$. Sketch the isotherms on the p-v plane so that the relative positions associated with the various descriptive models are shown.
2. Consider the state of water vapor where it is well described as an ideal gas. For example $T = 1500$ K and some as yet unspecified pressure. This gas is expanded isentropically through a temperature ratio of 3. Since the gas is triatomic the specific heat ratio γ may be assumed to be 4/3 (see value of C_p at 298 K). Find the pressure ratio and the enthalpy drop per unit mass for the expansion using a constant C_p (i.e., constant γ). Compare your results to those obtained using another logical value of γ—that derived from the value of C_p (1500 K). Tabulate your results for comparison with other calculations below.

   Using the JANAF tables, find the same quantities using a variable C_p and ideal gas model calculation.

   Finally consult the steam tables and determine the true results for an initial pressure you specify. Note that state 1 may not be in the tables and in particular that state 2 may be close to the saturated vapor state. Find the initial pressure, p_1, which leads to just reaching the saturation state and the value that is safe (say 10% error) for use of an ideal gas model.

3. Examine a gas with an alternate equation of state in the manner of Section 4.2.3. Consult a text on classical thermodynamics for various forms which may be used for this purpose.
4. A nonreacting mixture of 50% each by mole of CO and CO_2 is expanded from 1500 K through a pressure ratio of 20. Find the final temperature, assuming constant C_p and therefore γ at the initial condition, and compare this result to that obtained using the gas property tables.
5. A gas is found to have a linear variation of C_p between two temperatures. Find an analytic expression for Δh and Δs for a state change between T_1 and T_2. Compare your answers to that obtained for a gas that has such a specific heat variation. See Appendix C.

BIBLIOGRAPHY

Sears, *An Introduction to Thermodynamics, the Kinetic Theory of Gases and Statistical Mechanics*, Addison-Wesley, Cambridge, Mass., 1952.

REFERENCES

4–1. Bolz, R. E., Tuve, G. L., *Handbook of Tables for Applied Engineering Science*, The Chemical Rubber Co., Cleveland, Ohio, 1970.

5

THERMOCHEMISTRY: REACTIONS, EQUILIBRIUM, AND KINETICS

This chapter primarily concerns combustion and the consequences of gas physics and chemistry that have an impact on the way energy from fuels is converted and thus on the way engines must be designed. In addition, some more generally useful relations suitable for energy conversion not involving combustion are also developed.

Fossil fuels are chemically combined, or "burned," with oxygen or air to release the chemical energy associated with the formation of CO_2 and H_2O from the hydrocarbon fuel (Sections 1.7 and 1.8). Since most combustion processes take place in conversion devices at constant pressure, emphasis will be on this method. However, constant volume combustion as it takes place in the internal combustion engine is the single most important alternative way of burning hydrocarbon fuels and is discussed briefly in Section 5.5. The discussion includes the bookkeeping methodology required to determine the heat release and the temperature of the combustion products given the reactant conditions. The procedure will make use of the JANAF thermochemical data tables (Appendix C, ref. 5–1) and similar notation will be used here.

5.1. CHEMISTRY OF COMBUSTION WITH SPECIFIED PRODUCT COMPOSITION

If one *assumes* that the fuel is burned completely in air, then the combustion process may be described by:

$$C_nH_mO_p + a\ O_2 + 3.76a\ N_2 \rightarrow b_1\ CO_2 + b_2\ H_2O + b_3\ O_2 + 3.76a\ N_2 + \text{heat} \tag{5-1}$$

The fuel is characterized by an average molecule having *n* carbon, *m* hydrogen, and *p* oxygen atoms. The *p* oxygen atoms are present in solid fuels such as coal and in alcohols for example. Air contains approximately the equivalent (including trace components) of 3.76 moles of nitrogen for each mole of oxygen (see Appendix B). Certain combustion processes may be carried out with O_2-enriched air in order to achieve higher temperatures in the combustion process; see, for example, Chapter 11.

The approximation made initially is that only CO_2 and H_2O are produced and there are no incomplete combustion products such as CO, H_2 and minor products such as NO, NO_2, etc. By this method, one calculates the highest possible heat release.

Species conservation statements for the atoms of C, H, O give values of b_1, b_2, and b_3 for a given amount of O_2 supplied, a:

$$\text{C}: n = b_1$$

$$\text{H}: m = 2b_2$$

$$\text{O}: p + 2a = 2b_1 + b_2 + 2b_3$$

or

$$b_3 = \frac{1}{2}\left\{p + 2a - 2n - \frac{m}{2}\right\} \tag{5–2}$$

If a is chosen such that $b_3 = 0$, then the heat released is shared with the smallest mass of combustion products, and hence, the temperature of these products for

$$a = n + \frac{m}{4} - \frac{p}{2} = a_s$$

should be greater than for $a > a_s$. If $a < a_s$, the C and H components cannot be burned completely to CO_2 and H_2O and the temperature of the products decreases as a is decreased below a_s. Thus a maximum temperature is achieved for a near a_s. This statement is true only within the approximation that minor products are neglected. More realistic calculations to be described subsequently show this conclusion to be very good for many problems of engineering interest.

One refers to $a = a_s$ as stoichiometric combustion, $a < a_s$ as fuel-rich and $a > a_s$ as fuel-lean or air-rich combustion. The ratio a_s/a is defined as the fuel/air equivalence ratio, ϕ. $\phi > 1$ is for fuel rich combustion.

The calculation of heat released in a chemical reaction requires knowing how much material reacted. In the following, the product gas composition is assumed to be known. It may have been determined experimentally or the assumption of complete combustion may be used for cases where excess oxygen is available, $\phi \leq 1$.

The calculation method to be developed assumes that the reactants and products are gaseous. The reactants may be liquid or solid fuels which undergo a pyrolysis and phase-change reaction where heat is used to break up the complex and heavy molecules into lighter gaseous ones.

In comparison to the chemical heat release, pyrolysis heat is not large for most circumstances, but must be accounted for if accurate results are sought. In some circumstances, where simple fuels are involved, the heat associated with the change of phase from liquid to gas is the most important heat involved in the fuel gasification process. Figure 5–1 summarizes the elements of this process. Examples where the effects of fuel processing heat are important include the burning of wet wood, of water/alcohol mixtures, and so on.

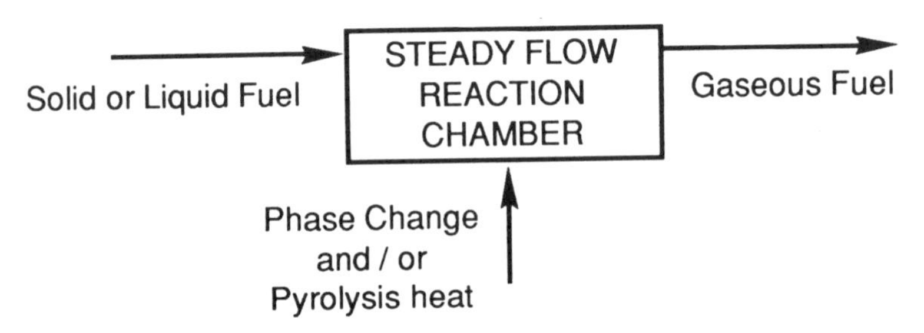

Fig. 5–1. Conversion of a fuel to gaseous reactants.

5.2. HEAT OF FORMATION

According to the First Law of Thermodynamics for a flow process with constant pressure, the heat released in a chemical reaction at constant pressure is an enthalpy difference between two thermodynamic states (eq. 4–13)

$$d'q = dh - v\,dp$$

or

$$q_{1-2}(p = \text{constant}) = h_2 - h_1$$

so that the absolute value of the enthalpy in either state is not of interest and a reference may be chosen for convenience. It is customary to assign a zero value of the enthalpy to a particular chemical state at a chosen temperature. For the work here, the thermodynamic standard state is where the chemical elements are most abundantly found at standard temperatures and pressure (STP, $p = 1$ atm, $T = 298$ K). Hence the heats for formation are zero by definition for gaseous H_2, N_2, O_2, solid C, etc. The heat of formation of any compound or other phases of these elements is the heat required to form 1 *mol* of the product from the elements in the thermodynamic standard state.

Examples:

$$\begin{aligned}
&1.\ C(s) + O_2 + \mathbf{\Delta H_f}(CO_2) \rightarrow CO_2(g)\\
&2.\ C(s) + \mathbf{\Delta H_f}(C(g)) \rightarrow C(g)\\
&3.\ H_2 + \tfrac{1}{2}O_2 + \mathbf{\Delta H_f}(H_2O) \rightarrow H_2O(g)\\
&4.\ H_2(g) + \mathbf{\Delta H_f}(H_2(l)) \rightarrow H_2(l)
\end{aligned} \tag{5–3}$$

Here s, l, and g refer to solid, liquid and gaseous phases. Table 5–1 gives the heats of formation for substances of interest in combustion problems. Specifically, ΔH_f is the heat that must be supplied to a constant pressure, steady-flow reaction chamber where reactants enter and products exit, both at the reference temperature (see Fig. 5–2).

Table 5–1. Heats of formation in kcal/gm-mol at $T_{ref} = 298$ K

Compound	ΔH_f	**Compound**	ΔH_f
O_2	0	CH_4	−17.895
H_2	0	C_3H_8	−24.82
H_2 (l)	−1.92	H_2O (g)	−57.798
C (s)	0	H_2O (l)	−68.31
C (g)	170.890	$CH_{1.93}$ (JP-4)	−5.89
CO	−26.417	$CH_{1.55}$ (JP-3)	−3.61
CO_2	−94.054	O	59.559
H	52.100	N	112.965
OH	9.432	NO	21.58

Note that reactions that release heat are exothermic ($\Delta H_f < 0$), whereas endothermic reactions require heat input. Exothermic reactions generally result in the formation of stable products.

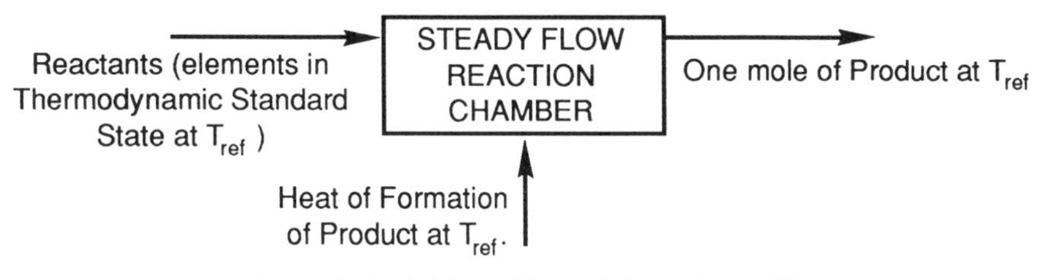

FIG. 5–2. Definition of heat of formation, ΔH_f.

5.3. HEAT OF REACTION

For chemical reactions where standard-state elements are not involved, the difference between the heats of formation for the products and the reactants is called the *heat of reaction*

$$H_{RP} = \sum \Delta H_{f,P} - \sum \Delta H_{f,R} \tag{5–4}$$

where $\sum \Delta H_f$ is the number of moles times the enthalpy per mole (of reactants or products) summed over all mixture components (see Fig. 5–3).

Example:

Reactants: CH_4 and O_2.
The elements in the standard state are: C(s), H_2 and O_2.
The products after reaction are (so specified): CO_2 and H_2O.

The heat of reaction is, like most heats of formation, negative for combustion processes and a measure of the heat released per mole of fuel. For the (fuel-lean, $a \geq 2$) combustion of methane (natural gas),

$$CH_4 + a\,O_2 \rightarrow 2H_2O + CO_2 + (a - 2)O_2$$

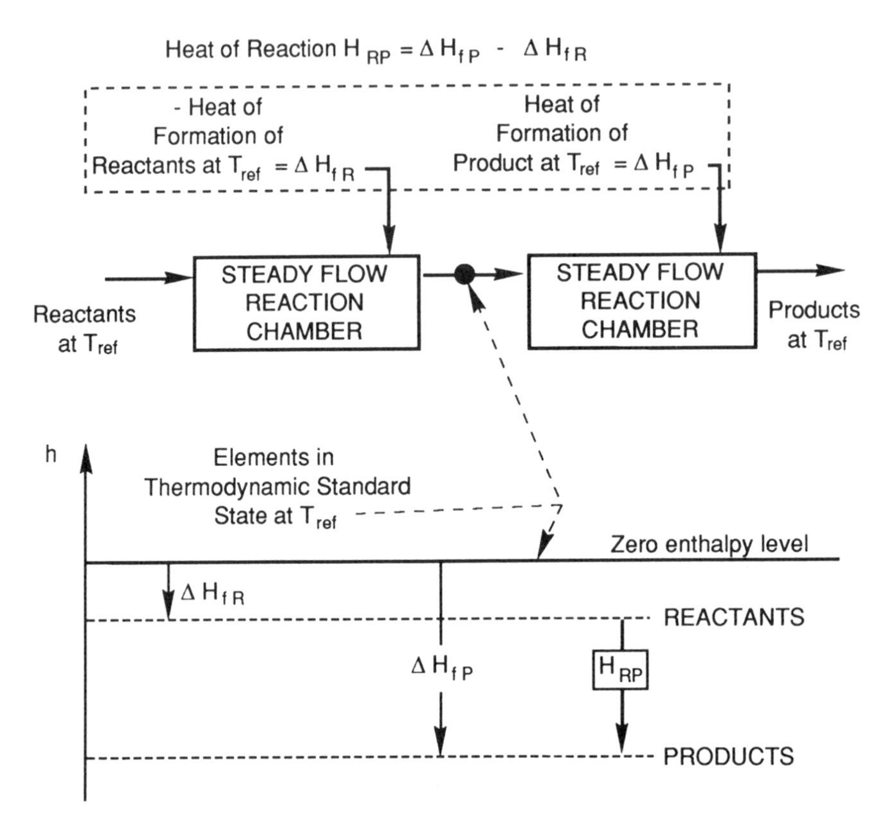

FIG. 5–3. Heat of reaction, H_{RP}, from the heats of formation of reactants and products.

the heat of reaction is:

$$
\begin{aligned}
H_{RP} &= 2\,\Delta H_f(H_2O) + \Delta H_f(CO_2) + (a-2)\,\Delta H_f(O_2) - [\Delta H_f(CH_4) + a\,\Delta H_f(O_2)] \\
&= 2(-57.8) + (-94.05) - (1)(-17.9) - (a-2-a)(0) \\
&= -191.75 \text{ kcal/mole } CH_4 \qquad (5\text{–}5)
\end{aligned}
$$

Note that 1 mole of CH_4 weighs 16 g or 35.3×10^{-3} lb and with 0.252 kcal/BTU this H_{RP} corresponds to a heating value of 21,500 BTU/lb (of CH_4), with the H_2O assumed to remain in gaseous phase. If the H_2O is allowed to condense, additional heat evolves and the quantity is described as the higher heating value (HHV) in contrast to the lower value (LHV) calculated here (see Table 1–3). Table 5–1 shows that the difference in heating values amounts to 10.5 (=68.3 – 57.8) kcal per mole H_2O (=21 kcal/mole CH_4) or about 10% of the H_{RP}.

In this example the heat of formation of CH_4 is also about 10% (17.9/172) of the LHV, which indicates the magnitude of the penalty one must pay for having the fuel packaged as CH_4 rather than solid C and gaseous H_2. Considering the mechanical problems associated with storage and transport

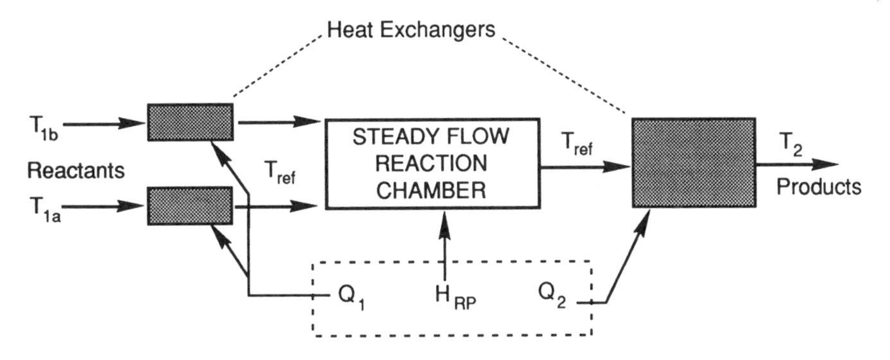

FIG. 5–4. Heat balance for reactants and products *not* at the reference temperature. Heat exchangers are required to change materials temperature between T_{ref} and T.

which are eased by having the fuel as CH_4, this appears to be an acceptable penalty.

5.4. ADIABATIC FLAME TEMPERATURE

In general, neither the reactants nor the products of a combustion reaction are involved at the reference temperatures. One might, for example, carry out the combustion with very small heat losses to the external environment, and one may ask: What temperature can be achieved under such circumstances? The combustion process may be imagined to be carried out in the manner illustrated in Fig. 5–4, where reactants enter the process at temperature T_1 (which may be different for fuel and air). One or two heat exchangers are used to add Q_1 to raise (or lower) the temperature of the reactants to T_{ref}.

The reaction chamber converts *reactants* to *products* at T_{ref}. A second heat exchanger "raises" the product's temperature to T_2. The heats involved are shown in Fig. 5–4, and their sum (dashed box) is the loss to the environment. If this loss is zero, the resultant product temperature, T_2, is the *adiabatic flame temperature for the specific reactants at their entry temperatures.* These processes are illustrated in a plot of enthalpy versus temperature in Fig. 5–5. Shown in the figure are the enthalpies of products and reactants. The differing slopes of the curves reflect the differing and varying values of C_p of the gas mixtures.

As an illustration of the procedure for determining the adiabatic flame temperature T_2, consider the following:

> Methyl alcohol liquid at 298 K is burned stoichiometrically with pure oxygen preheated to 500 K. Assume that the combustion products consist only of CO_2 and H_2O. (Note that this assumption is unrealistic since the temperature will turn out to be so large that CO_2 and H_2O will be dissociated. A more correct calculation is described in Section 5.12.)

$$CH_3OH + \tfrac{3}{2}O_2 \rightarrow CO_2 + 2\,H_2O \tag{5–6}$$

From Fig. 5–4

$$Q_1 = -H^*(CH_3OH, 298) - \tfrac{3}{2}H^*(O_2, 500) \tag{5–7}$$

where $H^*(i, T)$ is the enthalpy difference $H_i(T) - H_i(298)$ for component i is written this way because this difference is given in the JANAF tables. Thus

$$Q_1 = -0 - (1.5)(1.455) \text{ kcal/mole } CH_3OH$$

$$\begin{aligned} H_{RP} &= \Delta H_f(CO_2) + 2\,\Delta H_f(H_2O) - \Delta H_f(CH_3OH) \\ &= -94.05 + 2(-57.80) - (-57.04) \\ &= -152.61 \text{ kcal/mole } CH_3OH \end{aligned} \tag{5–8}$$

For an adiabatic process,

$$\begin{aligned} Q_2 &= -Q_1 - H_{RP} = 154.8 \text{ kcal/mole } CH_3OH \\ &= H^*(CO_2, T_2) + 2H^*(H_2O, T_2) \end{aligned} \tag{5–9}$$

which may be tabulated as a function of T_2.

T_2 (K)	$H^*(H_2O)$	$H^*(CO_2)$	Q_2
1000	6.209	7.98	20.4
2000	17.37	21.85	56.59
3000	30.20	36.53	96.93
4000	43.80	51.54	139.14
5000	57.83	66.75	182.41

Interpolating between 4000 and 5000 gives $T_2 = 4363$ K. Interpolation again between 4300 and 4400 K where Q_2 is 152.04 and 156.35, respectively, gives

$$T_2 = 4300 + 100\,\frac{154.8 - 152.0}{156.4 - 152.0} = 4365 \text{ K}$$

The closeness of these two estimates is related to the linearity of the functions involved.

If the gas were to be cooled to 3200 K by the addition of N_2 at 298 K, find the amount of N_2 required per mole CH_3OH.

This problem does not require iteration because T_2 is known. The reaction is

$$CH_3OH + \tfrac{3}{2}O_2 + x\,N_2 \rightarrow CO_2 + 2H_2O + x\,N_2 \tag{5–10}$$

Since the temperature of N_2 is 298 K the enthalpy contribution to Q_1, on the

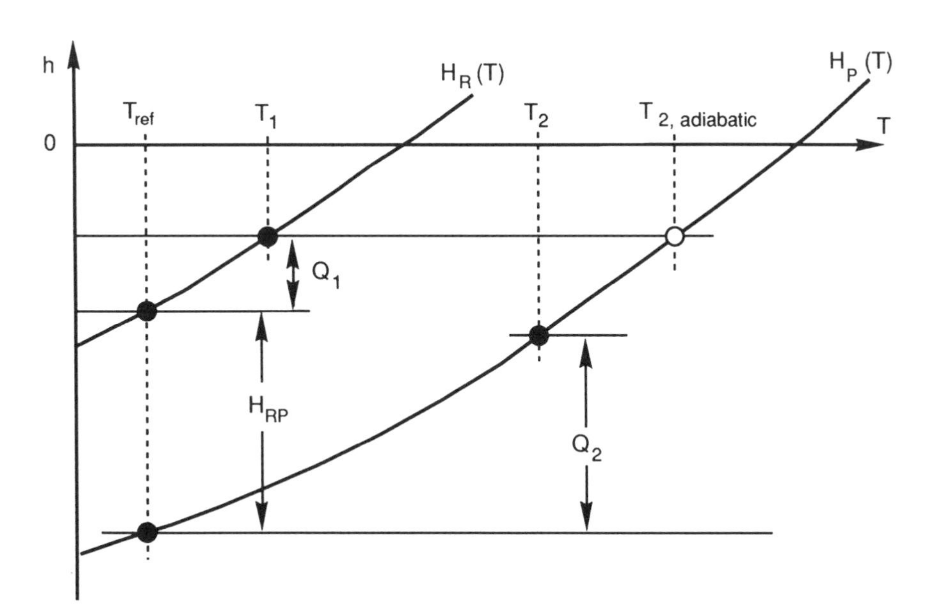

FIG. 5–5. Temperature enthalpy relations during the combustion process.

reactant side, is zero. N_2 is not involved in the chemical reaction, hence H_{RP} is not altered. Only Q_2 involves an additional x moles of N_2

$$-Q_1 - H_{RP} = 154.8 = H^*(CO_2, 3200) + 2\,H^*(H_2O, 3200) + xH^*(N_2, 3200)$$

$$154.8 = 39.51 + 2(32.88) + 23.94x \tag{5–11}$$

whence x = 2.07 moles of N_2 are required per mole of CH_3OH.

5.5. CONSTANT VOLUME COMBUSTION

Constant volume combustion processes, like those occurring in the Otto cycle engine, can be approximated by a short adiabatic process wherein no work is removed from the gas during combustion. The First Law then states that the heat released equals the change in internal energy of the gas, or

$$\Delta U + Q = 0 \tag{5–12}$$

Again this statement is divided into three steps to account for reactants and products not being at the reference conditions. Thus,

$$(U_{R,\,ref} - U_{R1}) + U_{RP} + (U_{P2} - U_{P,\,ref}) = 0 \tag{5–13}$$

for adiabatic reactions.

The JANAF tables list enthalpies of gases from which the internal energies required can be readily generated. For any substance the internal energy, written as $U(T) - U(T_{ref})$, can be found from the definition of $H(= U + pV)$, which, for an ideal gas, may be written

$$H = U + R_u T \tag{5–14}$$

Hence

$$U(T) - U(T_{ref}) = H(T) - H(T_{ref}) - R_u(T - T_{ref}) \tag{5–15}$$

Expressions of this kind may be used to calculate the internal energy differences $U_{R1} - U_{R,ref}$ and $U_{P2} - U_{P,ref}$. The heat released in the reaction of transforming reactants to products, both at T_{ref}, is obtained from:

$$\begin{aligned} U_P - U_R &= \sum_{i,P} n_i U_i - \sum_{i,R} n_i U_i \\ &= \sum_{i,P} (u_i + R_u T_f) - \sum_{j,R} (u_j + R_u T_f) + U_P - U_R + R_u T_f\left(\sum_{i,P} n_i - \sum_{j,R} n_j\right) \end{aligned} \tag{5–16}$$

5.6. EQUILIBRIUM AND ENTROPY

It is known from elementary thermodynamics that the entropy of an isolated system cannot decrease. This is as much information as can be derived from the Second Law. There is, however, an additional principle applied to possible state changes which can be stated in the form: "The entropy of isolated systems tends to increase." This statement means that the entropy of an isolated system will increase if conditions permit it, and will continue to increase until a definite state of equilibrium is reached.

The principle of increasing entropy bears some resemblance to a theorem in mechanics: the potential energy of a mechanical system decreases if conditions permit and continues to decrease (with friction present) until a definite state of equilibrium is reached. The mechanical theorem is applied in statics to determine the conditions of mechanical static equilibrium. If V is the potential energy of a mechanical system, the work done in displacing it slightly from some initial condition is δV. If δV is calculated and found to be positive for all possible displacements that satisfy the constraints of the system, then the initial condition is one of stable equilibrium (i.e., V must be a minimum). In most cases, the condition for equilibrium is $\delta V = 0$ to the first order displacements, whereas the test for stability requires calculation of the higher-order terms.

An argument similar to that described above can be used to establish a criterion for chemical equilibrium of an isolated system in thermodynamics (see Fig. 5–6). Suppose one imagines a small change to take place in the isolated

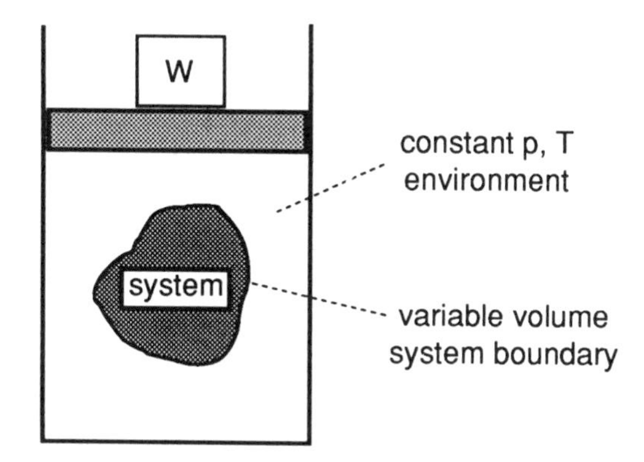

Fig. 5–6. Reaction control volume for chemical changes: constant pressure and temperature system and surroundings

system so that no constraint is violated (e.g., the energy remains constant and the number of atoms involved is invariant). For example, there is an increase in one of the molecular constituents. Let the change in entropy of the system be δS, a quantity that can be calculated even though it may not correspond to a real change. If, for all such displacements of the isolated system, one finds that $\delta S \leq 0$, then the original condition must be a state of equilibrium. This is seen to be correct because the entropy of an isolated system cannot decrease, and if all possible displacements of the system require a decrease of entropy, then no displacement can actually take place and the system is in equilibrium.

The condition for equilibrium is satisfactory for isolated systems, but one is more often interested in systems that have some connection with their surroundings. Before determining the criteria in these cases it is convenient to involve the Gibbs free energy, a property defined in Section 4.2, eq. 4–15.

5.7. GIBBS FREE ENERGY IN CHEMICAL SYSTEMS

The work done by a system in a reversible process is, according to the First Law and the definition of entropy ($T\,dS = dQ_{rev}$)

$$dW = T\,dS - dU \tag{5–17}$$

If the system is imagined to be immersed in a medium at pressure p, the "useful" work is

$$dW' = T\,dS - dU - p\,dV \tag{5–18}$$

since work $p\,dV$ is expended in the medium and only the net is available as useful work. If the pressure and temperature are held constant, then

$$(dW')_{T,p} = -d(U + pV - TS) = -d(H - TS) = -dG \tag{5–19}$$

and the work done by a system in this reversible change is the decrease in the quantity $G = H - TS$. This combination of properties occurs frequently in connection with systems undergoing changes with constant p and T and is the Gibbs free energy which is easily interpreted from eq. 5–19. On the basis of a unit mass, $g = h - Ts$.

Consider again the criterion for equilibrium. In the following, the conditions for equilibrium for certain types of nonisolated systems will be established. Suppose one has a finite heterogeneous system—that is, one consisting of various phases and chemical constituents that may possibly react. Imagine the surroundings of the system to consist of a very extensive body of homogeneous substance at a uniform pressure and temperature. The only interactions between the system and the surroundings will be the result of work done as the volume of the system changes and/or heat flows across the boundary of the system, and it is assumed that the system and surroundings are in mechanical and thermal equilibrium at p and T. The combination of system and surroundings forms an isolated system, and the condition for equilibrium of the combination is that

$$\delta S + \delta S_0 \leq 0 \qquad (5\text{–}20)$$

for all possible (i.e., virtual) changes. Here δS is the increase of entropy of the system and δS_0 is that of the surroundings.

In the change which is imagined to take place, the pressure and temperature will remain constant because the surroundings are very large. Hence the increase of entropy of surroundings is $\delta S_0 = \delta Q/T$, where δQ is the heat flow from the system into the surroundings. The increase of entropy of the system cannot, in general, be represented by $\delta Q/T$ because phase changes and chemical reactions may be involved, and these contribute to the entropy change even when T is constant. The heat transferred to the surroundings can be written as a gain of $-\delta Q$ to the system so that the criterion for equilibrium for the system can be stated as

$$\delta S - \frac{\delta Q}{T} \leq 0 \qquad \text{or} \qquad T\,\delta S - \delta Q \leq 0 \qquad (5\text{–}21)$$

for all possible displacements. From the first law of thermodynamics, the heat flow δQ can be replaced by $\delta Q = \delta U + \mathrm{p}\,\delta V$ where δU is the increase of energy for the system and $\mathrm{p}\,\delta V$ is the work done by the system on the surroundings. Hence the criterion for equilibrium becomes (with $T > 0$)

$$\begin{aligned} &T\,\delta S - \delta U - p\,\delta V \leq 0 \\ \text{or} \quad &\delta(TS) - S\,\delta T - \delta U - \delta(\mathrm{p}V) + V\,\delta\mathrm{p} \leq 0 \\ \text{or} \quad &\delta(H - TS) + S\,\delta T - V\,\delta p \geq 0 \qquad (5\text{–}22) \end{aligned}$$

for all virtual changes.

The criterion above can be expressed in terms of a property when changes are restricted to the temperature and pressure remaining constant:

$$\delta G \geq 0 \tag{5–23}$$

The inequality sign is the condition for stability. Usually the condition for equilibrium alone is desired for which the equality sign applies. According to eq. 5–19, this may be interpreted as: All state changes away from equilibrium require external work input.

5.8. THE LAW OF MASS ACTION FOR IDEAL GAS MIXTURES

The Gibbs free energy for a mixture of ideal gases is written in terms of the enthalpy and the entropy of the system. Recognizing the system to be the gas mixture, it is appropriate to use h and s defined on a "per mole" basis.

$$G = H - TS = \sum n_j h_j - T \sum n_j s_j \tag{5–24}$$

where the sum is over the components of the mixture. For each component j one can write

$$h_j = \int_{T_0}^{T} C_{pj}\, dT + h_{0,j} \qquad \text{where } h_{0,j} = h_j(T_0) \tag{5–25}$$

and

$$s_j = \int_{T_0}^{T} C_{pj} \frac{dT}{T} - R \ln \frac{p_j}{p_0} + s_{0,j} \qquad \text{where } s_{0,j} = s_j(p_0, T_0) \tag{5–26}$$

where T_0 and p_0 are the reference temperature and pressure (usually 298 K and 1 atm). The enthalpy and entropy are per mole of component j referenced to T_0, p_0.

Thus the Gibbs free energy may be written as the sum of terms each of which depends only on temperature and pressure:

$$G = \sum n_j \left\{ g_j^0(T) + RT \ln \frac{p_i}{p_0} \right\} \tag{5–27}$$

where the function,

$$g_j^0(T) \equiv h_{0,j} + \int_{T_0}^{T} C_{pj}\, dT - T s_{0,j} - T \int_{T_0}^{T} C_{pj} \frac{dT}{T} \tag{5–28}$$

is the temperature-dependent portion of the Gibbs free energy for each gas component, j.

The mole fraction is $x_j = p_j/p$ (eq. 4–60) where p is the pressure of the mixture. The partial pressure terms are

$$\frac{p_j}{p_0} = x_j \frac{p}{p_0} \qquad \text{or} \qquad \ln \frac{p_j}{p_0} = \ln x_j + \ln \frac{p}{p_0} \tag{5–29}$$

and the mixture's Gibbs free energy change is

$$\delta G = \sum \delta n_j \left\{ g_j^0(T) + RT \ln \frac{p_j}{p_0} \right\} + \sum n_j RT \frac{\delta x_j}{x_j} \tag{5–30}$$

Since $n_j = x_j n$, the second summation equals $\sum nRT\,\delta x_j = p \sum \delta x_j = 0$ since $\sum x_j = 1$ (eq. 4–65), and it follows that

$$\sum \delta n_j \left\{ g_j^0(T) + RT \ln \frac{p_j}{p_0} \right\} = 0$$

The differentials δn_j must be proportional to the stoichiometric coefficient of component j since this coefficient determines in what *relative* proportion the reacting components interact. Denoting these coefficients by ν_j, one has

$$\delta G = 0 = \sum \nu_j \left\{ g_j^0(T) + RT \ln \frac{p_j}{p_0} \right\} \tag{5–31}$$

EXAMPLE

As an example, consider the reaction

$$CO_2 \Leftrightarrow CO + \tfrac{1}{2}O_2$$

written with all terms on the left hand side

$$CO_2 - CO_2 - \tfrac{1}{2}O_2 \Leftrightarrow 0$$

One has $j = 1(CO_2)$, $\nu_1 = 1$; $j = 2(CO)$, $\nu_2 = -1$; and $j = 3(O_2)$, $\nu_3 = -\frac{1}{2}$. Equation 5–32 may be rewritten as

$$\sum \ln \left(\frac{p_j}{p_0} \right)^{\nu_j} = - \frac{\sum \nu_j g_j^0}{RT}$$

or, taking antilogs,

$$\prod_j p_j^{\nu_j} = \prod_j p_0^{\nu_j} \exp - \frac{\sum \nu_j g_j^0}{RT} \equiv K_p(T) \tag{5–32}$$

Here $\prod$ is the product over j (viz., $\prod_{j=1}^{j=4} A_j = A_1 A_2 A_3 A_4$).

This is the *law of mass action* written in terms of partial pressures. $K_p(T)$ is the equilibrium "constant" (at a given T). This form is convenient for many purposes because K_p is independent of the pressure and is thus easily tabulated as in the JANAF tables (see Appendix C). Note that K_p has dimensions of pressure given by $(1\ \text{atm})^{\Sigma v_j}$. This coefficient sum is zero when the number of molecules of reactants equals that of the products. An example of such an (equimolal) "shift" reaction is:

$$H_2O + CO \Leftrightarrow H_2 + CO_2$$

Equimolal reactions are therefore independent of pressure. The sum $\sum v_j$ is not zero for the reaction:

$$N_2 \Leftrightarrow 2N$$

Figure 5–7 is a plot of a number of reaction K_p's of interest in combustion problems. Note the very strong temperature dependence. The triatomic molecules have large K_p at low temperature whereas the monatomic species have very small values in this range.

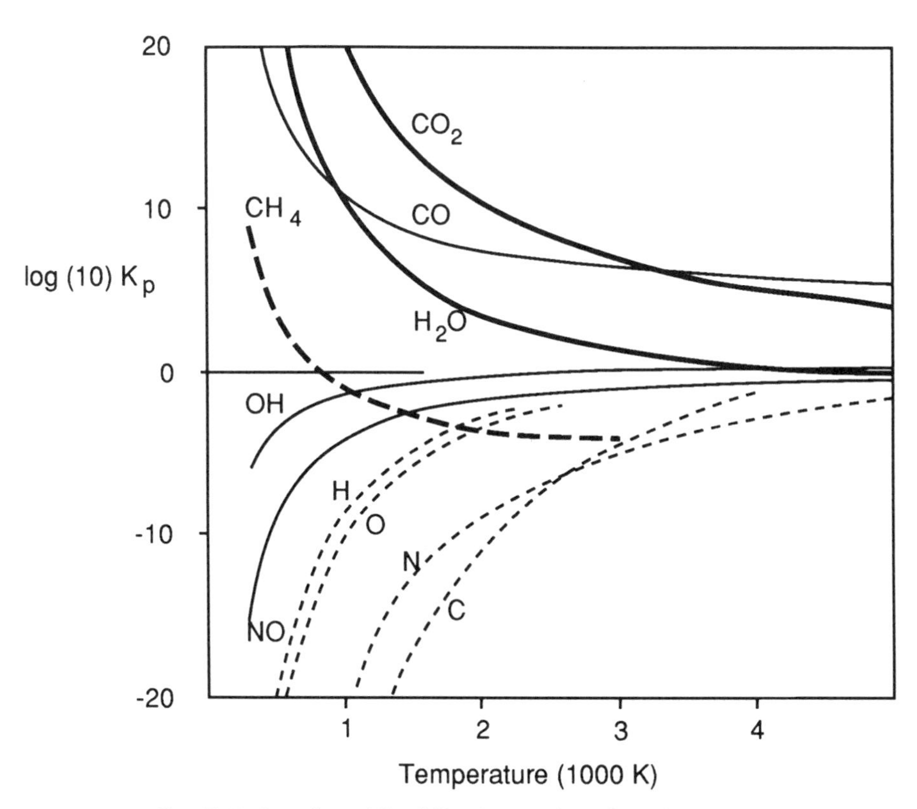

FIG. 5–7. Log (base 10) of K_p of a number of product gases.

The form of the law of mass action written as eq. 5–33 shows that the equilibrium constant is equal to

$$K_p = C \exp - \frac{\Delta G_{RP}}{RT} \qquad \text{where } C = \prod_j p_0^{\nu_j} = 1 \qquad \text{if } p_0 = 1 \text{ atm}$$

ΔG_{RP} is given by eq. 5–28 as

$$\Delta G_{RP} = \sum_{j,\,\text{products}} \nu_j g_j^0 - \sum_{j,\,\text{reactants}} \nu_j g_j^0$$

For gases, g^0 is given by eq. 5–28. The integral of the log T (last term) is such a weak function of temperature that it may be considered a constant. Thus, the last three terms of eq. 5–28 contain factors approximately proportional to T so that this portion of $\Delta G_{RP}/RT$ is itself a constant for all contributions. The principal element describing the temperature dependence of the equilibrium constant is therefore of the form

$$K_p \cong \exp - \left\{ \frac{\sum_{j,P} \nu_j h_{0j} - \sum_{j,R} \nu_j h_{0j}}{RT} + \text{constant} \right\} = (\text{constant}) \exp - \left(\frac{H_{RP}}{RT} \right) \tag{5–33}$$

The validity of the above discussion can be verified by a check of the variation of the logarithm of K_p with $1/T$ as shown in Fig. 5–8 for the formation reaction of several compounds. Appendix C gives the curve fit constants for the various K_p's of interest in the H/C/N/O combustion system.

Another form is obtained from eq. 5–32 when $p_j/p_0 = x_j p/p_0$ is written as in the beginning of the development. Then

$$\prod x_j^{\nu_j} = K_c(p, T) = \left[\exp - \frac{\sum \nu_j g_j^0}{RT} \right] \left(\frac{p}{p_0} \right)^{\Sigma \nu_i} \tag{5–34}$$

This is the law of mass action written in terms of concentration. Note that K_c is a function of pressure, hence more awkward to tabulate.

5.9. THE EQUILIBRIUM CONSTANT FOR SPECIFIC REACTIONS

The law of mass action (eq. 5–33) may be written in an infinite number of ways depending on the form of the reaction statement. Since the oxygen dissociation reactions

$$O_2 \Leftrightarrow 2O, \; 2O \Leftrightarrow O_2, \qquad \text{and} \qquad \tfrac{1}{2}O_2 \Leftrightarrow O$$

are all entirely equivalent, with each involving various assignments of ν_j, it

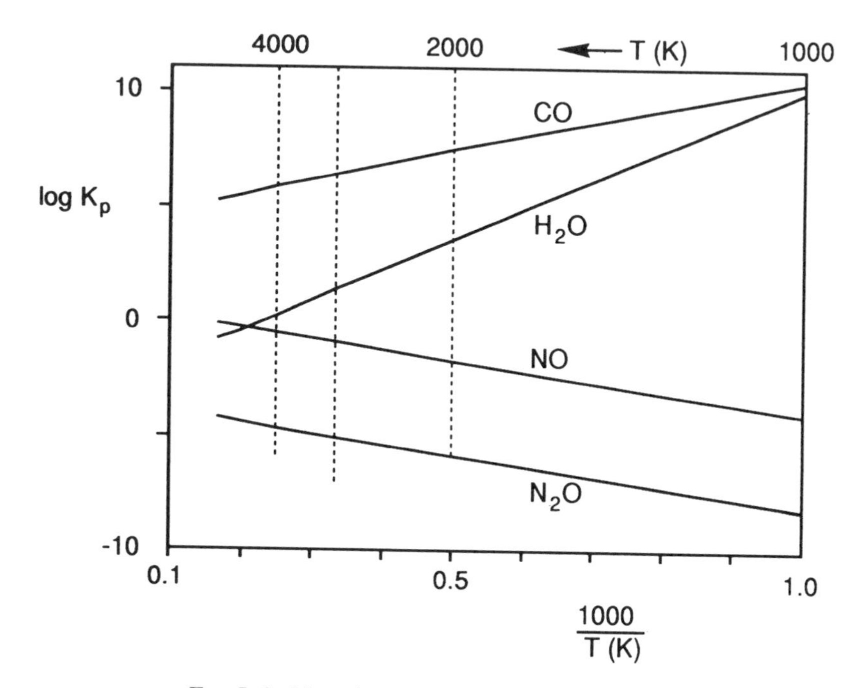

FIG. 5–8. Linearity relation check on eq. 5–34.

follows that the K_p that corresponds to each of the above will be different. Caution must therefore be exercised in interpreting numerical values of K_p.

The JANAF tables present K_p data in a standardized, unambiguous way. The tables are generated for each substance, and hence K_p is logically determined for the reaction that yields the 1 *mole* of the substances from the elements in their thermodynamic standard state. In tables for CH_4, CO_2, H_2O, N_2, N one finds the logarithm (base 10) of K_p for the following reactions

$$\begin{aligned} CH_4 &\Rightarrow 2H_2 + C(s) \\ CO_2 &\Leftrightarrow C(s) + O_2 \\ H_2O &\Leftrightarrow H_2 + \tfrac{1}{2}O_2 \\ N_2 &\Leftrightarrow N_2 \\ N &\Leftrightarrow \tfrac{1}{2}N_2 \end{aligned} \tag{5–35}$$

where the diatomic gases and solid carbon are the standard-state quantities involved. N_2 is already in the standard state, hence $K_p = 1$.

In summary: the JANAF data tables give

$$\log_{10} K_p = \log_{10} \frac{p^1(\text{partial p. of substance})}{\prod_j p_j^{\nu_j}(\text{of the elem. in std. state})} \tag{5–36}$$

with all partial pressures in atmospheres.

The values of K_p for reactions not involving the standard state are obtained by appropriate product of K_p's for the elementary formation reactions.

EXAMPLE 1

Consider the determination of K_p for the reaction:

$$CO_2 \Leftrightarrow CO + \tfrac{1}{2}O_2$$

$$K_p \text{ (this reaction)} = \frac{p_{CO}\sqrt{p_{O_2}}}{p_{CO_2}} = \frac{K_p(CO)}{K_p(CO_2)}$$

where

$$K_p(CO_2) = \frac{p_{CO_2}}{p_{C(s)}p_{O_2}} \quad \text{and} \quad K_p(CO) = \frac{p_{CO}}{p_{C(s)}\sqrt{p_{O_2}}} \tag{5–37}$$

Although $p_{C(s)}$ canceled out of this reaction (it should, since it is not present), $p_{C(s)}$ is 1 atm in this formulation. An example of a reaction involving solid carbon is the water gas reaction:

$$C + H_2O \Leftrightarrow H_2 + CO$$

K_p for this reaction is

$$K_p = \frac{p_C p_{H_2O}}{p_{H_2}p_{CO}} = \frac{p_{H_2O}}{p_{H_2}p_{CO}} = \frac{K_p(H_2O)}{K_p(CO)} \tag{5–38}$$

The pressure of a gas mixture may influence the equilibrium composition.

EXAMPLE 2

Consider the dissociation reaction of nitrogen. The problem is to find the composition of a mass of nitrogen heated to a temperature sufficiently high for atomic N to be present.

$$N_2 \rightarrow aN_2 + bN_2 + \text{(nothing else, an assumption)}$$

Conservation of N atoms states

$$2 = 2a + b \quad \text{or} \quad b = 2(1 - a)$$

and the total number of moles is

$$a + b = 2 - a$$

The partial pressures of the mixture components, in terms of the total mixture pressure, are:

$$p_N = \frac{b}{2-a} p = \frac{2(1-a)}{2-a} p \qquad \text{and} \qquad p_{N_2} = \frac{a}{2-a} p$$

The law of mass action gives ($N \Leftrightarrow \frac{1}{2}N_2$)

$$K_p = \frac{p_N}{\sqrt{p_{N_2}}} = \frac{\frac{2(1-a)}{2-a} p}{\sqrt{\frac{a}{2-a} p}} = \frac{2(1-a)}{\sqrt{a}} \frac{1}{\sqrt{2-a}} \sqrt{p}$$

Note that K_p carries units of $(\text{atm})^{1/2}$. By writing this equation as

$$\frac{K_p(T)}{\sqrt{p}} = f(a) \tag{5–39}$$

One may conclude that a, and therefore the mole fraction of N that is $a/(2-a)$, depends on p and T. The table for K_p gives

T(K)	log K_p
300	−79.3
1000	−21.5
3000	−4.86
6000	−0.62

For general values of $K_p/\sqrt{p}$ the solution of eq. 5–39 gives a. A large K_p (very large T) requires that a be very small. For small values of the K_p parameter the function $f(a)$ is dominated by the $(1-a)$ term. In that limit a approaches unity, and little dissociation is to be expected.

$$\text{Large } (\log K_p > 0)\text{: } a \cong \frac{2p}{K_p^2} \ll 1$$

$$\text{Large } (\log K_p < 0)\text{: } a \cong 1 - \frac{K_p}{2\sqrt{p}} \approx 1$$

Thus it is clear that as temperature decreases, $a \to 1$ and therefore $b \to 0$.
The form of the equation for $f(a)$ also reveals that increasing T and hence K_p has the same consequence on a as does lowering the pressure. Either change results in increased dissociation.

5.10. THE IONIZATION REACTION

Vapors of the alkali metals may be made to undergo the reaction

$$A \Leftrightarrow A^+ + e^-$$

for the production of current-carrying charges. A gas mixture containing an equal number of positive and negative charges is called a *plasma*. The equilibrium statement is

$$\frac{p_e p_{a^+}}{p_a} = K_p(T)$$

In this example, the form of K_p is particularly simple because of the similarity between the A and A^+ molecules and the simplicity of the new fragment, the electron. Recall that

$$\ln K_p = \frac{\sum_i \nu_i g_i^0}{RT}$$

where g_i^0 is a summation (or an integral when the states may be assumed not to be quantized) over all energy states available to the various reactants. In fact, one can write the equilibrium statement as

$$\prod_i p_i^{\nu_i} = \prod_i Z_i^{\nu_i}$$

where Z_i is the partition function (see for example, ref. 5–2). Thus

$$\frac{p_e p_{a^+}}{p_a} = \frac{Z_e Z_{a^+}}{Z_a}$$

For the alkali metal atoms, where the chemically active electron is alone in the outermost shell, the energies available to A, A^+ molecules are "identical" except for the energy required to remove the outer electron (the ionization energy). Thus since

$$K_p \sim \exp - \left(\frac{\sum \nu_i g_i^0}{RT}\right) \sim \exp - \left(\frac{\varepsilon_i}{kT}\right)$$

Here ε_i is the ionization potential (=3.89 ev for Cs, =4.54 ev for K, for examples). Note the conversion: 1 ev = 11,600°K.

The only energy states available to the electron are in translation where each of the three freedom axes gives $\frac{1}{2}kT$. A thorough development gives the constants so that one obtains for the equilibrium equation (with $p = nkT$)

$$\frac{n_e n_{a^+}}{n_a} = \left(\frac{2g_i}{g_0}\right)\left(\frac{2\pi m_e kT}{h^2}\right)^{3/2} \exp - \left(\frac{\varepsilon_i}{kT}\right) \tag{5-40}$$

which is known as the *Saha* equation. Note (g_i/g_0) is the spin degeneracy ratio ($=1$ for Cs, K; the choice of g as a symbol is customary and unrelated to the Gibbs free energy). Values for Planck's constant (h) and Boltzmann's constant (k) are given in Appendix A.

EXAMPLE 3

For a gas mixture consisting of neutrals, ions, and electrons and whose total density of ionizable species is known,

$$n_{at} = n_a + n^+$$

one may define a "degree of ionization" $= \alpha \equiv n_e/n_{at}$. The number density of electrons and ions must be equal if the atoms are only single ionized. Thus

$$n_e = n^+$$

For the quantity, α, the *Saha* equation reads

$$\frac{\alpha^2}{1 - \alpha} = \frac{S(T)}{n_{at}} \tag{5-41}$$

with $S(T)$ defined as the right hand side of eq. 5–40. When plotted against the function of T and p, α is shown in Fig. 5–9. Single species gases (plasmas) for which S/n is larger than about 100 are said to be fully ionized while smaller values describe partially ionized plasmas.

5.11. EQUILIBRIUM COMBUSTION CHEMISTRY

In the development on combustion described above, the product gas composition is assumed known. For example, in the combustion of 1 mole H_2 with 1.2 moles O_2, the desired product gas is

$$H_2 + 1.2\,O_2 \rightarrow H_2O + 0.7\,O_2$$

plus the heat associated with the formation of the one mole of H_2O. Such a statement is an approximation since the conditions of the gas, namely its

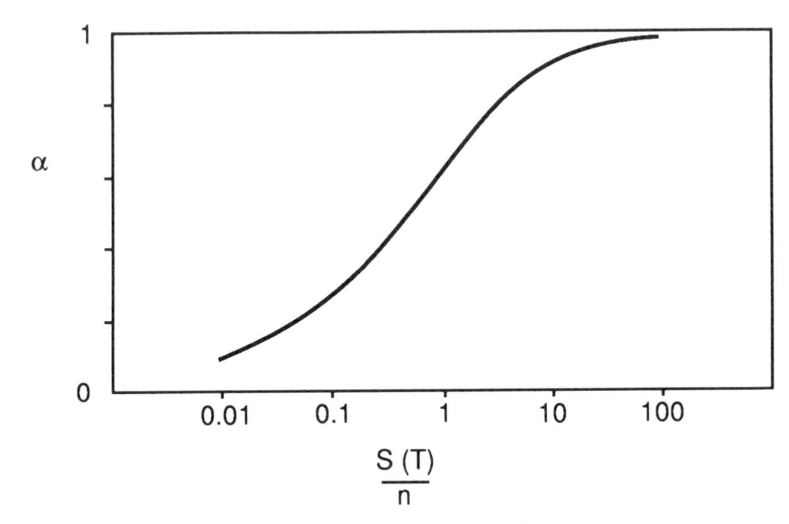

FIG. 5–9. Ionization fraction of a single gas species for various temperatures and pressures. The parameter plotted on the abscissa is defined in eq. 5–41.

pressure and temperature, are such that other product components are, in fact, present. Such products may include

$$H, O, OH, H_2O_2, H^+, e^-$$

from a large virtual list. Experience and knowledge that some combinations of H and O just do not exist or do exist in such small quantities under conditions of interest, allow the list to be bounded.

Species conservation statements permit the writing of relations among the product components that reflect the number of atoms supplied as reactants. Consider for example

$$H_2 + 1.2\,O_2 \rightarrow a\,H_2O + b\,O_2 + c\,H + d\,O + e\,OH + f\,H_2 \tag{5–42}$$

One may state that for the six unknown coefficients, a–f, (a and b may be reasonably close to 1 and 0.7 respectively) species conservation requires

$$H: 2 = 2a + c + e + 2f \tag{5–43}$$

$$O: 2.4 = a + 2b + d + e \tag{5–44}$$

as relations among the coefficients. The number of unknowns (6, here) exceeds the number of species conservation equations (2, here) by the difference between (1) the number of product species and (2) the number of elements supplied as reactants.

To determine the distribution of molecular species within the product mixture, one must ask how these various species can interact. Then, as a consequence of the possible interactions one must determine that distribution that is most likely to be observed. In thermodynamic terms, the equilibrium equations give the partial pressures or composition for that state.

How are the required equilibrium equations set up? The $6 - 2 = 4$ independent relations one must devise are, for example: H_2O can be created by the combination of 2 atoms of H and 1 of O, or 1 molecule of H_2 and 1 atom of O, or 2 molecules of H_2O may be obtained from 2 molecules of H_2 and 2 atoms of O, or 2 molecules of H_2O may be obtained from 2 molecules of H_2 and 1 or O_2. These may be written as:

$$H_2O \Leftrightarrow 2H + O \qquad (5\text{–}45 \text{ or A})$$

$$H_2O \Leftrightarrow H_2 + O \qquad (5\text{–}46 \text{ or B})$$

$$H_2O \Leftrightarrow H_2 + \tfrac{1}{2}O_2 \qquad (5\text{–}47 \text{ or C})$$

and more are possible.

The test concerning the validity of these statements as being the entire set required to describe the possible interaction in the product gas components is:

1. Their number must equal the number of equations required (4, here).
2. They cannot be indeterminate
3. They must include each possible component at least once.

In the example above, the equations A, B, and C are *not* multiples of one another and include all components but OH. Hence a statement like

$$OH \Leftrightarrow O + H \qquad (5\text{–}48 \text{ or D})$$

is all that is necessary to complete the number of interactions. An additional statement such as

$$2OH \Leftrightarrow H_2 + O_2$$

is superfluous since it can be obtained from 2D – (A – B) – (2B – C). Furthermore, the dissociation reactions

$$O_2 \Leftrightarrow 2O \qquad \text{and} \qquad H_2 \Leftrightarrow 2H$$

are obtainable from (2B – C) and (A – B), respectively.

Thus for the set of product gas components chosen, the 4 most convenient equilibrium equations are:

$$H_2O \Leftrightarrow H_2 + \tfrac{1}{2}O_2$$

$$OH \Leftrightarrow \tfrac{1}{2}H_2 + \tfrac{1}{2}O_2$$

and

$$H \Leftrightarrow \tfrac{1}{2}H_2$$
$$O \Leftrightarrow \tfrac{1}{2}O_2$$

The partial pressures are given by relations like:

$$p_{H_2O} = \frac{a}{a + b + c + d + e + f}(p) \text{ etc.}$$

and these are needed for the equilibrium equations:

$$K_p(H_2O) = \frac{p_{H_2O}}{p_{H_2}\sqrt{p_{O_2}}} \qquad K_p(OH) = \frac{p_{OH}}{\sqrt{p_{H_2}}\sqrt{p_{O_2}}}$$

$$K_p(H) = \frac{p_H}{\sqrt{p_{H_2}}} \qquad K_p(O) = \frac{p_O}{\sqrt{p_{O_2}}}$$

Thus in each interaction *between product gas components*, an equation involving various K_p's results. For example, in the problem $H_2 + 1.2O_2 \rightarrow 6$ products, four equations were formulated in addition to the two species conservation statements for O and H.

This forms the set of six algebraic equations to be solved. Obtaining a solution to this set in no way guarantees that the right six products were chosen. In practice, either judgment or a thorough justification that the other components' contributions are negligible is required. A knowledge of the expected result, such as particular mole fractions being small is, of course, useful in the execution of an iterative solution to the algebraic equations.

For equivalence ratios $\phi > 1$ (fuel-rich combustion), the products of combustion include CO, H_2, as well as fuel fragments, all of which represent unrealized heat energy. For energy conversion processes in heat engines, this situation is not generally of interest. Suffice it to say that in determining the products of fuel-rich combustion, one generally finds that the atomic species are not present and that the principal products, in addition to CO_2 and H_2O are lighter-weight fuel fragments, C_nH_m. One may note, however, in the discussion of the mechanics of NO_x (i.e., a mixture of NO and NO_2) production (Sections 5.13 and 5.14), fuel-rich combustion to suppress NO_x formation followed by further combustion at lower temperature has been considered in the Internal Combustion Engine (ICE) and in complex cycles aimed at reduced emissions of NO_x. The reader is referred to the literature for a discussion of fuel fragment productions for $\phi > 1$. For example, ref. 5–3, among many excellent others and ref. 5–4 are examples of discussions of engineering approaches to solving the combustions NO_x emissions problem at the source.

5.12. HEAT RELEASE AND FLAME TEMPERATURE WITH EQUILIBRIUM PRODUCT COMPOSITION

In the determination of species concentration in a fuel-lean combustion problem like that given by eq. 5–42, one assumed the temperature of the resultant product gas to be known, enabling the values of K_p to be gleaned from the thermochemical tables. This is usually not possible since the temperature determines the composition and the composition, in turn, determines the heat released which determines the temperature.

Consider the following example of a fuel-rich combustion process:

$$CO + 0.4(O_2 + 3.76N_2) \rightarrow a\,CO_2 + b\,CO + c\,O_2 + 1.52N_2 \quad (5\text{–}49)$$

with CO and the air at 600 K. The adiabatic flame temperature T_2 and the composition are desired at, say, 1 atm pressure. The fuel-air equivalence ratio is clearly greater than unity because 0.5 moles O_2 are required for stoichiometric combustion. The species balances yield

$$C\colon 1 = a + b$$

$$O\colon 1.8 = 2a + b + 2c$$

$$\text{or} \quad b = 1 - a \qquad \text{and} \qquad c = 0.4 - \frac{a}{2}$$

One may assume that N_2 does not participate in the reaction, hence $(0.4) \times 3.76 = 1.52$ moles N_2 are present in the product gas.

The heat brought in by reactants, $H(600) - H(298)$, is

Species	kcal per mole	# of moles	Q_1
CO	2.187	1.0	−2.137
O_2	2.210	0.4	−0.884
N_2	2.125	1.52	−3.188
			−6.209 $\dfrac{\text{kcal}}{\text{mole CO brought in}}$

The heat of reaction is

$$\begin{aligned} H_{RP} &= \sum \Delta H_{fP} - \sum \Delta H_{fR} \\ &= b\,H_f(CO) + a\,H_f(CO_2) - 1\,H_f(CO) \\ &= (b - 1)(-26.42) + a(-94.05) = -a(67.63) \text{ using the C balance.} \end{aligned}$$

The heat required to raise the products to the unknown T_2 [using H^*

(CO_2, T) to mean $H(T) - H(298)$ for CO_2 etc. as used above] is:

$$Q_2 = a\,H^*(\mathrm{CO_2}, T) + b\,H^*(\mathrm{CO}, T) + c\,H^*(\mathrm{O_2}, T) + 1.50\,H^*(\mathrm{N_2}, T)$$

$$\text{or} \quad = a\,H^*(\mathrm{CO_2}, T) + (1 - a)H^*(\mathrm{CO}, T) + \frac{0.8 - a}{2} H^*(\mathrm{O_2}, T) + 1.50\,H^*(\mathrm{N_2}, T)$$

The energy balance that gives the flame temperature T is therefore:

$$Q_1 + H_{\mathrm{RP}} + Q_2(T, a) = 0 \tag{5–50}$$

The equilibrium statement for the reaction

$$\mathrm{CO_2} \Leftrightarrow \mathrm{CO} + \tfrac{1}{2}\mathrm{O_2}$$

is K_p (for this reaction) $= \dfrac{K_p(\mathrm{CO_2}, T)}{K_p(\mathrm{CO}, T)} = \dfrac{p_{\mathrm{CO_2}}}{p_{\mathrm{CO}}\sqrt{p_{\mathrm{O_2}}}} = \dfrac{a}{1-a}\sqrt{\dfrac{1.92 - \dfrac{a}{2}}{0.4 - \dfrac{a}{2}}}\dfrac{1}{\sqrt{p}}$ (5–51)

where $\quad p_{\mathrm{CO_2}} = \dfrac{ap}{z},\ p_{\mathrm{CO}} = \dfrac{bp}{z},\ p_{\mathrm{O_2}} = \dfrac{cp}{z}$

and z is the total number of moles $= a + b + c + 1.52 = 1.92 - a/2$. p is the mixture pressure. Equations 5–50 and 5–51 allow solution for the two unknowns $T = T_2$ and a. A method for solution is to choose a temperature, find K_p, solve eq. 5–51 for a and see if eq. 5–50 is satisfied.

Typical results from such equilibrium considerations are given in Figs. 5–10 through 5–13 for the combustion of CH_4 with air, reproduced from ref. 5–5, with permission. Figure 5–10 shows the equilibrium composition for stoichiometric reactants and for products at 1 atm and the temperature given. Note the general appearance of monatomic species at high temperatures, and the decrease in other species. The effect of pressure is noted in Fig. 5–11. Generally, higher pressure at a given temperature suppresses dissociation. Figure 5–12 shows the role of equivalence ratio and reactant temperature on the adiabatic flame temperature. A peak temperature is experienced near stoichiometric conditions, as one would expect. Figure 5–13 is the resultant variation of the equilibrium concentration (mole fraction) of an undesired product, NO, as the temperature of the product gas and the equivalence ratio are altered.

5.13. CHEMICAL KINETICS AND NO FORMATION IN COMBUSTION

To this point, the procedures necessary to calculate properties of fluids that have undergone chemical changes are described with an eye toward developing

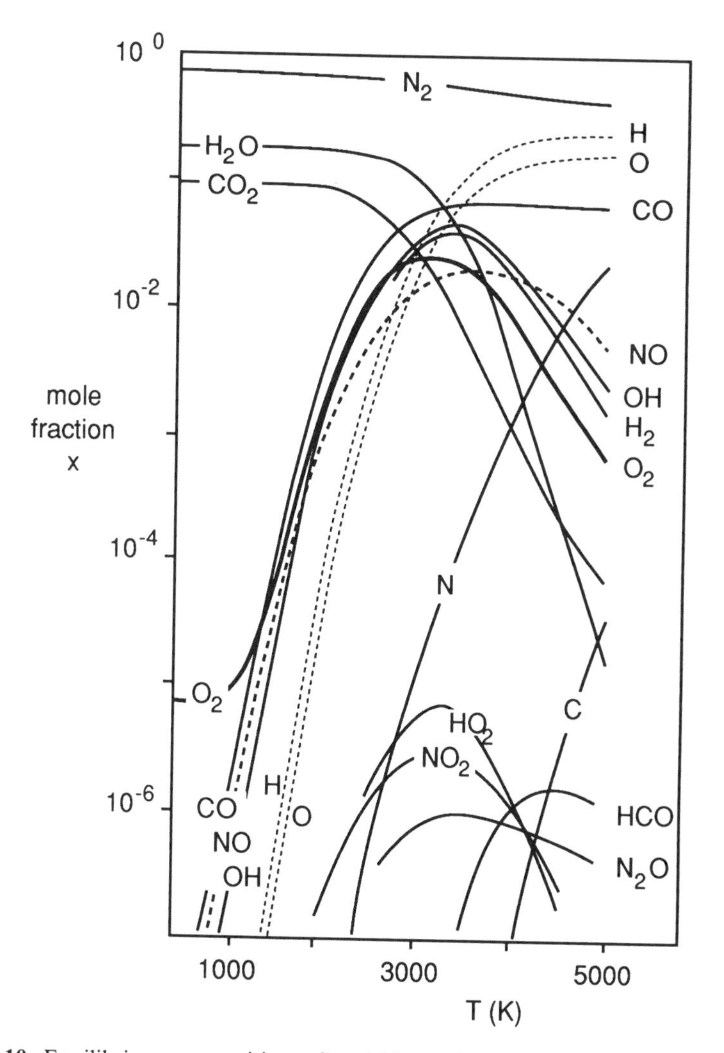

Fig. 5–10. Equilibrium composition of stoichiometric methane-air mixture at 1 atm pressure. (from ref. 5–5.)

an understanding of the process for generating the heat required in a heat engine. This will be useful for the details of the conversion process as well as the determination of the efficiency that is naturally bounded by the temperature attainable in the conversion. Sufficient time has also been assumed to be available for all chemical reactions to proceed to equilibrium.

For the generation of specific gas components, the time required may not be available because of constraints imposed by the design of the machinery wherein the combustion process takes place. For example, in an internal combustion engine, the time allowed for the combustion process is constrained to the period of time when the piston is near top dead center. The "completeness" of the reaction process is governed by the kinetics of the heat release reactions and the fluid mechanics of the energy exchange process.

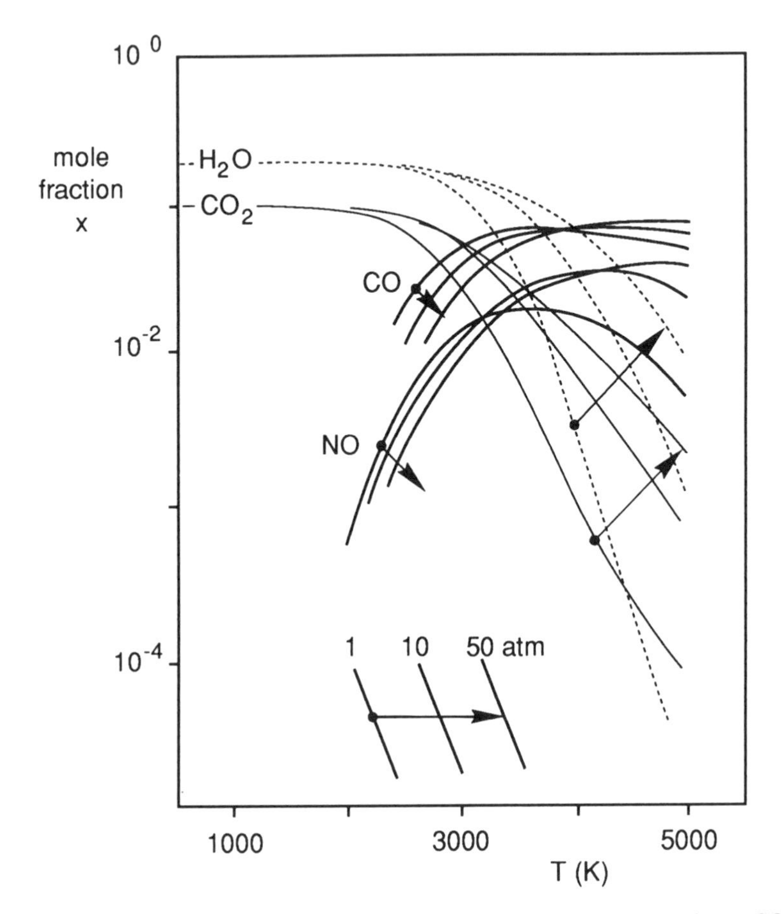

FIG. 5–11. Effect of pressure on the fraction of selected combustion products, CO and NO in particular. Conditions described in Fig. 5–10 apply. Note that increasing p and T affect the results, sometimes in competing ways. (from ref. 5–5.)

Furthermore, there usually is a requirement for a high mechanical speed so that the engine's power may be sufficiently high. In other types of heat engines such as those involving steady combustion, a similar requirement for compactness bounds the time available. Fortunately, the heat-release reactions are sufficiently rapid so that practical devices may be built. The kinetics of heat releasing chemical reactions are fast compared to the kinetics of the production of an unwanted by-product of the combustion process: nitrogen oxide, NO_x. The kinetics of the mixture of NO and NO_2 will be described because it plays an important role in real combustion systems and because it is a good introductory example of the topic for more general purposes.

The production of some combustion gas components is such that the temperature-time history is important in determining how much or how little of the product is produced. One such component is nitric oxide, NO, which is an element of photochemical smog and fortunately not an important component determining heat release. In what follows, the time-dependent

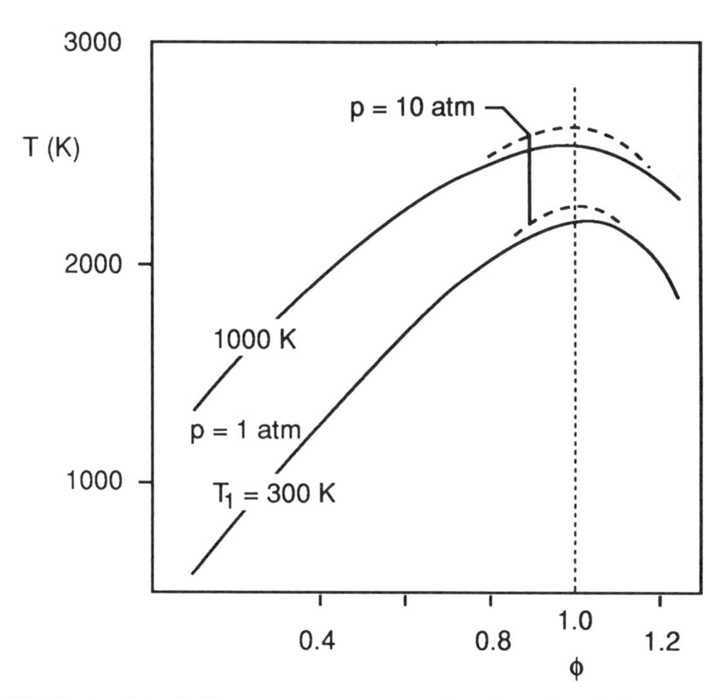

FIG. 5–12. Ideal adiabatic flame temperature as a function of equivalence ratio, initial temperature and pressure. Peak temperature occurs approximately at stoichiometric conditions. Higher pressure suppresses dissociation and the energy it requires, hence higher peak temperatures. (from ref. 5–5.)

behavior of NO formation is described to illustrate the relationship of the combustion process to the achievement of efficient conversion to work. The relation between NO production and the temperature history of fluid elements undergoing combustion is used as an example of gaseous chemical kinetics as an area of study. The related photochemical smog problem is reviewed in Chapter 18.

The equilibrium concentration of NO in combustion gas is controlled by the value of K_p for the reaction producing it and the concentration (i.e. partial pressures) of the reactants involved. In what follows, it is more useful to speak in terms of molecular concentrations (number of a chemical species/volume) rather than partial pressures.

The law of mass action may be written as (eq. 5–33)

$$\frac{p_{NO}}{\sqrt{p_{N_2}}\sqrt{p_{O_2}}} = K_p(T) \cong \exp - \Delta H/RT = \frac{(NO)_e}{\sqrt{(N_2)_e(O_2)_e}} \qquad (5\text{–}52)$$

where $\Delta H = 21.4$ kcal/mole is the heat release in the reaction (eq. 5–33). Here (X) is defined as the number density of X related to partial pressure by the ideal gas equation of state [e.g., $p_{NO} = (NO)kT$]. The variation of any species

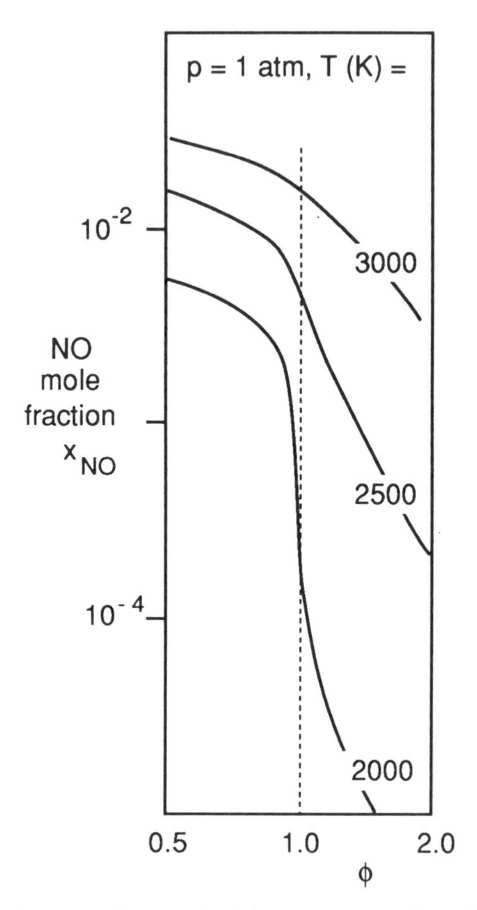

FIG. 5–13. Effect of combustion product temperature and equivalence ratio on the equilibrium mole fraction of NO. (from ref. 5–5.)

concentration involved in a chemical reaction subjected to a change in p and T thus proceeds from one value to another.

As far as the production of NO is concerned, one can show that one may model the combustion process as a step rise in temperature and a consistent change in the chemical composition. In other words, the rate of change of NO concentration, (NO), is slow compared to the combustion processes.

To illustrate how the kinetics of chemical composition evolves with time, consider a situation where a gas sample may be heated very suddenly. One way of doing this in practice is to subject the gas to the passage of a shock wave that will cause the temperature and pressure of the gas to rise very rapidly.

To make this analysis as simple as possible, consider diatomic oxygen molecules heated sufficiently to dissociate by a reaction such as

$$O_2 \rightarrow a\, O_2 + b\, O$$

where the equilibrium of O_2 and O concentrations are related by

$$\frac{(\mathrm{O})_e}{\sqrt{(\mathrm{O}_2)_e}} = K_c(p, T) \tag{5–53}$$

Prior to reaching equilibrium, and in particular during the early stages, when the O concentration is very small, one can say that the rate of O production must be proportional to the number of times that an O_2 molecule encounters another because some of those collisions will result in a dissociation of one of the molecules. The collision partner could be molecules of any other kind present in the mixture. The collision frequency between two kinds of molecules is evidently proportional to their number densities.

$$\frac{d(\mathrm{O})}{dt} \text{ is proportional to } (\mathrm{O}_2)(\mathrm{M})$$

Here (M) is the number density of the collision partners. The proportionality constant is called a (forward) rate constant and is expected to depend on temperature because it determines the energy of the collision partners and on the kind of molecules involved. Thus

$$\frac{d(\mathrm{O})}{dt} = k_f(\mathrm{O}_2)(\mathrm{M}) \tag{5–54}$$

When a finite amount of O is present, the collisions between O atoms may lead to recombination:

$$\mathrm{M} + \mathrm{O} + \mathrm{O} \rightarrow \mathrm{O}_2 + \mathrm{M}$$

Here a third body is required to satisfy the laws of mechanics applied to the collision process. A proof of this statement involves the requirement of satisfying the momentum and energy equations applied to the three-body collision system. Figure 5–14 shows the situation before and after the collision and the algebraic proof that the third body must be present for the energy released, E, to be positive.

$$\text{Conservation of momentum: } \mathbf{V}_1 + \mathbf{V}_2 = 2\mathbf{V}_f$$

$$\text{Conservation of energy: } \tfrac{1}{2}V_1^2 + \tfrac{1}{2}V_2^2 + \frac{E}{m} = 2\tfrac{1}{2}V_f^2$$

The conservation relations may be combined to read

$$V_f^2 = -\frac{E}{m} + \mathbf{V}_1 \cdot \mathbf{V}_2,$$

which is an improbable result when the recombination energy is large, as is typically the case.

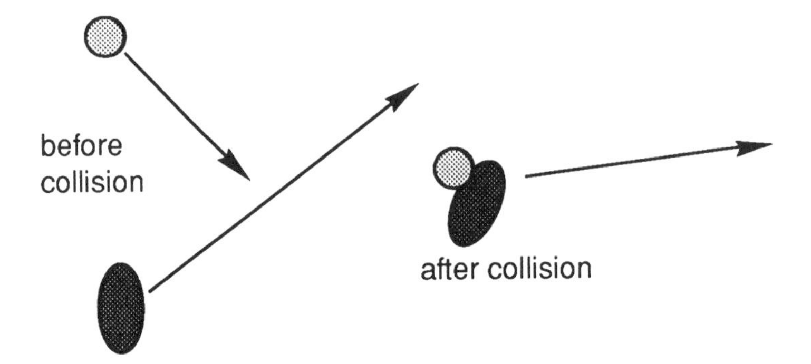

FIG. **5–14.** Collision dynamics showing that a third body is required because the conditions without it are improbable.

The recombination reaction thus makes a negative contribution to the net production rate of (O):

$$\frac{d(\mathrm{O})}{dt} = k_f(\mathrm{O}_2)(\mathrm{M}) - k_b(\mathrm{O})^2(\mathrm{M}) \tag{5–55}$$

Here k_b is the backward rate constant. The presence of M (which may also be O_2 or O) may be factored into the rate constants.

When equilibrium is reached, $d/dt = 0$ and this equation must yield an equilibrium relation between O and O_2 concentrations, namely

$$(O)_e^2 = \frac{k_f}{k_b}(\mathrm{O}_2)_e \tag{5–56}$$

A comparison of this relation and eq. 5–53 shows that the ratio of rate constants is directly related to the equilibrium constant. In an experimental situation, one therefore needs to only determine either k_f or k_b to be able to write the complete rate equation (eq. 5–55). The expected history of O and O_2 fraction in an O_2 sample heated suddenly is shown in Fig. 5–15. The variation of NO also shown in an air sample is discussed subsequently.

If several reactions lead to the possible production of a series, then one may say that the total production is the additive contribution from the separate reactions since the mechanics of the reacting collisions are independent. Thus

$$\left.\frac{d(\text{species})}{dt}\right)_{\text{total}} = \left.\frac{d(\ )}{dt}\right)_{\text{reaction 1}} + \left.\frac{d(\ )}{dt}\right)_{\text{reaction 2}}$$

$$= k_{f1}(\ )(\ ) - k_{b1}(\ )(\ ) + k_{f2}(\ )(\ ) - k_{b2}(\ )(\ ) \tag{5–57}$$

5.13.1. *NO Production in Heated Air*

Zel'dovich and Semenov (see ref. 5–6) suggest that NO is produced by the following reactions. At high temperature equilibrium is established

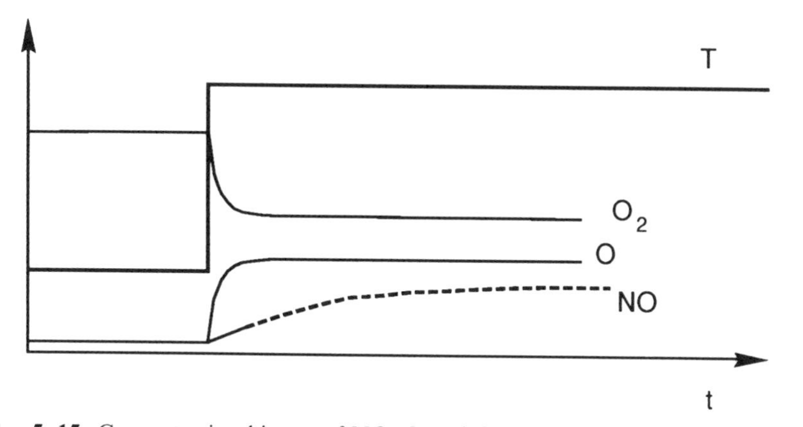

FIG. 5–15. Concentration history of NO, O and O_2 in air that is suddenly heated.

between O, O_2

$$O_2 \Leftrightarrow 2O \tag{5–58}$$

The O atom is then available in a "shuttle" reaction in which it acts as a catalyst:

$$O + N_2 + 75.5\ \text{kcal} \underset{k_3}{\overset{k_1}{\Leftrightarrow}} NO + N \tag{5–59}$$

This is an endothermic reaction, where k_1 is the forward rate constant. The liberated N atom then reacts with O_2 in an exothermic reaction:

$$N + O_2 \underset{k_4}{\overset{k_2}{\Leftrightarrow}} NO + O + 32.5\ \text{kcal} \tag{5–60}$$

which produces more NO and releases the O catalyst atom. The equilibrium constants are closely related to the energy releases as indicated in eq. 5–52.

For these two reactions, one may write the rate of NO production as

$$\frac{d(NO)}{dt} = k_1(O)(N_2) - k_3(NO)(N) + k_2(N)(O_2) - k_4(O)(NO) \tag{5–61}$$

Similarly for O one has:

$$\frac{d(O)}{dt} = -\frac{d(N)}{dt} = k_3(NO)(N) - k_1(O)(N_2) + k_2(N)(O_2) - k_4(NO)(O) \tag{5–62}$$

Experimental measurement of the relevant rate constants shows that (O) adjusts very rapidly so that O is vitually in equilibrium, or $d(O)/dt = 0$. This implies that (N) is also in equilibrium. Physically this means that the time scale for (O) equilibration is very short compared to that of (NO) which is the quantity of interest here. Figure 5–15 illustrates the variation of some air components (NO in particular) when suddenly heated.

5.13.2. *NO Relaxation Time Constant*

With $d(\mathrm{O})/dt = 0$, eq. 5–62 may be used to solve for the concentration (N). Substituting (N) into eq. 5–61, one obtains

$$\frac{d(\mathrm{NO})}{dt} = 2\frac{(\mathrm{O})}{k_2(\mathrm{O_2}) + k_3(\mathrm{NO})}\{k_1k_2(\mathrm{O_2})(\mathrm{N_2}) - k_3k_4(\mathrm{NO})^2\} \qquad (5\text{–}63)$$

This equation looks very much like a classical rate equation involving the difference between a forward rate and a backward rate since the ratio $(\mathrm{NO})^2/(\mathrm{N_2})(\mathrm{O_2})$ has the form of an equilibrium statement for the reaction:

$$2(\mathrm{NO}) \Leftrightarrow \mathrm{N_2} + \mathrm{O_2} \qquad (5\text{–}64)$$

Evidently, the K_p for this reaction is

$$K_{\mathrm{pNO}}^2 = \frac{k_1k_2}{k_3k_4} = \frac{(\mathrm{NO})_e^2}{(\mathrm{N_2})_e(\mathrm{O_2})_e} \qquad (5\text{–}65)$$

which may be shown to be approximated by

$$K_{\mathrm{pNO}}^2 \cong \frac{64}{3}\exp\left(-\frac{43.3}{R_uT}\right) \qquad (5\text{–}66)$$

The equilibrium constants for the two shuttle reactions and the oxygen dissociation reaction are

$$\frac{k_1}{k_3} = \frac{(\mathrm{NO})_e(\mathrm{N})_e}{(\mathrm{O})_e(\mathrm{N_2})_e} = \frac{32}{9}\exp\left(-\frac{75.5}{R_uT}\right) \qquad (5\text{–}67)$$

$$\frac{k_2}{k_4} = \frac{(\mathrm{NO})_e(\mathrm{O})_e}{(\mathrm{N})_e(\mathrm{O_2})_e} = 6\exp\left(+\frac{32.2}{R_uT}\right) \qquad (5\text{–}68)$$

$$K_{\mathrm{pO}} = \frac{(\mathrm{O})_e}{\sqrt{(\mathrm{O_2})_e}} = 6.6 \times 10^{-2}\exp\left(-\frac{61}{R_uT}\right) \qquad (5\text{–}69)$$

Experimentally, the rate constant k_1 is found to be

$$k_1 = 8.3 \times 10^{-11}\exp\left(-\frac{74}{R_uT}\right) \qquad (5\text{–}70)$$

the term in front of the bracket of eq. 5–63 has dimensions of reciprocal time so that this term must be examined to deduce the time scale on which adjustment to equilibrium takes place. One may rewrite eq. 5–63 as

$$\frac{d(\mathrm{NO})}{dt} = 2\frac{k_3 k_4(\mathrm{O})}{k_2(\mathrm{O}_2) + k_3(NO)}\{(\mathrm{NO})_e^2 - (\mathrm{NO})^2\} \tag{5–71}$$

using eq. 5–65 and the very good approximations $(N_2) \cong (N_2)_e$ and $(O_2)_e \cong (O_2)$.

The rate constants k_2 and k_3 are associated with the very similar reactions:

$$k_2\text{: N} + \mathrm{O}_2$$

$$k_3\text{: N} + \mathrm{NO}$$

since the nitrogen atom in NO is at least similar in mass to the O atom in O_2. It is reasonable therefore to expect that k_2 and k_3 be of the same order. In combustion situations where there is excess O_2 or in heated air problems one would expect that $k_2\,(O_2) \gg k_3(NO)$. Under such circumstances, eq. 5–71 reads

$$\frac{d(\mathrm{NO})}{dt} \cong 2\frac{k_3 k_4}{k_2}\frac{(\mathrm{O})}{(\mathrm{O}_2)}\{(\mathrm{NO})_e^2 - (\mathrm{NO})^2\}$$

or with eq. 5–69

$$\frac{1}{(\mathrm{NO})_e}\frac{d(\mathrm{NO})}{dt} \cong (\mathrm{NO})_e\frac{2k_3k_4}{k_2}\frac{K_{p\mathrm{O}}}{\sqrt{(\mathrm{O}_2)}}\left(1 - \frac{(\mathrm{NO})^2}{(\mathrm{NO})_e^2}\right) \tag{5–72}$$

or again with eq. 5–65

$$\frac{1}{(\mathrm{NO})_e}\frac{d(\mathrm{NO})}{dt} = 2k_1\sqrt{(\mathrm{N}_2)}\frac{K_{p\mathrm{O}}}{K_{p\mathrm{NO}}}\left(1 - \frac{(\mathrm{NO})^2}{(\mathrm{NO})_e^2}\right) \tag{5–73}$$

This equation has the virtue that it gives an easy mathematical solution if the heating takes place in a stepwise fashion so that all the terms k_1, K_{pO}, K_{pNO} and $(NO)_e$ are truly constant because the temperature rises to a new fixed value. It is then logical to define a ratio α, to measure the degree of equilibrium achievement (similar to that used with the Saha equation, eq. 5–41) as

$$\alpha = \frac{(\mathrm{NO})}{(\mathrm{NO})_e} \tag{5–74}$$

which approaches 1 asymptotically. Then with $1/2\tau$ equal to the term in front of the bracket of eq. 5–73, one obtains

$$\frac{d\alpha}{dt} = \frac{1}{2\tau}(1 - \alpha^2) \tag{5–75}$$

which has a $\tanh^{-1}$ solution. For α close to 1, that is,

$$\varepsilon = \pm(1 - \alpha) \qquad \text{or} \qquad 1 - \alpha^2 = +2\varepsilon$$

Equation 5–75 then becomes

$$\frac{1}{\varepsilon}\frac{d\varepsilon}{dt} = -\frac{1}{\tau} \tag{5–76}$$

which states that the final approach is exponential in character with a relaxation time, τ. The numerical value may be obtained from the values given, namely

$$\tau(\text{sec}) = 2.06 \times 10^{-3} \frac{1}{\sqrt{(N_2)}} \exp\left(\frac{113.5}{R_u T}\right)$$

for an atmospheric pressure value of the number density of nitrogen ($= 2.1 \times 10^{25}$ m^{-3}), the values of τ are given in Table 5–2. Note the extremely slow rate at low temperature and the extremely rapid rate at high values. (A year is 3×10^7 sec!).

For stoichiometric or fuel-rich combustion, the $k_2(O_2)$ term in eq. 5–71 may be small or comparable to $k_3(NO)$. Under these circumstances, computation of the relaxation time constant becomes more complex because it requires knowledge of the (O) concentration which, in turn, depends on the heat-release reactions. Table 5–3 as well as the data of Figs. 5–16 through 5–19 are from ref. 5–5 (with permission) to illustrate the complexity of the computation problem from merely a chemical point of view. Figure 5–16 shows the variation of temperature of the mixture for 1 and 10 atm, respectively, whereas Fig. 5–17 illustrates the corresponding composition at 1 atm pressure and 1000 K initial

Table 5–2. NO rise time constant for heated air

$T(K)$	τ(sec)
1000	2×10^{12}
1700	140
2000	1
2300	2.3×10^{-2}
3000	7.8×10^{-5}
4000	7.2×10^{-7}

Table 5–3. Twelve reaction scheme for CH_4/air system

Reaction	A	B	C	D	E	F
1. $H_2 + O \rightarrow H + OH$	4.0 + 13	0	−4.7 + 3	1.69 + 1	3.68 + 2	−9.1 + 2
2. $H + H_2O \rightarrow H_2 + OH$	1.0 + 14	0	−1.0 + 4	6.90 + 1	−0.31	−8.0 + 3
3. $OH + CO \rightarrow H + CO_2$	3.1 + 11	0	−3.0 + 2	2.25 − 7	1.25	1.3 + 4
4. $N + O_2 \rightarrow O + NO$	1.3 + 10	1	−2.1 + 4	1.20 + 1	−0.11	1.5 + 4
5. $O + N_2 \rightarrow N + NO$	7.0 + 13	0	−3.8 + 4	2.1	8.90 − 2	−3.7 + 4
6. $NO \rightarrow N + O + M$	4.0 + 20	−1.5	−7.5 + 4	8.9 + 1	−0.34	−7.6 + 4
7. $H_2 + O_2 \rightarrow 2OH$	0	0	−1.96 + 4	3.1 + 4	−0.81	−1.0 + 4
8. $H + O_2 \rightarrow O + OH$	2.2 + 14	0	−8.3 − 13	3.9 + 2	−0.40	−8.6 + 3
9. $O + H_2O \rightarrow 2OH$	8.4 + 13	0	−9.12 + 3	1.04 + 1	−0.28	−8.93 − 3
10. $H + OH \rightarrow H_2O + M$	4.0 + 19	−1	0	2.3 − 3	−0.38	−6.0 + 5
11. $2H \rightarrow H_2 + M$	5.0 + 17	−1	0	1.6 − 1	5.2 − 2	5.2 + 4
12. $2O \rightarrow O_2 + M$	3.0 + 17	−1	0	2.3 − 4	−0.61	6.0 + 4

Note: k_f is the rate constant for the reaction as indicated written as

$$k_f = AT^B \exp\left(\frac{C}{T}\right) \quad \text{and} \quad K_p = DT^E \exp\left(\frac{F}{T}\right)$$

M is a third body with molecular weight of the gas mixture. For shorthand, a number, such as 9.2×10^{-6} is written as 9.2 − 6. (From ref. 5–5.)

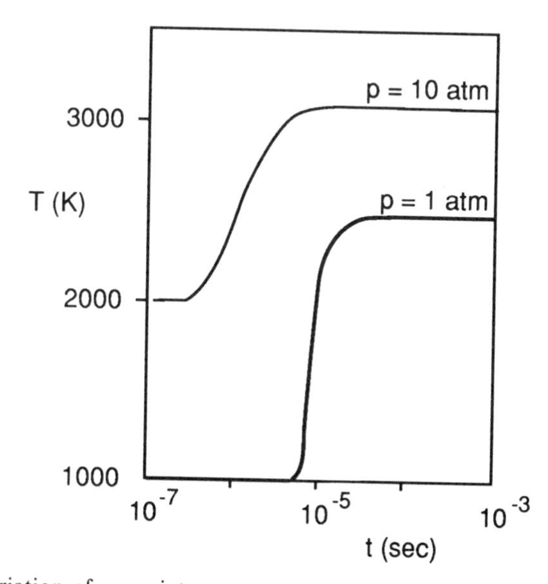

Fig. 5–16. Variation of gas mixture temperature after initial reaction. Two conditions are shown which correspond to (steady flow) combustion in an engine and in a heater at atmospheric pressure. (From ref. 5–5.)

temperature. The process involved is to initiate combustion of CH_4 and air given initial reactant temperature and a constant pressure using accepted rate data for the reactions. The author examined 12 (Table 5–3) and 15 reactions to describe the chemical evolution of the gas mixture. Shown in Fig. 5–17 is the concentration time history of the components considered. Note the

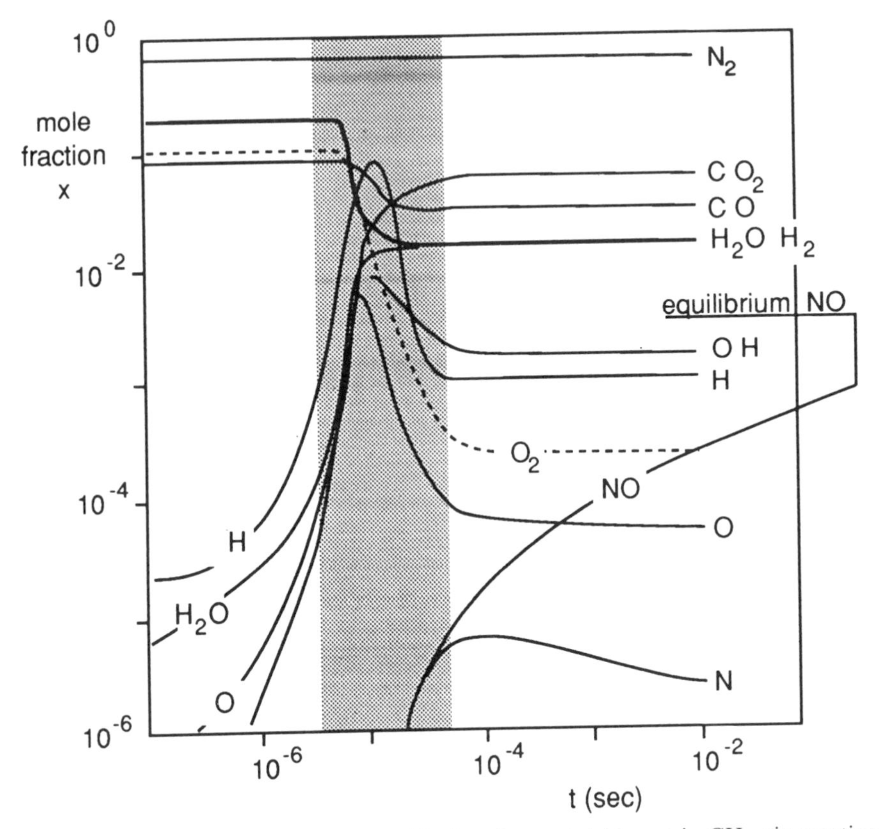

FIG. 5–17. Variation of composition with time for a stoichiometric CH_4-air reaction at $T_1 = 1000$ K. Pressure is 10 atm and adiabatic flame temperature is 2477 K. (From ref. 5–5.)

comparatively slow NO variation with time on a logarithmic scale. At higher inlet temperatures, the mixture temperature rise starts somewhat earlier, and the equilibrium concentration for most species is reached sooner. The effect of reactant temperature and equivalence ratio on the NO production evolution is illustrated in Fig. 5–18.

5.14. PRACTICAL COMBUSTION DEVICES

This chapter has concentrated on the thermochemistry that underlies the function of practical combustors. The heaviest emphasis was on steady flow devices with approximately constant pressure. In practice, a number of combustor types are available to carry out the conversion of fuel thermal energy to heat. The design of such devices is strongly controlled by the kind of application where combustion is required.

One distinguishes between combustors where the hot gas is used to heat another fluid such as the liquid in a Rankine cycle engine and where the hot

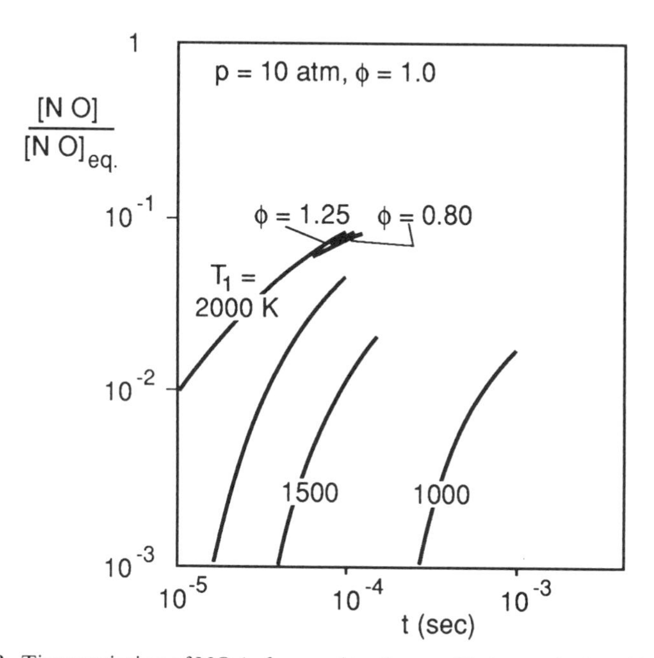

FIG. 5–18. Time variation of NO (referenced to the equilibrium value) at 10 atm pressure and at the initial temperature given. Effect of equivalence ratio is noted. (From ref. 5–5.)

gas participates in the cycle as the working fluid. An example of such a situation is the gas turbine. The goals of most combustors include

- Combustion efficiency (low fuel waste)
- Reliable operation
- Low heat loss to regions other than those intended (gas temperature should be close to the adiabatic flame temperature)
- Low cost and long life burner
- Low maintenance requirements
- Low emissions of undesirable "pollutants"
- Low weight and volume
- Adaptability to various fuel types, if necessary

The principal methods of carrying out combustion are the flame type burners, fluidized bed reactors, and surface catalytic reactors. In the power industry, the flame combustors include those typified by the gas turbine wherein a hot combustion "eddy" is maintained which continuously ignites fresh fuel-air mixtures. Flame combustors usually use gaseous or liquid fuels, with preheating if the fuel has a high average molecular weight. Such a burner for an aircraft engine is shown schematically in Figs. 7–16 and 12–40. The reader may wish to consult an article by Blazowski in ref. 5–7 for a thorough discussion of such combustors. The performance of flame combustors is dominated by the fluid

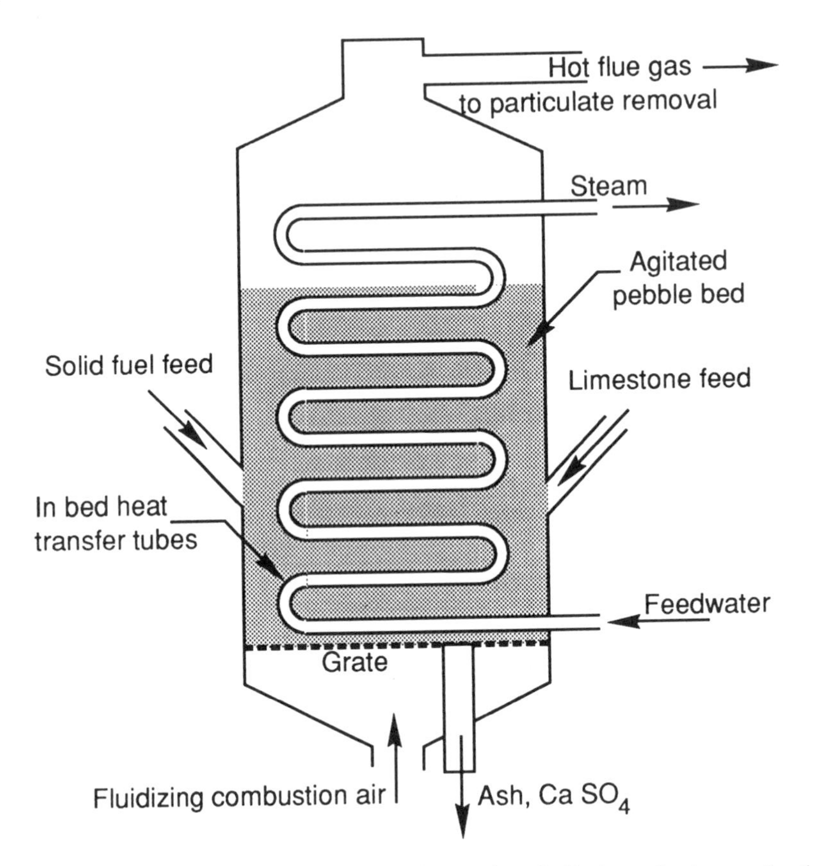

Fig. 5–19. Schematic of a fluidized bed combustor in a boiler/superheater application. Limestone is for sulfur oxides removal (see Chapter 18). Direct use in an engine cycle requires particulate removal.

mechanics of the reacting mixture because that controls flame stability, product gas uniformity, emissions, and heat transfer to the walls of the burner. The dynamics of the reacting gas also allows for ways to control emissions.

Figure 5–19 (after ref. 5–8) shows a fluidized bed reactor for heater/boiler/superheater for water. The fuel for such combustors is generally solid, although fuel oils and gas are used. In this kind of burner, the combustion air is forced through a grate to lift and set in motion a bed of pebbles that agitate the fuel particles and retain them long enough for combustion to be complete. The pebbles (made of silica, alumina, etc.) serve to control the temperature of the gas, so that its variation is slow, by prolonging the residence of the gas and fuel. As a result, the fluidized bed burner type tends to have a temperature distribution that is less "peaked" than a flame type burner and leads to reduced formation of pollutants like NO_x.

An advantage of the fluidized bed is that chemicals for the control of emissions such as sulfur oxides (see Section 18.4.4) are readily incorporated into the combustor. The figure shows limestone used for that purpose. When the fluidized bed reactor is pressurized as it would have to be for use in a gas turbine, significant effort at particulate removal from the hot gas stream would have to be expended.

PROBLEMS

1. CH_4 is burned with 50% excess oxygen (both at 298 K) to products at 1000 K. Assuming the products do not contain any CO, estimate the amount of CH_4 in the product mixture. Show that the percentage CO is indeed small.
2. An equilibrium mixture of only H_2O, H_2 and O_2 exists at 3000 K. This mixture is to be heated to 3500 K in a constant pressure process. Determine the heat required per unit mass. Are atomic species negligible?
3. Write a rate equation for the production of NH_3 when a mixture of N_2 and H_2 is suddenly heated.
4. Show that in a mixture of two ionizable species, the degree of ionization of species a is suppressed by the presence of b in comparison to the situation where b is absent altogether.
5. Determine the adiabatic flame temperature of CO burning with O_2 at a fuel-rich equivalence ratio of 1.2. Reactant temperatures are 298 K and $p = 1$ atm.
6. A reacting mixture of X and X_2 is expanded isothermally from $p = 10$ to $p = 1$. $K_p = 1$ at the temperature. In terms of the component enthalpies as listed in thermochemical tables, determine the enthalpy change for this process. Is heat liberated in the process?
7. H_2 is heated to 5000 K (in an arcjet for example) and expanded from $p = 1$ atm to $p = 0.01$ atm. Calculate the final temperature using a nonreacting, perfect gas assumption and compare this result to an equilibrium expansion of the real (imperfect gas) and a frozen flow case with imperfect gas.
8. Calculate the Gibbs free energy of the reactants involved in the dissociation reaction of H_2O and estimate the value of the equilibrium constant at 2000 K. Compare to the value given in the JANAF tables.
9. On a p-v and/or T-s diagram draw the lines where a substance (H_2O or NH_3 for example) is x% dissociated and the line where the specific volume given by the ideal gas equation of state differs from the actual value by x%. If x is 1 or 5 these lines identify the region where the ideal nonreacting gas assumption is reasonable.

BIBLIOGRAPHY

Hill, P. G., Peterson, C. R., *Mechanics and Thermodynamics of Propulsion*, Addison-Wesley, Reading, Mass., 1965 and 1991.

REFERENCES

5–1. JANAF Thermochemical Tables, 2nd ed., Office of Standard Reference Data, National Bureau of Standards, Washington, D.C., 1970.

5–2. Fay, J. A., *Molecular Thermodynamics*, Addison-Wesley, Reading, Mass., 1965.

5–3. Heywood, J. B., *Internal Combustion Engine Fundamentals*, McGraw-Hill, New York, 1988.

5–4. Decher, R. Hertzberg, A., Corlett, R. C., "A Compound Cycle Employing Two Stage Combustion to Reduce Automotive Air Pollution," SAE paper 720736, Natonal Combined Farm, Construction and Industrial Machinery and Powerplant Meeting, Milwaukee, September 1972.

5–5. Marteney, P. J., "Analytical Study of the Kinetics of Nitrogen Oxide Formation in Hydrocarbon-Air Combustion," *Combustion Science & Technology*, **1**, 461–9, 1970.

5–6. Zel'dovich, Ya. B., Razier, Yu. P., *Physics of Shock Waves and High Temperature Hydrodynamic Phenomena*, Academic Press, New York, 1966.

5–7. Oates, G. C., Ed., *Aerothermodynamics of Aircraft Engine Components*, AIAA, New York, 1985.

5–8. Fennelly, P. F., "Fluidized Bed Combustion," *American Scientist*, **72**, 254–61, May–June 1984.

6

COUNTERFLOW GAS HEAT EXCHANGERS

In many energy conversion applications, the heat generated by combustion, from a reactor or from other heat sources, is carried by a fluid and must be transferred to another fluid in a steady flow manner. The heat exchanger allows such a transfer of heat between dissimilar fluids at different pressures and temperatures.

6.1. HEAT EXCHANGER ARRANGEMENTS

The fluid media carrying thermal energy may be liquids, gases, or multiphase mixtures. Typically, the two fluids flow through ducts that have a common boundary which allows the transfer of heat. A simple geometry is shown in Fig. 6–1, where a tube carrying one fluid is surrounded by the flow of the second fluid. The fluids may flow in the same direction (parallel flow, Fig. 6–2a), in opposing directions (counterflow, Fig. 6–2b) or at some angle near 90 degrees to each other (cross flow, Fig. 6–2c). A common example of a cross flow heat exchanger is the automobile radiator in which water flows up or down and air moves rearward.

Cross flow heat exchangers may be rigged to approximate counterflow (multiple pass crossflow) as shown in Fig. 6–2d. Two implementations of such devices are shown in Fig. 6–3. Figure 6–4 shows the temperature profiles of a hot flow losing heat to a cold flow in a cross geometry. Evidently the portion of the cold flow nearest the hot flow entry will be exposed to a higher temperature and thus reaches a greater temperature than the streamtube nearer the hot outlet end. The resultant nonuniformity in temperature represents an irreversibility when mixing occurs to remove it. In practical situations this irreversibility penalty may be offset by lower cost, weight, and/or volume.

In heat engines designed for processing a working fluid, heat exchange is sometimes carried out in a rotating arrangement as illustrated in Fig. 6–5. In the figure is seen a rotating "cake" of tubular elements which allow flow passage. By rotating, any one element is exposed to hot gas at which time the element walls heat up and later transfer that heat to the cold flow when it flows through the tubes. When the pressure of the cold flow is elevated, seals must be provided. Note also that an irreversibility is incurred as a result of the physical carryover of the trapped gas volume. In practice, the hot and cold fluid carrying tubes are D-shaped, each to cover half the circle available. The effectiveness of storing heat in the solid material of a regenerator is discussed in Section 10.6 in connection with its use in the Stirling cycle engine.

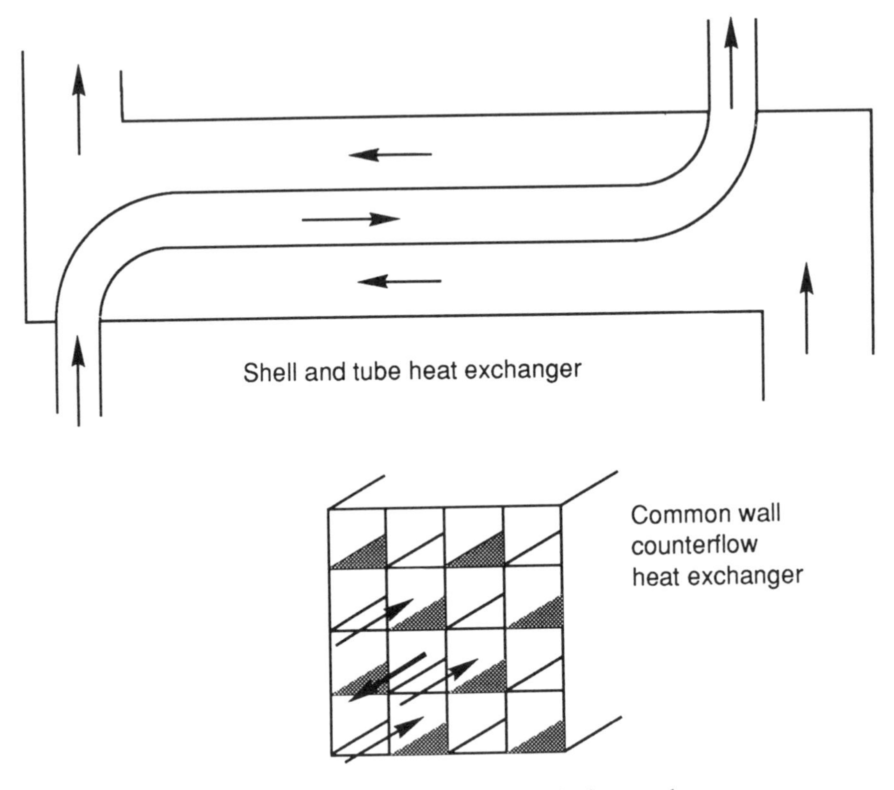

FIG. 6–1. Common wall and shell-and-tube heat exchangers.

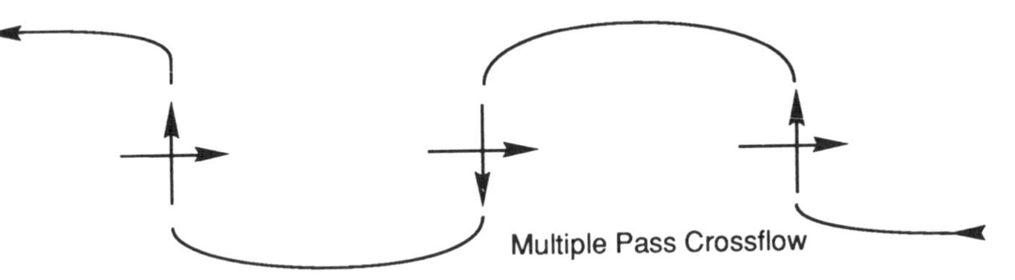

FIG. 6–2. Flow arrangements in heat exchangers.

6.2. TEMPERATURE DISTRIBUTION IN HEAT EXCHANGERS

If the two fluids flow in opposite directions, the hot and the cold flows' temperatures decrease and increase along the length of the exchanger as shown in Fig. 6–6 (left). By contrast, a parallel flow heat exchanger brings the temperatures to values between the initial hot and cold fluid temperatures (Fig. 6–6, right). The parallel flow design is advantageous if a small fraction of heat available is to be transferred in a short device. However, the process is inefficient because of the large and varying temperature drop across which heat flows from one fluid to the other. The cross flow exchanger achieves

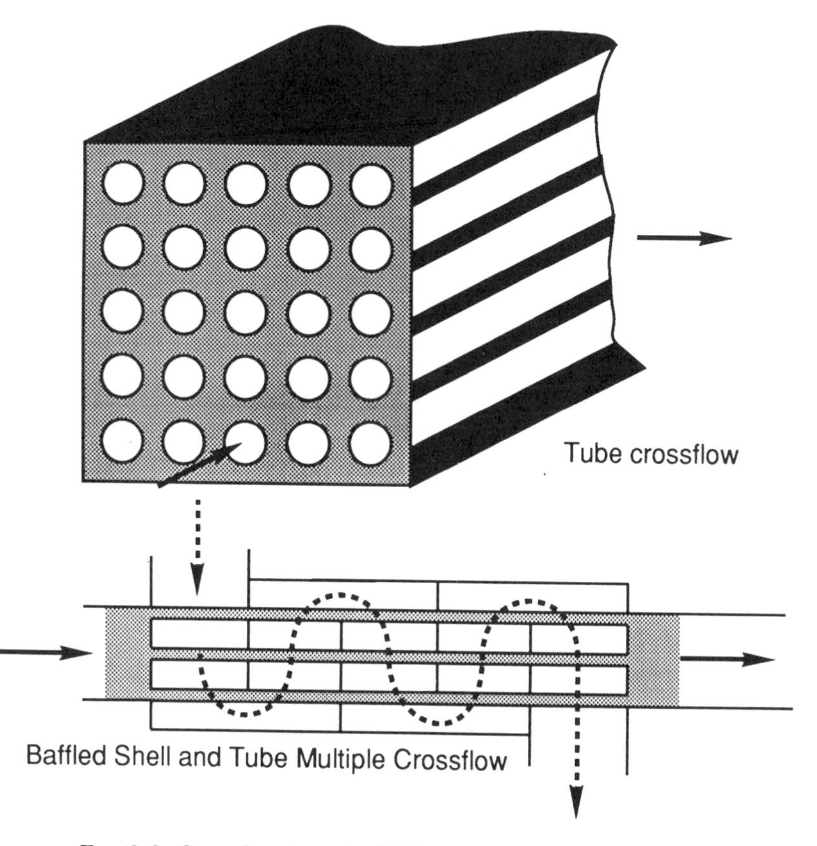

FIG. 6–3. Cross flow heat (multiple pass at bottom) exchangers.

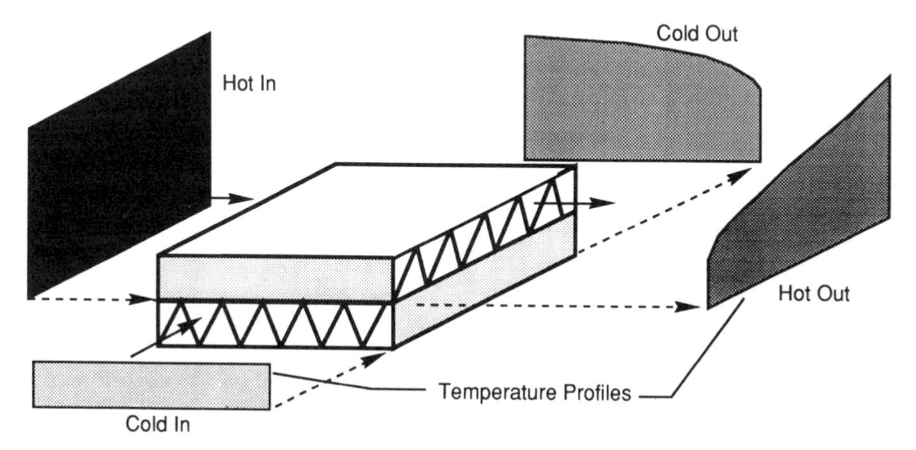

FIG. 6–4. Temperature profiles in a plate-fin crossflow heat exchanger. Note non-uniformity of the discharges.

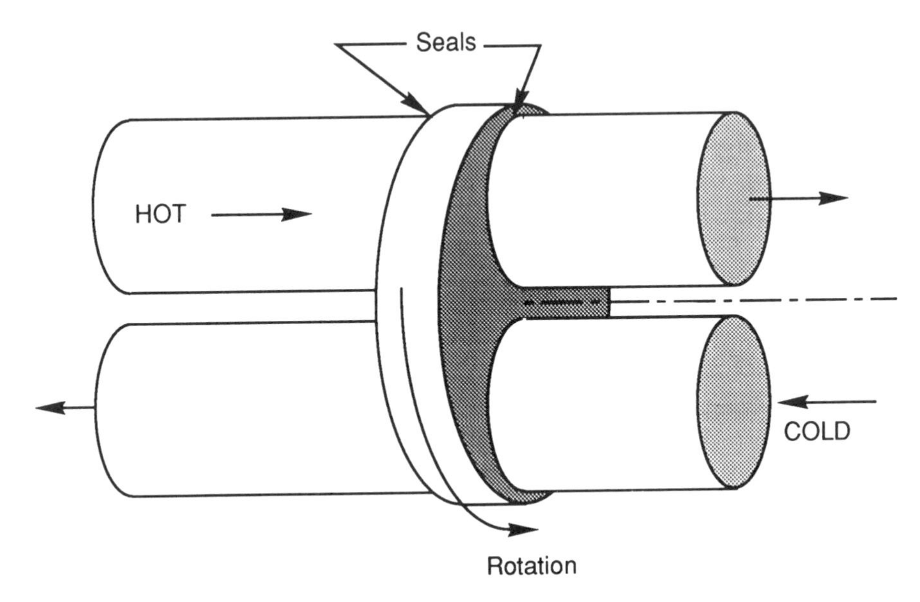

FIG. 6–5. Rotating automotive gas turbine regenerator.

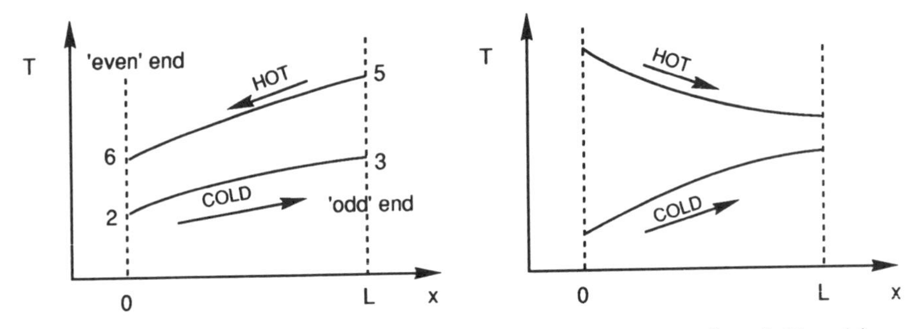

FIG. 6–6. Temperature variation along the fluid path in a counter flow (left) and in a parallel flow (right) heat exchanger

performance levels between that of counter and that of the parallel flow heat exchangers. Because of the superior thermodynamic performance of the counterflow exchanger, attention here is restricted to that device. For purposes of the analysis to follow, the duct through which the fluid passes is assumed square. For duct shapes other than square, one may take it to be characterized by an equivalent square duct of side length, D, also termed hydraulic diameter, defined by

$$D = 4\,\frac{\text{Flow Area}}{\text{Perimeter}} \tag{6–1}$$

The tubes are also assumed to be sufficiently long to allow one to neglect the entrance region where boundary layers develop and eventually merge to form a fully developed flow.

In the following, a limitation of approximate modeling accuracy is accepted in order to develop a physical understanding for the relationships involved in heat exchanger design and performance. The reader should ensure that this level of descriptive accuracy is warranted by the application considered. Specifically excluded from consideration are extreme geometric situations, liquid and two-phase flows, and performance analysis where the goal is to achieve high numerical accuracy.

The phenomenological description of heat transfer from a wall to a fluid is given by

$$dq = h(T_{\text{wall}} - T)\, dA_{\text{w}} \tag{6–2}$$

where

dq = increment of heat added to the flow,
the temperatures are of the wall and of the fluid, respectively
dA_{w} = differential element of heat transfer surface area
= wetted perimeter × dx (=4D dx, as shown in Fig. 6–7)
h = fluid to wall heat transfer coefficient.

This important descriptive variable for heat transfer can be written in terms of the nondimensional Stanton Number, N_{ST}, defined by $h = \rho u C_{\text{p}} N_{\text{ST}}$. Use of the symbol h is traditional and should not cause confusion with enthalpy.

In gases, the similarity between the mechanism of heat transport (by kinetic energy of the molecules, $1/2\, m\, v^2$) and that of momentum transport (by the momentum of the molecules, mv) allows one to reason that there ought to be a simple relationship between the nondimensional parameters that describe friction (momentum transport to the wall) and heat transfer (energy transport). This similarity is indeed found in the laboratory and is termed the Reynolds Analogy. This analogy states that there is an approximately proportional relationship (ref. 6–1) between the transport of momentum and of heat *in gases.* In liquids, the analogy is not valid because of the important intermolecular forces that are associated with the small distance between molecules. As a result of these forces, the mechanisms for energy and momentum transport differ in significant ways. In gases, the Reynolds Analogy is also limited in validity, and the reader should consult a text on heat transfer for better correlations if higher accuracy demands it. In this chapter, the purpose is limited to the development of a mathematically simple model for a qualitative description of the key parameters in heat exchangers as they affect engine cycle design. Mathematically, the analogy states

$$N_{\text{ST}} = \tfrac{1}{2} f \tag{6–3}$$

where the friction factor, f, is defined as:

$$f = \frac{\text{wall shear stress}}{\tfrac{1}{2}\rho u^2} \tag{6–4}$$

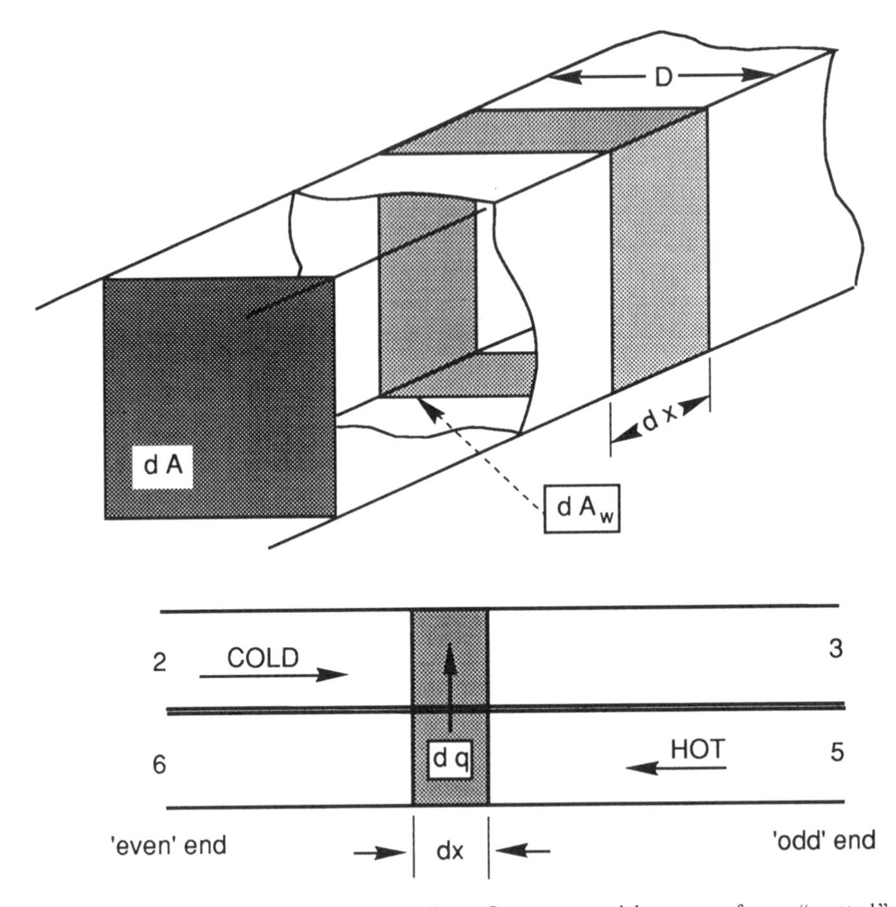

FIG. 6–7. Heat transfer to a flowing medium: flow area and heat transfer or "wetted" area.

The Reynolds number dependence of f is shown in Fig. 6–8 for laminar and turbulent flow. This plot of

$$4f = \frac{\Delta p}{\frac{1}{2}\rho V^2 \frac{L}{D}}$$

is termed a Moody diagram and the figure is after ref. 6–2. For smooth flow the approximation

$$f = 0.04\, Re^{-0.16} \tag{6–5}$$

is valid in the range $4 \times 10^3 < Re_D < 2 \times 10^7$ (ref. 6–3). This is but one of a number of useful approximations which are used in practice. In this text, other expressions are also used. Further,

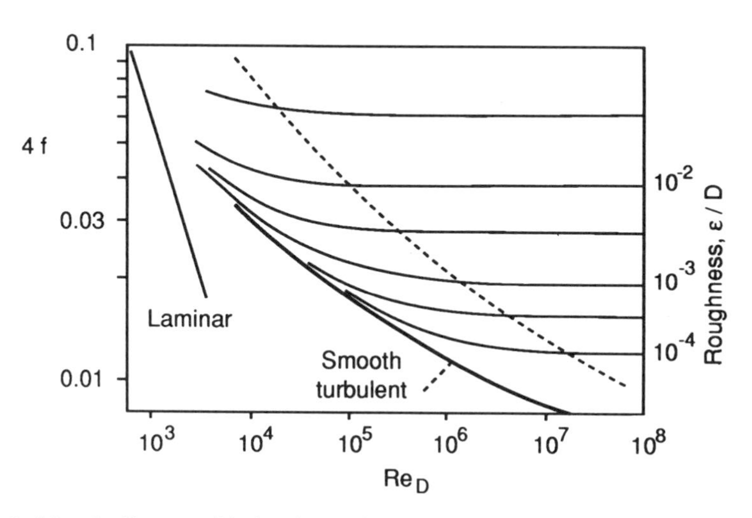

FIG. 6–8. Moody diagram: friction factor for fully developed flow in a circular tube of diameter D and roughness ε.

$$\frac{1}{\sqrt{f}} = 1.74 + 2\log\left(\frac{\varepsilon}{D}\right) \tag{6–6}$$

applies (ref. 6–4) to a fully rough duct where f becomes independent of Reynolds number as seen in Fig. 6–8. ε is the length scale ratio of the roughness to tube size.

For the entire heat exchanger, the extreme temperatures are as noted in Fig. 6–7. The numbering system is made to be consistent with that used for engines using a heat exchanger. Note that there are two ends, one with "odd" numbers and the other with "even" numbers. These are termed the "odd" and "even" ends. Further, the sign of dx is different for the two flows because their directions differ. At any station, x, the steady heat balance between fluids reads:

$$\dot{m}_{\mathrm{C}} C_{p\mathrm{C}}\, dT_{\mathrm{C}} = \dot{m}_{\mathrm{H}} C_{p\mathrm{H}}\, dT_{\mathrm{H}} \tag{6–7}$$

where $\dot{m}$ for the hot (H) or cold (C) flows is given by

$$\dot{m} = \rho V A \tag{6–8}$$

where V is a mean flow velocity through the tube. It follows that:

$$\alpha\, dT_{\mathrm{H}} = dT_{\mathrm{C}} \tag{6–9}$$

where

$$\alpha = \frac{\dot{m}_{\mathrm{H}} C_{p\mathrm{H}}}{\dot{m}_{\mathrm{C}} C_{p\mathrm{C}}}$$

The constant α is the capacity rate ratio. From integration of eq. 6–9 one has

$$T_3 - T_2 = \alpha(T_5 - T_6) \tag{6–10}$$

This equation relates the end temperatures of the heat exchanger. To obtain an additional relation between T_3 and T_6, given T_2 and T_5, the details of the heat transfer must be examined. The heat transfer rate between fluid and wall is given by eq. 6–2. This equation may be generalized to the problem of heat transfer between two fluids by writing eq. 6–2 in terms of an overall thermal conductance, U:

$$dq = -U(T - T_w)\, dA_w \tag{6–11}$$

From one fluid to another, the thermal conductance (= the reciprocal of thermal resistance) is the sum of fluid film conductances, $1/h$, and a solid material resistance, t/k. Here t is the material thickness and k is the thermal conductivity:

$$\left(\frac{1}{U}\right)_{\text{overall}} = \frac{1}{h_H} + \frac{1}{h_C} + \frac{t}{k} \approx \frac{2}{h} \tag{6–12}$$

Typically, the wall resistance is small (due to the large k of metals) and thus can be neglected. Figure 6–9 shows the temperature profile of the hot and cold fluids: The high thermal conductivity leads to a nearly uniform metal wall temperature. The heat balance at any station, x, is

$$\dot{m}_C\, dq_C = \dot{m}_C C_{pC}\, dT_C = -h_C(T_C - T_w)\, dA_w;\; T_w > T_C \tag{6–13}$$

and

$$\dot{m}_H\, dq_H = \dot{m}_H C_{pH}\, dT_H = +\, h_H(T_H - T_w)\, dA_w;\; T_H > T_w \tag{6–14}$$

From eq. 6–7, these heat flow rates are equal so that the wall temperature becomes with eqs. 6–13 and 6–14:

$$T_w = \frac{T_C + \beta T_H}{1 + \beta} \tag{6–15}$$

The ratio of film heat transfer coefficients (typically close to unity) is defined by

$$\beta = \frac{h_H}{h_C} \tag{6–16}$$

With the definition of a nondimensional axial length, $\xi = \dfrac{x}{L}$, the differential wetted surface area dA_w is (Fig. 6–7)

$$dA_w = A_w\, d\xi = 4DL\, d\xi \qquad 0 \leq \xi \leq 1 \tag{6–17}$$

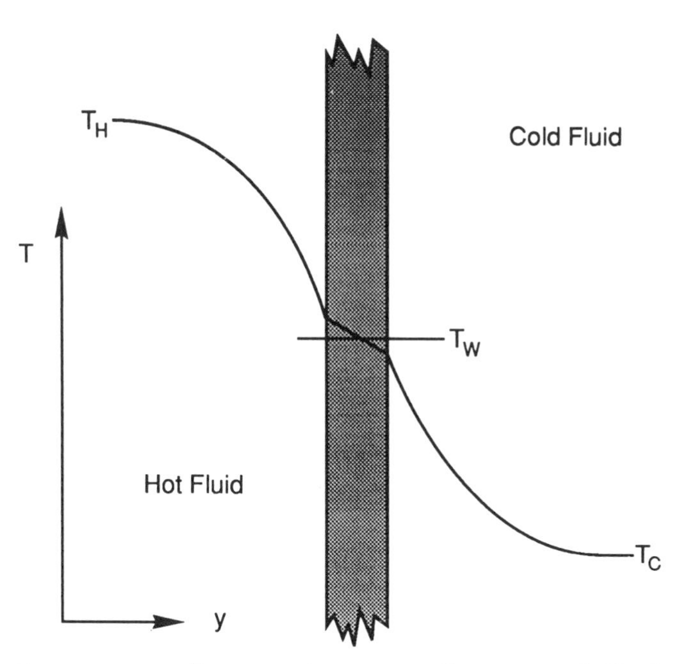

FIG. 6–9. Temperature profile near and across a wall across two fluids that exchange heat.

Here it is implicit that the wetted areas for the two sides of the heat transfer surface are equal. This implies that the hydraulic diameters of the two flows must be equal if they share a common flow length as assumed here. Thus no distinction between Ds needs to be made. Equation 6–13 then becomes, on substituting for T_w:

$$dT_C = \frac{h_C A_C}{\dot{m}_C C_{pC}} \frac{A_w}{A_C} \frac{\beta}{1+\beta} (T_H - T_C)\, d\xi \qquad (6\text{–}18)$$

where A_C is the flow area of the cold fluid conduit. With the definition of the Stanton number

$$N_{ST} = \frac{hA}{\dot{m}C_p} = \frac{h}{\rho u C_p} = N_{STC} \quad \text{or} \quad N_{STH} \qquad (6\text{–}19)$$

for the cold and hot flows, respectively. Equations 6–18 and (similarly) 6–14 become, respectively:

$$\left.\begin{aligned} dT_C &= -\,N_{STC}\frac{4L}{D}\frac{\beta}{1+\beta}(T_C - T_H)\, d\xi \\ dT_H &= -\,N_{STH}\frac{4L}{D}\frac{1}{1+\beta}(T_C - T_H)\, d\xi \end{aligned}\right\} \qquad (6\text{–}20)$$

Substracting the two equations 6–20 yields

$$d(T_C - T_H) = -(T_C - T_H)\frac{4L}{D}\frac{N_{STH}}{1+\beta}\left\{\beta\frac{N_{STC}}{N_{STH}}\frac{A_H}{A_C} - 1\right\}d\xi \qquad (6\text{–}21)$$

The combination in the braces reduces to

$$\frac{h_H}{h_C}\frac{\left(\dfrac{hA}{\dot{m}C_p}\right)_C}{\left(\dfrac{hA}{\dot{m}C_p}\right)_H}\frac{A_H}{A_C} - 1 = \alpha - 1 \qquad (6\text{–}22)$$

as defined by eq. 6–9. Before proceeding, it is useful to introduce a new parameter, ϕ:

$$\left.\begin{aligned} &\frac{4L}{D}N_{STH}\frac{\alpha - 1}{1+\beta} = \phi(\alpha - 1) \\ &\text{where} \\ &\phi = \frac{4L}{D}N_{STH}\frac{1}{1+\beta} = \frac{2}{1+\beta}\frac{fL}{D} \end{aligned}\right\} \qquad (6\text{–}23)$$

so that an integration of eq. 6–21 gives

$$\frac{T_H - T_C}{T_6 - T_2} = \exp[(1-\alpha)\phi\xi] \qquad (6\text{–}24)$$

The difference $\Delta T = T_H - T_C$ thus varies exponentially with distance and the heat capacity rate ratio, α. This ratio determines whether ΔT increases or decreases, as seen in Fig. 6–10. The parameter ϕ is seen to be the length-scaling parameter for the transfer of the heat: A large L/D and N_{ST} are desirable for a short heat exchanger. The value of ΔT at the $\xi = 1$ or "odd" end relative to the value at the "even" end is then given by

$$\frac{T_5 - T_3}{T_6 - T_2} = \exp[(1-\alpha)\phi] \equiv e \qquad (6\text{–}25)$$

This equation together with eq. 6–10 allows calculation of outlet temperatures if the inlet values T_5 and T_2 are known. These values are:

$$\left.\begin{aligned} T_3 &= \frac{\alpha(e-1)T_5 + (\alpha - 1)T_2}{\alpha e - 1} \\ T_6 &= \frac{e(1-\alpha)T_5 + (1-e)T_2}{1 - e\alpha} \end{aligned}\right\} \qquad (6\text{–}26)$$

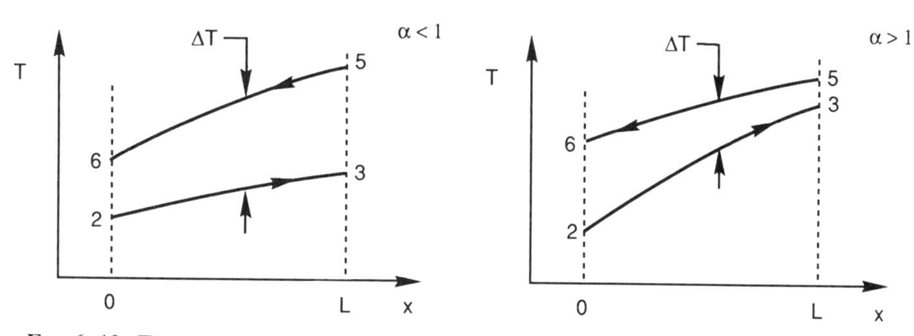

FIG. 6–10. Temperature profile in a counterflow heat exchanger with $\alpha < 1$ (left) and $\alpha > 1$ (right).

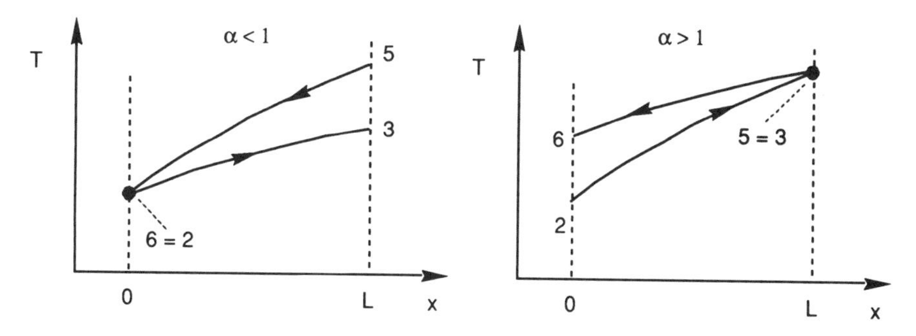

FIG. 6–11. Counterflow heat exchanger with large ϕ.

where e is used as an abbreviation for the term in eq. 6–25. Two special cases allow interpretation of these algebraic results.

6.2.1. *Limit of* L *Very Large* ($\alpha \neq 1$)

From eq. 6–26 and $\phi \to \infty$, it is clear that the exponential term is either zero or infinite:

$$\alpha > 1\text{: } T_3 \text{ approaches } T_5$$

$$\alpha < 1\text{: } T_6 \text{ approaches } T_2$$

This leads to the temperature profiles shown in Fig. 6–11. For $\alpha > 1$, eq. 6–10 gives with $T_3 = T_5 = T_{\text{odd}}$

$$T_6 = \frac{T_2}{\alpha} + \frac{\alpha - 1}{\alpha} T_{\text{odd}}$$

Similarly for $\alpha < 1$

$$T_3 = (1 - \alpha) T_{\text{even}} + \alpha T_5 \tag{6–27}$$

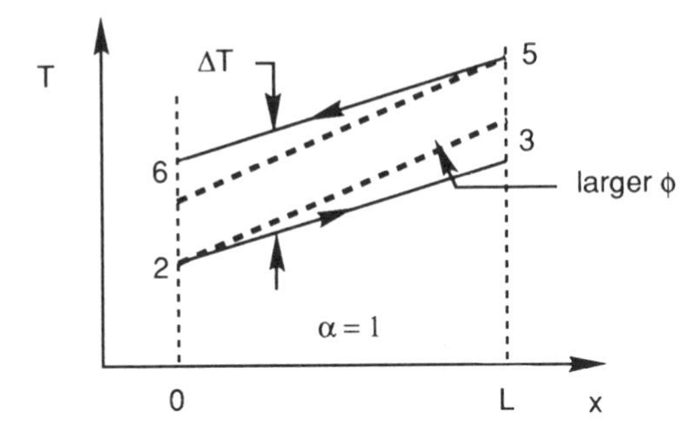

FIG. 6–12. Matched capacity rate counterflow heat exchanger temperature profile, $\alpha = 1$. Larger ϕ leads to smaller ΔT.

6.2.2. *Limit of* $\alpha = 1$, *Matched Capacity Rate*

In heat exchangers where the same fluid flows in the two sides of the exchangers with merely differing temperatures and pressures (e.g., in a closed cycle gas turbine regenerator), α is unity. Expressions for the various temperatures can be obtained mathematically by expanding the exponential terms for small argument:

$$\exp[(1-\alpha)\phi] = 1 + (1-\alpha)\phi + \text{higher order terms} \tag{6–28}$$

Equations 6–26 then simplify (with $\alpha = 1$) to:

$$\left.\begin{aligned} T_3 &= \frac{\phi}{1+\phi} T_5 + \frac{1}{1+\phi} T_2 \\ & \frac{1}{1+\phi} T_5 + \frac{\phi}{1+\phi} T_2 \end{aligned}\right\} \tag{6–29}$$

which shows that in the limit of large ϕ (i.e., L/D), T_3 approaches T_5, and T_6 approaches T_2, so that a smaller ΔT is required to derive heat from one medium to the other as the heat transfer area is made larger. The ΔT across which the heat flows from one fluid to another is given by $T_5 - T_3$ on the "odd" end and $T_6 - T_2$ on the "even" end. These can be shown to be equal from eqs. 6–29 so that the variation of temperature along the tube must be linear in x (see eqs. 6–21 and 6–22 for $\alpha = 1$) for both hot and cold flows, as shown in Fig. 6–12.

6.3. HEAT EXCHANGER EFFECTIVENESS

A convenient efficiency for the effectiveness of a heat exchanger with $\alpha = 1$ is:

$$\eta_x = \frac{\text{heat transferred}}{\text{maximum heat transferable}}$$

$$= \frac{T_5 - T_6}{T_5 - T_2} = 1 - \frac{\Delta T}{T_5 - T_2} \tag{6–30}$$

where ΔT is the temperature drop each unit of heat experiences as the flow proceeds along the tube. The heat exchanger effectiveness can be expressed as a value of this ΔT written as a ratio with the maximum temperature difference available $T_5 - T_2$.

$$\frac{\Delta T}{T_5 - T_2} = \frac{1}{1 + \phi} \qquad \text{for } \alpha = 1. \tag{6–31}$$

or, in terms of ϕ, (again, $\alpha = 1$):

$$\eta_x = \frac{\phi}{1 + \phi} \tag{6–32}$$

Evidently, ϕ is a good measure of effectiveness. In terms of η_x, the temperatures T_3 and T_6 are given by:

$$\begin{aligned} T_3 &= (1 - \eta_x)T_2 + \eta_x T_5 \\ T_6 &= (1 - \eta_x)T_5 + \eta_x T_2 \end{aligned} \tag{6–33}$$

These expression are useful for the Brayton cycle analysis described in Chapter 9.

6.4. PRESSURE LOSSES

The process of heat transfer from the fluid medium to the wall is necessarily accompanied by viscous friction so that a pressure drop penalty is incurred. For the purposes of estimating this loss one may assume:

1. Fluid flows in a channel of uniform hydraulic diameter
2. An algebraic mean of shears at the two ends describes the mean shear

A method circumventing this last assumption is discussed in Section 6.4.1. Here the goal is to determine the total pressure change a fluid experiences in terms of the variables of interest for heat transfer.

The momentum equation relates pressure changes to dynamic changes and to forces experienced by the fluid. The one-dimensional form of this statement reads:

$$\rho u \frac{\partial u}{\partial x} + \frac{\partial p}{\partial x} = \frac{\text{shear force}}{\text{volume}} \tag{6–34}$$

The average shear force may be estimated from the wall shear stress, τ:

$$\tfrac{1}{2}(\tau_1 + \tau_2)A_w$$

where the subscripts describe the states at the inlet and exit. The average shear force per unit volume is therefore

$$\tfrac{1}{2}(\tau_1 + \tau_2)\frac{\pi LD}{\pi LD^2/4} = \tfrac{1}{2}(\tau_1 + \tau_2)\frac{4}{D}$$

After integrating, for a constant area (ρu = constant) compressible flow, the momentum equation becomes

$$(p_1 + \rho_1 u_1^2) - (p_2 + \rho_2 u_2^2) = \tfrac{1}{2}(\tau_1 + \tau_2)\frac{4L}{D}$$

or

$$p_1 + \rho_1 u_1^2(1 - \phi_1) = p_2 + \rho_2 u_2^2(1 + \phi_2) \tag{6–35}$$

where the following relations have been used:

$$\frac{\tau_i}{\frac{1}{2}\rho_i u_i^2} = f_i, \frac{A_w}{A} = \frac{4L}{D}, \quad \text{and} \quad \phi_i = \frac{f_i L}{D}$$

(see eq. 6–23 and note that β is taken to be unity as a reasonable value). With $\rho u^2 = \gamma p M^2$ and the definition of stagnation pressure (see eq. 12–9):

$$p_t = p\left(1 + \frac{\gamma - 1}{2}M^2\right)^{\gamma/(\gamma - 1)} \tag{6–36}$$

one obtains, when p_2/p_1 is eliminated,

$$\frac{p_{t2}}{p_{t1}} = \frac{1 + \gamma M_1^2(1 - \phi_1)}{1 + \gamma M_2^2(1 + \phi_2)}\left(\frac{1 + \frac{\gamma - 1}{2}M_2^2}{1 + \frac{\gamma - 1}{2}M_1^2}\right)^{\gamma/(\gamma - 1)} \tag{6–37}$$

which is a function of M_1^2 and M_2^2.

The ratio M_2/M_1 may be written in terms of stagnation property changes. The Mach numbers are defined from (see eq. 12–7):

$$\frac{M_2^2}{M_1^2} = \frac{u_2^2}{u_1^2}\frac{T_1}{T_2} \tag{6–38}$$

With constant area flow ($\rho_1 u_1 = \rho_2 u_2$) and the ideal gas state equation, this ratio becomes:

$$\frac{M_2^2}{M_1^2} = \frac{p_1^2}{p_2^2}\frac{T_2}{T_1} \tag{6–39}$$

which may be written in terms of stagnation conditions:

$$\frac{M_2^2}{M_1^2} = \frac{p_{t1}^2}{p_{t2}^2}\frac{T_{t2}}{T_{t1}}\left(\frac{1+\dfrac{\gamma-1}{2}M_2^2}{1+\dfrac{\gamma-1}{2}M_1^2}\right)^{(\gamma+1)/(\gamma-1)} \tag{6–40}$$

By eliminating M_2 between them, the combination of eqs. 6–37 and 6–40 is the following functional relationship:

$$\frac{p_{t2}}{p_{t1}} = fcn\left(\frac{T_{t2}}{T_{t1}}, M_1, \phi \text{ (and, of course } \gamma)\right) \tag{6–41}$$

which is not of general interest since Mach numbers are small for most heat exchanger flows. In that limit, equation 6–37 becomes ($M^2 \ll 1$):

$$\frac{p_{t2}}{p_{t1}} = 1 + \gamma M_1^2(\tfrac{1}{2} - \phi_1) - \gamma M_2^2(\tfrac{1}{2} + \phi_2) \tag{6–42}$$

To the same order, eq. 6–40 reads

$$\frac{M_2^2}{M_1^2} = \frac{T_{t2}}{T_{t1}} \tag{6–43}$$

Combining eqs. 6–42 and 6–43 one obtains:

$$\frac{p_{t2}}{p_{t1}} = 1 - \gamma M_1^2\psi; \qquad \text{where } \psi = \frac{1}{2}\left(\frac{T_{t2}}{T_{t1}} - 1\right) + \left(\frac{T_{t2}}{T_{t1}}\frac{\phi_2}{\phi_1} + 1\right)\phi_1 \tag{6–44}$$

ψ is seen to account for two effects:

1. Pressure loss associated with the amount of heat transferred (a direct result from the entropy change) and
2. Friction effects governed by $\phi(=fL/D)$.

Evidently keeping M_1 small minimizes the total pressure loss.

6.4.1. *General Description of a Flow through a Long Tube with Heat Transfer*

The assumption that the average shear acting on the flow is the mean between inlet and exit is approximate when there is heat transfer to the fluid. This is true because the inlet and exit conditions may be different. A better way to approach the problem of obtaining a more realistic calculation and to check on the quality of the approximate calculation is to integrate the one-dimensional, steady equations of fluid motion. This approach ignores the entrance effects and is described by ref. 6–2.

The fully developed pipe flow friction factor depends primarily on the Reynolds number which is almost constant in area flow because ρu is invariant and the viscosity is generally a weak function of temperature. With the friction factor, f, known and the variation of total temperature specified by the physical problem, the variation of Mach number is given by

$$\frac{dM^2}{M^2} = \frac{1 + \frac{\gamma - 1}{2} M^2}{1 - M^2} \left\{ (1 + \gamma M^2) \frac{dT_t}{T_t\, d\xi} + \gamma M^2 \frac{4fL}{D} \right\} d\xi \qquad (6\text{–}45)$$

The variation of total temperature derivative for a matched capacity rate, counterflow heat exchanger, as an example, is given by

$$\frac{dT_t}{T_t\, d\xi} = \frac{\left(\frac{T_{t2}}{T_{t1}} - 1\right)}{1 + \left(\frac{T_{t2}}{T_{t1}} - 1\right)\xi} \qquad (6\text{–}46)$$

since the T_t variation is linear (see Fig. 6–12). The variation of Mach number, $M(\xi)$, may be obtained by numerical integration of eq. 6–45 with initial conditions. Further, the variation of other thermofluid mechanical variables may be obtained from the conservation equations. The variation of total pressure is of interest and is given by (ref. 6–2):

$$\frac{dp_t}{p_t} = -\gamma M^2 \left(\frac{dT_t}{T_t\, d\xi} + \frac{4fL}{D} \right) d\xi \qquad (6\text{–}47)$$

Integration gives the total pressure drop as a function of the following variables

$$\frac{p_{t2}}{p_{t1}} = fcn\left(M_1^2, \frac{fL}{D}, \frac{T_{t2}}{T_{t1}} \right) \qquad (6\text{–}48)$$

The results of such a calculation are shown in Figs. 6–13 and 6–14.

6.4.2. *Special Case: Further Interpretation of $\alpha = 1$ Case*

For the $\alpha = 1$ case, eq. 6–44 can be manipulated into a form that involves the extremal temperatures T_5 and T_2 and thus reveals the relationship between the two terms in that equation. From eq. 6–29 for the cold flow:

$$\frac{T_{t2}}{T_{t1}} - 1 = \frac{T_3 - T_2}{T_2} = \frac{T_5 - T_2}{T_2} \frac{1}{1 + \phi}$$

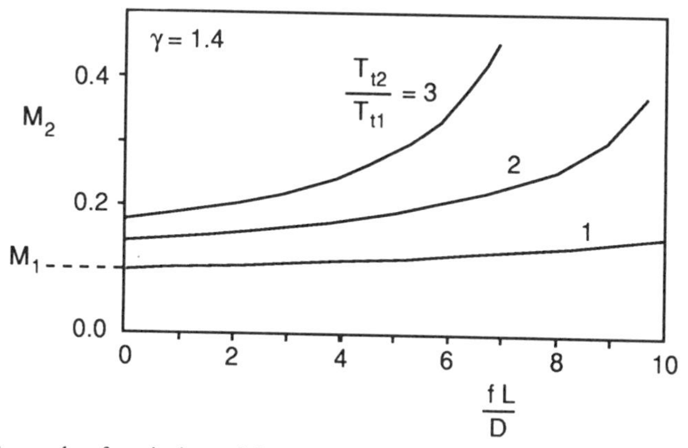

FIG. 6–13. Example of variation of downstream Mach number following heat addition. Integration of eq. 6–45. $M_0 = 0.1$ and $\gamma = 1.4$.

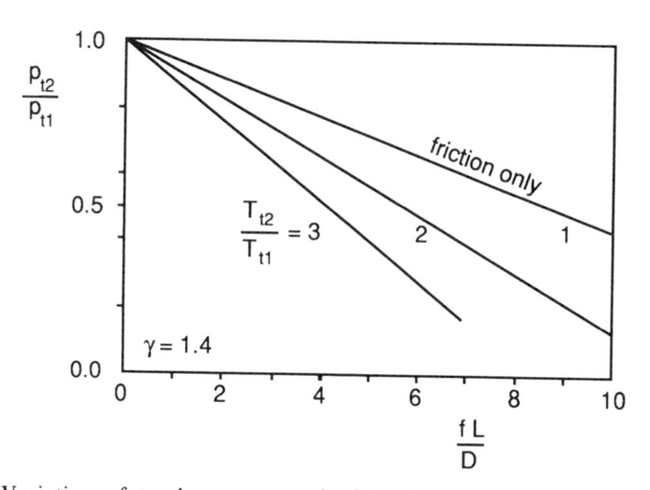

FIG. 6–14. Variation of total pressure ratio following heat addition. Integration of eq. 6–47. Conditions correspond to those noted in Fig. 6–15.

Further, assuming that ϕ's for the two flow are the same, it follows:

$$\frac{T_{t2}}{T_{t1}} + 1 = \frac{T_3 + T_2}{T_2} = \frac{\phi T_5 + (2 + \phi) T_2}{(1 + \phi) T_2}$$

Combining these to form the ψ term in eq. 6–44 gives

$$\psi = \frac{(1 + 2\phi^2)\dfrac{T_5}{T_2} + 2\phi^2 + 4\phi - 1}{2(1 + \phi)} \tag{6–49}$$

which is plotted in Fig. 6–15 and suggests that M_1 must be chosen so that $\psi M_1^2 \ll 1$ for the total pressure loss to be small.

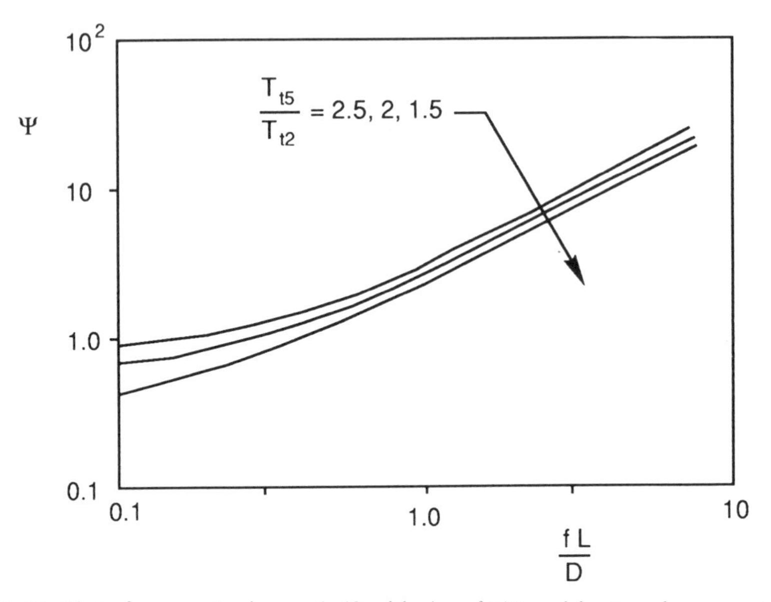

FIG. 6–15. Plot of parameter in eq. 6–49 with $\phi = fL/D$ and heat exchanger extreme temperature ratio.

6.5. PHYSICAL CHARACTERISTICS OF A MATCHED CAPACITY RATE ($\alpha = 1$) HEAT EXCHANGER

The heat exchanger is assumed to be as shown in Fig. 6–16, with an overall length L and "square," (i.e., planar symmetric) cross section. The cross section overall dimension is w. Square tubes of dimension D carry one of the fluids (say the hot one); the other fluid travels outside these tubes in the opposite direction. The specific device envisioned here is a recuperator where $\alpha = 1$ and C_p is the same for the two flows. This allows a simple determination of the volume and surface area required to illustrate the role of the performance parameters. Let the hot flow area fraction of the total be ε and let the number of tubes be n^2. Thus

$$\frac{A_H}{A_H + A_C} = \varepsilon \qquad \text{and} \qquad w^2 = A_H + A_C \text{ from Fig. 6–16}$$

It follows that

$$A_H = (nD)^2 = \varepsilon w^2 \qquad \text{or} \qquad n = \sqrt{\varepsilon}\,\frac{w}{D} \tag{6–50}$$

For a level of mass flow rate $\dot{m}$ and known flow conditions, the flow area required is obtained (for low Mach number) from the 1-D continuity equation:

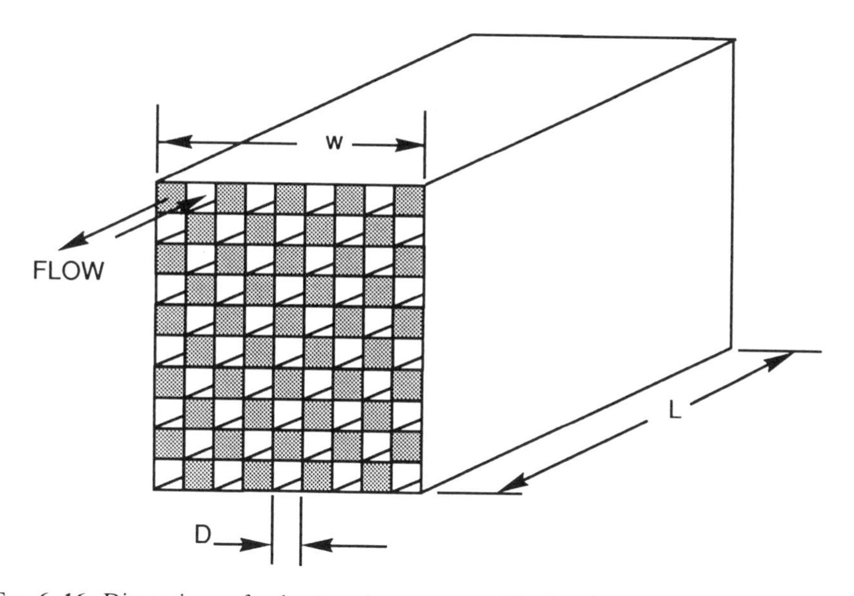

FIG. 6–16. Dimensions of a heat exchanger assembly for the development of scaling relationships.

$$A_H \approx \frac{\dot{m}\sqrt{RT_t}}{p_t} \frac{1}{M_H} \tag{6–51}$$

which can be written in terms of a throat area (A^*) required to bring this flow to sonic conditions.

$$A_H = A_H^* M_H^{-1} \frac{p^*}{p_t} \sqrt{\frac{T_t}{T^*}} = A_H^* M_H^{-1} \Gamma \qquad \text{where } \Gamma \equiv \left(\frac{\gamma + 1}{2}\right)^{-(\gamma+1)/[2(\gamma-1)]} \tag{6–52}$$

This allows determination of the ratio ε as

$$\varepsilon = \frac{A_H}{A_H + A_C} = \left[1 + \frac{A_C^*}{A_H^*}\frac{M_H}{M_C}\right]^{-1} = \left[1 + \frac{p_{tH}}{p_{tC}}\sqrt{\frac{T_{tC}}{T_{tH}}}\frac{M_H}{M_C}\right]^{-1} \tag{6–53}$$

with the condition that the mass flow rates are equal. An additional parameter describing the exchanger is its aspect ratio, $\lambda = L/w$ which must be specified. The volume is

$$\text{Vol} = w^2 L = L(A_H + A_C) = \lambda(A_H + A_C)^{3/2} = \lambda\left[\Gamma\left(\frac{A_H^*}{M_H} + \frac{A_C^*}{M_C}\right)\right]^{3/2} \tag{6–54}$$

It is noteworthy that small length parameter λ is desirable for a small volume which, in practice, is achieved at the expense of having complex ends. The volume is also kept minimal by having A^*/M small (i.e., design for large values of the stagnation pressure), presuming that the stagnation temperature is determined by cycle considerations.

An important parameter in some applications is the weight of the exchanger. For the same $\alpha = 1$ device, the weight is given by the product n^2 (tubes) times $2D$ (half the perimeter since the walls are shared) times L (length) times t (material thickness) times ρ_m (material density), thus

$$\frac{W}{\rho_m} = 2n^2DLt = 2n^2D^2\left(\frac{t}{D}\right)L = 2\varepsilon(w^2L)\left(\frac{t}{D}\right) \tag{6–55}$$

With eq. 6–50, this material volume parameter in eq. 6–55 becomes:

$$\frac{W}{\rho_m} = 2\frac{A_H}{D}Lt = 2\varepsilon\left(\frac{t}{D}\right) \text{(overall heat exchanger volume)} \tag{6–56}$$

The thickness ratio t/D would be fixed by structural considerations that either a pressure tensile stress or a bending stress be within specified limits. In either case the pressure difference across the device is the applied load:

1. Pressure tension (round tube). The maximum stress level is σ_m which must equal

$$\sigma_m = (p_{t2} - p_{t1})\frac{D}{t} \quad \text{or} \quad \frac{t}{D} = \frac{\Delta p_t}{\sigma_m}$$

2. Bending stress (square tube).

$$\sigma_m = \frac{\text{bending stress}}{\text{moment of inertia}}\frac{2}{t} = cst\frac{\Delta p_t D^2 Lt}{Lt^3}$$

or

$$\frac{t}{D} = cst\sqrt{\frac{\Delta p_t}{\sigma_m}} \tag{6–57}$$

Hence t is directly related to D and nongeometrical design parameters. This allows the conclusion that the weight of a heat exchanger is proportional to the parameters that determine overall volume: M, λ, and the tube configuration. Section 9.11 expands on these ideas in connection with the Brayton cycle.

PROBLEMS

1. By linearizing eq. 6–37, show that eq. 6–42 follows. Show also that eq. 6–43 is valid for the same approximation.

2. Estimate the magnitude of the conductive and convective terms in eq. 6–12 for typical conditions of heat exchanger flow, to show that the conductive term is indeed small.
3. Develop an expression for the hydraulic diameter for a duct of rectangular cross section, noting in particular the limits of large and unit aspect ratio.
4. Write the equations governing the momentum and energy in a boundary layer or fully developed flow to arrive at the conclusion that the Reynolds analogy should be physically valid, especially for the case of laminar flow.
5. Show that in the limit $\alpha = 1$, the temperature distributions in a counterflow heat exchanger are linear under the restrictions stated in Section 6.2. Is this true even when the heat transfer coefficients, h, on the hot and cold sides are unequal?
6. Show by examination of eq. 6–45 that a heated, viscous flow experiences a Mach number increase, which if conditions allow will result in a limiting $M = 1$, viscous or thermal choking, depending on the relative importance of the two effects. Integrate numerically the equation for M for separate friction and heating effects, the latter by choosing a T_t profile that reflects a thermal power input.
7. Show the validity of eq. 6–57 by examining a square duct under pressure.
8. By writing the boundary layer equations with pressure gradient, estimate the duct length required to arrive at fully developed flow where the boundary layer theory will no longer be valid.
9. A heat exchanger flow tube is found to operate at a $Re_D = 10^5$ for some conditions that involved flowing a certain *fixed* mass flow through the tube. If the diameter is made smaller, is the Reynolds number increased or decreased? i.e., since Re_D can be written as a reference Reynolds number times D^n, what is n?
10. A matched capacity rate heat exchanger is to operate close to ideal. Show that L/D can be written as

$$\frac{L}{D} \approx \frac{\text{constant}}{1 - \eta_x}$$

and estimate the constant.

BIBLIOGRAPHY

Kays, W. M., London, A. L., *Compact Heat Exchangers*, The National Press, Palo Alto, Cal., 1955.

Hodge, B. K., *Analysis and Design of Energy Systems*, Prentice-Hall, Englewood Cliffs N.J., 1985.

REFERENCES

6–1. Shapiro, A. H., *Dynamics and Thermodynamics of Compressible Fluid Flow*, The Ronald Press Co., New York, 1953.

6–2. Moody, L. F., "Friction Factors for Pipe Flow," *Trans. ASME*, **66**: 671–84, Nov. 1944.

6–3. Prandtl, L., *Strömungslehre*, Fried. Vieweg & Sohn, Braunschweig, 1949.

6–4. Genereaux, R. P., "Fluid Flow Design Methods," *Industrial Engineering Chemistry*, **29**(4): 385–8.

7

HEAT ENGINE PROCESS MODELING

The description of engine cycles requires a model for the elementary processes making up the cycle. These include those that involve the transfer of heat and those that involve work interactions between the thermodynamic system and the environment. A realistic performance assessment requires identification and quantification of the important sources of irreversibilities so that their influence may be assessed. This chapter is a discussion of important irreversible processes relevant to engine performance analysis.

7.1. DEFINITIONS AND INTRODUCTION TO CYCLE ANALYSIS

A heat engine is a device wherein a combination of components or processes convert heat to mechanical power. The heat engine works by causing a mass of working fluid to undergo a series of state changes resulting from heat and work interactions such that the paths drawn in the p-V or T-s planes are closed.

Three important aspects of a heat engine must be considered in order to successfully obtain a descriptive model of the engine. These are

1. The cycle
2. The irreversibilities of the elementary processes
3. The choice of working fluid

The work calculated by such an analysis of a cycle is the work that would be produced in the absence of mechanical friction losses.

Attention will be restricted to systems where the thermodynamic state is specified by two variables—a simple system. The word "system" is used to describe a unit mass of the working fluid. Figure 7–1 illustrates the heat engine, its components or component processes, and the interactions it experiences with the external world.

7.1.1. *Cyclic Operation of a System*

A closed cyclic path in the p-V plane indicates that net work is absorbed or produced because

$$\int p \, dV = W_{\text{rev}} \qquad (7\text{–}1)$$

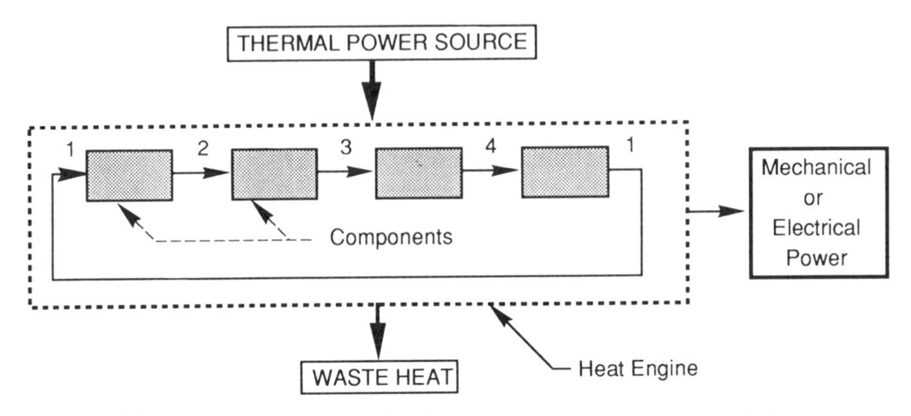

Fig. 7–1. Schematic of a heat engine (and its components) interacting with the external world.

The convention is that dV is positive, work is done *by* the system. The irreversible work is a conversion to heat. A closed cyclic path in the T-s plane shows that a net amount of heat is provided *to* the system. This heat may be added reversibly (from a heat source at T) or irreversibly (through friction or from a heat source at $T_s > T$). In general, the heat input is given by

$$Q \leq \int T\,ds \tag{7–2}$$

With the inequality sign, the area under a path on the T-s plane can be interpreted only when additional information is available. For example, if the process is an adiabatic work interaction, the area under the path is an irreversible conversion of work to heat. For a reversible heat interaction, the equality sign in eq. 7–2 applies and the interpretation is direct: the area represents the heat involved. Thus in much of the work to be done in cycle modeling, the T-s plane proves useful: the reversibility of adiabatic processes is clearly displayed, and the area under a path is related to heat input and its reversibility.

When the cyclic process is performed, the paths in the T-s and p-V planes are closed, so that state variables (u in particular) return to the starting values. Under these circumstances, the First Law states that the net heat added is the work produced (or vice versa) so that heat is converted to work and the areas in the planes mentioned may be interpreted as limiting values of heat or work involved. Thus

$$\oint du = 0 \qquad \text{gives} \qquad \oint d'q = \oint d'w \tag{7–3}$$

Figure 7–2 shows the two state planes and the closed cycle paths of interest.

Since constant pressure and constant volume processes are of interest for many types of heat engines, the mathematical form of such process lines must

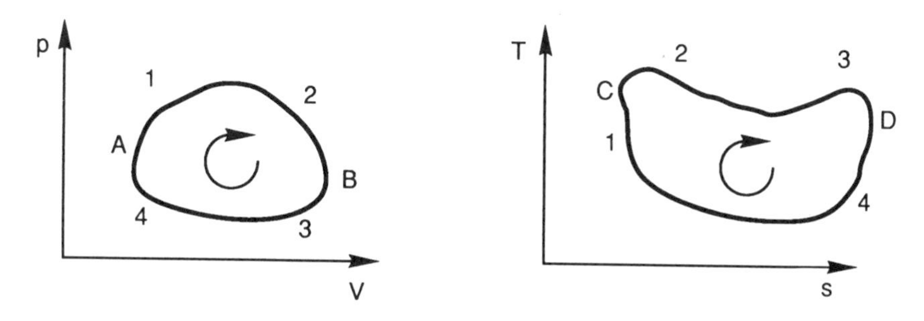

FIG. 7–2. State path diagrams showing net work output from system at left (clockwise path) and net heat input to system at right (also clockwise path).

be developed in the T-s plane. This is particularly simple for the ideal and perfect gas because the equations are analytic.

A process line is described by a relation between two of the three variables p, T, and s. From the combined first and second laws (eq. 4–14) applied to a perfect gas,

$$T\,ds = C_p\,dT - v\,dp \tag{7–4}$$

using the ideal gas equation of state for v (eq. 4–2),

$$ds = C_p \frac{dT}{T} - R \frac{dp}{p} \tag{7–5}$$

For a gas with constant C_p the equation of a line of constant pressure (isobar) is obtained from integration of eq. 7–5:

$$T = \exp\left(\frac{s}{C_p}\right) + T_0; \qquad \text{for } p = \text{constant} \tag{7–6}$$

A similar expression applies to the line of constant (specific) volume. From eq. 7–4 and the equation of state, the differential pressure may be eliminated so that

$$T\,ds = C_v\,dT + p\,dv$$

It follows that with $dv = 0$:

$$T = \exp\left(\frac{s}{C_v}\right) + T_0; \qquad \text{for } v = \text{constant} \tag{7–7}$$

Thus both isolines are exponentials and since $C_p > C_v$, eqs. 7–6 and 7–7 show that lines of constant v in the T-s plane are steeper than lines of constant p as shown in Fig. 7–3.

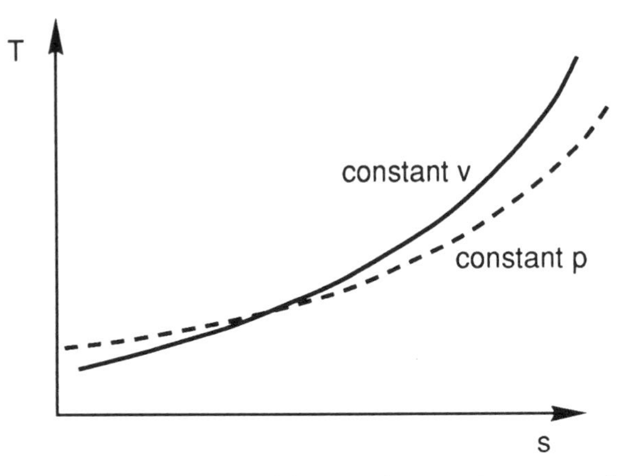

FIG. 7–3. Constant v and p exponential curves for a perfect gas on the T-s plane.

Before proceeding to discuss cycles and the thermodynamic paths involved, the adequacy of the state point as a description of the system must be addressed.

7.1.2. *Description by Use of a State Point*

The events that are made to occur in the working fluid as it proceeds from one state to another necessarily occur in finite time and with the fluid in contact with the surrounding boundaries. The finite process time violates the requirement for infinite time available for the adjustment of conditions to achievement of true equilibrium. This would appear to be an undesirable situation but, in actuality, the reverse is true: a short process adjustment time is desirable. The reason is that the working fluid is never spatially uniform so that it may not be described by a simple set of state variables. The transport of mass, momentum, and energy takes place whenever conditions are nonuniform which, in turn, leads to the transport of energy out of the working fluid to the surroundings, resulting in a loss. This is to say that state process changes must take place sufficiently rapidly that the losses do not play an important role in affecting the events one would wish the working fluid to experience. Thus in many descriptions of processes, there is a multiplicity of important time scales. In the description of a combustion process, for example, one can list the following scales:

1. Time between collisions of molecules: collision frequency
2. Chemical reaction time scales: chemical kinetics time constant
3. Flow or process time scales: machine angular speed or flow velocity/characteristic physical length
4. Transport or diffusion time scale: species, momentum, or thermal diffusivity

The process of successful modeling is seen to require an understanding of the correct hierarchical relationship between the relevant times scales and the use of approximations based on this relationship. In modeling of heat engine processes, the first order of approximation is that state points are descriptive of the process: The working fluid is uniform and processes are temporally or spatially distinct. This approach will be seen to be successful in describing steady flow machines and less so in describing reciprocating heat engines, particularly those where the transport of heat is by conduction, such as is the case in the Stirling engine.

7.2. TYPES OF HEAT ENGINES

In discussing heat engines it is useful to distinguish among engines that differ in important ways. Engines can be either steady or intermittent, and they can be closed or open.

7.2.1. *Steady and Intermittent Process Engines*

The steady flow engines are built around an array of components through which the fluid flows, never building up thermal energy or mass within the component. Among such components are compressors, expanders, and heat exchangers. Two important steady flow cycles are the Rankine and Brayton cycles, for which the component arrangements are shown in Fig. 7–4.

Intermittent operation engines, on the other hand, force the fluid system to undergo the processes required by keeping it confined by the same material boundaries and sequentially altering the environment by movement of the boundaries. The automotive Otto cycle engine is probably the best example of such an engine where timing provided by cranks, valve gear, and fuel system determine what happens to the system (the air in the cylinder). As far as the thermodynamics is concerned, it matters little whether the processes are carried out in a steady flow engine or in an intermittent engine, the result is the same: The system experiences a changing set of conditions that produce work. The rate at which these processes are carried out determines the power of the engine. Figure 7–5 shows the history of an intermittent-type engine and how events at the boundary control the events to which the gas in the cylinder volume is subjected.

7.2.2. *Closed and Open Cycles*

Closed cycle engines are those that use a given mass of working fluid in a continuing way. Such engines are built to take advantage of particular fluid properties and often use fluids of high purity. Such cycles can be operated at any pressure level chosen to suit the power density requirements of the machinery, consistent with strength and operating life of the components. They require the use of heat exchangers to transfer of heat between the system and the heat reservoir. Chapters 9 and 11 discuss steady flow cycles (Brayton and

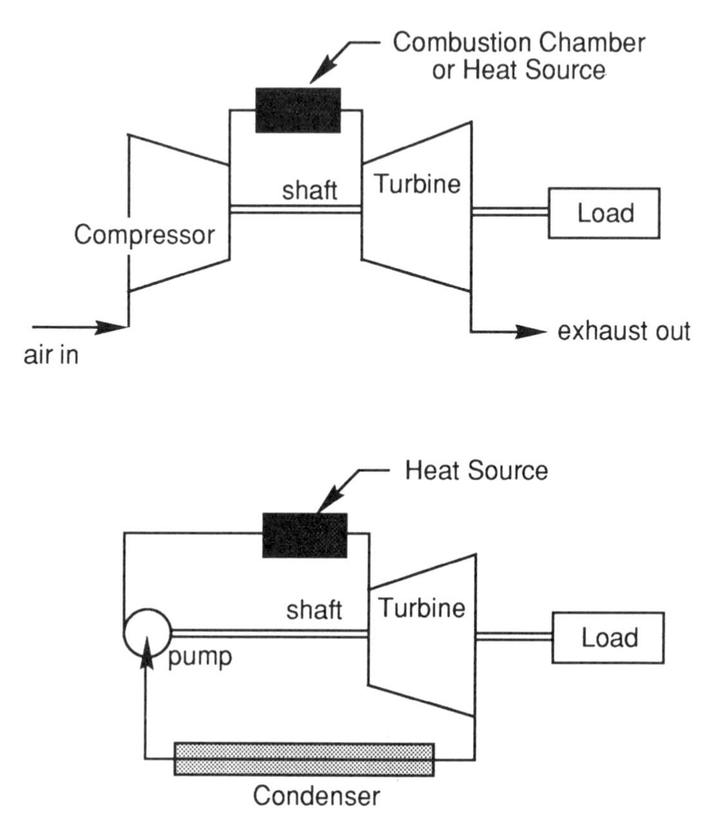

FIG. 7–4. Schematic of two steady flow heat engines: Brayton (top) and Rankine (bottom) cycles.

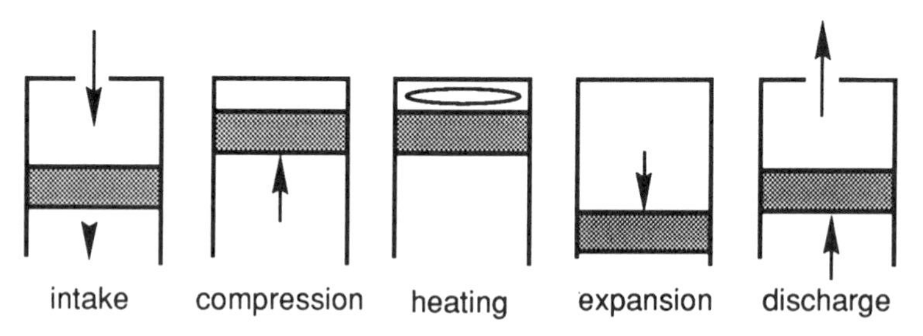

FIG. 7–5. Intermittent or batch process engine: an example, the Otto cycle.

Rankine, respectively). Figures 9–1 and 9–2 contrast closed and open cycles schematically.

Open cycles, on the other hand, employ a dischargeable substance (such as the water that the steam locomotives used, air, or combustion product gas) as a working fluid. The advantage is that the heavy and costly heat rejection heat exchanger(s) may be eliminated. When the properties of the working fluid are altered chemically, as they are in the combustion process, it requires that the

FIG. 7–6. A mid-twentieth century railroad locomotive uses an open Rankine cycle engine. (Photo by P. Illert).

Table 7–1. Design characteristics of open and closed cycles

Closed	Open	Consequences
Can choose working fluid	Must use air (or water)	Affects max. T, losses, corrosion
Can choose density	Must use ambient density	Affects compactness
Requires input heat exchanger	Can use combustor with air	Cost, compactness, component life
Requires rejection heat exchanger	Simple discharge	Cost

processed fluid be discarded. These characteristics make the use of open cycles very attractive for applications where power is desired for propulsion of a transport vehicle. Such a vehicle needs to carry only the fuel required for the combustion process and uses the air available locally. The steam locomotive (Fig. 7–6) also carries the working fluid that is condensed during expansion and in the cool environmental air as visible droplets. Open cycle engines include the Otto, Brayton, Diesel, and Rankine cycles which provide a major portion of the world's motive power. Table 7–1 summarizes the major differences between open and closed cycles and their impact on practical engines.

7.2.3. *The Environment as an Engine Component*

The open cycle operates in a cyclic manner when the air environment is included as a component in the process. The heat rejection process may be viewed as cooling to ambient temperature followed by dilution to very nearly initial

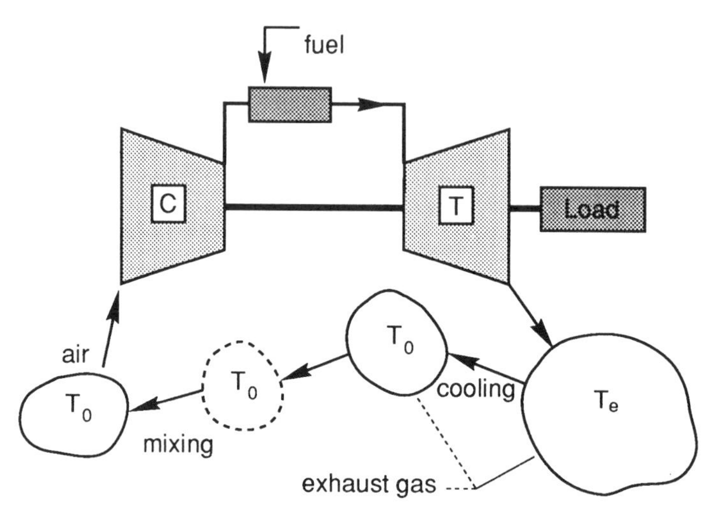

FIG. 7–7. The atmosphere as an engine component in a steady flow heat engine. Combustion gas is cooled and diluted to approximate fresh air.

chemical composition. Figure 7–7 illustrates the process of heat rejection in an open cycle engine. For engineering purposes, this process is often adequately modeled by considering the process to be a constant pressure flow process, ignoring the chemical composition aspects.

7.3. WORKING FLUIDS

Any phase and/or combination of phases of a substance or mixtures of substances may be employed in a heat engine. The requirement that volume changes be significant so that a large quantity of work is produced usually dictates the use of the gaseous phase or combinations of the gas and liquid phases of a substance as a working fluid. The mathematical ease with which the working fluid may be described depends on the proximity of the state point of interest to the "saturation dome" of the substance (Section 4.2.6).

7.3.1. *Gases*

The simplest description is of the "ideal gas" for which

$$pV = RT \tag{7–8}$$

applies (see eq. 4–2). This approximation requires physically that the molecules spend little of their time in the proximity of other molecules, as shown in Fig. 4–1.

The notion of a "perfect gas" is embodied in the approximation that the internal modes of energy storage are uniformly accessible when the temperature is changed. As a result, internal energy and enthalpy are directly proportional

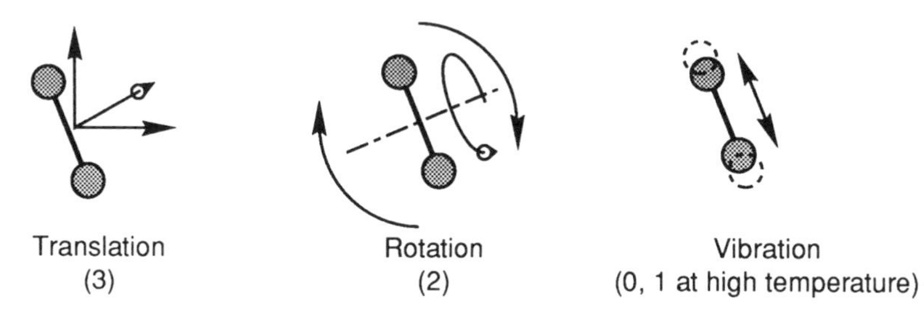

FIG. 7–8. Specific heat is related to the number of ways in which a molecule can store and surrender energy by collisions.

to temperature. Stated in other words: For a (calorically) perfect gas, the specific heats, C_p and C_v, are constant (Section 4.2.2).

In actuality, when sizable temperature changes are considered, internal energy modes are accessible to varying degrees and the result is that the specific heats are temperature dependent. Thus for an

$$\text{Imperfect and ideal gas: } C_p \quad \text{and} \quad C_v = \text{functions of } (T) \qquad (7\text{–}9)$$

Consult the thermochemical tables in Appendix C for an appreciation of the variation of C_p. Note the specific heat of a diatomic gas like N_2 and a more complex one like CO_2. Tables for monatomic gases like helium are not provided because they are indeed perfect, having no internal degrees of freedom such as rotation or vibration that can be excited by collisions with other molecules (Fig. 7–8). The validity of assuming the specific heats are constant depends on the range of temperature changes experienced in a process and the need for accuracy (see Chapter 4).

The following hierarchy of approximations applies to modeling of gases in cycle analysis (in order of increasing accuracy):

1. Ideal gas, calorically perfect gas: R, C_p, C_v are constant.
2. Ideal gas, piecewise perfect gas: R is constant; C_p and C_v have different constant values depending on gas temperature and chemical composition.
3. Piecewise ideal gas: R is constant at value associated with the (temperature dependent) molecular weight of the gas composition; C_p and C_v are temperature dependent.
4. Nonideal gas or vapor: used for states near the saturation dome. Property tables or elaborate mathematical descriptions must be employed.

Figure 7–9 illustrates the descriptive regimes of a substance in the p-v plane.

7.3.2. *Two Phase Cycle Fluids*

Condensable fluids are used in cycles operating to take advantage of liquid properties at low temperature and vapor or gas properties at high temperature.

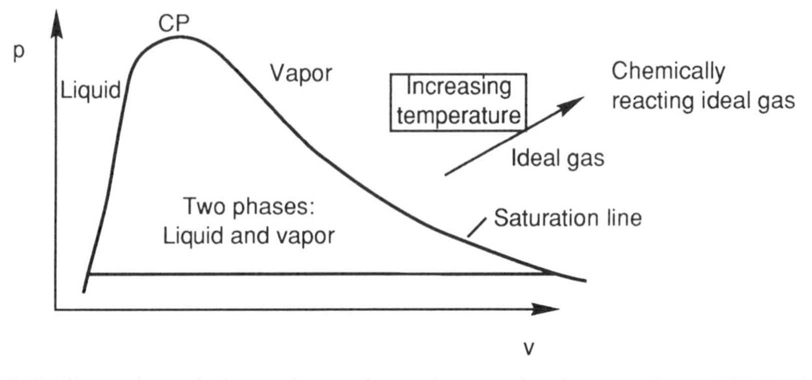

FIG. 7–9. Illustration of the regimes of a substance in the *p-v* plane. CP is the critical point.

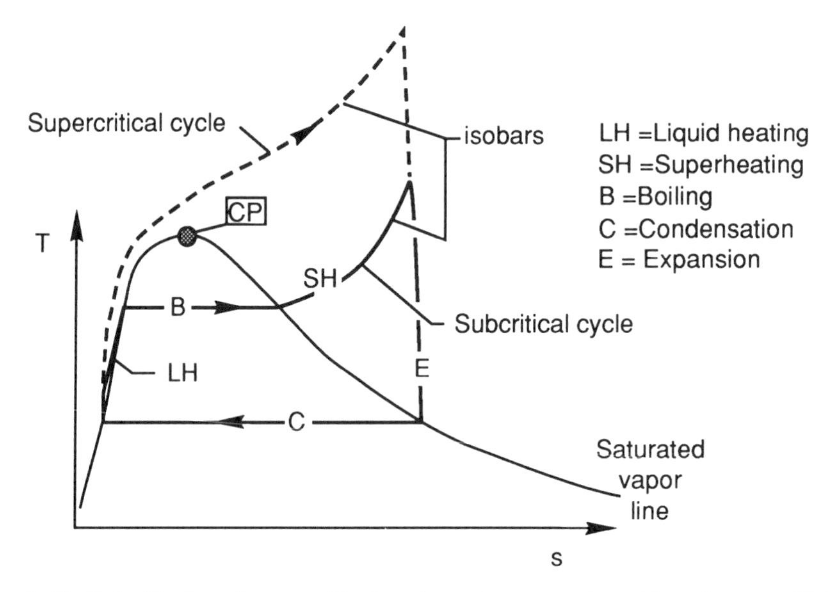

FIG. 7–10. Subcritical and supercritical cycles using a condensable substance. Note compression and expansion work processes.

Typically the critical temperature lies between these limits. Modeling of such cycles is complicated by the awkward description of the fluid properties. The unique advantage of using a condensable substance is that the work of compression is very small owing to the small and nearly constant specific volume of a liquid. (See Section 7.6.2 where this notion is quantified.) Figure 7–10 shows the state path of two cycles on the *T-s* plane, one (solid line) involving liquid heating, boiling, and vapor heating (termed superheating in the literature). The other (dashed line) involves a continuous single-phase change from liquid to vapor. In the subcritical cycle, the heating process line crosses under the critical state, whereas in the supercritical heating process, the process path does not cross saturation lines.

7.4. HEATS INTERACTIONS BETWEEN SYSTEM AND RESERVOIRS

7.4.1. *Heat Reservoir*

The materials from which the system, or the heat engine, gains heat and to which it loses heat are called heat reservoirs. The high temperature reservoir supplies heat at a specific temperature, and the system cannot be made to rise to a temperature higher than this value. This nearly obvious statement is, in fact, a form of the Second Law of Thermodynamics. Similarly, the low temperature reservoir limits the lowest temperature of a system that must reject heat. The existence of these reservoirs therefore imposes limits to the range of temperatures in which a cycle operates.

One may visualize a heat reservoir as a material mass with which the system (the working fluid) comes into contact. In this mass, some form of heat energy is released, and as a consequence, the system is heated. This mass could be the core of a nuclear reactor, the tubes of a heat exchanger, or the gas participating in the combustion process. In this last, but common, case, one may visualize the nitrogen component of air as the system that is heated by the reacting oxygen and fuel molecules. One can say further that the oxygen behaves as if it participates in this heating process even though it altered chemically. Thus when one models combustion as a heating process, one is neglecting the mass addition of fuel. For many purposes, this is a valid approximation, although proper modeling of this mass addition is straightforward (see Section 7.5).

A balance between the generation of heat within the reservoir and the removal rate by the system determines the temperature of the reservoir and must be such that neither the reservoir mass nor the material confining the working fluid is compromised if the engine is to have a practical lifetime. This maximum temperature as well as the minimum temperature available for the system are the most important parameters governing the performance of the cycle (see Fig. 7–11).

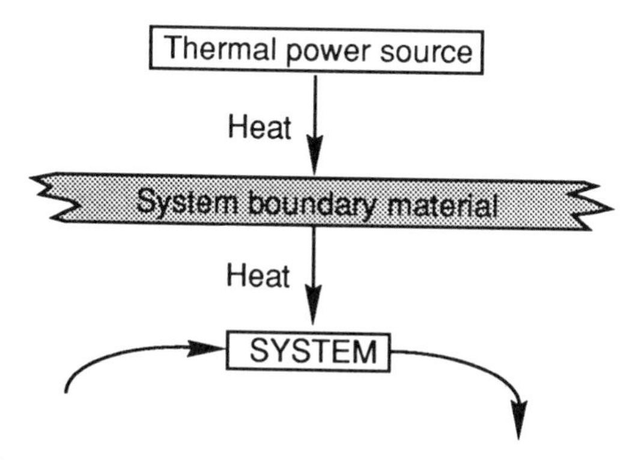

Fig. 7–11. Effective reservoir temperature determined by a balance between input rate from power source and removal rate by the system.

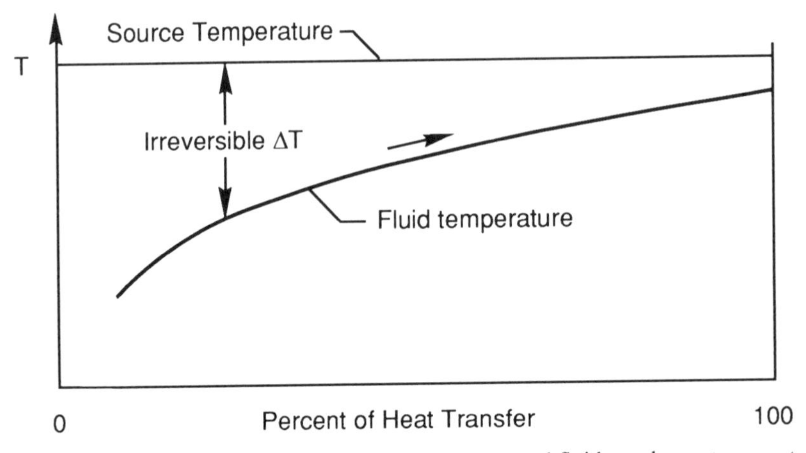

Fig. 7–12. Heat transfer irreversibility between a source and fluid at a lower temperature.

7.4.2. *Reversible and Irreversible Heat Transfer*

The definition of entropy is by the relation (eq. 2–11):

$$ds = \frac{dq_{\text{rev}}}{T} \tag{7–10}$$

where the heat transfer process is reversible as noted. This means that if one had a system at temperature T and a heat reservoir at a very slightly higher temperature $T + dT$, then heat flows nearly reversibly from the reservoir to the system. This process is reversible when dT is made to approach zero. Irrespective of the reservoir temperature, the integral of $T\,ds$ represents the heat that flowed to the system. From a practical point of view, the area required to transfer a finite amount of heat across a very small temperature difference is either very large or requires a very long time. Either requirement conflicts with the desire for high power density measured in terms of power/volume. Thus, in practice, considerable irreversibility is tolerated because it leads to a significant reduction of machinery volume and cost. Thus the process of heat transfer in practical devices may be visualized as shown in Fig. 7–12. The temperature rise of the system from its lowest value to the highest, which is the source temperature, is a measure of the irreversibility suffered in the process.

7.4.3. *Practical Methods of Heat Transfer*

The most common heat interactions between reservoir and system are at constant volume and at constant pressure. The Otto and Stirling cycle engines are the best examples of constant volume heating. Steady flow engines such as the Brayton and Rankine have nearly constant pressure heating. The Diesel cycle engine employs a combination of these processes. None of these processes

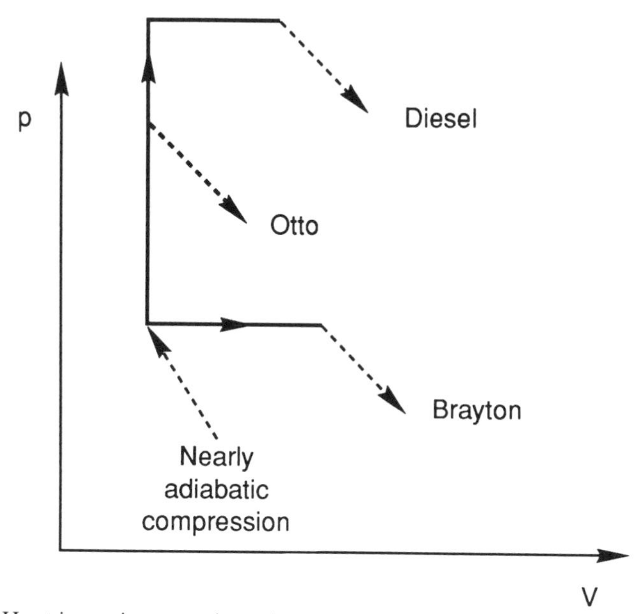

Fig. 7–13. Heat input in a number of gas cycles, normalized to the same state at the start of heat input.

are reversible because part of the heat is added at a lower temperature than is another part. This irreversibility leads to a significant performance difference between the real and the reversible cycle. Figure 7–13 shows the ideal heating processes in the p-V plane for various engine cycles noted. In the T-s plane, heating lines are described by eqs. 7–6 and 7–7 and illustrated in Fig. 7–3.

7.4.4. *Constant Volume Heat Addition by Combustion*

In Otto, and to some extent in Diesel cycle engines, the constant volume combustion process is carried out by forcing the rate of heat addition to be very fast in relation to the volume changes associated with the piston motion. This is done by releasing the heat rapidly to the air in the cylinder volume. In the Otto or spark ignition engine, the fuel is a premixed component of the compressed working fluid and is ignited at an optimum time. In the Diesel engine, fuel is rapidly injected into the compressed air at the appropriate time. In either case, the time scale for the combustion and fluid mixing processes is usually sufficiently rapid that the combustion may be modeled as occurring at constant volume (Fig. 7–14). The major irreversibility associated with this process is that the equivalent reservoir temperature exceeds the system temperature at all times (see Fig. 7–12) and, for a time, by a significant ΔT. Another important source of irreversibility is associated with the heat transferred to the walls by virtue of their relatively low temperature. This heat transferred represents a lost opportunity for conversion to work, at least by the system within the cylinder volume.

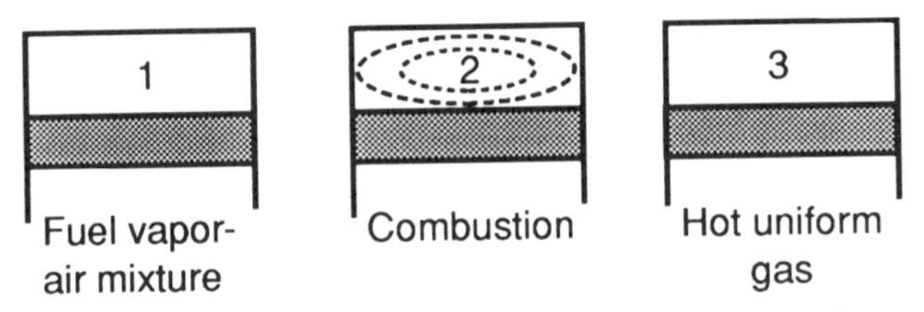

FIG. 7–14. Rapid combustion at top dead center (constant volume) in an Otto cycle engine.

The process of heat addition by the slower conduction process is carried out in the Stirling engine. Chapter 10 is a discussion of this cycle and some of the special engine characteristics which arise from this mode of heat transfer.

7.4.5. *Constant Pressure Heat Addition*

In steady-flow heat-transfer processes, the fluid has a necessarily finite convection speed through the device. In such a device, the fluid experiences two unavoidable irreversibilities. These are associated with the friction of fluid moving past a stationary wall and the fact that heat transfer to a moving fluid occurs to the fluid at the source temperature which must exceed the fluid's stagnation temperature (see eq. 6–44). This stagnation temperature is necessarily larger than the static temperature by a factor related to the Mach number of the flow. For duct flow with heat and momentum transfer, there are total pressure decreases (see Section 6.4) associated with both of these effects. Both contributions to the total pressure loss depend on the Mach number (M^2) of the flow and are thus minimized by designing the flow area to be large so that M is small. For a given mass flow rate, the relationship between area and Mach number is given by (eq. 12–10):

$$\dot{m} = \rho u A = \frac{p_t}{\sqrt{RT_t}} \sqrt{\gamma}\, MA \left(1 + \frac{\gamma - 1}{2} M^2\right)^{(\gamma+1)/(2(\gamma-1))} \qquad (7\text{–}11)$$

$$\approx \frac{p_t}{\sqrt{RT_t}} \sqrt{\gamma}\, MA \qquad \text{for } M^2 \ll 1 \qquad (7\text{–}12)$$

Since pressure and temperature are presumably fixed by thermodynamic and physical constraints, the losses are minimized by balancing compactness (as measured by the flow area, A) against thermodynamic performance.

In any case, the so-called constant-pressure heat addition process in steady flow is actually one where the total pressure necessarily falls (see eq. 6–51), albeit as slightly as desired. Figure 7–15 shows the ideal and actual heat addition processes that are analytically described in Section 6.4.

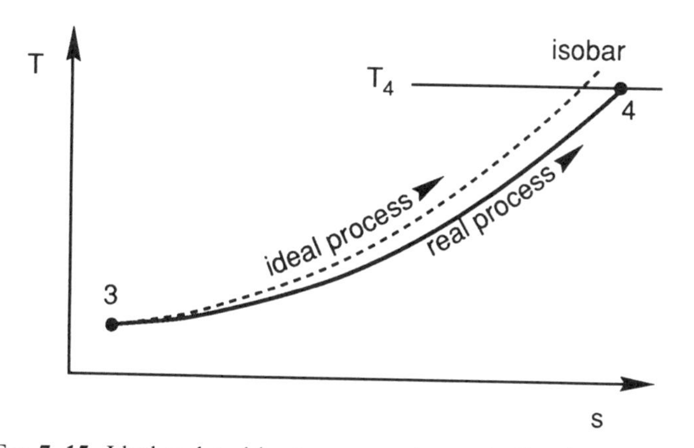

FIG. 7–15. Ideal and real heat processes in steady flow heat addition.

7.5. SPECIAL FEATURES OF STEADY-FLOW HEAT ADDITION BY COMBUSTION

7.5.1. *Chemical Aspects*

The composition of the fuel determines to a large extent the value of the fuel as a heat producer. In the following, one may estimate the temperature rise to be expected in a constant pressure combustor.

Consider a fuel whose hydrogen to carbon ratio is 2, typical of a liquid petroleum fuel, that is, $(CH_2)_n$. Here n is arbitary, hence one may choose $n = 1$. Other values of n would be manifest in the fuel's heat of formation and its physical properties. In the combustion with air, 1.5 moles of oxygen are required to burn the fuel to only CO_2 and H_2O, which bring 1.5×3.76 moles of N_2. The balance reads

$$CH_2 + 1.5O_2 + 1.5 \times 3.76N_2 \rightarrow CO_2 + H_2O + 5.64N_2 \qquad (7\text{–}13)$$

With the formation of 1 mole each of CO_2 and H_2O, the heating value of the fuel, H (see Section 5.3 and Table 1.8) is determined.

For a steady flow combustor processing a mass flow rate $\dot{m}_3$ of air with $\dot{m}_f$ of fuel (Fig. 7–16), the mass balance reads

$$\left.\begin{aligned} &\dot{m}_4 = \dot{m}_3 + \dot{m}_f \\ &\text{or} \\ &\dot{m}_4 = \dot{m}_3(1 + f); \qquad \text{where } f = \frac{\dot{m}_f}{\dot{m}_3} \end{aligned}\right\} \qquad (7\text{–}14)$$

and the thermal balance reads

$$\dot{m}_4 h_4 = \dot{m}_3 h_3 + \dot{m}_f H$$

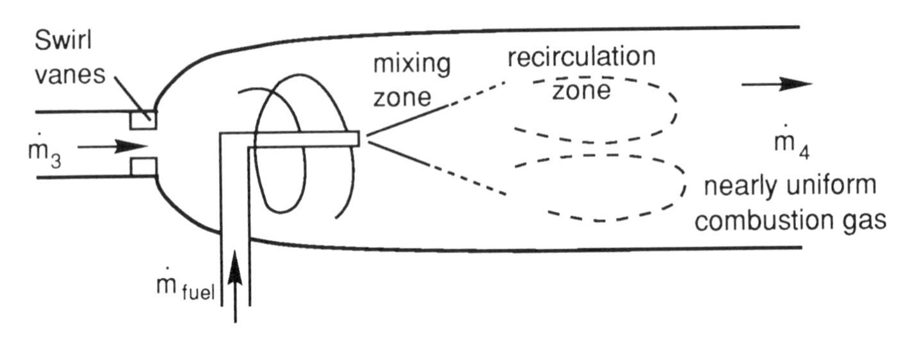

Fig. 7–16. A steady flow combustor. The state point numbering system corresponds to that used in the Brayton cycle, Chapter 9.

or

$$h_4 = \frac{h_3 + fH}{1 + f} \tag{7–15}$$

For a piecewise perfect gas, this becomes

$$T_4 = \frac{C_{p3}}{C_{p4}} \frac{T_3 + (H/C_{p3})f}{1 + f} \tag{7–16}$$

Here Hf/C_{p3} is a temperature rise parameter. To estimate this parameter, consider the stoichiometric combustion of the fuel described above CH_2. The fraction, f, for this case is

$$f = \frac{12 + 2}{(1.5 \times 32) + (5.64 \times 28)} = 0.068$$

and H/C_{p3} is of the order of 45000 K (!) so that the temperature rise, fH/C_{p3}, is of the order of 3000 K. Nondimensionally, one may define $\theta_i = T_i/T_0$ and $\Delta\tau_b = fH/(C_{p3}T_0)$, a temperature rise parameter, so that eq 7–16 can be written as:

$$\theta_4 = \theta_3 + \Delta\tau_b \tag{7–17}$$

where $\Delta\tau_b \sim 10$, the approximation $f \ll 1$ has been used in eq. 7–15 or 7–16, and $C_{p4} = C_{p3}$ is assumed.

Naturally $\Delta\tau_b$ can be any value less than this maximum value if there is a reduction in the fuel used relative to the amount of air supplied, that is, f can be smaller than the value given.

If the combustion process is carried out at constant volume (see Section 2.1) the principal difference is that the temperature rise parameter is larger by the ratio of $\gamma = C_p/C_v$, because the first law for constant volume combustion must be written in terms of internal energy (eq. 2–4) rather than enthalpy (eq. 4–13).

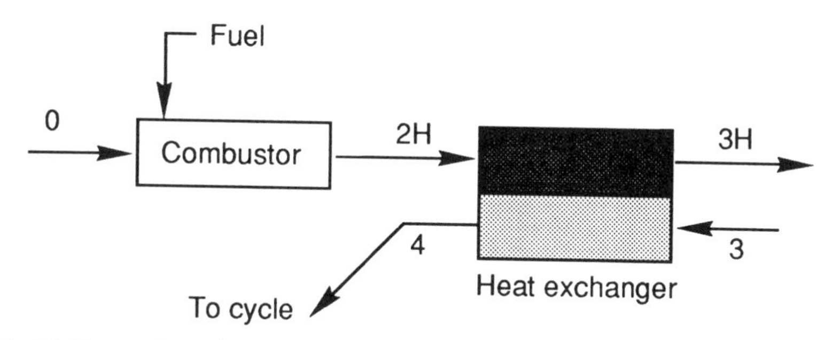

FIG. 7–17. External combustion scheme for the transfer of heat to a steady flow cycle.

7.5.2. *External Combustion*

In Chapter 6 the performance of heat exchangers was considered. The performance of such devices plays an important role in cycles where combustion must take place outside the cycle. In the discussion of Brayton or Stirling cycles, the working fluid may be (often must be) different from the combustion gas. In such instances, combustion may take place in an atmospheric pressure burner using ambient air to generate heat to be transferred to the working fluid. Using a steady flow model, one may consider the cycle to require the working fluid temperature to be increased from T_3 to T_4.

Figure 7–17 shows an arrangement for doing this. For this discussion, $\dot{m}$ is the air mass flow rate and $\dot{m}(1 + f)$ is the amount of combustion gas processed. The cycle on the other hand processes $\dot{m}'$.

Three specific heats (perfect gas is assumed) characterize the gases involved

1. Air: C_p,
2. Combustion gas: $C_p(1 + x)$ where x depends somewhat on the fuel air ratio, f

and

3. Cycle gas: C'_p.

x is a small factor to account for the variation in C_p. Consider first the relatively poor arrangement shown in Fig. 7–17. The combustion gas T_{3H} exits at relatively high temperature if the heating $\Delta T = T_4 - T_3$ is small. This represents a waste of the energy in the waste exhaust gas which can be quantified by a fuel utilization efficiency, η_f, which is determined below.

$$\eta_{fu} = \frac{\dot{m}' C'_p (T_4 - T_3)}{\dot{m} f H} \tag{7–18}$$

The performance of this heat transfer technique is relatively poor even when the fL/D parameter of the heat exchanger is very large. The resulting temperature profile is as shown in Fig. 7–18 with temperatures 4 and $2H$

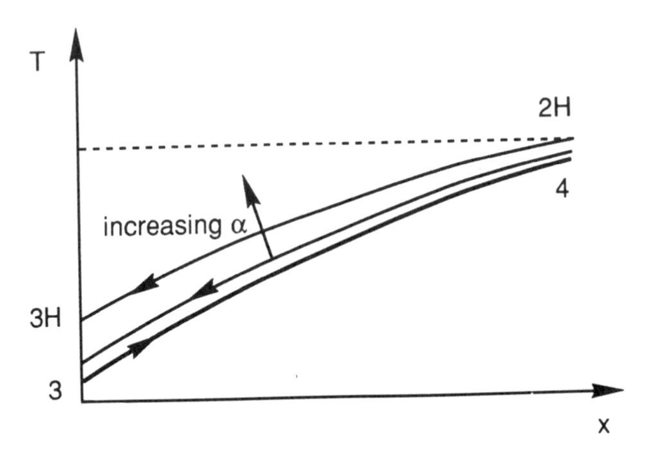

FIG. 7–18. Temperature profiles in heat exchanger for various values of α.

identical. Three lines are shown with the capacity rate ratio, α defined as

$$\alpha = \frac{\dot{m}(1+f)C_p(1+x)}{\dot{m}'C_p'} \tag{7–19}$$

having a unit value increasing to very large. Evidently, T_{3H} is lowest for $\alpha = 1$, hence this is the case of interest.

Combining eqs. 7–17 with 7–18 and 7–19, one obtains,

$$\eta_{fu} = \frac{(T_4 - T_3)}{T_0}\frac{1}{\Delta\tau_b}\frac{(1+x)(1+f)}{\alpha} \tag{7–20}$$

which shows that α should be as small as possible, that is, $\alpha = 1$.

The combustor energy balance gives $\Delta\tau_b$:

$$\left.\begin{aligned} \dot{m}C_pT_0 + \dot{m}_fH &= (1+f)(1+x)\dot{m}C_pT_{2H} \\ \text{or} \qquad 1 + \Delta\tau_b &= (1+f)(1+x)\frac{T_{2H}}{T_0} \end{aligned}\right\} \tag{7–21}$$

By eliminating $\Delta\tau_b$ it follows that

$$\eta_{fu} = \frac{\theta_4 - \theta_3}{\theta_{2H} - \dfrac{1}{(1+f)(1+x)}} \qquad \text{where } \theta_i = \frac{T_i}{T_0} \tag{7–22}$$

The magnitude of this efficiency can be estimated recognizing that f and x are small compared to unity and T_{2H} will be close to T_4. For a Brayton cycle designed for maximum specific work and $\theta_4 = 6$, η_{fu} is of the order of 0.7 and the small f and x term contributions raise η_{fu}. Note how η_{fu} becomes small when the temperature difference θ_4–θ_3 is itself small. This has serious consequences for regnerated cycles where the goal is to add all external heat at *constant* and relatively high temperature.

A fuel efficiency for the cycle may be defined as the ratio of specific work to fuel energy added to the external combustor. Thus

$$\eta_{th,f} = \frac{w}{\Delta\tau_b} \tag{7–23}$$

is maximized for a Brayton cycle under the same conditions as is the specific work since the term $\Delta\tau_b$ is given by θ_4 (see eq. 7–21 with $\theta_4 = \theta_{2H}$). When a less ideal heat exchanger ($T_4 < T_{2H}$) is considered along with the minor effects due to specific heats (x) and fuel mass fraction (f), this conclusion is altered somewhat. Thus with external combustion (in contrast to internal combustion) the point of maximum specific work and maximum fuel utilization are expected to lie much closer together when cycle parameters, such as pressure ratio, are chosen to optimize the cycle.

Air Preheating

Much improved performance can be obtained if the heat from the combustion gas is used in an air preheater as shown in Fig. 7–19. Optimally, as in the discussion above, the capacity rate ratio for the heat exchanger to the cycle should be unity. That parameter cannot be forced to unity for the preheater because the properties and amounts traversing the two sides are fixed and differ. Hence the capacity rate ratio for the preheater, α_{PH}, is

$$\alpha_{PH} = (1 + f)(1 + x); \left[\frac{\text{hot}}{\text{cold}}\right] \tag{7–24}$$

This parameter is greater than 1 so that the temperature profiles for the preheater are as shown on the right side of Fig. 7–20. Also shown on the left side are the temperature profiles in the heat exchanger that transfers heat to the cycle. This configuration is quite practical and the real performance losses

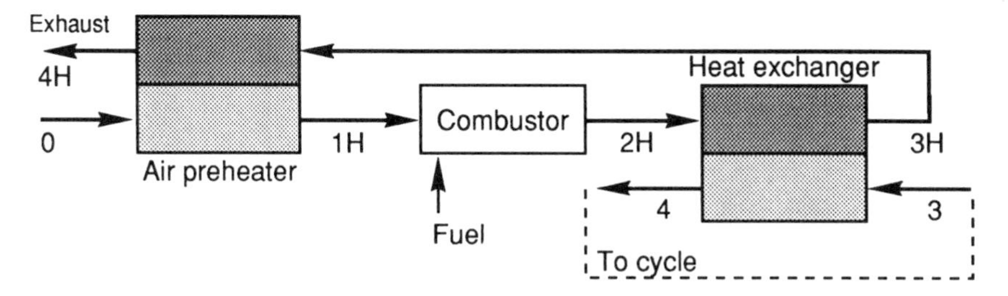

FIG. 7–19. External combustion for a steady flow cycle with air preheating.

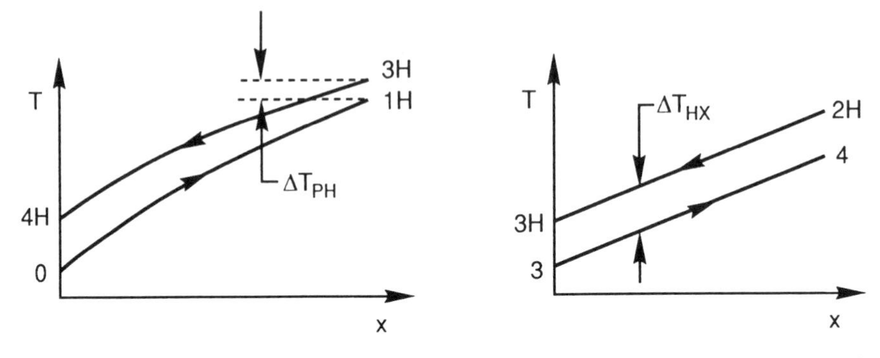

Fig. 7–20. Temperature profiles in the two heat exchangers shown in Fig. 7–19. Cycle heat exchanger on left and preheater on right.

are well modeled by consideration of the ΔTs. Thus one may define non-dimensional measure of ΔT as:

$$\delta_{\mathrm{PH}} = \frac{\Delta T_{\mathrm{PH}}}{T_0} \quad \text{and} \quad \delta_{\mathrm{HX}} = \frac{\Delta T_{\mathrm{HX}}}{T_0} \tag{7–25}$$

where the temperature differences are indicated in Fig. 7–20. The combustor energy balance now reads

$$\left.\begin{aligned} \dot{m}C_{\mathrm{p}}T_{1\mathrm{H}} + \dot{m}_{\mathrm{f}}H &= (1+f)(1+x)\dot{m}C_{\mathrm{p}}T_{2\mathrm{H}} \\ \text{or} \qquad & \\ \Delta\tau_{\mathrm{b}} &= (1+f)(1+x)\theta_{2\mathrm{H}} - \theta_{1\mathrm{H}} \end{aligned}\right\} \tag{7–26}$$

Here $\Delta\tau_{\mathrm{b}}$ is defined as in eq. 7–17, but without the assumption that f and x are small. The temperature ratios are obtained in terms of cycle temperatures from

$$\theta_{2\mathrm{H}} = \theta_4 + \delta_{\mathrm{HX}} \quad \text{and} \quad \theta_{1\mathrm{H}} = \theta_3 - \delta_{\mathrm{PH}} + \delta_{\mathrm{HX}} \tag{7–27}$$

so that the fuel utilization efficiency is

$$\eta_{\mathrm{fu}} = \frac{\alpha_{\mathrm{PH}}(\theta_4 - \theta_3)}{\alpha_{\mathrm{PH}}\theta_4 - \theta_3 + (\alpha_{\mathrm{PH}} - 1)\delta_{\mathrm{HX}} + \delta_{\mathrm{PH}}} \quad \text{where } \alpha_{\mathrm{PH}} = (1+f)(1+x) \tag{7–28}$$

This efficiency is unity when the δs are zero and $\alpha_{\mathrm{PH}} = 1$. Since α_{PH} is generally larger than unity, the efficiency can be better than 100%. This is due to the larger mass flow and specific heat on the heating side. Finite ΔT's naturally reduce the efficiency.

7.6. WORKS INTERACTIONS

The execution of work interactions between system and environment involve the displacement of the system control surface in such a way that the volume is reduced or increased. A reduction of the volume leads to an increase in density and is termed a compression process. The inverse process involving the system doing work on the environment is an expansion process.

7.6.1. *Compression and Expansion*

Compression or expansion processes may be carried out by a number of means including displacement of a piston or the pressure forces resulting from flow over aerodynamic blading. A compressor of the latter type is referred to as a turbocompressor, whereas the expander is a turbine.

A pure work interaction is adiabatic. In most kinds of machinery, the process of compression or expansion may be modeled as adiabatic if the process time is short compared to the time required for heat transfer processes to take place (Fig. 7–21). Even if the process is adiabatic, real processes are not reversible, owing to the friction associated with the motion of the fluid relative to the wall.

An adiabatic work process for a steady flow process is easy to describe. Since much of this work is concerned with steady flow cycles, it is instructive to determine the work required to compress (i.e., raise the density or, equivalently, the pressure of) a substance (Fig. 7–22) and conversely to determine the

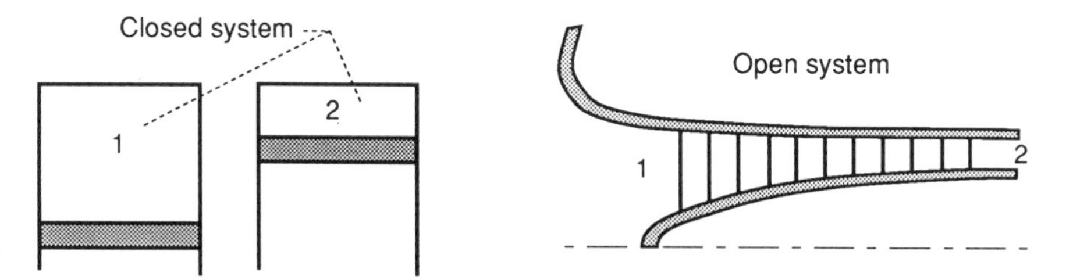

FIG. 7–21. Work processes can be made nearly adiabatic through a short process time.

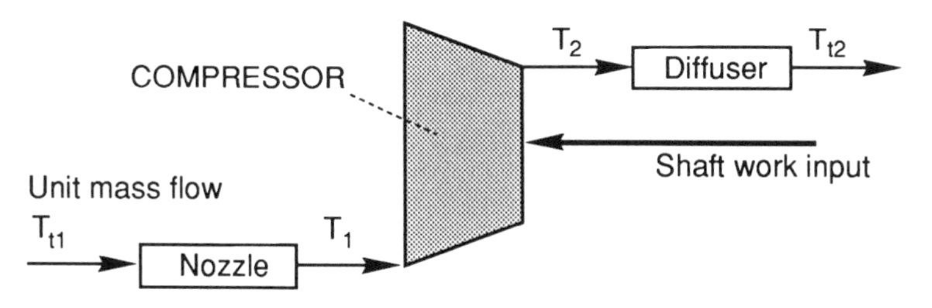

FIG. 7–22. Steady flow aerodynamic compressor showing flow devices that are part of the compressor to connect it to stagnation input and output states.

work that can be obtained from expansion by means of adiabatic processes. The work interaction changes the total enthalpy, eq. 2–6:

$$\dot{W}_{ext} = \dot{m}(h_{t2} - h_{t1}) \qquad \text{or} \qquad w = (h_{t2} - h_{t1})$$

or for a perfect gas,

$$w = C_p T_{t1}\left(\frac{T_{t2}}{T_{t1}} - 1\right) \tag{7–29}$$

Here w is the work per unit mass required between states 1 and 2. This equation shows that the work of compression for a chosen total temperature ratio is low when the initial state temperature is low. The temperature ratio is the key parameter that describes the compression work. This temperature ratio may be related to other property ratios, such as the pressure ratio, when the degree of reversibility is known. Specifically, if the process is reversible, the isentropic relations may be used. For example for an ideal and perfect gas,

$$\frac{p_{t2}}{p_{t1}} = \left(\frac{T_{t2}}{T_{t1}}\right)^{\gamma/(\gamma-1)} \tag{7–30}$$

In displacement compressors, the piston motion is continuous with a number of port or valving events scheduled to process the gas as shown in the p-V diagram of Fig. 7–23. The ideal process starts with the cylinder volume nearly zero, and continues with the low pressure (say atmospheric air) gas taken into the full displacement volume V_1. The mass of the gas processed is

$$m = \rho_0 V_1 = \rho_1 V_1 \tag{7–31}$$

By reducing the gas volume to V_2 adiabatically, reversible compression takes

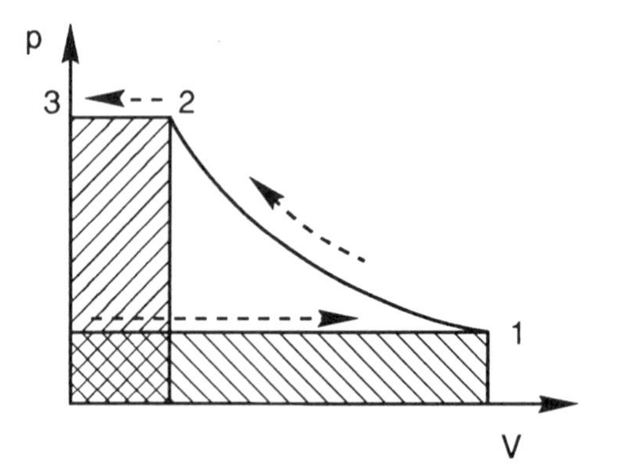

FIG. 7–23. Elements of compression of a closed system: intake, compression, and discharge.

place. From the first law, the compression work is

$$w_{1,2} = \Delta u = C_v(T_2 - T_1) \qquad \text{where} \quad w_{1,2} = \int_{V_1}^{V_2} p\, dV \tag{7–32}$$

The gas is then discharged at constant pressure (reversible process) from state point 2 to 3 and the cylinder volume reduced to zero. Note that for this discharge process, the system volume, (that of the mass of gas processed) remains constant: The control volume is open for this part of the process.

The total work required to carry out this process is the sum of the works for each of the three processes:

$$\text{Intake work} = -p_0(V_1 - V_0) = -p_1V_1 = -RT_1$$
$$\text{Compression work} = C_v(T_2 - T_1)$$
$$\text{Discharge work} = p_2(V_2 - V_3) = p_2V_2 = RT_2$$

so that the total work is

$$w = C_v(T_2 - T_1) + R(T_2 - T_1) \tag{7–33}$$

where the first term represents the $p\,dV$ work of the closed system whereas the second is the displacement work associated with transferring the fluid into and out of the cylinder (control) volume. This last term is the "flow work" in steady flow processes. Equation 7–29 gives the work

$$w_{1,2} = \Delta h = C_p(T_2 - T_1) \tag{7–34}$$

which is the same as that given by eq. 7–33 when the gas is ideal and perfect. This relation shows that a displacement compressor may be viewed as a steady flow compressor when the time scale is long enough to smooth out the delivery pulses.

In a similar way, the (positive) work of expansion (Fig. 7–24) may be

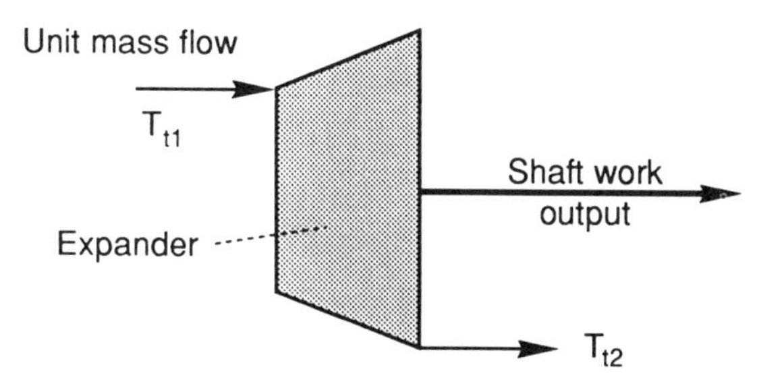

Fig. 7–24. Steady flow work extraction. Nozzle and diffuser, if present, are part of the expander.

calculated from

$$w = C_p T_{t1}\left(1 - \frac{T_{t2}}{T_{t1}}\right) \tag{7–35}$$

Note that this equation is also written so that the entry conditions appear explicitly. As in reversible compression, the expansion process total temperature ratio is related to the total pressure ratio. Thus, if the pressures are fixed, so is the exit temperature for a given inlet temperature. It follows therefore that the larger T_{t1}, the greater the work available from the expander.

7.6.2. *Liquid Compression*

The work of "compression" for a liquid is of interest in the study of cycles with a condensable fluid medium. The useful characteristic is that the specific volume v of a liquid is practically constant, independent of pressure. Furthermore, it is very small compared to that of the vapor. The steady-flow work of compression is therefore

$$w = h_2 - h_1 = u_2 - u_1 + (p_2 v_2 - p_1 v_1) \tag{7–36}$$

where the internal energy difference is negligible and the specific volumes of the two states are nearly equal. Thus

$$w_{\text{liquid}} \cong v(p_2 - p_1) \tag{7–37}$$

The small compression work is a key feature of the Rankine cycle for which the "recirculating" work (i.e., the work required from the expansion process to realize compression) is therefore small. Figure 7–25 shows the T-s diagram for the liquid compression and heating process where a useful approximation is to

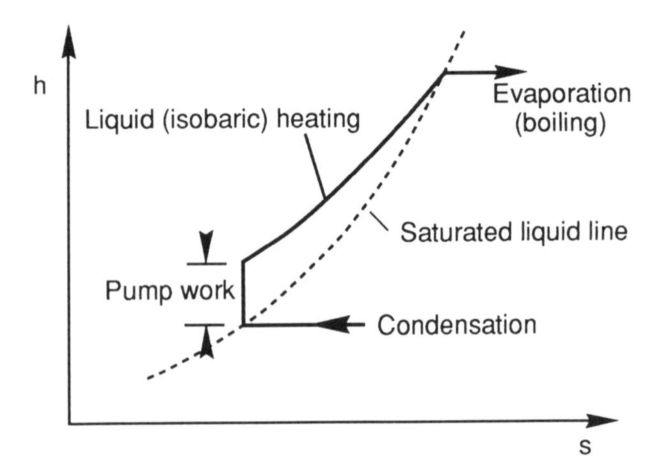

Fig. 7–25. T-s diagram of the pumping and liquid heating processes.

combine the two processes into a heating process and neglect the small compression work.

7.6.3. *Irreversible Adiabatic Work Processes*

A steady-flow work process that is adiabatic and irreversible may be characterized by an adiabatic efficiency.

Compressor Adiabatic Efficiency

For a compressor, one defines adiabatic compressor efficiency as the ratio of the work required to compress a mass of fluid from some initial state to a final pressure reversibly to the actual work required. This formulation ensures that the efficiency is less than unity.

$$\eta_c = \frac{\text{reversible work}}{\textbf{actual work} \text{ for the same pressure ratio}} = \frac{h_{2,s} - h_1}{h_2 - h_1} \leq 1$$

If the gas is perfect:

$$\eta_c = \frac{T_{2,s} - T_1}{T_2 - T_1} = \frac{\left(\dfrac{p_2}{p_1}\right)^{(\gamma - 1)/\gamma} - 1}{\dfrac{T_2}{T_1} - 1} \tag{7–38}$$

For compressors where there is a significant kinetic (or other) energy component, the temperatures and pressures must be *total*, rather than static, so that this part of the energy is properly counted (see Fig. 7–22).

Expander Adiabatic Efficiency

For an expander, the adiabatic efficiency is similarly defined as the ratio of actual work derived from the expander to the work a reversible expander would have provided for the same mass processed and the same pressure ratio.

$$\eta_E = \frac{\textbf{actual work}}{\text{reversible work for the same pressure ratio}} = \frac{h_1 - h_2}{h_1 - h_{2,s}} \leq 1$$

Again, if the gas is perfect:

$$\eta_E = \frac{T_1 - T_2}{T_1 - T_{2,s}} = \frac{1 - \dfrac{T_2}{T_1}}{1 - \left(\dfrac{p_2}{p_1}\right)^{(\gamma - 1)/\gamma}} \tag{7–39}$$

Note that the definitions of these adiabatic efficiencies are inverted from one another. This is to ensure that the efficiency is less than unity for an irreversible process in both cases.

For nonadiabatic work processes, other means are used to describe the degree of heat transfer. The adiabatic process is for a perfect gas described by

$$pV^{\gamma} = \text{constant} \tag{7–40}$$

In practice, one may find that, particularly in displacement processes, the following relation describes the process well:

$$pV^{n} = \text{constant} \tag{7–41}$$

Here n is called the *polytropic exponent* and the process is also called *polytropic.* It follows that for $n \neq \gamma$ the process is not adiabatic.

One may describe a polytropic efficiency by

$$e \equiv \eta\left(\lim \frac{p_2}{p_1} \to 1\right) \tag{7–42}$$

The adiabatic efficiency for a differentially small change in pressure can be obtained for a compressor or an expander by writing the temperature ratio and pressure ratio for small changes in state and substituting into the definitions of adiabatic efficiency.

$$\text{for } \frac{p_2}{p_1} \text{ close to 1; } \frac{p_2}{p_1} = 1 + \frac{\Delta p}{p_1} \quad \text{and} \quad \frac{T_2}{T_1} = 1 + \frac{\Delta T}{T_1}$$

so that it follows

$$\eta_c = \frac{\left(\dfrac{p_2}{p_1}\right)^{(\gamma-1)/\gamma} - 1}{\dfrac{T_2}{T_1} - 1} \Rightarrow e_c = \frac{\gamma - 1}{\gamma}\frac{\Delta p}{p}\frac{T}{\Delta T} = \frac{\gamma - 1}{\gamma}\frac{d \ln p}{d \ln T} \tag{7–43}$$

The last step is valid when the small quantities are made differentially small. By involving the state equation, one can show that the polytropic process (eq. 7–41) is described by a characteristic exponent, n_c, defined by:

$$\frac{d \ln p}{d \ln T} = \frac{n_c}{n_c - 1} \tag{7–44}$$

or with eq. 7–43,

$$e_c = \frac{\gamma - 1}{\gamma} \frac{n_c}{n_c - 1} \tag{7–45}$$

which can be solved for n_c,

$$n_c = \left(1 - \frac{\gamma - 1}{\gamma} \frac{1}{e_c}\right)^{-1} \cong \gamma[1 + (\gamma - 1)(1 - e_c)];\ (>\gamma) \text{ for } e_c \text{ close to 1.} \tag{7–46}$$

Similarly, for the expander:

$$e_E = \frac{\gamma}{\gamma - 1} \frac{d \ln T}{d \ln p} = \frac{\gamma}{\gamma - 1} \frac{n_E - 1}{n_E} \tag{7–47}$$

The polytropic process allows writing the relationship between the total temperature and total pressure from integration of eqs. 7–44 and 7–47 with n or e constant. The results are

$$\left.\begin{aligned} \text{compressor: } \frac{p_{t2}}{p_{t1}} &= \left(\frac{T_{t2}}{T_{t1}}\right)^{n_c/(n_c - 1)} = \left(\frac{T_{t2}}{T_{t1}}\right)^{e_c\gamma/(\gamma - 1)} \\ \text{expander: } \frac{T_{t2}}{T_{t1}} &= \left(\frac{p_{t2}}{p_{t1}}\right)^{(n_E - 1)/n_E} = \left(\frac{p_{t2}}{p_{t1}}\right)^{e_E(\gamma - 1)/\gamma} \end{aligned}\right\} \tag{7–48}$$

Substituting these relations into the definitions for adiabatic efficiencies (eqs. 7–38 and 7–39) gives the relation between the polytropic and adiabatic efficiencies.

$$\eta_c = \frac{\left(\frac{p_{t2}}{p_{t1}}\right)^{(\gamma - 1)/\gamma} - 1}{\left(\frac{p_{t2}}{p_{t1}}\right)^{(\gamma - 1)/\gamma e_c} - 1} \qquad \eta_E = \frac{1 - \left(\frac{p_{t2}}{p_{t1}}\right)^{e_E(\gamma - 1)/\gamma}}{1 - \left(\frac{p_{t2}}{p_{t1}}\right)^{(\gamma - 1)/\gamma}} \tag{7–49}$$

For the compressor, the adiabatic efficiency is always less than the polytropic efficiency, whereas for the expander the reverse is true. This reflects the increased work involved in the compressor processing the gas in the later portion of the process which has been heated by earlier irreversibility. The expander gas, on the other hand, can partially convert to work the heat generated earlier in the process. Figures 7–26 and 7–27, show the relation between these efficiencies and the differences between compressors and expanders.

Since polytropic and adiabatic efficiencies are related, the question arises: When does one use which? The adiabatic measure is used for situations where the performance of a particular work component is to be quantified. The

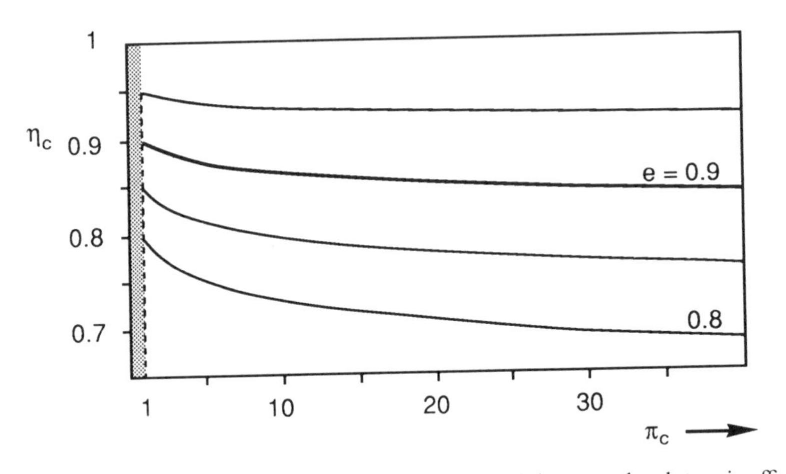

Fig. 7–26. Relation between a compressor adiabatic efficiency and polytropic efficiency for various pressure ratios. Polytropic efficiency has the value shown at unit pressure ratio.

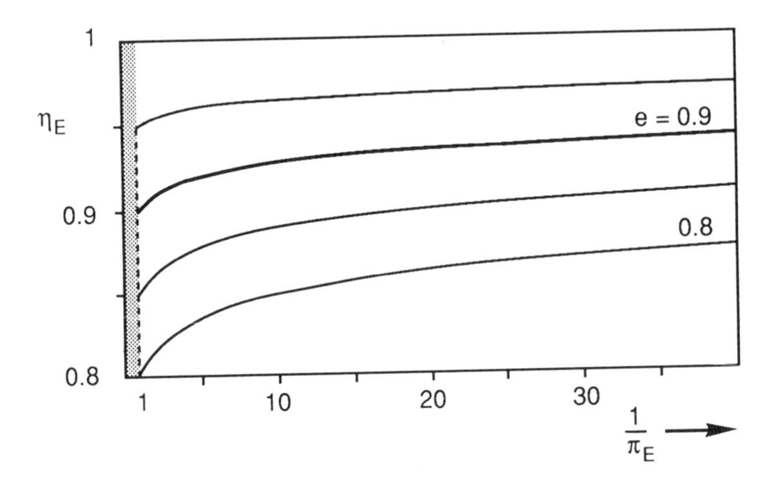

Fig. 7–27. Relation between adiabatic efficiency and polytropic efficiency for a turbine operating with various pressure ratios. Since the turbine pressure ratio is less than unity, its reciprocal is plotted for comparison to Fig. 7–26. Polytropic and adiabatic efficiencies are equal at unit pressure ratio.

polytropic one, on the other hand, is more useful in the design process of identifying the optimum cycle pressure ratio where an efficiency must be used that applies to members of a family of work components. Presumably all members of the family will, in differential sense, perform equally well.

For such a family of compressors, the value of the polytropic efficiency, e, can be assumed constant, and the differential relation between pressure and temperature can be integrated to represent the locus of end states of this family of compressors each with the same differential efficiency. Equation 7–48 is an analytical representation of such end states. This locus may be represented on

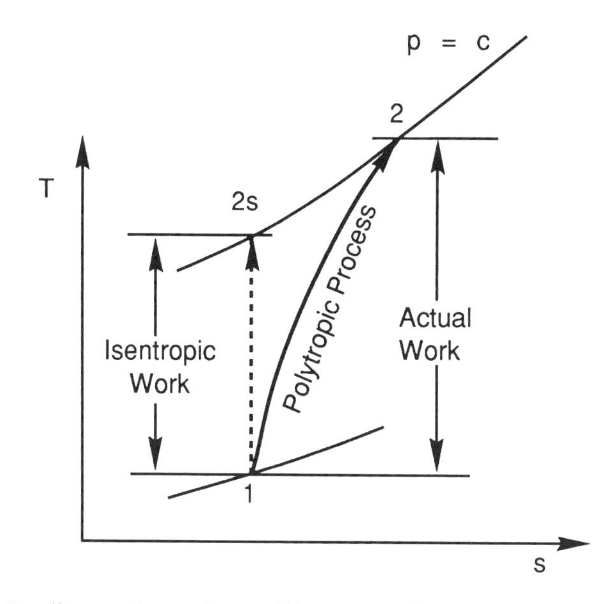

FIG. 7–28. *T-s* diagram for an irreversible compression process characterized by constant polytropic efficiency.

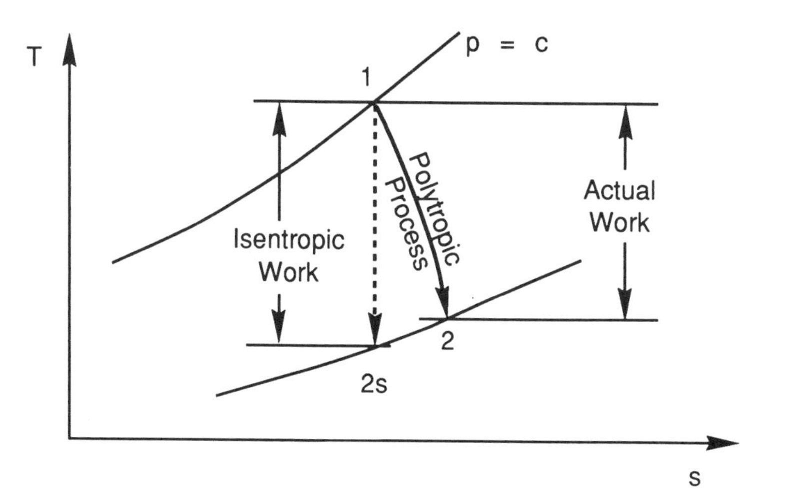

FIG. 7–29. *T-s* diagram for an irreversible expansion process, constant polytropic efficiency.

a *T-s* plane as shown in Fig. 7–28. Note both expansion and compression processes have paths, which, for increasing pressure ratio, lead to increasing entropy (Fig. 7–29). In Fig. 7–28 or 7–29, a process characterized by an adiabatic efficiency cannot be shown except for its end points.

7.6.4. *Total to Static Efficiency*

In some compression and expansion processes involving flow dynamics, there is often a need to separate the processes in rotating components from those in stationary elements. For example, the rotor of a radial flow compressor produces a very high speed flow which is then slowed in a diffuser. It is therefore appropriate to discuss a rotor's ability to produce a jet, in terms of an efficiency that does not involve a stagnation end state. A so-called total to static efficiency is discussed in detail in Section 14.5.4 where the specifics of the radial flow compressor are described.

7.6.5. *Heat Transfer in a Polytropic Process*

From eq. 7-44, the polytropic process in terms of p, T is:

$$p = \text{constant}\ (T)^{n/(n-1)} \tag{7–50}$$

For any process

$$dq = dh - v\,dp = C_p\,dT - \frac{RT}{p}\,dp \tag{7–51}$$

and using the differential relation between p and T for the polytropic process, one obtains:

$$\frac{dq}{R} = dT\left(\frac{\gamma}{\gamma - 1} - \frac{n}{n - 1}\right) = dT\frac{\gamma}{\gamma - 1}(1 - e_c) \tag{7–52}$$

which means that for $n > \gamma$ or $e_c < 1$, it follows that a heat gain is experienced ($dq > 0$) in a compression process ($dT > 0$). This is equivalent to saying that part of the compressor work has been irreversibly converted into heat. Thus one can model irreversible adiabatic processes with the *polytropic* exponents n_c and n_E and expect that

$$\text{Compression: } n_c > \gamma \text{ (equivalent heat gain)}$$

$$\text{Expansion: } n_E < \gamma \text{ (equivalent heat loss)}$$

7.6.6. *Cyclic Heat Storage in Reciprocating Work Machines*

In all piston/cyclinder machines, the working fluid undergoes processes that result in changes of temperature and pressure. These changes in conditions result in transfer of heat between the walls of the cylinder and the top of the piston. During the intake portion of a compression cycle, the air is cool relative to the walls. Thus it is reasonable to expect that the walls will accommodate to a mean between the temperatures of the compressed charge and the ambient air temperature. The weighting is related to the heat transfer coefficient and the time spent at the various conditions. The following is an examination of the role of cyclic wall heat storage in determining the efficiency of a compressor.

Let π_c be the total pressure ratio imposed on the working fluid during

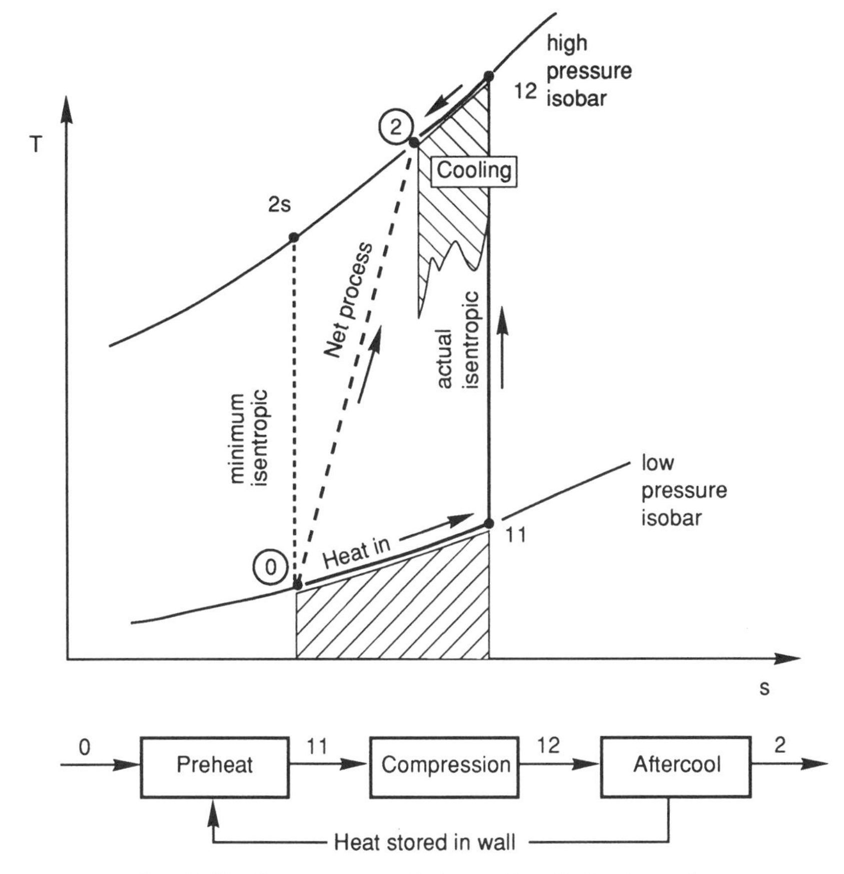

FIG. 7–30. Compression with internal cyclic heat transfer.

compression. Ignoring the exact physics of the wall/gas heat transfer process, one may estimate the mean surface temperature to be the average between T_0 and T_2 where

$$T_2 = T_0(\pi_c)^{(\gamma - 1)\gamma}$$

One can assume that a quantity of heat Q, is stored in the wall, which is returned to the incoming fresh charge. This process may be modeled (ref. 7–1) in a simple way as an overall adiabatic process with preheating, compression, and aftercooling. Figure 7–30 shows the physical processes and the states in the T-s diagram. Such simplified modeling describes the role of the relevant parameters. References 7–2 and 7–3 discuss the general problem in greater detail.

The net process 0–2 is evidently irreversible and the work of compression is larger than the thermodynamic minimum work for the same pressure ratio. The process 0–2s is the ideal isentropic process, whereas the real process consists of 0–11 heating, 11–12 isentropic compression, and 12–2 cooling. The process of internal heat transfer may thus be quantified in terms of an effective adiabatic efficiency defined as

$$\eta_c = \frac{T_{2s} - T_0}{T_{12} - T_{11}} \tag{7–53}$$

Heat balances establish the relationship between the various temperatures:

$$\frac{T_{11}}{T_0} = 1 + q \qquad \text{and} \qquad \frac{T_2}{T_{12}} = 1 - q\frac{T_0}{T_{12}} \qquad \text{where } q \equiv \frac{Q}{mC_P T_0} \tag{7–54}$$

For the isentropic processes

$$\frac{T_{12}}{T_{11}} = \frac{T_{2s}}{T_0} = \tau_{ci} = \pi_c^{(\gamma-1)/\gamma} \tag{7–55}$$

Thus the efficiency becomes:

$$\eta_c = \frac{\tau_{ci} - 1}{\dfrac{T_{11}}{T_0}(\tau_{ci} - 1)} = \frac{1}{1 + q} \tag{7–56}$$

which is independent of the pressure ratio and valid when q is small. As will be seen subsequently, q can, under certain circumstances, be of order 1 or even greater. In such circumstances, the analysis is likely to be wrong quantitatively but it will correctly indicate that the efficiency penalty is serious. A similar analysis may be carried out for an expander and the result is

$$\eta_E = 1 - q \tag{7–57}$$

By invoking a descriptive model of the heat transfer process such as the Reynolds analogy and describing a mean heat transfer coefficient in terms of local processes involved, one can show that q is a time weighted and spatially averaged integral of density, wall-to-gas temperature difference, velocity, and equivalent friction factor:

$$q = \int \frac{A_w}{V} u \frac{\Delta T}{T_0} f\, dt \tag{7–58}$$

which is very difficult to evaluate. Here A_w is the heat-transferring surface area and V the volume, both of which are time dependent. The friction factor, f, is primarily dependent on the Reynolds number which is also time dependent: $Re \sim \rho u l \sim \rho \omega l^2$ and shows the principal factor that aids in reducing q: large characteristic size l.

7.7. COMBINED HEAT AND WORK INTERACTIONS: ISOTHERMAL PROCESSES

There are two important circumstances under which one considers constant temperature processes. The first is the heat interaction associated with the change of state of a substance from liquid (A) to vapor (B) or vice versa (Fig. 7–31), and the second is in connection with isothermal compression or expansion. Both of these kinds of processes play an important role in the conceptualization of a reversible heat engine and are therefore important to the design of real heat engines.

The steady flow cycle permits the design of an isothermal work process. The true isothermal compression process is obtained as a mathematical limit of operating between temperatures T_0 and $T_0 + dT$, first by compressing adiabatically which raises the temperature by dT and then cooling the compressed gas by dT at constant pressure. A large number of such processes will lead to nearly isothermal compression. Similarly, isothermal expansion can be carried out by sequentially expanding and reheating the expanded gas (Fig. 7–32).

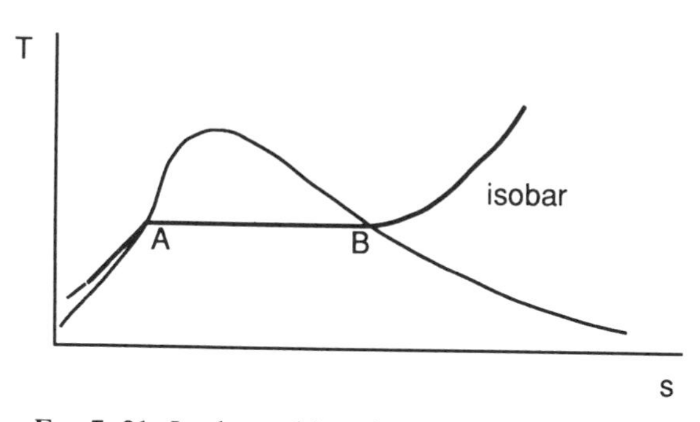

FIG. 7–31. Isothermal heat interaction from A to B.

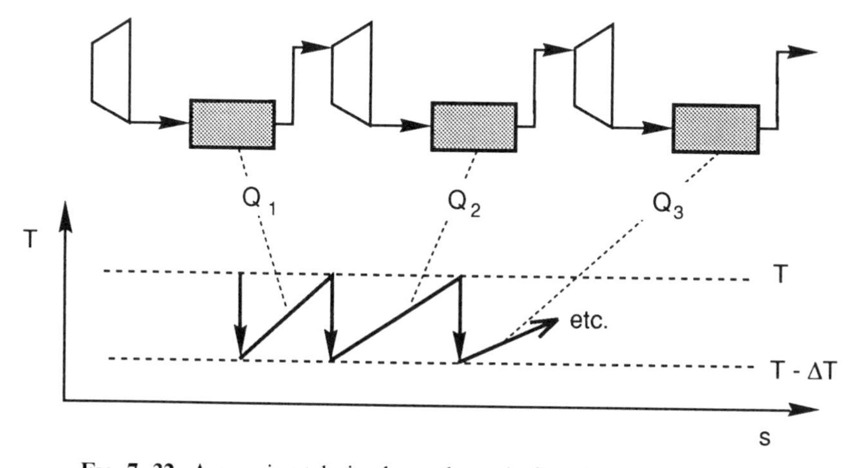

FIG. 7–32. Approximately isothermal steady flow (expansion) process.

7.8. THE CARNOT OR REVERSIBLE CYCLE

The cycle involving reversible adiabatic compression, isothermal heat addition, reversible adiabatic expansion and finally isothermal expansion is called the Carnot cycle (Fig. 7–33). Its unique characteristic is that an engine using it as a work producing cycle has the highest possible thermal efficiency as defined in Chapter 2. This efficiency is of obvious interest when the conversion of heat to work is carried out by any other cycle. The heat input to the Carnot cycle is

$$Q_{\text{in}} = T_{\max}(\Delta s_{\max \mathrm{T}}) \tag{7–59}$$

and the heat output is

$$Q_{\text{out}} = T_{\min}(\Delta s_{\min \mathrm{T}}) \tag{7–60}$$

so that the net work produced by the cycle is

$$W = Q_{\text{in}} - Q_{\text{out}} \tag{7–61}$$

from the First Law of Thermodynamics applied to a cyclic process. It follows that the efficiency of the engine is

$$\text{thermal efficiency} = 1 - \frac{T_{\min}}{T_{\max}} \frac{\Delta s_{\min \mathrm{T}}}{\Delta s_{\max \mathrm{T}}} = 1 - \frac{T_{\min}}{T_{\max}} = \eta_{\text{Carnot}} \tag{7–62}$$

provided that Δs is the same for both minimum and maximum temperature sides of the cycle, which is possible only if the adiabatic processes joining them are reversible. The thermal efficiency defined by eq. 7–62 is the so-called Carnot efficiency which clearly identifies the importance of the temperatures of the heat reservoirs, or more precisely, the temperatures to which the fluid in the cycle can be brought. Can the Carnot cycle be implemented in practice?

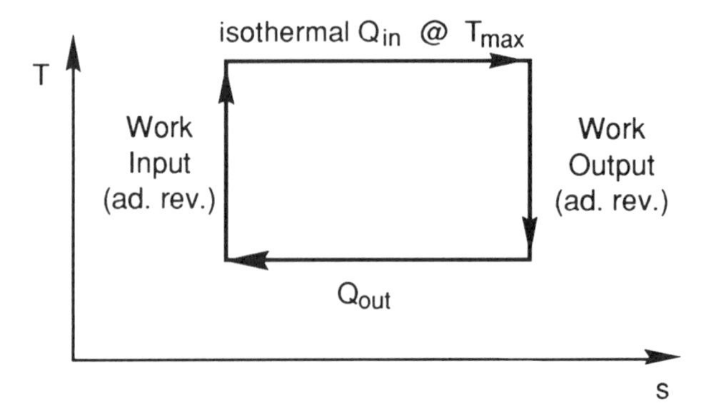

FIG. 7–33. The reversible or Carnot cycle consisting of two isotherms and two adiabatic work processes

7.8.1. *Near Impossibility of Using a Two Phase Fluid in a Carnot Cycle*

The addition of heat to an evaporating liquid can easily be carried out isothermally because the pressure and temperature are uniquely related during the phase change and the pressure can easily be controlled. The fact that pressure is constant during this process makes a practical flow process possible. Thus, one might imagine that a combinaton of an isothermal heat input and a heat rejection might lead to a useful reversible heat engine. Unfortunately, these two processes must be joined by adiabatic reversible ones and these necessarily involve a two-phase mixture. These compression and expansion processes are nearly impossible to carry out reversibly and adiabatically:

1. If the process is carried out slowly to ensure equilibrium between the two phases, then it is difficult to provide the necessary insulation required for the process to be adiabatic.
2. On a faster time scale, the process might be adiabatic, but the lack of equilibrium between the two phases leads from the starting temperature and pressure to a higher pressure but with each of the two phases reaching different temperatures (see Fig. 7–34). This leads to an irreversible transfer of heat from one phase to the other.

The previous section and Fig. 7–32 shows, however, that using a *gas* as a working fluid does permit the design of a Carnot cycle. It is merely (sic) very involved and expensive in practice. Section 9.8 explores the performance of this so-called Ericsson cycle, whereas Chapter 10 describes the Stirling cycle.

7.8.2. *Carnot Efficiency of an Air Combustion Engine*

The conversion of chemical energy to heat which is then converted to work is limited by the Second Law of Thermodynamics. In principle, this limitation is not very serious, even for the problem of burning a fuel with air.

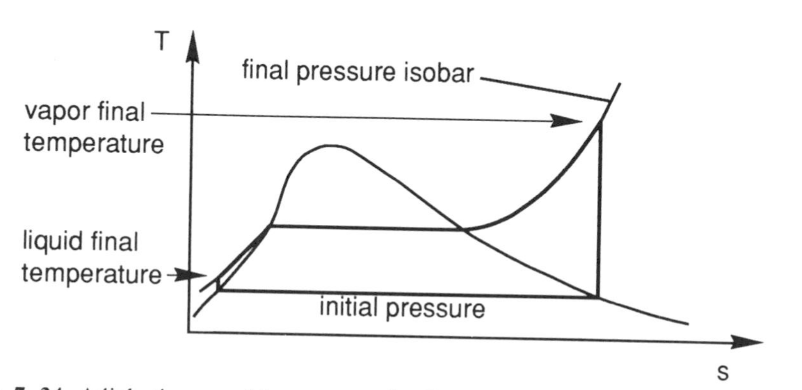

FIG. 7–34. Adiabatic reversible processes for liquid and vapor between same pressure extremes.

Consider, as shown in Section 7.5, combustion of a fuel with the following composition:

$$CH_{2n} + \left(1 + \frac{n}{2}\right)(O_2 + 3.76\ N_2) \rightarrow CO_2 + nH_2O + 3.76\left(1 + \frac{n}{2}\right)N_2$$

One can estimate the amount of work producible by this reaction by assuming that a Carnot engine can be devised operating between the reference temperature (298 K) and the temperature of the combustion products. Assuming the temperatures of fuel and air are 298 K, one has for the product enthalpy Q_2 (see Section 5.4)

$$Q_2 = \Delta H_{\mathrm{RPf}} - Q_1(=0) \tag{7–63}$$

Further, if one assumes that the fuel heat of formation is negligible, then

$$\Delta H_{\mathrm{RPf}} = -94.05 - n\ 57.8\ kcal{*}mol \tag{7–64}$$

If the product gas is perfect, then the temperature T_2 is obtained from

$$Q_2 = \left[3.76\left(1 + \frac{n}{2}\right)C_{\mathrm{p2}} + (1 + n)C_{\mathrm{p3}}\right](T_2 - T_{\mathrm{f}}) \tag{7–65}$$

where C_{p2} and C_{p3} are the molar specific heats of the diatomic and triatomic molecules.

$$C_{\mathrm{p2}} = 3.5\ R_{\mathrm{u}} \qquad (\gamma = 7/5)$$

$$C_{\mathrm{p3}} = 4\ R_{\mathrm{u}} \qquad (\gamma = 4/3)$$

$$R_{\mathrm{u}} = 1.987 \qquad \mathrm{cal/mol\ K}$$

This gives

$$T_2 = T_{\mathrm{ref}} + \frac{1 + 0.615\,n}{1 + 0.430\,n}2758 \tag{7–66}$$

The resulting T_2 varies between 3000 K for $n = 0$ (pure H_2) and over 4000 K for large n (pure C). The corresponding Carnot efficiency $\eta_{\mathrm{co}} = 1 - T_{\mathrm{f}}/T_2$ therefore depends somewhat on the fuel composition through n. More importantly the value is very high. In practice, however, the irreversibilities of real cycles and temperature limitations on real materials preclude achievement of these efficiency levels.

7.9. CYCLE PERFORMANCE PARAMETERS

7.9.1. *Thermal Efficiency*

Thermal efficiency has already been discussed in Chapter 2 as a significant parameter for the heat engine. Its importance lies in the fact that the heat resource represents a cost one must pay for the delivery of work. The impact of this cost on the system is described in Chapter 3.

7.9.2. *Specific Work*

Another important parameter that governs the "quality" of an energy conversion device is the amount of work the cycle produces per unit mass processed. This parameter is a measure of the compactness of the engine because for a specified level of flow- or speed-related losses the flow area or volume dictates the rate at which the fluid can be handled and, thus, the heat input as well as the work output rates. This parameter (w) is the

$$\text{specific work} = \text{work output per unit mass or}$$
$$= \text{power output per unit mass process rate}$$

This parameter is related (but not necessarily equal) to the area enclosed by the cycle process lines drawn in either the T-s plane of the p-V plane. Figure 7–35 illustrates the fact that the ratio of maximum to minimum pressure (i.e., the cycle pressure ratio in a cycle like the Brayton shown) plays an important role in determining the area enclosed in the state variable planes and thus the specific work.

7.9.3. *Nondimensionalization of Specific Work and mep*

It is often convenient to nondimensionalize the specific work. This will generally be done by using the enthalpy of the fluid at the minimum cycle temperature which is usually a fixed parameter for a particular investigation. An exception to this is in the case of space power systems where the peak cycle temperature is more likely to be fixed and the minimum cycle temperature is chosen for an optimum radiator size (see Section 9.10). In this text, specific

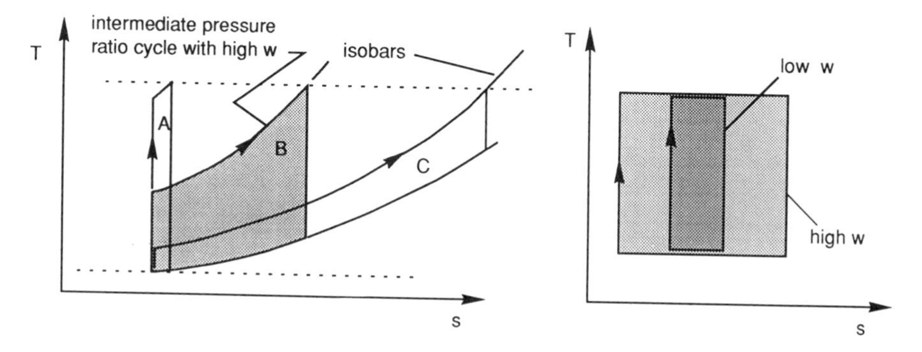

FIG. 7–35. Examples of various cycles with varying w, Brayton (left) and Carnot (right) cycles.

work will be nondimensional and given by:

$$w \equiv \frac{W}{mC_pT_0} = \frac{\dot{W}}{\dot{m}C_pT_0} \tag{7–67}$$

In the displacement heat engine community, this parameter is written to reflect the idea that engine size is conveniently measured by piston displacement or cylinder volume. Thus work per unit volume carries the units of pressure and is called the "mean effective pressure," or "mep" for short. These quantities are related through:

$$w \equiv \frac{\gamma - 1}{\gamma}\frac{W}{p_0V_0} \quad \text{and} \quad \text{mep} = \frac{W}{V_0} \tag{7–68}$$

In connection with the Otto cycle described in Chapter 8, the definition of specific work is based on C_v rather than on C_p. This convenience introduces a different factor of γ into eq. 7–68. Further, by using the swept volume rather than the initial air volume, a small variation in this relation is introduced. These issues account for the differences seen in this simple relation between specific work and mep and the more precise ones developed in Chapter 8.

7.9.4. *Part Load Characteristics*

Most heat engines must be able to provide power levels that are less than the peak value for some fraction of the time. An important issue in evaluating cycles is to understand the part load requirements of the end user and the performance of the cycle at this part load condition, see Chapter 3. In the cycle analysis, two types of examination will be carried out to examine the performance of an engine. The first is the so-called design-point optimization arrived at by conceptually allowing some parameter(s) to vary until the design point objective has been met. For example, in Fig. 7–35, the *T-s* paths for three cycles are shown. These differ in pressure ratio but have the same maximum cycle temperature, T_4. One such engine design might meet the goals set for efficiency and specific work. This particular engine concept might now be asked to operate at a power output level that is different from (usually less than) the design value, by reducing T_4, for example. The performance of this engine at this new condition is termed "off-design" (or often, part load since the full load maximum power point is usually the most challenging to achieve). These conceptual analysis approaches are illustrated in Fig. 7–36. Typical user characteristics are shown in Figs. 3–8 and 3–9.

Naturally, the heat engine should have a high value of efficiency integrated over the relevant time period as described in Section 3.3.4. Figure 7–37 shows the characteristics of several engine types at part load (ref. 7–4).

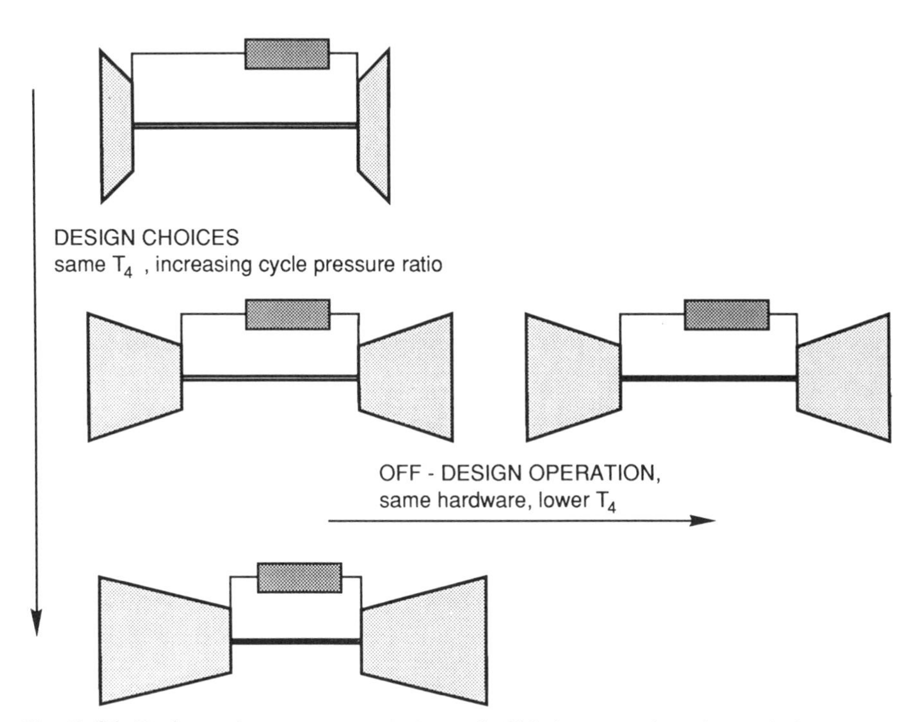

FIG. 7–36. Design cycle parameter variation and off-design operation of a particular design.

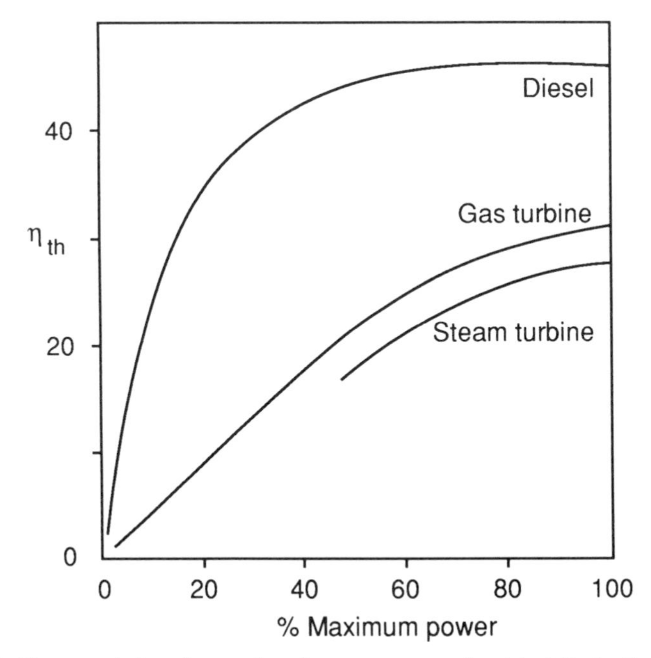

FIG. 7–37. Characteristics of several engine types at part load (ref. 7–4). Caution should be used with figures such as this to be sure that performance is stated for a common set of constraints.

7.9.5. *Cost of Conversion Machinery*

In the analysis of cycles, one deals primarily with the problem of choosing a cycle configuration and parameters that achieve a design goal. To a large extent, the complexity and the mass of materials involved in the components determine the costs of the engine (see Chapter 3). It is usually very difficult to estimate costs directly. For this purpose, experience, correlations with mass or other physical parameters, or a detailed cost buildup may be used. In preliminary design, however, the physical parameters themselves are often used as measures to identify system costs. These may be heat transfer area for a heat exchanger, pressure ratio or volume ratio for work components, pressure level for a system as a whole, and so on.

7.9.6. *Environmental Impact*

The cycle designer must deal with issues relating to the discharge of chemical constituents and low-grade waste heat from the point of view of local impact as well as from a larger, even global, perspective. These issues are considered in detail in Chapter 18.

PROBLEMS

1. Obtain an expression for n_E in terms of the polytropic efficiency, similar to eq. 7–46.
2. Show the validity of eq. 7–57.

REFERENCES

7–1. Decher, R., "Wall Heat Storage in Displacement Compressors," *International Journal of Energy Research*, **12**: 379–86, 1988.
7–2. Rios, P. A., "An Approximate Solution to the Shuttle-Heat Transfer in a Reciprocating Engine," *Transactions of the ASME, J. of Engineering for Power*, 177–82., April 1971.
7–3. Lee, K. P., Smith, J. L., "Influence of the Cyclic Wall to Gas Heat Transfer in the Cylinder of a Valved Hot Gas Engine," 13th IECEC, SAE P-75, 1798–1804, San Diego, Cal., 1978.
7–4. Ahlquist, I., "Choosing the Prime Mover for Cogeneration," *Cogeneration World*, **5**(6) Nov.–Dec. 1986.

8

ENERGY LIMITED CYCLES: OTTO AND DIESEL ENGINES

The Otto and Diesel cycle engines are in widespread use as the motive power for a large variety of applications whose development continues (ref. 8–1). There appears no end to the permutations with which these basic cycles appear attractive for one reason or another (e.g., refs. 8–2, 8–3, among a wide body of literature). These piston/cylinder engines use open cycles with distillate fuels as a source of heat. Both cycles are intermittent in operation which permits stoichiometric combustion conditions because the combustion process is short and followed by rapid expansion and internal cooling of the enclosing cylinder walls. Because of the large cycle temperature extremes the thermodynamic consequence is that the Carnot efficiencies, and therefore the real efficiencies, are high. Other features that contribute to the usefulness of the cycles are good part-load performance and high power-to-weight ratio.

In this chapter the discussion is limited to the four-stroke cycles. The advantage of the two-stroke engine is greater power for a given engine rotational speed because there are twice as many power strokes per unit time. The disadvantage is the carryover to the exhaust of unburned fuel products with its detrimental impact on the environment which has prevented a significant modern use of two-stroke engines except in very small engines. Exhaust gas treatment of such engines may make it possible to use the cycle in larger engines for future light weight powerplants. From a modeling point of view, the exhaust gas carryover involves greater descriptive complexity of the elementary processes. For example the mixing of fresh and spent charges is a dynamic process which is not easily described in a general way.

In Section 7.5, the chemistry of near stoichiometric combustion is examined. To first order, the fuel mass involved is small and the temperature rise is large. Further, since the products of combustion are a relatively small constituent of the product gas, the thermochemical properties of the heated gas are approximately equal to those of N_2 or air. Although the combustion process is complex and has much to do with the real details of a working engine, such as pollutant emissions as well as providing a potential for future improvements (ref. 8–4), the first-order effect is the heating of the air working fluid. A cycle analysis using these working fluid property assumptions in termed "air standard." This chapter is limited to discussion of such a fluid description. The principal objectives are to examine the role of irreversibilities and to describe analytically the performance of the basic cycles and their variations at part-load.

The perfect gas approximation is used here for the purpose of describing the performance of these two important engine cycles, which differ from the steady flow cycles in important ways. The use of this approximation leads to error in numerical values of performance parameters, a fact that may have to be corrected, with better gas model and loss description to increase the accuracy of the results sought. Work process irreversibilities are emphasized here because their inclusion is important and algebraically direct. Further, if development of adiabatic engines proceeds successfully, these irreversibilities will increase in relative importance in determining engine performance (refs. 8–5, 8–6). For the purposes of cycle modeling, the assumption is made that distinct, rather than continuous, processes are involved in executing the cycles and that the control volume contains a gas that can be described as uniform. The important pressure loss associated with entry into and exit from the cylinder volume is speed dependent (see Section 14.7) and in order to reduce the problem to examination of the most important variables it is assumed that the processes occur slowly enough for the flow losses past valves to be negligibly small. While this is unrealistic for practical engines, the assumption does provide an understanding of the theoretical capability of the cycle.

The temperature rise that results from the process of constant pressure and constant volume heat addition may be determined from the First Law of Thermodynamics (see eqs. 2–4, 2–6, and 7–17):

$$\text{constant pressure: } T_3 - T_2 = \frac{Hf}{C_p} = T_1 \Delta\tau_b \tag{8–1}$$

$$\text{constant volume: } T_3 - T_2 = T_1 \Delta\tau_b \gamma \tag{8–2}$$

If the process is carried out at constant volume, the temperature rise parameter is larger by a factor of γ. Note $\Delta\tau_b$ (a number of the order of 10, less any heat loss during combustion) can be smaller than some maximum value depending on the fuel/air mixture ratio, consistent with ignition limits. In this chapter, the Otto and a model Diesel cycle are examined in detail. The modeling simplification of the Diesel involves the assumption that the constant volume portion of the cycle is absent. Its inclusion requires specification of an additional parameter that would obscure the overall behavior sought while improving the accuracy. Figure 8–1 shows the T-s diagrams of the cycles in question.

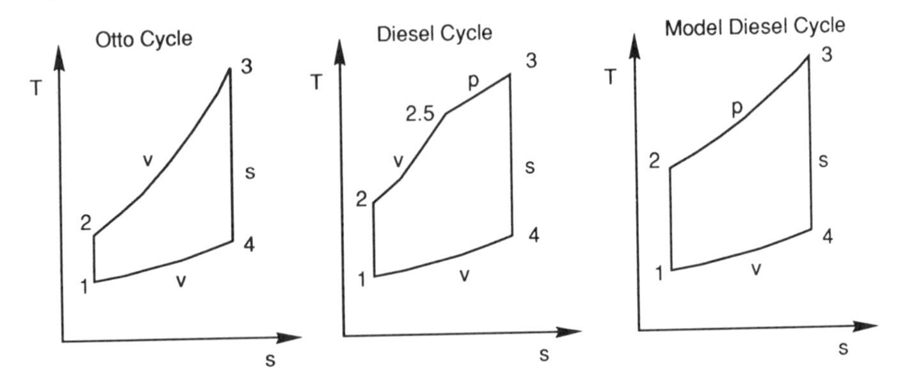

FIG. 8–1. T-s diagrams for the idealized Otto, Diesel and model Diesel engine cycles. Letters denote property that is constant during the ideal process.

8.1. THE OTTO CYCLE

The T-s diagram for the reversible Otto cycle is shown in Fig. 8–1. In this ideal description, two isentropic processes and two constant volume processes form the cycle. For a more realistic cycle, the work processes can be characterized by an efficiency. The heat rejection process takes place in the atmosphere. The nondimensional specific work, $w = W/(mC_vT_1)$ or its equivalent, the mean effective pressure, as well as the thermal efficiency for this cycle as an approximation of performance of an engine built around this cycle are of interest. Section 8.2 seeks similar calculation of performance for the cycle at part-load when the realities of its implemention in a real engine are more evident.

8.1.1. *Air Standard Analysis of an Irreversible Otto Cycle*

The nondimensionalization of W is more appropriately by C_v, rather than C_p. For purposes of determining work output and efficiency the heat interactions are needed. From the first law (eq. 8–1)

$$\text{Heat input to cycle} = \frac{Q_i}{mC_vT_1} = \frac{T_3}{T_1} - \frac{T_2}{T_1} = \gamma\,\Delta\tau_b \tag{8–3}$$

$$\text{Heat rejected by cycle} = \frac{Q_o}{mC_vT_1} = \frac{T_4}{T_1} - 1 \tag{8–4}$$

The compression and expansion processes are modeled as polytropic. One may define a compression temperature ratio $\tau_c \equiv T_2/T_1$. This ratio, together with the polytropic exponent, gives the piston volume ratio, which is the conventional parameter used for reciprocating engine description:

$$\tau_c \equiv \frac{T_2}{T_1} = \left(\frac{V_1}{V_2}\right)^{n_c - 1} \equiv R_v^{n_c - 1} \tag{8–5}$$

The "c" subscript on n is to distinguish between the compression (c) and expansion (E) processes. Combining eqs. 8–5 and 8–3 gives

$$\frac{T_3}{T_1} = R_v^{n_c - 1} + \gamma\Delta\tau_b \tag{8–6}$$

For the expansion process, with $V_4 = V_1$ and $V_3 = V_2$, the expansion temperature ratio is

$$\frac{T_4}{T_3} = \left(\frac{V_3}{V_4}\right)^{n_E - 1} = \left(\frac{V_2}{V_1}\right)^{n_E - 1} = R_v^{-(n_E - 1)} \tag{8–7}$$

so that the heat rejected is

$$\frac{T_4}{T_1} - 1 = \frac{T_4}{T_3}\frac{T_3}{T_1} - 1 = R_v^{n_c - n_E} - 1 + \gamma \Delta\tau_b R_v^{-(n_E - 1)} \tag{8–8}$$

From the first law, the net work produced is $(Q_i - Q_0)/(mC_v T_1) =$

$$w = \gamma \Delta\tau_b (1 - R_v^{-(n_E - 1)}) - (R_v^{n_c - n_E} - 1) \tag{8–9}$$

or in terms of τ_c

$$w = \gamma \Delta\tau_b (1 - \tau_c^{-\alpha}) - (\tau_c^{(1-\alpha)} - 1) \tag{8–10}$$

where α near unity is introduced for convenience:

$$\alpha \equiv \frac{(n_E - 1)}{(n_c - 1)} \text{ so that } 1 - \alpha = \frac{(n_c - n_E)}{(n_c - 1)}$$

$$[\text{for small } e\text{'s } (1 - \alpha) \approx \gamma\{(1 - e_c) + (1 - e_E)\}] \tag{8–11}$$

The expression for the work per unit mass (eq. 8–10) is comparable to expressions obtained for other cycles. In Section 8.2 a similar result is obtained for work as calculated from an integral of $p\,dV$. In the ideal case $n_c = n_E = \gamma$, eq. 8–10 reduces to

$$w_i = \gamma \Delta\tau_b \left(1 - \frac{1}{\tau_c}\right) \tag{8–12}$$

The mean effective pressure is defined by (compare to eq. 7–68)

$$\text{mep} \equiv \frac{W}{V_1 - V_2} = \frac{w}{1 - \dfrac{V_2}{V_1}} \frac{mC_v T_1}{V_1} = \frac{1}{\gamma - 1} p_1 \frac{w}{1 - \tau_c^{-1(n_c - 1)}}$$

Combining this last equation and eq. 8–10, one obtains:

$$\frac{\text{mep}}{p_1} = \frac{1}{\gamma - 1} \frac{\gamma \Delta\tau_b (1 - \tau_c^{-\alpha}) - (\tau_c^{(1-\alpha)} - 1)}{1 - \tau_c^{-1/(n_c - 1)}} \tag{8–13}$$

or ideally

$$\frac{\text{mep}_i}{p_1} = \Delta\tau_b \frac{\gamma}{\gamma - 1} \frac{1 - \tau_c^{-1}}{1 - \tau_c^{-1/(\gamma - 1)}} \tag{8–14}$$

The thermal efficiency and its ideal value are calculated from the First Law

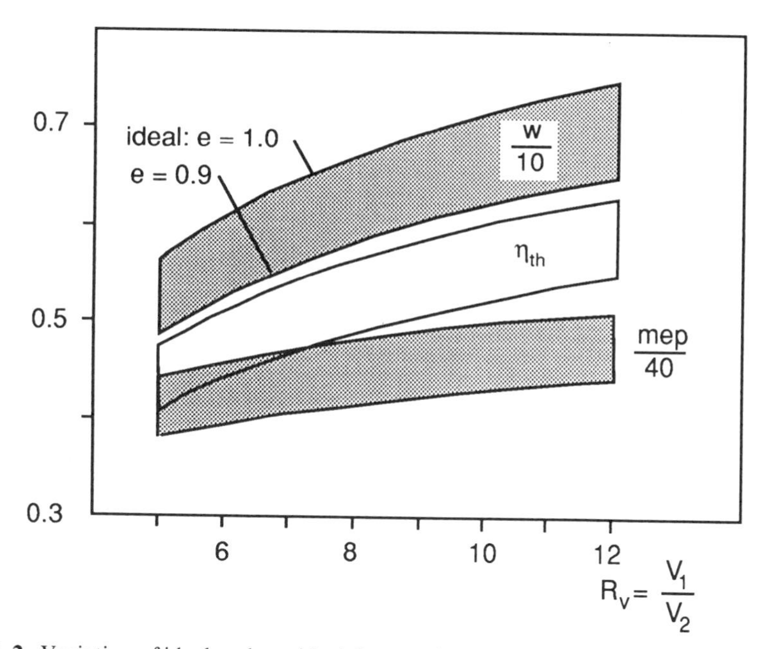

FIG. 8–2. Variation of ideal and nonideal Otto cycle performance parameters, $\Delta\tau_b = 8.5$, $\gamma = 1.4$. For nonideal cycle, polytropic efficiencies = 0.9.

using eqs. 8–3 and 8–10;

$$\eta_{th} = \frac{w}{\gamma\,\Delta\tau_b} = 1 - \tau_c^{-\alpha} - \frac{\tau_c^{(1-\alpha)} - 1}{\gamma\,\Delta\tau_b} \quad \text{and} \quad \eta_{th,i} = 1 - \frac{1}{\tau_c} \tag{8–15}$$

Note the dependence of the ideal cycle on the *compression temperature ratio* which is a common characteristic of all heat engine cycles. The full power performance of the Otto cycle is shown in Fig. 8–2 for the ideal and nonideal cycles. The design variable R_v (defined as V_1/V_2 and termed the "compression" ratio) is varied. The heat input is fixed by specification of $\Delta\tau_b = 8.5$. The lower boundary on a plot is for both polytropic efficiencies $e = 0.9$ whereas the upper is for $e = 1.0$. These plots are obtained using eqs. 8–9, 8–13, and 8–15. Note the increasing, but relatively weak, dependence of performance on R_v. The highest practical value of R_v is fixed by the requirement that the fuel-air mixture should not ignite prematurely as it is compressed. Typical values of R_v range between 6 and 12 and are sensitive to the fuel's ignition characteristics as measured by its octane number. Comparing the actual to ideal cycles' performances, it appears that a 90% polytropic efficiency in the work processes results in less than 10 percentage points loss in thermal efficiency and a similar reduction in specific work, w. These characteristics provided the motivation to employ higher compression ratio in automotive engines in the 1950s and 60s until fuel cost became an issue.

8.2. PART-LOAD PERFORMANCE OF THE OTTO CYCLE

The application of the Otto cycle engine in practice is usually dominated by its ability to operate well at part power. In order to operate at less than full power, the Otto cycle engine is equipped with a throttle whose function is to reduce the pressure (and therefore the density) of the incoming charge. If the cycle thermodynamic cycle characteristics did not change as a result of the altered inlet pressure, then the power output of a given engine is reduced because the mass of air (and therefore fuel) is reduced. The fuel system's function is to keep the air mass and fuel mass proportional to each other, deviating from that task only to ensure good combustion with low emissions. Thus from eq. 8–9,

$$\frac{W}{W_{max}} = \frac{m}{m_{max}} \frac{w}{w_{max}} \tag{8–16}$$

where the largest influence on W is due to the variation of the mass processed

$$\frac{m}{m_{max}} = \frac{\rho_1}{\rho_{1\,max}} = \frac{p_1}{p_{1\,max}} \tag{8–17}$$

In actuality, the cycle is also altered which changes w so that the ratio w/w_{max} decreases from unity. This variation is examined first, followed in Section 8.2.2, by a discussion of the effect of varying mass.

8.2.1. *Thermodynamic Path Description*

In this section, the changes to the cycle are described. Figures 8–3 and 8–4 show the *p-V* and *T-s* diagram of the ideal Otto cycle at part power. Both of these representations are used to describe the cycle. Care must be used, however, because the two diagrams show different aspects of the same cycle. The *p-V* diagram shows the volume in the cylinder space whereas the *T-s* shows the thermodynamic state of the *system*. The system is the mass of fluid being processed which may or may not fully occupy the cylinder volume. It may also occupy part of the space in the intake or exhaust manifolds.

Consider first the *p-V* diagram in Fig. 8–3. Air enters at atmospheric pressure at p_0 and T_0. The throttling process is at constant enthalpy (T), and is irreversible. At point 1 the Otto cycle is executed to 2, 3, and 4. The volumes V_4 and V_1 are identical, and hence points 1 and 4 lie on a constant volume line. The pressure p_4 is larger than atmospheric, hence another irreversible expansion takes place to point 5. At this point the system occupies part of the exhaust manifold where the pressure is close to atmospheric. The constant pressure process 5-0 is the heat rejection process. Note that at full power with no throttling at the inlet, all of the noisy throttling takes place at the end of the cycle, 4–5. In a real Otto cycle engine, the heat rejection process takes place (1) as the warm (T_5) gas is physically discharged from the cylinder volume and (2) to the cylinder walls, a part of the process which is neglected here.

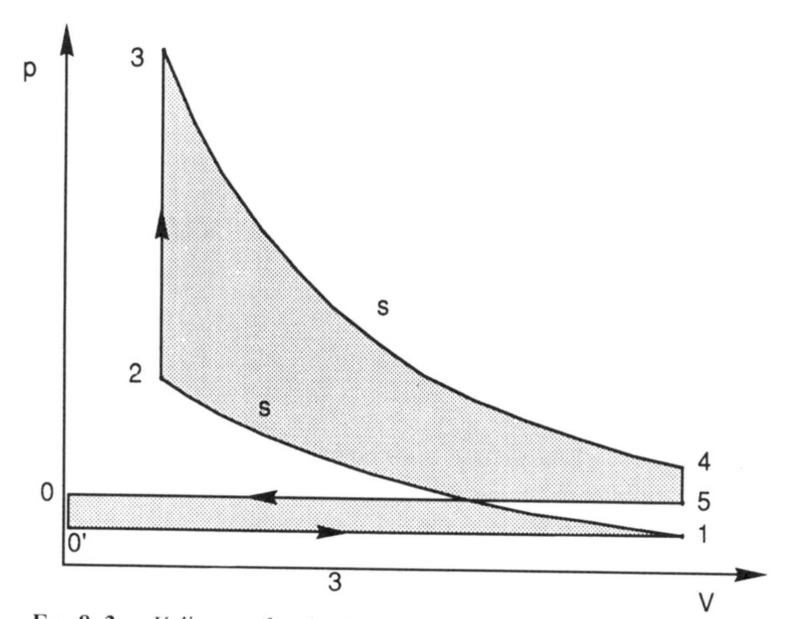

FIG. 8–3. *p-V* diagram for the (zero clearance) Otto cycle at partpower.

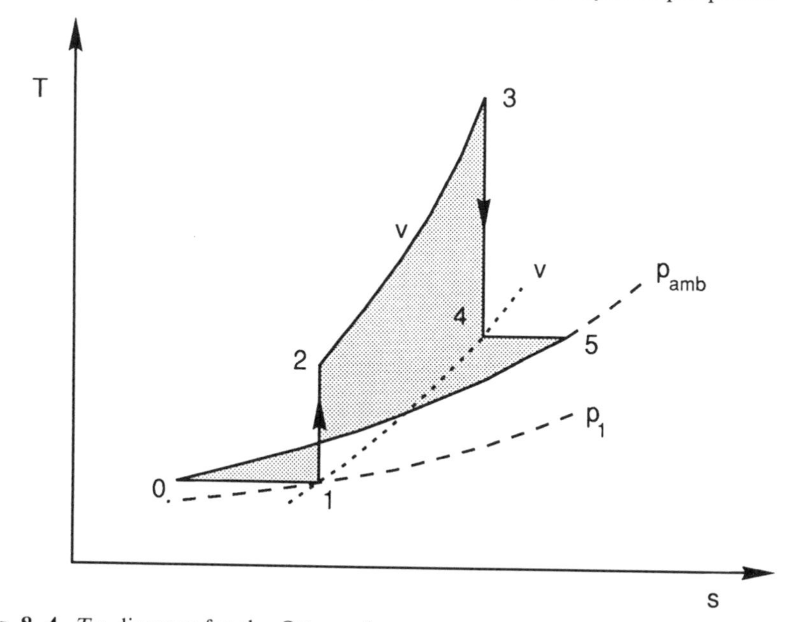

FIG. 8–4. *T-s* diagram for the Otto cycle at part power; 5–6 is an isobar and 2–3 is a constant (specific) volume or isometric line.

The state of the processed fluid is described more accurately by the *T-s* diagram in Fig. 8–4. One may imagine the family of processes described by varying throttle settings by noting the changes brought about when the isobars p_0 and p_1 change their relation with constant volume ratio,

$$\frac{v_1}{v_2} = \frac{V_1}{V_2} = R_v.$$

The real engine's piston is constrained to move to a minimum, nonzero, volume V_2 where combustion takes place. In the intake process, one rotation of a four-stroke engine later, this volume retains some of this exhaust gas which mixes with the fresh incoming charge. This aspect of the process is ignored in the analysis here but is included in Sections 8.3 and 8.7 to refine the model.

8.2.2. *Performance Analysis*

The following is the analysis of the part-load performance. To first order, $\Delta\tau_b$ remains the same since the fuel–air ratio is maintained, and thus eq. 8–3 remains valid. This analysis is carried out assuming unity polytropic efficiencies so that the approach may be illustrated without the algebraic complexity. Incorporation of losses in the work processes through specification of the polytropic efficiencies is readily incorporated into a numerical calculation procedure. From Fig. 8–3 the work in the four processes is:

$$\text{3–4:} \quad \int p\,dV = \frac{1}{\gamma - 1} p_3 V_2 (1 - \tau_c^{-1})$$

$$\text{1–2:} \quad = \frac{1}{\gamma - 1} p_1 V_1 (\tau_c - 1)$$

$$\text{0–1:} \quad = p_1 V_1$$

$$\text{5–6:} \quad = -p_0 V_1$$

Here τ_c is defined in eq. 8–5. One may define a throttling pressure ratio, $p_1/p_0 = x \leq 1$, so that the total work is

$$W = \left[\frac{\gamma}{\gamma - 1} x\{1 - \tau_c^{-1}\}\,\Delta\tau_b - (1 - x)\right] p_0 V_1$$

or nondimensionally

$$w = \gamma\,\Delta\tau_b\left(1 - \frac{1}{\tau_c}\right)x - (1 - x)(\gamma - 1) \tag{8–18}$$

which is consistent (with $x = 1$) with the full power case (eq. 8–11). The heat input is $mC_v T_0 \gamma\,\Delta\tau_b$ where $m = x p_0 V_1/RT_0$. The thermal efficiency is therefore:

$$\eta_{th} = \left(1 - \frac{1}{\tau_c}\right) - \frac{1 - x}{x}\,\frac{\gamma - 1}{\gamma}\,\frac{1}{\Delta\tau_b} \tag{8–19}$$

A plot of specific work and efficiency (eqs. 8–18 and 8–19) is shown in

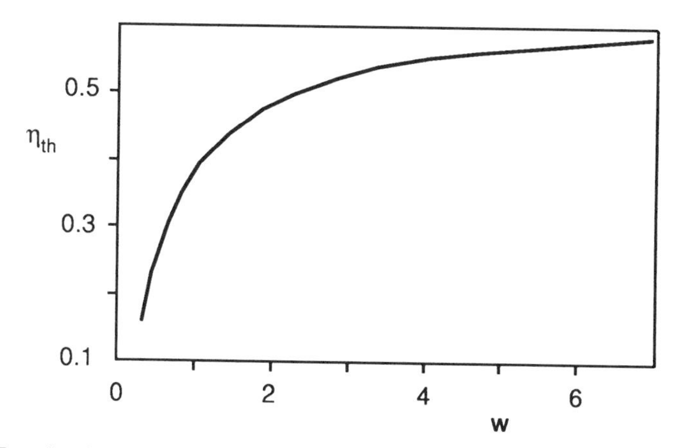

Fig. 8–5. Part-load efficiency for the ideal Otto cycle. $\Delta\tau_b = 8.5$, $\gamma = 1.4$, $\tau_c = 2.4$. w and η_{th} given by eqs. 8–18 and 8–19. w is of the order of 6 and mep of the order of 18 (atm).

Fig. 8–5 for $\Delta\tau_b \sim 8.5$, $\gamma = 1.4$, and $\tau_c = 2.4$ corresponding to $R_v = 9$. Note the weak dependence of efficiency away from the maximum specific work point. For a real engine's part load performance, a number of additional factors play important roles, among these are:

1. Mixing of spent gas in the clearance volume
2. Heat transfer modifying the polytropic exponents
3. Real gas effects
4. The flow irreversibilities associated with gas entry and exit

The first of these is examined in Section 8.3.

The important conclusion, however, is that from a thermodynamic point of view,

1. The specific work for the Otto cycle is large.
2. The Otto cycle is efficient.
3. The part-load efficiency falls slowly with decreasing work output.

8.3. CHARGE DILUTION: CLEARANCE VOLUME

In practical Otto cycle engines, the exhaust stroke is limited to that of the compression stroke for mechanical reasons: Unequal strokes for the cylinder during compression and discharge are difficult to implement in practice. This means that the process 5–0 in Fig. 8–3 should be terminated at $V = V_2$ rather than $V = 0$. Thus when the volume is V_2, the mass of trapped gas that has the temperature T_4 will prevent the full volume of fresh air from being processed. The ratio of mass processed per cycle, to that which could be processed if the

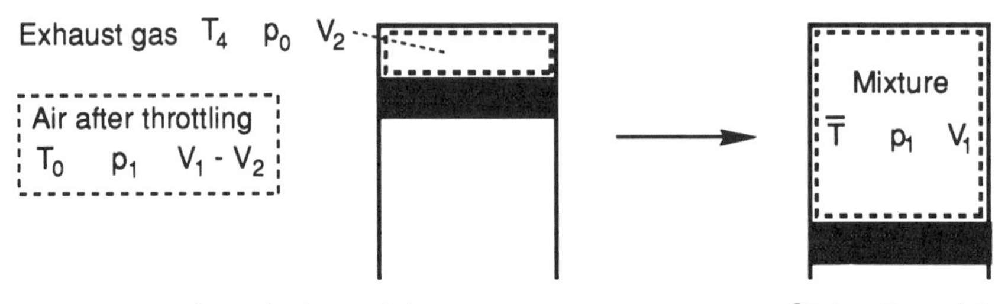

FIG. 8–6. Initial and final states after mixing of new fresh air charge with exhaust trapped in the clearance volume.

clearance were zero, is defined as the *volumetric efficiency*:

$$\eta_v = \frac{\text{actual mass per cycle}}{\text{ideal mass per cycle}} \tag{8–20}$$

Furthermore, the temperature of the gas mixture to be compressed is increased by the mixing of the hot spent gas and the fresh air. To determine this temperature, consider the state of affairs before and after the mixing process illustrated in Fig. 8–6. From the figure, the *masses* of the three gas volumes are:

$$\text{Exhaust gas: } \frac{p_0 V_2}{R T_4} \tag{8–21}$$

$$\text{Mixture gas: } \frac{p_1 V_1}{R \overline{T}} \tag{8–22}$$

$$\text{New air: } \frac{p_1 V_1}{R \overline{T}} - \frac{p_0 V_2}{R T_4} = \frac{p_1 V_1}{R T_0}\left\{\frac{T_0}{\overline{T}} - \frac{p_0 V_2}{p_1 V_1}\frac{T_0}{T_4}\right\} \tag{8–23}$$

The internal energies of these masses are calculated from $U = mC_v T$. Thus

$$\text{Exhaust gas: } \frac{p_0 V_2}{R T_4} C_v T_4 = \frac{C_v}{R} p_0 V_2 \tag{8–24}$$

$$\text{New air: } \frac{p_1 V_1}{R T_0}\left\{\frac{T_0}{\overline{T}} - \frac{p_0 V_2}{p_1 V_1}\frac{T_0}{T_4}\right\} C_v T_0 = \frac{C_v}{R} p_1 V_1 \left\{\frac{T_0}{\overline{T}} - \frac{p_0 V_2}{p_1 V_1}\frac{T_0}{T_4}\right\} \tag{8–25}$$

$$\text{Mixture gas: } \frac{p_1 V_1}{R \overline{T}} C_v \overline{T} = \frac{C_v}{R} p_1 V_1 \tag{8–26}$$

Equating the internal energies before and after the mixing process gives the mixture gas temperature, $\bar{T}$ (with heat loss is neglected):

$$\frac{\bar{T}}{T_0} = \left\{1 - \left(\frac{p_0}{p_1}\frac{V_2}{V_1}\right)\left(1 - \frac{T_0}{T_4}\right)\right\}^{-1} \tag{8–27}$$

Since this temperature ratio is greater than unity, the cycle operates with an increased minimum cycle temperature and thus a reduced thermal efficiency. This temperature ratio is a function of the compression ratio, V_1/V_2. It also depends on the load condition through p_1/p_0. The ratio T_4/T_0 (see Fig. 8–4) is a constant to first order because the fuel–air ratio is maintained. The importance of the dilution effect is determined by $p_0 V_2/(p_1 V_1)$ which is small at full power but becomes of order 1 at part power (i.e. low p_1). Since the ideal mass processed is $p_1 V_1/(R T_0)$, the volumetric efficiency (see eq. 8–20) is thus found to be:

$$\eta_v = \frac{T_0}{\bar{T}} - \frac{p_0}{p_1}\frac{V_2}{V_1}\frac{T_0}{T_4} \quad \text{or with eq. 8–27} \quad \eta_v = 1 - \frac{p_0 V_2}{p_1 V_1} = 1 - \frac{1}{x R_v} \tag{8–28}$$

which shows that how the volumetric efficiency is reduced at part-load. At full load, the volumetric efficiency is close to unity.

8.4. THE DIESEL CYCLE

In the Diesel cycle engine, the compression ratio is made sufficiently high so that fuel injected into the compressed air will burn spontaneously. This imposes special requirements on the physical characteristics of the fuel but has the advantage that the fuel injection rate can be used to control the conditions in the combustion chamber. The equivalence ratio (see Section 5.1) in Diesel engines is *fuel lean* to ensure that the last amount of fuel injected has sufficient oxygen available to burn as completely as possible.

In contrast to the Otto cycle engine, the Diesel engine has no throttle. The air process rate therefore depends only on the engine's rotational speed. The amount of fuel injected determines the heat input and therefore the mechanical power output. In practical terms, this is accomplished by allowing the length of time over which the injection takes place to vary: a short injection squirt during low power and a longer one at full power, with the additional constraint that the pressure in the combustion chamber does not exceed limits.

The Diesel cycle can be modeled as a combination of constant volume combustion during the initial portion of the combustion process while the later portions are better described as constant pressure, although the real process is a smooth transition between them (Fig. 8–7). It is a straightforward procedure to include both types of combustion: it entails specification of a parameter, such as maximum pressure allows. This complication of the model is omitted here in order to emphasize the thermodynamics. In practice, the strength of the piston and cylinder materials and the allowable heat transfer rate to the walls

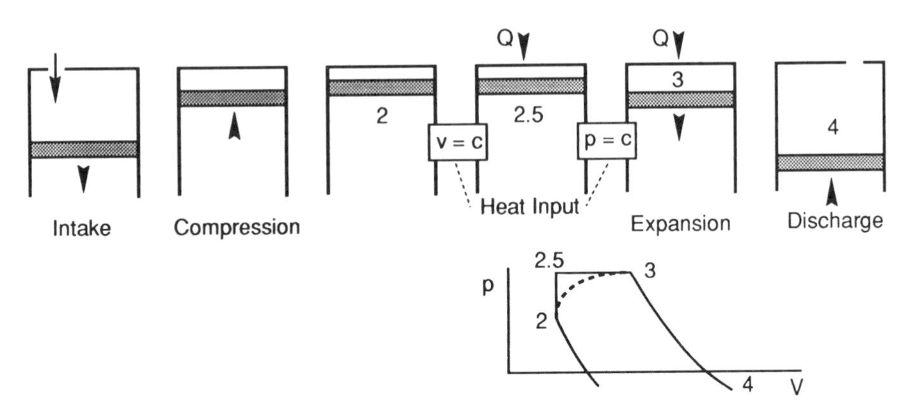

FIG. 8–7. Processes in the Diesel cycle. Solid line 2-3-4 is an approximation to the real process shown dashed.

limit the pressure: typical pressure limits lie near 70–80 atm. To describe this cycle in a simple manner, the constant volume portion of the process is neglected, and the analysis is of the cycle whose *T-s* diagram is shown in Fig. 8–1c. Typical volume compression ratios in Diesel engines are in the range of 12–25. This brings the gas to 30–90 atm for an adiabatic process with $\gamma = 1.4$. In reality, most of the heat addition is at constant pressure, particularly at high power output. Heat losses are partly equivalent to a direct loss of fuel heat energy in the combustion process itself and are thus neglected here. A simple accounting for heat loss is described in Section 8.5.

Figure 8–8 shows the approximate Diesel cycle represented on the *T-s* plane. The cycle is analyzed in a manner similar to that used for the Otto cycle. The (constant pressure) heat input to the cycle is

$$q_{\text{in}} = \frac{Q_{\text{i}}}{mC_{\text{p}}T_1} = \frac{T_4}{T_1} - \frac{T_2}{T_1} = \Delta\tau_{\text{b}} \tag{8–29}$$

where eq. 8–5 can be used to invoke the same nondimensional parameters as used in the Otto cycle description.

The (constant volume) heat rejected is

$$q_0 = \frac{Q_0}{mC_{\text{p}}T_1} = \left(\frac{T_5}{T_1} - 1\right)\frac{1}{\gamma} \tag{8–30}$$

The temperature ratio T_5/T_1 can be written with eq. 8–29:

$$\frac{T_5}{T_1} = \frac{T_5}{T_4}\frac{T_4}{T_1} = \frac{T_5}{T_4}(\tau_{\text{c}} + \Delta\tau_{\text{b}}) \tag{8–31}$$

where the ratio T_5/T_4 can be written for a polytropic process:

$$\left(\frac{T_5}{T_4}\right)^{n_{\text{E}}/(n_{\text{E}}-1)} = \frac{p_5}{p_4} = \frac{p_5}{p_1}\frac{p_1}{p_2}\frac{p_2}{p_4} \tag{8–32}$$

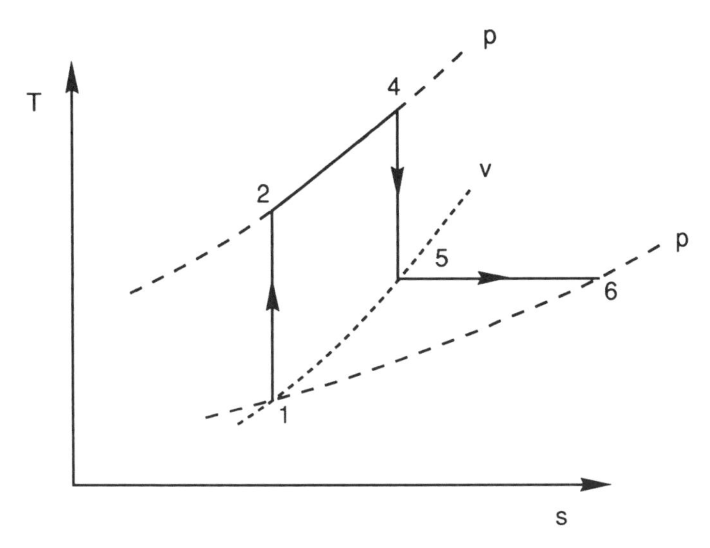

FIG. 8–8. *T-s* diagram for a Diesel cycle approximated with constant pressure combustion.

These pressure ratios are given by

$$\frac{p_4}{p_2} = 1 \quad \text{and} \quad \frac{p_5}{p_1} = \frac{T_5}{T_1} \quad \text{since} \frac{V_5}{V_1} = 1 \quad \text{and} \quad \frac{p_2}{p_1} = \tau_c^{n_c/(n_c - 1)} \tag{8–33}$$

Combining these into eq. 8–32 yields

$$\left(\frac{T_5}{T_4}\right)^{n_E/(n_E - 1)} = \frac{T_5}{T_1} \tau_c^{-n_c/(n_c - 1)} \tag{8–34}$$

Equations 8–31 and 8–34 combine so that the temperature ratio T_5/T_1 can be obtained

$$\frac{T_5}{T_1} = (\tau_c + \Delta\tau_b)^{n_E} \tau_c^{-n_c\alpha} \tag{8–35}$$

Here α is defined by eq. 8–11. The ideal value of T_5/T_1 is (with $n_E = n_c = \gamma$)

$$\left(\frac{T_5}{T_1}\right)_{id} = \left(1 + \frac{\Delta\tau_b}{\tau_c}\right)^{\gamma}$$

Equation 8–35 allows calculation of the rejected heat using eq. 8–30. The net work and its ideal value are

$$w = q_i - q_0 = \Delta\tau_b - \frac{1}{\gamma}\left\{\frac{T_5}{T_1} - 1\right\} \qquad w_{id} = \Delta\tau_b - \frac{1}{\gamma}\left[\left(1 + \frac{\Delta\tau_b}{\tau_c}\right)^{\gamma} - 1\right] \tag{8–36}$$

The work output is evidently maximized for high compression ratio. Equivalently, the mean effective pressure is similar to that in eq. 8–12, except that w is nondimensionalized by C_p rather than C_v:

$$\frac{\text{mep}}{p_1} = \frac{\gamma}{\gamma - 1} \frac{w}{1 - \tau_c^{-1/(n_c - 1)}} \tag{8–37}$$

Finally, the thermal efficiency is given by

$$\eta_{th} = \frac{w}{q_i} = 1 - \frac{\frac{T_5}{T_1} - 1}{\gamma\, \Delta\tau_b} \quad \text{and} \quad \eta_{th,id} = 1 - \frac{1}{\gamma\, \Delta\tau_b}\left(\left(1 + \frac{\Delta\tau_b}{\tau_c}\right)^{\gamma} - 1\right) \tag{8–38}$$

Note here that τ_c large is desirable, but increasing $\Delta\tau_b$ leads to decreasing thermal efficiency. This can be seen by looking at the limit of small $\Delta\tau_b$. Linearizing the last term in brackets in eq. 8–38,

$$\left.\begin{aligned} &\left(1 + \frac{\Delta\tau_b}{\tau_c}\right)^{\gamma} \cong 1 + \gamma\frac{\Delta\tau_b}{\tau_c} \\ \text{leads to} \quad & \\ &\eta_{th,id} \cong 1 - \frac{1}{\tau_c} \qquad \text{for } \Delta\tau_b \to 0 \end{aligned}\right\} \tag{8–39}$$

which means that the ideal cycle approaches the Brayton cycle limit (see Fig. 7–35 and Chapter 9) when the heat input is made vanishingly small. This is indeed a high efficiency for a large value of τ_c. The second-order term in the linearization of eq. 8–39, is negative, showing that the efficiency of the cycle decreases as $\Delta\tau_b$ is increased.

If the ideal and nonideal equations are used to calculate the variation of w, mep, and η with the volume ratio V_1/V_2 as a design variable and $\Delta\tau_b \sim 8.5$ as done previously for the Otto cycle, the characteristics are shown in Fig. 8–9. Note the similar behavior of all three of these variables and their relatively slow rate of increase with increasing V_1/V_2. Taking the polytropic efficiencies to be 0.9 for both compression and expansion, one sees a reduction in these performance parameters of approximately 15%.

The values of w, mep/p_1, and η are of the order of 4.5, 12, and 0.5, respectively. These numbers will be useful to compare to the steady flow cycle, such as the Brayton to be considered in Chapter 9.

The part load efficiency can be calculated from eqs. 8–34 and 8–32. The approach is to decrease $\Delta\tau_b$ from the maximum value (8.5 in the case here), calculate η and w, and plot these against each other for a chosen design volume ratio (25 here). The result is shown in Fig. 8–10 for polytropic efficiency = 1 (ideal) and 0.90. Note the approach to Carnot efficiency at small heat input (and work output). For the more realistic cycle, note that the efficiency is nearly flat from 100% power to 40%.

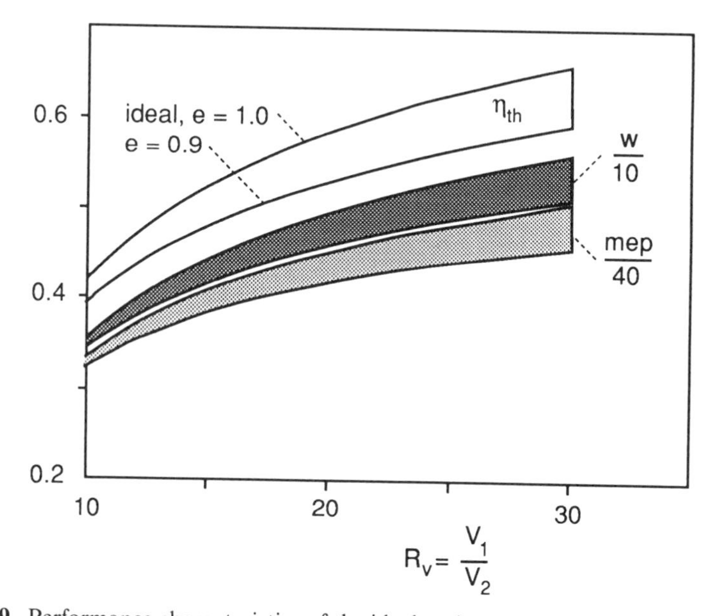

FIG. 8–9. Performance characteristics of the ideal and nonideal Diesel cycle. Polytropic efficiencies for compression and expansion = 1.0 and 0.90, respectively. Compare with Fig. 8–5 for the Otto cycle.

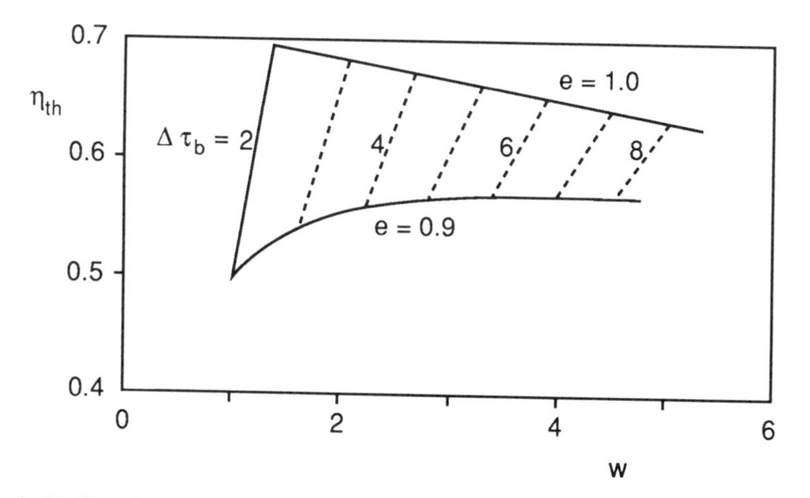

FIG. 8–10. Part-load performance characteristics of the ideal and nonideal Diesel cycles. Values of heat input, specified as $\Delta\tau_b$, are noted.

8.5. DIESEL CYCLE HEAT LOSS

The conditions of the combustion gas lead to a large heat transfer rate from the gas to the cylinder wall. These losses are important and have been the subject of major development efforts aimed at their reduction. The term "adiabatic Diesel" is applied to such engines, refs. 8–5 and 8–6. A first-order way of modeling this loss is to say that the loss represents an increase in the heat input required. This loss may be argued to be proportional to a difference between the mean wall temperature and that of the environment:

$$q_L = \frac{Q_{lost}}{mC_pT_1} = \frac{cst}{T_1}\left(\frac{T_2 + T_3}{2} - T_1\right) = L\left(\frac{\Delta\tau_b}{2} + \tau_c - 1\right) \qquad (8\text{–}40)$$

One may take the efficiency of the "real" engine (i.e., the one with the heat loss) to be equal to that of an ideal (loss free) divided by a factor reflecting the loss. Thus

$$\eta_{th} = \frac{\eta(L = 0)}{1 + L\dfrac{(\frac{1}{2}\Delta\tau_b + \tau_c - 1)}{\Delta\tau_b}}$$

An estimate of L may be made with actual performance data and then the constant L may be assumed independent of $\Delta\tau_b$. If the term is defined as q_L at maximum $\Delta\tau_b$, that is,

$$L\left(\frac{\Delta\tau_{b,\max}}{2} + \tau_c - 1\right) \equiv q_L$$

then a plot of η_{th} vs $\Delta\tau_b$ with L fixed so that the efficiency penalty is an assumed 10 percentage points at full power is shown in Fig. 8–11. Evidently, the heat loss serves to reduce η_{th} by a relatively uniform increment over the load range.

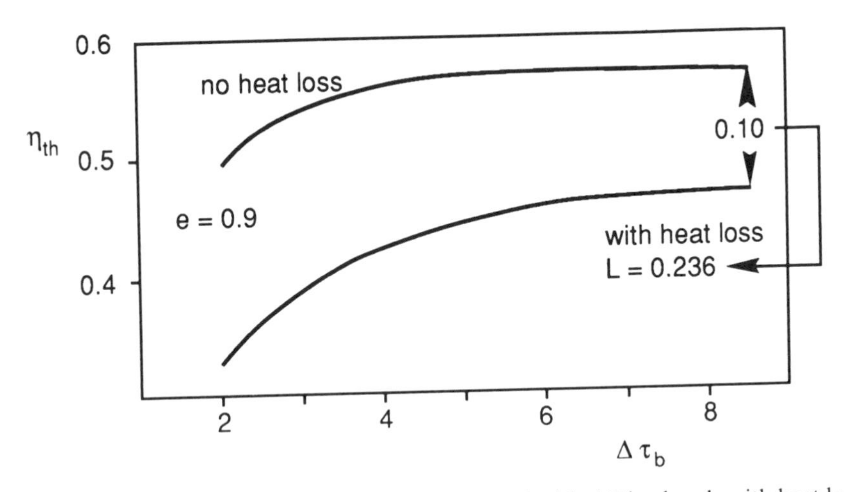

FIG. 8–11. Part-load performance characteristics of the ideal Diesel cycle with heat loss. $R_v = 25$. The constant L in eq. 8–40 is fixed by the loss at maximum power as noted.

Both the Otto and the Diesel cycles are remarkably similar from the points of view of thermodynamic performance. Efficiency and specific work have similar values. They differ in that the Diesel operates with higher compression ratio than the Otto cycle, which should lead to better performance. The fact that heat addition is carried out at constant pressure rather than constant volume leads to a smaller temperature rise for the Diesel and thus lower Carnot efficiency. The net effect is that both cycles appear to reach similar peak cycle temperatures. In practice, the irreversibilities play an important role in both cycles.

It is worthy of note that induction flow losses, which are governed by filters and induction system design as well as losses associated with valves, are important in that they serve to reduce the initial pressure of the processed charge and thus its density. This influences the power output per unit displacement (or mep), and this reduction depends on engine rotational speed (see also Section 14.7).

8.6. ENGINE SPEED EFFECTS

The work per unit mass for a thermodynamic cycle is determined by the state parameters. The power of an engine is therefore determined by the mass process rate per unit time. For displacement engines such as the Otto and Disel cycles, this is directly proportional to the size (volume) of the cylinders that are employed. This total volume is the displacement of the engine and is most commonly measured in liters. The number of engine revolutions per unit time thus affects the power output most directly. A concern for determining performance is the influence of this engine speed on various processes, primarily the irreversibilities. The combustion process that occurs in the Otto cycle at constant volume usually takes place rapidly enough that the penalty of having the heat released during the piston motion is relatively small.

An irreversibility that occurs in the real engine is heat transfer from the cylinder wall during compression and into the cylinder wall during expansion. As the time available becomes shorter at higher speed for a given rate of heat transfer, less heat is transferred compared to the low-speed situation. This is generally true even though the rate is influenced by the gas motion. In this sense, operation at higher speed leads to a more nearly adiabatic work process.

On the other hand, the flow processes into and out of the cylinder past valves generate greater pressure differences across the valve when the engine rotational speed is higher. This leads in particular to flow losses during the times when the piston demands mass transfer and its speed is high (i.e., during the middle of the intake and the expansion strokes of a four-stroke engine). Such losses make operation at high rotational speed undesirable from an efficiency point of view. An examination of flow loss effects is described in Section 14.7.

The effect of increasing loss with speed is compounded by the increasing frictional work that has to be expended by the sliding ring seals and other

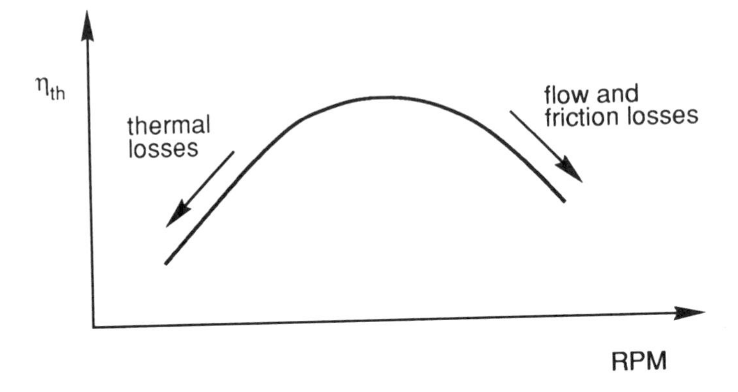

Fig. 8–12. Schematic variation of efficiency with rotational speed.

rotating components as well as by the acceleration of reciprocating elements. While this has a relatively small effect on the thermodynamic performance, it reduces the real output of the engine by demanding a fraction of the mechanical power produced. The excess over such requirements is the net power output.

As a consequence of such considerations, one would expect the efficiency at full power output of a displacement engine to display a maximum at an intermediate speed, with low-speed performance influenced by thermal losses and high-speed performance diminished by friction and valve flow losses (Fig. 8–12).

8.7. POWER IN THE EXHAUST STREAM

The nature of displacement engines is that it is impractical to have the mechanical arrangement execute different strokes for the compression and expansion processes. A practical consequence is that when the Otto or Diesel cycle engine is operated at any level higher than minimal, the temperature and power of the exhaust gas are such that useful power could be obtained from this stream. The internal combustion engine (ICE) executing either the Otto or the Diesel cycle can then be thought of as a pump delivering a gas at an elevated enthalpy level. A design question is how to take advantage of this power.

The most obvious possibility is to use an expander that couples its mechanical power directly with the output from the engine as illustrated in Fig. 8–13. This scheme requires matching the speeds of the two shafts, which may be difficult because the ICE shaft speed is dictated by piston speed limitations and the expander might be a turbine where the shaft rotational speed is high in the size required. The ratio of these two speeds may be of the order of 30, which implies that gear ratios of this order are required. Such gears involve losses and may be heavy. On the other hand, if the expander is of a displacement type, the speed match is significantly better. The displacement expander has the

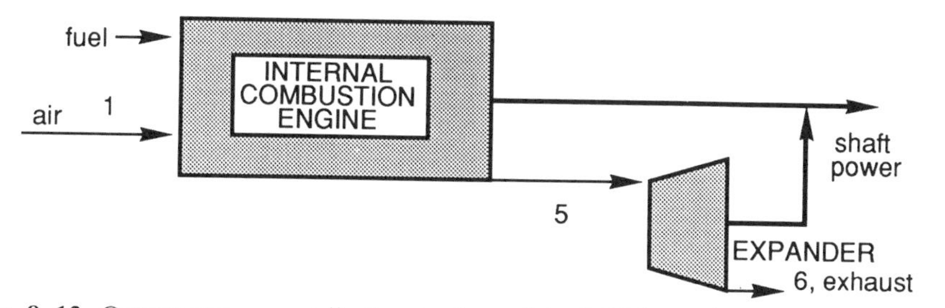

Fig. 8–13. Output power contribution to that of the ICE from the exhaust by means of an expander.

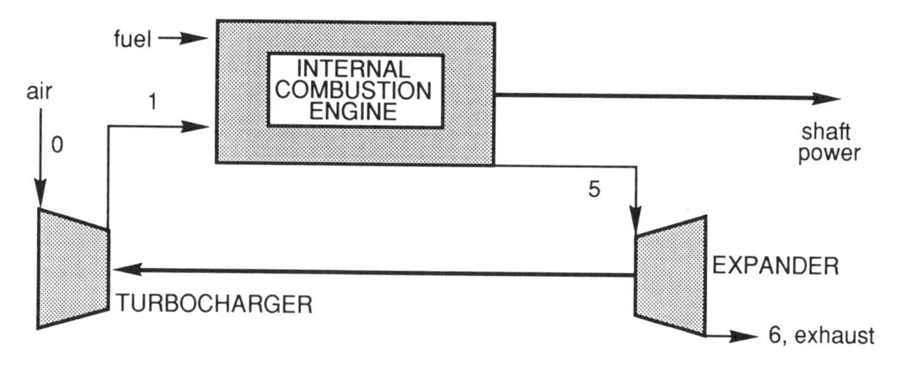

Fig. 8–14. The turbocharged ICE.

disadvantage of being limited in the expansion pressure ratio range it can process efficiently.

The power per unit displacement volume of the ICE is dictated by the rotational speed which is limited by friction considerations and by the density of the fresh charge gas. Increasing this charge density by compressing the gas reduces the size required for a given power output. Modern turbocharged engines utilize the expander power available to drive a compressor. Since both the turbine and the compressor may be aerodynamic flow devices, the high shaft speed required for both components allows them to be easily matched on a separate shaft as shown in Fig. 8–14. Serious efforts to develop an engine incorporating both turbocharging and exhaust gas energy recovery have been undertaken, ref. 8–7, among others.

An additional density increase may be obtained by locating an intercooler ahead of the ICE to reduce the temperature of the fresh charge. The use of the intercooler has the added advantage of lowering temperature levels which leads to increased service life of the engine.

Some of the advantages of turbocharging may also be obtained through supercharging where a compressor is driven from the ICE shaft power as shown in Fig. 8–15. This does not affect thermal efficiency very much since the waste stream power is not utilized. The largest effect is the allowable reduction in the displacement and potentially the size and weight of the engine. Supercharged engines may also be intercooled as shown.

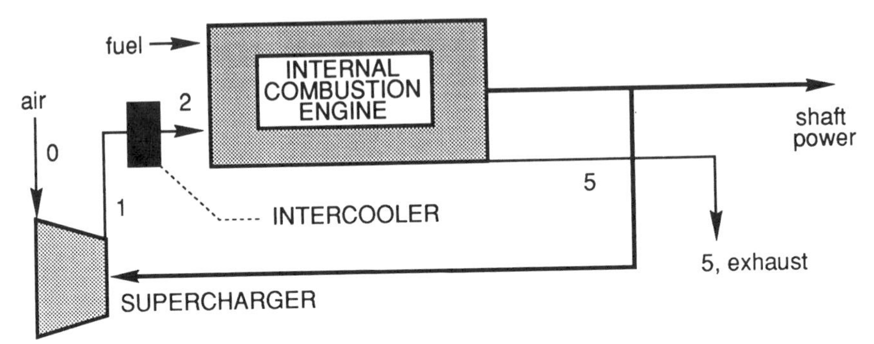

Fig. 8–15. The supercharged ICE with intercooling.

The ICE is used in many applications where part-load performance is important. For this reason, the part-load efficiency is examined for the Otto cycle and the Diesel cycle engines with external work component enhancements. The thermodynamics of the two cycles differ considerably so that they are considered separately. The goal is to obtain the efficiency as a function of fractional power output for an engine processing a unit mass of air. In particular, the role played by the important design parameters is described. Cost, complexity, weight and volume are not addressed here even though these are important cost and performance issues.

8.8. THE TURBOCHARGED OTTO CYCLE ENGINE

The performance of the ICE is modeled through the previously developed polytropic processes describing compression and expansion; and to make the estimates realistic, a fraction of the heat released is assumed lost to the cylinder walls. Entry and exit processes past valves are assumed loss free, and the throttling process is assumed adiabatic. The compression of the fresh charge is diluted by the residual waste gas. This influences the volumetric efficiency that is therefore examined. The system model is shown in Fig. 8–16 and is quite general. Intercooling and/or supercharging can easily be handled in the same model. The state points and component descriptions are noted.

The following parameters describe the various components or processes:

Supercharger (S):	total temperature ratio τ_s, adiabatic efficiency η_s
Throttle:	pressure ratio x, adiabatic
Mixer:	trapped mass per unit processed mass y
Compression (C):	volume ratio, $R_v > 1$, polytropic exponent n_c
Combustion (B):	heat release parameter $\Delta\tau_b$
Expansion (E):	polytropic exponent n_E
Final expansion (F):	adiabatic efficiency η_f.

All the above parameters, except the throttle pressure ratio, are taken as fixed

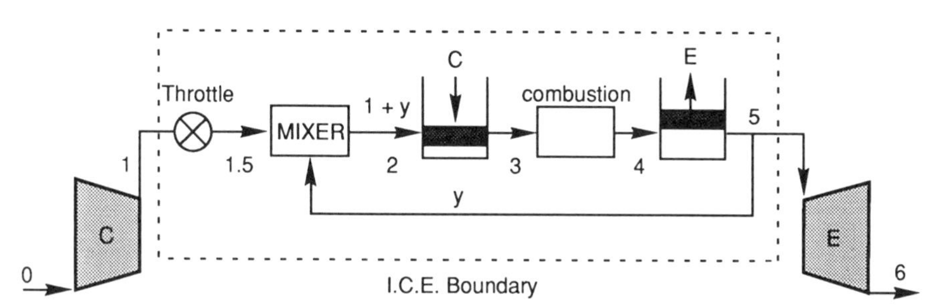

Fig. 8–16. Process elements of the ICE with exhaust work extraction and supercharging.

for a specific design. A thermodynamic analysis requires a number of total temperature ratios. These are

Air charging compressor:

$$\tau_s \equiv \frac{T_1}{T_0}; \qquad \pi_s \equiv \frac{p_1}{p_0} \qquad \text{and} \qquad \eta_s = \frac{\pi_s^{\frac{\gamma-1}{\gamma}} - 1}{\tau_s - 1} \tag{8–42}$$

Compression:

$$\frac{T_3}{T_2} \equiv \tau_c = (R_v)^{1/(n_c - 1)} \tag{8–43}$$

Expansion:

$$\frac{T_5}{T_4} = (R_v)^{-(n_E - 1)} = (\tau_c)^{-\alpha} \qquad \text{where } \alpha \equiv \frac{n_E - 1}{n_c - 1} \tag{8–44}$$

Final Expansion:

$$\tau_f \equiv \frac{T_6}{T_5}; \qquad \pi_f \equiv \frac{p_0}{p_5} \qquad \text{and} \qquad \eta_f = \frac{1 - \tau_f}{1 - \pi_f^{\frac{\gamma-1}{\gamma}}} \tag{8–45}$$

Equations 7–16, 7–17, or 8–1 give the temperature rise through the combustion process:

$$\frac{T_4}{T_0} - \frac{T_3}{T_0} = \frac{\gamma \Delta \tau_b}{1 + y} \qquad \text{and} \qquad \tau_b \equiv \frac{T_4}{T_3} \tag{8–46}$$

The $1 + y$ term accounts for the reduction in fresh charge mass flow due to the recirculation of trapped exhaust. The combustion temperature ratio is defined for convenience.

These relations allow the determination of the temperature and pressure ratio across the ICE processes as

$$\frac{T_5}{T_2} = \frac{T_5}{T_4}\frac{T_4}{T_3}\frac{T_3}{T_2} = (\tau_c)^{1-\alpha}\,\tau_b \tag{8–47}$$

and

$$\frac{p_5}{p_2} = \frac{p_5}{p_4}\frac{p_4}{p_3}\frac{p_3}{p_2} = (\tau_c)^{1-\alpha}\,\tau_b \tag{8–48}$$

α is defined by eq. 8–11 and equal to unity for the ideal fluid case. Thus both "pumping" parameters of the ICE scale directly as the combustion temperature ratio, τ_b.

Following an approach similar to that outlined in Section 8.3, the mass of the new and recirculated charges are:

$$m_{\text{new}} = \frac{p_2 V_3}{T_2}(R_v - 1) \qquad \text{and} \qquad m_{\text{recirc}} = \frac{p_2 V_3}{T_2} \tag{8–49}$$

From these, the fraction y is

$$y = \frac{1}{R_v - 1} \tag{8–50}$$

The independence of this ratio of exit temperature effects is due to the fact that both temperature and pressure at state point 5 scale with τ_b. The temperature of the mixture to be compressed may be determined from an energy conservation statement:

$$\frac{T_2}{T_1} = \frac{R_v - 1}{R_v - \tau_b(\tau_c)^{1-\alpha}} \tag{8–51}$$

The conventional volumetric efficiency (defined as the ratio of new mass to mass which could be accommodated without dilution) follows:

$$\eta_v = \frac{\dfrac{p_2 V_3}{T_2}(R_v - 1)}{\dfrac{p_2 V_3}{T_1} R_v} = 1 - \frac{\tau_b(\tau_c)^{1-\alpha}}{R_v} \tag{8–52}$$

The combustion temperature ratio may be determined from the parameters of the cycle. From its definition one has:

$$\tau_b \equiv \frac{T_4}{T_3} = 1 + \frac{y\Delta\tau_b}{1+y}\frac{T_1}{T_2}\frac{1}{\tau_s\tau_c} \tag{8–53}$$

The combustion temperature ratio may be determined from the parameters of the cycle. From its definition one has:

$$\tau^{b} \equiv \frac{T_4}{T_3} = 1 + \frac{\gamma\Delta\tau_b}{1+y}\frac{T_1}{T_2}\frac{1}{\tau_s\tau^c} \tag{8–53}$$

which yields with y and T_2/T_1 given as above,

$$\tau_b = \frac{1 + \dfrac{\gamma\Delta\tau_b}{\tau_s\tau_c}}{1 + \dfrac{\gamma\Delta\tau_b}{\tau_s\tau_c^{\alpha}}\dfrac{1}{R_v}} \tag{8–54}$$

For $R_v = 9$, $\gamma\Delta\tau_b = 12$, $\tau_s\tau_c = 2.5$ and $\alpha \sim 1$, the temperature ratio $\tau_b = 3.8$. The temperature and pressure ratios are approximately:

$$\frac{T_5}{T_2} \cong \frac{p_5}{p_2} \cong 3.8 \qquad \text{and} \qquad \frac{T_2}{T_1} \cong 1.5$$

These values make it evident that considerable power is available from the exhaust stream which can be readily determined. The pressure ratio and the temperature of the exhaust stream are

$$\frac{p_5}{p_0} = \frac{p_5}{p_2}\frac{p_2}{p_{1.5}}\frac{p_{1.5}}{p_1}\frac{p_1}{p_0} = (\tau_c^{1-\alpha}\tau_b)(1)(x)(\pi_s) \tag{8–55}$$

and

$$\frac{T_5}{T_0} = \frac{T_5}{T_2}\frac{T_2}{T_1}\frac{T_1}{T_0} = (\tau_c^{1-\alpha}\tau_b)\left(\frac{R_v - 1}{R_v - \tau_c^{1-\alpha}\tau_b}\right)(\tau_s) \tag{8–56}$$

The power per unit mass flow rate (divided by C_pT_0) is

$$w_E = \frac{\dot{W}}{\dot{m}C_pT_0} = \eta_f\frac{T_5}{T_0}\left(1 - \left(\frac{p_5}{p_0}\right)^{-(\gamma-1)/\gamma}\right) \tag{8–57}$$

For the numerical case cited above with a 1.4 pressure ratio supercharger ($\tau_s = 1.1$), the work (divided by $\dot{m}C_pT_0$) from the exhaust is shown in Fig. 8–17. Also shown is the small work required by the compressor (horizonal line). Evidently, if the processes are reversible as in this calculation, there is considerable work available from the exhaust. In practice, there are significant pressure losses associated with the entry into and exit from the cylinder as well as heat losses during and after combustion. These effects reduce this performance level significantly. The magnitude of the work available should be compared with that from the Otto cycle engine (eq. 8–18 and Fig. 8–5). The work obtainable ignoring flow losses across the valves is seen to represent about 30% of the engine's output. The practical reality that it is difficult to

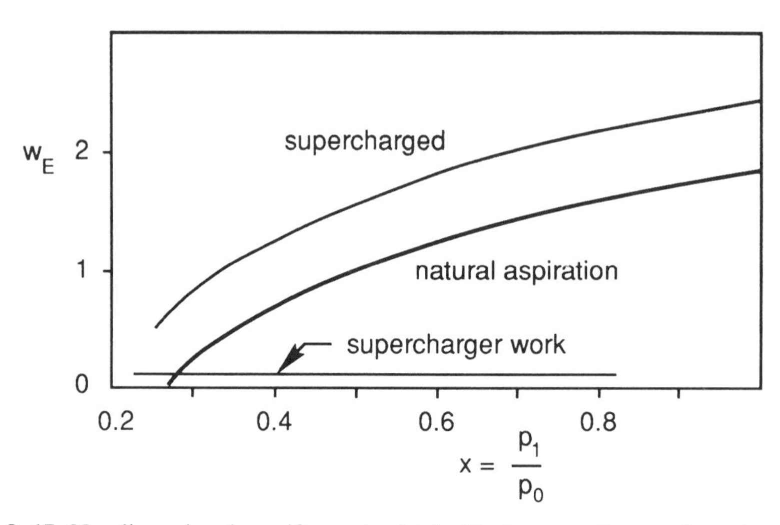

Fig. 8–17. Nondimensional specific work obtainable from an Otto cycle engine with zero internal pressure losses other than the primary throttle. Also shown is the (small) work required by the turbocharger. Parameters are noted in the text.

take advantage of this power by means other than driving the compressor, which does not require much. Clearly, the supercharged engine which operates at elevated exhaust temperature and pressure is capable of greater final expansion work. Intercooling will reduce the entry temperature and thus the power of the expander somewhat. The advantage of intercooling lies in the lower heat load to the ICE and in a reduction in displacement required in comparison to the engine not employing intercooling.

8.9. TURBOCHARGING OF THE DIESEL CYCLE

The Diesel cycle may be analyzed by a method similar to that of the Otto cycle engine. The Diesel cycle is simpler to describe in that the charge dilution effects are closer to negligible because the volume ratio is so large (25 rather than about 8 for the Otto cycle), and there is no throttle. Equation 8–35 gives the temperature ratio across the Diesel engine as

$$\frac{T_5}{T_2} = (\tau_c + \Delta\tau_b)^{n_E}\, \tau_c^{1-\alpha} = \frac{p_5}{p_2} \qquad (8\text{–}58)$$

which is also the pressure ratio since the volume at the end of expansion equals the volume of the compressed charge prior to compression. Here the exponent α is unity if the polytropic exponent for compression and expansion are equal. The station numbering system is similar to that used for the Otto cycle engine and is illustrated in Fig. 8–16.

These values of the state description allow one to calculate the power that can be extracted by a turbine in the exhaust stream. The pressure ratio for the turbine is p_5/p_0 and the inlet temperature is T_5. Thus the nondimensional work is given by eq. 8–57. Here the temperature ratio and pressure ratio across the supercharger are required and are the same ratios as given by eq. 8–42. Combining these gives

$$w = \eta_f\left(\tau_s\left(1+\frac{\Delta\tau_b}{\tau_c}\right)^{n_E}\tau_c^{(1-\alpha)} - \frac{1}{1+\eta_s(\tau_s-1)}\left(1+\frac{\Delta\tau_b}{\tau_c}\right)^{n_E/\gamma}\tau_c^{(1-\alpha)/\gamma}\right) \quad (8\text{–}59)$$

where η_f is the turbine effiency. This expression simplifies for all polytropic exponents being equal to γ to

$$w_{E,i} = \eta_f\left[\tau_s\left(1+\frac{\Delta\tau_b}{\tau_c}\right)^{\gamma} - \frac{1}{1+\eta_s(\tau_s-1)}\left(1+\frac{\Delta\tau_b}{\tau_c}\right)\right] \quad (8\text{–}60)$$

If there is an intercooler, the first τ_s may be as small as unity. Thus the following three cases may be compared

1. No supercharger, $\tau_s = 1$ in both terms
2. Supercharger present, τ_s is as specified
3. Supercharger and intercooler, $\tau_s = 1$ in the first term only

These three cases are shown in Fig. 8–18 where $\Delta\tau_b/\tau_c$ ranges from zero to 4 and efficiencies are unity. The supercharger total temperature ratio is 1.15 ($\pi_s = 1.6$).

A comparison with Fig. 8–17 shows that the power available from the Diesel cycle appears larger than for the Otto cycle and certainly larger than the

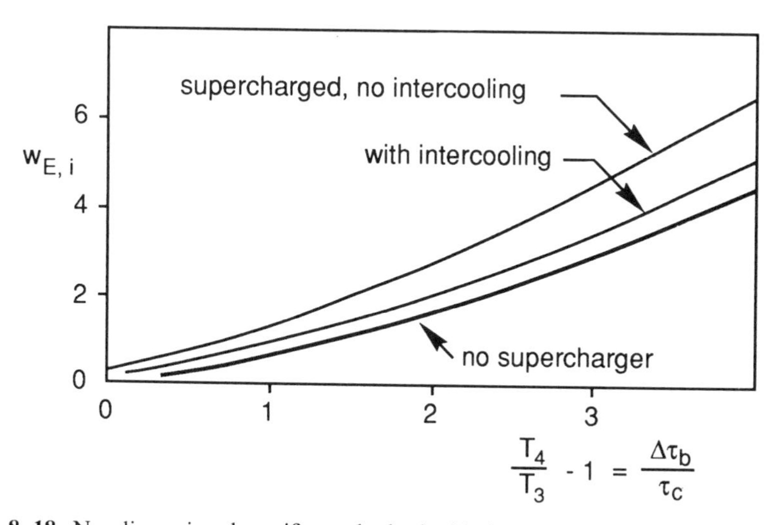

Fig. 8–18. Nondimensional specific work obtainable from the exhaust stream of a Diesel engine in three configurations. Parameters noted in the text.

compressor work required. The excess work in both cycles may be so large in comparison to what is needed or tolerated by the piston engine that a waste gate is provided to dump excess exhaust gas ahead of the turbine. This is particularly appropriate on aircraft Otto cycle engines where the function of the turbocharger is to compensate for the lower ambient density at higher altitude.

8.10. WAVE COMPRESSORS/EXPANDERS IN ICE'S

The task of supercharging the Otto or Diesel cycle engine from the energy in the exhaust stream may be carried out by a class of devices that use unsteady energy transfer (refs. 8–8, 8–9). A quantitative assessment of the performance of such a charger requires description in terms of the wave dynamics in a tube. This is given in Appendix D with the specific application to the inlet charging process. The mechanics of internal combustion engine charging is described here.

Figure 8–19 shows an internal combustion engine and two reservoirs, one of which is the inlet gas "manifold," whereas the other is the exhaust gas stream. These two reservoirs are connected by a wave energy exchanger consisting of a series of tubes that form a cylindrical ring. This ring of tubes is mechanically

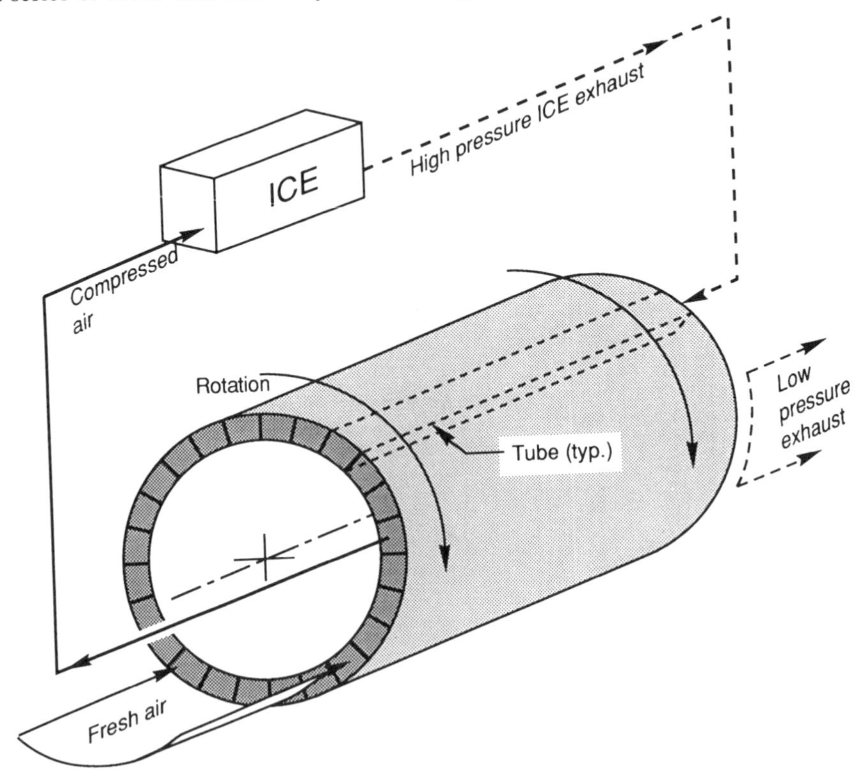

Fig. 8–19. Schematic of an ICE charged with a wave energy exchanger.

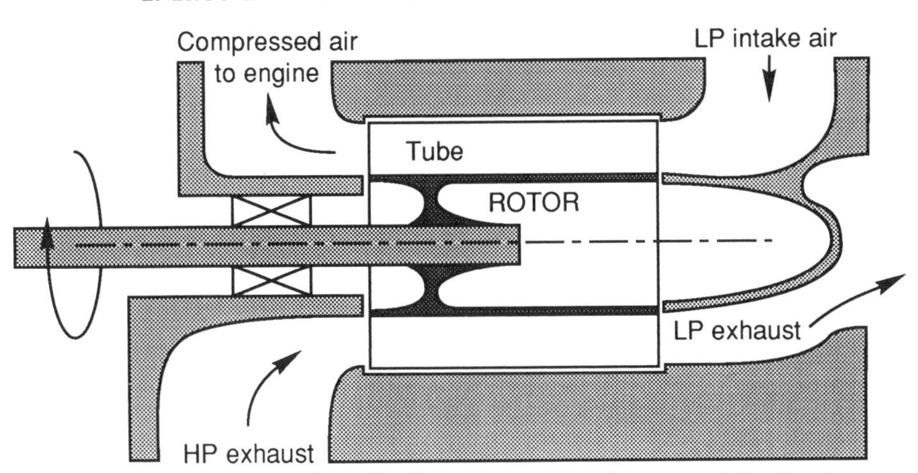

Fig. 8–20. Cross-sectional schematic of a wave energy exchanger used for charging an internal combustion engine. Engine-driven rotor and the gas flow connections are shown.

driven by the engine. The tubes have a sufficiently large length to diameter ratio so that gas motion in the tube is close to one-dimensional. The ends of the tubes are also connected to two other reservoirs at various times: a fresh air inlet and a waste gas duct. Figure 8–20 shows a cross sectional schematic of a wave energy exchanger, in particular, the rotor and the connections to the four reservoirs. Figure 8–21 shows the developed (i.e., unwrapped) rotor tubes. The left and right sides of the figure are the stationary connections.

The figure shows the sequence of events as the tube filled with fresh low pressure air is exposed to the high pressure exhaust gas reservoir. A compression (shock) wave (#1, heavy solid line) is formed which travels (to the left) into the low pressure air and compresses it. For illustrative purposes, this wave is shown as a wave front which, in reality, is a series of one-dimensional waves normal to the tube cross section. At the same time, the contact surface separating the two gases travels more slowly than the compression wave in the same direction. An unsteady expansion wave (not shown) travels into the high pressure reservoir so that the pressure there is diminished, that is, transferred to the gas in the tube. The left traveling shock wave (#1) reflects from the wall and/or the compressed gas space and now travels to the right, raising the pressure of the air a second time (shock wave #2). The air continues to travel to the right into the compressed gas space, albeit more slowly. Eventually the contact surface approaches the air outlet where the process must be terminated so that no exhaust gas is carried over into the compressed gas. The timing of this event is determined by the closing of the high pressure access to the tube on the right which forces an expansion wave into the gas column (expansion wave #3, shown as a double line). That expansion wave meets the compressed air discharge side which then closes the left end of the rotor tube. The compressed air leaves to the left into a manifold where it is used as inlet air for the ICE.

A buffer space to the tube's left end brings the contact surface to rest for a time to allow the right end to be exposed to the low-pressure atmospheric

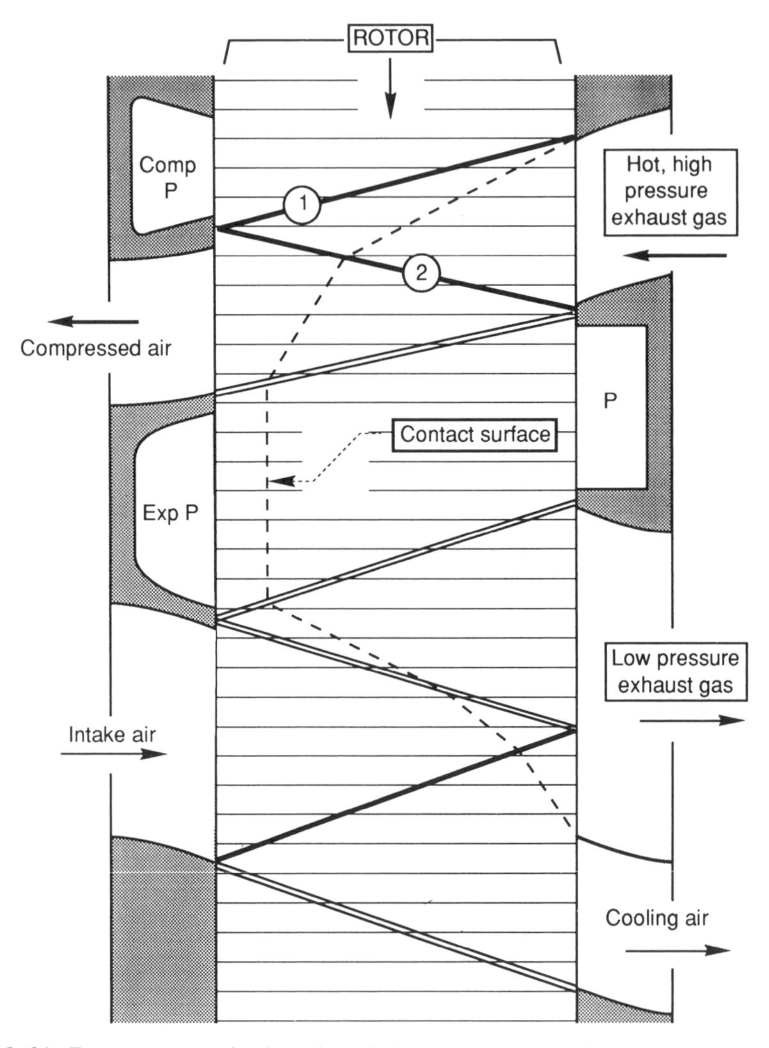

FIG. 8–21. Event sequence in the tubes of the wave energy exchanger rotor using hot, high-pressure exhaust gas to compress the incoming fresh air charge. Heavy lines are compression waves, whereas double lines are expansion waves. Contact surface is dashed. The oblique lines shown are meant to suggest the location of the wave that is normal to the tube wall. "P" denotes end pocket for wave control.

environment. This exposure results in an expansion wave traveling to the left and in motion of the waste gas to the right and out. This same expansion wave also results in motion of fresh air to the left. The exhaust opening is sufficiently long to allow the contact surface to exit and fresh charge to fill the tube.

At this point the functional process is complete. The low pressure intake port may, however, be held open for a longer time than that required to merely refill the tube. Figure 8–21 shows a portion of the fresh air charge carried through the tube without compression. This air is useful to serve as a coolant to keep wall temperatures acceptably low for long service life and may allow operation at gas temperatures above the wall material capability. This is in contrast to the steady flow turbocharging process where the turbine suffers steady exposure to a temperature close to the stagnation temperature.

PROBLEMS

1. Heat addition takes place in a confined volume that changes by a factor of two during the heating process. Find the state after heating is completed. Solution requires a modeling assumption about the relationship between heat added and volume.
2. Calculate the specific work and thermal efficiency of an approximate Diesel cycle that has $R_v = 20$, $e_c = e_E = 0.90$, $\Delta\tau_b = 8$, with constant pressure combustion until the pressure is 50 atm (from 1 atm no pressure losses), and constant pressure thereafter. Devise (and implement) a calculation procedure to obtain part-load performance.
3. Internal combustion (Otto cycle) engines in aircraft are severely limited by the decreasing atmospheric density at higher altitude. Turbochargers are a way of "fooling" the engine into thinking it is flying at sea level. Noting that peak pressure in the cylinder needs to be less than some design value at all times, devise an operating model for a turbocharged engine which is then operated at varying altitude and at a full power setting.
4. Estimate the benefit of intercooling a turbocharged automotive engine in terms of the reduction in displacement this affords over the nonintercooled engine when both engines produce the same peak power output. Assume turbo boost pressure ratio is 1.5 and make reasonable assumptions about the intercooler performance.

BIBLIOGRAPHY

Taylor, C. F., *The Internal Combustion Engine in Theory and Practice*, Vols. I & II, MIT Press, Cambridge, Mass., 1966.

Heywood, J. B., *Internal Combustion Engine Fundamentals*, McGraw-Hill, New York, 1988.

Ramos, J. I., *Internal Combustion Engine Modeling*, Hemisphere Publishing, New York, 1989.

Haddad, S., Watson, N., *Design and Applications in Diesel Engineering*, Ellis Horwood, Ltd., Chichester, U.K., 1984.

Ferguson, C. R., *Internal Combustion Engines: Applied Thermosciences*, John Wiley & Sons, New York, 1986.

REFERENCES

8-1. Heywood, J. B., "Future Engine Technology—Lessons from the 1980's and 1990's," *ASME J. of Engineering for Gas Turbines and Power*, **113**(3), July 1991.

8–2. Wyczalek, F. A., "Two-Stroke Engine Technology in the 1990s," SAE paper 910663 or SP 849, 1991.
8–3. Gerace, R., Gerace, A., "The Gerace Engine," *Mechanical Engineering*, January 1984.
8–4. Dale, J. D. Oppenheim, A. K., A rationale for Advances in the Technology of I.C. Engines," SAE 820047, 1982.
8–5. "The Adiabatic Diesel Engine," SAE SP-543, 1983.
8–6. "Adiabatic Engines, a Worldwide Review," SAE SP-571, 1984.
8–7. Hope, J. I., Johnston, R. D., "A New Concept for Reduced Fuel Consumption in Internal Combustion Engines," IECEC Paper 719051, Society of Automotive Engineers, 1971.
8–8. Seippel, C. "Pressure Exchanger," U.S. Patent No. 2,399,394, issued 1946.
8–9. Kirchhofer, H., "Comprex Supercharging of Diesel Engines for Vehicles,," BBC Publication No. CH-T123 143 D, also in Automobil Industrie, **1**, 1977. In German.

9

TEMPERATURE LIMITED CYCLES: BRAYTON, ERICSSON, AND SPACE POWER CYCLES

The performance of engines using the Brayton cycle is dominated by the temperature limits to which the engine can be subjected: the heat source, the sink, or the turbine. Modern engines that use this cycle employ steady flow. The simple Brayton cycle consists of two adiabatic work processes joined by constant pressure heat interactions. The Ericsson cycle is a modification of the simple cycle designed to achieve high efficiency.

A discussion of space power generation cycles is included because the design of these cycles is dominated by consideration of the temperature extremes. Because of its close relationship to the Brayton cycle used for the same purpose, a simple model of the Rankine cycle is also described to highlight the essential thermodynamic aspects of space power generation.

9.1. ELEMENTARY BRAYTON CYCLE ANALYSIS

The simple Brayton cycle is easily implemented, using a gas as a working fluid, in a steady flow process by configuring a compressor and an expander or turbine with a heat source in an open cycle as shown in Fig. 9–1 or in a closed cycle with heat exchangers as shown in Fig. 9–2. Figure 9–1 also shows two shafting arrangements whose performance difference is primarily of interest for part-load operation as detailed in Chapter 16. The compressor uses a fraction of the shaft power output from the expander. The remainder is the useful power transferred to the load.

The component processes can be modeled with a simple description of the working fluid: the gas behaves as ideal and (calorically) perfect. The *T*-*s* diagram (Fig. 9–3) shows the ideal cycle state points. In the following, specific work and thermal efficiency are determined.

From the First Law (eq. 2–10) for an adiabatic process, the work of compression is

$$\dot{W}_c = \dot{m}C_pT_0\left(\frac{T_2}{T_0} - 1\right) \tag{9–1}$$

and the work of expansion (as through an expander such as a turbine) is

$$\dot{W}_t = \dot{m}C_PT_4\left(1 - \frac{T_2}{T_4}\right) \tag{9–2}$$

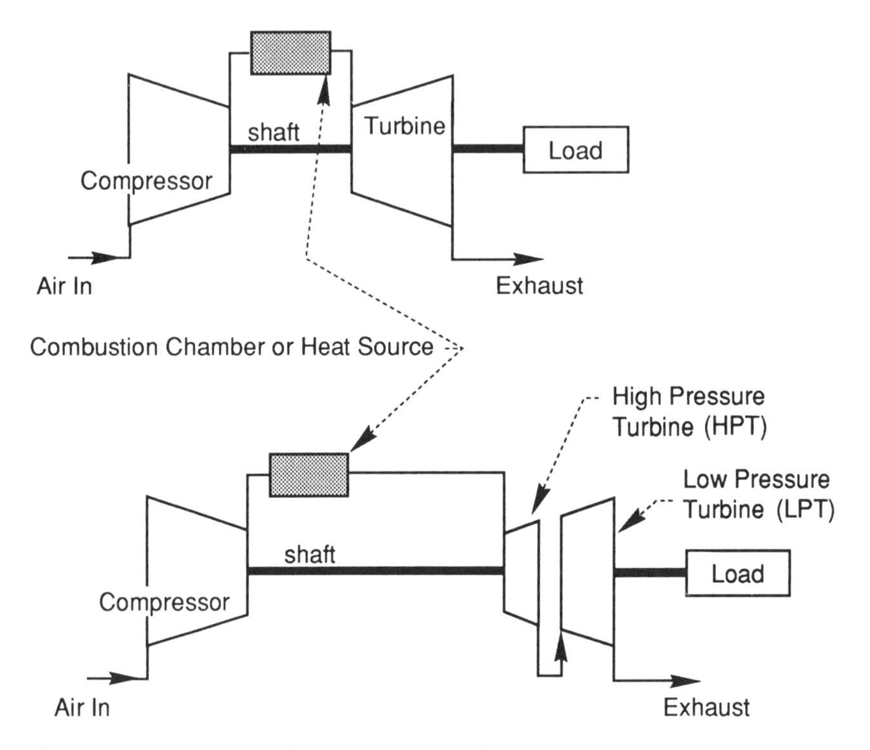

FIG. 9–1. Open Brayton cycle engines with single (top) or dual shaft (bottom) arrangements.

The peak and minimum cycle temperatures are fixed by constraints. The peak temperature is limited by the materials and the turbine blade cooling technique when the heat is supplied by combustion or by limitations of the heat source itself. The environmental temperature is the lowest temperature to which the cycle can be driven. With process 0–0 and 4–5 isentropic and for a perfect gas, the ideal relation between pressures and temperatures are

$$\frac{p_2}{p_0} = \left(\frac{T_2}{T_0}\right)^{\gamma/(\gamma-1)} \tag{9–3}$$

and

$$\frac{p_4}{p_5} = \left(\frac{T_4}{T_5}\right)^{\gamma/(\gamma-1)} \tag{9–4}$$

With no pressure losses in the heat interactions, $p_2 = p_4$ and $p_5 = p_0$, the temperature ratios may be related:

$$\frac{T_5}{T_4} = \left(\frac{T_2}{T_0}\right)^{-1} \tag{9–5}$$

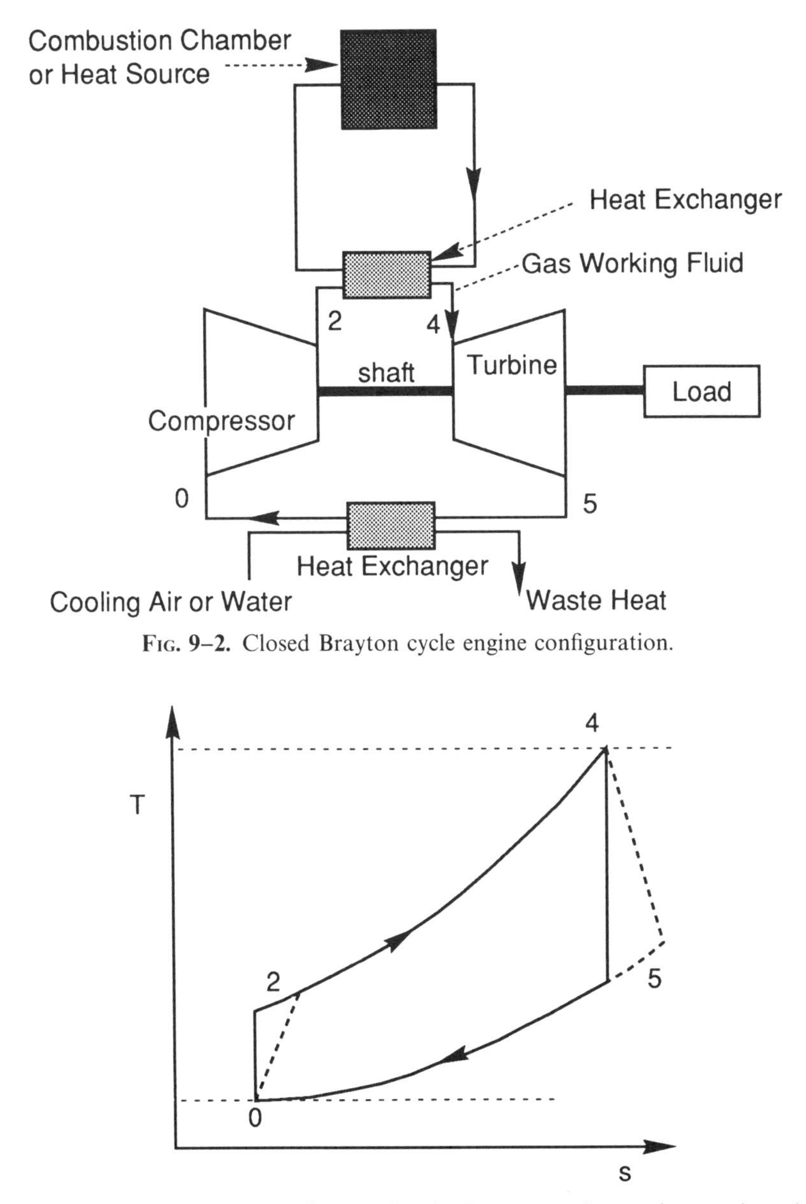

FIG. 9–2. Closed Brayton cycle engine configuration.

FIG. 9–3. Temperature-entropy diagram for the Brayton cycle: two isentropic and two isobaric processes. Dashed lines depict nonideal work processes, and dotted lines are the temperature extremes allowed.

These temperature ratios determine the work involved in the compressor and the expander. They are determined by the pressure ratios for the processes that can be chosen freely within limits. The most basic design problem is to identify the performance of a member of a family of similar engines that differ only in the value of the pressure ratio. This family is chosen to operate at the largest temperature ratio T_4/T_0 possible because that value is of greatest interest. Such an operating point is often referred to as a "design point" which is then used as a reference for determining the performance of the engine operated at "part load" or "off-design" at lower values of T_4/T_0. In this chapter, only the role of design parameters is explored. Discussion of part-load or off-design performance requires component characteristics which are developed in Chapters 14 and 15.

One may designate the compressor temperature ratio to be a measure of the chosen pressure ratio because of the isentropic relationship. Let

$$\tau_c = \frac{T_2}{T_0} \qquad \text{and} \qquad \theta_4 = \frac{T_4}{T_0} \tag{9–6}$$

Here θ_4 is a constant for a specified technology level. These quantities allow the nondimensionalization of the works involved:

$$\text{Compression:} \ \frac{\dot{W}_c}{\dot{m}C_pT_0} \equiv w_c = \tau_c - 1 \tag{9–7}$$

and

$$\text{Expansion:} \ \frac{\dot{W}_t}{\dot{m}C_pT_0} \equiv w_t = \theta_4\left(1 - \frac{1}{\tau_c}\right) \tag{9–8}$$

The net work is given by

$$\frac{\dot{W}_{net}}{\dot{m}C_pT_0} = w = w_t - w_c = \left(\frac{\theta_4}{\tau_c} - 1\right)(\tau_c - 1) \tag{9–9}$$

while the heat input is given by

$$\dot{Q} = \dot{m}C_p(T_4 - T_2) \qquad \text{or} \qquad q = \frac{\dot{Q}}{\dot{m}C_pT_0} = \theta_4 - \tau_c \tag{9–10}$$

from which the thermal efficiency may be determined:

$$\eta_{th} \equiv \frac{\dot{W}}{\dot{Q}} = \frac{w}{q} = 1 - \frac{1}{\tau_c} \tag{9–11}$$

Note the similarity between eq. 9–11 for the ideal simple Brayton cycle and eq. 8–15 for the ideal Otto cycle: efficiency is determined by the compression temperature ratio. For the simple Brayton cycle specific work and efficiency are plotted in Fig. 9–4 for two values of θ_4. The compressor temperature ratio can be made to achieve a value as high as θ_4, but very little fuel (see eq. 9–10) can be added under such circumstances so that the specific work becomes vanishingly small. However, this little heat is added near the maximum cycle

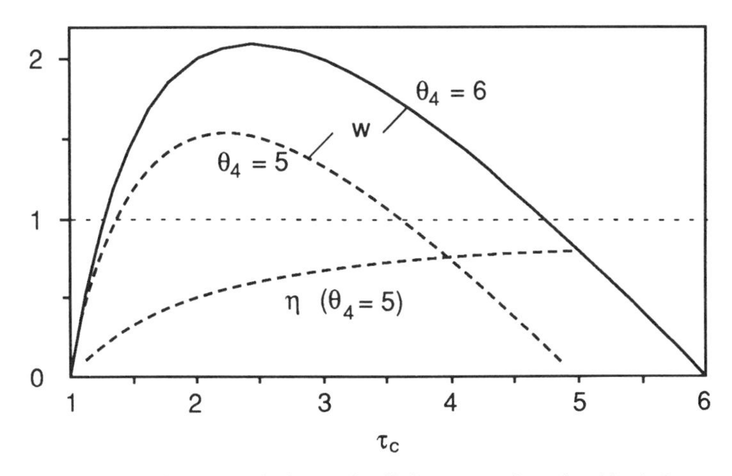

FIG. 9–4. Specific work, w, and thermal efficiency, η, for the ideal Brayton cycle plotted with varying values of compressor temperature ratio. The cycle temperature ratio is noted.

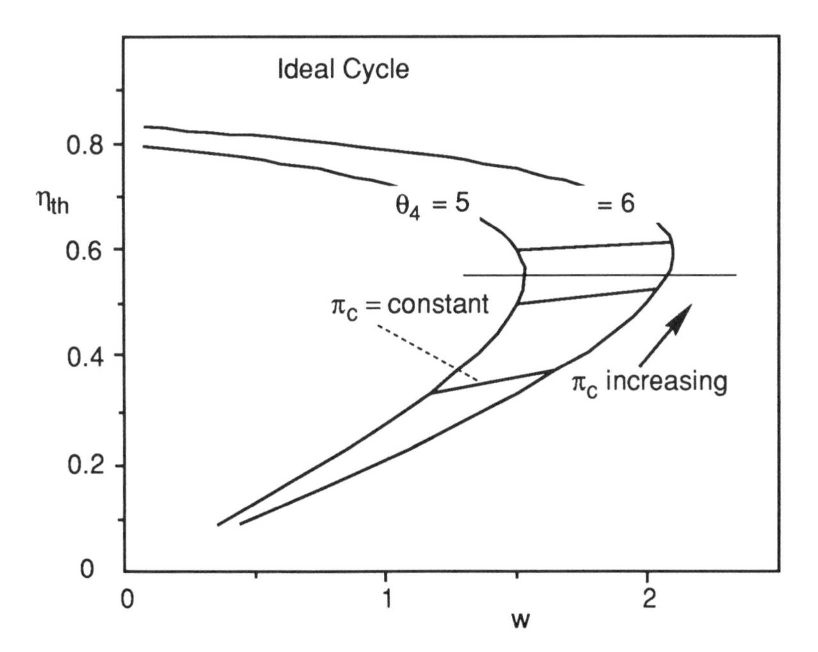

FIG. 9–5. Thermal efficiency, η, crossplotted against specific work, w, for the ideal Brayton cycle.

temperature and the waste is rejected near the minimum temperature, so that near-Carnot efficiency is realized. In a practical engine, one would generally choose the τ_c value to be larger than the value for maximum specific work, where the efficiency is still rising.

An interesting way to present this information is to plot efficiency vs. specific work with τ_c changing along the curve. Figure 9–5 is such a plot for the two cases on the previous plots. Note that the achievement of a larger value of θ_4 increases specific work and the increase in efficiency is small. As a point of reference, $\theta_4 = 5$ and 6 corresponds to 1440 and 1730 K, respectively, for T_0 = standard temperature (288 K). A unit value of specific work for air corresponds to 289 kw sec/kg, 176 HP per lb/sec, or 0.224 HP/SCFM depending on one's choice of units. In conventional units this power corresponds to 5.9 kw at a process rate of 1000 liters of standard density air per minute.

The temperature ratio τ_c, for maximum specific work, τ_{cw}, is obtained by differentiation of w (eq. 9–9) with respect to τ_c and setting the derivative to zero. The result is

$$\tau_{cw} = \sqrt{\theta_4} \tag{9–12}$$

The value of the maximum specific work and the corresponding efficiency are:

$$w_{max} = (\sqrt{\theta_4} - 1)^2 \qquad \text{and} \qquad \eta_{max\,w} = 1 - \frac{1}{\sqrt{\theta_4}} \tag{9–13}$$

which, for this ideal cycle, are seen to depend only on the cycle temperature ratio.

9.2. REGENERATION

Brayton cycles with very low pressure ratio (and therefore τ_c) have a low specific work, but worse, they also have a low efficiency. A cycle with a low pressure ratio accepts heat with the system rising in temperature from T_2 to T_4 and rejects it while falling from T_5 to T_0. Evidently with $T_2 < T_5$, some of the heat required as heat input could be supplied from the rejected heat. Figure 9–6 shows the T-s diagram for a low-pressure cycle where heat is transferred from the low pressure side of the cycle to the high-pressure side prior to heating by the external heat source. The physical arrangement is shown in Fig. 9–7 with the state points noted. With $T_3 = T_5$, as is possible with a good (matched capacity rate) counterflow heat exchanger (see Chapter 6), the heat required to raise the gas temperature from T_2 to T_3 is regenerated heat taken from the low-pressure working fluid as it falls from T_5 to T_6. The heat added to the cycle is that which raises the working fluid temperature from T_3 to T_4 and the rejection process is from 6 to 0. The net work is given by eq. 9–9 exactly as for the simple (not regenerated) cycle. The heat input, however, is (with $T_5 = T_3$)

$$\frac{\dot{Q}}{\dot{m}C_P T_0} = \frac{T_4}{T_0} - \frac{T_3}{T_0} = \frac{T_4}{T_0} - \frac{T_5}{T_4}\frac{T_4}{T_0} = \theta_4\left(1 - \frac{1}{\tau_c}\right) \tag{9–14}$$

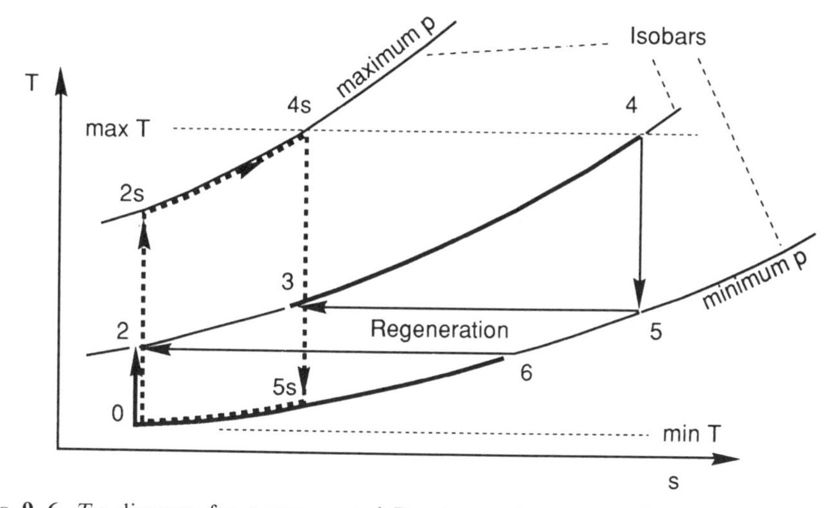

FIG. 9–6. T-s diagram for a regenerated Brayton cycle, contrasted to a high-pressure ratio simple cycle where regeneration is not possible.

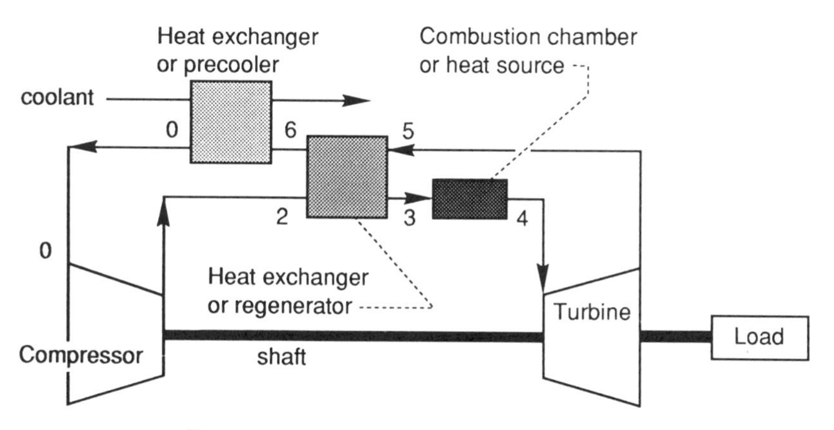

FIG. 9–7. Brayton cycle with regeneration.

and the efficiency is

$$\eta_{th} = 1 - \frac{\tau_c}{\theta_4} \tag{9–15}$$

which equals the Carnot efficiency in the limit as τ_c approaches unity since

$$\eta_{th,\,Carnot} = 1 - \frac{1}{\theta_4}$$

Note that for the Brayton cycle with regeneration two (or three) heat exchangers are used, and the pressure ratios across the two work components are the reciprocals of one another.

9.3. CYCLE PERFORMANCE WITH NONIDEAL WORK COMPONENTS AND REGENERATOR

The performance of various irreversible Brayton cycle configurations is examined here using specified values of the adiabatic efficiencies (η_c and $\eta_t \leq 1$) because simple algebraic expressions result which clearly display the role of these important irreversibilities. Note that for cycle optimization (rather than analysis) purposes, it is usually better to use polytropic efficiencies for the reasons cited in Section 7.6.3.

In Section 6.3, the performance of a heat exchanger was characterized by an effectiveness which is a measure of fL/D or size. This effectiveness was defined by eqs. 6–30 and 6–32:

$$\eta_x = \frac{T_3 - T_2}{T_5 - T_2} \tag{9–16}$$

The compressor pressure ratio, π_c, is used as the principal design variable. Because it appears in the equatons in the form $\pi_c^{(\gamma-1)/\gamma}$, use will be made of a shorthand parameter defined as

$$\tau_{cs} \equiv \pi_c^{(\gamma-1)/\gamma} \tag{9–17}$$

This isentropic temperature ratio is an equivalent measure of π_c. From the definitions of adiabatic efficiency, eq. 7–38, one has

$$\eta_c = \frac{\tau_{cs} - 1}{\tau_c - 1} \qquad \text{and} \qquad \eta_t = \frac{1 - \tau_t}{1 - \tau_{ts}} \tag{9–18}$$

where $\tau_c = T_2/T_0$ and $\tau_t = T_5/T_4$ are the compressor and expander (turbine) temperature ratios respectively. The isentropic ratios carry an additional subscript, s. The net cycle work output is

$$w = \theta_4(1 - \tau_t) - (\tau_c - 1) \qquad \text{or} \qquad w = \eta_t\theta_4(1 - \tau_{ts}) - \frac{1}{\eta_c}(\tau_{cs} - 1) \tag{9–19}$$

in terms of τ_{cs} and τ_{ts}. The isobars on the high and low sides of the cycle imply that the temperature ratios are related through $\tau_{cs} = \tau_{ts}^{-1}$ since zero pressure loss through the heat addition and removal processes is assumed in this model. The heat input to the cycle is

$$q = \theta_4 - \frac{T_3}{T_0} \tag{9–20}$$

From the definition of η_x (eq. 9–16), one can write

$$\frac{T_3}{T_0} = \frac{T_2}{T_0}(1 - \eta_x) + \eta_x \frac{T_4}{T_0}\frac{T_5}{T_4} = \tau_c(1 - \eta_x) + \eta_x \theta_4 \tau_t \tag{9–21}$$

Using the adiabatic efficiencies (eq. 9–18), the heat input, q, becomes:

$$q = (1 - \eta_x)\left[\theta_4 - \left(1 + \frac{\tau_{cs} - 1}{\eta_c}\right)\right] + \eta_x \eta_t \theta_4\left(1 - \frac{1}{\tau_{cs}}\right) \tag{9–22}$$

and the thermal efficiency follows from w and q:

$$\eta_{th} = \frac{\theta_4 \eta_t\left(1 - \frac{1}{\tau_{cs}}\right) - \frac{(\tau_{cs} - 1)}{\eta_c}}{(1 - \eta_x)\left[\theta_4 - \left(1 + \frac{\tau_{cs} - 1}{\eta_c}\right)\right] + \eta_x \eta_t \theta_4\left(1 - \frac{1}{\tau_{cs}}\right)} \tag{9–23}$$

Several interesting limits may be examined. For $\eta_x = 0$, one obtains the simple Brayton cycle result (see eq. 9–11):

$$\eta_{th}(\text{simple cycle}) = \frac{\frac{\eta_c \eta_t \theta_4}{\tau_{cs}} - 1}{\frac{\eta_c(\theta_4 - 1)}{\tau_{cs} - 1} - 1} \tag{9–24a}$$

Ideal regeneration results are obtained for $\eta_x = 1$:

$$\eta_{th}(\text{regenerated cycle}) = 1 - \frac{\tau_{cs}}{\eta_c \eta_t \theta_4} \tag{9–24b}$$

For both cycles, it is evident that the role of η_c and η_t is to reduce the effective cycle temperature ratio. In other words, the irreversibilities have an influence on the efficiency which is as important as the temperature ratio, θ_4. The specific work for these two cases is the same and is given by eq. 9–19. One may determine the "pressure" ratio (see eq. 9–17) which gives maximum specific work by

$$\frac{dw}{d\tau_{cs}} = 0 \qquad \text{and the result is } \tau_{cs,w} = \sqrt{\eta_c \eta_t \theta_4} \tag{9–25}$$

for which the maximum work is given by

$$w_{max} = \frac{1}{\eta_c}(\sqrt{\eta_c \eta_t \theta_4} - 1)^2 \tag{9–26}$$

The cycle temperature ratio giving the maximum efficiency for the *simple* (non-regenerated) cycle is given by a null differentiation of the expression for efficiency:

$$\left.\begin{aligned} \tau_{cs,\eta} &= \frac{\tau_{cs,w}^2}{C}\left(\sqrt{1 + C\frac{\theta_4}{\tau_{cs,w}^2}} - 1\right) \qquad \text{where } C \equiv \theta_4 - \tau_{cs,w}^2 - 1 \\ &\text{or} \\ \tau_{cs,\eta} &\approx \frac{\theta_4}{1 + \sqrt{(1-\eta_c)(1-\eta_t)\theta_4}} \qquad \text{for } \eta\text{'s close to 1} \end{aligned}\right\} \tag{9–27}$$

These results are more general than those obtained in eqns. 9–9 and 9–11. Figure 9–8 shows the variation of w for the conditions noted. The value of τ_{cs} at maximum w is about 2.

The thermal efficiencies for the same conditions depend on η_x and are plotted in Fig. 9–9. A cross plot of η vs. w is shown in Fig. 9–10. Note that there is

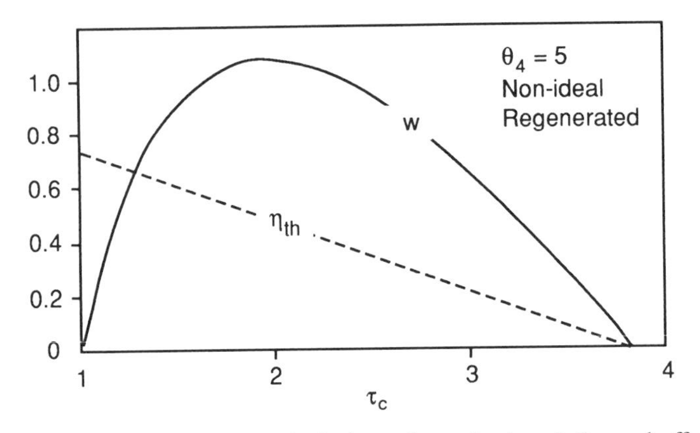

FIG. 9–8. Variation of specific work (independent of η_x) and thermal efficiency for a fully regenerated cycle ($\eta_x = 1$). $\eta_c = 0.85$, $\eta_t = 0.90$.

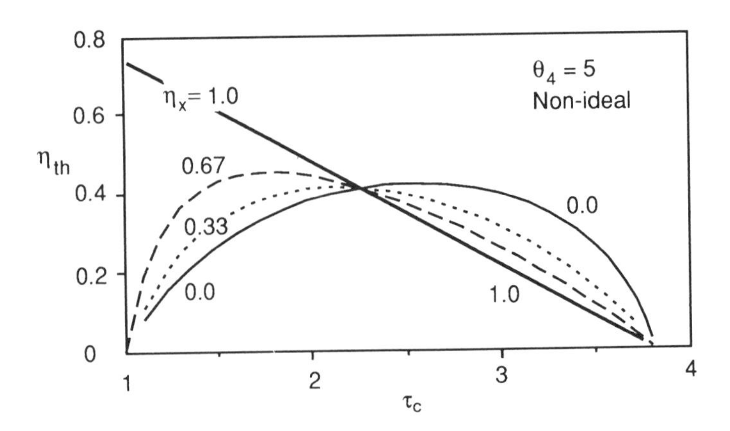

FIG. 9–9. Variation of thermal efficiency for specific regenerator effectiveness. $\eta_c = 0.85$, $\eta_t = 0.90$.

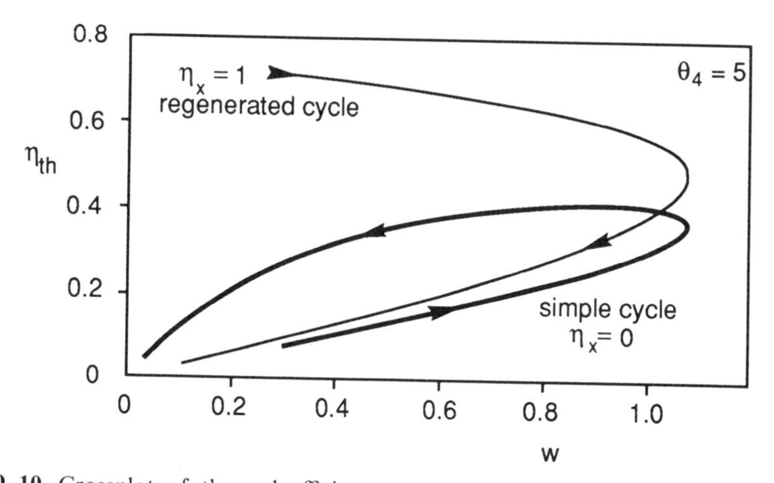

Fig. 9–10. Crossplot of thermal efficiency and specific work for simple cycle and regenerated cycle. Arrows show direction of increasing cycle pressure ratio. Parameters are identical to those in Fig. 9–9.

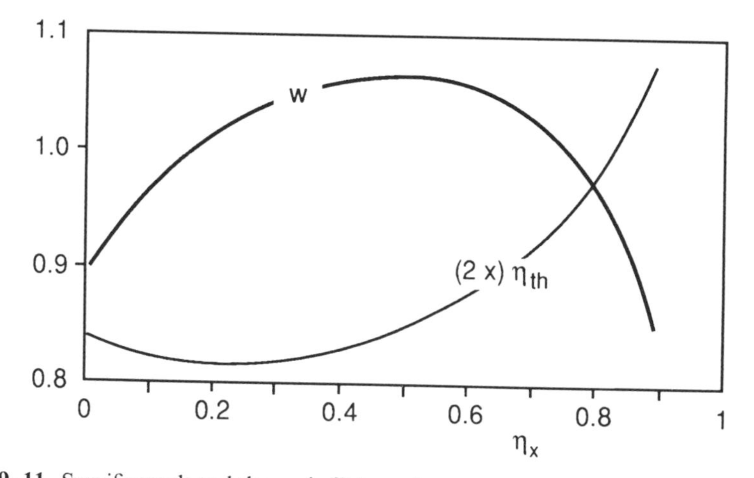

Fig. 9–11. Specific work and thermal efficiency for a regenerated Brayton cycle optimized in pressure ratio so that thermal efficiency is maximized. Heat exchanger effectiveness is varied. Parameters are identical to those given in Fig. 9–9.

nearly linear relationship between η and w for the intermediate values of η_x. For an effective regenerator a relatively low value of τ_{cs} is desirable, whereas for the simple cycle a higher pressure ratio gives high w and high efficiency. The regenerated cycle can always be designed to be superior to the simple cycle, although the practical considerations of the heat exchanger (size, weight, or cost) may limit its employment.

With η_x varying between its practical limits, one can plot the maximum thermal efficiency (larger for higher η_x) and the specific work at the maximum efficiency point which has the undesirable characteristic of falling rapidly when η_x approaches unity (Figs. 9–11 and 9–12). The pressure ratio required for maximum efficiency is shown in Fig. 9–13 for $\gamma = 1.4$.

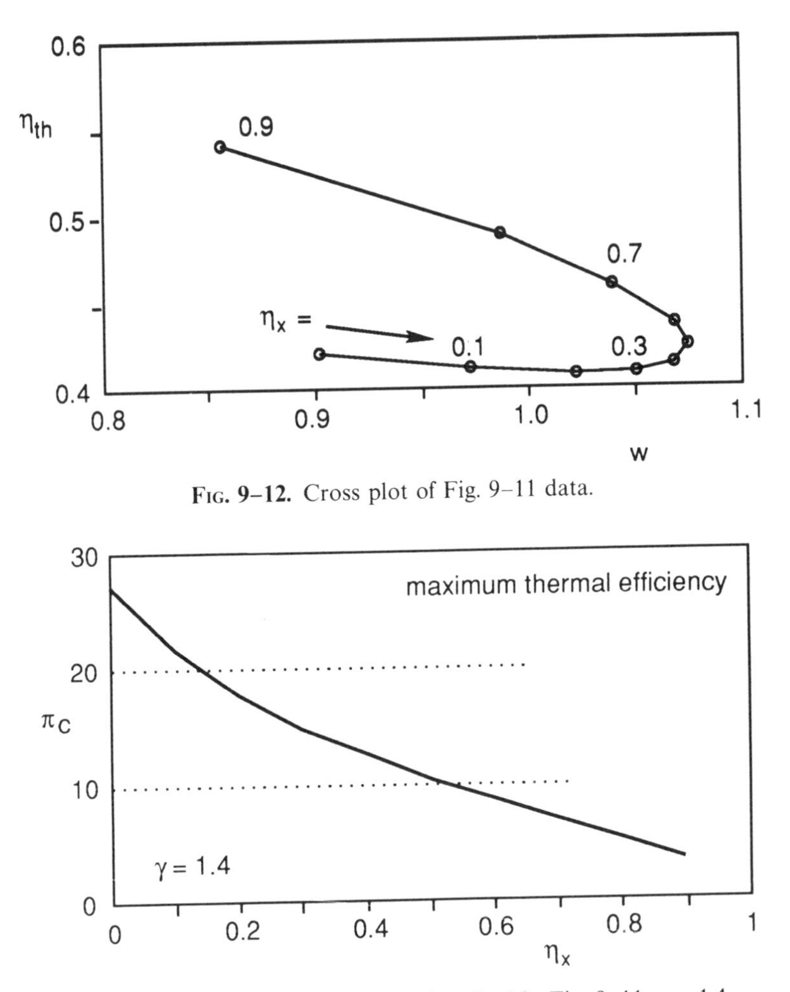

Fig. 9–12. Cross plot of Fig. 9–11 data.

Fig. 9–13. Pressure ratio for cycle described in Fig. 9–11. $\gamma = 1.4$.

9.4. ANALYSIS OF THE SIMPLE CYCLE WITH POLYTROPIC EFFICIENCIES

Polytropic efficiencies are better suited to describing families of compressors or compressors with a wide range of pressure ratio than the adiabatic efficiencies (see section 7.6). Thus this section addresses the design problem of choosing the proper values of the parameters. The combustor is again designed so that $p_4 = p_2$, in which case the cycle is modified as shown by the dotted lines in Fig. 9–3. When a polytropic efficiency characterizes the compression process, it is more convenient to use τ_c rather than τ_{cs} as a design variable. The temperature ratio is related to the pressure ratio (eq. 7–48) for constant e_c:

$$\frac{p_2}{p_0} = \tau_c = (\tau_c)^{\gamma e_c/(\gamma - 1)} \equiv (\tau_c)^{c_c} \tag{9–28}$$

where c_c and similar polytropic parameters are defined as follows:

$$c_c \equiv \frac{e_c\gamma}{\gamma - 1}; \qquad c_t \equiv \frac{1}{e_t}\frac{\gamma_t}{\gamma_t - 1}; \qquad \text{and} \qquad e_{ct} \equiv \frac{c_c}{c_t} = e_c e_t \frac{\gamma}{\gamma - 1}\frac{\gamma_t - 1}{\gamma_t} \tag{9–29}$$

The t subscript is used to describe the gas processed in the expansion process. The constants defined in eqs. 9–29 are convenient parameters for description of the compression and expansion processes. For ease of interpretation in the work to follow, note that c_c is of order 3–3.5 while e_{ct} is somewhat less than unity since it is approximately equal to the product of the two polytropic efficiencies.

With these definitions, the compressor and expander works are:

$$w_c = \tau_c - 1 \qquad \text{and} \qquad w_t = \theta_4(1 - \tau_c^{-e_{ct}}) \tag{9–30}$$

The net specific work, w, is then $w_t - w_c$. The value of τ_{cw} is then determined from

$$\frac{dw}{d\tau_c} = 0 \qquad \text{or} \qquad \tau_{cw} = (e_{ct}\theta_4)^{(1/(1+e_{ct}))} \tag{9–31}$$

and it follows that

$$w_{max} = \theta_4 - \left(1 + \frac{1}{e_{ct}}\right)(e_{ct}\theta_4)^{(1/(1+e_{ct}))} + 1 \tag{9–32}$$

Plots of specific work and efficiency for the case of equal compression and expansion polytropic efficiencies are shown in Figs. 9–14 and 9–15. The pressure ratio corresponding to the maximum w case is shown in Fig. 9–16 for

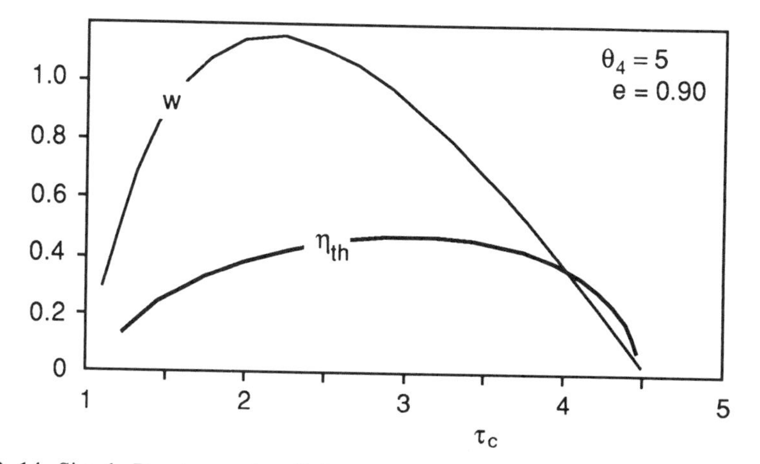

Fig. 9–14. Simple Brayton cycle: efficiency and specific work variation with *polytropic* work efficiencies used to characterize the component performance. Compare with Figs. 9–4, 9–8, and 9–9.

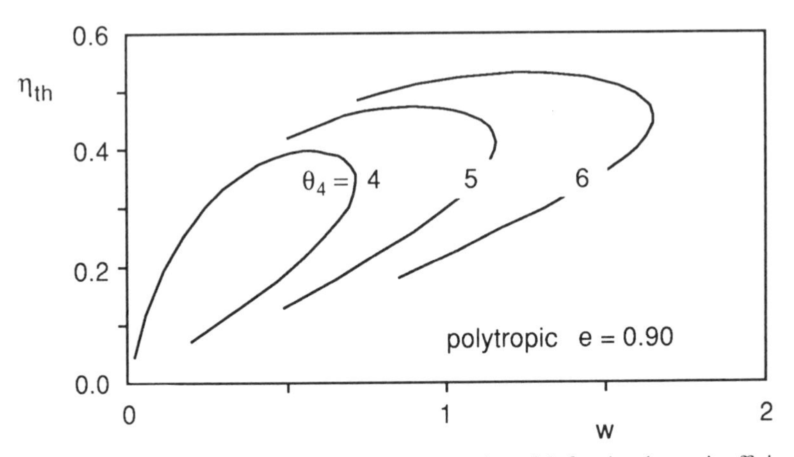

FIG. 9–15. Simple Brayton cycle performance crossplot with fixed polytropic efficiencies. Compare with Fig. 9–10.

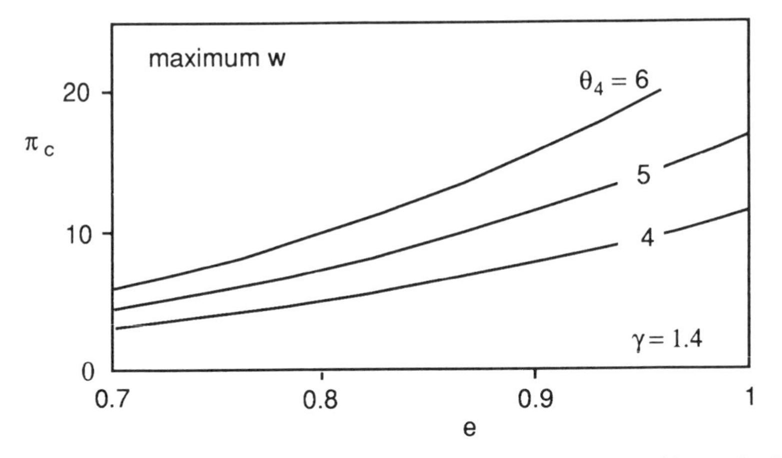

FIG. 9–16. Simple Brayton cycle pressure ratio for maximum specific work. Equal polytropic efficiencies.

the general and the ideal cases. The significant difference between these results and the ideal calculation is the reduced efficiency at large τ_c. The maximum efficiency is usually encountered at a higher τ_c than τ_{cw}. The values of specific work and efficiency that result from designing a cycle for maximum efficiency are shown in Fig. 9–17. Similarly the maximum specific work cycle result is also shown. Note at the lower component efficiencies, the distinction between the two maxima becomes less important.

The role of work component polytropic efficiencies is obviously important. The magnitude of the polytropic efficiencies determines not only the magnitude of the maximum specific work and efficiency, as seen in Fig. 9–17 but also the pressure ratio that must be chosen to obtain the maximum specific work shown in Fig. 9–16. For designing an engine, knowing the magnitude of the component efficiency is thus doubly important. This is illustrated in Fig. 9–18 where curves

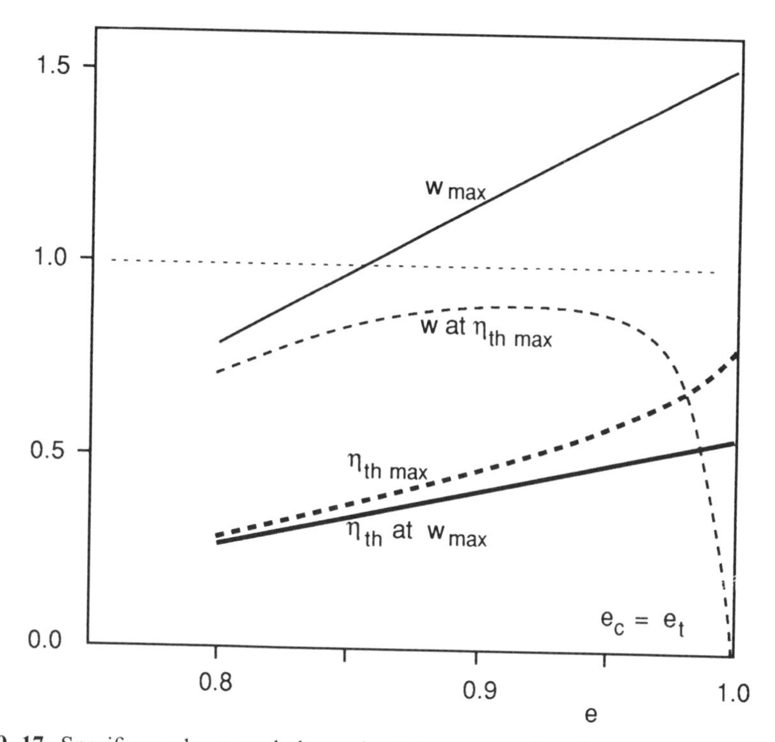

Fig. 9–17. Specific work, w, and thermal efficiency as a function of equal polytropic efficiencies (whose magnitude is plotted on the abscissa) for a simple cycle *designed for maximum efficiency*: dashed lines, and for a cycle *designed for maximum specific work*: solid lines. $\theta_4 = 5$.

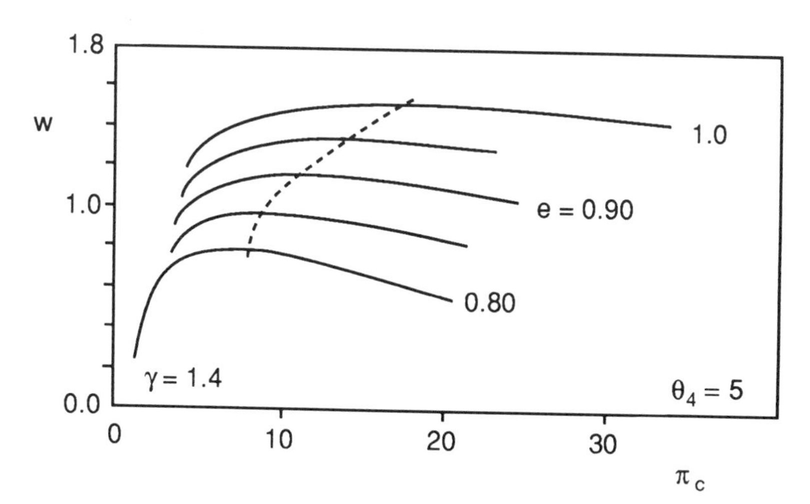

Fig. 9–18. Specific work as a function of pressure ratio showing the sensitivity to (equal) component efficiencies.

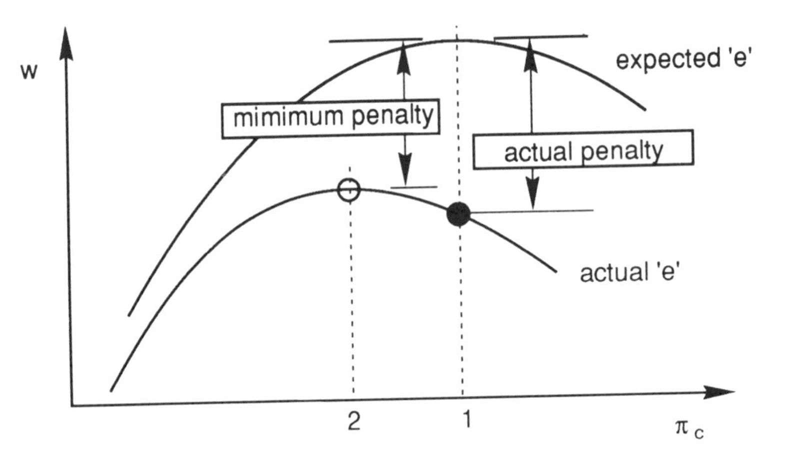

FIG. 9–19. Expansion of Fig. 9–18 showing the penalty of uncertainty in component efficiencies.

of specific work are shown as compressor total pressure ratio is varied for $e_c = e_t$. τ_c may be chosen on the expectation that $e_c = e_t = 0.95$ leads to a particular value of specific work. If the actual efficiencies were found to be 0.90 then w for the cycle would be lower than the value that would be obtained had the efficiencies been recognized to be 0.90. This sensitivity is quite high when the component efficiencies are high and a maximum thermal efficiency design is to be identified. Figure 9–19 shows an expansion of the data of Fig. 9–18 with an identification of the penalties associated with design of a cycle for an improperly chosen pressure ratio based on an invalid value of component efficiencies.

9.5. TURBINE COOLING WITH COMPRESSOR BLEED

The performance of the Brayton cycle obviously varies with expander inlet temperature in an important way: Both specific work and efficiency depend on the achievable value of the parameter θ_4. The question naturally arises: Could working gas from the compressor be used to cool the structure of the expander in order to raise the effective expander inlet temperature and thus improve cycle performance? The cycle considered is shown in Fig. 9–20 which illustrates the removal of bleed gas from the compressor that is used to cool the turbine in an aerodynamic expander. Let μ be the mass fraction of the bleed flow so that the remainder $(1-\mu)$ is processed by the combustor and expander in the normal way. For the modeling of this situation, three issues must be addressed:

1. What is the pressure of the bleed gas? If transpiration cooling is to be used, for example, then the total pressure of the bleed gas should be quite close to turbine inlet total pressure, that is, the bleed gas would have to be compressed through the full cycle pressure ratio, as shown in the sketch.

2. How much work can be extracted from the bleed flow gas after it has

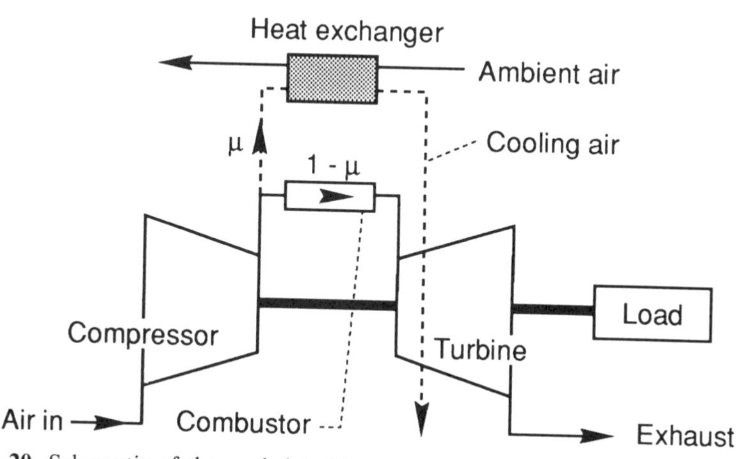

Fig. 9–20. Schematic of the cooled turbine engine, μ is the fraction of compressed air withdrawn from cycle for cooling.

done its cooling task and has been mixed back into the main turbine flow? For example, if the cooling gas experiences a small pressure loss and is used in only the first turbine stages, then some work might be recovered from later stages of the flow. On the other hand, if the cooling is by convection, then the pressure remaining in the flow might be quite low so that the work recovered is small. A conservative modeling approach is to neglect the possibility of work recovery from the bleed gas.

3. How much bleed gas is required to achieve a unit increase in turbine inlet temperature? The answer to this question depends on the temperature of the coolant gas that may be fixed at a value between ambient requiring the heat exchanger shown in Fig. 9–20 and the compressor outlet temperatures, in which case the exchanger is unnecessary. The amount of coolant gas required as a fraction of the turbine flow rate depends also on the design details of the cooling process. Figure 9–21 shows a typical coolant fraction for an open cycle turbine in terms of the parameters noted.

It is this last question that is the critical one when optimizing system performance. For that purpose, the required assumptions associated with questions 1 and 2 above are dealt with by assuming the following:

1. Bleed gas pressure ratio equals cycle pressure ratio
2. Zero work is recovered from the bleed gas.

A number of approaches to finding a relation between μ and θ_4 may be cited. These arise from experimental observation of the performance of various cooling techniques and/or from satisfaction of the relevant conservation laws. Any observed variation between coolant mass flow required to achieve an elevation in the effective turbine inlet temperature depends on the details of the design. Figure 9–21 shows a representative relation between μ and θ_4 for a specific design. This kind of relation can be used to estimate the size of parameters useful for the analysis.

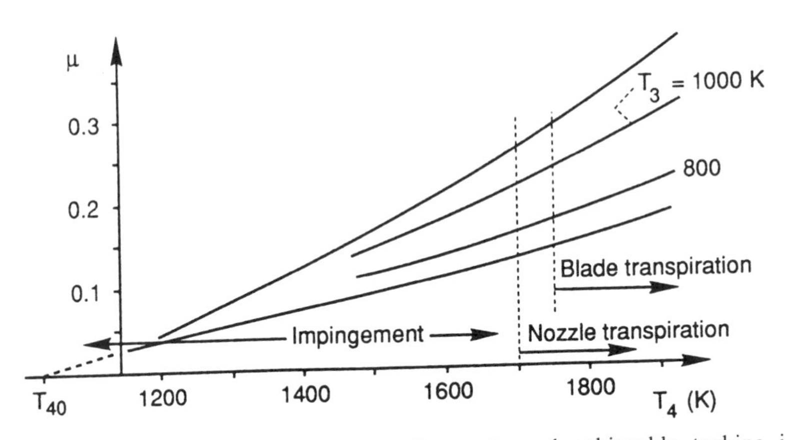

Fig. 9–21. Relation between coolant mass flow rate and achievable turbine inlet temperature in an aircraft gas turbine of a typical design. Temperature values are nominal since they change with technological development. The notation refers to the various cooling techniques which are necessary to achieve the performance level indicated.

For the purpose of examining the important effects one may assume a linear relation of the form:

$$\mu = a\theta_2(\theta_4 - \theta_{40}) \tag{9–33}$$

where a is a constant that must be determined from the experimental data, and θ_{40} is the nondimensional maximum temperature that can be experienced by the turbine blade metal. From Fig. 9–21, $a \sim 0.04$ and T_{40} is noted (~ 1050 K).

9.5.1. *Analysis*

For ideal components, the compressor work and turbine work are given by

$$w_c = \tau_c - 1 \qquad \text{and} \qquad w_t = (1 - \mu)\theta_4\left(1 - \frac{1}{\tau_c}\right) \tag{9–34}$$

The net work from the cycle is therefore (with eq. 9–33):

$$w = w_t - w_c = (1 - \mu)\left(\theta_{40} + \frac{\mu}{a\theta_2}\right)\left(1 - \frac{1}{\tau_c}\right) - (\tau_c - 1) \tag{9–35}$$

This expression can be differentiated with respect to μ to determine the value of μ which gives maximum work. The result is

$$\mu_w = \frac{1 - a\theta_2\theta_{40}}{2} \tag{9–36}$$

and the corresponding value of the maximum work is

$$w' = \frac{1}{a\theta_2}\left(\frac{1 + a\theta_2\theta_{40}}{2}\right)^2\left(1 - \frac{1}{\tau_c}\right) - (\tau_c - 1) \tag{9–37}$$

This relation is equivalent to the expression for the specific work of a Brayton cycle (eq. 9–9), except that the effective maximum cycle temperature is

$$\theta_{4\text{eff}} = \frac{1}{a\theta_2}\left(\frac{1 + a\theta_2\theta_{40}}{2}\right)^2 \tag{9–38}$$

which is greater than θ_{40} when

$$a\theta_2\theta_{40} \leq 1 \tag{9–39}$$

Thus the cooling technique employed must be good enough for a to have the smallest possible positive value that leads to satisfaction of the condition given by eq. 9–39.

9.5.2. *Coolant Gas Temperature*

The coolant gas may vary between ambient and the value exiting the compressor. In nondimensional terms, this means $1 \leq \theta_2 \leq \tau_c$ and the effectiveness of the cooling heat exchanger, η_x, may be used as a parameter. Thus

$$\theta_2 = \eta_x + (1 - \eta_x)\tau_c \tag{9–40}$$

gives $\eta_x = 0$ for no cooling of the compressed coolant flow and $\eta_x = 1$ for an extremely effective heat exchanger. In the following, these limited cases will be referred to as "warm" (to contrast with the "hot" turbine inlet temperature) and "ambient," respectively. With a chosen level of heat exchanger performance, the specific work, w', is a function of only the compressor temperature ratio, τ_c. The variation of w' (the specific work with the coolant flow chosen to maximize it) is shown in Fig. 9–22. The performance evidently depends on good heat exchanger effectiveness, and this improved performance requires a design with higher pressure ratio (τ_c).

9.5.3. *Performance*

The heat input to the cycle is

$$q = (\theta_4 - \tau_c)(1 - \mu) \tag{9–41}$$

from which the efficiency ($\eta_{th} = w'/q$) can be calculated. The result is shown in Fig. 9–23. Note that high thermal efficiency is obtained for $\eta_x = 0$. For $\eta_x = 1$, performance is also quite good, but somewhat lower, for intermediate values.

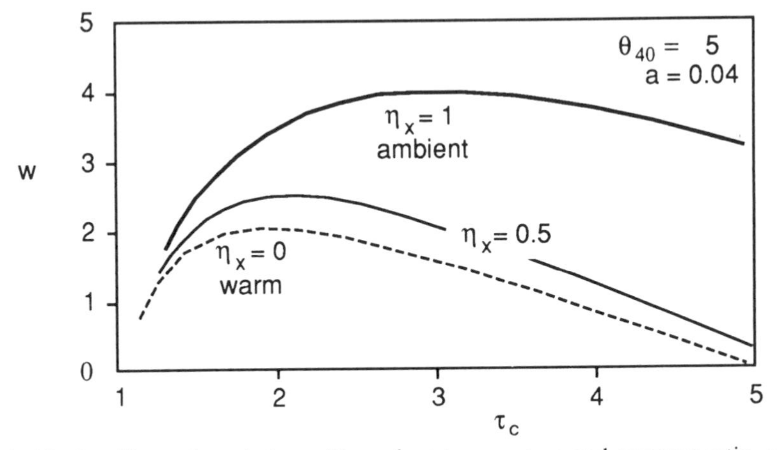

FIG. 9–22. Specific work variation with coolant temperature and pressure ratio. $\eta_x = 0$ corresponds to uncooled air used in the turbine (warm coolant), while $\eta_x = 1$ corresponds to air cooled to ambient temperature (ambient coolant). Coolant flow rate is optimized to maximize specific work. Ideal work components.

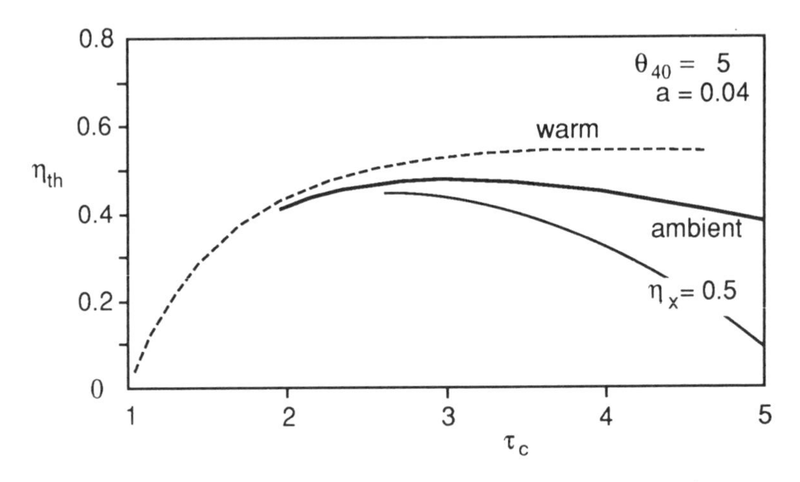

FIG. 9–23. Thermal efficiency variation with coolant temperature and pressure ratio. Limitations identified in Fig. 9–22 also apply here.

This is illustrated in the crossplot of Figs. 9–22 and 9–23, shown as Fig. 9–24. These results suggest that the colder the coolant, the more the design points for maximum efficiency and maximum specific work coincide. For the uncooled coolant ($\eta_x = 0$) the values of τ_c for the two maxima are more separated.

The effective turbine inlet temperature (eq. 9–38) and the mass flow rate (eq. 9–36) corresponding to the conditions above are shown in Figs. 9–25 and 9–26. Evidently the effective turbine inlet temperature is elevated significantly, depending on the coolant temperature and the cycle pressure ratio. The benefit to the warm coolant case decreases toward higher cycle pressure ratio. In practice, a limit on effective θ_4 limits the choice of options. For the

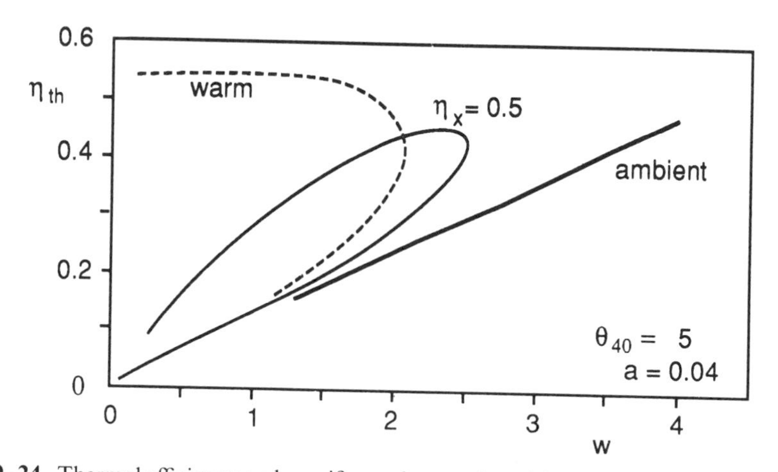

FIG. 9–24. Thermal efficiency and specific work crossplot of data in Figs. 9–22 and 9–23.

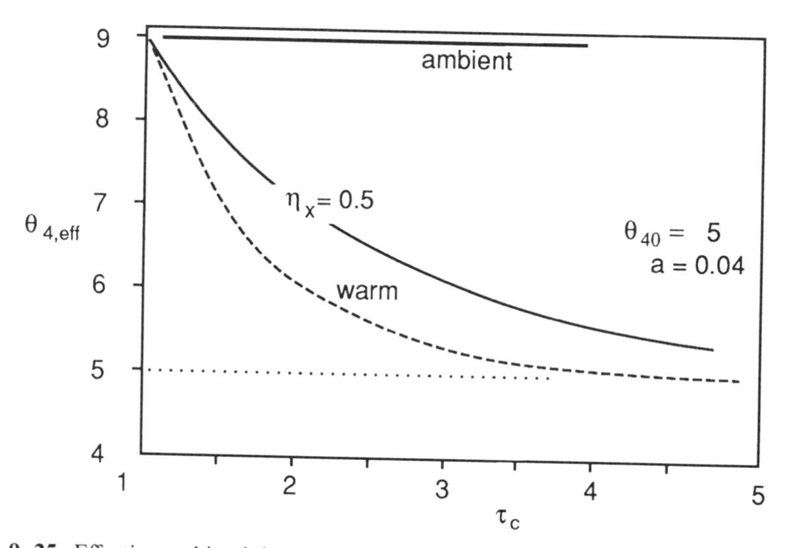

FIG. 9–25. Effective turbine inlet temperature. Limitations identified in Fig. 9–22 apply here.

ambient coolant case, large mass flows are required for this maximum work case. Irreversibilities normally present in such engines will alter these conclusions to some degree, especially for the higher pressure ratio designs.

9.5.4. *Limited Maximum Temperature*

Recognizing that there are limits to the peak cycle temperature that are independent of considerations of the turbine cooling (combustion chemistry, for example), the question may be asked how an upper limit on the temperature affects cycle performance. For stated levels of θ_4 and of θ_{40}, the bleed mass flow rate is fixed by the temperature of the coolant, as seen from

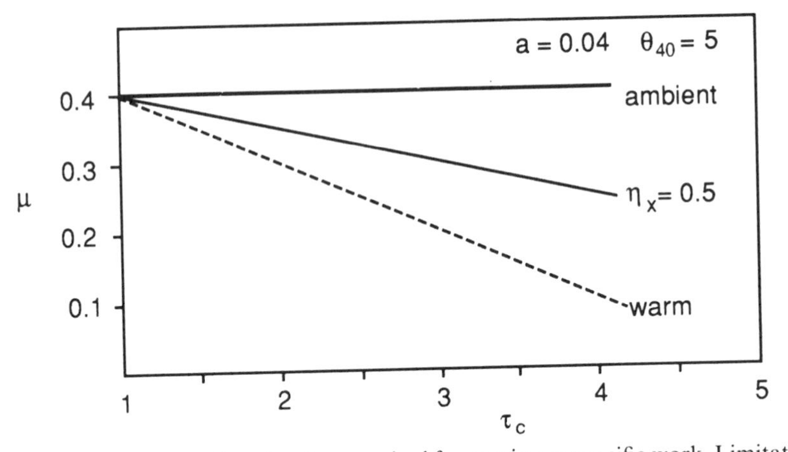

FIG. 9–26. Coolant mass flow fraction required for maximum specific work. Limitations identified in Fig. 9–22 also apply here.

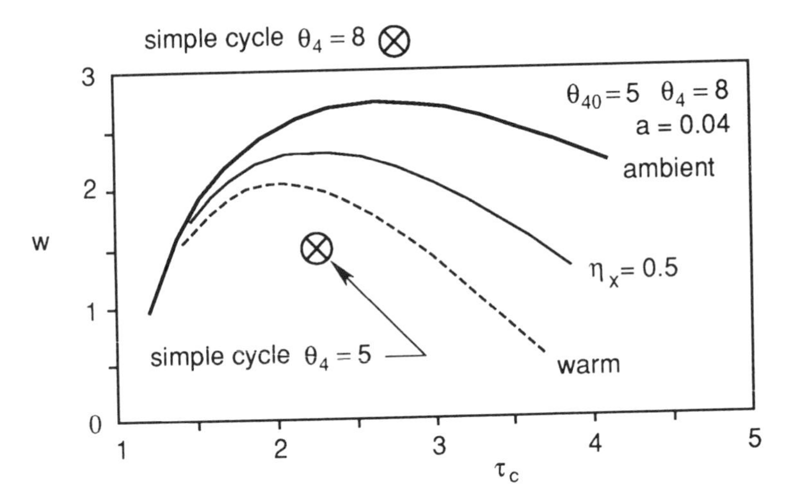

FIG. 9–27. Cycle performance with limited effective peak cycle temperature: specific work. Coolant mass flow fraction determined by temperature limits indicated. Ideal components. The points identify the performance of the simple cycle operated at the two-peak cycle temperatures: $\theta_{40} = 5$ and $\theta_4 = 8$.

eq. 9–33. Thus,

$$\mu = a(\theta_4 - \theta_{40})(\eta_x + (1 - \eta_x)\tau_c) \tag{9–42}$$

Substituting this into the equation for the specific work and efficiency gives the results shown in Figs. 9–27 to 9–30. Figure 9–27 shows that the "pressure ratio," τ_c, for maximum specific work varies with coolant temperature. Note that for the simple cycle, this pressure ratio is $\sqrt{5} = 2.24$ or $\sqrt{8} = 2.82$ for the two maximum cycle temperatures. Of real interest is that the value of the

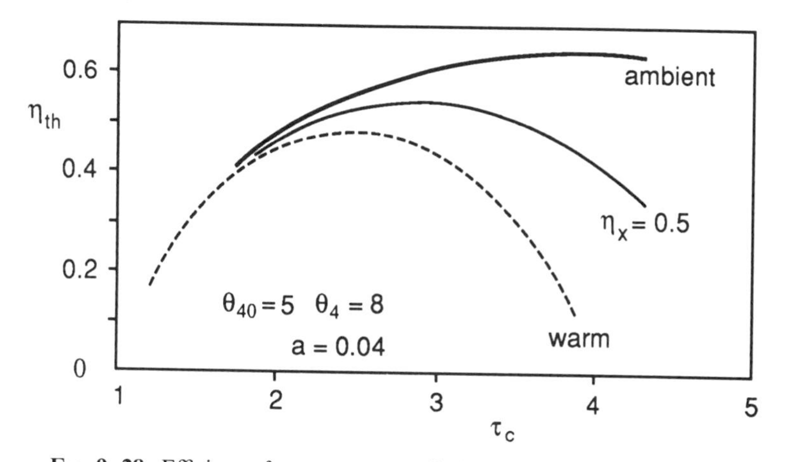

FIG. 9–28. Efficiency for temperature limits indicated in Fig. 9–27.

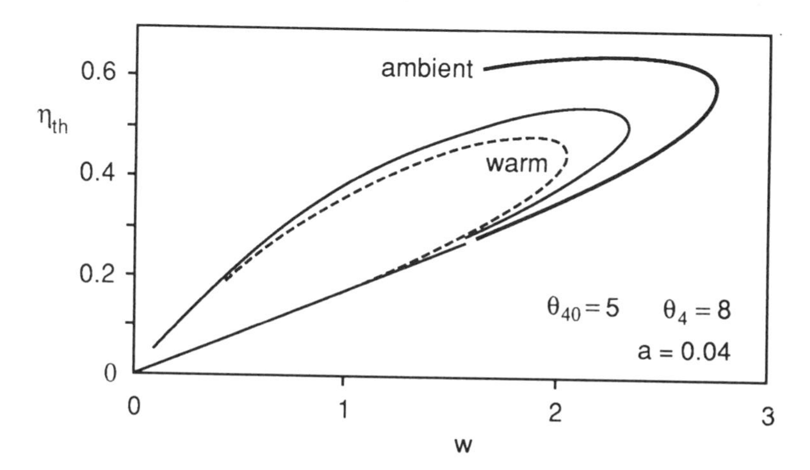

FIG. 9–29. Specific work-efficiency crossplot for temperature limits indicated in Fig. 9–27.

maximum specific work for the simple cycle is 1.52 for the lower θ_4 case, which is the one to which a comparison is fair. The points on Fig. 9–27 identify the performance of these two cases: the use of cooling flow allows one to approach much better performance than for the uncooled case.

The variation of thermal efficiency differs from that of the simple uncooled cycle (Fig. 9–4) by having a maximum, which the uncooled cycle has only in the Carnot limit of very high pressure ratio. Relative to the value of "pressure ratio" for maximum specific work, that for maximum efficiency is reached for somewhat higher values of τ_c as shown in Fig. 9–28. Figure 9–29 is a cross plot of the data in Figs. 9–27 and 9–28. Here, in contrast to the situation where peak cycle temperature was not constrained, the pressure ratios for maximum efficiency and maximum specific work are closer for the warm coolant case compared to the ambient coolant case.

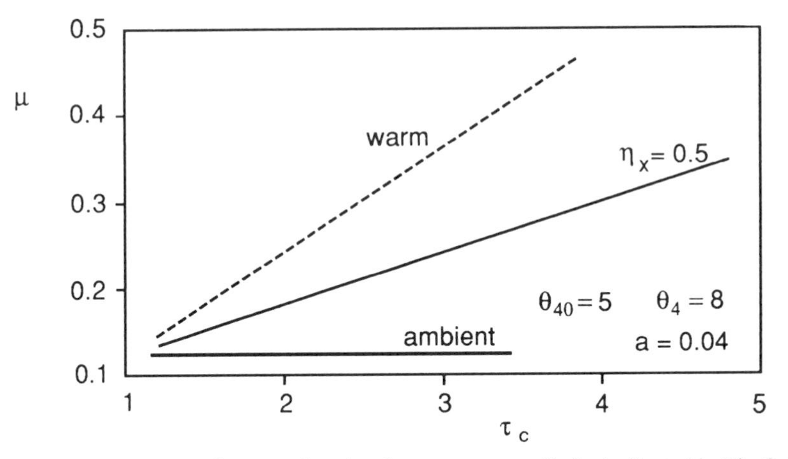

Fig. 9–30. Coolant mass flow rate fraction for temperature limits indicated in Fig. 9–27.

The fractional mass flow rate that is implicit in these results is shown in Fig. 9–30. Note that for the ambient coolant, the fraction required is independent of cycle pressure ratio, as expected, while for the warm coolant, more is required with increasing pressure ratio.

Note that also in the context of a limited effective peak cycle temperature, it pays to cool the cooling gas, which reduces the need for gas bleed and improves performance in terms of both specific work and efficiency. The pressure ratio for best performance is higher for the cooled coolant than it is for the uncooled. For the uncooled case, both maximum w and efficiency design points tend to lie closer together, making the identification of the best design rather easier.

For a realistic system, these results must be calculated for reasonable values of the work component efficiencies, but it is to be expected that significant performance, efficiency, and specific work may be realized by the use of coolant to increase the turbine inlet temperature. The analysis also shows the sensitivity to the constant a, which should ideally be as small as possible to minimize the coolant flow rate required.

9.6. NONADIABATIC WORK PROCESS

9.6.1. *Intercooling*

Since the simple Brayton cycle uses a high pressure ratio, one might wonder about the effectiveness of cooling the compressed gas during the compression process to reduce the work required. A two-step compression process involves a single *intercooling* stage. The process consists of compression to a fraction of the final pressure, cooling back to (near) ambient conditions by means of a heat exchanger and finally continued compression of the gas. The work of compression is reduced because the work for two equal intercooled pressure ratio stages is less than the work for a single-step compression. For a cycle with a given pressure ratio, the compression work per unit mass flow rate is given by (see eqs. 9–1 or 9–7):

$$w_{c1} = \tau_c - 1$$

For a cycle with two equal pressure ratio compression stages the pressure ratio (and therefore the temperature ratio) per stage is $\sqrt{\tau_c}$ so that the work required is

$$w_{c2} = 2(\sqrt{\tau_c} - 1) \tag{9–43}$$

The ratio of these two works is given by

$$\frac{w_{c2}}{w_{c1}} = \frac{2}{\sqrt{\tau_c} + 1} < 1$$

The (log)T-s diagram for this process is shown in Fig. 9–31. The cycle thus promises to produce greater work while suffering a penalty in size, complexity, and cost associated with the use of the heat exchanger.

Incorporating these ideas into a cycle analysis as carried out above, including the polytropic efficiencies to characterize the work components, results in the specific work versus efficiency curves shown in Figs. 9–32 and 9–33. To allow an assessment of the role of the intercooler, a comparison between two cycles is shown in the figures; one shows performance without intercooler (9–32) whereas the other includes an intercooler as the only cycle modification (9–33). In these figures the lines of constant $T_4/T_0 = \theta_4$ are like the ones shown in Fig. 9–13. Since pressure varies along these curves, points of a chosen cycle pressure ratio are connected and labeled. For the simple cycle, a temperature ratio $\theta_4 = 5$ or so would dictate choosing the pressure ratio to be near 20. For that same pressure ratio, with intercooling, the specific work

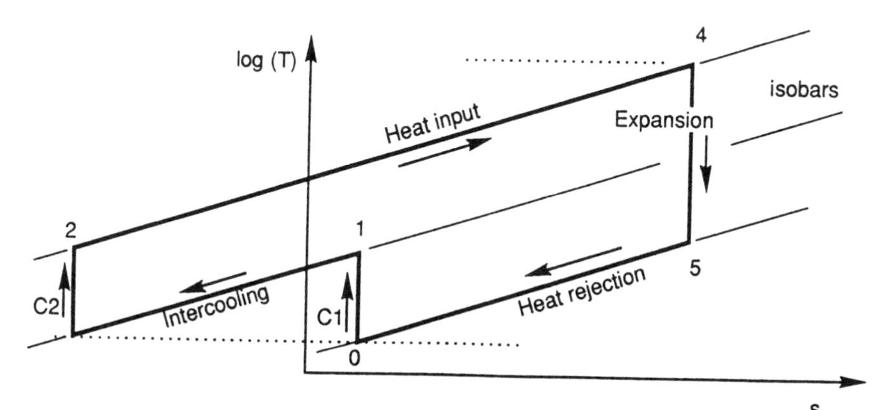

Fig. 9–31. log(T)-s diagram for a Brayton cycle with two (intercooled) compression stages.

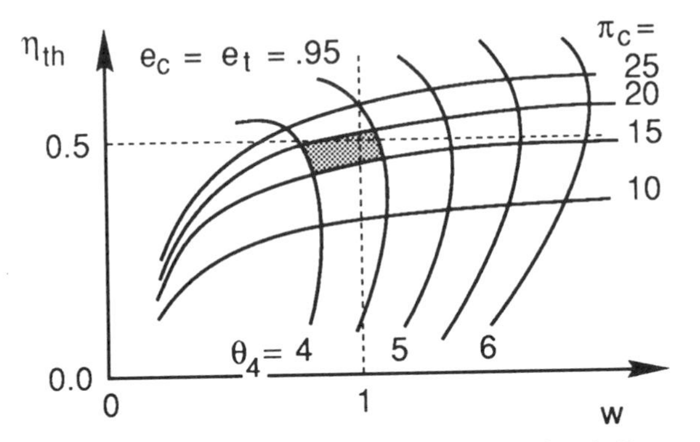

Fig. 9–32. Efficiency-specific work cross plot for a simple Brayton cycle.

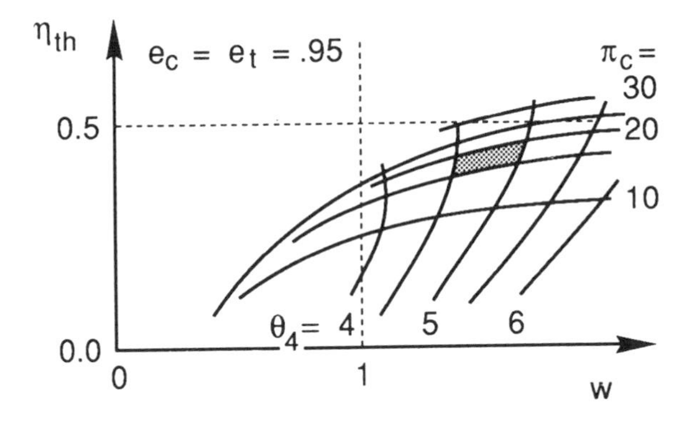

Fig. 9–33. Efficiency-specific work crossplot for a simple cycle with one intercooler.

increases but efficiency falls as shown in Fig. 9–33. The reason is that more heat can be added to the cycle fluid between the *lower* compressor outlet temperature and the fixed expander inlet level, θ_4. The real benefit of the intercooler is realized when the pressure ratio is raised to the vicinity of 30 or more, in which case a mild efficiency benefit is realized, but more importantly, one obtains a large increase in specific work.

9.6.2. *Reheat*

A process that is similar to intercooling may also be carried out at the heat input side of the cycle. Figure 9–34 shows the *T-s* diagram where the process of heat addition (3–4) is followed by partial expansion (4 to 4.1) which reduces the temperature allowing a further addition of heat (4.1 to 4.2) presumably again to the same maximum cycle temperature. The second expansion (4.2 to 5) completes the process. The process of reheating leads to a temperature of T_5, which is higher than it would be without reheating; this provides a larger temperature range available for regeneration. Figure 9–35 shows two cycles designed for the same overall pressure ratio and limited by the same overall

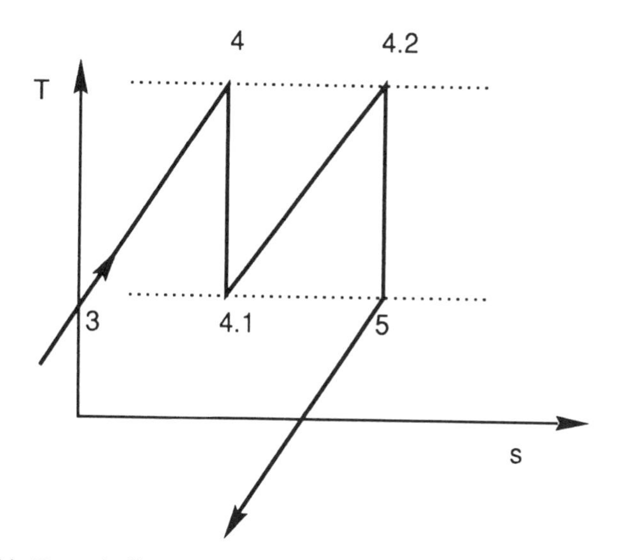

FIG. 9–34. *T-s* path diagram for the reheat process. Sloped lines are isobars.

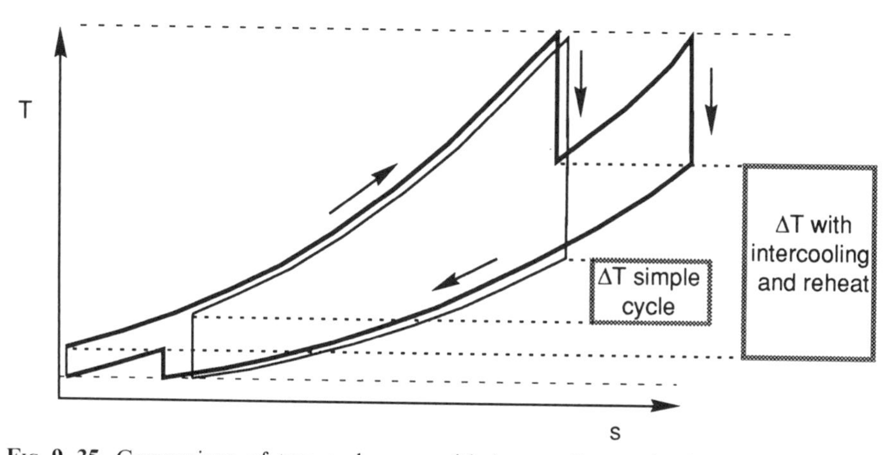

FIG. 9–35. Comparison of two cycles one with intercooling and reheat showing the larger ΔT across the regenerator available when using intercooling and/or reheat.

temperature ratio θ_4, one with no intercooling or reheat and another with one stage of each. From a Carnot point of view, the heat input is added at a relatively higher temperature with reheat and rejected at a lower temperature with intercooling so that the efficiency is better than for the simple cycle. Note the larger amount of heat (ΔT) is indicated) handled in the regeneration process in the case of the engine with intercooling and reheat. Such a comparison may be inappropriate since one would probably not choose two such cycles to have identical cycle pressure ratios. Nevertheless, this argument can be extended to a larger number of stages to realize an efficiency closer to a Carnot efficiency. In the limit, the heat addition and rejection

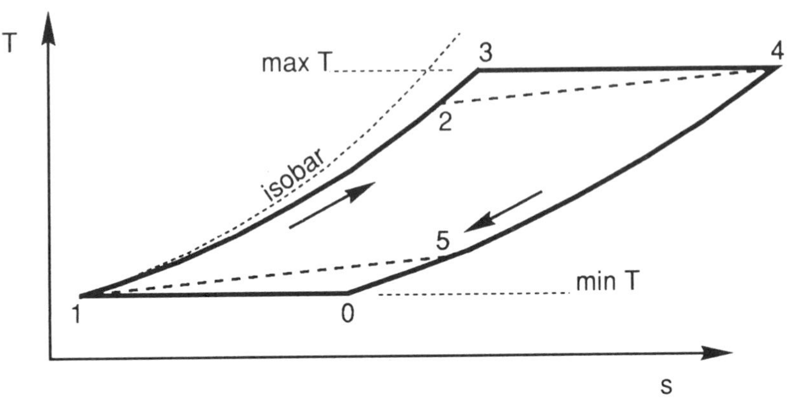

Fig. 9–36. The Ericsson cycle on a *T-s* diagram.

processes are carried out isothermally and the cycle is termed the Ericsson cycle shown in Fig. 9–36. Such cycles are generally interesting only when regeneration is employed.

9.7. WAVE ENERGY EXCHANGER

The wave energy exchanger described in section 8.10 for charging of the ICE may also be used in the Brayton cycle to circumvent peak cycle temperature limitations imposed by the steady flow turbine. Weatherston and Hertzberg (ref. 9–1) describe the wave energy exchanger as an energy conversion device and how it may be employed in a cycle to improve thermal efficiency. Based on their work efficiencies in excess of 50% were judged possible with (then) realistic conventional component performance assumptions.

9.8. THE ERICSSON CYCLE

In the limit of a large number of intercooling and reheat stages, the compression and expansion processes become isothermal, with heat removal during compression and input during expansion. Regeneration is obviously required. Furthermore, in a practical setting the regeneration process involves pressure loss and effectiveness which are appropriate for inclusion in the analysis. In a closed Ericsson cycle, the capacity rate ratio is unity (see eq. 6–9) so that the effectiveness given by eq. 9–16 or equivalently for the state point labels in Fig. 9–36 is:

$$\eta_x = \frac{T_2 - T_1}{T_4 - T_1} \tag{9–44}$$

Note that, in general, the pressure falls from 1 to 2 and 4 to 5.

The Ericsson cycle performance is evaluated by applying the First Law to the cycle elements and evaluating the magnitude of the heat interactions. For the isothermal process (3 to 4) the heat input is

$$Q_{34} = RT_3 \ln\left(\frac{p_3}{p_4}\right) \tag{9-45}$$

from the First Law written for a constant temperature process (eq. 4–12). Additionally, an input at approximately constant pressure from 2 to 3 is required:

$$Q_{23} = C_p(T_3 - T_2) = C_p\Delta T = C_pT_0(\theta_4 - 1)(1 - \eta_x) \tag{9-46}$$

where ΔT is associated with the ineffectiveness of the regenerator (see eq. 6–31). The total heat input is

$$Q_{in} = Q_{23} + Q_{34} = C_p\Delta T + RT_3 \ln\left(\frac{p_3}{p_4}\right)$$

Similarly the heat rejected is

$$Q_{out} = C_p\Delta T + RT_0 \ln\left(\frac{p_1}{p_0}\right) \tag{9-48}$$

Let the pressure ratio $p_1/p_0 = \pi_c$. The flow through the heat exchanger will cause the pressure to fall through the cold and hot flow passages. Thus

$$p_3 = p_1 - \Delta p_c \qquad \text{and} \qquad p_0 = p_4 - \Delta p_H \tag{9-49}$$

Combining these gives

$$\frac{p_3}{p_4} = \pi_c(1 - \delta) \tag{9-50}$$

where

$$\delta = 1 - \frac{1 - \dfrac{\Delta p_C}{p_1}}{1 + \dfrac{\Delta p_H}{p_0}} \sim \frac{\Delta p_c}{p_1} + \frac{\Delta p_H}{p_0} \tag{9-51}$$

The last form is valid for small pressure loss fractions. These pressure relations allow the heat input and output to be written nondimensionally as

$$q_i = \frac{\dot{Q}}{\dot{m}C_pT_0} = (\theta_4 - 1)(1 - \eta_x) + \frac{\gamma - 1}{\gamma}\theta_4 \ln\left[\pi_c(1 - \delta)\right] \tag{9-52}$$

and

$$q_0 = (\theta_4 - 1)(1 - \eta_x) + \frac{\gamma - 1}{\gamma} \ln \pi_c \tag{9–53}$$

From the First Law, the net work produced is

$$w = q_i - q_0 = \frac{\gamma - 1}{\gamma}\left[(\theta_4 - 1) \ln \pi_c - \ln\left(\frac{1}{1 - \delta}\right)\right] \tag{9–54}$$

Note that the regenerator effectiveness plays no role in determining the specific work. For small δ, the last term is approximately equal to δ. In the limit of no losses, the specific work becomes

$$w = (\theta_4 - 1) \ln \tau_c \tag{9–55}$$

where $\tau_c = \pi_c^{(\gamma-1)/\gamma}$, which suggests that, for large w, one should design such a cycle to have a high pressure ratio.

The thermal efficiency of the Ericsson cycle is given by:

$$\eta_{th} = \frac{w}{q_i} = 1 - \frac{1}{\theta_4}\left\{\frac{(\theta_4 - 1)(1 - \eta_x) + \ln \tau_c}{\left(\frac{\theta_4 - 1}{\theta_4}\right)(1 - \eta_x) + \ln \tau_c + \ln (1 - \delta)^{(\gamma-1)/\gamma}}\right\} \tag{9–56}$$

In the limit $\eta_x = 1$ and $\delta = 0$, the Carnot efficiency is recovered. Thus, for small losses (η_x near 1 and $\delta \ll 1$), one may linearize the term in brackets in eq. 9–56 around the $\ln \tau_c$ term. In this limit η_{th} becomes

$$\eta_{th} \sim 1 - \frac{1}{\theta_4}\left\{1 + \frac{\frac{\gamma - 1}{\gamma}\delta + (\theta_4 - 1)\left(1 - \frac{1}{\theta_4}\right)(1 - \eta_x)}{\ln \tau_c}\right\} \tag{9–57}$$

Clearly, the importance of the losses should be judged relative to the magnitude of the $\ln (\tau_c)$ term.

9.8.1. *Practical Implementation of the Ericsson Cycle*

A series of compression stages with intercooling is the practical way to execute an approximation to the isothermal process called for in the Ericsson cycle. In such a cycle, the temperature of the working fluid will fluctuate between T_0 and $T_0 + \Delta T_C$ in the compressor and an analogous range in the expander. The regenerator operates between $T_0 + \Delta T_C$ and $T_4 - \Delta T_H$. It is convenient to include the last element of the precooler in the accounting for the compression

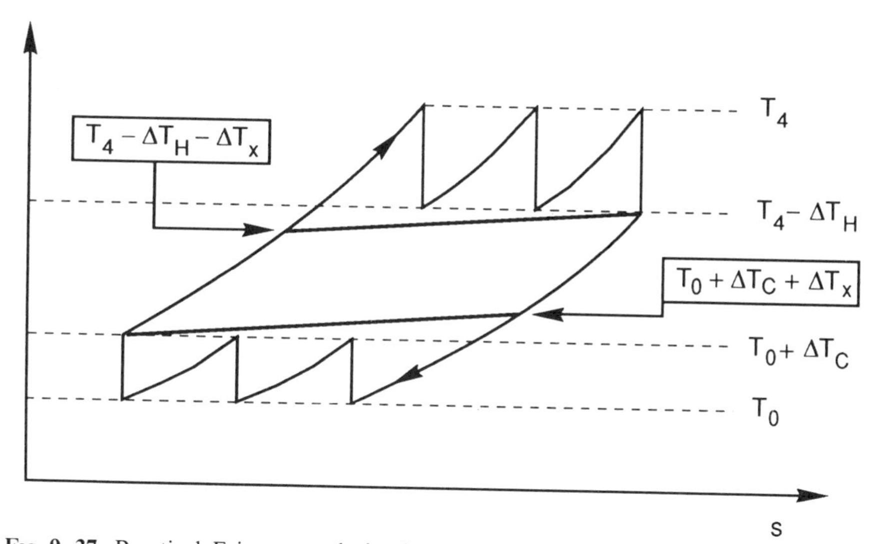

FIG. 9–37. Practical Ericsson cycle implemented with a finite number of reheat and intercooling stages.

process. Figure 9–37 shows the complete cycle and the important temperature levels. The three stages of compression are described in greater detail in Fig. 9–38 which shows the states between compression states, reflecting particularly the pressure loss between compressors.

Compression

The pressure ratio of the nonadiabatic compression process is

$$\pi_c = \left(\frac{p_2}{p_0}\right)_{\text{cycle}} = \frac{p'_4}{p'_0} = \frac{p'_4}{p_3}\frac{p'_3}{p_2} \cdots \frac{p'_1}{p_0} \times \frac{p_3}{p'_3} \cdots \frac{p_0}{p'_0} \tag{9–58}$$

where the first ratios are the increases in the adiabatic elements and the last are the heat exchange pressure losses. In the general case where there are N such processes, each work input pressure ratio is given by

$$\frac{p_{i'}}{p_{i-1}} = \left(\frac{T_1}{T_0}\right)^{c_c} = \left(1 + \frac{\Delta T_C}{T_0}\right)^{c_c} \tag{9–59}$$

Here c_c is as defined in eq. 9–29. Each loss is given by $p_i/p_{i'} = 1 - \delta$, so that the overall pressure ratio is therefore

$$\pi_c = \left[\left(1 + \frac{\Delta T_C}{T_0}\right)^{c_c}(1 - \delta)\right]^N \tag{9–60}$$

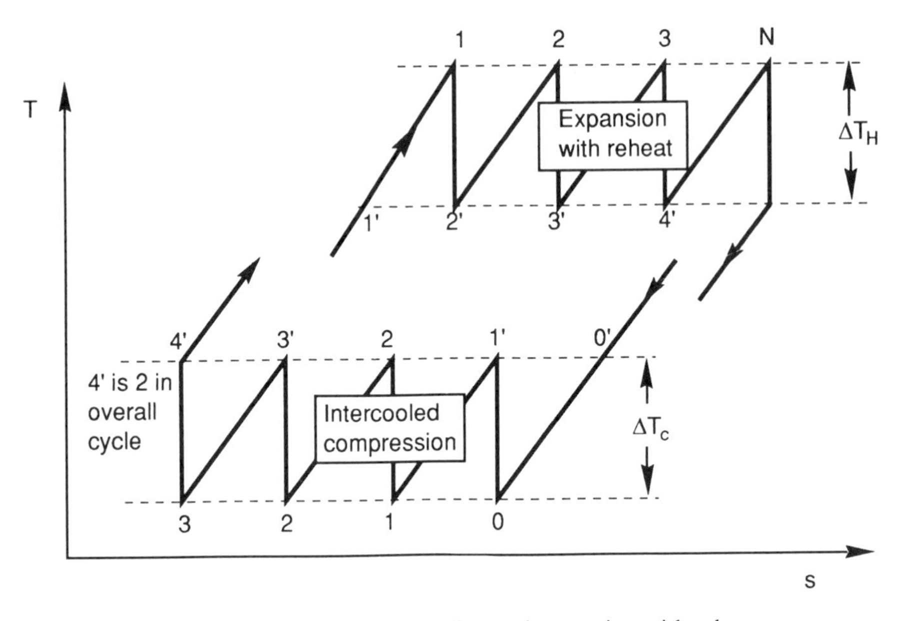

Fig. 9–38. The compression with intercooling and expansion with reheat processes.

from which

$$\frac{\Delta T_C}{T_0} = \left(\frac{(\pi_c)^{1/N}}{1-\delta}\right)^{1/c_c} - 1 \tag{9–61}$$

The total (specific) work compression required is given by

$$w_c = N\frac{\Delta T_c}{T_0} = N\left[\left(\frac{(\pi_c)^{1/N}}{1-\delta}\right)^{1/c_c} - 1\right] \tag{9–62}$$

which becomes, with $\tau_{ca} = (\pi_c)^{1/c_c}$ (the subscript a is meant to indicate "adiabatic" temperature ratio) and $\delta \ll 1$,

$$w_c \cong N\left((\tau_{ca})^{1/N}\left(1 + \frac{\delta}{c_c}\right) - 1\right) \tag{9–63}$$

This expression corresponds to eq. 9–43 derived for $N = 2$. These expressions may be compared with the isothermal work

$$w_c(\text{isothermal}) = \ln \tau_{ca}(\sim RT \ln \pi_c) \tag{9–64}$$

although a direct comparison of the two expressions is not meaningful because the optimum pressure ratios for maximum specific work or maximum efficiency are different for the two cycles. The parenthetical extension to eq. 9–64 carries units and is therefore dimensional.

Expansion

The expansion process can be described in an entirely analogous manner. Let $\pi_t = p_N/p_1(<1$, see Fig. 9–38), then the pressure losses during reheat are

$$\frac{p_i'}{p_i} = 1 + \delta \tag{9–65}$$

and the expansion pressure ratios are, with the definitions in eq. 9–29:

$$\frac{p_{i+1}'}{p_i} = \left(1 - \frac{\Delta T_H}{T_4}\right)^{\gamma_t/(\gamma_t - 1)e_t} = \left(1 - \frac{\Delta T_H}{T_4}\right)^{c_t} \tag{9–66}$$

the overall expansion pressure ratio becomes

$$\pi_t = \left(1 - \frac{\Delta T_H}{T_4}\right)^{Nc_t}\left(\frac{1}{1 + \delta}\right)^N \tag{9–67}$$

from which the expansion work for all N stages is

$$N\frac{\Delta T_H}{T_4} = N\frac{\Delta T_H}{T_4}\frac{T_4}{T_0} = N\theta_4\{1 - [\pi_t^1/^N(1 + \delta)]^{1/c_t}\} \tag{9–68}$$

which reads for $\delta \ll 1$ and $\tau_t = (\pi_t)^{1/c_t}$

$$w_t \cong N\theta_4\left[1 - \tau_t^{1/N}\left(1 + \frac{\delta}{c_t}\right)\right] \tag{9–69}$$

The (nondimensional) net work is given by (eqs. 9–63 and 9–69):

$$w = w_t - w_c = N_E\theta_4\left[1 - \tau_t^{1/N_E}\left(1 + \frac{\delta}{c_t}\right)\right] - N_C\left[\tau_{ca}^{1/N_c}\left(1 + \frac{\delta}{c_c}\right) - 1\right] \tag{9–70}$$

Note that the number of intercool (c) and reheat (E = expansion) stages may differ, hence the subscripts on N. The heat input to the cycle is

$$q = \frac{N_E\Delta T_H + \Delta T_x}{T_0} \tag{9–71}$$

One can readily show that the efficiency determined by these expressions increases with increasing N's.

A variation of this cycle is of interest in connection with cycles that attempt to complement the simple Brayton cycle, by converting to work the heat rejected during the low and varying temperature heat rejection process. This discussion is presented in Section 11.4.2: Performance of an Ideal Match Supercritical Cycle.

9.9. PRESSURE LOSS DURING HEAT ADDITION

When a significant pressure loss occurs during a heat interaction between a steady flow control volume and the environment, the approximation that pressure is constant is not appropriate. This approximation was used in specifying the relationship between heat input and enthalpy change using the First Law (eq. 4–12):

$$dq = dh - v\,dp \tag{9–72}$$

To quantify the influence of this effect an ideal Brayton cycle whose only nonideal phenomenon is a falling pressure during heat input is considered. The component could be a burner or a heat exchanger. Attention is limited here to the ideal cycle, and one must recognize that, in the limit of very small loss, the optimum value of pressure ratio for maximum efficiency tends toward a large value given by $(\theta_4)^{\gamma/(\gamma-1)}$.

The heat addition portion of this cycle is shown in Fig. 7–15 with decreasing (total = static, $M \sim 0$) pressure during the heat input process. With the usual definitions

$$\pi_c = \frac{p_2}{p_0} \qquad \tau_c = \frac{T_2}{T_0} \tag{9–73}$$

and $p_5 = p_0$ for the heat rejection process, the expansion pressure ratio is

$$\pi_t = \frac{p_5}{p_4} = \frac{p_0}{p_2 - \Delta p} = \frac{1}{\pi_c(1-\delta)} \tag{9–74}$$

the expander work is

$$w_t = \theta_4\left(1 - \frac{T_5}{T_4}\right) = \theta_4[1 - \pi_c^{-(\gamma-1)/\gamma}(1-\delta)^{-(\gamma-1)/\gamma}] \tag{9–75}$$

and the net work follows as

$$w = w_t - w_c = \theta_4\left[1 - \frac{1}{\tau_c(1-\delta)^{(\gamma-1)/\gamma}}\right] - (\tau_c - 1) \tag{9–76}$$

The calculation of efficiency requires the cycle heat input (from eq. 9–72), assuming the gas is ideal:

$$q = \frac{1}{C_p T_0}\left(\int_2^4 T\,ds\right) = \frac{1}{C_p T_0}\left[\int_2^4 C_p\,dT - R\int_2^4 T\,d(\ln p)\right] \tag{9–77}$$

The last term of this expression can be evaluated if a relation between p and

T is known. A plausible relation is a linear one of the form:

$$\frac{p}{p_2} = 1 - \delta\frac{T - T_2}{T_4 - T_2} \qquad \text{from which} \qquad \frac{dp}{p_2} = -\delta\frac{dT}{T_4 - T_2} \tag{9-78}$$

and

$$RT\,d\ln p = RT\frac{-\delta\dfrac{dT}{T_4 - T_2}}{1 - \delta\dfrac{T - T_2}{T_4 - T_2}} \tag{9-79}$$

Note that the δ term may be associated with the Mach number of flow suffering the heating. From the differential form of eq. 6–45 or eq. 6–51, $\delta \sim \gamma M^2/2$. With a change in variable

$$x = \frac{T - T_2}{T_4 - T_2}$$

the second integral in eq. 9–77 may be evaluated. The result is

$$\int RT\,d\ln p = -\delta RT_2\left[\left(\frac{T_4}{T_2} - 1\right)\int_0^1 \frac{x\,dx}{1 - \delta x} + \int_0^1 \frac{dx}{1 - \delta x}\right] \tag{9-80}$$

With δ small, the entire term is small so that the δ's in the integrals are of second order and therefore negligible. Thus

$$q_i = \frac{(T_4 - T_2)}{T_0} + \delta\frac{\gamma - 1}{\gamma}\left[\frac{T_4 + T_2}{2T_0}\right] \approx (\theta_4 - \tau_c)\left(1 + \delta\frac{\gamma - 1}{\gamma}\frac{\theta_4 + \tau_c}{\theta_4 - \tau_c}\right) \quad \text{for } \delta \ll 1 \tag{9-81}$$

With the small δ approximation, the specific work expression also simplifies to

$$w \cong \left\{\theta_4\left[1 - \frac{1}{\tau_c}\left(1 + \frac{\gamma - 1}{\gamma}\delta\right)\right] - (\tau_c - 1)\right\} \tag{9-82}$$

and the efficiency (for small δ) becomes,

$$\eta_{th} = \left(1 - \frac{1}{\tau_c}\right)\left(1 - \left[\frac{\gamma - 1}{\gamma}\delta\right]\left\{\frac{\theta_4 + \left(\dfrac{\tau_c + \theta_4}{2}\right)(\tau_c - 1)}{(\theta_4 - \tau_c)(\tau_c - 1)}\right\}\right) \tag{9-83}$$

where $(1 - 1/\tau_c)$ is the ideal efficiency of the Brayton cycle. Note that the pressure loss factor appears as $((\gamma - 1)/\gamma)\,\delta$. The term in wavy brackets multiplying this factor depends only on the cycle parameters τ_c and θ_4. To

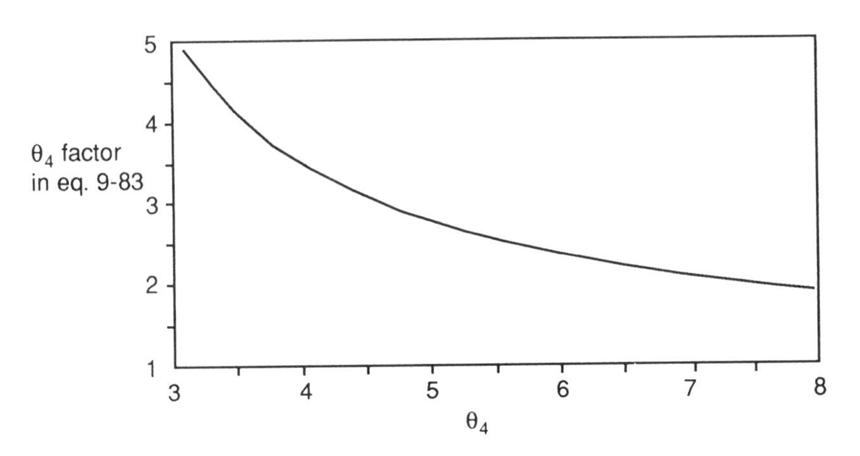

FIG. 9–39. Efficiency reduction factor which multiples $[(\gamma - 1)/\gamma]\delta$ in eq. 9–83 with τ_c for maximum w (eq. 9–84).

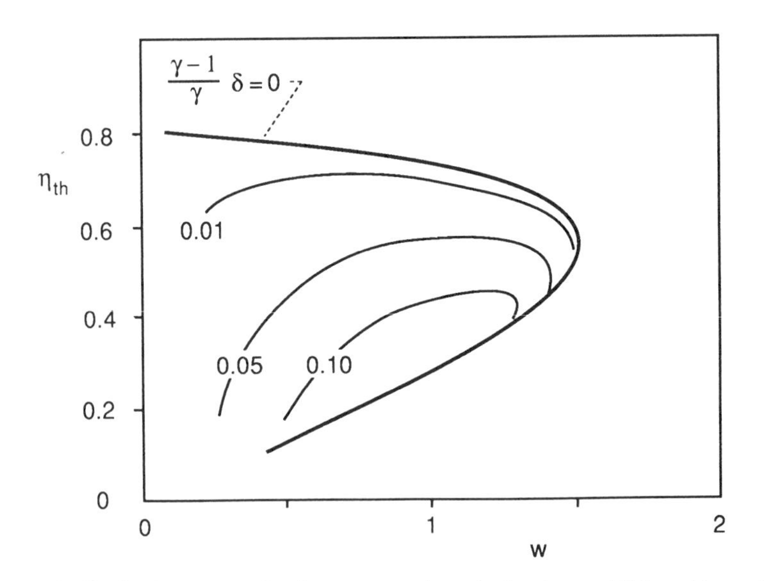

FIG. 9–40. Performance plot for pressure loss during heat addition. ($\Delta\tau_c = 0.1$).

estimate the magnitude of this term, note that τ_c for maximum w (eq. 9–82) is

$$\tau_{cw} = \sqrt{\theta_4\left(1 + \frac{\gamma - 1}{\gamma}\delta\right)} \cong \sqrt{\theta_4} \text{ to first order} \tag{9–84}$$

Using this value, the multiplying factor { } in eq. 9–83 varies between 3.5 for $\theta_4 = 4$ and 2.5 for $\theta_4 = 6$. Figure 9–39 shows its variation with θ_4. The variation of w and η_{th} for various values of $[(\gamma - 1)/\gamma]\delta$ are shown in Fig. 9–40. Note that a maximum efficiency is reached at pressure ratio less than $(\theta_4)^{\gamma/(\gamma-1)}$.

9.10. STEADY FLOW HEAT ENGINES FOR SPACE POWER

The space environment is unique in that the reservoir temperature to which heat may be rejected is very low (of the order of 3K, unless earthshine is an important contribution to the thermal background). This implies that the thermal efficiency of a cycle may be made quite high since θ_4 could be very large. A practical system using a low rejection temperature must reject the heat by (true) radiation, and the rate of heat rejection *per unit area* for a surface at a temperature T_0 varies (approximately, neglecting absorption of heat from the space background) as:

$$q_0 = \varepsilon\sigma T_0^4 \tag{9–85}$$

Here σ is the Stefan-Boltzmann constant and ε ($\leqslant 1$) is an emissivity to account for the fact that real surfaces are "gray" rather than "black" ($\varepsilon = 1$). A lower temperature radiator therefore requires a greater area to emit a given quantity of heat. For the following, ε is taken to be unity.

A low value of T_0 implies a large radiator area which, in turn, implies a massive power system, typically dominated by the radiator. There appears to exist an optimum rejection temperature which balances the decreasing efficiency and decreasing mass of a system whose design rejection temperature is increased. For space power systems, the mass must be minimized for lowest cost and/or maximum acceleration of the vehicle with a finite thermal power available.

The leading contenders for such (dynamic flow) systems are the Rankine and Brayton cycles, with the heat being supplied by a nuclear reactor, and the heat being rejected by a true radiator. The schematic diagram of a Rankine system is shown in Fig. 9–41. The Rankine cycle is considered here because

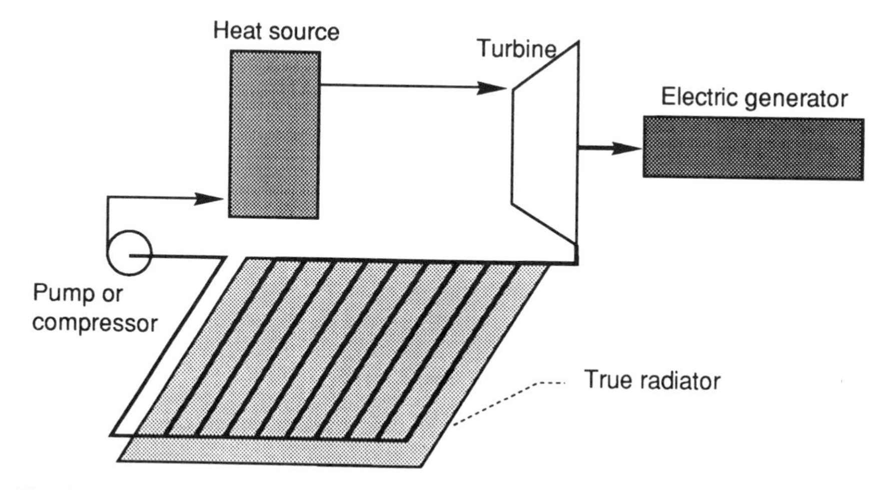

Fig. 9–41. Schematic of a dynamic space power system based on either the Rankine or the Brayton cycles.

of a close relationship exists between it and the Brayton cycle, especially when the Rankine cycle is modeled as Carnot-like.

When detailed calculations are performed, it becomes readily apparent that the radiator mass of such a vehicle can become the dominant portion of the mass of the vehicle. Obviously then, an effort must be made to develop a system that utilizes high peak cycle temperatures so that the *rejection* temperature itself is also high.

Consider the analysis of a space power system with Rankine or Brayton cycle engines. It is assumed that the optimal system will be one for which the mass of the power supply system is a minimum. In order to keep the analysis tractable, some (hopefully realistic) simplifying approximations are made. These are:

1. All parts other than the radiator have a mass proportional to the thermal power (Q_s) supplied:

$$\text{Mass of (pump + reactor + turbine + piping)} \equiv m_s = \frac{Q_s}{k_s} \qquad (9\text{–}86)$$

2. The radiator mass, m_r, is proportional to the radiator area:

$$m_r = k_r A_r \qquad (9\text{–}87)$$

Here k_r and k_s are proportionality constants characterizing the mass per unit power (sometimes referred to as specific mass) of the components. Ideally one would wish k_r to be small and k_s to be large. In the following, the role of these constants is examined for two cycles of interest in dynamic energy conversion engines in space.

9.10.1. *Simple Rankine Cycle*

Consider first, the simple subcritical Rankine cycle because it has fewer parameters. As far as this cycle is concerned, one need not necessarily consider water as the working substance because, at the high temperatures considered, the corresponding fluid pressures are large. The alkali metals (sodium, potassium, lithium, or mixtures thereof, etc.) may be more suitable working substances for this cycle.

The *T-s* diagram is shown in Fig. 9–42. The expander work $(h_4 - h_5)$ is reduced by the pump work $(h_2 - h_0)$, while heat input is $h_3 - h_2$. The efficiency of the cycle is:

$$\eta_R = \frac{(h_4 - h_5) - (h_2 - h_0)}{(h_4 - h_2)} \qquad (9\text{–}88)$$

Here the subscript on the thermal efficiency is "R" when the value is restricted to Rankine cycle, whereas the subscript "th" is used in expressions that are valid for all cycles. It is convenient to use the similarity between this and the Carnot

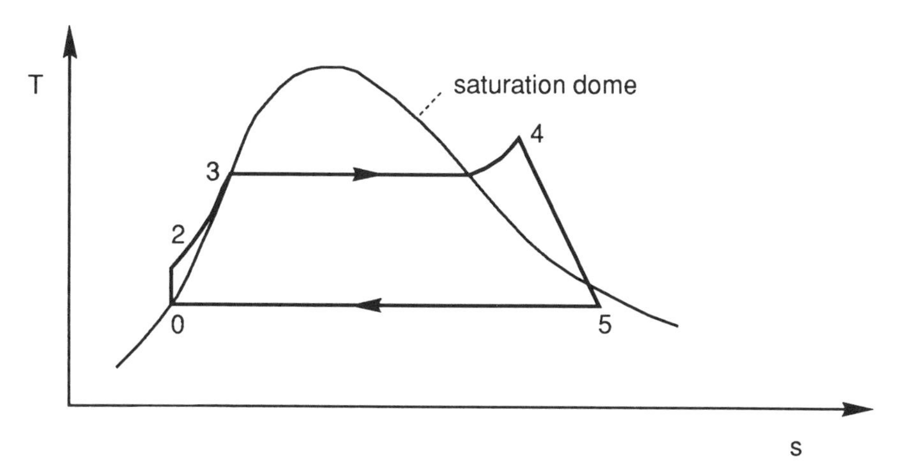

FIG. 9–42. *T-s* diagram for a condensable working fluid cycle.

cycle to write:

$$T_3 \cong T_4 = T_s, \qquad T_0 = T_5 = T_r \tag{9–89}$$

where T_s and T_r are effective source and radiator temperatures. The cycle performance may be modeled by saying that the efficiency equals a fraction, ϕ, of the Carnot efficiency. Figure 11–9 shows that, when water is used as the working fluid, ϕ is of the order ~ 0.5. The efficiency is

$$\left.\begin{aligned} \eta_R &= \phi\left(1 - \frac{T_r}{T_s}\right) \\ &\text{or} \\ T_r &= T_s\left(1 - \frac{\eta_R}{\phi}\right) \end{aligned}\right\} \tag{9–90}$$

Usually the maximum supply temperature, T_4 (or T_s in this case), is limited by material considerations so that this temperature may be treated as a constant. The efficiency for minimum system mass is to be determined.

Radiator Mass

The mass of the radiator follows from eq. 9–87 to give,

$$m_r = k_r \frac{\text{thermal power rejected}}{\text{thermal power per unit area}} = k_r \frac{Q_r}{\sigma T_0^4} \tag{9–91}$$

where Q_r is the heat radiated. It is desirable to obtain this expression in terms of the source temperature, efficiency, and the work output W. It follows that,

$$\left.\begin{aligned} \eta_{th} &= \frac{W}{Q_s} = \frac{W}{W + Q_r} \\ \text{from which} \\ Q_r &= W\left(\frac{1}{\eta_{th}} - 1\right) \end{aligned}\right\} \tag{9–92}$$

Utilizing eqs. 9–91 and 9–92, one obtains

$$m_r = \frac{k_r}{\sigma T_4^4} W\left(\frac{1 - \eta_R}{\eta_R}\right)\left(1 - \frac{\eta_R}{\phi}\right)^{-4} \tag{9–93}$$

Equations 9–86 and 9–93 then give for the total mass of the system,

$$\left.\begin{aligned} m_{tot} &= m_r + m_s = \frac{k_r}{\sigma T_4^4} W\left(\frac{1 - \eta_R}{\eta_R}\right)\left(1 - \frac{\eta_R}{\phi}\right)^4 + \frac{W}{k_s}\frac{1}{\eta_R} \\ \text{or} \\ \frac{k_s m_{tot}}{W} &= \frac{1}{\eta_R}\left\{1 + \frac{1}{\Gamma}\frac{1 - \eta_R}{\left(1 - \frac{\eta_R}{\phi}\right)^4}\right\} \end{aligned}\right\} \tag{9–94}$$

where $\Gamma = \sigma(T_4)^4/k_r k_s$ is a system parameter that should ideally be large. One may also write the (nondimensional) power density as

$$\frac{W}{k_s m_{tot}} = \eta_R\left[1 + \frac{(1 - \eta_R)}{\Gamma\left(1 - \frac{\eta_R}{\phi}\right)^4}\right]^{-1} \tag{9–95}$$

which is plotted in Fig. 9–43 for $\Gamma = 1$, a value that may be attainable. Note the maxima at various values of the cycle quality parameter, ϕ.

Analytical expressions describing these maxima may be obtained. Rather than use eq. 9–95 directly, it is easier to kind the maximum of the natural logarithm of the expression that occurs at the same value of η_R. After some algebra, it then follows that

$$\frac{\partial \ln\left(\frac{W}{m_{tot}}\right)}{\partial \eta_R} = \frac{1}{\eta_R} - \frac{\frac{4}{\phi}}{1 - \frac{\eta_R}{\phi}} - \frac{4\Gamma\left(-\frac{1}{\phi}\right)\left(1 - \frac{\eta_R}{\phi}\right)^3 - 1}{\Gamma\left(1 - \frac{\eta_R}{\phi}\right)^4 + (1 - \eta_R)} = 0 \text{ at the maximum.}$$

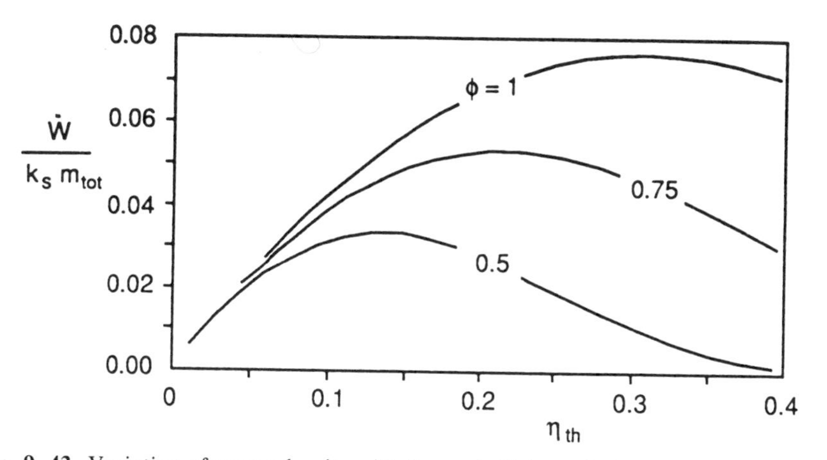

Fig. 9–43. Variation of power density with thermal efficiency for $\Gamma = 1$. Plot of eq. 9–95.

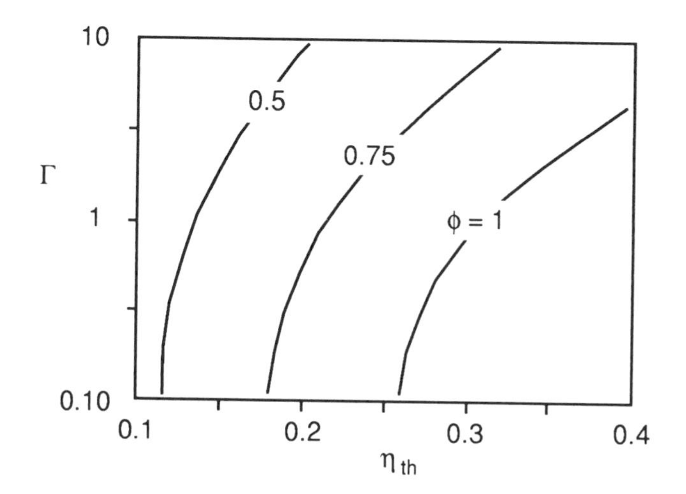

Fig. 9–44. Variation of thermal efficiency required for maximum power density for specified value of Γ. Plot of eq. 9–96.

Here ϕ is assumed not to vary with η_R, which, though not strictly true, is reasonable approximation. This equation can be rewritten as

$$\Gamma = \frac{\phi^4}{(\phi - \eta_R)^4}\left[\frac{4\eta_R(1 - \eta_R)}{\phi - \eta_R} - 1\right] \text{at maximum } \frac{W}{m_{tot}} \tag{9–96}$$

and gives the value of the efficiency that must be chosen to give maximum system power density. A numerical inversion of the variables yields the plot shown in Fig. 9–44. Evidently, a better cycle (larger ϕ) at given value of Γ gives a better efficiency (η_R) when the cycle is designed to give minimum system mass.

The value of maximum $W/k_s m_{tot}$ is then obtained from eq. 9–95. A plot, not shown, of $W/(k_s m_{tot})$, and efficiency for maximizing this specific power shows that larger values of Γ are desirable. This will be algebraically evident from the small $\Gamma \ll 1$ limiting case developed below. Naturally, a detailed analysis of the system is required so that realistic values of the constants k_r and k_s can be identified and for those values, the critical cycle characteristics such as the source temperature T_4.

Special Case: Radiator Mass Dominant

One special limiting case of interest occurs when the situation is considered where the radiator mass is dominant. In such a case, the proportionality constant k_r (eq. 9–86) becomes very large so that the Γ group becomes very small. The limit of such a process is Γ approaching zero. In this case the algebraic difficulties associated with finding the value of η_R for maximum W and the corresponding value of W are overcome because eq. 9–96 becomes

$$4\eta_R(1-\eta_R) = \phi - \eta_R \qquad \text{or} \qquad \eta_{R,\text{opt}} = \frac{5}{8}\left\{1 - \sqrt{1 - \frac{16}{25}\phi}\right\} \tag{9–97}$$

This is the efficiency giving maximum power density for a system with mass of radiator dominant. The value of the corresponding maximum is shown in Fig. 9–45. Note that when Γ is small, the maximum power density as shown in Fig. 9–45 is proportional to Γ. The abscissa of this plot can be shown to be the ratio of output power per unit $(A_r\sigma(T_4)^4)$. From eq. 9–97 it follows that if the Rankine cycle is very close to a Carnot cycle, so that $\phi \approx 1$, the optimum efficiency is:

$$\eta_{R,\text{opt}} = 0.25$$

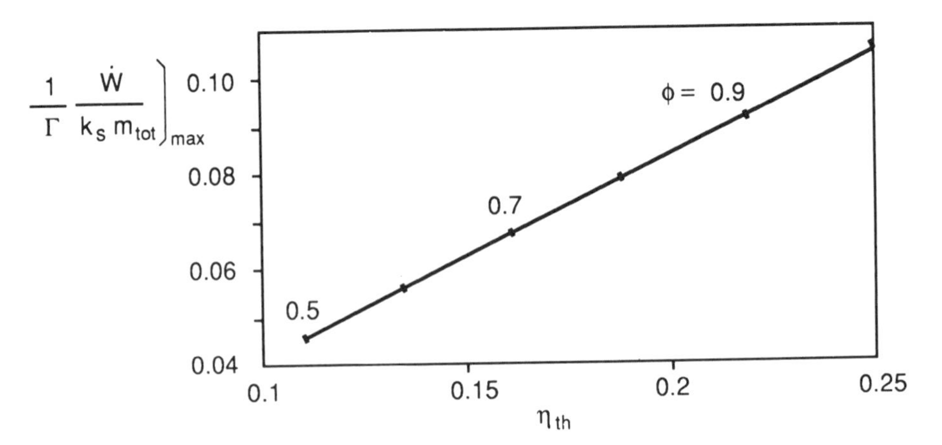

Fig. 9–45. Maximum power density for systems with dominant radiator mass. Note proportionality to Γ.

For $\phi = 0.5$ the optimum value of η_R is 0.11. This somewhat surprising result serves to emphasize what is to be optimized. The low efficiency leads to maximum power density because the resulting high rejection temperature gives a small radiator area.

It is also apparent that a higher efficiency will be optimum when the effects of the mass of reactor, pump, etc., are included, because these masses are proportional to the heat supplied. All these effects are present in the more complete eqs. 9–94 and 9–96.

9.10.2. *Simple Brayton Cycle*

A similar analysis of a space power system using a Brayton cycle may be carried out. The cycle configuration is chosen for maximum cycle specific work. This cycle has the state points 2 and 5 at the same temperature as shown in Fig. 9–46. This follows directly from eqs. 9–5 and 9–12.

The Brayton cycle differs from the Rankine cycle because the heat interaction between fluid and environment take place at varying temperature. This makes the determination of the radiator area required interesting and different from that involving an isothermal phase change.

Consider a differential area of the radiator that causes the working fluid to lose an amount of heat that is measured by a temperature decrease of dT. The area required is given by

$$\sigma T^4 \, dA = -\dot{m} C_p \, dT \qquad (9\text{–}98)$$

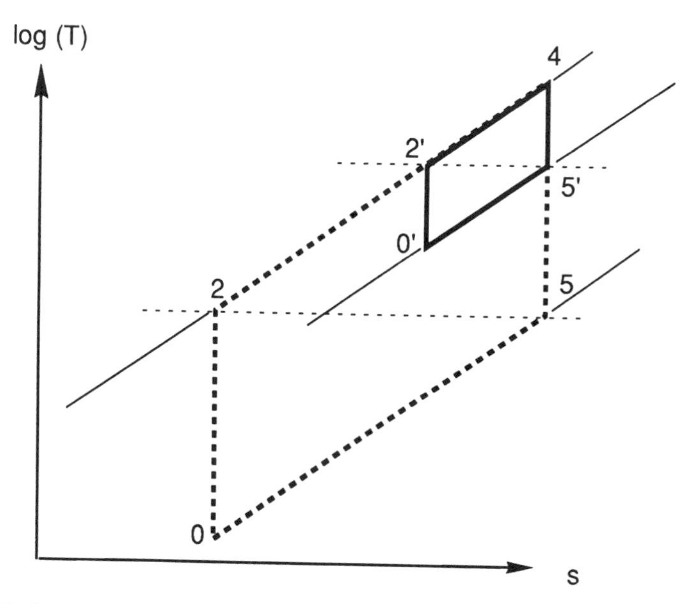

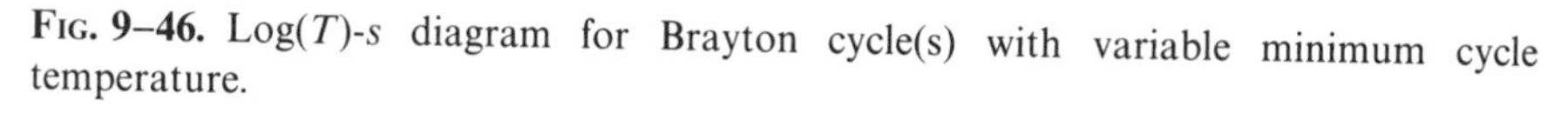
Fig. 9–46. Log(T)-s diagram for Brayton cycle(s) with variable minimum cycle temperature.

or by rearranging and integrating with a constant value of the product $\dot{m}C_p/\sigma$:

$$A = -\frac{\dot{m}C_p}{\sigma}\int_{T_5}^{T_0}\frac{dT}{T^4} = \frac{\dot{m}C_p}{3\sigma}\left(\frac{1}{T_0^3} - \frac{1}{T_5^3}\right) \tag{9–99}$$

The total heat rejected is $Q_r = \dot{m}C_p(T_5 - T_0)$ and the radiator mass is proportional to the area as given by eq. 9–87. Combining these relations gives the radiator mass per unit thermal power:

$$\frac{m_r}{Q_r} = \frac{k_r}{\sigma T_0^4}\left\{\frac{1}{3}\frac{1 - \left(\frac{T_0}{T_5}\right)^3}{\frac{T_5}{T_0} - 1}\right\} \tag{9–100}$$

Since $T_5/T_0 = T_4/T_2$ and T_4 is fixed, this parameter is essentially a measure of the compressor outlet temperature or, more importantly, the minimum cycle temperature T_0. This can be seen in Fig. 9–46, where two cycles are shown, both with $T_5 = T_2$ but one with $T_5/T_0 = T_4/T_2$ close to unity (0′, 2′, 4, 5′ cycle) and the other with T_4/T_2 much larger. One can show with the aid of eq. 9–12 that the condition for maximum specific work leads also to $T_2 = T_5$.

The ideal Brayton cycle has an efficiency given by (see eqs. 9–5 and 9–11)

$$\eta_B = 1 - \frac{T_2}{T_4} = 1 - \frac{T_0}{T_5} = \frac{W}{Q_r + W} \tag{9–101}$$

From this follow the general results

$$\frac{W}{Q_r} = \frac{\eta_B}{1 - \eta_B} \tag{9–102}$$

and

$$\frac{T_4}{T_2} = \frac{T_5}{T_0} = \frac{1}{1 - \eta_B} \tag{9–103}$$

Combining eqs. 9–100 and 9–102 gives, with $T_2 = T_5$

$$\frac{m_r}{W} = \frac{k_r}{3\sigma T_2^4}\,\frac{1 - \eta_B}{\eta_B}\,\frac{\left(\frac{T_4}{T_2}\right)^3 - 1}{1 - \frac{T_2}{T_4}}$$

Writing this in terms of T_4, the source temperature, as in the parameter Γ, one obtains with eq. 9–103:

$$\frac{m_r}{W} = \frac{k_r}{3\sigma T_4^4} \frac{1}{\eta_B^2(1-\eta_B)^3}\left(\frac{1}{(1-\eta_B)^3} - 1\right) \tag{9–104}$$

The mass of the remaining system, m_s, is again assumed proportional to Q_s (eq. 9–86) and the total system mass is:

$$\frac{k_s m_{tot}}{W} = \frac{k_s(m_s + m_r)}{W} = \frac{1}{\eta_B}\left\{1 + \frac{1}{\Gamma}\frac{1-(1-\eta_B)^3}{3\eta_B(1-\eta_B)^6}\right\} \tag{9–105}$$

This expression is equivalent to eq. 9–94 for the Rankine cycle. The parameter Γ consists of the same quantities except that numerical values of k_r, k_s and T_4 will be different because the physical characteristics are different for the two cycles. For comparison purposes, however, one may use Γ to examine the differences between the simplified Rankine (with non-zero ϕ) and the Brayton cycles.

Figure 9–47 is a plot of system mass per unit power $(m_{tot}k_s)/W)$ versus system efficiency for various values of Γ. This figure corresponds to Fig. 9–45 for the Rankine cycle except that the ordinate is inverted and various Γ's are shown here. The values of the minima are plotted as a function of Γ itself in Fig. 9–48. Also shown is the variation of the same performance parameter for the Rankine cycle for $\phi = 0.75$. In these terms, the two cycles are equivalent: The Brayton cycle is like a Carnot cycle with $\phi = 0.75$. Note that for both cycles a large value of Γ is desirable.

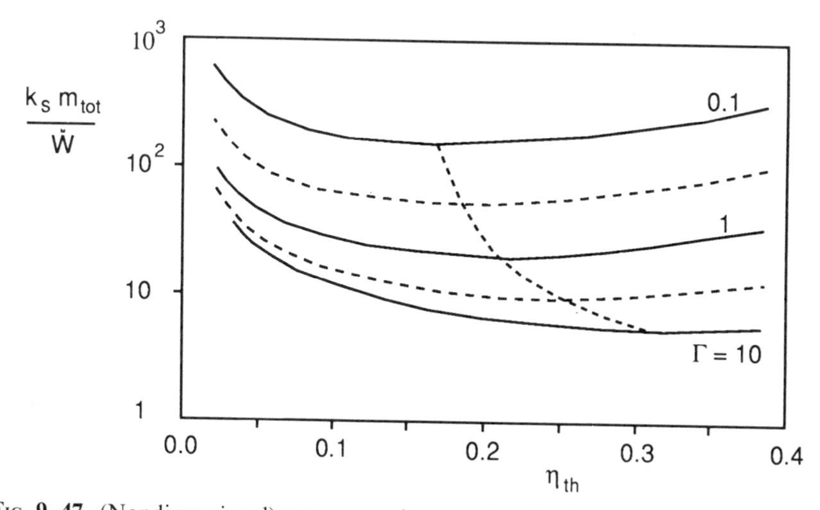

FIG. 9–47. (Nondimensional) mass per unit power (eq. 9–105) variation with efficiency, maximum specific work Brayton cycle with ideal components.

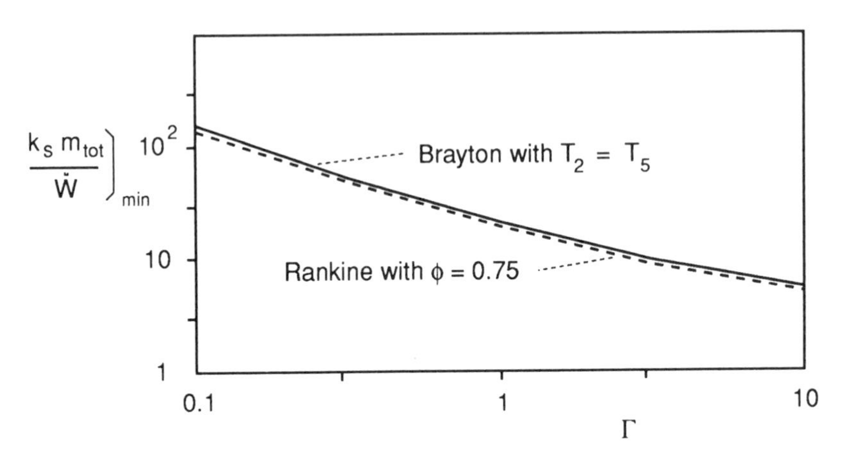

FIG. 9–48. Minimum mass per unit power for idealized Rankine and Brayton cycles.

9.11. CLOSED BRAYTON CYCLE ENGINES

The commercial electric power industry has had a continuing interest in the closed cycle (Fig. 9–2) gas turbine (CCGT) as prime movers for the generation of electric power. This interest was initiated in Europe by J. Ackeret and C. Keller at Escher Wyss results in a 1935 patent. References 9–2 and 9–3 are discussions of the salient features of the so-called A–K (after the originators) machines. The technology was considered to be an able competitor to existing Rankine cycle systems. Since the earliest times after the development of combustion turbines, a serious effort by a number of individuals and enterprises has been undertaken to investigate the possibility of using the closed cycle gas turbine as a prime power generator, both with combustion and with nuclear thermal power sources. In the United States the nuclear heat sources effort centered on the so-called high temperature gas cooled reactor (HTGR). The reader should consult the extensive technical literature on the systems that have been considered. For a variety of reasons, the closed cycle gas turbine technology has not been developed to the point of large-scale commercialization. An important one is that the capital investment in the conventional Rankine power plant machinery and, in particular, in the nuclear components had been large, and there has always been the concern that development of a new technology might not be economically successful and is therefore risky.

Further, the successful employment of the closed Brayton cycle requires development of a new class of higher temperature gas cooled (rather than liquid cooled) reactors. The reason is that the Brayton cycle demands the highest practical turbine inlet temperature. The Rankine cycle using steam, on the other hand, is used at temperature levels that are limited by the high pressure imposed by the pressure-temperature relation along the saturation line. The practical result is that a temperature limit of around 600–700 K is imposed on the cycle. Such a peak cycle temperature level is consistent with a Carnot efficiency of about 50% and real plants achieve about 35%, a good fraction of the thermodynamically possible efficiency level.

With some exceptions, the development of the required gas cooled reactor types has not been undertaken beyond the prototype development because of the cost. It may well be that, in the future, concern with fuel supplies, environmental and safety issues may be cause for reexamination of the need for the nuclear gas turbine. From a thermodynamic viewpoint, the motivation to compete with the Rankine cycle is that the closed cycle gas turbine has the potential for higher thermal efficiency and better part-load performance.

The following is an examination of some of the important and interesting points related to the thermodynamic and fluid dynamic design and operation of a closed cycle gas turbine. The design process must address the choice of working fluid and its operating pressure. The unique capability of a closed cycle to operate with a reduced inventory of working fluid gives it unique flexibility in operating at part load. This aspect of the cycle is discussed in Chapter 16.

9.11.1. *Choice of Working Fluid in the Closed Cycle Gas Turbine*

The greatest experience with working fluid in gas turbines has been with air in compressors and with steam and combustion gas in turbines. While such systems will not be producing power commercially in the near future, their fundamental advantages have not changed, and it is important to cite some of the technical issues involved to the student of such cycles.

In the closed cycle gas turbine, the ideal working fluid is helium. It is chemically inert and thus does not pose a threat to the turbine life by chemical corrosion. It obeys an ideal gas state equation and is thermally perfect at all conditions considered so that the mathematical description of the fluid and thus performance is particularly simple. From a nuclear viewpoint, helium is an excellent reactor coolant because it does not undergo absorption reactions with neutrons, and thus does not become radioactive.

In Table 9–1 the properties of helium are compared at 300 and 1000 K where similar cycles might operate. The pressure at the low and high temperature conditions are 1 and 30 atm, respectively. Note that the transport

Table 9–1. Properties of air and helium at conditions of interest in CCGTs

Property	p(atm)	T(K)	Air	Helium
Mol. Wt.			28.9	4
γ	1	300	1.4	1.67
	30	1000	1.36	1.67
C_p	1	300	1.005	5.20 kJ/kg K
	30	1000	1.142	5.20
k	1	300	0.028	0.15 w/m-K
	30	1000	0.068	0.36
ν	1	300	16	120×10^{-6} m^2/sec
	30	1000	3.4	28

properties ν (kinematic viscosity) and k (thermal conductivity) vary with temperature, even for helium. A number of parameters of interest are examined to assess the consequences of a change from air (where the industrial experience is considerable) to helium. The material in this section owes some of its origin to refs. 9–2, 9–3, and 9–4.

SPECIFIC HEAT. The specific heat of helium is five times larger than that for air so that for a chosen temperature increment, helium carries five times the power per unit mass.

PRESSURE RATIO. The cycle may be considered to operate between two chosen temperatures (T_2 and T_1). The pressure ratio is given by

$$\frac{p_2}{p_1} = \left(\frac{T_2}{T_1}\right)^{\gamma/(\gamma-1)} \tag{9–106}$$

where the exponent in eq. 9–106 equals 3.5 for air and 2.5 for He. This leads to a lower pressure ratio for the helium which is easier from the point of view of compressor design because the magnitude of the adverse pressure gradient in the compressor tends to be lower.

VOLUME RATIO. The design of the turbomachinery is simpler when the (geometric) expansion or compression volume ratio is smaller. In the limit of an incompressible fluid, the axial flow machinery may be built with little or no annular contraction. The volume ratio is given by:

$$\frac{v_2}{v_1} = \left(\frac{T_2}{T_1}\right)^{1/(\gamma-1)} \tag{9–107}$$

which is also smaller for helium compared to air. As an example, consider a temperature change from 300 K to 450 K. The pressure and volume ratios implied by eqs. 9–106 and 9–107 are summarized in Table 9–2.

Table 9–2. Characteristics of cycle parameters for a 1.5 temperature ratio in a compression process

	air	helium
pressure ratio (p_2/p_1)	4.4	2.66
volume ratio (v_2/v_1)	0.36	0.55
stage pressure ratio	1.2	1.03
number of compressor stages	8	33
flow velocity	1	2.3
flow area (high pressure side)	1	0.62
flow area (low pressure side)	1	0.37

STAGE PRESSURE RATIO. The Euler turbine equation (eq. 14–2) relates the work done by torques exerted with angular speed to the change in stagnation enthalpy:

$$C_p(T_{t2} - T_{t1}) = \omega(v_{\theta 2}r_2 - v_{\theta 1}r_1) \tag{9–108}$$

For an approximately isentropic flow with a small change in state conditions, the Euler turbine equation may be written for a compressor stage with axial flow ($r_1 = r_2 = r$):

$$\frac{T_{t2}}{T_{t1}} = 1 + \frac{(\omega r)^2}{C_p T_{t1}}\left(\left(\frac{v_\theta}{\omega r}\right)_2 - \left(\frac{v_\theta}{\omega r}\right)_1\right)$$

or, for a reversible process,

$$\frac{p_{t2}}{p_{t1}} = 1 + \frac{(\omega r)^2}{R_u T_1}\frac{T_1}{T_{t1}}\,[\mathrm{MW}]\ \mathrm{fcn\ (angles)} \tag{9–109}$$

which is developed from first principles as eq. 14–20. Here the ratios between velocities are grouped to reflect the fact that they are functions of angles in the blading only. The specific heat ratio has been written in terms of the molecular weight, MW (see eq. 4–5). The total to static temperature ratio is a function of Mach number and is close to unity. Thus for air, the pressure ratio is approximately 1.2 for a single axial compression stage, whereas that of the helium compressor is about 1.03 for the same wheel speed (ωr) and inlet temperature.

NUMBER OF COMPRESSOR STAGES. The overall pressure ratio given by eq. 9–106. This, the stage pressure ratio (eq. 9–109) allows the number of stages, N, for the same temperature rise to be determined:

$$\text{Overall pressure ratio} = (\text{stage pressure ratio})^{\mathrm{N}} \tag{9–110}$$

Table 9–2 gives N for air and helium using the numbers quoted above.

PRESSURE DROPS IN FLOW DUCTS. Specification of the fractional pressure drop dictates the magnitude of the fluid flow velocity. This has an implication for the compactness of the system as discussed at the conclusion of Section 9.11.1. Since the work lost to fluid friction in a duct is at the expense of that obtainable from an adjacent work component, consider the following arrangement: a turbine followed by a heat exchanger where the flow negotiates a tube with a specified L/D. The states of the fluid are as shown in Fig. 9–49. In the heat exchanger, a (small) pressure loss will be experienced:

$$\frac{p_3}{p_2} = 1 - \delta, \qquad \delta \ll 1 \tag{9–111}$$

If this loss is lumped into the turbine and an efficiency of the combination

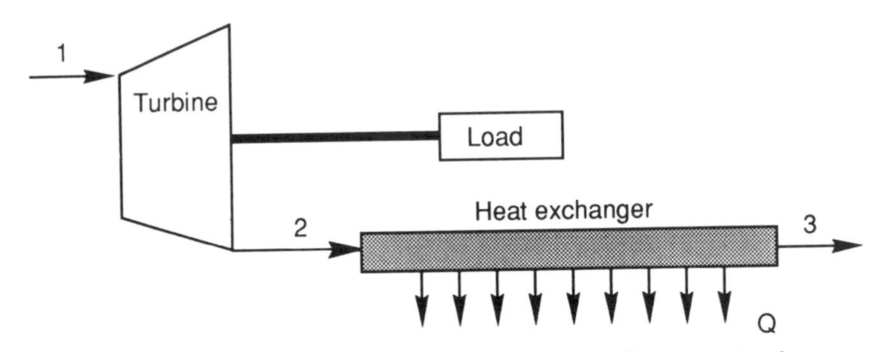

Fig. 9–49. Combination of turbine and heat exchanger to illustrate role of pressure drops in the latter.

process is defined, then one has (eq. 7–39):

$$\eta_t = \frac{1 - \dfrac{T_{t2}}{T_{t1}}}{1 - \left(\dfrac{p_{t3}}{p_{t1}}\right)^{(\gamma-1)/\gamma}} \tag{9–112}$$

or with $\dfrac{p_3}{p_1} = \left(\dfrac{p_3}{p_2}\right)\left(\dfrac{p_2}{p_1}\right)$ and assuming low flow Mach number ($p \sim p_t$),

$$\eta_t = \frac{1 - \dfrac{T_2}{T_1}}{1 - \left(\dfrac{p_2}{p_1}\right)^{(\gamma-1)/\gamma}\left(1 - \dfrac{\gamma - 1}{\gamma}\delta\right)} \tag{9–113}$$

By inspection of eq. 9–113 it follows that the same level of thermodynamic performance is obtained if the combination $((\gamma - 1)/\gamma)\,\delta$ is the same for the two flows. The loss factor, δ, may be written in terms of a friction factor as:

$$\delta = \frac{\Delta p}{p} = \frac{fL}{D}\frac{1}{2}\frac{\rho u^2}{p} \tag{9–114}$$

Combining these results allows one to state that the ratio of velocity allowable in helium to that in air is:

$$\frac{u_{He}}{u_{air}} = \sqrt{\frac{\left(\dfrac{\gamma - 1}{\gamma}\mathrm{MW}\right)_{air}}{\left(\dfrac{\gamma - 1}{\gamma}\mathrm{MW}\right)_{He}}} = 2.27 \tag{9–115}$$

where MW is the gas molecular weight. In practice, helium velocities of the order of 100 m/sec are nominally considered.

FLOW CROSS-SECTIONAL AREA. The flow cross-sectional area is given by the steady flow continuity equation

$$A = \frac{\dot{m}}{\rho u} \tag{9–116}$$

where the density is given by the state equation and the mass flow rate by an enthalpy flux:

$$\dot{m} \sim \frac{\text{power}}{C_p T} \tag{9–117}$$

with the velocity given by the "equal pressure loss effect" relation (eq. 9–115), and the pressures on the high pressure side taken as equal for the two cycles one obtains, the flow area ratio $A(\text{He})/A(\text{air})$ high and the low pressure side. The values are given in Table 9–2. Evidently, the use of helium leads to more compact flow devices compared to air when flow area is used as a comparative criterion.

9.11.2. *Effect of Pressure Level on the Helium Closed Cycle Gas Turbine Component Characteristics*

In principle, the efficiency of the closed cycle using helium depends on pressure *ratio* and the corresponding temperature ratio and not on the level of the pressure. On the other hand, the mass processed by a duct of a given diameter depends on the density (and therefore for a fixed temperature, on the pressure) of the fluid. Thus the power output of the engine is directly affected by the choice of pressure as are the physical characteristics of several of the engine's components. In this section, the impact on the following is examined:

1. Physical dimensions of the machinery
2. Reynolds number in the flow passages
3. Physical dimensions of heat exchangers

when helium is used at varying pressures. The approach is to consider an increase in the pressure of the fluid over some reference level according to

$$p = Z p_0 \tag{9–118}$$

so that increasing Z determines the sought after effect. The constraints are that the flow velocity (i.e., the Mach number) and the temperatures which the flow experiences are fixed by consideration of the cycle thermodynamics as outlined above. Reynolds number effects are ignored except where their influence is specifically examined. Geometric similarity is assumed, and a reference scale is denoted by a zero subscript. The approach and some of the results described here are based on refs. 9–3 and 9–5. These are generally useful for an examination of processing equipment in power and chemical plants in relation to the choice of fluid and its pressure.

9.11.3. *Physical Dimensions of the Machinery*

FLOW PASSAGE SIZE. The continuity equation gives $\rho u A$ = constant or, with the state equation,

$$pD^2 = \text{constant or } \frac{D}{D_0} = \frac{1}{\sqrt{Z}} \tag{9–119}$$

ROTATION SPEED. The force balance on an element of the structure yields the fact that the centripetal acceleration acting on the mass of the element is balanced by the stress times an effective area. Thus for a fixed (maximum) stress level, σ, the linear speed of the element is given by:

$$V = \sqrt{\frac{\sigma}{\rho_m}} \tag{9–120}$$

where ρ_m is the material density of the blade or wheel disk material. This result could also have been obtained from dimensional analysis. Denoting ω as the angular speed, one obtains

$$\omega D = \text{constant or } \frac{\omega}{\omega_0} = \sqrt{Z} \tag{9–121}$$

using the size scaling result above (eq. 9–119).

SHAFT TORQUE. Torque is given by power divided by angular speed. Thus

$$\text{Torque} = \frac{\text{power}}{\text{angular speed}} \sim \frac{1}{\omega} \sim \frac{1}{\sqrt{Z}} \tag{9–122}$$

MACHINERY SHAFT DIAMETER. The torsional stress level on a shaft is given by the torque and the polar moment of inertia, J. If the shaft diameter is d, then $J \sim d^4$. For a fixed stress level on the outermost shaft elements where it is largest, d is given by:

$$\text{stress} = \frac{d * \text{torque}}{J} \quad \text{or} \quad \frac{d}{d_0} = Z^{-1/6} \tag{9–123}$$

BLADE STRESS. The individual blades of a compressor or turbine rotor experience "lifting" forces which tend to deflect them like the bending experienced at the root of an airplane wing. The blade would be designed to operate at a fixed lift coefficient so that the load (force) per unit area on the blade is of order

$$c_1 \tfrac{1}{2}\rho u^2 \text{ varies as } [p] \text{ for fixed } u \text{ and } T.$$

From the idea of geometric similarity, it follows that the blade length, or span, is proportional to D. Combining these notions into a statement concerning the stress at the root due to the bending moment, one has

$$\text{stress, } \sigma \sim \frac{D^* \text{ bending moment}}{I} \sim Z \tag{9–124}$$

since the bending moment is proportional to lift ($\sim ZD^2$) times span ($\sim D$). The cross-sectional moment of inertia, I, is proportional to D^4. It follows that the stress level is proportional to pressure and thus limits the pressure.

DUCT WALL THICKNESS. Figure 9–50 shows a cut section of a duct carrying a pressurized fluid. The applied pressure load is balanced by a stress acting on the thin walls. Equating these forces yields a relation for the wall thickness, t:

$$2\sigma t L = pDL \qquad \text{or} \qquad t \sim pD \sim \sqrt{Z} \tag{9–125}$$

For fixed stress level, the wall thickness, t, increases with increasing pressure but only as the square root of pressure.

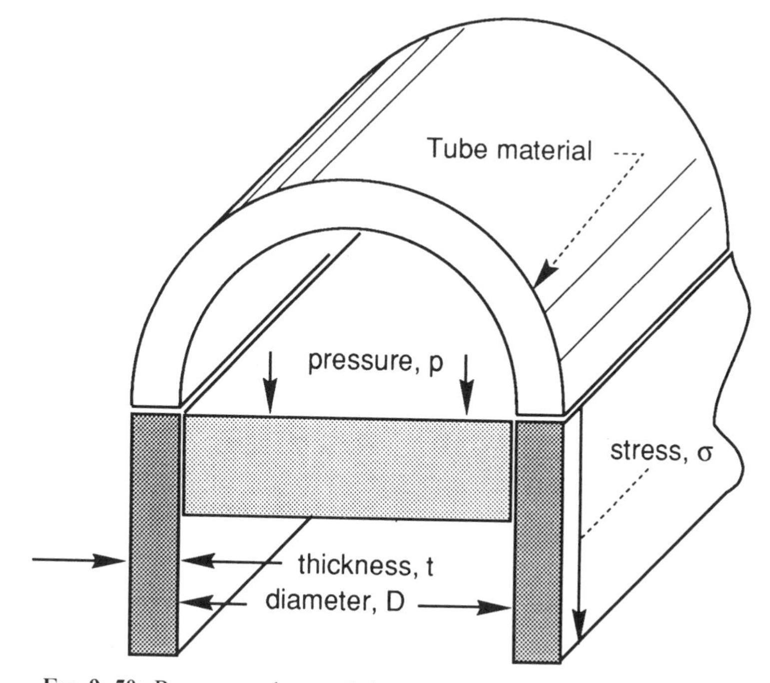

FIG. 9–50. Pressure and stress balance in a thin-walled cylindrical duct.

MASS OF DUCT MATERIAL PER UNIT LENGTH. The mass per unit length is proportional to the product $t \times D$ for a chosen material, Thus, with eqs. 9–119 and 9–125, this linear mass density is *independent* of pressure level.

9.11.4. *Fluid Mechanics in the Flow Passages*

Duct flow, as well as flow over surfaces, is strongly influenced by the flow Reynolds number, a nondimensional ratio of inertial to viscous forces. For example, Fig. 6–8 shows the variation of the friction factor for fully developed pipe flow. Similar variations describe the friction factor of boundary layer flow over flat plates and, with additional information, the flow over curved surfaces. In general, the variation of friction factors is such that they decrease with increasing Reynolds number, Re. Thus higher Reynolds number flows are generally less strongly affected by the fluid friction.

REYNOLDS NUMBER. With u and T fixed, the Reynolds number (Re) is given by its definition,

$$\mathrm{Re} = \frac{\rho u D}{\mu} \sim pD \tag{9–126}$$

With the variation of these quantities given by eqs. 9–118 and 9–119, this parameter varies as

$$\frac{\mathrm{Re}}{\mathrm{Re}_0} = \sqrt{Z} \tag{9–127}$$

which favors high pressure level operation to keep viscuous losses small, although heat transfer rate may be adversely affected.

9.11.5. *Heat Exchanger Dimensions*

Heat exchangers are devices wherein heat is transferred from one fluid to another across a wall that supports a pressure difference. These devices are complex, and careful attention must be paid to their design to optimize a system configuration. A key parameter governing the performance of the exchanger is the heat transfer coefficient h which controls the surface area required to transfer a given amount of heat when the temperatures of the wall and of the fluid are specified. Using the Reynolds analogy (eq. 6–2) and an approximate description for the friction coefficient in a turbulent duct flow (see also eq. 6–3a):

$$h = \rho u C_p \frac{f}{2} \qquad \text{or} \qquad h = (cst)\rho u C_p{}^* \,\mathrm{Re}^{-1/4} \tag{9–128}$$

and

$$\frac{\Delta p}{p} = \frac{fL}{D}\frac{\rho u^2}{p} = \frac{fL}{D}\frac{u^2}{RT} \tag{9–129}$$

Examination of the expression for the heat transfer coefficient (eq. 9–128) reveals that two variables may be manipulated to influence its magnitude: pressure, p, and velocity, u. These two variables must be related to each other if the influence of p is to be ascertained. To do this, two interesting options are available:

1. Hold Reynolds number constant (by letting D decrease when p is increased, see eq. 9–126); or
2. Keep D constant and let the flow length, L, change as required, holding the pressure loss fraction constant.

In both cases the mass flow rate and temperatures are fixed by design. The characteristics of the heat exchanger using these two options are examined below.

Case 1. Constant Reynolds Number

The constant Reynolds number specification implies that

$$puD = \text{constant} \tag{9–130}$$

the heat transferred through N tubes is $Q = NhLD\,\Delta T \sim N(puD)L$, or, for constant Re (eq. 6–130),

$$Q \sim NL \tag{9–131}$$

The total mass flow processed by the N tubes is fixed so that the mass flow rate per tube (puD^2) gives the relation between N and D as

$$\dot{m} \approx N(puD^2) = ND = \text{constant} \tag{9–132}$$

since the product (puD) is constant. Thus from eqs. 9–131 and 9–132, the ratio L/D is also constant. The fractional pressure drop is given by eq. 9–129, which states that

$$\frac{\Delta p}{p} \sim fu^2\left(\frac{L}{D}\right) \tag{9–133}$$

Here f, which is a function of Reynolds number, is constant. Thus for a fixed pressure loss percentage, the velocity, u, must therefore be constant. This implies that

pD is constant from eq. 9–130,

D and L (from eq. 9–130 with constant u) vary as $\sim(Z)^{-1}$, and

N varies as $\sim Z$(from eqs. 9–131 and 9–132).

The total flow area of the tubes varies as $ND^2[=(ND)D]$ which implies that this area varies as

$$\text{flow area} \approx \frac{1}{Z} \tag{9–135}$$

so that the end area of the heat exchanger is reduced by operation at increased pressure. The total (gas) volume of the heat exchanger varies as

$$\text{volume} = (\text{flow area}) \times (\text{length}) \sim Z^{-2} \tag{9–136}$$

which results in an important reduction in volume at higher pressure. The volume of the metal required in the exchanger is the product of material thickness, t, $\times$ (LD) $\times$ number of tubes. Thus

$$\text{metal volume} \sim NLD \sim Z^{-1} \tag{9–137}$$

since, for a specified stress level, the thickness varies as (pD) which is constant (see eq. 9–134). The ratio of metal to gas volumes in the heat exchanger varies as Z. These results are summarized in Table 9–3 and contrasted to the constant tube diameter case described below.

Table 9–3. Summary of pressure scaling relationships in heat exchangers

	Reynolds No., Re_D = constant	Tube Diameter D = constant
p, pressure	1	1
D, tube diameter	−1	0
L, tube length	−1	0.25
u, flow velocity	0	0
N, no. of tubes	1	−1
t, tube wall thickness	0	1
End (flow) area	−1	−1
Metal volume	−1	0.25
Gas volume	−2	−0.75
Ratio metal/gas vol.	1	1

Note: Shown is the exponent on Z for quantity listed.

Case 2. Constant Tube Diameter

This limit is of interest when the manufacturing cost of heat exchanger tubing is an issue. The total heat to be transferred is proportional to

$$Q = hNLD \tag{9–138}$$

where

$$h \sim puf \sim Zu/\mathrm{Re}^{1/4} \sim (Zu)^{3/4}$$

The heat condition (eq. 9–138) therefore reads:

$$\frac{(NLu^{3/4})}{(NLu^{3/4})_0} = Z^{-3/4} \tag{9–140}$$

The number of tubes is obtained from the mass flow condition: $\dot{m} \sim Npu =$ constant, or

$$\frac{(uN)}{(uN)_0} = \frac{1}{Z} \tag{9–141}$$

Lastly, with an invariant pressure loss fraction given by

$$\frac{\Delta p}{p} = \frac{fL}{D}\frac{\rho u^2}{p} \sim \frac{Lu^2}{\sqrt[4]{\mathrm{Re}}}$$

it follows that

$$\frac{(u^{1.75}L)}{(u^{1.75}L)_0} = Z^{0.25} \tag{9–142}$$

Combining eq. 9–140. 9–141, and 9–142, one arrives at the following set of relationships:

$$\frac{u}{u_0} = 1, \qquad \frac{N}{N_0} = \frac{1}{Z} \qquad \text{and} \qquad \frac{L}{L_0} = \sqrt[4]{Z}. \tag{9–143}$$

The conclusion is that as pressure is raised under the constant D constraint, the number of tubes decreases and their length increases weakly.

The heat exchanger flow area is ND^2, which varies as Z^{-1}. The wall thickness (t) varies as pD, (i.e., as Z); and finally, the metal volume or mass varies as $tNLD$ which varies as $Z^{1/4}$. Table 9–3 and Fig. 9–51 summarize and illustrate these ideas.

PROBLEMS

1. Consider a space power system radiator that is gray rather than black. When optimized as in the text, how does the gray radiator system differ from the black one? What if the gray material properties were such as to modify the emission law to that given by eq. 9–87 except that the exponent is smaller, say = 3?
2. Rather than minimize system mass in a space power system, it may be better to minimize the time to accelerate to a given speed. How does such an optimization differ from the minimum mass system?

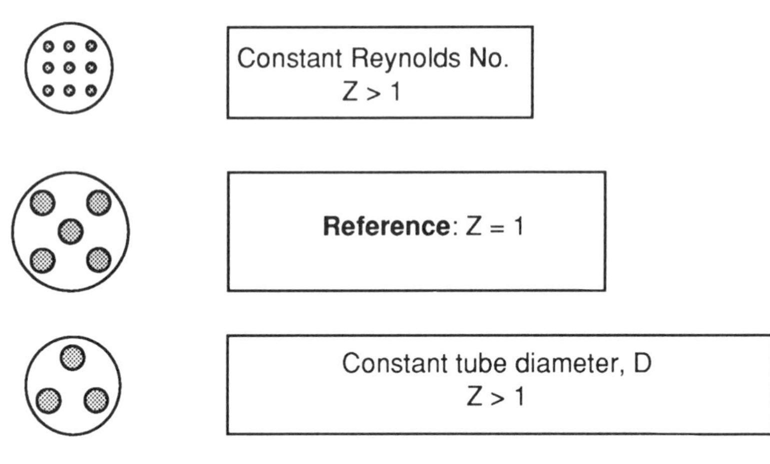

FIG. 9–51. Heat exchanger configurations described in Table 9–3.

3. From first principles and data regarding weight characteristics for space power reactors, obtain a numerical value for the parameter Γ as defined in eq. 9–97.

4. Show, with the aid of eq. 9–12, that the condition for maximum specific work in the simple Brayton cycle leads also to $T_2 = T_5$.

5. A constant pressure Diesel cycle engine as described in Chapter 8 is designed to produce the same power as a gas turbine, both at full power. Using reasonable cycle performance characteristics, what is the ratio of mass flow rates handled by the two engines?

6. Determine the pressure ratios for maximum specific work (or equivalently $\tau_{c,w}$) for an ideal cycle with m stages of intercooling and n stages of reheat. Take as examples $m = 1$, $n = 0$; $m = 0$, $n = 1$; and $m = n = 1$. Compare to $m = n = 0$ for which $\tau_{c,w} = \sqrt{\theta_4}$.

7. Intercooling is generally easier to implement than reheat. Determine the specific work and efficiency of a cycle employing isothermal compression with a single, irreversible expansion characterized by a polytropic efficiency. Find, for suitable numerical values, the pressure ratio for maximum specific work and maximum efficiency. Note that the use of regeneration is required for high thermal efficiency.

8. Following a development similar to that of Section 9.5.1 where the coolant mass flow fraction is determined for maximum specific work, find the system parameters that give maximum efficiency. Calculate and plot the thermal efficiency versus specific work variation for this case. Let $\eta_x = 0, 0.5$, and 1 and compare to the results of Fig. 9–24 (A direct attack to maximize efficiency may not be useful).

9. Extend the calculation in Section 9.5.3 to include irreversibilities in the work components. For realistic values of either polytropic or adiabatic efficiencies, the cases $\eta_x = 0$, 0.5, and 1, calculate and compare to the

results of Fig. 9–24. Comment on the relationship between these results and those for no intercooling and no reheat.

10. An intercooled compressor operates with minimum temperature for the first stage T_0 and for the second $T_0\ (1 + \varepsilon)$ because the heat exchanger is not able to return the compressed gas back to T_0. Finite ε results from a finite heat exchanger flow length. The compressor flow is reversible. If the overall compression ratio is π_c what should be the pressure ratio for each of the two stages if the compression work is to be minimized? Note the special limit of $\varepsilon = 0$. How is this conclusion altered if the compression processes are irreversible?
11. Consider the work required by a compressor with N stages of incooling. With the use of eq. 9–62, plot the work as a function of N for $\delta = 0$ and 0.01. Show that the $\delta = 0$ case approaches the result given by eq. 9–64 and explain the minimum seen for finite δ.
12. Equation 9–56 describes the thermal efficiency of the Ericsson cycle. Assume that the goal is to build such a cycle with a thermal efficiency of 5 percentage points less than Carnot efficiency ($\eta = \eta_{carnot} - 0.05$). Assume $\theta_4 = 4$. The cycle has zero pressure losses and the heat exchanger effectiveness is within 5% of ideal. Find the compressor *pressure* ratio required for such a cycle.
13. Show that for a Rankine space power system with small ϕ and small Γ, the efficiency for minimum mass is $\phi/5$.

BIBLIOGRAPHY

Wilson, D. G., *The Design of High Efficiency Turbomachinery and Gas Turbines*, MIT Press, Cambridge, Mass., 1984.

Helms, H. E., Lindgren, L. C., Heitman, P. W., Thrasher, S. R., *Ceramic Applications in Turbine Engines*, Noyes Publications, Park Ridge, N.J., 1986.

REFERENCES

9–1. Weatherston, R. C., Hertzberg, A., "The Energy Exchanger, a New Concept for High Efficiency Gas Turbine Cycles," ASME paper 66-GT 117 and Transactions of the ASME, J of Engineering for Power, 89, (2), pp. 217–28, April 1967.

9–2. Keller, C., "Die Aerodynamische Turbine im Vergleich zur Dampf and Gasturbine," Escher Wyss Mitteilungen, 1942/43.

9–3. Ackeret, J., Keller, C., "Aerodynamische Wärmekraftmaschine mit Geschlossenem Kreislauf," Escher Wyss Mitteilungen, 1942/43.

9–4. Keller, C., Schmidt, D., "Die Verwendung der Helium Gasturbine mit Geschlossenem Kreislauf für Kernkraftwerke," Escher Wyss Mitteilungen, 1966, No. 1.

9–5. Keller, C., Schmidt, D., "Die Verwendung der Helium Gasturbine mit Geschlossenem Kreislauf für Kernkraftwerke," Escher Wyss Mitteilungen, 1967, No. 3. Also ASME paper 67 GT-10, Houston, March 1967.

10

STIRLING CYCLE ENGINES

The Stirling cycle engine is an example of an engine cycle where the external thermal environment is manipulated to force the execution of a thermodynamic cycle by the system. As such, the Stirling cycle is representative of a kinematic engine. The system is a mass of gas bounded by material walls through which the heat travels and by which work is done. The conductive mode of heat transfer brings with it unique aspects of engine design.

The history of the Stirling engine is long, dating back to the early nineteenth century. The process of its development includes involvement by many individuals and corporate entities who were relentless in their pursuit of this engine as a prime mover. The reader is encouraged to review the well illustrated development history of the Stirling engine (among other engines) in the book by Kolin (ref. 10–1). To date, this engine has not met with large-scale commercial success in spite of a number of interesting features. The discussion focus in this chapter is on kinematic Stirling engines which employ pistons and cylinders to force the execution of a cycle. Nonkinematic engines include free piston (ref. 10–2) and liquid piston engines (ref. 10–3).

The discussion to follow is intended to highlight the basic issues associated with modeling this engine and thereby describing its performance. These issues will be seen to be complex enough to preclude reaching the goal of having a mathematical model that clearly identifies the development path to a viable engine configuration whose performance is competitive with existing engine types.

10.1. THE STIRLING CYCLE

The fact that the Carnot cycle could operate with a large thermal efficiency for a specific set of reservoir temperatures has led to the exploration of practical cycles that achieve this level of performance. In Chapter 7 the argument was made that the Carnot cycle with its two isotherms and two adiabats is not practical as an engine for the production of power. Two kinds of practical engine cycles can (theoretically) be made to reach the Carnot efficiency level using constant pressure or constant volume heat transfer processes to join the isothermal ones. The key feature is regeneration. Figure 10–1 shows the T-s diagrams of the three cycles of interest. In the Carnot engine (Fig. 10–1a), the processes 2–4 and 5–0 are individually adiabatic. By contrast to the Carnot cycle in which the temperature-changing processes are completely adiabatic,

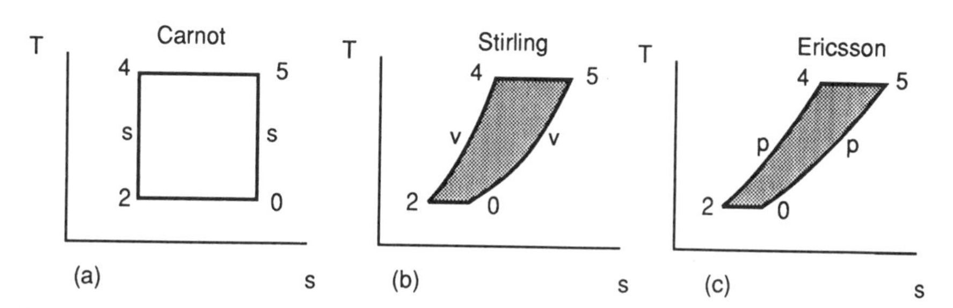

FIG. 10–1. T-s diagrams of cycles capable of Carnot efficiency: (a) the Carnot cycle, (b) the Stirling cycle, and (c) the Ericsson cycle. Cycles (b) and (c) require regeneration.

the cycle with constant volume (Fig. 10–1b) and constant pressure (Fig. 10–1c) processes between these states are externally adiabatic. That is, in the Stirling and Ericsson cycles there is an internal transfer of heat from one part of the cycle to another. Taking these two processes together, the cycle exchanges no heat with the environment. Cycles with other paths linking the isotherms may be devised, but these are of lesser interest because the processes of constant volume and constant pressure heat exchange can be approximately achieved in practice.

The constant pressure and volume heat transfer processes may be carried out by heat exchangers that are termed regenerators. In the case of the Ericsson cycle, the engine resembles a closed cycle gas turbine with an infinite number of intercooling and reheat stages. This cycle is examined in Chapter 9.

The ideal Stirling engine forces a gaseous working fluid to execute the cycle shown in Fig. 10–1b where the principal state points and processes are shown. The regeneration process takes place at constant volume so that piston-cylinder (i.e., kinematic motion), machines are required for the implementation of the Stirling engine. Figure 10–2 shows the physical succession of events that are implied by the state diagram. The engine shown is a single cylinder device in which a single mass of working fluid undergoes the required changes. In this kind of engine, the regenerated heat must be passed into and out of a solid material through the system boundaries. In the following section, the cycle operation is described. The effectiveness of a single cylinder regenerator is considered in Section 10.6.

10.1.1. *Physical Event Sequence*

Figure 10–2 shows a rectilinear arrangement of a high-temperature reservoir on the left, a regenerator, and a low temperature reservoir on the right. The working fluid, or system, is the gas between two pistons for which a mechanism is designed to execute the motion required. Whenever the working fluid moves to a region adjacent to the heat reservoirs, heat may flow as required through the bounding walls. The axial conduction of heat through the regenerator and out through the piston face surfaces may be made negligibly small by

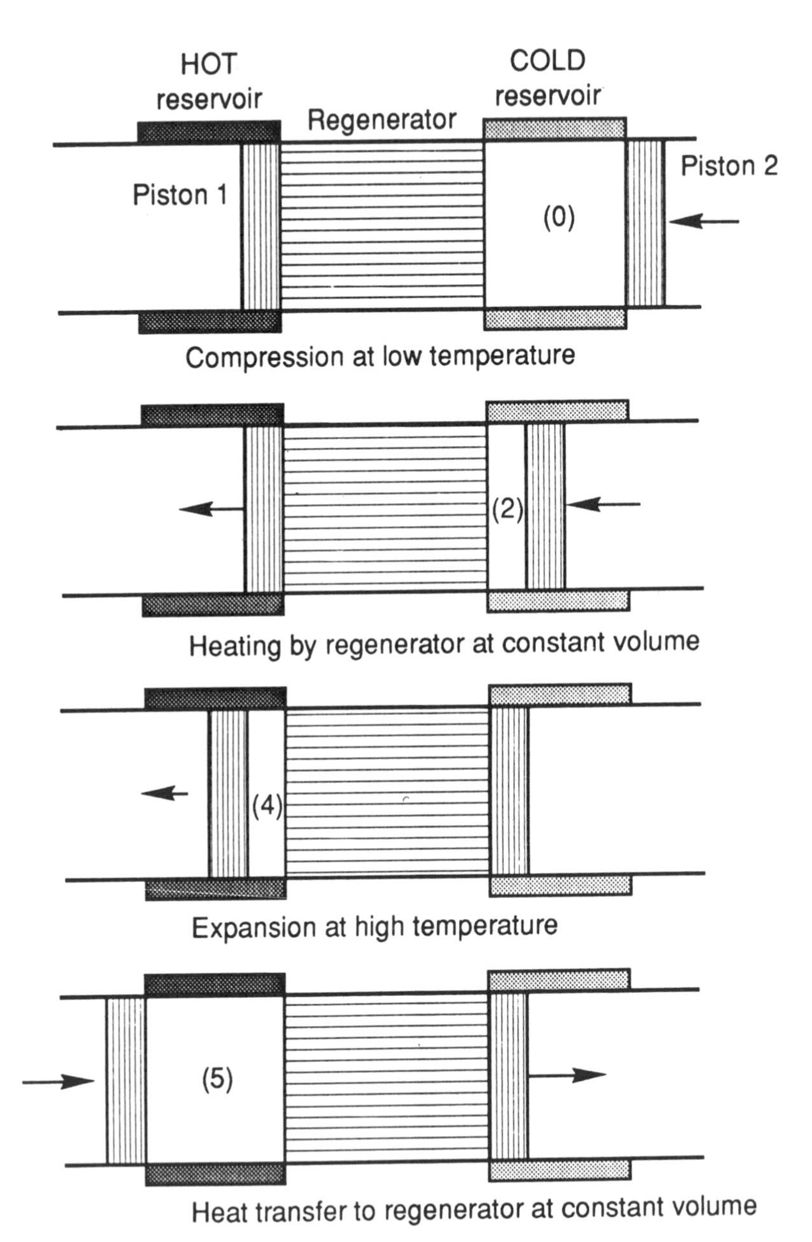

FIG. 10–2. Schematic of the elementary processes in a Stirling cycle. Note high- and low temperature reservoirs and the regenerator.

appropriate design and is therefore neglected. For the purpose of modeling the sequence of events, it may be assumed that the volume associated with the passages in the regenerator may be neglected.

The operating sequence of a Stirling engine begins (Fig. 10–2, top) with the system in equilibrium with the low-temperature reservoir (point 0). While the system remains in thermal contact with this reservoir, the right piston moves to compress the gas (isothermally). Heat is transferred to the low-temperature reservoir during this process.

At point 2 (Fig. 10–2), the compression and heat removal from the cycle is complete and the working fluid is then forced through the regenerator to capture the heat stored in it. When an ideal regenerator, the working fluid exits the regenerator with a temperature equal to that of the high-temperature reservoir (state point 4, Fig. 10–2) when both cylinders move simultaneously to the left. The requirement for the process to be constant volume is ensured by moving the two piston surfaces at the same speed.

Isothermal expansion from 4 to 5 is now carried out by allowing motion of the left piston. During this process heat is added to the system and the expansion work is realized. At state point 5 in Fig. 10–2, the volume reaches a maximum. State point 0 is restored by the simultaneous movement of both pistons returning the working fluid to the cold compression space while it surrenders its high-temperature heat to the regenerator.

In the single working space machine, such as the one shown here, this heat must come from the solid material making up the regenerator. In two (or more) cylinder machines (see Fig. 10–3), one may take advantage of the phase difference in the cyclic events to carry out a heat-transfer process that resembles that of a counterflow heat exchanger. This may be done by making the regenerator of very small heat capacity (which is not easy in practice, particularly for an engine with small power output) and arranging the two heat exchange processes to occur simultaneously. Although this is not simple, it is theoretically possible. This at least allows contemplation of the existence of an ideal regeneration process.

10.1.2. *Attractive Engine Features*

If one were able to carry out these events with completely effective regeneration, one would realize Carnot efficiency from the cycle. Unfortunately, it is not possible to do this in practice, and performance penalties are necessarily incurred. Before considering the nature and impact of the irreversibilities, one may note some of the characteristics of Stirling cycle engines. Comments relevant to many of these advantages are given below.

Near-Carnot efficiency is possible

Ability to use a wide range of heat sources, including low cost fuels in an external, steady flow combustion chamber

Low noise due to low engine rotational speeds

Long life designs may be possible

Power-to-weight ratio may be competitive with other engine types

Engines may be built in a wide range of sizes

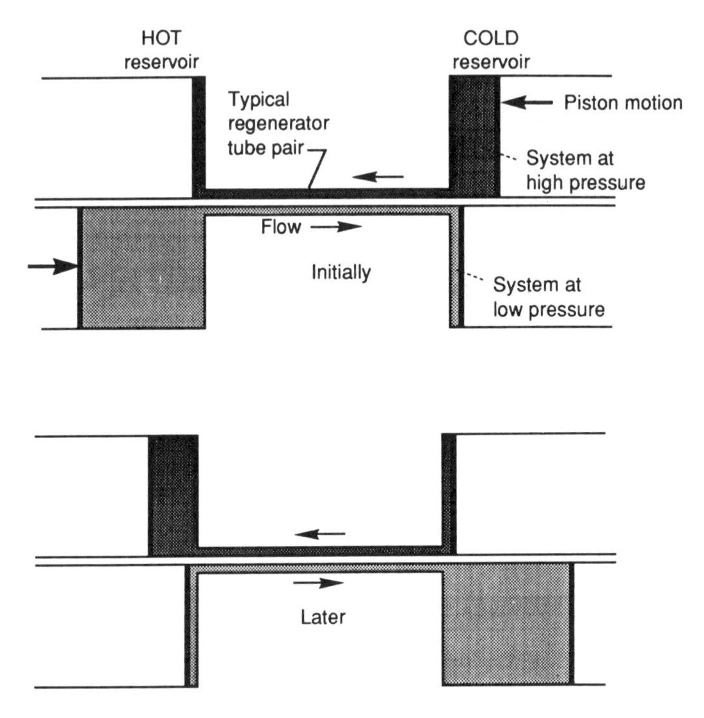

Fig. 10–3. Counterflow heat exchange in a multiple cylinder Stirling cycle engine.

CARNOT EFFICIENCY. The process of transferring heat occurs through material walls that have temperature capability limits. With ambient temperatures of the order of 300 K the maximum cycle temperature ratio that may be envisaged by such engines is of the order of 4 when attention is limited to realistic engine materials. This is particularly true since a design constraint may be to operate this engine at elevated pressure and thus the walls need to be strong as well. This temperature ratio implies that that Carnot efficiency is of the order of 75%. Work to date has shown that real Stirling engines appear to be able to realize about half of the Carnot efficiency (ref. 10–4). An engine with a conversion efficiency in this range is of great interest.

FUEL USAGE. The heat source is a constant temperature zone perhaps supplied by the combustion off a hydrocarbon as illustrated in Fig. 10–4. Under these low-temperature conditions, the fuel quality (as measured by its heating value) does not need to be high, enabling the designer to consider the use of low-cost fuels. One must realize that since the heat is provided to the cycle at the maximum cycle temperature, the enthalpy of the combustion products at this temperature is lost unless a special effort is made to capture this heat by means of another cycle or to use it for air-preheating (see discussion of external combustion, Section 7.5.2).

EMISSIONS. The simplest combustor serving the needs of this engine is an atmospheric pressure device using either a stabilized flame or a fluidized bed

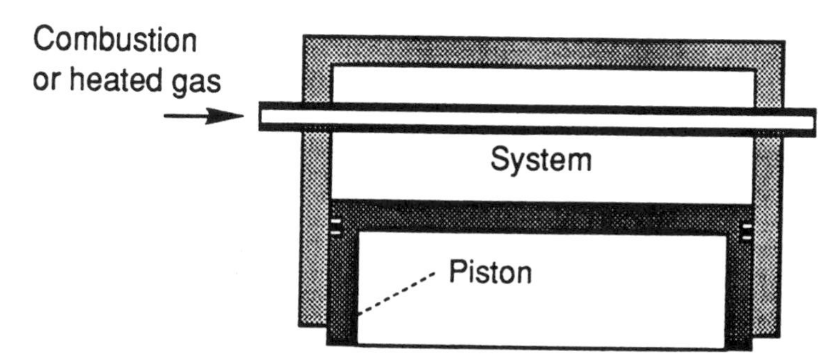

Fig. 10–4. Heat input to a Stirling cycle engine: transfer of heat from hot (combustion) gas to working fluid across a tube wall.

reactor. Regardless of how the combustion heat is provided, the potential for low emissions appears excellent because of the low combustion temperatures. This motivates the development of this engine for some applications.

ROTATION SPEED. The speed controlling process is the transfer of heat into and out of the working fluid. Since this time is usually long compared to cycle times determined by the effects of mechanical and fluid friction, the friction is generally manageable. Benefits associated with the low speeds are low mechanical noise, although the problems of controlling heat and mass leakage become more acute when process frequencies are low. Lastly, the power output is proportional to speed, which results in a relatively low power per unit engine mass.

10.1.3. *Physical Engine Arrangements*

The technical community distinguishes among a number of component arrangements all designed to approximate the events required by the Stirling cycle. The basic configurations are illustrated in Fig. 10–5 as "alpha," "beta," and "gamma" engines. They all share the same thermal components and differ primarily in the arrangement of piston and connecting rods as well as the location of the regenerator.

The "alpha" configuration is characterized by two separate piston/cylinder arrangements driven by two crankshafts. One piston is for compression whereas the other is for expansion of the working fluid. The "beta" configuration incorporates both compression and expansion spaces into the same cylinder and a common crankshaft. This engine type uses a displacer piston to force the fluid through the regenerator. An advantage is that the overlap in the piston and displacer stroke generally results in a reduction in the dead space of the system and thus allows higher compression ratios. "Gamma" configurations are classified as a variation on "beta" engines with a separate displacer cylinder provided.

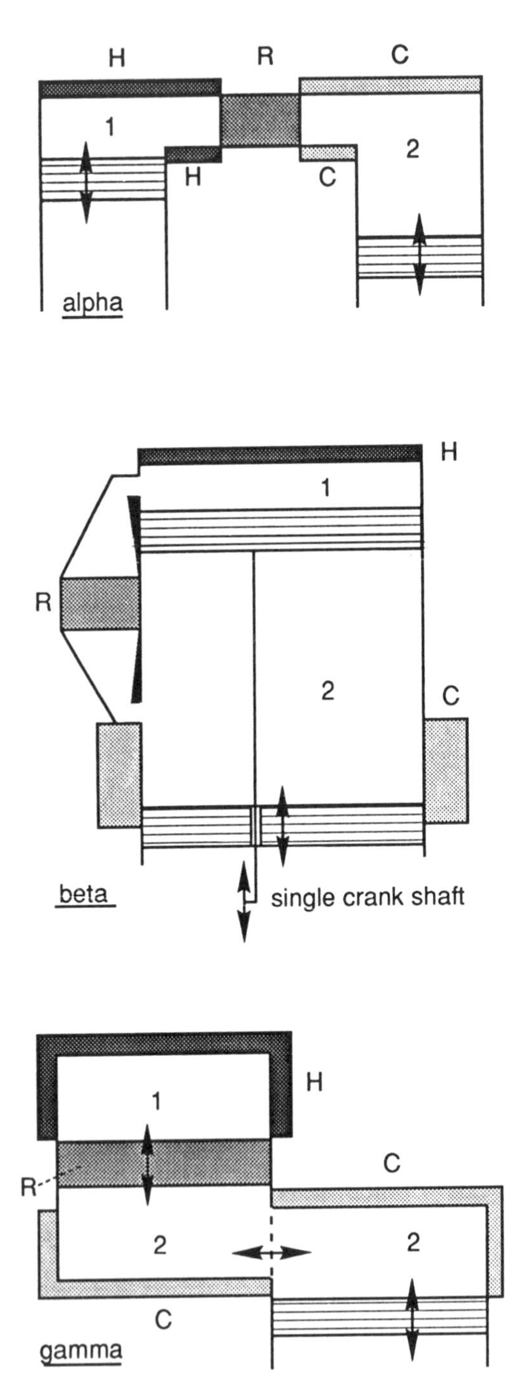

FIG. 10–5. Stirling engine configurations: "alpha", "beta", and "gamma" configurations.

10.2. NEARLY IDEAL CYCLE PERFORMANCE

The regenerator plays an important role in the performance of the Stirling cycle. It is instructive to examine the cycle that is ideal except for the regenerator effectiveness. Expressions for thermal efficiency and specific work for this model will lead to insight on the role of other parameters.

The state points for the Stirling cycle with a nonideal regenerator are shown in Fig. 10–6. On a log T versus s plot, lines of constant volume are linear as shown. The regenerator transfers heat to the working fluid between states 3 and 4 and accepts it between 5 and 6. The temperature drop $\Delta T = T_4 - T_3$ is a measure of the effectiveness of the regenerator (see Section 6.3). For the cyclic conservation of thermal energy in the regenerator:

$$C_v(T_3 - T_2) = C_v(T_5 - T_6) \tag{10–1}$$

The effectiveness of the regenerator is given as

$$\eta_x = \frac{\text{actual heat transferred}}{\text{heat transferable}} = \frac{T_3 - T_2}{T_4 - T_2} \tag{10–2}$$

using the First Law which states that for a constant volume process,

$$Q_{a,b} = C_v(T_b - T_a) \tag{10–3}$$

For the isothermal process 4 to 5, the heat transferred is given by an integration

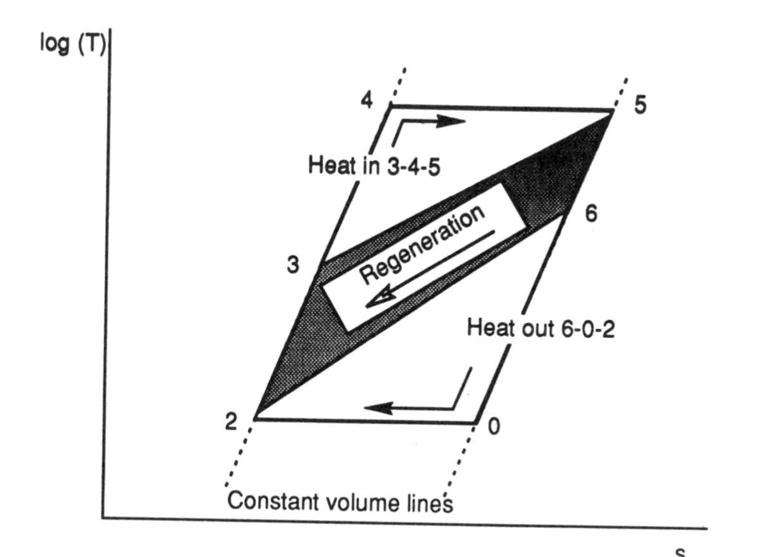

Fig. 10–6. Temperature (Log T)-entropy diagram for Stirling cycle with less than ideal regeneration.

of eq. 2–4

$$Q_{4,5}RT_4 \ln\left(\frac{V_5}{V_4}\right) = RT_4 \ln R_v = C_v T_4 \ln \tau_{ca} \tag{10–4}$$

Note that the combination $(\gamma - 1) \ln R_v$ is the logarithm of the adiabatic temperature ratio, τ_{ca}, the temperature ratio that would be experienced if the compression and expansion were adiabatic and reversible. Thus one may express the performance of this cycle in terms of the same parameters governing other cycles, namely a temperature ratio characteristic of the temperature limits, and one characteristic of the work component processes. For convenience, all energies and temperatures are nondimensionalized by C_v and/or T_0 as appropriate. A unit mass of fluid of the working fluid is assumed. Thus

$$\theta_4 = \frac{T_4}{T_0} \quad \text{and} \quad q = \frac{Q}{C_v T_0} \tag{10–5}$$

The heat input to the cycle is

$$q_H = (1 - \eta_x)(\theta_4 - 1) + \theta_4 \ln \tau_{ca} \tag{10–6}$$

while the heat output from the cycle is

$$q_L = (1 - \eta_x)(\theta_4 - 1) + \ln \tau_{ca} \tag{10–7}$$

The specific work is obtained from $w = q_H - q_L$ or

$$w = (\theta_4 - 1) \ln \tau_{ca} \tag{10–8}$$

which is independent of the regenerator effectiveness, making the power produced by the engine independent of this component. The thermal efficiency, on the other hand, involves η_x and is given by:

$$\eta_{th} = \frac{w}{q_H} = \frac{1 - \dfrac{1}{\theta_4}}{1 + \dfrac{(1 - \eta_x)}{\ln \tau_{ca}}\left(1 - \dfrac{1}{\theta_4}\right)} \tag{10–9}$$

Note that the efficiency approaches the Carnot value when R_v or τ_{ca} is large and that a large value of $\eta_x(\leqslant 1)$ is desirable. For conditions typical of a Stirling engine, the second term in the denominator is of order unity so that roughly half of the Carnot efficiency is attained (ref. 10–4). For chosen values of the regenerator effectiveness, the relationship between specific work and thermal efficiency is shown in Fig. 10–7 [in this figure specific work is defined here

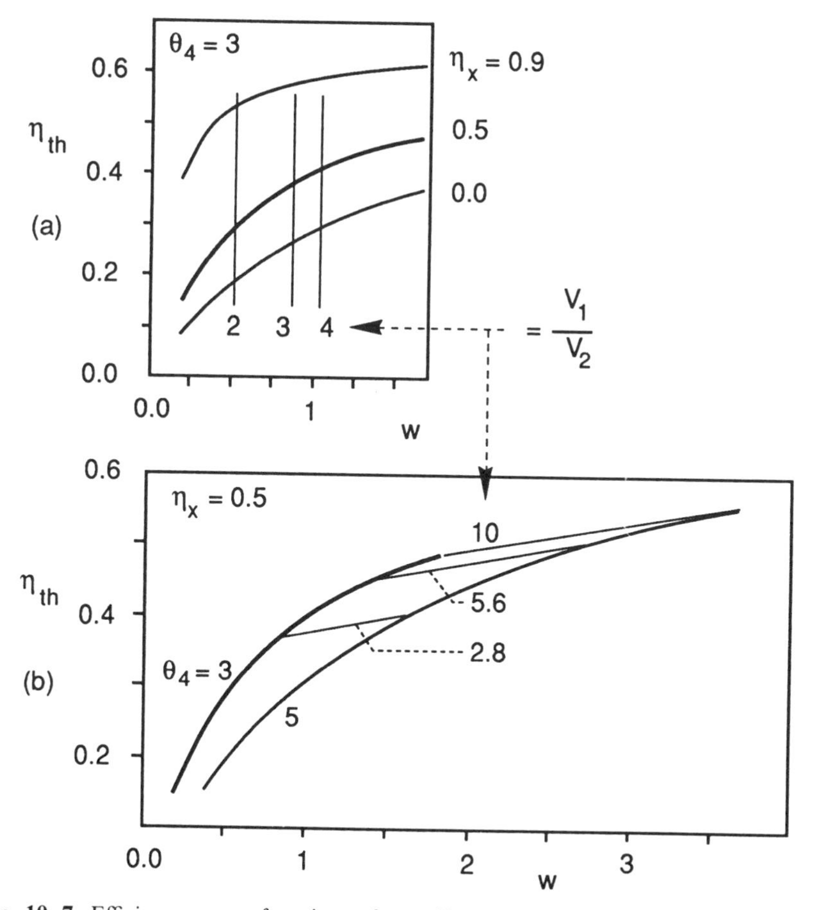

FIG. 10–7. Efficiency as a function of specific work for the Stirling cycle. Parametrization of heat exchanger effectiveness (top) and temperature ratio (bottom). The numbers associated with lines correspond to the volume ratio of the piston motion, V_1/V_2.

as $W/(C_p T_0)$ so that comparison with the Brayton cycle in particular is direct]. Figure 10–7a is for a temperature ratio of $\theta_4 = 3$ and various constant values of η_x. Figure 10–7b shows the rôle played by the temperature ratio. Evidently, under these circumstances, a large value of θ_4 is desired for large specific work, but this improvement comes at the expense of decreased efficiency. In practice, metallic material limits limit θ_4 to values between 3 and 4 (288 K ambient and perhaps 1100 K maximum temperatures), although higher temperatures may be possible with the use of ceramics.

From eq. 10–8, it is evident that the only remaining design variable, the volume ratio R_v, increases along the curves in Fig. 10–7. It is therefore desirable to have R_v as large as possible, provided that additional losses do not counter this trend. In practice, volume ratios near 2–3 are found most practical (i.e., this choice results in near maximum values for either or both w and thermal efficiency). This discrepancy is explained partially by the suggestion of Reader and others (refs. 10–5, 10–6) that this cycle description is not accurate. Section 10.4 is a review of the so-called pseudo-Stirling cycle which appears to be a better model for describing the cycle and thus for identifying optimal cycle parameters.

10.2.1. *Power Output*

The power output of any engine (including the Stirling engine) varies as

(specific work) × (mass per unit time processed).

The specific work depends on the cycle variables. With temperature limits fixed by materials and volume ratio by losses and the cycle temperature ratio, it follows that the performance of Stirling engines is proportional to the mass of working fluid in the system and the operating frequency. This power can also be written as

$$\dot{W} = B_0 f \bar{p} V_0$$

where the constant of proportionality, B_0, is called the Beale number. With f in hertz, p in bar (or atmospheres), and the displacement V_0 in cm^3, the Beale number is found to vary between 0.010 and 0.020 when the power is expressed in watts. Walker (ref. 10.4) suggests an intermediate value as a rough design number to enable scaling the size of a projected Stirling engine. A variant of the Beale formula (refs. 10–4 and 10–7) better accounting for temperature effects is found to be

$$\dot{W} = F f V_0 \bar{p} \frac{T_H - T_L}{T_H + T_L}$$

where $F = 2$ for the *indicated* power of the Stirling cycle. For the actual power, diminished by friction and losses, F falls to about 0.3 for practical engines.

10.3. PRACTICAL ENGINE DESIGN CONSIDERATIONS

Since the power output from a Stirling engine is proportional to the mass flow rate processed and to the specific heat of the working medium, the choice of working fluid and its mean pressure are design variables. The gas of choice is one with a high specific heat and therefore a low molecular weight. Hydrogen or helium is often chosen for Stirling engines, although if power density is not an issue, air may be employed because of its lower cost and greater ease of containment. The seals of the engine should be able to contain the working fluid indefinitely. Low molecular weight gases unfortunately leak readily both to the outside world and past seals of the piston cylinder arrangement. The leakage path to the outside is important when the working pressure is

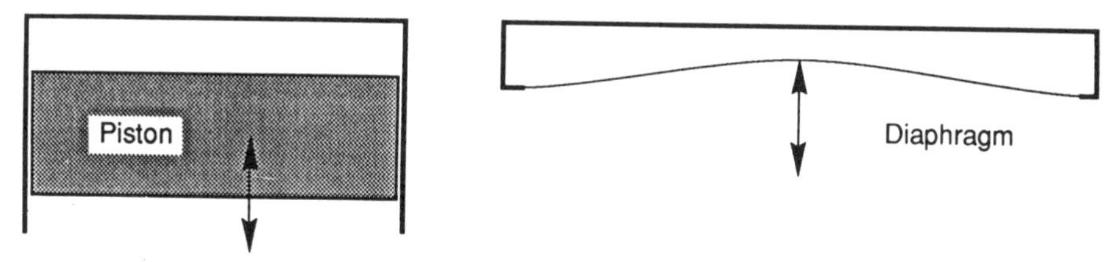

Fig. 10–8. Piston/cylinder and diaphragm displacement machines.

considered. The path past seals is important primarily in the design of the piston seals where mechanical friction power loss is an issue. In order to avoid a case where leakage becomes important in limiting the unattended operational lifetime, some Stirling engines have been designed with the reciprocating boundary motion controlled by a diaphragm that is firmly attached to the cylinder wall. This arrangement allows welding of the seal, which is superior to assembled arrangements such as gaskets and the like (Fig. 10–8). In these geometries the effective use of the swept volume is relatively poor.

The cycle efficiency depends on the compression temperature ratio, or equivalently, the physical volume *ratio*. One is therefore free to choose the mean pressure *level* in the working fluid of the engine to maximize power per unit engine size. Higher mean pressures obviously result in higher density so that the mass of the fluid is increased when one operates at high mean pressure. This makes for compact engines and helps in increasing the Reynolds number of the various flow processes which is, in turn, helpful in reducing the relative importance of fluid friction effects. A consideration in specifying the pressure level is the buffer pressure of the engine. This issue is discussed in greater detail in Section 10.7.

The relation between thermal efficiency and specific power of a real Stirling engine operated with different working fluids at a mean pressure of 10.7 MPa is shown in Fig. 10–9 from ref. 10–6, or the original ref. 10–8. The power output correlates with the higher specific heats. Efficiencies are in a generally similar range, and depend on speed because flow friction losses as well as heat transfer rates are affected.

Figure 10–10 (from ref. 10–6) shows the variation of torque as a function of rotational speed for various working fluids. At low speed, leakage is important while friction plays a limiting role at high speeds. One would expect the torque to be proportional to pressure for a chosen working fluid and this is shown in Fig. 10–11.

10.3.1. *Rotational Speed Issues*

Ideally the Stirling cycle operates at a rotational speed which is slow enough for the heat transfer process to be completed. This time scale is dictated by the thermal diffusivity of the gas since that of the solid materials involved is significantly shorter. Further, gas transfer processes through regenerators should be completed sufficiently slowly for the pressure losses to be small. These

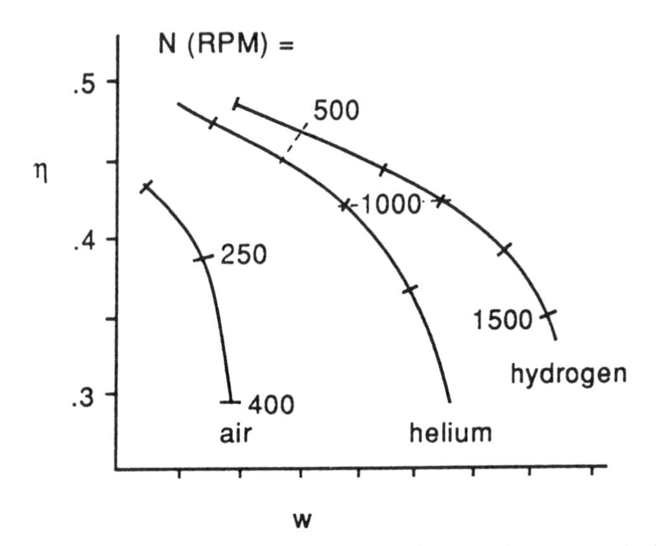

Fig. 10–9. Efficiency as a function of specific power of one engine operated with various working fluids. Engine speed in RPM is indicated.

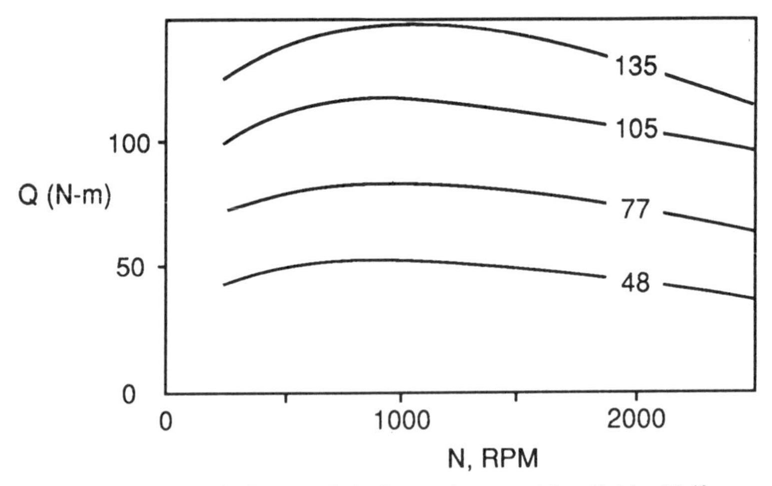

Fig. 10–10. Torque-speed characteristic for various working fluids. Uniform working fluid pressure.

requirements conflict with the requirement for a high cycle frequency to obtain high power from a given device.

The time scales for heat transfer can be shortened by ensuring the flow to be turbulent to provide good mixing of the volume of gas to be heated or cooled and good thermal contact with the wall. Permitting some level of pressure loss in the various flow processes typically leads to an improvement in specific work performance.

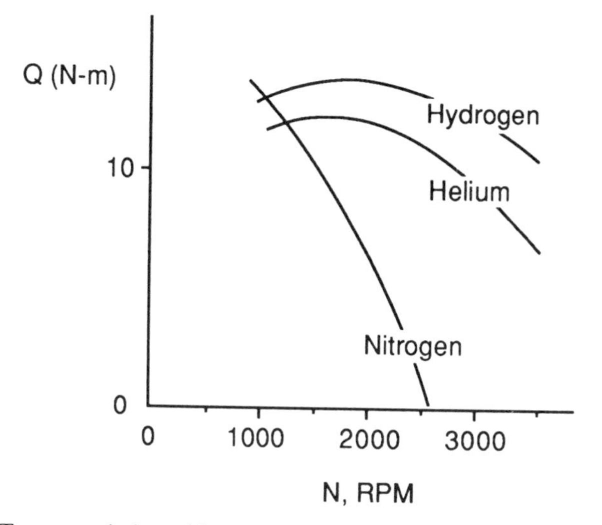

Fig. 10–11. Torque variation with speed for varying maximum pressure (in atmospheres) of the working fluid.

10.4. THE PSEUDO-STIRLING CYCLE

The cycle analysis discussed in Sections 10.1 and 10.2 fails to describe the cycle performance in an accurate manner. This analysis thus fails to be useful for finding the best cycle design for a particular purpose. A number of investigators (ref. 10–5) have examined ways to overcome the model limitations in order to arrive at a better cycle description. One such cycle is the so-called pseudo-Stirling cycle which recognizes that the compression and expansion processes meant to be isothermal are actually carried out so quickly that a better way to describe the compression process is to say that it consists of an adiabatic compression followed by a constant volume heat interaction which brings it twice to near the minimum cycle temperature.

Figure 10–12 shows the *T-s* diagram path for this cycle. The standard Stirling cycle is given by 1-2-3-4-1, which is not shown explicitly here. The pseudo-Stirling processes are

- 1-2′ adiabatic compression
- 2′-2 constant volume heat rejection to the environment
- 2-2″ constant volume heat transfer *from* the regenerator
- 2″-3 external heat input to the cycle
- 3-4′ adiabatic expansion
- 4′-4 second phase of constant volume heat input to the cycle
- 4-4″ constant volume heat transfer *to* the regenerator
- 4″-1 second-phase constant-volume heat rejection to the environment

Figure 10–13 shows the physical sequence of the events describing the cycle.

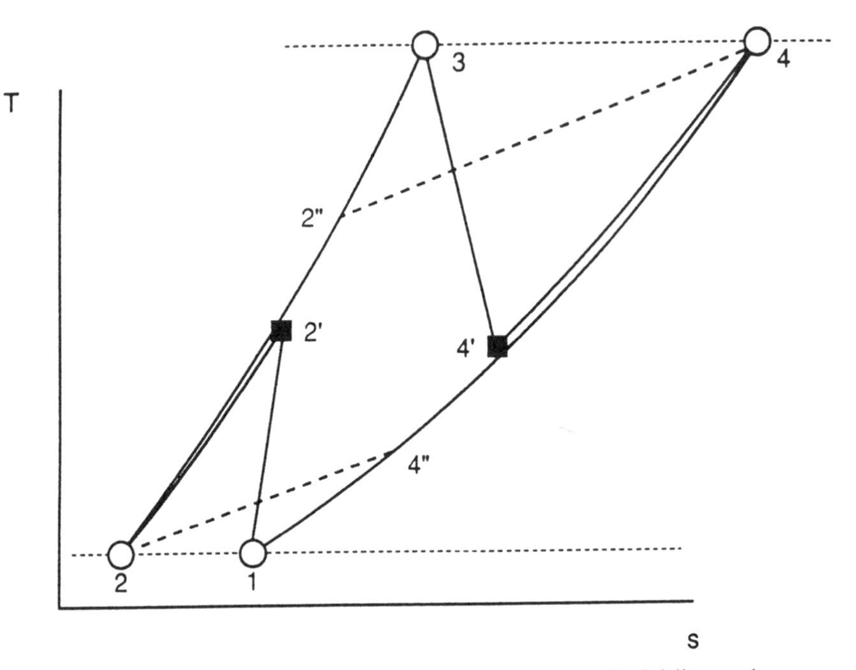

FIG. 10–12. Temperature-entropy diagram of the pseudo-Stirling cycle.

For the purpose of calculating efficiency, the heat input to the cycle is given by

$$\frac{Q_{in}}{C_v} = (T_3 - T_{2''}) + (T_4 - T_{4'}) \tag{10–10}$$

The effectiveness of the regenerator is the ratio of heat transferred to the heat transferable or

$$\eta_x = 1 - \frac{(T_4 - T_{2''})}{(T_4 - T_1)} \tag{10–11}$$

In the limit of the perfect regenerator. ($\eta_x = 1$) the temperature $T_{2''}$ approaches T_3, and $T_{4''}$ approaches T_1. The common method for analyzing this cycle is to use the cycle volume ratio, $R_V = V_1/V_2$ as the independent variable. Assuming the compression process to be reversible, this volume ratio is related to the temperature ratio by

$$\tau_c = \left(\frac{V_1}{V_2}\right)^{\gamma - 1} \tag{10–12}$$

which introduces the specific heat ratio, γ, into the analysis. A better way is to use the temperature ratio as the independent design variable and write the performance equations in terms of this parameter. This avoids a requirement to

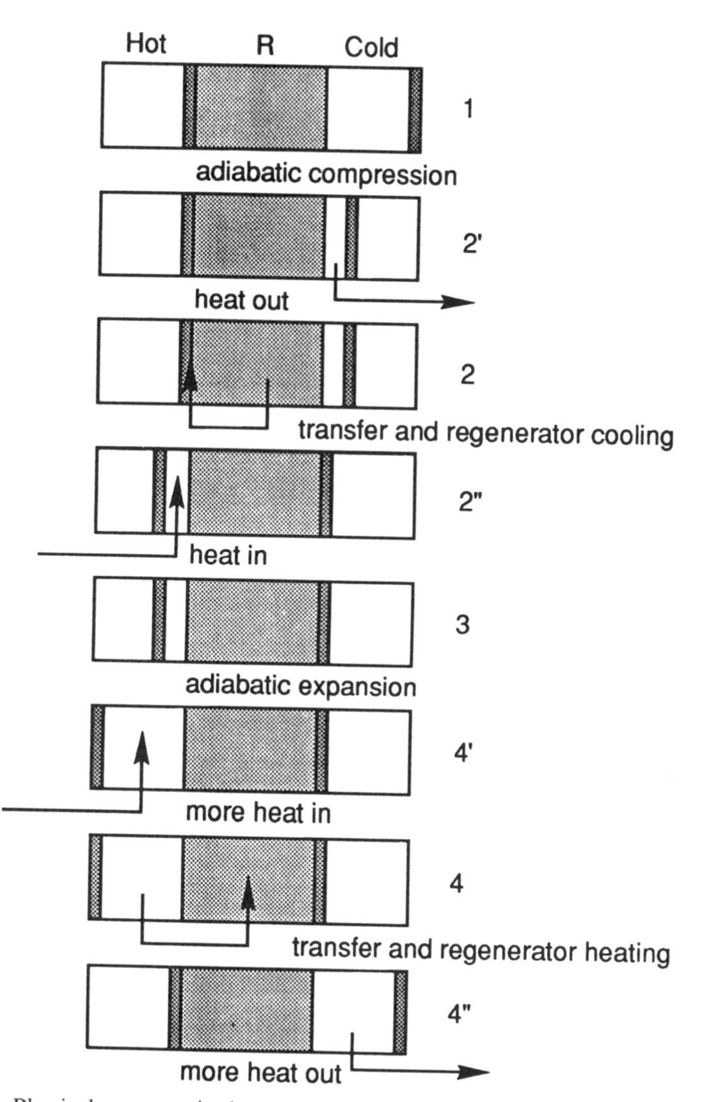

FIG. 10–13. Physical processes in the pseudo-Stirling cycle. The states shown correspond to those shown in Fig. 10–12.

specify γ directly. The relation between volume and temperature ratio as given by eq. 10–12 is shown in Fig. 10–14. It will be seen that the temperature ratio range of interest is 1–2 and that the volume ratio range varies between 2 and 4, depending on γ for the working fluid. The nondimensional specific work is given by the expander work less the compression work.

$$w = \frac{W}{mC_v T_1} = \theta_4\left(1 - \frac{1}{\tau_c}\right) - (\tau_c - 1) \tag{10–13}$$

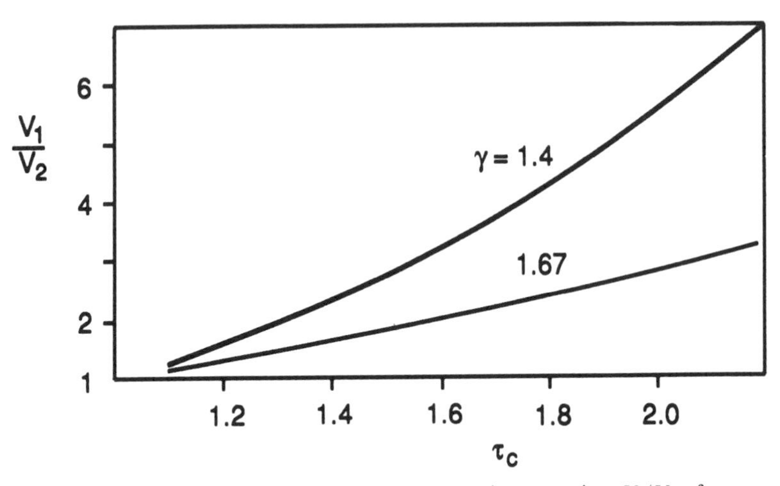

FIG. 10–14. Influence of specific heat ratio on volume ratio, V_1/V_2, for specified compression temperature ratio, τ_c.

Note the similarity of the equations describing the Brayton cycle in Section 9.1 and in particular eq. 9–9. The specific work defined here differs from that of the Brayton cycle by a factor γ and w for the pseudo-Stirling cycle is correspondingly larger. The temperature parameter θ_4, with $T_3 = T_4$, was defined as in Section 10.2 to parallel the nomenclature developed in Sections 9.1, 10.1, and 10.2 describing the ideal Stirling cycle. The value of $\theta_4 = 3$ may be a realistic value and is therefore chosen for numerical examples. The ratio of the work and heat input (both per unit mass) gives the thermal efficiency:

$$\eta_{\text{th}} = \frac{\theta_4\left(1 - \dfrac{1}{\tau_c}\right) - (\tau_c - 1)}{(\theta_4 - 1)(1 - \eta_x) + \theta_4\left(1 - \dfrac{1}{\tau_c}\right)} \tag{10–14}$$

The heat input is obtained from eq. 10–10. The variation of thermal efficiency with compression temperature ratio (τ_c) is shown in Fig. 10–15 for a number of values of the regenerator efficiency. A perfect regenerator allows operation at the Carnot efficiency, although the value of the compression ratio required is unity. This cycle design is seen to produce very little (zero, in fact) work from eq. 10–13. For a good regenerator (0.85), the best performance is significantly below the Carnot value. An ineffective regenerator gives an uninteresting performance, even with the ideal component efficiencies assumed here.

Since the specific work is independent of the heat exchanger effectiveness, the maximum value of this parameter depends only on the cycle temperature ratio. Thus, following the approach of Section 9.1, one has for the pseudo-Stirling cycle,

$$\tau_{c,\max w} = \sqrt{\theta_4} \tag{10–15}$$

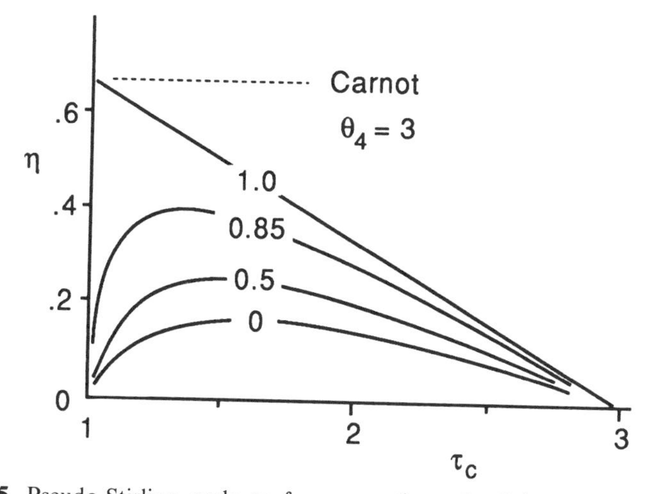

Fig. 10–15. Pseudo-Stirling cycle performance: thermal efficiency as a function of compression temperature ratio. Heat exchanger effectiveness noted.

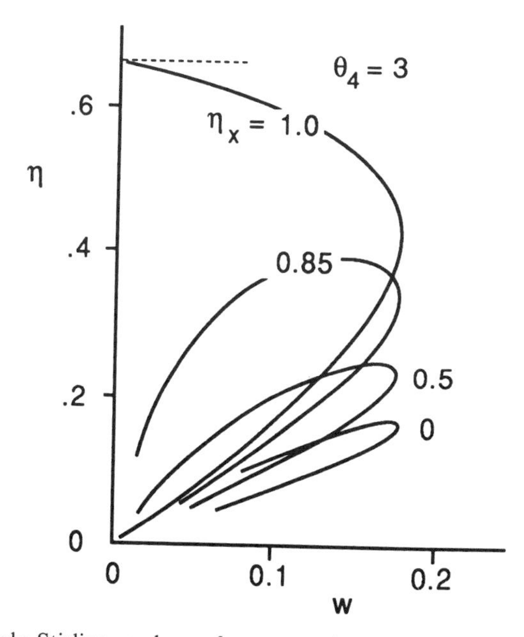

Fig. 10–16. Pseudo-Stirling cycle performance: thermal efficiency as a function of specific work. Heat exchanger effectiveness noted.

A cross plot of specific work and thermal efficiency is given in Fig. 10–16. Note that τ_c for maximimum efficiency can be calculed from eq. 10–14 which yields

$$\tau_{c,\,\max\eta} = \frac{1 + \sqrt{1 + (\beta + 1)(\theta_4 \beta - 1)}}{(\beta + 1)} \qquad \text{where } \beta = \left(1 - \frac{1}{\theta_4}\right)(1 - \eta_x) \tag{10–16}$$

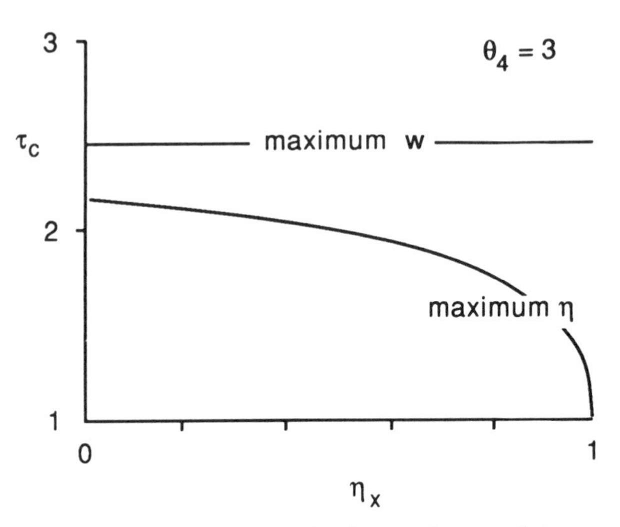

FIG. **10–17.** Variation of temperature ratio for maximum efficiency and that for maximum specific work with regenerator effectiveness.

This value, together with that for maximum efficiency is shown in Fig. 10–17. Note that this cycle, unlike the Brayton cycle, requires a higher τ_c for maximum specific work than for maximum thermal efficiency.

In the Stirling cycle community, there is a custom of using the work per unit displacement as a measure of engine compactness. This is probably associated with the development history of the other reciprocating cycles and is equivalent to the mean effective pressure. Reader (ref. 10–5) defines a dimensionless work "transfer" (from the cycle) as

$$w_0 = \frac{W}{p_1(V_1 - V_2)} = \frac{W}{p_1 V_1(1 - R_v^{-1})} = \frac{W}{mC_v T_1(1 - \tau_c^{-1/(\gamma-1)})} \frac{C_v}{R} \tag{10–17}$$

or, in terms of the conventional specific work, w:

$$w_0 = \frac{1}{\gamma - 1} \frac{w}{1 - \tau_c^{-1/(\gamma-1)}} \tag{10–18}$$

This work per unit displacement does depend on the gas specific heat ratio, γ, as shown in Fig. 10–18 where this parameter is shown for a diatomic working fluid ($\gamma = 7/5$) and a monatomic one ($\gamma = 5/3$) as well as the work per unit mass. Note that maximizing w_0 favors low values of the compression ratio in comparison to maximizing work per unit mass. For $\gamma = 7/5$, the maximum in w_0 is quite flat whereas for $\gamma = 5/3$ gas there is no a maximum which leads to an interesting choice for τ_c. This w_0 parameter is of interest in situations where a design goal is to minimize the total piston movement in the cylinder.

The thermal efficiency is that achieved by selecting the τ_c for maximum efficiency and is a function of only the regenerator effectiveness. This thermal

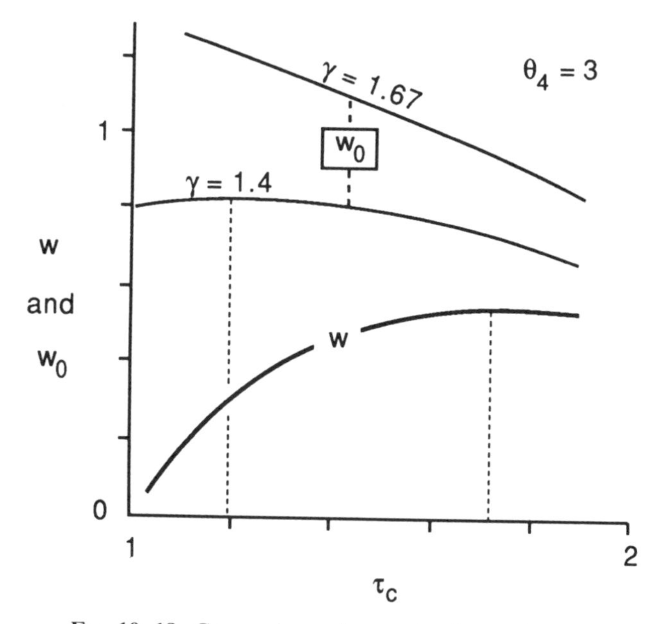

FIG. 10–18. Comparison of w_0 and w in eq. 10–18.

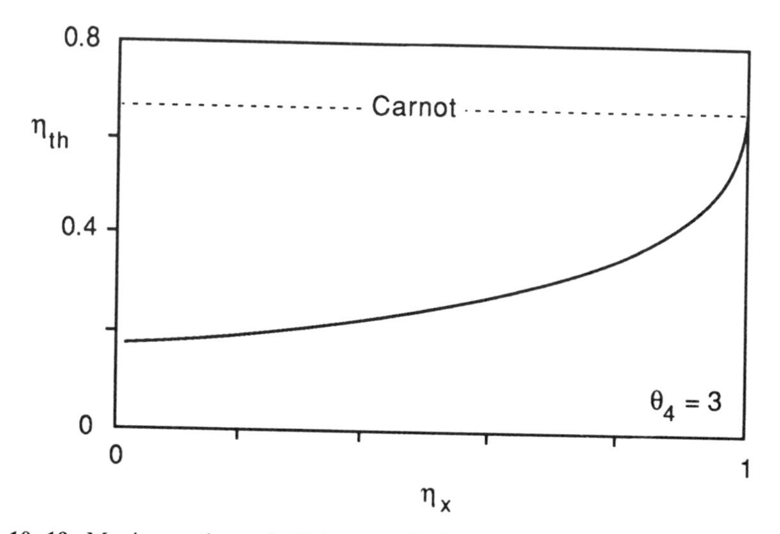

FIG. 10–19. Maximum thermal efficiency variation with heat exchanger effectiveness.

efficiency is shown in Fig. 10–19. Evidently it is very important to have an effective regenerator in the cycle.

10.5. IRREVERSIBILITIES IN STIRLING CYCLE ENGINES

The cycle is implemented in practice by an arrangement of cranks, connecting rods, and similar mechanisms to ensure that the working fluid volume is as required in the p-V diagram. Since no practical mechanism can execute a discontinuous motion, the engine designer must be content with an approximation to the desired motion. Classical approaches to dealing with this fact include writing the volumes of the working fluid spaces as harmonic functions with appropriate mechanical relations and analyzing the work ($p\,dV$ integrations) and heat interactions as the system proceeds through one cycle.

10.5.1. *Cycle Form in the* T-s *Plane*

A simple thermodynamic approach is to recognize that the path in the T-s diagram is a figure inscribed in the ideal path. The ideal path in the plane of (ln T)-s is a rectangle. Thus one might inscribe a super ellipse (exponents $m > 2$ in the expression below) to determine the severity of a nonideal path. The super ellipse is described by an equation of the form

$$\left(\frac{x}{a}\right)^m + \left(\frac{y}{b}\right)^m = 1 \tag{10–19}$$

where $m = 2$ constitutes an ordinary ellipse. Values for $m = 4$, 8 and larger values will cause the path to approach the ideal path. Without showing the details of how one might do this, the T-s plane path is shown for $m = 2, 4, 8$ in Fig. 10–20. A cycle with 50% regenerator effectiveness can be analyzed and the results are shown in Fig. 10–21. The conclusion is that for modest "rounding" of the cycle, the performance penalty is not large.

10.5.2. *Schmidt Analysis*

A more complex approach is the classical analysis originally put forth by Schmidt (1871, ref. 10–10). His analysis involves recognition of the fact that the

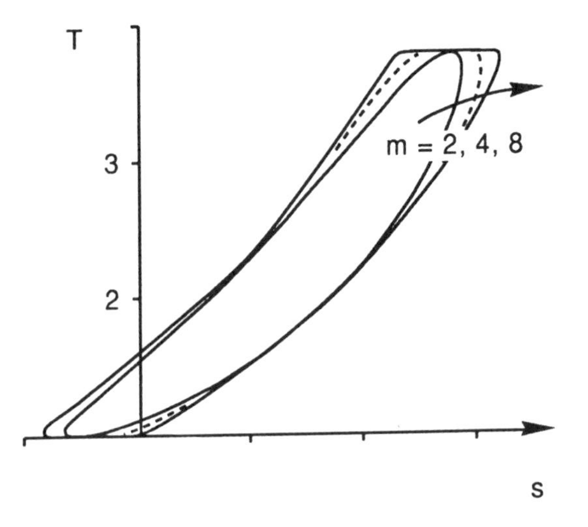

Fig. 10–20. Superelliptic Stirling T-s diagram. Value of the exponent m is indicated.

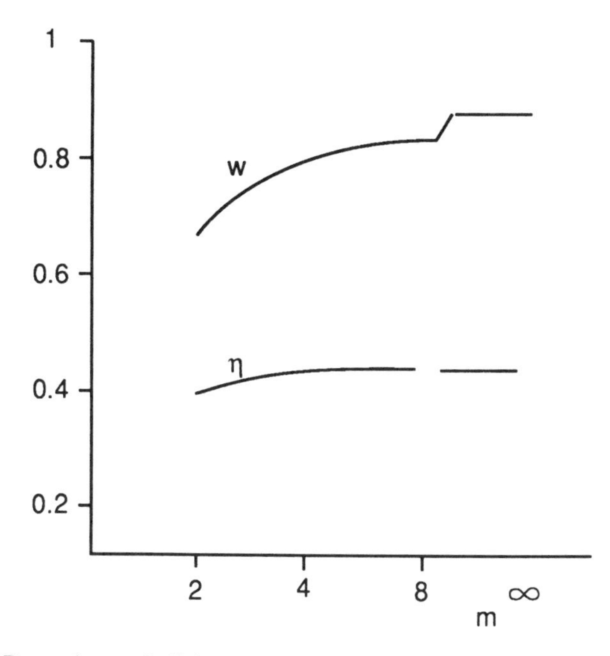

FIG. 10–21. Dependence of efficiency and specific work of the superelliptic Stirling cycle on the exponent, m.

piston motion is near sinusoidal as the length of the connecting rod is made large compared to the crack radius. The work of compression and expansion is obtained in such an analysis as an integration of $p\,dV$ over the engine shaft's angular motion. The heat budget is determined from constant volume and isothermal heat interactions. The work of Walker (1980, ref. 10–6) summarizes work associated with the Schmidt analysis and the assumptions that are made to carry it out. The simplified analysis is still complex, but it is algebraic, lending itself well to machine computation for an approximate determination of cycle performance. The reader is also referred to refs. 10–11 to 10–15 for further discussion.

10.5.3. *Modern Analysis Methods*

The failure of simple analysis methods to predict accurately the performance of Stirling engines suggests that many physical effects must be correctly included in any analysis. The reasons are that a significant amount of the total heat budget is stored and released in solid materials. This means that accurate means must be employed to describe the unsteady heat flow phenomena. The behavior of the solid is rather straightforward through the use of thermal diffusivity and conductivity. For the gas, however, the transport is harder to describe because it involves turbulent motion, which implies that mixing proceeds at a finite rate. The most difficult part of the description is the interface of these two materials where unsteady boundary layers of varying scales play important roles. The modeling difficulty is the lumping of all the phenomena into input/output "nodes" to a sufficient degree of accuracy.

In order to overcome such difficulties, finite element codes have been developed for understanding the results of experimental engine tests and to direct the designer to better configurations. These codes are characterized by the number of nodes (i.e., elements involved in the heating and working of the engine). Modern computers handle large numbers of such nodes with ease, and the analysis problem is reduced to identifying the significant nodes and their roles in the engine. Conservation equations are then written for mass, energy, etc., and the engine is run through a cycle. After periodicity is satisfied, the quantities of interest such as work output and heat input are determined. The success of these models is at the present time not as good as it should be to enable one to draw general conclusions regarding the best way to design a generic Stirling cycle engine. One important reason is that these nodal analyses depend strongly on the geometry of the engine.

The Stirling cycle analysis is confounded by the fact that the gas temperatures are limited to values that are low compared to say, a Brayton cycle engine or even an internal combustion engine. This means that a significant fraction of the heat can be transported away from its desired location, constituting a loss.

10.6. SOLID REGENERATOR PERFORMANCE

A single cylinder Stirling cycle engine relies on the storage of heat from the high-temperature side of the cycle to be transferred to the low-temperature side in an unsteady periodic fashion by having the working fluid "slosh" back and forth through the regenerator. The following is an examination of the performance of a solid material regenerator as a thermodynamic heat transfer device. The goal is to store heat at the highest and lowest temperatures of the cycle and thus achieve ideal effectiveness. This goal is shown to be unattainable and depends on the physical characteristics of the regenerator material as well as its design. For a review of a recent numerical investigation, see ref. 10–16.

10.6.1. *Model*

Consider the cold compressed gas on the right side of the regenerator shown in Fig. 10–22. This fluid is to be transferred to the left side of the regenerator at constant volume while being heated. This heating process is not describable on a state diagram because the working fluid is nonuniform until the process is completed and the fluid is well mixed.

Initially the regenerator is at a high temperature, and the cold fluid is about to enter the long flow tube through the regenerator. The first element of that fluid discharged to the left side will be hot because the regenerator walls are hot. These walls will cool down as the process continues and the last element will be colder. Finally, the mixed out temperature of the gas on the left side is an

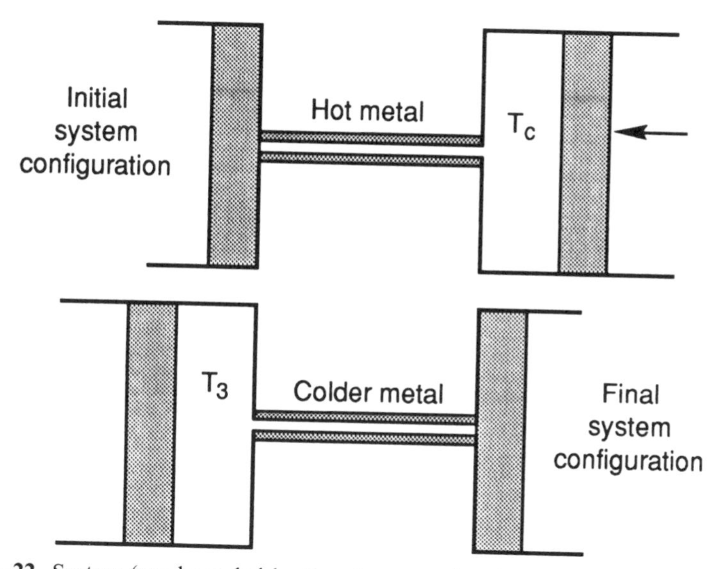

Fig. 10–22. System (gas bounded by the pistons and walls) description during heating of cold compression gas.

average of these temperatures which is necessarily lower than the highest (desirable) temperature T_4. Thus the effectiveness must be less than unity. The temperatures of the regenerator walls and of the fluid depend on the heat transfer process which therefore controls the regenerator performance.

For the purpose of analyzing the flow and heat transfer, one may assume that the piston motion is such that it provides a steady flow through the tubes of the regenerator. The time scale for flow through the regenerator is therefore short compared to the time associated with the complete fluid transfer. This allows one to assume steady flow through the tubes.

The regenerator is modeled as a series of long parallel tubes which bring the gas in contact with the material of the regenerator. Figure 10–22 shows one such tube and the fluid that flows through it. Since may materials of interest have a thermal diffusivity which is high when compared to the physical dimensions and the flow time scale, this material can be assumed to be capable of smearing out the temperature gradients both along the axis of the tube and in the radial direction. Thus for analysis purposes, the temperature of the tube material is assumed to be spatially, but not temporally, uniform. The time variation of the material temperature is assumed to be slow compared to the flow time. This allows one to say that during the transit of a fluid element, the wall temperature is constant. In effect, the assumption made is that the volume allowed to the gas in the regenerator is small compared to the volume of fluid forced through the regenerator. Extending this work to allow an axial variation of material temperature is possible, but requires more extensive numerical calculations

The goal of the present calculation is to estimate the average hot space temperature (T_3) at the end of the regenerator heat input process. Figure 10–23

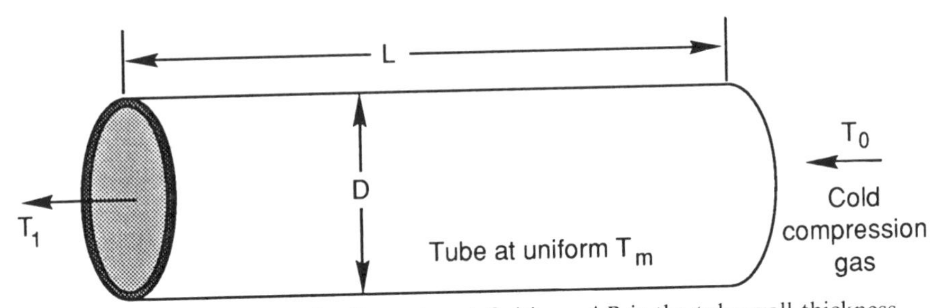

FIG. 10–23. Regenerator tube dimension definitions. ΔR is the tube wall thickness.

illustrates the physical characteristics of the flow tube. The dimensions are indicated: length L, diameter D, and wall thickness ΔR The initial and uniform temperature of the incoming fluid is T_c, whereas the slowly varying outflow temperature is T_1. The mass of the gas transferred is m, whereas τ denotes the time over which this process is to take place. Thus the transfer time is

$$\tau = \frac{m}{\rho u \frac{\pi}{4} D^2} \tag{10–20}$$

The requirement that

$$u\tau \gg L \tag{10–21}$$

justifies the assumption of steady flow through the tube since the tube length must be short compared to the total flow column length $u\tau$. At any time, the metal in the tube wall is a uniform T_m.

10.6.2. *Fluid and Regenerator Material Temperatures*

The steady flow heat transfer relation that describes the axial distribution of gas temperature is

$$\dot{m}C_p \frac{dT}{dx} = \rho u C_p \frac{\pi D^2}{4} \frac{dT}{dx} = h(T_m - T)\pi D \tag{10–22}$$

where h is the heat transfer coefficient which, for simplicity, may be assumed to be proportional to the local pipe flow fraction coefficient by the Reynolds Analogy (eq. 6–2):

$$h = \rho u C_p \frac{f}{2} \tag{10–23}$$

Denoting the nondimensional distance along the tube as $x/L = \xi$, eq. 10–22 can be written as

$$\frac{dT}{T_m - T} = 2\frac{fL}{D}\,d\xi = 2\phi\,d\xi \tag{10–24}$$

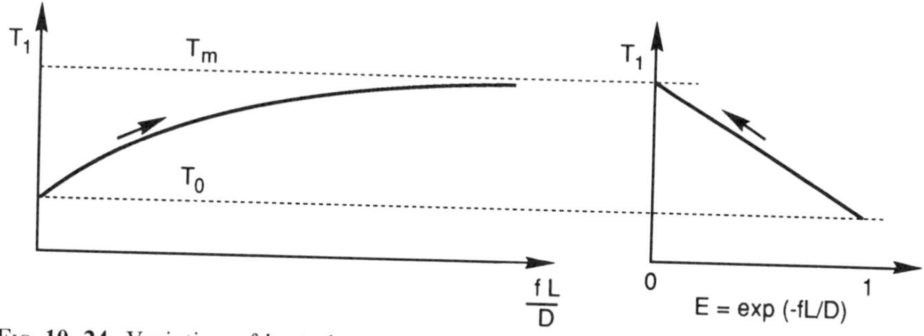

Fig. 10–24. Variation of heated gas temperature with changing effective flow lengths.

where ϕ is shorthand for fL/D as introduced in Chapter 6. This equation can be integrated to a point at ξ

$$\frac{T_m - T(\xi)}{T_m - T_c} = \exp(-2\phi\xi)$$

or at $\xi = 1$,

$$\frac{T_m - T_1}{T_m - T_c} = \exp(-2\phi) \equiv E < 1 \qquad \text{or} \qquad T_1 = T_m - (T_m - T_c)E \quad (10\text{–}25)$$

where T_c is the cold entry temperature, whereas T_1 is the temperature of the exiting heated gas. The term $E = \exp(-2\phi)$ is introduced for algebraic simplicity. Figure 10–24 shows the variation of the outflow temperature as the design (as characterized by fL/D) is varied. The metal temperature T_m is constant during the time interval between entry and exit of the fluid element. The friction factor is assumed constant because it depends only on the flow Reynolds number, which is approximately constant during the flow transit because both Reynolds number and mass flow rate are proportional to ρu. A similar approach yields a relation between the hot entry temperature and the outflow temperature during the regenerator heating process:

$$\frac{T_{1'} - T_m}{T_H - T_m} = \exp(-2\phi) = E \qquad (10\text{–}26)$$

During the flow cycle time, the metal temperature will change as a result of heat input or output. In time dt the heat lost by the metal, equals that gained by the gas. The heat lost to the gas is the integral of the nonuniform temperature difference between gas and wall which is

$$\int_0^L h(T(x) - T_m)\pi D\, dx$$

The temperature difference is given by the general x form of eq. 10–25. With the Reynolds analogy, this integral then reads

$$-(T_m - T_c)\rho u C_p \frac{f}{2}\pi D L \int_0^1 \exp(-2\phi\xi)\,d\xi = -\rho u \frac{\pi D^2}{4} C_p(T_m - T_c) \times (1 - \exp(-2\phi))$$

Thus the heat balance on the tube material reads,

$$\left.\begin{aligned} \rho_m(\pi\,\Delta R D L)C_{pm}\frac{dT_m}{dt} &= -\rho u\frac{\pi D^2}{4}C_p(T_m - T_c)(1-E) \\ \text{or}\qquad m_m C_{pm}\frac{dT_m}{dt} &= -\rho u\frac{\pi D^2}{4}C_p(T_m - T_c)(1-E) \end{aligned}\right\} \quad (10\text{–}27)$$

The total mass processed, m, may be introduced though eq. 10–20 so that the tube material temperature may be determined from eq. 10–27:

$$\frac{dT_m}{(T_m - T_c)} = -(1-E)\alpha\frac{dt}{\tau} \qquad \text{where } \alpha = \frac{mC_p}{m_m C_{pm}} \quad (10\text{–}28)$$

The ratio α (defined in a manner similar to that in eq. 6–9) is the heat capacity ratio of the gas processed to that of the tube material mass. Equation 10–28 shows that the metal temperature varies exponentially in time and identifies the characteristic time scale. Integrating this equation over the process time period, τ, one obtains an expression for the minimum and maximum metal temperatures. The result is

$$\frac{T_{m,\,min} - T_c}{T_{m,\,max} - T_c} = \exp(-\alpha(1-E)) \equiv \psi \quad (10\text{–}29)$$

The right side of this equation controls the tube material temperature variation, and the new parameter, ψ, embodies the two physical aspects of the problem:

1. The heat capacity ratio, α
2. The nondimensional heat transfer area, $E = \exp(-2fL/D)$

By an analogous procedure, the metal heating process may be described. If the term fL/D for the hot flow is the same as that of the cold flow, then no special distinction needs to be made between the hot and cold flows. To first order, one would expect that only the temperature dependence of the fluid viscosity will change the Reynolds number and thus the friction factor. The mass flow rate and the mass flow rate per unit area should be

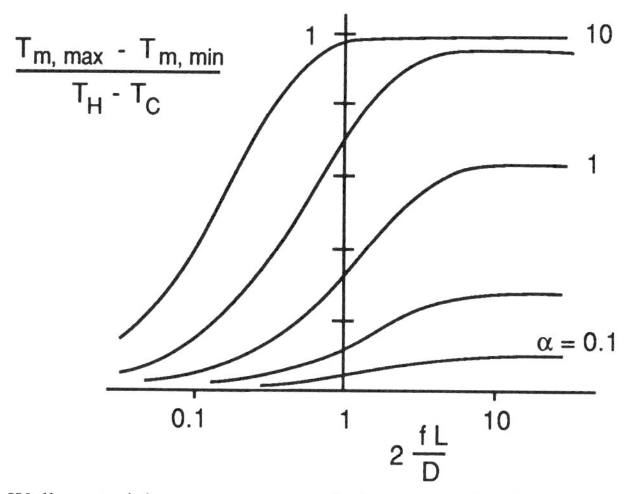

Fig. 10–25. Wall material temperature variation as a fraction of maximum fluid temperature difference.

close to equal for the two cases,

$$\frac{T_H - T_{m,\max}}{T_H - T_{m,\min}} = \exp - \alpha(1 - E) = \psi \tag{10–30}$$

Solving eqs. 10–29 and 10–30 for the extremal metal temperatures, one obtains:

$$T_{m,\max} = \frac{T_H + \psi T_C}{1 + \psi} \quad \text{and} \quad T_{m,\min} = \frac{\psi T_H + T_C}{1 + \psi} \tag{10–31}$$

This set of equations may be used to gain insight into the role of the physical parameters. For example, the difference between extremal metal temperatures may be found from eq. 10–31, namely:

$$\frac{T_{m,\max} - T_{m,\min}}{T_H - T_C} = 1 - \psi \tag{10–32}$$

which is plotted in Fig. 10–25 as a function of fL/D. The metal temperature fluctuations are large when the metal mass is small (large α). For small fL/D, the heat transfer is ineffective and the temperature fluctuations are therefore small.

Using eqs. 10–29 and 10–30, the variations of these temperatures may be plotted as functions of ψ as shown in Fig. 10–26. Evidently for small α, the regenerator is passive in the sense that it operates at constant wall temperature, half-way between the cycle extremes.

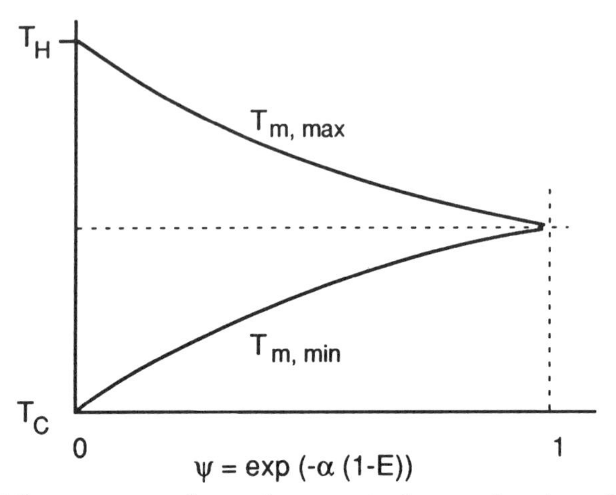

FIG. 10–26. Wall temperature fluctuation magnitude as a function of heat transfer parameters.

10.6.3. *Temperature at State Point* 3

A thermodynamic analysis of the regenerated cycle, such as that shown in Fig. 10–6, requires knowledge of the conditions at state point 3. The temperature at this point is intimately associated with the effectiveness of the regenerator (see eq. 10–2).

The integration of eq. 10–28 gives the temporal variation of the metal temperature:

$$\frac{T_m(t) - T_C}{T_{m,\min} - T_C} = \exp\left(-\alpha(1 - E)\frac{t}{\tau}\right) \tag{10–33}$$

The gas outflow temperature is related to the metal temperature by eq. 10–25, so that

$$T_1(t) = (1 - E)(T_m(t) - T_C) + T_C \tag{10–34}$$

Since the flow is steady in time, the effective mixed out temperature of the fluid heated by the regenerator is an average of the temperature T_1. A combination of eqs. 10–33 and 10–34 allows writing the equation for $T_1(t)$ in terms of constants of the problem. The temperature difference between the heated gas exiting the regenerator and the cold supply is therefore

$$T_1(t) - T_C = (1 - E)(T_{m,\min} - T_C)\exp\left(-\alpha(1 - E)\frac{t}{\tau}\right) \tag{10–35}$$

which should be as large as possible. Two aspects of the design compete for this end result:

1. E small (i.e., large fL/D equivalent to good thermal contact)
2. large heat transfer driving ΔT which is obtained for large E and small α (see Fig. 10–25)

The largest value of the ΔT is obtained from

$$T_{\mathrm{m,min}} - T_0 = \frac{\psi}{1+\psi}(T_{\mathrm{H}} - T_{\mathrm{C}}) \qquad \text{which is} = \tfrac{1}{2}(T_{\mathrm{H}} - T_{\mathrm{C}}) \qquad \text{for } \psi \to 1 \tag{10–35}$$

The mass averaged temperature for the exiting gas is:

$$T_3 - T_{\mathrm{C}} = \frac{1}{\tau}\int_0^{\tau}(T_1(t) - T_{\mathrm{C}})\,dt = \frac{\psi(1-\psi)}{(1+\psi)}(T_{\mathrm{H}} - T_{\mathrm{C}}) \tag{10–36}$$

where ψ is as defined as the exponential terms in eq. 10–30. The ratio of the temperature differences in eq. 10–36 is the regenerator effectiveness:

$$\eta_{\mathrm{x}} = \frac{T_3 - T_{\mathrm{C}}}{T_{\mathrm{H}} - T_{\mathrm{C}}}$$

This quantity is plotted in Fig. 10–27 as a function of the heat transfer parameter, fL/D. One may conclude that for large heat capacity in the regenerator, the regenerator effectiveness is always small. It reaches its maximum value of 0.5 for a low heat capacity in the regenerator and for large fL/D when the thermal contact between gas and regenerator is good. This performance

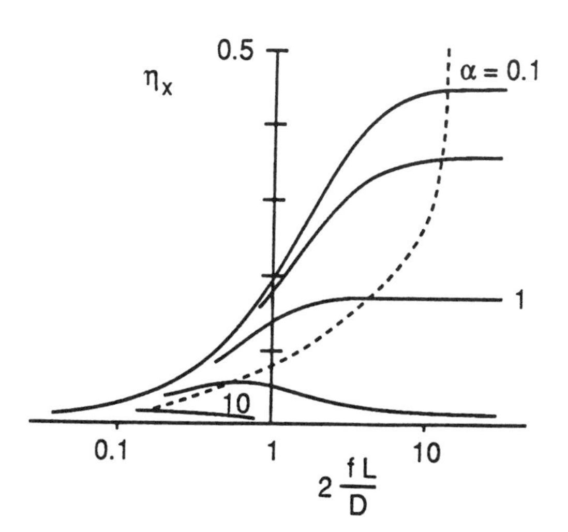

Fig. 10–27. Regenerator effectiveness for a regenerator wherein heat is stored in the solid wall material.

limitation is eliminated when multiple cylinders are employed and the heat is transferred directly from one fluid to another.

10.7. MECHANICAL PERFORMANCE OF KINEMATIC ENGINES

Kinematic heat engines alternately accept and produce mechanical power by involvement with energy storage devices. The storage mechanisms include the flywheel and the buffer gas which exerts a force on the rear side of the pistons. The buffer may be visualized as the volume of space around the crank and connecting rods of an internal combustion engine. This space may be a closed vessel in which case the pressure level is a design variable or it may be (vented to) the atmosphere as in most common internal combustion engines. Senft (ref. 10–17) describes the effect of buffer gas pressure on the efficiency characteristics of kinematic heat engines. The following is a review of some of his work.

The flywheel stores the energy necessary for carrying out the compression process after the work producing expansion is completed. In a multicylinder engine the requirement for such storage is reduced because the cylinders are arranged to transfer power between cylinders so that they produce power "smoothly" over the complete engine revolution. The buffer gas can store energy during the time period when the working gas is expanding and the piston moves to reduce the volume of the buffer gas. The point in the working fluid's cycle where the net force on the piston is zero depends on the buffer pressure. The buffer gas pressure may also vary but is close to constant because the volume is generally large compared to that of that of the working fluid.

Figure 10–28 shows the physical components of a kinematic engine. The energy transfer processes between the components are described in Fig. 10–29.

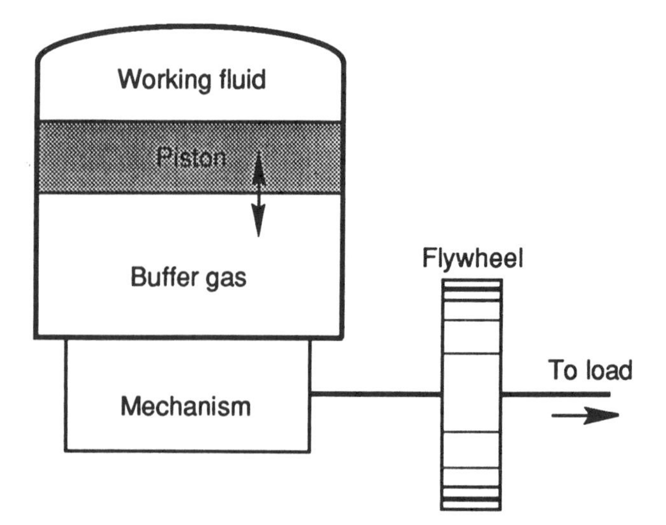

FIG. 10–28. Physical arrangement of a kinematic heat engine.

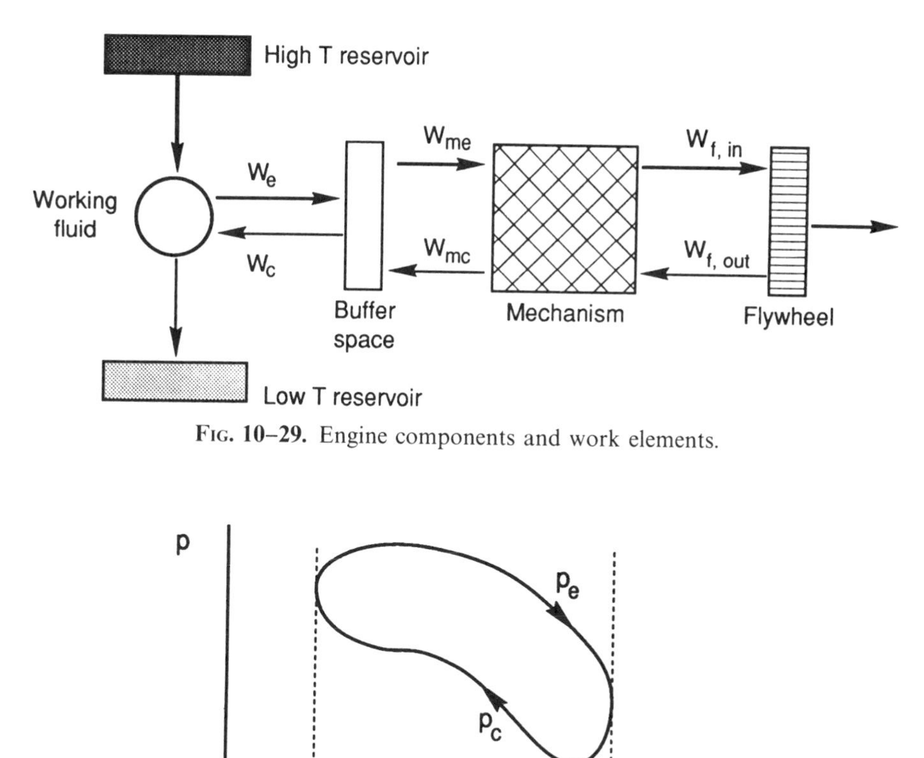

FIG. 10–29. Engine components and work elements.

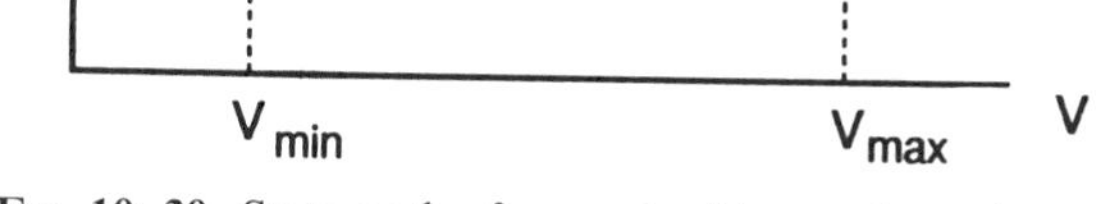

FIG. 10–30. State path of a regular kinematic engine.

Shown are the thermal reservoirs, the working fluid, the material in the buffer space, the physical mechanism involved in making the required volume changes, and the flywheel. The magnitudes of the works flowing to and from the various components during a cycle are indicated.

A cycle that is described by a path in the *p-V* plane that does not cross itself is a *regular* cycle. For these purposes, only regular cycles such as those described in Fig. 10–30 are considered. The cycle will cause the system volume to undergo changes including a maximum and a minimum that are denoted as V_{max} and V_{min}, respectively. The cycle executes a clockwise path to produce net work. An expansion produces work equal to the area under the curve labeled p_e in Fig. 10–30, whereas the compression work is the area under p_c. The path integral

$$W = \oint p\, dV = \int_{V_{min}}^{V_{max}} (p_e - p_c)\, dV \tag{10–37}$$

gives the net work produced when the pressure behind the piston is negligibly small (as in a vacuum). The magnitudes of the expansion and compression works are, respectively,

$$W_e = \int_{V_{min}}^{V_{max}} p_e \, dV \qquad \text{and} \qquad W_c = \int_{V_{min}}^{V_{max}} p_c \, dV \tag{10–38}$$

10.7.1. *Buffer Gas Pressure*

If the piston is buffered, that is, has a finite pressure on the backside (= say a constant p_0), then the work is:

$$W = \oint (p - p_0) \, dV \tag{10–39}$$

which is identical to eq. 10–37 when p_0 is constant. The works of compression and expansion (consistent with the external pressure) may be written as:

$$W_{0e} = \int_{V_{min}}^{V_{max}} (p_e - p_0) \, dV \qquad \text{and} \qquad W_{0c} = \int_{V_{min}}^{V_{max}} (p_c - p_0) \, dV \tag{10–40}$$

W_{0e} and W_{0c} may be positive or negative depending on the signs of $(p - p_0)$ and of dV. Figure 10–31 shows the buffer level and the cycle pressures. Evidently at points 1, 2, 3, and 4, the sign of the contribution to the work integral changes. The various contributions to the total integral are labeled as areas *A-E*.

The cyclic indicated work is given by

$$W_0 = W_e - W_c = W_{0e} - W_{0c} \tag{10–41}$$

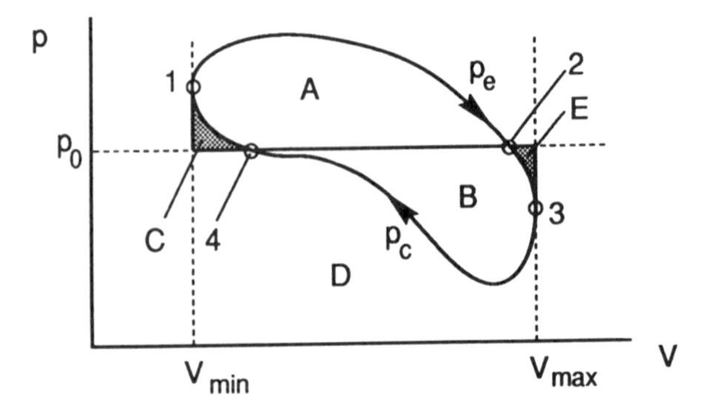

Fig. 10–31. Work elements of a kinematic heat engine. Dashed pressure level is the buffer pressure.

10.7.2. *Mechanical Efficiency*

Since the mechanism is a potential source for irreversibilities, the work that proceeds to and from that component is of interest. One may define a mechanism efficiency as the ratio of output to input works. Such a ratio may be a function of piston position, piston velocity, and its design. For the cycle, an average value of this efficiency is to be determined to find the best buffer pressure level. The mechanical efficiency is defined as

$$\eta_m = \frac{W_s}{W_0} = \frac{\text{shaft work}}{\int p\, dV \,\text{work}} \tag{10–42}$$

Following the notation of Senft, the various $p\,dV$ integrals of interest are shown in Fig. 10–31. The so-called *indicated* work is the sum of areas A and B. Hence

$$W_0 = W_A + W_B \tag{10–43}$$

This work is the *net of the working gas expansion work* (=area under p_e curve) which is

$$W_e = W_A + W_B + W_C + W_D \tag{10–44}$$

and the *net of the working gas compression work* (=area under p_c curve) which is

$$W_c = W_C + W_D \tag{10–45}$$

The $p\,dV$ integrals illustrated in Fig. 10–31 are denoted as subscripted W's. The work delivered *to* the mechanism during expansion is:

$$W_{me} = W_A + W_C \tag{10–46}$$

while the (positive) work delivered during compression is

$$W_{mc} = W_B + W_E \tag{10–47}$$

The total work delivered *to* the mechanism is therefore the sum of these two contributions:

$$W_{m,\text{out}} = W_A + W_B + W_C + W_E \tag{10–48}$$

The work delivered *by* the mechanism is the work performed during the counterclockwise or negative portions of the cycle, C and E. Thus

$$W_{m,\text{in}} = W_C + W_E \tag{10–49}$$

The work available to the flywheel is less than the work available, $W_{m,\text{out}}$

by a factor that is an effectiveness ($e < 1$). The work stored in the flywheel is therefore:

$$W_{f,in} = eW_{m,out} = e(W_A + W_B + W_C + W_E) \tag{10–50}$$

and the work returned to the mechanism by the flywheel is

$$W_{f,out} = \frac{W_{m,in}}{e} = \frac{(W_C + W_E)}{e} \tag{10–51}$$

At the most basic level, one assumes that a single effectiveness parameter governs the departure from ideal behavior of the processes at play in the mechanism. The net shaft work is the sum of the flywheel works:

$$W_s = W_{f,in} - W_{f,out} = eW_0 - \left(\frac{1}{e} - e\right)(W_C + W_E) \tag{10–52}$$

so that Senft's formula for the mechanical efficiency is

$$\eta_m = e - \left(\frac{1}{e} - e\right)\frac{(W_C + W_E)}{W_0} \tag{10–53}$$

Evidently, the mechanical efficiency can be no greater than the process effectiveness parameter. This minimum value is obtained when the sum $W_C + W_E$ is zero. From Fig. 10–31 it becomes apparent that for a particular value of e, the best value of the buffer pressure is such as shown. Higher or lower values increase the area C or E. This combination of areas can be made zero only for cycles that have equal pressures at V_{min} and V_{max} (see e.g., Fig. 10–32).

10.7.3. *Optimum Buffer Pressure in an Ideal Stirling Engine*

Consider a simple ideal Stirling cycle as shown in Fig. 10–33. The sloped lines 4–5 and 0–2 are isotherms which are hyperbolas in the T-s plane. The heavy lined cycle is a reference cycle with $p_5 = p_2$. The dashed cycle has a larger cycle volume ratio while the small cycle has smaller θ_4. Both cycles have $p_5 < p_2$ and will be penalized because the last portion of the mechanical efficiency term as written in eq. 10–53 is nonzero.

For the ideal cycle, the relevant pressures are given by:

$$\frac{p_2}{p_0} = R_v\left(\equiv \frac{V_0}{V_2}\right) \quad \text{and} \quad \frac{p_5}{p_0} = \theta_4 \tag{10–54}$$

Thus for $R_v > \theta_4$ the ratio $(W_E + W_C)/W_0$ is of interest. Senft gives both the

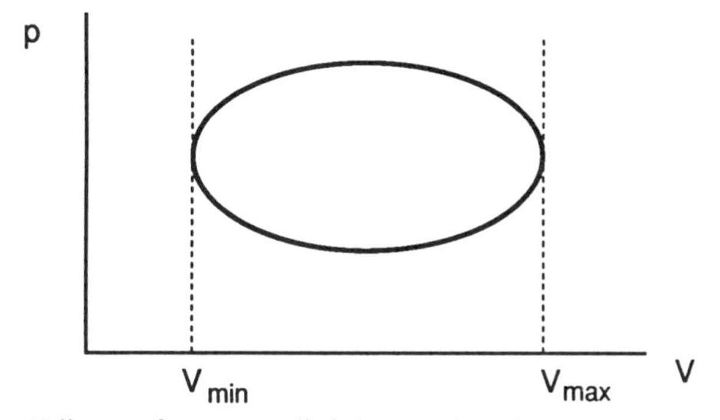

FIG. 10–32. p-V diagram for an unrealistic heat engine with high mechanical efficiency.

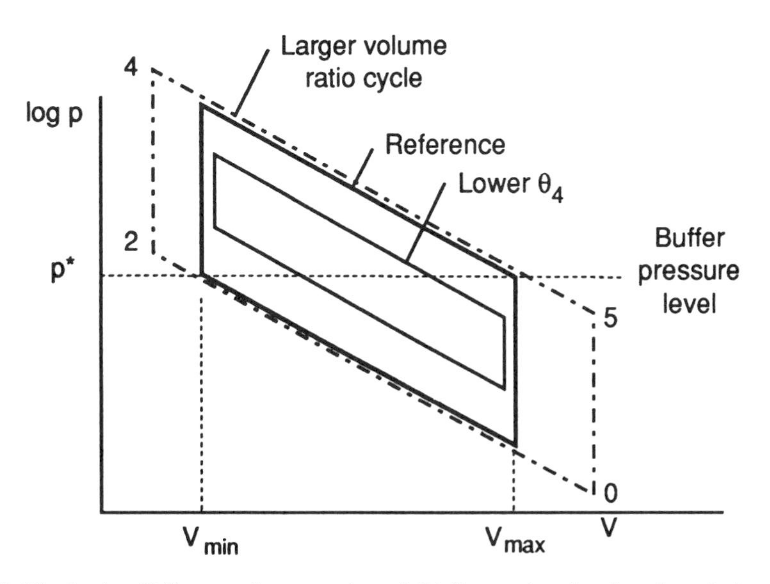

FIG. 10–33. (log) p-V diagram for a number of Stirling cycles showing the relationship to the buffer pressure.

optimum buffer pressure given by

$$\frac{p^*}{p_0} = (\theta_4 + 1)\frac{R_v}{R_v + 1} \tag{10–55}$$

and the work ratio as

$$S = \frac{W_E + W_C}{W_0} = \left(\frac{-1}{\theta_4 - 1}\right)\left\{\frac{\ln \theta_4 + (\theta_4 + 1)\ln\left(\frac{(1 + 1)/\theta_4}{1 + R_v}\right)}{\ln R_v} + \theta_4\right\} \tag{10–56}$$

These expressions are valid for $R_v > \theta_4$. As shown in ref. 10–17, eqs. 10–56 and 10–53 taken together constitute a universal limit on mechanical efficiency to which all reciprocating engines are subject. No engine with the same values of the parameters e, R_v, and θ_4 can have a higher mechanical efficiency. Using the last cited equations, one can show that for an engine with $\theta_4 = 2$, $R_v = 10$, and a mechanism effectiveness $e = 0.8$, the mechanical efficiency cannot exceed 0.67. This sort of calculation provides a rational scale against which one may assess the performance of real engines.

When $R_v < 4$, $S = 0$ and with optimum buffer pressure, the mechanical efficiency of the ideal Stirling cycle is equal to the effectiveness of the mechanism, a limit on the performance as shown by eq. 10–53. This is usually the case for Stirling engines designed to operate from high-temperature sources, $\theta_4 \sim 3$ and having adequate internal heat exchangers to operate at high speed, typically with R_v of order 2. However, in cases where optimal buffering is not possible, such as when the buffer pressure is atmospheric, an engine can suffer severe mechanical efficiency penalties (ref. 10–18), even to the point of not being able to achieve self-sustaining operation. This has important consequences for supercharging as shown in ref. 10–19.

Equations 10–53 and 10–56 have especially important consequences for engines designed to operate with low-temperature heat resources such as industrial waste, and geothermal and passive solar energy. Reference 10–18 shows that when $e^2\theta_4 < 1$, there is a limited range of compression ratio R_v under which an engine can run at all, no matter what its cycle. This range becomes more limited when θ_4 approaches unity. The analysis described has been extended to the adiabatic pseudo-Stirling cycle (refs. 10–20, 10–21) which shows even greater limitations on compression ratio and specific work.

BIBLIOGRAPHY

Wheatly, J. C., Swift, G. W., Migliori, A., "The Natural Heat Engine," Los Alamos Science, Fall 1986.

West, C. D., *Principles and Applications of Stirling Engines*, Van Nostrand Reinhold, New York, 1986.

Senft, J. R., *Ringbom Stirling Engines*, Oxford University Press, New York, 1993.

Walker, G., *Stirling Engines*, Clarendon Press, Oxford, 1973.

Hargreaves, C. M., *The Philips Stirling Engine*, Elsevier, New York, 1991

Stirling Engines—Progress toward Reality, I Mech E Publication 1982–2, London 1982

REFERENCES

10–1. Kolin, I., *Evolution of the Heat Engine*, Longman Group Ltd., London, 1972.

10–2. Walker, G. and Senft, J. R., *Free Piston Stirling Engines*, Springer, 1985.

10–3. West, C. D., *Liquid Piston Stirling Engines*, Van Nostrand, 1983.

10–4. Walker, G., "Elementary Design Guidelines for Stirling Engine," IECEC 7999230, 1979.

10–5. Reader, G. T., "The Pseudo Stirling Cycle—A suitable Performance Criterion?", IECEC 789116, Am. Chem. Soc., 1978.

10–6. Walker, G., *Stirling Engines*, Clarendon Press, 1980.

10–7. Senft, J. R., "A Simple Derivation of the Generalized Beale Formula," IECEC paper 829273, IEEE, 1982.

10–8. Meijer, R. J., Prospects of the Stirling Engine for Vehicular Propulsion, *Phillips Tech Review*, 31: 168–85, 1970.

10–9. Percival, W. H., NASA report 121097, Washington D. C., 1974.

10–10. Schmidt, G., Theorie der Lehmannschen Calorischen Maschine, *ZVDI*, 15(1), 1871.

10–11. Urieli, I., "A Review of Stirling Cycle Machine Analysis," IECEC 799236, 1979.

10–12. Berchowitz, D. M., Rallis, C. J., "A Computer and Experimental Simulation of Stirling Cycle Machines, IECEC 789111, 1978.

10–13. Reader, G. T., Hooper, C., *Stirling Engines*, E. & F. N. Spon, London, 1983.

10–14. Martini, W. R., "A Simple Method of Calculating Stirling Engines for Engine Design Optimization," IECEC 789115, 1978.

10–15. Martini, W. R., "A Simple Method of Calculating Stirling Engines for Engine Design Optimization," IECEC 789115, 1978.

10–16. Bergmann, C., Paris, J. A. D. R., "Numerical Prediction of the Instantaneous Regenerator and in-Cylinder Heat Transfer of a Stirling Engine," *Int. J. of Energy Research*, 15: 623–35, 1991.

10–17. Senft, J. R., "Mechanical Efficiency of Kinematic Heat Engines," *Journal of the Franklin Institute*, 324: 273–90, 1987.

10–18. Senft, J. R., "Limits on the Mechanical Efficiency of Heat Engines," IECEC 879071, 1987.

10–19. Senft, J. R., "Pressurization Effects in Kinematic Heat Engines," *Journal of the Franklin Institute*, 328: 255–79, 1991.

10–20. Senft, J. R., "Analysis of the Brake Potential of Crossley-Stirling Engines," IECEC 910314, 1991.

10–21. Senft, J. R., "Mechanical Efficiency Considerations in the Design of Ultra Low Temperature Differential Stirling Engines," IECEC 929024, 1992.

11

FLUID PROPERTY-LIMITED CYCLES: RANKINE AND CYCLE COMBINATIONS

The choice of an optimal cycle configuration is sometimes dominated by considerations of the working fluid's state properties, rather than temperatures achievable or energy available in the fuels used. Most substances operated near the saturation dome are far from ideal or perfect gas and impose performance limitations on practical cycles. Here a number of cycles and cycle combinations are discussed to illuminate the advantages and limitations of working fluid properties and their limitations.

11.1. VAPOR CYCLES

Cycles that employ a condensable working fluid realize a net work output which is a large fraction of the expansion work because the compression work of the fluid in the liquid phase is small. Section 7–6 describes the work required for compression of a liquid in contrast to that required by a gaseous substance. The work required by the compression portion of the cycle is termed the "recirculating work." Figure 11–1, which shows the elements of a vapor cycle, also illustrates this idea. Evidently, in cycles where the efficiencies of expansion work production, compression, and/or transfer between components are low, the possibility of building a successful heat engine is limited. Fortunately, the Rankine cycle is one where nearly all of the expander work is useful. This explains why the steam engine was the first heat engine to enjoy successful and widespread use and why the Brayton cycle, with its larger recirculating work, took longer to develop.

Compression of a substance in the liquid state and expansion in the gaseous state demands that the cycle operate with state points near the saturation dome of the substance. Such cycles are termed "vapor" cycles. Their description cannot make use of simple algebraic relations to describe the state changes and therefore requires use of state tables (e.g., ref. 11–1) or their equivalent. This makes it difficult to obtain expressions for specific work and efficiency in terms of physically interesting parameters such as temperature and pressure. Algebraic results similar to those obtained for the Brayton cycle (Section 9.3) using an ideal and thermally perfect gas description are not available but numerical results are readily obtained.

It will be evident from the discussion in this chapter that the phase characteristics (such as critical pressure and temperature) play an important role in determining the performance of vapor cycles when these are constrained

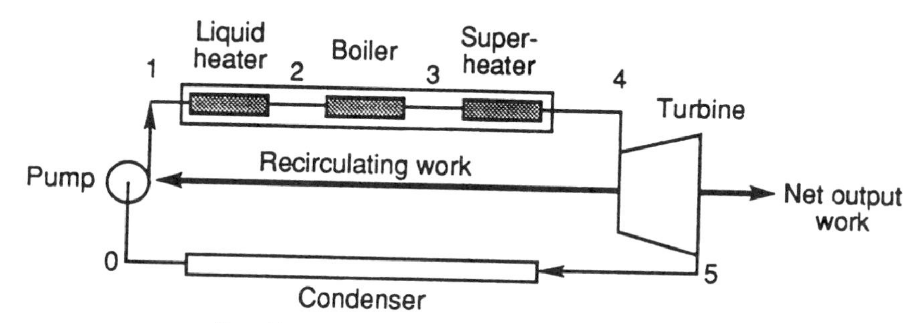

FIG. 11–1. Vapor cycle and recirculating work.

to working with temperatures and pressures that are imposed by the heat source or by the materials that may be employed in the manufacture of the components. The mismatch between these characteristics and engineering constraints generally requires the modification of the cycle to improve its performance, such as connection with other cycles or use of complex component arrangements. This chapter takes up the problem of cycle performance calculation using available tabular state data in connection with two kinds of simple cycles. In one of these, the path stays under the critical point and the cycle is called a "subcritical" cycle. In the other, the state path does not include a portion where the liquid and gaseous phases coexist in the heating segment. Such a cycle is termed "supercritical." Figure 7–10 illustrates the two cycle types on the same fluid and with reversible expansion.

The figure shows a saturation dome that is representative for a substance like water or mercury. The reader may wish to consult a text such as Wood (ref. 11–2) for saturation characteristics of other substances of interest to energy conversion. One may conclude that for substances with a large number of atoms per molecule, the trend is for both saturation lines to have generally positive slopes in the T-s plane. Further, some fluids may not have a critical point because thermal dissociation takes place at the higher temperatures.

In the discussion to follow, water will be used as an example to illustrate the points to be made. Although it may be of interest to determine the role of fluid properties in limiting performance to aid in the search for better working fluids for energy conversion, the important and practical fluids are those available in nature. For this purpose, water is by far the most commonly used substance. The critical point of water is 647 K and 218 atm ($=22.090 \times 10^3$ kPa) pressure. In connection with water, it is often convenient and common to use Centigrade degrees for the measurement of temperature. This intimate connection is exploited here to the extent it is appropriate.

11.2. CRITICAL CYCLE

The limiting case between sub- and supercritical cycles lies in a cycle that follows the saturation line on the compression and heat input side while it follows the vapor line on the expansion side. Heat is also added during expansion and

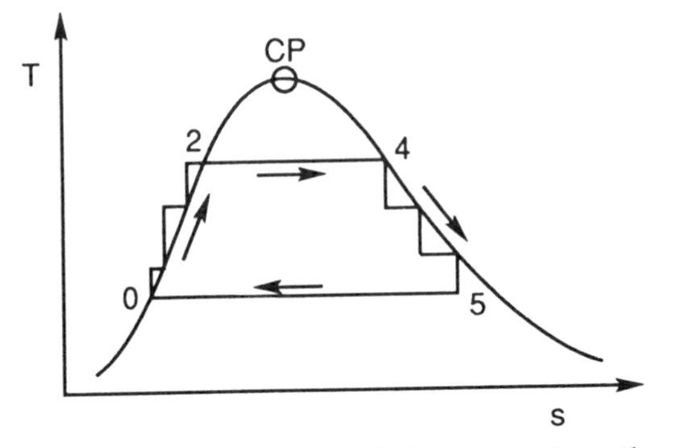

FIG. 11–2. A critical cycle in the limit of small departures from the saturation lines.

rejected in the two-phase condensation process. Figure 11–2 shows such a cycle. It should be apparent that such a cycle is impractical from an engineering viewpoint, because it would have to be implemented through differential or small adiabatic and isobaric steps which approximate the process along the saturation lines. A few such steps are shown schematically in Fig. 11–2. This cycle, nonetheless, illustrates the performance potential of vapor cycles as well as a sense of what characteristics of the saturation dome (i.e. the two saturation lines taken together) properties are desirable in a working fluid of a heat engine. Among these is that the saturation lines should be steep between the operating temperatures (say 1200 K and 300 K) so that little of the heat is added at less than the peak cycle temperature.

The heat input process for a cycle using a substance whose saturation dome is as illustrated in Fig. 11–2 may be broken into three (increasing s) segments: the liquid-phase heating, the two-phase heating (boiling), and the gas-phase heating. These are

$$q_{in} = q_{02} + h_{fg} + q_{45} = \int_0^2 T\,ds_f + h_{fg}(T_4) + \int_4^5 T\,ds_g \qquad (11\text{–}1)$$

Here h_{fg} is the phase change enthalpy at the peak cycle temperature. The integrals may be evaluated using the tabular saturation line data. The integrals may be written as

$$\int_0^2 T\,ds = \int_{TP}^2 T\,ds - \int_{TP}^0 T\,ds \qquad (11\text{–}2)$$

where the two-phase data are developed from the triple point (TP) data to any state point under the saturation dome. These integral functions for the liquid and vapor lines are as shown for water in Fig. 11–3. The units of q are in kJ/kg. The heat input integrals are shown as lines for an example where the cycle operates between 100°C and 200°C. q_{45} and q_{02} are shown as the differences in the functions on the left and right sides of the plot.

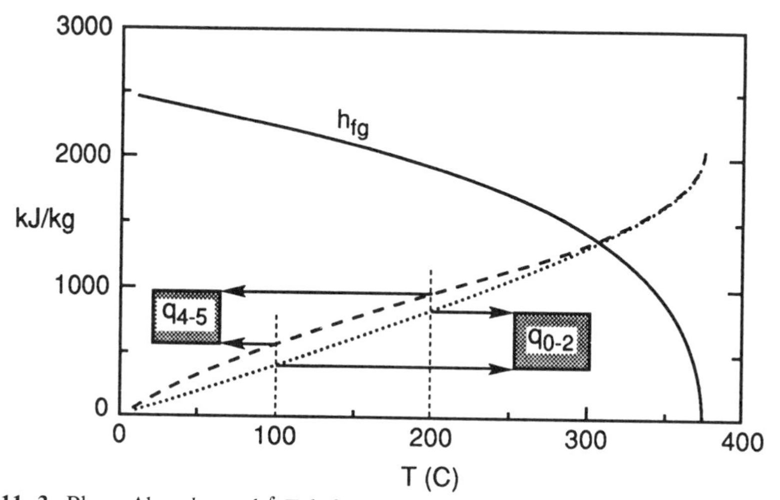

FIG. 11–3. Phase $\Delta h = h_{fg}$ and $\int T\,ds$ from triple point to T for liquid (q_{0-2}) and vapor (q_{4-5}) phases of water. Units are kJ/kg.

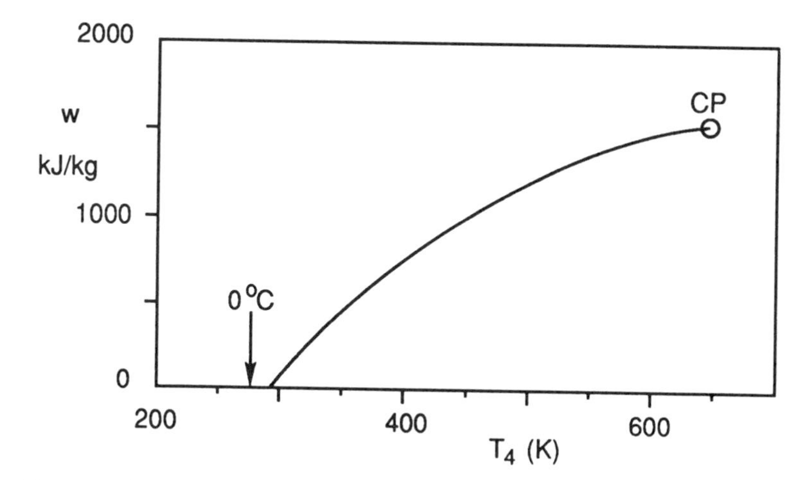

FIG. 11–4. Specific work (w, in kJ/kg) for the critical cycle using water at T_{max} given and with $T_{min} = 17.5°C$.

The rejected heat from the cycle is $h_{fg}(T_0)$ so that the work and efficiency follow.

$$w = q_{in} - h_{fg}(T_0) \qquad \text{and} \qquad \eta_R = \frac{w}{q_{in}} \tag{11–3}$$

The specific work performance of the critical cycle using water is shown in Fig. 11–4 for a minimum cycle temperature of 17.5°C. The plot is terminated at the critical point. The corresponding efficiency is shown in Fig. 11–5 together with the Carnot efficiency obtained for the temperatures chosen.

Here, in contrast to Chapters 8 and 9, w will be carried with dimensions

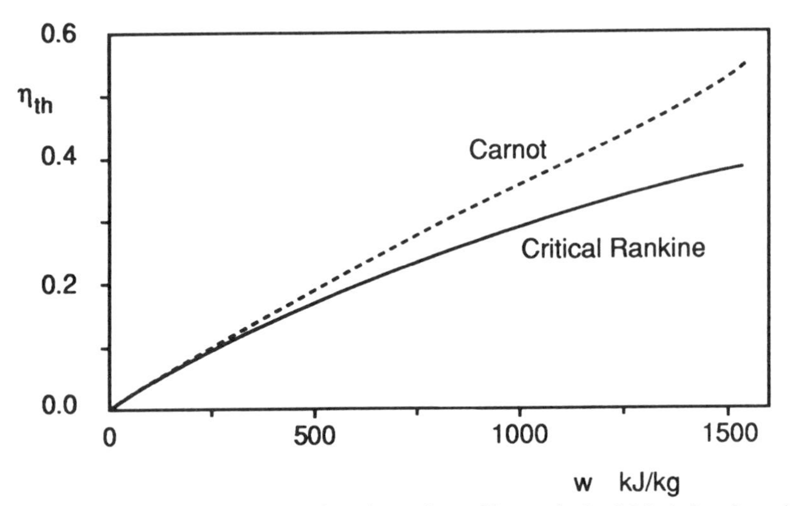

FIG. 11–5. Thermal efficiency as a function of specific work (in kJ/kg) for the critical cycle (water with $T_{min} = 17.5°C$). Carnot efficiency is for the temperature corresponding to that given in Fig. 11–4.

because use of dimensionless ratios is neither common nor practical. Any discussion of a cycle is generally about a specific working fluid whose state parameters are known. For comparison, however, division by $C_p T_0$ could be carried out, and this constant is 1230 and 1560 kJ/kg at 20°C and 100°C, respectively. For the lower temperature, the nondimensional w parameter is seen to be of the order of unity for the cycle operated to the critical point, less at lower peak cycle temperatures. The value may be compared with eq. 9–26 for the Brayton, eq. 8–10 for the Otto, and 8–36 for the Diesel cycles.

11.3. SUBCRITICAL CYCLES

Figure 11–6 shows the T-s diagram of a simple Rankine cycle and Fig. 11–1 shows the process elements as they occur around the cycle. The high value of the critical pressure and the relatively low value of the critical temperature of water (in comparison to heat exchanger material capability) often mandate use of the subcritical cycle for this fluid. The cycle is of great practical interest because higher temperatures allow the use of a single gaseous phase in steady flow machinery such as a turbine (4–5). Briefly, a pump compresses the liquid and heating takes place in three steps: liquid heating (1–2), boiling (2–3), and vapor heating (3–4), also called superheating. Heat is rejected in condensation (5–0). The recirculating work to the pump will be shown to be very small.

The thermal efficiency of the ideal cycle is lower than that of the Carnot value because a significant temperature drop is seen to be required to heat the working fluid from the source temperature T_4 to the lower fluid temperature along the heating process. The real Rankine engine also suffers flow irreversibilities in the expansion process not unlike a real gas turbine. This loss

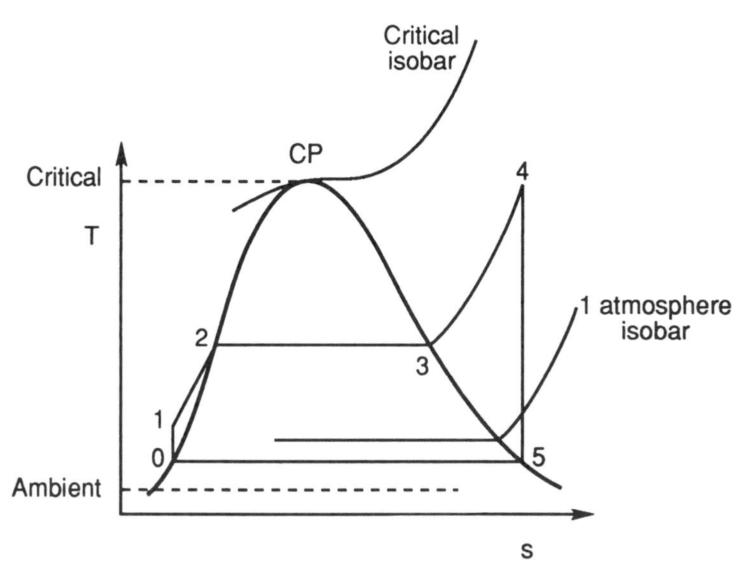

FIG. 11–6. Subcritical cycle (using water) on T-s diagram showing three isobars of interest.

can be characterized by an adiabatic efficiency, which is assumed to be unity in the present discussion.

When the vapor is at a temperature larger than the saturation value at a given pressure it is said to be *superheated*. The expansion of the superheated vapor by means of a turbine requires that the exit state be such that the formation of condensation droplets is avoided to prevent erosion of the turbine blades due to droplet impact. This limitation is not involved if the expansion is by means of piston displacement. The visible white exhaust from steam locomotives (Fig. 7–6), which use displacement expanders, consists of the condensed water droplets (together with the combustion gas products).

In open cycle engines, the exit pressure must be near atmospheric. For a closed Rankine cycle engine the pressure can be as low as the ambient temperature of the coolant (air or a body of water) allows across the surfaces of a heat rejecting condenser. One would expect that closed cycle engines can be made more efficient because the minimum temperature may be made lower than the (sea level) 100°C limit for open cycle water engines. The use of water in a low-temperature condenser results in low pressures experienced in that device and a large surface area required which brings with it challenges to achieve low-cost construction. Leaks here occur inward, and care must be used to avoid contamination by impure water or gases—oxygen, in particular.

11.3.1. *Performance of the Subcritical Rankine Cycle with Water*

The performance of the subcritical cycle is determined by the peak pressure and the temperature reached in the cycle for a specified condenser temperature. Since the expansion end point is generally fixed from a design point of view

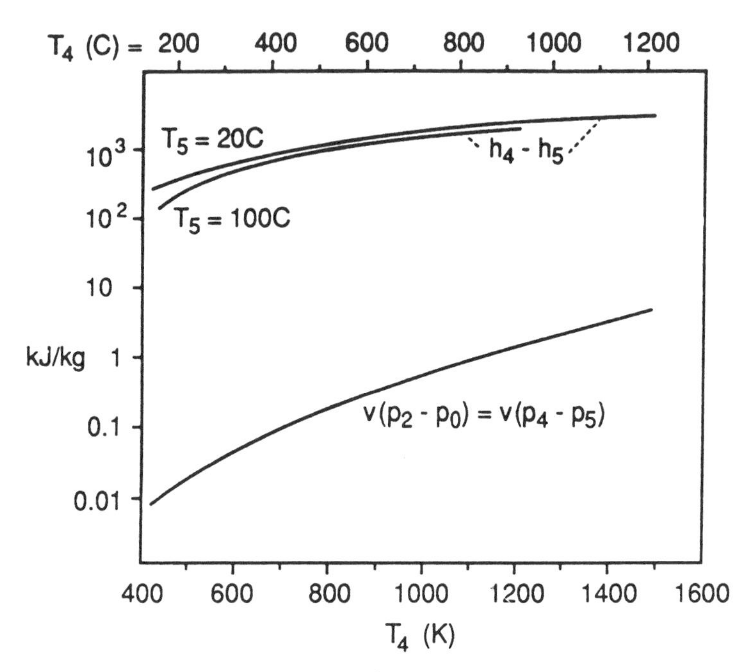

FIG. 11–7. Expander work ($h_4 - h_5$) for two final temperatures (and associated pressures) and pump work (for $T_0 = 20°C$) for the simple Rankine cycle. Units are kJ/kg. The expander work is close to linear with temperature.

(i.e., $T_5 = T_0$) is specified, greater work is obtained from the expansion when the pressure, temperature, and therefore the enthalpy of the fluid at the expander inlet is increased. Thus one may estimate the performance of the simple Rankine cycle by assuming the turbine to be reversible and the pump work to be negligible. The conditions at state point 4, namely h and p are determined by the specified T_4 and the condition $s_4 = s_0$. The calculation technique to determine these conditions is similar to that described in Section 11.4.1 for the supercritical cycle. The pump and expander work (per unit mass) are given by:

$$w_{\text{pump}} = v_f(p_2 - p_0) \quad \text{and} \quad w_{\text{exp}} = (h_4 - h_5) \tag{11–4}$$

whereas the heat input is

$$q_{\text{in}} = h_4 - h_1 \tag{11–5}$$

from which the thermal efficiency follows. These quantities are plotted in Figs. 11–7 and 11–8 for $T_5 = 20°$ and 100°C characteristic of closed and open cycles. Note that the pump work is indeed negligible in relation to the expander work.

Figure 11–8 shows the thermal efficiencies plotted against specific work for the simple Rankine cycle and for the Carnot cycle at the corresponding temperatures. Note that the simple Rankine cycle has a T_0 dependence that is counter to the expected result where lower T_0 would be expected to have higher efficiency.

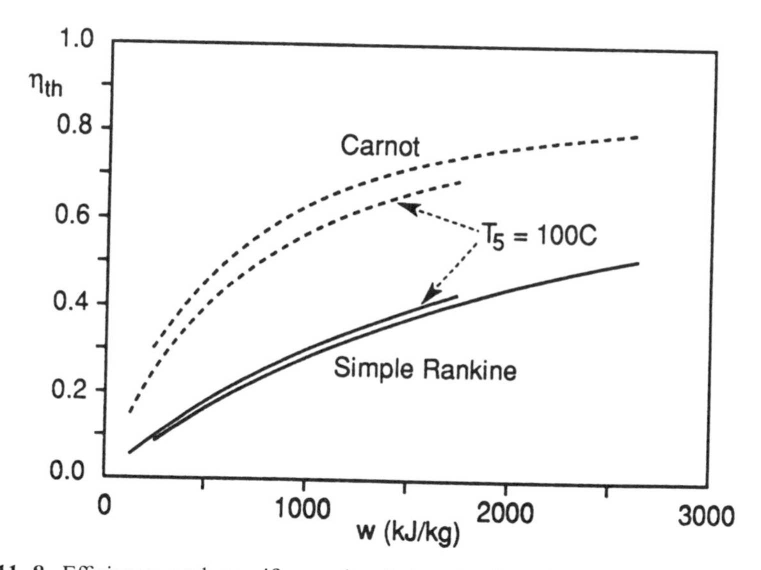

Fig. 11–8. Efficiency and specific work of the simple subcritical Rankine cycle at $T_{min} = 20°C$ and 100°C (dashed lines). Carnot efficiency corresponds to the temperature extremes of the Rankine cycle. Note that the simple Rankine cycle efficiency is about half the Carnot value.

Here the temperature and pressure are increasing in the increasing efficiency direction. This curve should be contrasted with Fig. 9–10 for the Brayton cycle. The ideal, simple, and subcritical Rankine cycle performance is relatively independent of T_0 and yields about half of the Carnot efficiency. Recall the modeling of the space power Rankine cycle in Section 9.10.1 where the fraction of Carnot efficiency, ϕ, is used parametrically.

For two cycles operating at the same maximum temperature, T_4, the higher rejection temperature requires operation at higher pressure as shown in Fig. 11–9.

A comparison of the $\eta(w)$ characteristic (for the subcritical cycle, Fig. 11–8) with Fig. 11–5 for the critical cycle shows that the critical and subcritical cycles closely overlie one another to the maximum value for the critical cycle. Thus specific work and efficiency are closely related: Improved w also results in improved η_{th}. Using efficiency as a way of comparing critical and subcritical cycles, one may conclude that the subcritical cycle has a larger potential for both improved specific work and improved efficiency: The subcritical cycle will want to be operated at higher temperature and/or at lower pressure. Engineering considerations thus determine the extent to which performance and cost may be compromised between these two kinds of cycles. See Figs. 11–10 and 11–11 for such a comparison. In this calculation the minimum temperatures are nearly identical (17.5°C for the critical and 20°C for the subcritical cycles).

Evidently from Fig. 11–11, the heaviest engineering development demands on the subcritical cycle are on temperature capability whereas on the the critical cycle it is on pressure capability.

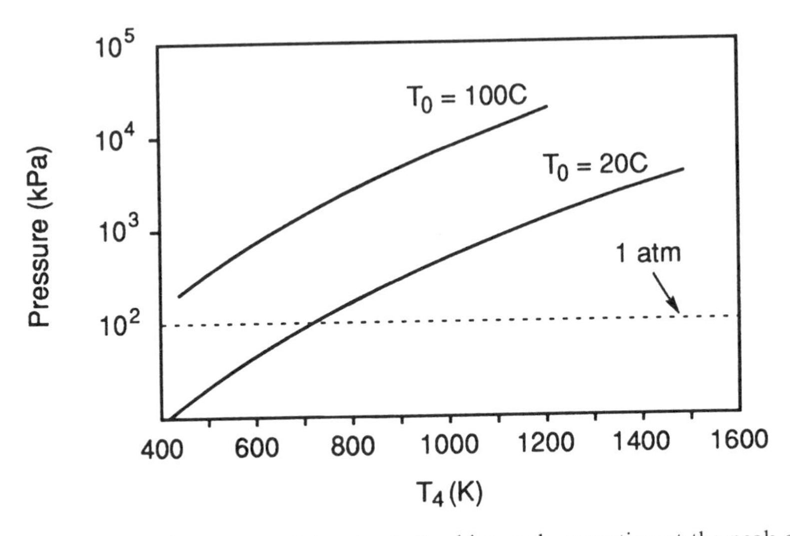

FIG. 11–9. Peak cycle pressure of the simple Rankine cycle operating at the peak cycle temperature given.

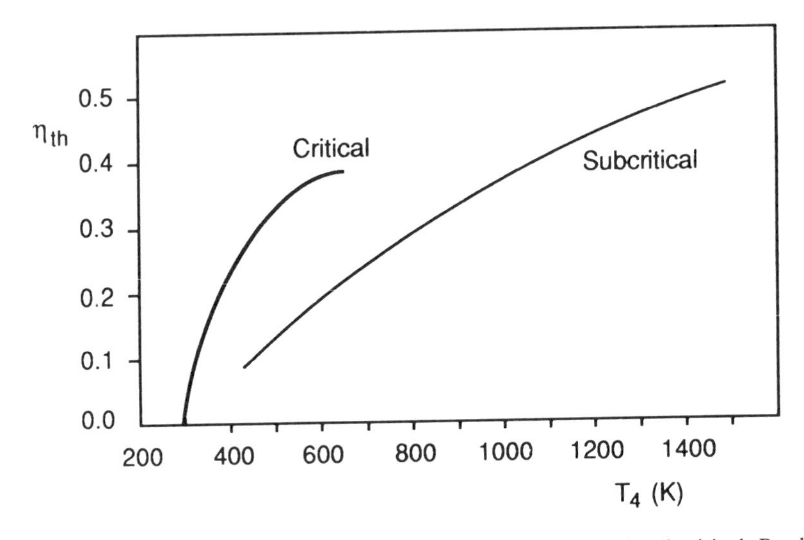

FIG. 11–10. Peak cycle temperature comparison of critical and subcritical Rankine cycles using water. Minimum cycle temperature ~20°C, see text.

11.3.2. *Subcritical Cycle with Reheat*

Intermediate performance between the critical and subcritical cycles may be achieved by allowing the expansion process of the subcritical cycle to proceed in a number of partial expansions with intermediate reheat. Figure 11–12 shows such a cycle with two-step expansion, that is, a single reheat stage.

Figure 11–13 shows a *T-s* diagram comparison of the simple and the reheat cycles operating at the same extremal pressures. The peak temperature (T_4) experienced in the cycle is reduced by this process. It should be evident that

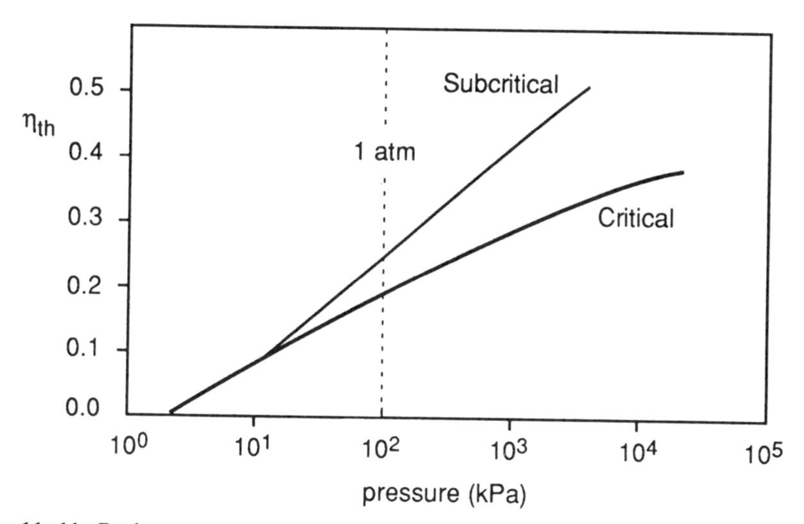

FIG. 11–11. Peak pressure comparison of critical and subcritical Rankine cycles using water. Minimum cycle temperature ~20°C, see text.

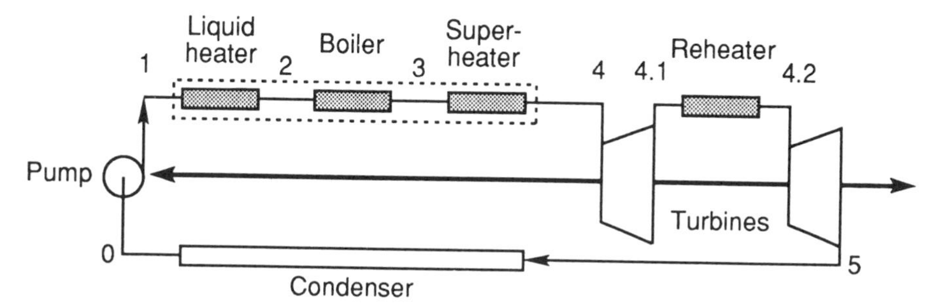

FIG. 11–12. Component arrangement for the reheated subcritical Rankine cycle.

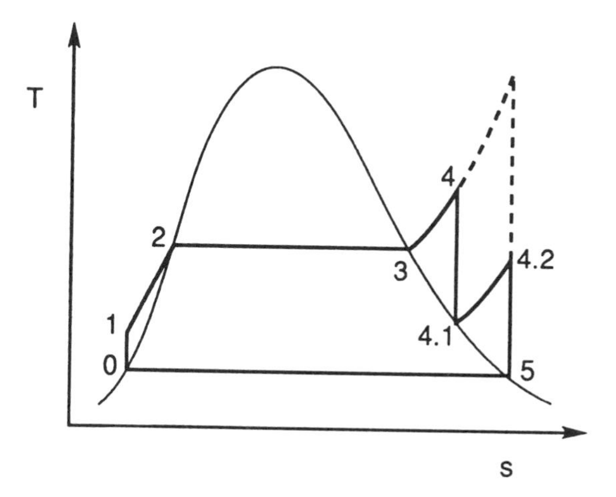

FIG. 11–13. *T-s* diagram for the reheat cycle (solid) contrasted to the simple cycle (dashed) at the same pressure.

reheat is a means for practical implementation of a compromise between the critical and simple subcritical cycles. Performance between these two limiting cases is therefore to be expected. Choice of the optimal intermediate pressure is an interesting design exercise because the fluid properties are strongly involved, as is the degree of condensation allowed in the last stage of the expansion processes.

The many stage reheat process approximates the following of the saturation line in the expansion process. In practice, feedwater heating may be (and often is) used to approximate a similar process on the liquid heating side.

Regeneration similar to that available to the Brayton cycle is not possible in the Rankine cycle because the temperature of the expanded fluid is at the minimum cycle temperature. Feedwater heating, however, is often carried out with low-pressure steam. This is, in effect, a form of regeneration using mass transfer. The description and analysis of such processes is beyond the scope of this work. The reader is referred to texts emphasizing the details of power plant design (e.g., Hill, ref. 11–3) or to the specialist literature.

It should be noted that thermal efficiencies near 40% have been achieved in wide service for Rankine plants using (low cost) *coal* combustion since the 1960s. This is a remarkably large fraction of the possible efficiency as can be seen in Fig. 11–10. Nuclear–Rankine power plants deliver electric power with efficiency in the 30–35% range. This is limited primarily by the lower peak cycle temperature to which the reactor material can be subjected. Using natural gas as a fuel combined cycle (Section 11.5), plants today guarantee delivery of electric power with efficiencies in the low-50 percent (52–54%) range. The utilization of coal in integrated gasification and power plants promises to deliver electric power using coal as a primary resource at efficiencies which may approach 60% in the next decade!

11.3.3. *Practical Rankine Cycles Operating with Gaseous Flow Heat Sources*

The heat typically supplied to a Rankine cycle is supplied at decreasing temperature as the heat transfer process proceeds. This is true in constant-pressure combustors of pure Rankine cycles or in heat exchangers used to transfer the heat from a gas turbine in a combined cycle. The fact that boiling (i.e., the phase change from liquid to vapor) takes place at constant temperature because the pressure is constant presents a challenging design problem. This is true because the thermodynamics requires a minimum ΔT between the heat supply and the system. Consider the heat transfer process between a hot gas and, say, water at saturated liquid state. A single pass heat exchanger (boiler and superheater) results in temperature profiles as shown in Fig. 11–14, left side. Here and in the discussion to follow, the heat-transfer process is idealized in that the minimum ΔTs between the two fluids are taken as zero, in spite of the fact that this presents practical difficulties (see Section 11.5.4).

The large average ΔT represents an irreversibility that should be minimized. This can be done by splitting the water path into two: heating a fraction of the flow at low pressure and the remainder at a higher pressure. The higher pressure water is boiled at a higher temperature so that the heat exchange process is as

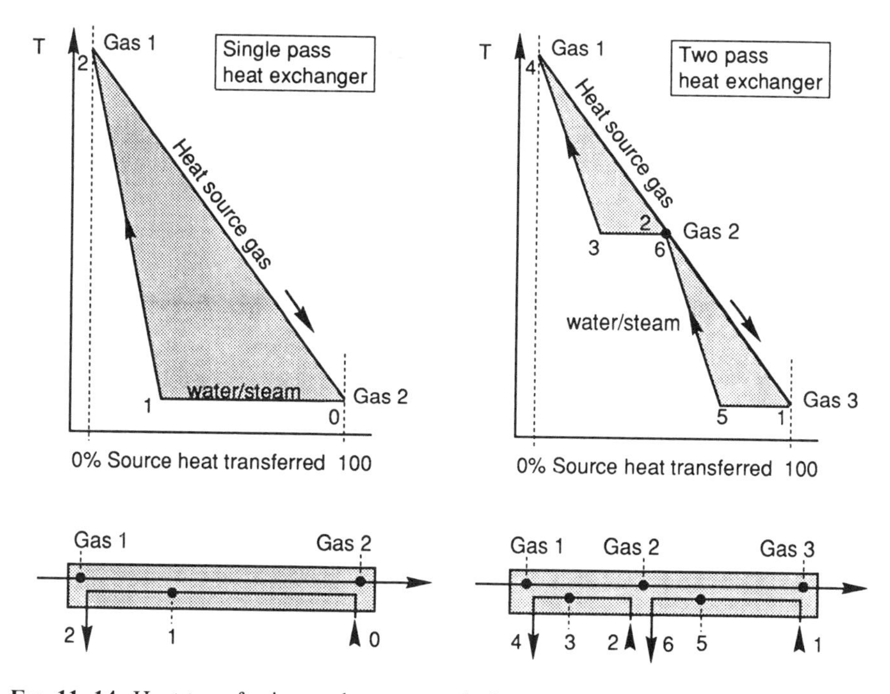

FIG. 11–14. Heat transferring exchanger to a boiler and superheater. Single pass on left and double pass on right. State points correspond to those identified in Fig. 11–15. ΔTs are minimal. Gas flow is to the right, water to the left. Water enters the boiler in saturated liquid state. See combined cycle layout in Fig. 11–21. Shaded area is heat transfer ΔT to be minimized.

shown in Fig. 11–14, right side. [Such a two-loop process is illustrated in the combined cycle shown in Fig. 11–21.] The state points correspond to those shown in the heat exchanger sketch. Both flows enter the boilers in the saturated liquid state. From the sketch it is evident that the temperature of the lower-pressure superheat (T_6) should equal the two-phase temperature of the high-pressure side (T_2) and these are ideally equal to the gas temperature at that point in the gas side (Fig. 11–14). The pressure of the high-pressure fraction is determined by the turbine inlet temperature and its efficiency. For example, if the turbine expansion process consists of two elements, one from 4 to 6 and another of the total flow from 6 to 7, the pressure at state point 4 is determined by s_7 and T_4. The T-s diagram is shown in Fig. 11–15 with mass flow fraction x at low pressure and $1-x$ at high pressure.

The component arrangement is shown in Fig. 11–16. For the present ignore the dashed processes and consider the mass flow fractions, y_L and y_H, to equal zero. Shown are the two turbines, the two pumps (labeled LPP and HPP), two boilers (B) and superheaters (SH), and one condenser. The numbers in square brackets ([x] etc.) denote the mass flow rate relative to that in the condenser. The numbers correspond to the state point identified in Fig. 11–15.

The pump outlets states are, for analysis purposes, the same as its input except that the pressure of the liquid is elevated. The pump work is negligibly

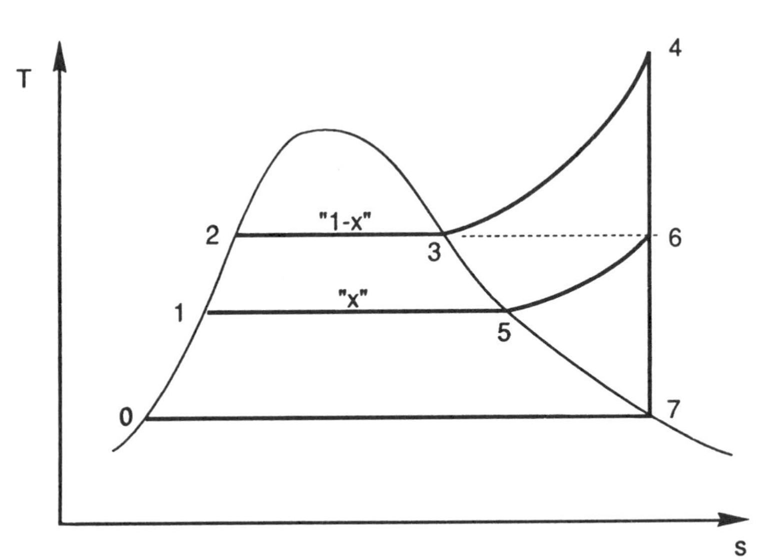

FIG. 11–15. *T-s* diagram for the two loops of a Rankine cycle.

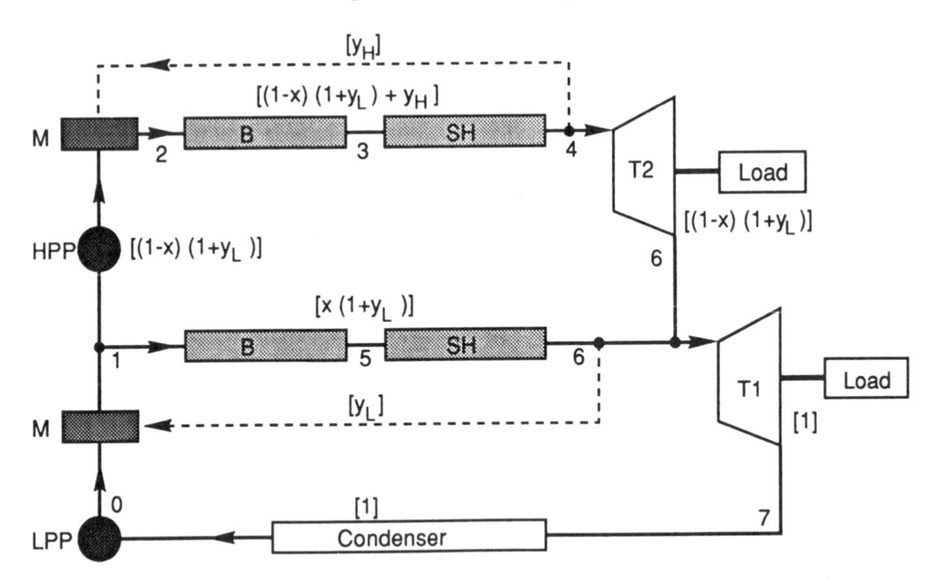

FIG. 11–16. Component arrangement for the two-loop Rankine cycle. Dashed lines are steam returns to feed water heaters which elevate the compressed liquid to the saturated liquid state. Quantities in brackets are relative mass flows. State points corresponding to Fig. 11–15. Mixers (M) are the feedwater heaters.

small. In order for the liquid to reach the saturated liquid state, heat must be provided. This heat should ideally be provided from the lowest temperature heat available. A practical way of providing it is to return the superheat steam *at the same pressure* to the pump outlet flow and mix these two flows. These are the feedwater heating return flows shown in Fig. 11–16 as dashed lines. The mass fraction of steam returned is given by y_i and determined by the mixer enthalpy balances. For the LPP outlet flow at p_1, the mixer outlet enthalpy is required to be h_1. Thus

$$h_0 + y_L h_6 = (1 + y_L) h_1 \qquad \text{or} \qquad y_L = \frac{h_1 - h_0}{h_6 - h_1} \tag{11-6}$$

For the HPP outlet flow, a similar relation is, for the as yet undetermined mass fraction x:

$$\left.\begin{aligned} &(1 + y_L)(1 - x)h_1 + y_H h_4 = ((1 + y_L)(1 - x) + y_H)h_2 \\ &\text{or} \\ &y_H = (1 + y_L)(1 - x)\frac{h_2 - h_1}{h_4 - h_2} \end{aligned}\right\} \tag{11-7}$$

For $T_2 = T_6 = T_{gas}$ the heat exchanger heat balance reads for the low water pressure segment:

$$\frac{[\dot{m}C_p]_{gas}}{\dot{m}_{steam}}(T_6 - T_1) = (x(1 + y_L))(h_6 - h_1) \tag{11-8}$$

and for the high-water pressure segment:

$$\frac{[\dot{m}C_p]_{gas}}{\dot{m}_{steam}}(T_4 - T_6) = ((1 - x)(1 + y_L) + y_H)(h_4 - h_2) \tag{11-9}$$

The ratio of eqs. 11–8 and 11–9 may be obtained to eliminate the gas parameters and thus the mass fraction x is determined. The ratio is

$$x = \frac{1 + y_L + y_H}{1 + y_L}\left(1 + \frac{T_4 - T_6}{T_6 - T_1}\frac{h_6 - h_1}{h_4 - h_2}\right)^{-1} \tag{11-10}$$

The three eqs. 11–6, 11–7 and 11–10 for a set from which the three unknowns x, y_L and y_H may be determined. It should be expected that the values of the y's should be relatively small, whereas x is nearer to one half.

With the mass flows known the performance of the lower cycle may be determined. Of particular interest is the work output by the two turbines given by

$$w = ((1 - x)(1 + y_L))(h_4 - h_6) + (h_6 - h_7) \tag{11-11}$$

The heat input is of interest if this Rankine cycle stands alone (i.e., there is no upper cycle which also produces work from the heat it receives). The heat input is

$$q = ((1 - x)(1 + y_L) + y_H)(h_4 - h_2) + x(1 + y_L)(h_6 - h_1) \tag{11-12}$$

The ratio of eqs. 11–11 and 11–12 gives the efficiency of the Rankine cycle.

The incorporation of a finite adiabatic efficiency for the turbine(s) impacts the calculation in that the work output is reduced and the pressures p_1 and p_2 are higher for a given T_4 and T_0 as can be deduced from an examination of Fig. 11–15.

11.4. SUPERCRITICAL CYCLES

The supercritical cycle is implemented with the same components as the subcritical cycle except that heating proceeds without phase change from a liquid to a superheated vapor at pressures above the critical value. A realistic cycle with its characteristic "triangular" shape is shown in the T-s plane as Figs. 7–10 and 11–17, the latter for an irreversible expansion process. It is this shape that makes the supercritical cycle interesting when heat is transferred from a gas at constant pressure, such as the Brayton cycle. See, for example, Fig. 9–3. Before discussing the integration of the supercritical with another cycle, it is appropriate to develop the procedure to calculate its performance.

11.4.1. *Supercritical Cycle Performance Calculation Methodology*

Consider the state points identified in Fig. 11–17. The heat source fixes the temperature T_4 and the heat rejection environment limits T_0. The turbine adiabatic efficiency is assumed known and the final expansion point is assumed to be on the saturation line to avoid droplet condensation in the turbine. For an irreversible turbine, the isentropic process necessarily leads into the two-phase region if the irreversible path ends on the saturation line. With this parameter specification, the procedure for calculating work output, and efficiency for a supercritical cycle employing a given substance is as follows:

1. Choose T_4, T_0, and a provisional estimate of p_4.
2. T_0 fixes the value of p_0 and s_0 from the saturation data of the working substance.

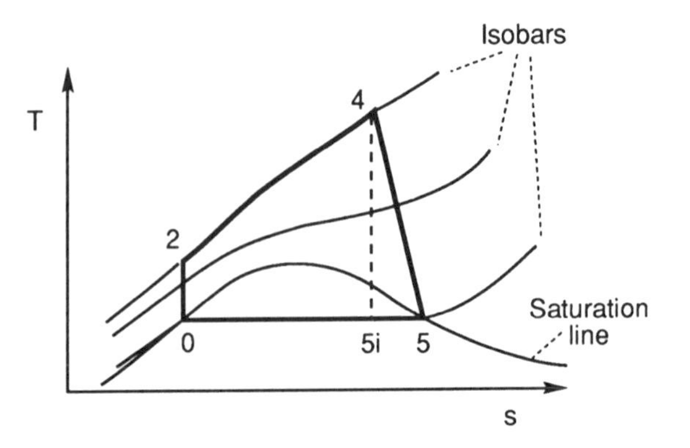

FIG. 11–17. T-s diagram for a supercritical Rankine cycle with irreversible expansion.

3. The pump work is given by: $w_c = v_f(p_4 - p_0)$.
4. Superheated data tables give s_4 and h_4 (from T_4 and p_4).
5. The isentropic expansion state point, 5i, is inside the saturation dome, hence the quality x_{5i} is given by:

$$s_{5i} = s_4 = s_f(T_0) + x_{5i}s_{fg}(T_0) \qquad \text{or} \qquad x_{5i} = \frac{s_4 - s_f(T_0)}{s_{fg}(T_0)}$$

6. The enthalpy at the point 5i is

$$h_{5i} = h_f(T_0) + x_{5i}h_{fg}(T_0)$$

7. The definition of adiabatic efficiency (eq. 7–39) allows determination of the final expansion point on the saturation line:

$$\eta_t = \frac{h_4 - h_5}{h_4 - h_{5i}} \tag{11–13}$$

8. h_5 will equal $h_g(T_0)$ if the pressure p_4 was correctly chosen, otherwise p_4 must be adjusted until this condition is met to a satisfactory degree of accuracy.
9. The net work output is $w = (h_4 - h_5) - w_c$.
10. The heat output is $q_{out} = h_{fg}(T_0)$.
11. From the First Law, the heat input is $q_{in} = q_{out} + w$.
12. And finally, the thermal efficiency is $\eta = w/q_{in}$.

Generally, when this cycle is used in conjunction with another, only the additional work output is of interest since it is added to the output of the first cycle. Under these circumstances, the efficiency of the supercritical cycle is of secondary concern.

One would expect a triangular cycle on the T-s plane such as this to be quite low in thermal efficiency. Hence a consideration of this cycle is most appropriate in connection with integrated operation of a higher temperature cycle (see Section 11.5.1).

11.4.2. *Performance of an Ideal Match Supercritical Cycle*

In use as a cycle to take advantage of waste heat from a Brayton cycle, the supercritical cycle should have an isobar that is like an isobar of a perfect gas. In that limit, the heat addition process can be calculated analytically. Further, if the working fluid is entirely an ideal gas made to execute the cycle described by the supercritical cycle, the limiting performance of this engine may be obtained because the Brayton cycle becomes a Carnot cycle, at least on the low-temperature end. In practice, this cycle would have to be built with a large

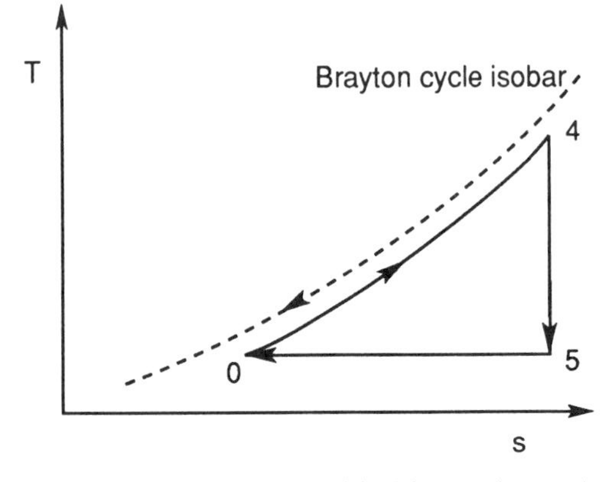

Fig. 11–18. Ideal gas supercritical bottoming cycle.

number of intercooled compression states as described in Section 9.8. For this cycle shown in Fig. 11–18, the heat input and rejected heat are

$$\frac{Q_{\text{in}}}{C_p T_0} = \frac{T_4}{T_0} - 1 \quad \text{and} \quad \frac{Q_{\text{rej}}}{C_p T_0} = \frac{s_5 - s_0}{C_p} = \frac{s_4 - s_0}{C_p} \tag{11–14}$$

Along an isobar, the entropy change can be related to T changes as

$$\frac{T_4}{T_0} = \exp\left(\frac{s_4 - s_0}{C_p}\right) \tag{11–15}$$

so that work and efficiency can be expressed simply in terms of the temperature ratio. Thus it follows directly

$$\eta = 1 - \frac{\ln \theta_4}{\theta_4 - 1} \quad \text{where } \theta_4 = \frac{T_4}{T_0} \tag{11–16}$$

This efficiency, shown in Fig. 11–19, is the standard against which real supercritical cycles operating with the same temperature extremes must be measured. Normally, this cycle will operate with an efficiency that is low compared to that of a primary cycle.

11.4.3. *The Liquid Fueled Rocket Engine*

When the useful power of a cycle is a jet employed to produce a propulsive thrust, the engine is a termed a *rocket*. Rockets are necessary for vehicular propulsion in space and the propellant used must be carried in the vehicle. Since the product gas is discharged into the environment, the cycle is necessarily *open*. Such an engine using liquid propellants employs a cycle which is approximately described by the Rankine cycle and is therefore described briefly here. An accurate description as a cycle requires consideration of the energy

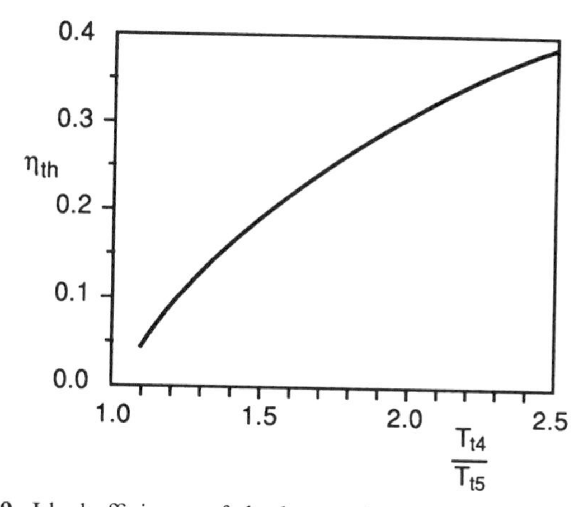

Fig. 11–19. Ideal efficiency of the bottoming cycle shown in Fig. 11–18.

required to create the propellants as storable liquids but this is not of direct interest in rocket vehicles because this process can be carried out separately before use.

The components of a (liquid propellant) rocket are two pumps, a combustion chamber, a thrust producing nozzle, and a turbine to drive the pumps. As in any cycle employing a gaseous working fluid, the total pressure and enthalpy determined the work available on expansion. The thrust of the engine is also determined by these parameters. The design and performance of the nozzle used to convert the enthalpy to jet kinetic energy and thrust is described in Section 12.4.

The similarity to the Rankine cycle is evident through the use of compression of the working fluid (fuel and oxidizer) as a liquid and expansion as a gas. In addition to serving as a working fluid in the expansion, the chemical combination of these liquids also serves as the heat source through combustion. The pressure of the gas in the combustion chamber is determined by the liquid propellant (injection) pressure. The temperature reached is determined by the choice of reactants through their chemical reaction energy. The gas temperature achieved sets the limit on the specific impulse (see Section 2.4) or, equivalently, on the efficiency of the reactants as propellants. High temperatures are typically desirable for this reason. Thus, in most high performance rocket engines the design point is near the point where the enthalpy per unit mass is largest because such a fluid can be expanded to the highest velocity.

The heat release is usually dominant in determining the product gas properties and the gas typically reaches conditions where it is ideal (see Chapter 3). The expansion through the nozzle is usually over such a large temperature range that a perfect gas description is not accurate, although it may be used for estimating the performance (see Sections 4.5 and 12.4).

The rocket uses two pumps to raise the propellant pressure (eq. 7–36). A common method for obtaining the required power is to bleed part of the propellant gas from the combustion chamber, dilute it with additional fuel to reduce temperature to an acceptable level, and expand this gas through a turbine (eq. 9–2) which drives the pumps. The pressure ratio for the turbine may be as large as the nozzle pressure ratio of the primary nozzle, and for a smaller pressure ratio, some thrust may be obtained from the turbine discharge.

There are a number of variations possible to configure this kind of engine. The reader is referred to the specialist literature for other possibilities and greater details. Problem 9 in this chapter is structured to develop an understanding of this application of the Rankine cycle

11.5. CYCLE COMBINATIONS

Cycle combinations can be classified according to whether the cycles involved interchange heat or mass (carrying heat). In heat-interchanging cycles, heat exchangers are used, whereas in cycles with mass interchange, the working fluid is sequentially processed by more than one cycle.

Heat interchanging cycle combinations are those where a heat exchanger or similar device is used to transfer say waste heat from a higher temperature (or *topping*) cycle as input to a lower temperature (or *bottoming*) cycle. The notion of a topping cycle is used when the principal cycle is augmented on the high-temperature end. Similarly, a bottoming cycle is a low-temperature augmentation of a cycle to take advantage of its waste heat. The term "combined cycles" is generally used in connection with a combination of cycles of equal importance and more specifically the Brayton and Rankine cycles operating as a mass interchanging combination. This combination is discussed in greater detail in Section 11.5.2.

11.5.1. *Efficiency Considerations*

Figure 11–20 shows two cycles operating between overall temperature extremes. The heats involved are

1. Input to the top cycle: Q_H
2. Output from the top cycle and input to the bottom cycle: Q_M
3. Output from the bottom cycle: Q_L

The efficiencies of the two cycles are

$$\eta_H = 1 - \frac{Q_M}{Q_H} \quad \text{and} \quad \eta_L = 1 - \frac{Q_L}{Q_M} \tag{11–17}$$

and the overall efficiency of the combination is

$$\eta_0 = 1 - \frac{Q_L}{Q_H} \tag{11–18}$$

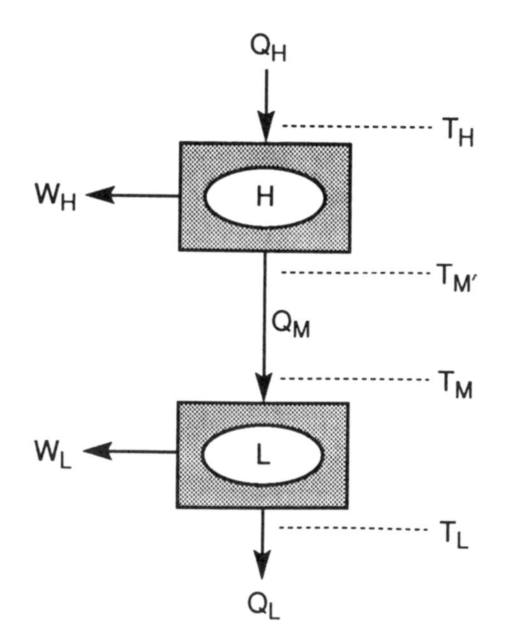

Fig. 11–20. Heat interchanging cycles with temperature drop from $T_{M'}$ to T_M.

or written in term of the efficiencies of the other cycles (eqs. 11–17),

$$\eta_0 = \eta_H + (1 - \eta_H)\eta_L \tag{11–19}$$

The cycles shown in Fig. 11–20 share an intermediate temperature that is near the rejection temperature of the top cycle and the input temperature for the bottom cycle. The question is, What temperature is optimal for this level? This question may be useful to see how far away from an optimum a design is if the working fluid temperatures of the two cycles are constrained. Consider that both, top and bottom, cycles are quasi-Carnot cycles whose efficiency is given by

$$\eta_H = \phi\left(1 - \frac{T_{M'}}{T_H}\right) \quad \text{and} \quad \eta_L = \phi_L\left(1 - \frac{T_L}{T_M}\right) \tag{11–20}$$

where ϕ less than unity is the totality of the measure of the irreversibilities. Further, let one of the cycles be slightly better than the other by having its ϕ larger by a small amount and let there be a finite temperature difference between the two cycles. Thus

$$\phi_L = \phi(1 + \varepsilon) \quad \text{and} \quad T_{M'} = T_M(1 + \delta) \tag{11–21}$$

Here T_M is the high temperature for the lower cycle, whereas $T_{M'} = T_M\,(1 + \delta)$ is the rejection temperature of the upper cycle. If one substitutes eqs. 11–21 into the expressions for the upper and lower cycle efficiencies (eq. 11–20) and then forms the overall efficiency using eq. 11–19, this expression can be maximized with respect to T_M. The result for optimum T_M, with ε and δ small so that the result can be linearized, is

$$T_{M,\max\eta} = \sqrt{T_L T_H}\left(1 + \frac{\varepsilon}{2(1-\phi)}\right)\left(1 - \frac{\delta}{2}\right) \tag{11–22}$$

Note that ε is zero when $\phi = 1$, the system consists of two Carnot cycles. For $\delta = 0$, the optimum temperature is the geometric mean of the extremes. When $\phi < 1$, a better bottom cycle ($\varepsilon > 0$) suggests a higher T_M. A larger temperature difference between the cycles favors the more efficient upper cycle by lowering T_M.

11.5.2. *Examples*

The best practical example of a heat interchanging cycle is the so-called combined cycle operation of the gas turbine and the Rankine cycle to take advantage of the waste heat. Figure 11–21 shows a simple combined cycle plant and the heat transfer process between the two cycles carried out in a two-loop

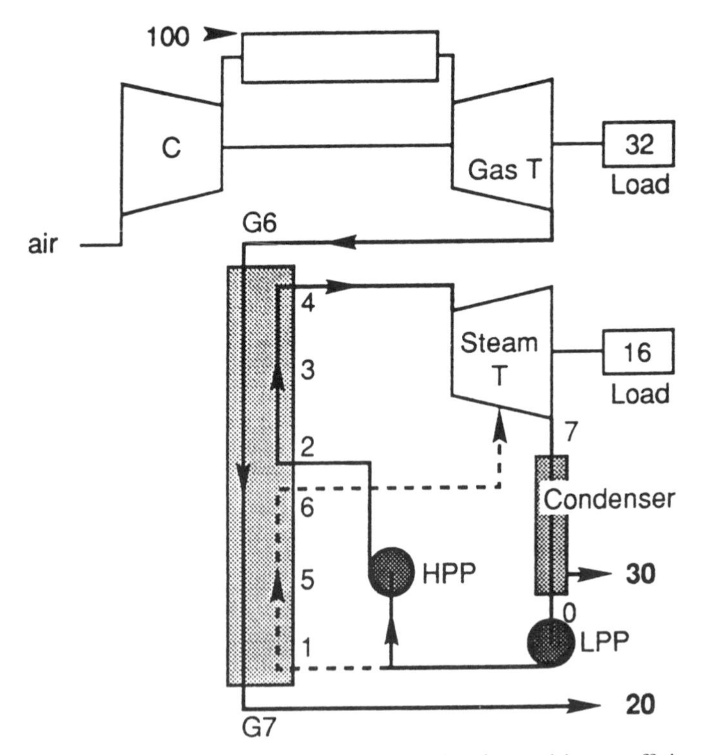

FIG. 11–21. Schematic of a typical combined cycle plant with net efficiency of 41% indicated. Pump work and losses account for balance of energy budget. In modern, more complex systems, this efficiency has reached around 53%. State points correspond to those in Fig. 11–16.

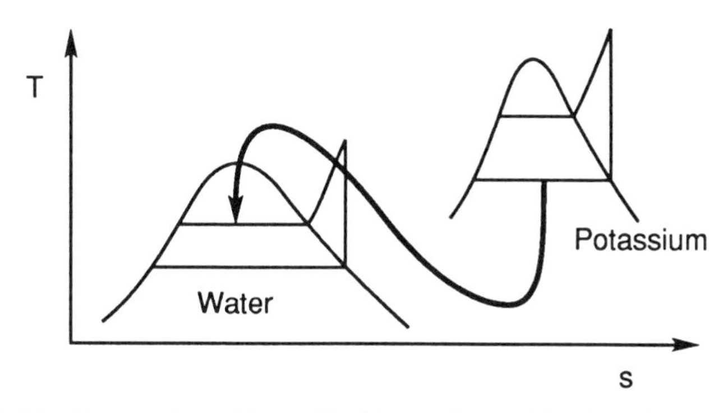

FIG. 11–22. T-s diagram for a binary Rankine cycle combination. Arrow shows heat transfer between cycles.

heat exchanger. The figure shows a typical energy budget. These plants can achieve high efficiency (>50%) with methane or distillate fuels and are in relatively common usage in systems of 1–100 MW, particularly when there is a need for heat (hot water) as a by-product and the need to be able to operate at part load. This last aspect can be achieved by discontinuing operation of one of the cycles. The performance analysis is a direct application of the methodology developed for the two individual cycles.

Other cycles have been proposed and sometimes built, but are not in common usage owing to the high capital cost. Such systems include liquid metal topping, organic bottoming of the water Rankine cycle, and organic Rankine bottoming of the Brayton cycle. A number of organic fluids have critical parameters that are low enough to work with input temperatures of a few hundred degrees Centigrade. These are listed and discussed below.

The topping cycle provides a good example to show how the mass flow rates through two heat interchanging cycles must be related. Consider the combination of a liquid potassium metal (K) cycle which provides some of the condensation heat to a water boiler with a some finite ΔT. If both cycles are subcritical as shown in Fig. 11–22, then the superheating must be done by means that allow the temperature demanded to be reached. This may be done with the primary heat source(s) as shown in Fig. 11–23.

11.5.3. *Mass Flow Matching*

The heat balance between the cycles is established by the condition that links the cycles, namely,

$$\dot{m}_{\mathrm{K}} h_{\mathrm{fg,K}}(T_{\mathrm{min,K}}) = \dot{m}_{\mathrm{w}} h_{\mathrm{fg,w}}(T_{\mathrm{boil,w}}) \tag{11–23}$$

where the temperatures are the potassium condensing and the water boiling values, related by $T_{\mathrm{boil,w}} = T_{\mathrm{min,K}} - \Delta T$. This relation establishes the mass flow rate ratio for the two cycles.

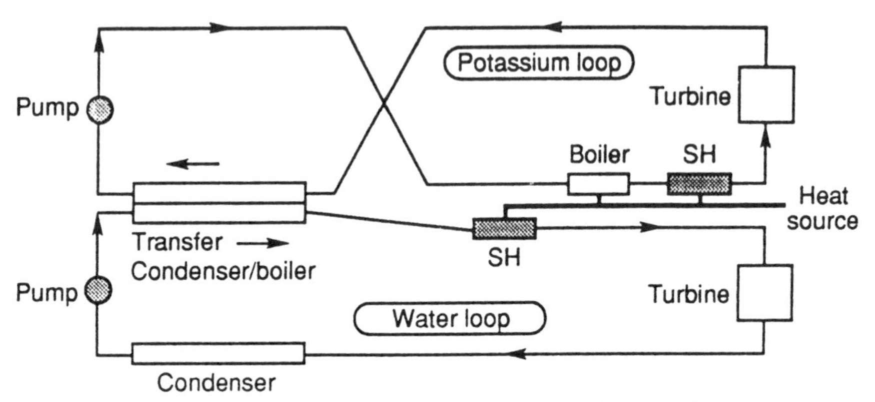

FIG. 11–23. Liquid metal binary Rankine cycle schematic showing the arrangement of components, and the superheater(s) in particular, denoted by SH.

11.5.4. *Pinch Point*

Gas cycles such as the Brayton cycle have a significant thermal (sensible) energy in the exhaust stream as discussed in Section 11.4. This heat may be used as heat input to a Rankine bottoming cycle. The temperatures on the two sides of the heat exchanger are shown in Fig. 11–24. Both subcritical and supercritical bottoming cycles are shown. The temperature difference will, due to the properties of the Rankine fluid, be nonuniform with a minimum at a so-called "pinch point." This point is important because the small ΔT results in a requirement for a correspondingly large heat transfer area. The total area required is, from the phenomenological heat transfer law (eq. 6–9),

$$\text{Area (wetted)} = \int_{\text{inlet}}^{\text{outlet}} \frac{dq}{h_c \Delta T} \tag{11–24}$$

where h_c is the convective heat transfer rate coefficient which depends on the heating mechanism, (i.e., single phase heating or boiling). Matching the bottoming cycle to the Brayton cycle is carried out by adjusting the temperature at the Rankine side inlet and by tailoring the mass flow rate in the bottom cycle to that of the upper cycle. In effect, changing the mass flow rate of one of the fluids relative to that of the other cycle changes the slope of the process line in the T-s plane (since s is per unit mass) and thereby changing the ΔT as shown in Fig. 11–24. This allows minimization of the integral in eq. 11–24, resulting in a lowest cost heat exchanger.

It follows from the characteristics of the supercritical cycle that there is a better match between this cycle and the Brayton cycle isobar: The irreversibility is smaller as is the area required. A number of working fluids (and their properties) which have been considered (e.g., ref. 11–4) in connection with a closed Brayton cycle with nuclear heat input is given in table 11–1.

Table 11–1. Properties of some materials for supercritical Rankine bottoming of the Brayton cycle

Substance	MW	p_{crit}(atm)	T_{crit}(K)
Ammonia NH_3	17	111.3	405
Isobutane C_4H_{10}	58	36.0	408
Propylene C_3H_6	42	45.4	365
Freon 12 CCl_2F_2	121	40.6	385
Propane C_3H_8	44	42.1	370

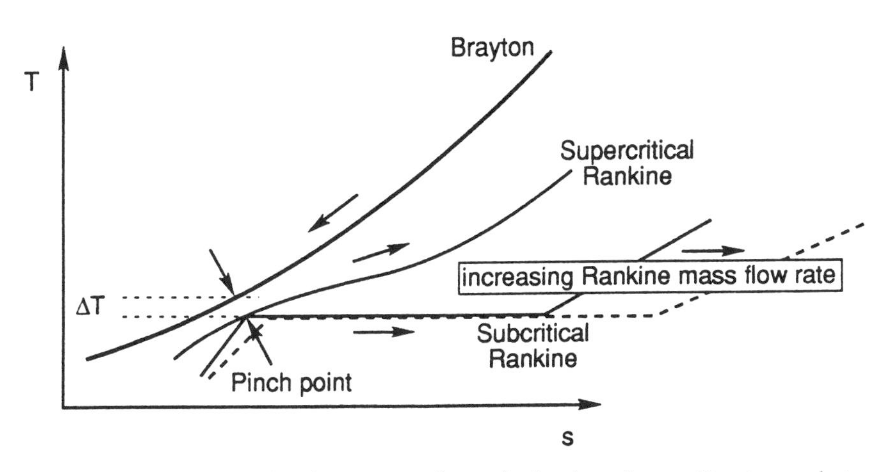

FIG. 11–24. *T-s* diagram for the process of transferring heat from a Brayton cycle to a Rankine cycle in a bottoming configuration. ΔT at the *pinch point* is the minimum temperature across the heat exchanger. Other arrows show process direction. The two subcritical process lines are for differing mass flow rates.

These substances vary in their relative values of the heat transfer coefficient in the precooler, of the condensing coefficient, and of the turbine flow area ratio, as well as in the critical properties. All are plausible candidates for use in the bottoming of the Brayton cycle between temperatures of 450–500 K on the hot end rejecting heat near 300 K.

11.5.5. *Mass Interchanging Cycles*

An open cycle gas turbine processes not only the gas delivered by the compressor but also the liquid fuel that is fed in to provide the heat. One could say that the fuel mass undergoes a Rankine cycle, albeit through a common expansion process as the Brayton cycle. In this same way one can imagine that steam generated from waste heat injected into the turbine stream may execute an expansion that depends on the properties of the mixed flow. The analysis that applies is a direct application of the methodology derived for the elementary processes (ref. 11–5).

A well-known attempt at achieving high efficiency through complexity is the coupling of the gas turbine with the Diesel engine. The Diesel engine operated near its maximum efficiency is fuel lean and the remaining oxygen can be combined in a steady flow combustor to drive a turbine. This is augmented with additional air as required by the temperature limits on the turbine. An engine based on such a cycle has been built, but evidently was concluded to be too complex to be economically interesting (ref. 11–6).

11.6. INTEGRATION OF THE MHD GENERATOR INTO A POWER SYSTEM

The magnetohydrodynamic (MHD) generator processes a compressible gas that must operate with a high temperature through the generator (ref. 11–7). This is true because the electromagnetic force interaction of the gas with its environment is small unless temperature in excess of 2000 K can be obtained in the generator. The MHD generator power extraction process parallels that of a turbine, which imparts to the cycle that utilizes it some of the characteristics of the Brayton cycle. There are, however, important differences that distinguish its performance from that of gas turbine cycles and their derivatives. The most important ones are, first, that the "turbine" inlet temperature to the MHD generator is not fundamentally limiting. Rather, the temperature at its *exit* must be sufficiently high to retain a short interaction length, and, at the same time, the temperature at the hot end of a follow-on heat exchanger such as the air-preheater must be lower than its material capability. Lastly, the peak cycle temperature may be made as high as allowed by the heat release from the fuel combustion. These features result in unique cycle design results.

The cycle using an MHD generator must consist of the generator, a compressor, heat exchangers and the cycle elements that allow the conversion of thermal power not utilized in the MHD generator. The cycle that naturally suggests itself is Brayton-like, where the MHD generator is used as a high-temperature turbine and a compressor is used to charge the generator.

The MHD interaction requires temperatures that are generally in excess of those achievable with combustion of fossil fuels, coal in particular, with ambient or near ambient air. An air preheater using the higher temperature of the generator exhaust is therefore required. The thermal ionization of the seed material in the combustion gas is also promoted when the pressure is low (Section 5.10). The exhaust pressure of the MHD generator should therefore be near atmospheric pressure, the lowest practical pressure of this combustion-driven, open cycle. From the viewpoint of thermodynamic analysis, it is easier to examine the Brayton cycle than the Rankine cycle. The latter is a likelier system to be used for the utilization of the MHD topping cycle "waste" heat because of the experience available and because the Rankine cycle is less sensitive to the use of solid contaminated gas. In this discussion, therefore, the Brayton cycle is used as a focus to illustrate the characteristics of the integrated system. Conclusions regarding overall efficiency performance are generally valid, and detailed analyses have been carried out for the Rankine cycle (refs. 11–8 and 11–9, among others).

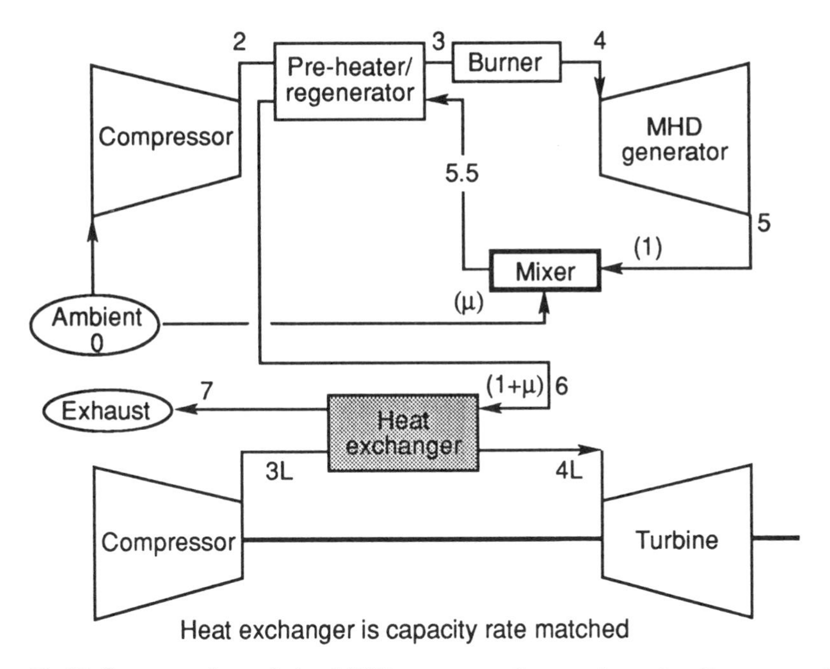

FIG. **11–25.** Incorporation of the MHD generator for topping of a Brayton cycle. Bottom cycle accepts all preheater exit heat: system 1. Unit mass is processed by the MHD generator.

The exhaust gases from the MHD generator are hotter than an air preheater can tolerate so that its entry temperature must be held to an acceptable level. Atmospheric pressure air is mixed with the MHD exhaust gas, and this hot gas may be used to preheat the combustion air. At this point, two options become available, one is to use the entire mixed gas mass to preheat the combustion gas (Fig. 11–25) or only an amount that allows the capacity rate of the air preheater to be matched (Fig. 11–26). The heat exchanger temperature profiles for these two cases are shown in Fig. 11–27 (ref. 11–10 and Section 6.2.2).

The objective of this cycle analysis is to determine the advantages and disadvantages of the two configurations shown and, for the better one, to determine the contribution of the MHD generator to the work output and overall thermal efficiency levels. Further, an examination of the sensitivity of performance to the magnitude of the critical parameters is a goal of this section.

11.6.1. *Cycle Analysis*

A cycle analysis requires a model of the principal components of the system. The goal is to describe clearly and simply the particular characteristics of the MHD generator and to connect this element with other components modeled to a similar level of accuracy. The assumption will be made that the scale of the system is sufficiently large such that losses that depend on surface-to-volume ratio, such as friction and heat transfer, are negligible.

Heat exchangers may be made to have performance limits that are close to

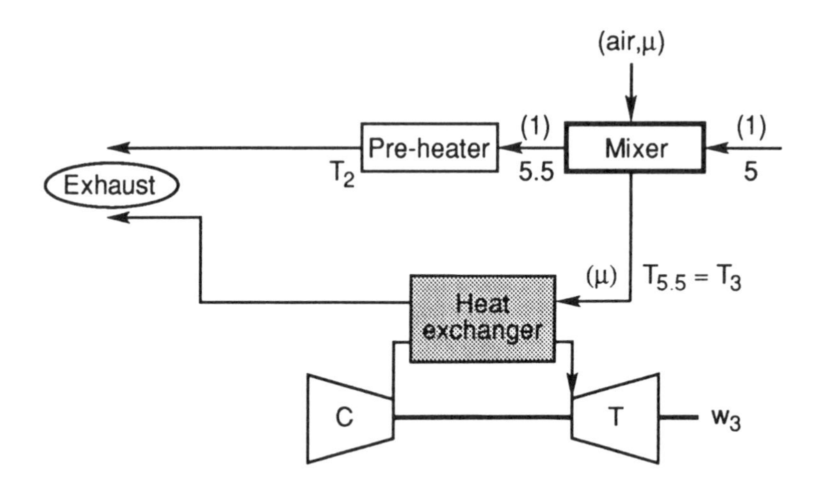

FIG. 11–26. Brayton cycle bottoming with two exhaust streams: system 2. Connection to MHD generator is at 5 and preheater connection is as in Fig. 11–25. Numbers in parentheses show relative mass flow rates.

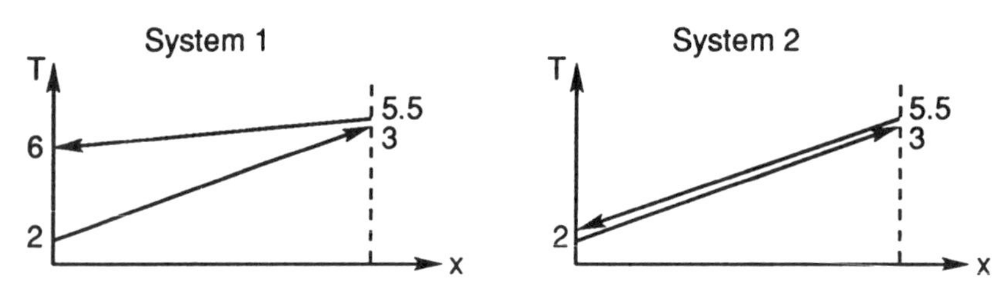

FIG. 11–27. Temperature profiles for the two systems' air preheater exchangers, x is the dimension along the flow.

ideal because that achievement is an economic rather than a thermodynamic limitation. Additional parameters such as temperature drop and pressure losses may easily be incorporated into a more detailed study. Here such inclusion is omitted to reveal the cycle limits as clearly as possible. The heat exchangers must therefore be of the counterflow configuration (Section 6.2).

The configuration with a single waste stream is termed "system 1" as in Fig. 11–25 whereas "system 2" describes the alternative with its hotter but smaller exhaust stream. The two configurations may be compared, assuming that the topping (MHD) cycle parameters are fixed. These parameters are examined first.

11.6.2. *MHD Topping Cycle*

The MHD generator is a turbinelike device that operates with a *minimum* (i.e., exit) temperature on the order of 2000 K or more. Its entry temperature is controlled by the details of the combustion process, in particular, by the fuel's heating value and the temperature of the combustion air. The fundamental measure of the MHD generator's reversibility as a work component is the polytropic efficiency of the expansion process. This polytropic efficiency is necessarily less than unity because of the Joule heating associated with current flow. The specification of the efficiency allows one to calculate cycle pressure ratio (Section 9.3).

For the analysis to follow, the working fluid is modeled as ideal and perfect air ($\gamma = 1.4$). Gas dynamic components such as a diffuser, burner, etc., are assumed ideal in that total pressure ratios are unity. The total temperatures are nondimensionalized by ambient temperature and denoted by θ, whereas total temperature ratios across components are denoted by τ. The numerical subscripts used to describe states are identified in Fig. 11–25. Thus

$$\left.\begin{aligned} &\tau_c = \frac{T_{t2}}{T_0} \qquad \text{and} \qquad \tau_m = \frac{T_{t5}}{T_{t4}} \\ \text{and} \\ &\theta_i = \frac{T_{ti}}{T_0};\ i = 3,\ 5,\ 5.5,\ 6 \end{aligned}\right\} \tag{11–25}$$

With ambient temperatures at 300 K, typical values for θ_5 (MHD generator exit) are of the order of 7 (2100 K), whereas θ_3 may be near 3 (900 K, a conservative value for the air preheater temperature capability). These values are used here, and the sensitivity to their choice is examined.

The independent parameter of interest is the enthalpy extraction in the MHD generator. This is measured by $\tau_m < 1$. The work (per unit mass) from the MHD generator is given by

$$w_m = \theta_4(1 - \tau_m) = \theta_5\left(\frac{1}{\tau_m} - 1\right) \tag{11–26}$$

where $1 - \tau_m$ is the fraction of the enthalpy extracted. For a polytropic expansion process, the cycle pressure ratio is related to the (total) temperature change by

$$\tau_m = (\pi_c)^{-k(\gamma-1)/\gamma} = (\tau_c)^{-ke_c} \tag{11–27}$$

Here k is the so-called load factor (typically 0.7–0.8) for the generator, and the polytropic efficiency (e_c) characterizes the compressor. τ_c is the compressor temperature ratio (for the MHD cycle). Pressure losses around the cycle are assumed negligible. The compressor work (per unit mass) is

$$w_c = \tau_c - 1 \tag{11–28}$$

The enthalpy extraction and topping cycle pressure ratio are shown in Fig. 11–28 for the conditions noted. Evidently for $(1 - \tau_m)$ of order 0.25,

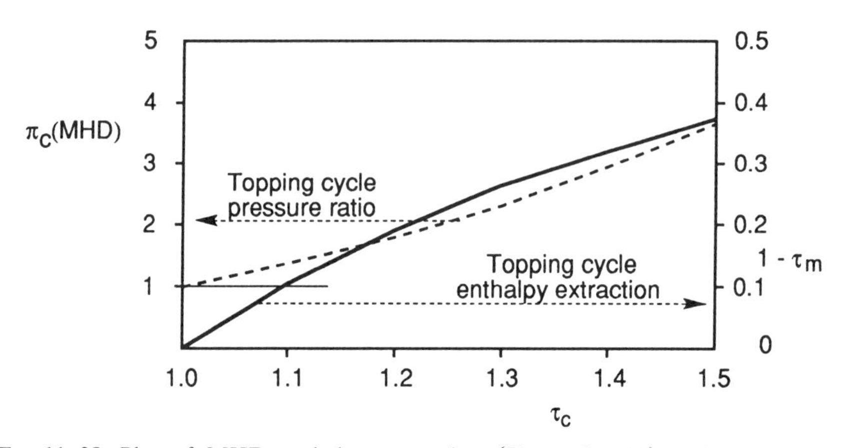

FIG. 11–28. Plot of MHD enthalpy extraction $\{[1 - \tau_m] \times 10\}$ and topping cycle pressure ratio for $k = 0.8$ and $e_c = 0.92$.

a τ_c ratio of the order of 1.3 is required, and the corresponding pressure ratio is around 2.2.

The temperature ratio θ_4 describes the generator entry temperature and θ_5 that at the exit flow which must be held to a value higher than some minimum. For an ideal (large L/D) preheater the exit air temperature equals the hot-gas entry temperature (Section 6.2). These two temperatures are limited to a practical maximum value, so that

$$\theta_3 = \theta_{5.5} \tag{11–29}$$

The net work for the MHD top cycle is obtained from eqs. 11–26 to 11–28. This work is a monotonically increasing function of $(1 - \tau_m)$.

The heat input to the cycle is given by the combustor energy balance for the fixed generator exit temperature and the work extracted through the generator:

$$q = \theta_4 - \theta_3 = \frac{\theta_5}{\tau_m} - \theta_3 \leq \Delta\tau_b \equiv \frac{\dot{m}_{\text{fuel}} H}{\dot{m}_{\text{air}} C_p T_0} \tag{11–30}$$

The temperature rise parameter $\Delta\tau_b$ is a nondimensional ratio of heating value per unit air mass flow and is numerically of the order of 8–10 (Section 7.5). Note that the air-preheater's function is to recycle exhaust heat in order to raise the compressed air temperature from the value τ_c to θ_3. A thermal efficiency may be calculated for only the topping cycle as

$$\eta_{\text{MHD}} = \frac{w_m - w_c}{q} = \frac{\theta_5\left(\dfrac{1}{\tau_m} - 1\right) - \left(\left[\dfrac{1}{\tau_m}\right]^{1/ke_c} - 1\right)}{\dfrac{\theta_5}{\tau_m} - \theta_3} \tag{11–31}$$

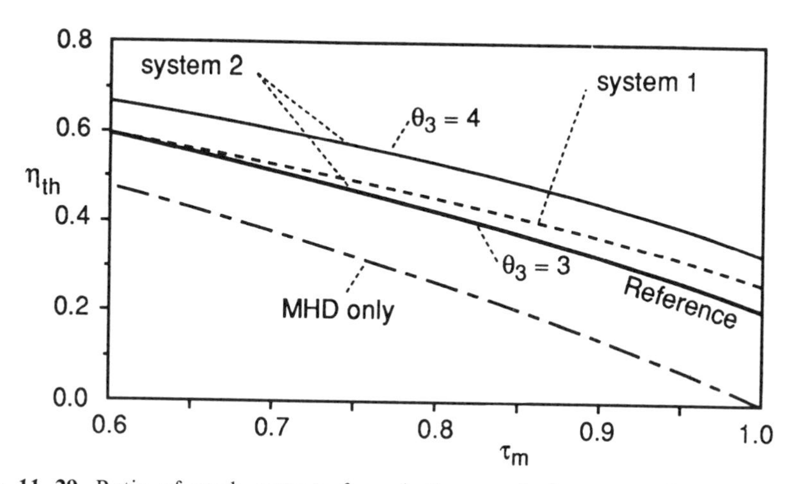

Fig. 11–29. Ratio of work outputs from bottom cycle for two cycle configurations showing relative advantage of system 2 over system 1.

This quantity is shown in Fig. 11–29 and is discussed subsequently although this efficiency is not of direct interest. Of much greater importance is the efficiency that includes the work contribution from the bottoming cycle in the numerator of this expression. Two values of θ_3 (=3 and 4, conservative and advanced values, respectively) are noted as well as the MHD generator exist temperature, $\theta_5 = 7$.

11.6.3. *Brayton Bottoming Cycle*

The heat input to the complete cycle is fixed by the topping cycle input, and the configuration determines how much work can be obtained from the bottom. The objective of the bottoming cycle design must be to maximize the contribution to the output work. Thus the configuration must be designed to have a pressure ratio to maximize specific work. For a Brayton cycle, this work is largest when the compressor temperature ratio is given by (Section 9.3):

$$\tau_{cL} = \sqrt{\theta_{4L}} \tag{11–32}$$

and this largest work output per unit mass is

$$w_{max,L} = (\sqrt{\theta_{4L}} - 1)^2 \tag{11–33}$$

The subscript "L" denotes the Brayton cycle's role as a *lower* or bottoming cycle. The maximum specific work condition requires equal turbine exit and compressor exit temperatures which make regeneration impossible.

The air flow rate required for diluting the hot exhaust needs no compression (hence no work) because both flows are at atmospheric pressure. The heat balance on the mixer with the hot side temperature given by θ_3 gives the amount of diluent air as a multiple μ of the mass flow rate through the MHD generator:

$$\mu = \frac{\theta_5 - \theta_3}{\theta_3 - \theta_0} = \frac{\theta_5 - \theta_3}{\theta_3 - 1} \tag{11-34}$$

The value of this parameter is 2 for the $\theta_3 = 3$ case and falls to 1 for $\theta_3 = 4$. In either case, the amount of diluent air required is significant.

System 1

This arrangement provides $(1 + \mu)$ units of mass to the lower cycle with a temperature of θ_6. This temperature is given by the thermal balance in the air preheater or,

$$\theta_6 = \frac{\tau_c + \mu\theta_3}{1 + \mu} \tag{11-35}$$

This temperature is the peak value of the lower cycle if no additional combustion is carried out. τ_c is the compressor exit temperature on the MHD topping cycle and, closely related to the enthalpy extraction τ_m as noted in eq. 11–27. The work from the lower cycle is therefore

$$w_{1,L} = (1 + \mu)(\sqrt{\theta_6} - 1)^2 \tag{11-36}$$

System 2

When the preheater is designed to be capacity rate matched, then the waste stream will consist of one mass unit at temperature τ_c and μ units at temperature θ_3, as shown in Figs. 11–27 and 11–28. When a maximum work Brayton cycle is used for the high temperature stream and the other is waste heat, the work realized is

$$w_{2,L} = \mu(\sqrt{\theta_3} - 1)^2 \tag{11-37}$$

The ratio of the two work outputs $w_{2,L}/w_{1,L}$ is shown in Fig. 11–30. This ratio is always greater than unity because the bottoming cycle operates at higher temperature in system 2. This advantage is improved when θ_3 is increased from the value of 3 to 4 as shown.

The overall system efficiency may be calculated using the MHD generator and bottoming cycle work outputs. Figure 11–29 shows this result with the enthalpy extraction as the independent variable. Nominally, the value of $\theta_3 = 3$ *and the system* 2 are used as a reference, heavy line. The system 1 is noted with a dashed line. In terms of efficiency, the two systems are seen to be close in performance. The thin solid line shows the improvement associated with operating system 2 with a better heat exchanger. Overall, it is evident that about 20% enthalpy extraction is required of the MHD system. At this point, the

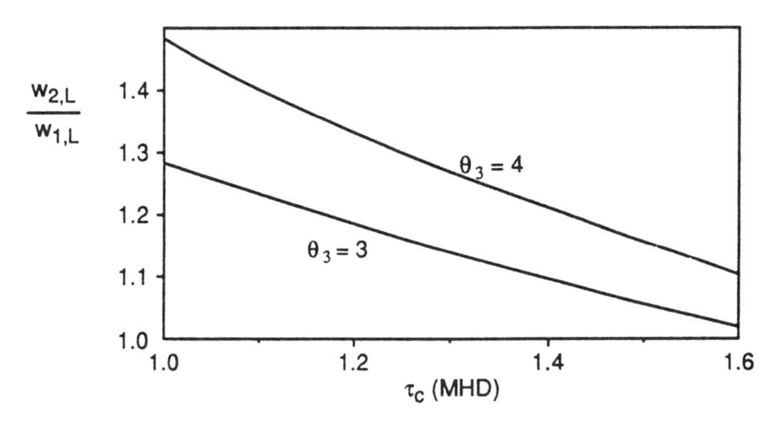

FIG. 11–30. Plot of overall thermal efficiency versus MHD total temperature ratio, $k = 0.8$ and $e_c = 0.92$. Heavy solid line is a system 2 reference case: $\theta_5 = 7$, $\theta_3 = 3$. Thin solid line is for $\theta_3 = 4$. Dashed line applies to system 1 with $\theta_3 = 3$.

bottoming cycle contributes enough to raise overall efficiency above 40%. It is noteworthy that increasing the enthalpy extraction (lower τ_m) leads to increased peak cycle temperature and therefore an improved efficiency and in a mildly increased waste heat temperature.

The parameter τ_m may not be reduced indefinitely. The temperature entering the MHD generator is equal to θ_5/τ_m which rises as τ_m is reduced. At some value, a maximum temperature will be reached that cannot be exceeded because of fuel thermal energy limitations. From eq. 11–22, for $\Delta\tau_b = 9$, and the parameters used here, minimum τ_m is of the order of 0.6. In practice, this situation would be circumvented by oxygen enrichment of the combustion air.

11.6.4. *Reversible Bottoming Cycle*

In order to see what fraction of the ideal this performance level represents, one may examine the performance of a reversible (Carnot) engine operating between θ_6 or θ_3 and the ambient temperature. The preheater exit air temperature is given above and the work per unit mass of the top cycle is

$$w_{1,\text{Rev}} = [(1+\mu)(\theta_6 - 1)]\left\{1 - \frac{1}{\theta_6}\right\} \quad \text{and} \quad w_{2,\text{Rev}} = [\mu(\theta_3 - 1)]\left\{1 - \frac{1}{\theta_3}\right\} \tag{11–38}$$

for the two systems. Here the first two quantities, [ ], are the heat fluxes and the last, { }, are the Carnot thermal efficiencies (Section 7.8). The ratio of bottom cycle work to reversible work for the two system configurations is shown in Fig. 11–31 to illustrate the potential of bottom cycle improvements. For the poorer system 1, the temperature ratio θ_6 depends on the MHD interaction whereas for the system 2 it depends only on θ_3 (horizontal lines). Two curves are shown for each case, one for $\theta_3 = 3$ and

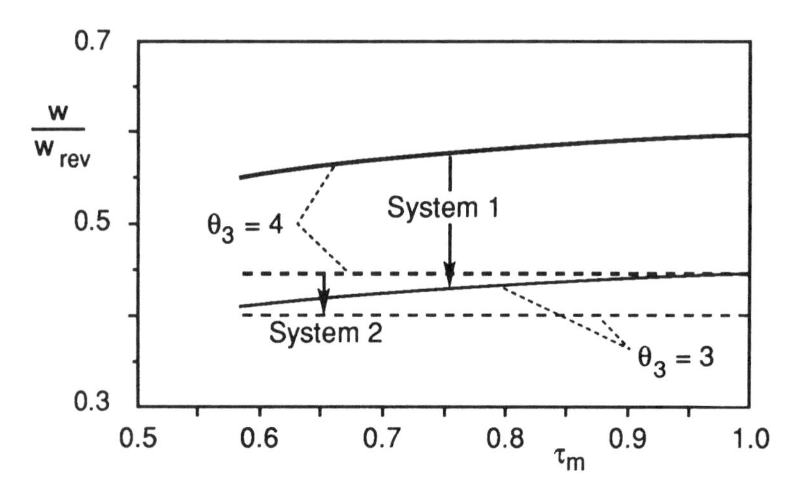

FIG. 11–31. Fraction of reversible bottom cycle work realized through use of a Brayton cycle. Upper lines are for $\theta_3 = 4$ whereas the lower (at end of arrows) is for $\theta_3 = 3$.

the other = 4. The directional arrow shows the result of using an improved heat exchanger. Evidently, the ideal Brayton cycle realizes about half (40–55%) of the work available, and a well-developed cycle should be able to reach performance between these two limits. In other words, in Fig. 11–29, an increase of another 10 percentage points over the combined performance should be possible with a more sophisticated cycle than a simple Brayton bottoming even though component irreversibilities in that cycle have been neglected.

The sensitivity to the input parameters to the chosen parameters may be examined using the relations developed above. In short, the sensitivity of thermal efficiency to component efficiencies is relatively small, as it is to the choice of the MHD generator exit temperature. On the other hand, raising θ_3 (the regenerator hot end temperature and turbine inlet temperature) from 3 to 4 results in 5–10 percentage points improvement in overall thermal efficiency. Increasing the heat exchanger capability decreases mass flow while increasing temperature. An improvement in θ_3 leads to a direct increase in the work contributed by the bottom cycle.

In summary, the efficiency of an MHD topped cycle improves with the increasing enthalpy extraction from the topping cycle. When the enthalpy fraction removed in the MHD cycle exceeds 20%, the potential for realistic cycles having thermal efficiencies greater than 40% is evident. The performance level is potentially good compared to today's efficiency for the generation of electric power.

The importance of θ_3 as a performance parameter is clear since it controls the work output from the lower cycle. The better cycle configuration is that where the temperature of the bottom cycle is highest, and its relative mass flow rate is correspondingly small. Under such circumstances, the performance level capabilities of the two heat exchangers, as measured by θ_3, are equally limiting.

11.7. UNUSUAL CYCLES

11.7.1. *The Dissociating Vapor Cycle*

A (closed) Brayton cycle that uses a working fluid where the specific heat changes from a low value during the cold portion of the cycle to a high value at the high temperature end may realize a performance gain within the constraints of fixed cycle temperatures because the works of compression and expansion are proportional to C_p. The dissociating vapor cycle has been proposed to achieve this. Balancing the advantages is the fact that phase change energy is involved in the work processes which makes them nonadiabatic. A complicating feature in the work processes is that the changing pressure tends to counter the trend associated with increasing temperature. Without going into the performance calculation, it turns out that in practice good working fluids are few and they are chemically corrosive. No attempts to implement such a cycle has been successful to date. The fluids considered for such cycles must have characteristic composition-change temperatures that lie between the limits where the cycle is to be operated. The reactions considered have been the dissociation of Al_2O_6 and N_2O_4.

11.7.2. *Unsteady Processes*

The characteristics of steady flow cycles are examined here and in Chapter 9. Unsteady processes occur in the Diesel and Otto cycles described in Chapter 8 and in the Stirling cycle described in Chapter 10. The Otto cycle combustion process is certainly unsteady, proceeding from spark plug to all corners of the combustion chamber. However, on a sufficiently long time scale, as one assumes in an air-standard analysis, the violent mixing makes the gas system uniform and the unsteady aspects become much less important in determining performance than do the average properties. A class of heat engines can be devised that utilize unsteady processes to gain a thermodynamic advantage. Such possibilities are limited by the characteristics of the working fluid in that its specific properties control wave processes (see ref. 9–1). Cycles that use such phenomena are not specifically discussed here, but their consideration requires a description of fundamentals of unsteady 1-D flow covered in Appendix D. The integration of wave compression into internal combustion engines is described briefly in Section 8.10.

11.7.3. *Kalina Cycle*

Mixtures of ammonia and water may be used in a Rankine power production cycle in the reverse sense that similar mixtures are used in regenerative ammonia refrigeration. Such cycles are termed "Adjustable Proportion Fluid Mixture" cycles; and of these, the Kalina cycle is the best known (refs. 11–11 and 11–12). The cycle utilizes a mixture that has the property that over a range of temperatures the equilibrium fraction of vapor and liquid can be varied. In effect, volume changes for the two phase mixture can be driven by temperature changes. The overall characteristic of the mixture is that the vaporization isobar on the T-s diagram increases to the right, unlike a single substance whose isobar is level. In other words, for the two-component mixture vaporization occurs at an increasing temperature rather than being isothermal as it is for the single component, two-phase evaporation process.

The shape of the isobar makes it attractive as a bottoming cycle for a gas turbine upper cycle. The reason is that the pinch point heat transfer process is not as restrictive, and there is as a result a more uniform and consequently smaller ΔT in the transfer process. The net result is that an approximately 5-percentage-point efficiency improvement may be realized over cycles with conventional bottoming.

PROBLEMS

1. An expansion process for a substance is designed to end with a specific degree of superheat or wetness. Show that the amount (as measured by a temperature increase) of superheat required in the heater is directly related to the steepness of the vapor saturation line on the T-s plane.
2. Consider the cycle shown in Fig. 11–13. Steam is used as a working fluid between a pressure of 100 atm and a temperature of 20°C. Calculate the reversible work available for
   a. A single expansion to the saturated state. What is T_{max}?
   b. Two equal pressure ratio expansions, reheat again ending at the saturation state. What should be the intermediate pressure and T_{max}?
   c. Is there an intermediate pressure where the work output is maximized?
3. By examining the properties off liquid potassium (or ammonia), develop plots similar to Figs. 11–11 and 11–12 for this substance.
4. Consider the ideal supercritical cycle shown in Fig. 11–18. Derive an expression of the specific work and the efficiency of this cycle when irreversibilities associated with finite ΔT between source cycle and in the turbine are included. Upper cycle temperatures and minimum cycle temperature are fixed. Find the influence coefficients for w on ΔT and on the adiabatic turbine efficiency.
5. Liquid water (rain?) is injected into the air at the inlet of a Brayton cycle engine. Describe with appropriate equations the performance of this engine that is operating at its design point. For this purpose, assume the water is in equilibrium with the air at its temperature and the compressor aerodynamics are unaffected to first order. Is the performance improved?
6. Consider the simple combined cycle illustrated in Fig. 11–21. Consider the components to be ideal and the gas turbine working fluid to be ideal air with $\gamma = 1.4$. The Brayton cycle is designed for maximum specific work. Calculate the state variables in the Rankine cycle and the mass flow split. Assume that the water condenser temperature is

20°C and θ_4 (Brayton cycle) = 5. Compare the efficiency to that given in Fig. 11–21. A more realistic calculation is for finite polytropic efficiencies and finite ΔTs in the heat exchanger.

7. A simple, one-pass heater Rankine cycle using water is to be used as a bottoming cycle for an MHD generator as in Fig. 11–25. By considering the capability of boiler/superheater materials, determine the pressures of the principal isobars in the bottom cycle. What would be mass flow ratio for the two cycles. Use numerical values of the parameters in the text, if necessary.

8. In the two-pass Rankine cycle discussion of Section 11.3.3, show that the expressions for the mass fractions are

$$y_L = \frac{h_1 - h_0}{h_6 - h_1}; \; x = \frac{h_4 - h_1}{h_4 - h_2}\left(1 + A_{h/T} + \frac{h_2 - h_1}{h_4 - h_2}\right)^{-1}$$

and

$$y_H = (x(1 + A_{h/T}) - 1)(1 + y_L)$$

where

$$A_{h/T} \equiv \left(\frac{T_4 - T_6}{T_6 - T_1}\right)\left(\frac{h_6 - h_1}{h_4 - h_2}\right)$$

Obtain numerical values for all cycle parameters (thermodynamic and fractional mass flow rates) when $T_4 = 600$°C and $T_0 = 20$°C, including specific work and thermal efficiency. For a general calculation, it may be helpful to plot the enthalpy, h (i.e. h_4 and h_6), as a function of T for the isentropic (or constant polytropic efficiency) line through the saturated vapor line at 20°C. With such a plot and h along the saturated liquid line, the differences in h required are easily determined.

9. Consider a rocket engine consisting principally of two turbopumps, a combustion chamber and a nozzle. The propellants are $H_2 + 0.25\ O_2$ at their respective triple point temperatures. These are burned to give only H_2 and H_2O. Assume a perfect gas. (Parts of this problem stated in square brackets require material covered in Chapter 12.) The turbines are driven by combustion gas from the main combustion chamber (i.e., at the same pressure) but at a temperature reduced by the dilution of additional fuel (H_2).

   a. Sketch the cycle components noting carefully the mass flow rates in all lines. Note the difficulties associated with drawing a T-s diagram because *two* reactants are processed in the cold part of the cycle.

   b. Determine the adiabatic flame temperature of the combustion gas [and the jet velocity on expansion to zero pressure].

   c. The combustion chamber operates at 7.5 MPa (and $M^2 \ll 1$) and the nozzle exit and ambient pressures are 0.1 MPa. Find the H_2/O_2

ratio for the turbine-drive gas so that the turbine inlet temperature is 1300 K.

d. Assuming that the turbine and the pumps are reversible, determine the turbine power requirements to provide for 1 kg/sec of primary propellant. State answer in kJ/kg.

e. Find the ratio of mass flow rate through turbine to that through the primary nozzle assuming both expansion components operate with the same pressure ratio.

f. [Find the thrust produced per kg of propellant used (both turbine and nozzle) assuming no thrust is obtained from the turbine flow.]

g. [Find the jet exit Mach number and the nozzle exit flow area per unit mass flow rate (m^2/(kg/sec))].

h. Obtain the same answers as above avoiding the use of the perfect gas assumption.

BIBLIOGRAPHY

Horlock, J. H., *Cogeneration: Combined Heat and Power*, Pergamon Press, Elmsford, N.Y., 1987.

Hodge, B. K., *Analysis and Design of Energy Systems*, Prentice-Hall, Englewood Cliffs, N.J., 1985.

Oates, G. C., *Aerothermodynamics of Gas Turbines and Rockets*, AIAA, New York, 1984.

REFERENCES

11–1. Keenan, J. H., Keyes, F. G., *Thermodynamic Properties of Steam*, John Wiley and Sons, New York, 1965.

11–2. Wood, B. D., *Applications of Thermodynamics*, Addison-Wesley, Reading, Mass., 1969.

11–3. Hill, P. G., *Power Generation*, Resources, Hazards, Technology and Costs, MIT Press, Cambridge, Mass., 1977.

11–4. Schuster, J. R., Vrable, D. L., Huntsinger, J. P., "Binary Plant Cycle Studies for the Gas Turbine HTGR," ASME paper 76-GT-39, 1976.

11–5. Decher, R., Bickford, P., "Steam Injection in High Pressure Ratio Brayton Cycle Engines," Proceedings Energy, Power and Environmental Systems Conference, May, 1985.

11–6. Sammons, H., Chatterton, E., "The Napier Nomad Aircraft Diesel Engine," SAE Trans., **63**, 107, 1955.

11–7. Sutton, G. W., Sherman, A., *Engineering Magnetohydrodynamics*, McGraw-Hill, New York, 1965.

11–8. Kessler, R., Hals, F. "Coal Burning Magnetohydrodynamic Power Generation," in *Magnetohydrodynamic Power Generation*, ASME Publication AES-23, October 1991.

11–9. Kessler, R., "Economic, Environmental and Engineering Aspects of Magnetohydrodynamic Power Generation," in Tester, J. W., Wood, D. O., Ferrari, N. A., eds., *Energy and Environment in the 21st Century*, MIT Press, Cambridge, Mass., 1991.

11–10. Kays, W. M., London, A. L., *Compact Heat Exchangers*, The National Press, Palo Alto, Calif., 1955.
11–11. Marston, C. H., "Development off the Adjustable Proportion Fluid Mixture Cycle," *Mechanical Engineering*, **114**(9), September 1992.
11–12. Marston, C. H., "Parametric Analysis of the Kalina Cycle," *J. of Engineering for Gas Turbines and Power*, **112**, 107–16, January 1990.

12

GAS KINETIC ENERGY: NOZZLES

Power machines of many kinds take advantage of fluids to transport and convert energy. Fluid kinetic energy is an important form that is useful for the purpose of propulsion, for generating shaft power in a turbine, for compressing a fluid in a steady flow process, and so on. Thermal energy is the most important energy form that may be interchanged with kinetic energy. The thermal energy is of interest because of its relation to the pressure that exerts forces on boundaries. This chapter is a discussion of the energy interchange between kinetic and thermal energies for which important examples are in so-called internal (in contrast to unbounded) flows. The emphasis here is on nozzles. Diffusers are discussed only briefly because their performance is dominated by viscous flow effects, and the necessary description of viscous flow phenomena is beyond the scope of this book.

12.1. INTERNAL FLOW DEVICES WITH AREA CHANGES: NOZZLES AND DIFFUSERS

The conversion of fluid thermal energy to kinetic energy or vice versa is easily accomplished by forcing the fluid to experience a static pressure change. Devices that are adiabatic, or very nearly so, and where the fluid experiences no work interaction are termed *nozzles* or *diffusers* depending on whether the fluid pressure falls or rises. Under such conditions, the steady flow energy equation for an element of fluid states that the total enthalpy must be conserved. Equation 2–6 is the steady flow energy equation for an open control volume. Restated for the conditions stated above, it reads

$$\dot{m}(h_{t2} - h_{t1}) = 0 \qquad \text{or} \qquad h_t \equiv h + \tfrac{1}{2}V^2 = \text{constant} \tag{12–1}$$

Here the possibility of changes in potential energy is neglected. Figure 12–1 shows the conversion process. Since the enthalpy, h, is closely related to the temperature, one can say that for processes between states 1 and 2 where C_p is approximately constant (Section 4.2),

$$C_pT_1 + \tfrac{1}{2}V_1^2 = C_pT_2 + \tfrac{1}{2}V_2^2 \tag{12–2}$$

Evidently the temperature decreases and velocity increases in *nozzles* whereas the reverse occurs in *diffusers*. Since such devices generally operate adiabatically

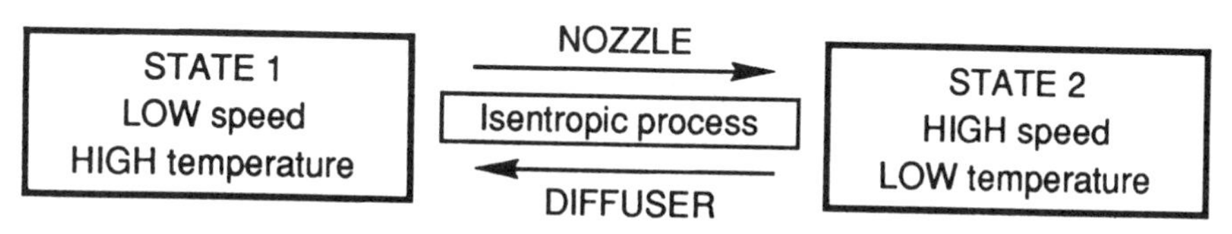

FIG. 12–1. State changes for processes without heat or work interactions.

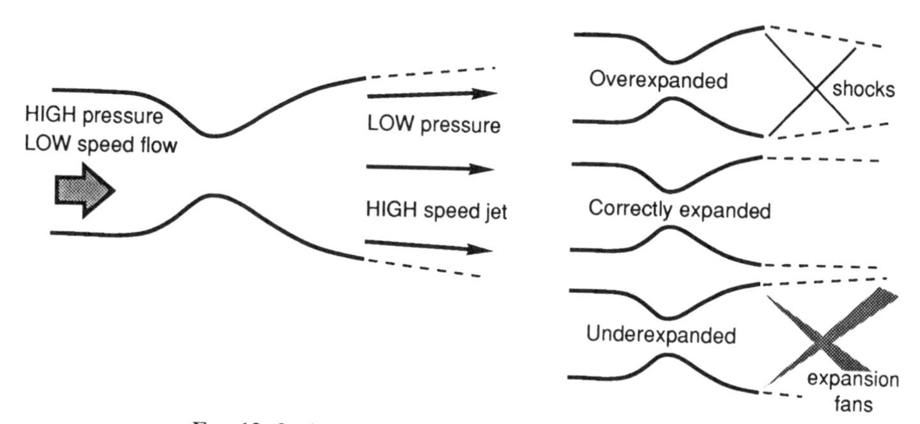

FIG. 12–2. A convergent-divergent propulsion nozzle.

and irreversibilities from friction and heat transfer are limited to the boundaries, the bulk of the fluid undergoes a reversible process. Thus entropy is conserved, and the relations between temperature changes and other state quantities such as pressure can be exploited. For example, for a process where C_p is constant, the isentropic relation can be written (eq. 4–79) as:

$$\frac{T_2}{T_1} = \left(\frac{p_2}{p_1}\right)^{(\gamma-1)/\gamma} \tag{12–3}$$

Thus a nozzle is used to accelerate a high-pressure fluid to high speed, such as in a jet into a lower-pressure environment. Conversely, a high-speed flow is slowed by a diffuser where pressure and temperature rise. Practical use of these devices as primary components includes use in aircraft and rocket propulsion. For example, nozzles are used for the production of a high-momentum jet propulsive thrust (Fig. 12–2), whereas a diffuser is used to match the initially high flow speed of incoming air to a lower speed required by the turbomachinery (Fig. 12–3). The dividing streamline shown is the boundary between flow that is processed by the turbomachinery and the external flow around the aircraft. Note the variation of the flow area as the flow proceeds from one state to another. Here part of the diffusion process takes place ahead of the inlet lip or "highlight," and the remainder takes place in the bounded duct.

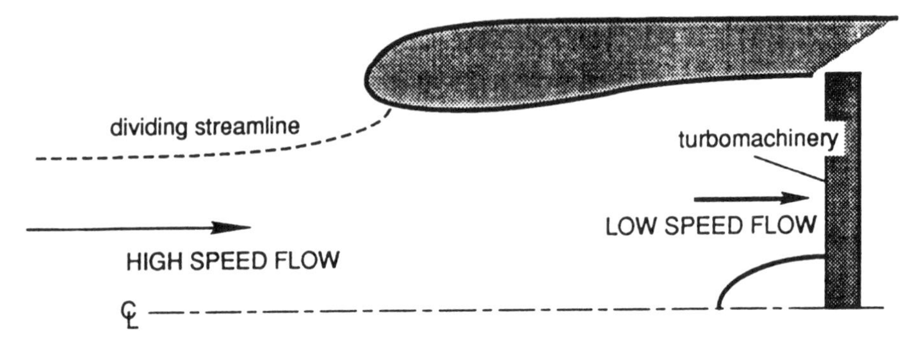

FIG. 12–3. An air-breathing gas turbine engine inlet (diffuser) which typically reduces flight Mach number from ~0.85 to ~0.55 at the fan inlet plane.

12.2. ONE-DIMENSIONAL COMPRESSIBLE FLOW GAS DYNAMICS

12.2.1. *Flow Area*

The following is a brief review of one-dimensional (1-D) gas dynamics, aimed at identifying the important descriptive variables in such flows; see Bibliography, where other texts with varying emphases are cited. The term 1-D is meant to convey that the flow properties in a duct can be described by a single spatial variable, say x. The flow is assumed uniform in the y and z directions. Viscous effects in flows concerned with the conversion of kinetic energy generally play a minor role in the description of such flows hence are neglected here. In situations where they cannot be neglected, their effects will be discussed. The 1-D and steady continuity equation states that mass does not build up in a control volume and can be written

$$\text{mass flow rate} = \dot{m} = \rho u A \tag{12–4}$$

In gives the value of ρu at any station where a cross-sectional flow area, A, can be identified. In the following, this equation is recast in a form that specifically gives the dependence on the more physically interesting parameters, such as temperature and pressure. The consequences of no heat or work interaction with the flow and of reversibility allow identification of constants of the motion which are useful and physically meaningful.

For an adiabatic flow, eq. 12–2 suggests that one such constant of the fluid motion is a temperature that measures the total energy of the flow. This temperature is the total or stagnation temperature of the flow. Further, since the pressure and temperature of an adiabatic reversible flow are related through an equation such as 12–3, it follows that a stagnation (or total) pressure exists and that it, too, remains constant. It is therefore appropriate to write the equation for mass flow rate in terms of these stagnation quantities.

The continuity equation can be combined with the state equation (to introduce the more physically interesting pressure and temperature, p and T, in lieu of ρ),

$$\rho = \frac{p}{RT} \tag{12–5}$$

and velocity is more conveniently expressed in terms of a nondimensional parameter involving the speed of sound which, for a perfect gas, is given by

$$a = \sqrt{\left(\frac{\partial p}{\partial \rho}\right)_s} = \sqrt{\gamma RT} \tag{12–6}$$

Thus one may write a nondimensional measure of velocity, the Mach number, M, defined by

$$M \equiv \frac{u}{a} \tag{12–7}$$

With C_p and R expressed in terms of γ, the energy (or enthalpy) equation (eq. 12–2) can be written as

$$T_{t2} = T_{t1} \qquad \text{where } T_t \equiv T\left(1 + \frac{\gamma - 1}{2} M^2\right) \tag{12–8}$$

The total temperature, T_t, is a flow property that may be measured physically when the flow is brought to rest (where M or $u = 0$) adiabatically. When this state is reached by means of a reversible process, the process obtained is the total or stagnation pressure which is obtained from eq. 12–3 as

$$\frac{p_t}{p} = \left(\frac{T_t}{T}\right)^{\gamma/(\gamma - 1)} = \left(1 + \frac{\gamma - 1}{2} M^2\right)^{\gamma/(\gamma - 1)} \tag{12–9}$$

using eq. 12–8. Here the static state, or the values measured by a moving observer at rest with respect to the fluid are unsubscripted. Flows through nozzles are generally reversible because shock waves are avoided through smooth and gently turning wall design. Further, the portion of the flow affected by friction effects, the boundary layer, grows on these walls and is also subject to accelerating forces. This allows the fluid in the boundary layer to behave in a manner that does not lead to flow separation. Hence the bulk of the flow through nozzles can be assumed reversible and conforming with the flow area imposed by the shape of the walls. Small corrections for the irreversibilities suffered on the walls are usually adequate to accurately estimate nozzle performance. This situation is unlike that in the diffuser where flow separation may be very important and certainly its consideration dominates the design of a diffuser.

The mass flow rate as given by eq. 12–4 can be written in terms of the total pressure and temperature by combining eqs. 12–5 through 12–9 as

$$\dot{m} = \frac{p_t A}{\sqrt{RT_t}} M\sqrt{\gamma}\left(1 + \frac{\gamma - 1}{2} M^2\right)^{-(\gamma+1)/[2(\gamma-1)]} \equiv \frac{p_t A}{\sqrt{RT_t}} f(M, \gamma) \quad (12\text{–}10)$$

This form of the continuity equation is useful because for adiabatic reversible flows, total temperature and total pressure are constant. The relation thus shows how the flow area and the Mach number are related. The functional dependence $f(M)$ defined by eq. 12–10 can be seen to have a maximum at $M = 1$ by differentiation ($df/dM = 0$). This Mach number has the special significance that the mass flow rate per unit area (i.e., $\dot{m}/A = \rho u$) has a maximum. Thus $M = 1$ for a flow with uniform total properties can be found only where the area is a *minimum*. Such an area is termed a *throat*, and the area is termed A^*. The Mach number at the throat fixes the ratio of local static pressure to stagnation pressure (eq. 12–9). For sonic flow, the static pressure is given by:

$$\left(\frac{p^*}{p_t}\right)^{(\gamma-1)/\gamma} = \frac{2}{\gamma + 1} \qquad \text{for } M^* = 1$$

For subsonic flow the pressure at the nozzle exit matches the ambient pressure ($p = p_0$ = environment pressure) and thus determines the mass flow rate. For supersonic flow in the section downstream of the throat, $M^* = 1$, and the mass flow rate *cannot* be influenced by (small) changes in pressure at the exit where the pressure is p_0. From eq. 12–10 it follows from conserved mass flow rate, the ratio of area where the Mach number is M to that where it is unity (A^*) is given by

$$\frac{A}{A^*} = \frac{1}{M}\left(\frac{1 + \dfrac{\gamma - 1}{2} M^2}{\dfrac{\gamma + 1}{2}}\right)^{(\gamma+1)/(2(\gamma-1))} \quad (12\text{–}11)$$

This function is plotted in Fig. 12–4 for various γ's and shows the minimum at $M = 1$. A physical interpretation of $M = 1$ in a throat may be obtained by examining the momentum equation for an isentropic 1-D flow without body forces

$$\rho u \, du + dp = \rho u \, du + a^2 \, d\rho = 0 \quad (12\text{–}12)$$

where in the second form, the speed of sound, a, is introduced through eq. 12–6. Combining this with the continuity equation (a differential form of eq. 12–4) by elimination of $d\rho$, yields

$$\frac{du}{u} = -\frac{dA}{A}\left(\frac{1}{M^2 - 1}\right) \quad (12\text{–}13)$$

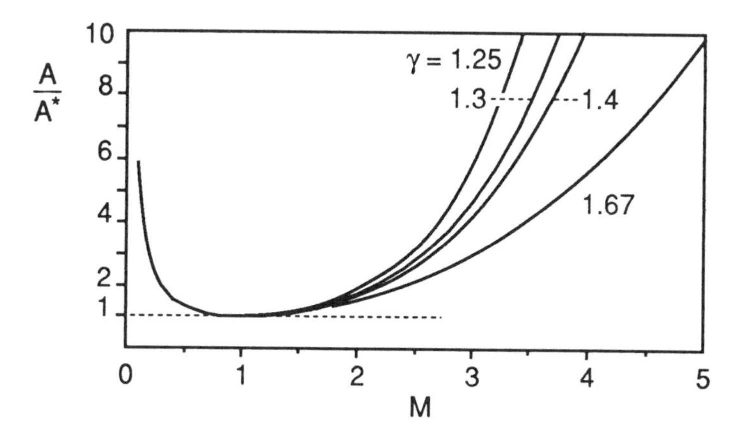

FIG. 12–4. Area ratio as function of Mach number. Plot of eq. 12–14 for a number of γ's.

This relation shows that the velocity can increase ($du > 0$) through $M = 1$ only when $dA = 0$ (i.e., at a throat).

These same equations can be rearranged to give the change in density associated with a velocity change as

$$\frac{d\rho}{\rho} = -M^2 \frac{du}{u} \tag{12–14}$$

Thus $M \ll 1$ implies small ρ changes as a result of velocity changes and hence nearly incompressible flow.

12.2.2. *Velocity and Static Pressure*

The pressure that leads to the particular Mach number is given by the fixed total pressure and the Mach number through eq. 12–9. Figure 12–5 shows that low pressures are required to reach high flow Mach number. The velocity is related to the Mach number through the speed of sound. One can define a total or stagnation speed of sound through eq. 12–6 using the stagnation temperature. This total sound speed is a constant in adiabatic flow and thus is a logical velocity with which to nondimensionalize velocity. Thus

$$\frac{u}{a_t} = \frac{u}{a}\sqrt{\frac{T}{T_t}} = \frac{M}{\sqrt{1 + \frac{\gamma - 1}{2} M^2}} \qquad \text{where } a_t = \sqrt{\gamma R T_t} \tag{12–15}$$

Figures 12–5, 12–6 and 12–7 are plots of eqs. 12–9 and 12–15, respectively, for $\gamma = 1.4$ and 1.667.

Forcing the flow to low pressure raises its velocity (i.e., its kinetic energy) to a value consistent with the thermal energy available. The asymptotic value

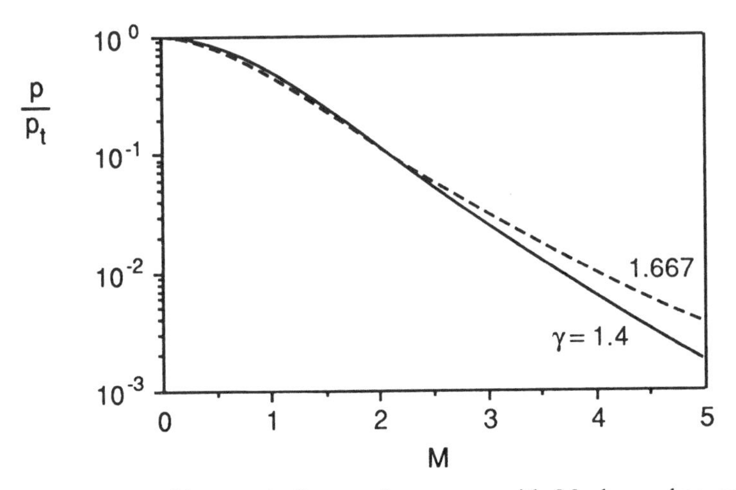

FIG. 12–5. Variation of isentropic flow static pressure with Mach number, eq. 12–9. The value of the static to total pressure ratio at $M = 1$ is 0.5283 (for $\gamma = 1.4$) and 0.4871 (1.667).

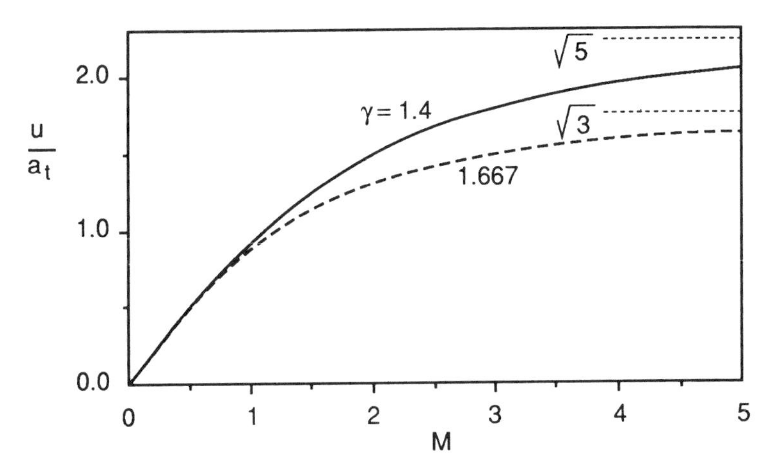

FIG. 12–6. Variation of flow velocity with Mach number, eq. 12–15.

of u, when all thermal energy is converted to kinetic energy, is

$$u_{\max} = \sqrt{2C_p T_t} = \sqrt{\frac{2}{\gamma - 1}}\, a_t \tag{12–16}$$

The constant is $\sqrt{5}$ ($\gamma = 1.4$) *or* $\sqrt{3}$ ($\gamma = 1.667$), as noted in Fig. 12–7. This equation shows that the realization of a jet of large kinetic energy (or large momentum) is obtained for a flow with a large total temperature, T_t.

Equations 12–15 and 12–16 can be combined to define a (first law or energy)

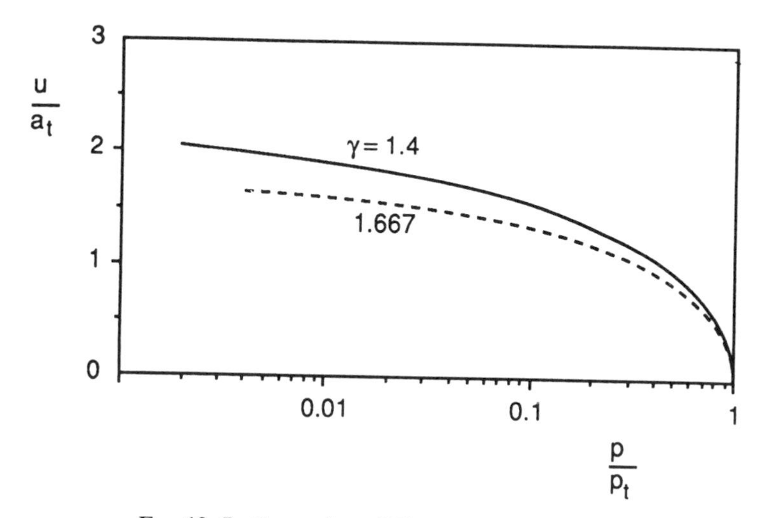

FIG. 12–7. Cross plot of Figs. 12–5 and 12–6.

nozzle kinetic energy efficiency

$$\eta_{KE} = \left(\frac{u}{u_{max}}\right)^2 = \frac{\dfrac{\gamma - 1}{2} M^2}{1 + \dfrac{\gamma - 1}{2} M^2} \tag{12–17}$$

which shows that a unit efficiency is obtained for very large M. Note the low value of this efficiency for sonic flow (17% for $\gamma = 1.4$). A second law efficiency addresses the reversibility of the flow process and is measured by the total pressure ratio for this adiabatic process. For real nozzles, a useful parameter is the jet thrust divided by the value obtained for the reversible process. This measure of efficiency is related to the *velocity* or *thrust* coefficient discussed in Section 12.5.5.

12.3. NOZZLES AND NOZZLE PRESSURE RATIO

In order to convert the thermal energy of a steady gas flow to jet kinetic energy, the flow must proceed from a high-pressure to a lower-pressure environment. Since the environmental pressure is generally fixed by the local atmospheric pressure, or since there is a realistic limit in area (from a size or weight point of view) to which a flow is expanded, it follows that the supply gas must be delivered at some high pressure to give an expansion *pressure ratio* as shown in Figs. 12–5 or 12–7. Limits on this upstream stagnation pressure are imposed either by the ability of the thermodynamic heat engine (such as a gas turbine) to produce it or by limitations on the strength of the duct (as in a rocket engine). It follows that the primary parameter controlling the design or performance of a nozzle is the *nozzle pressure ratio* (NPR) defined as

$$\mathrm{NPR} \equiv \frac{\text{Upstream total pressure}}{\text{Downstream environment pressure}} = \frac{p_t}{p_0} \qquad (12\text{–}18)$$

12.4. THRUST FROM PROPULSION NOZZLES

Many nozzles are used for the generation of a high-momentum jet for propulsion of a vehicle. Since most of the interesting problems are associated with the design of flight vehicle propulsion nozzles, this section is a discussion of the design and operation of such nozzles.

The approach is to treat the nozzle as a flow-through device which accepts a gas at specified total temperature and pressure at very low Mach number. The momentum associated with the entering flow can therefore be taken as zero and pressure forces may be arranged to sum to zero by symmetry. All forces on the control volume can then be attributed to the exiting flow. Figure 12–8 shows a control volume for a nozzle and the forces relevant to it. For present purposes, one may neglect external aerodynamic forces and take the control volume boundaries to be so far from the engine that they lie where the pressure is p_0. For a thorough discussion of aircraft thrust and drag forces, the reader is directed to refs. 12–1 and 12–2. Under such circumstances the (1-D) thrust is a reaction to pressure forces across the exit area A_e and momentum created within the control volume exiting through A_e. Thus,

$$F_g = (p_e - p_0)A_e + \dot{m}u_e \qquad (12\text{–}19)$$

This thrust is termed the *gross thrust* of the nozzle. The momentum term is by far

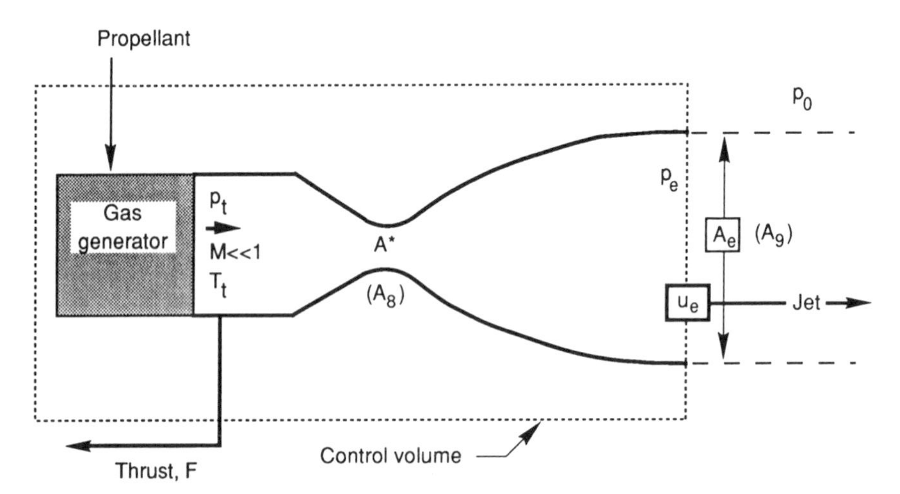

FIG. 12–8. Control volume and descriptive parameters of a nozzle.

the most important for a well-designed nozzle. Standard nomenclature for jet engines involves a station or state numbering system that uses "9" instead of the "e" for exit used above. Since this work is primarily applicable to the analysis of jet engines, the exit condition is noted as "9."

For the following, the view is taken that the gross thrust can be written with a nozzle expansion efficiency as

$$F_g = \eta_{ne}\dot{m}u_{9i} \qquad \text{where } \eta_{ne} \equiv \frac{u_9}{u_{9i}} + \frac{A_9(p_9 - p_0)}{\dot{m}u_{9i}} \tag{12–20}$$

Since the flow through the nozzle is isentropic, one may relate pressure changes to temperature changes from the stagnation state to the exit static state. It is convenient, in the expressions to follow, to define a parameter

$$s_0 \equiv \left(\frac{p_0}{p_{t9}}\right)^{(\gamma-1)/\gamma} \equiv \mathrm{NPR}^{-(\gamma-1)/\gamma} \tag{12–21}$$

The choice of the symbol s follows from the notion that this parameter is easily identified with the entropy. Physically, the quantity s_0 is the isentropic temperature ratio that would be experienced by the flow through the nozzle when the nozzle is properly expanded. It follows from eqs. 12–2 and 12–9 that the velocity for isentropic expansion to p_0, is

$$u_{9i} \equiv \sqrt{2C_pT_{t9}}\sqrt{\left(1 - \left(\frac{p_0}{p_{t9}}\right)^{(\gamma-1)/\gamma}\right)} = \sqrt{2C_pT_{t9}}\sqrt{1 - s_0} \tag{12–22}$$

The nozzle expansion efficiency, η_{ne}, depends on the nozzle pressure ratio and the nozzle area ratio as shown in Section 12.4.1. By using the equation for the choked flow ($M = 1$ in eq. 12–10) through the area A_8 (which is A^* in Fig. 12–8), the mass flow is given by

$$\dot{m} = \frac{p_{t8}A_8}{\sqrt{RT_{t8}}}\sqrt{\gamma}\left(\frac{\gamma+1}{2}\right)^{-(\gamma+1)/2(\gamma-1)} \tag{12–23}$$

The gross thrust is thus given by

$$\left.\begin{aligned} F_g &= p_{t8}A_8\pi_n\eta_{ne}f(\gamma)\sqrt{(1 - s_0)} \\ \text{where} \quad & \\ f(\gamma) &\equiv \left\{\gamma\left(\frac{\gamma+1}{2}\right)^{-(\gamma+1)/2(\gamma-1)}\sqrt{\frac{2}{\gamma-1}}\right\} \end{aligned}\right\} \tag{12–24}$$

Here π_n is a total pressure ratio ($=p_{t9}/p_{t8} < 1$) to account for possible irreversibilities in the nozzle flow. The variation of the function $f(\gamma)$ is shown in Fig. 12–9. Numerical values are 1.811 and 1.878 for $\gamma = 1.4$ and 1.35, respectively.

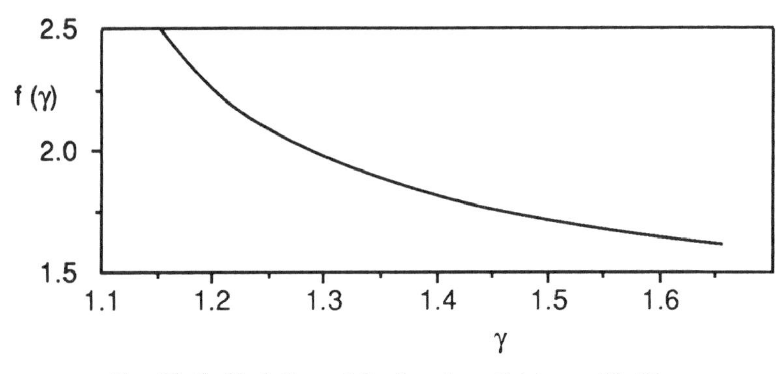

Fig. 12–9. Variation of the function $f(\gamma)$ in eq. 12–23.

12.4.1. *Nozzle Expansion Efficiency*

Nozzles operate either choked or unchoked. When they are unchoked, i.e., operating with a pressure ratio sufficiently low so that sonic conditions are not reached, the pressure of the fluid exiting the nozzle adjusts to the ambient pressure and the nozzle expansion efficiency is identically unity.

For choked flow, the flow pressure at the exit plane A_9 is determined by the stagnation pressure and the flow Mach number from eq. 12–11. Physically, this pressure can be different from the ambient pressure because communication between the two regions is by means of waves that can support steady pressure differences. The ideal situation is reached when the η_{ne} is unity by providing an exit area ratio A_9/A_8 that allows the pressure of the fluid exactly to reach ambient pressure.

From the definition of the nozzle expansion efficiency (eq. 12–18), one requires, in addition to u_{9i} given by eq. 12–22, the actual jet velocity,

$$\left.\begin{aligned} u_9 &\equiv \sqrt{2C_pT_{t9}}\sqrt{1-\left(\frac{p_9}{p_{t9}}\right)^{-(\gamma-1)/\gamma}} = \sqrt{2C_pT_{t9}}\sqrt{1-\frac{1}{\mu_9}} \\ \text{where} \quad & \\ \mu_9 &\equiv 1+\frac{\gamma-1}{2}M_9^2 = \left(\frac{p_{t9}}{p_9}\right)^{(\gamma-1)/\gamma} \end{aligned}\right\} \qquad (12\text{–}25)$$

The jet temperature ratio μ_9 is a function of the flow Mach number. One may characterize flow irreversibility by π_n and assume that the flow is adiabatic. The terms in the nozzle expansion efficiency may be combined using eqs. 12–20 to 12–25 to yield

$$\eta_{ne} = \sqrt{\frac{1-\dfrac{1}{\mu_9}}{1-s_0}\left\{1+\frac{\gamma-1}{2\gamma}\pi_n\frac{\left(1-(s_0\mu_9)\dfrac{\gamma}{\gamma-1}\right)}{(\mu_9-1)}\right\}} \qquad (12\text{–}26)$$

The ratio μ_9 is fixed by the geometric area ratio and s_0 is strictly a function of

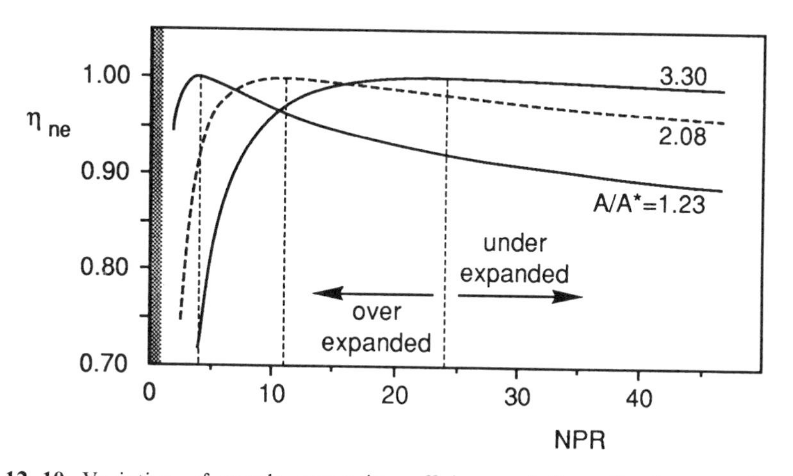

FIG. 12–10. Variation of nozzle expansion efficiency of three fixed-area ratio convergent– divergent nozzle configurations. The design jet Mach numbers for these three nozzles are 2.74 (for area ratio = 3.30, 2.24, and 1.56, respectively). $\gamma = 1.4$.

the operation condition of the flow, namely the NPR. Figure 12–10 shows the variation of this efficiency for three nozzle geometries. The efficiency is unity when $s_0 \mu_9 = 1$ (i.e., the nozzle is properly expanded), while it is less than ideal for both under-expansion ($p_9 > p_0$ or NPR > ideal) and over-expansion (reversed inequalities). For under-expanded flow, the penalty in efficiency is modest over a large NPR range (see Fig. 12–2).

Another way of writing the nozzle expansion efficiency is in terms of the pressure mismatch ratio p_9/p_0, or

$$\eta_{\text{ne}} = \sqrt{\frac{\dfrac{1}{s_0} - \left(\dfrac{p_0}{p_9}\right)^{-(\gamma-1)/\gamma}}{\dfrac{1}{s_0} - 1} \left\{1 + \frac{\gamma - 1}{2\gamma} \pi_{\text{n}} \frac{\left(1 - \dfrac{p_0}{p_9}\right)}{\left(\dfrac{1}{s_0}\left(\dfrac{p_0}{p_9}\right)^{(\gamma-1)/\gamma} - 1\right)}\right\}} \tag{12–27}$$

Note that the ideal result is obtained for $p_9 = p_0$.

12.5. CONVERGENT AND CONVERGENT–DIVERGENT NOZZLES

Convergent nozzles are ducts with a converging flow area to accelerate a subsonic flow ($M < 1$) to a higher subsonic or sonic condition. Figure 12–11 shows a convergent nozzle with sonic flow at the exit plane.

12.5.1. *Unchoked Flow in Convergent Nozzles*

For NPR less than that required to reach sonic flow, that is,

$$s_0 \geq \left(\frac{p_0}{p_{t9}} \quad \text{for } M = 1\right)^{(\gamma-1)/\gamma} = \frac{2}{\gamma + 1} \tag{12–28}$$

Table 12–1. Calculation procedure for convergent nozzle thrust

	Unchoked	Choked
Exit to ambient static pressure ratio	$x = \dfrac{p_9}{p_0} = 1$	$x = \dfrac{p_9}{p_0} > 1$
	$\pi_j = \dfrac{p_{t9}}{p_9} = \text{NPR}$	$= (x)\text{NPR}$
Jet temperature ratio, μ_9	$= (\pi_j)^{(\gamma-1)/\gamma}$	$= \dfrac{\gamma+1}{2}$
Exit Mach number, M_9	$= \sqrt{\dfrac{2}{\gamma-1}\{\mu_9 - 1\}}$	$= 1$
Jet velocity, $\dfrac{u_9}{\sqrt{RT_{t9}}}$	$= \dfrac{\sqrt{\gamma}\,M_9}{\sqrt{1+\dfrac{\gamma-1}{2}M_9^2}}$	$= \sqrt{\dfrac{2\gamma}{\gamma+1}}$
Mass flow rate, $\dfrac{\dot{m}\sqrt{RT_{t9}}}{p_{t9}A_8}$	$= M_9\sqrt{\gamma}(\mu_9)^{-(\gamma+1)/2(\gamma-1)}$ $\approx M_9\sqrt{\gamma}$	$= \sqrt{\gamma}\left(\dfrac{\gamma+1}{2}\right)^{-(\gamma+1)/2(\gamma-1)}$
Term $\dfrac{\dot{m}u_9}{p_0 A_8}$	$= \gamma M_9^2$	$= \gamma$
Term $\dfrac{A_9(p_9 - p_0)}{A_8 p_0}$	$= 0$	$= x - 1$
Thrust $\dfrac{F}{p_0 A_8}$	$= \gamma M_9^2$	$= x + \gamma - 1$
Expansion efficiency, η_{ne}	$= 1$	given by eq. 12–26

Note: Assumed known: p_{t9}, p_0, $A_8 = A_9$, RT_{t9}, γ. Calculate: $\text{NPR} = p_{t9}/p_0$.

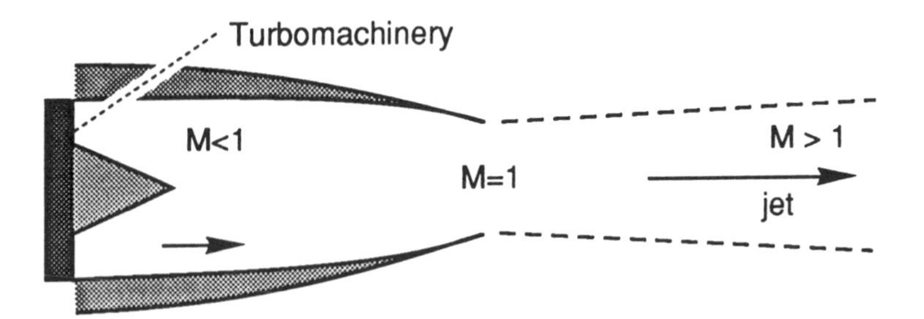

Fig. 12–11. Converging nozzle with supersonic jet flow.

the pressure at the exit plane adjusts to the ambient pressure and the nozzle expansion efficiency is unity. The thrust is then given by eq. 12–24 with $\eta_{ne} = 1$. The resulting equations for thrust and mass flow rate are summarized in the middle column of Table 12–1.

12.5.2. *Choked Flow in Convergent Nozzles*

For conditions in violation of eq. 12–28 (i.e., $s_0 < 2/(\gamma + 1)$), the flow is choked and the mass flow rate is given by eq. 12–23. The pressure at the nozzle exit, p_9, is fixed by p_t with $M = 1$. The exit or jet velocity is given by eq. 12–15 with $M = 1$. The pressure term in the thrust equation is

$$(p_9 - p_0)A_9 = p_{t9}A_8\left(\left(\frac{\gamma + 1}{2}\right)^{-\gamma/(\gamma - 1)} - \frac{p_0}{p_{t9}}\right) \tag{12–29}$$

so that the complete expression for the thrust includes the terms listed in Table 12–1. This table is a summary of the calculation procedures necessary for the determination of thrust and mass flow rate for a convergent nozzle operating at any pressure ratio.

12.5.3. *Convergent-Divergent Nozzles*

If the pressure ratio is substantially larger than that required for sonic flow, a diverging duct may be used to reduce the pressure further and thereby increase the jet velocity. Such a nozzle is a convergent-divergent nozzle, an example of which is shown in Fig. 12–2.

Such nozzles are rarely operated at unchoked conditions so that only the choked case is examined. The equations developed above apply to this case with the expectation that the jet flow will be supersonic. In such flows, wave systems consisting of shocks or expansion fans (Fig. 12–2), allow the exit static pressure (p_9) to be mismatched to the environment pressure (p_0). A calculation procedure using known geometric and operating parameters to calculate mass flow rate and thrust is outlined in Table 12–2 using the equations developed.

12.5.4. *Deviations from Ideal and One-dimensional Performance*

Boundary Layers

To the extent that flow is nonuniform across a flow area under study, the performance parameters of the nozzle must be corrected. Nonuniformities arise from the effects of friction resulting in boundary layers as discussed in Section 12.6. In general, boundary layers reduce the available flow area in the mass flow determining area, A_8. If the boundary layer is thin relative to the radius (of an axisymmetric nozzle, the reduced area is approximately given by

$$A_{8,\text{eff}} = A_{8,\text{geom}} - 2\pi R\delta_8^* = A_{8,\text{geom}}\left(1 - \frac{2\delta_8^*}{R_8}\right) \tag{12–30}$$

Table 12–2. Calculation procedure for thrust of a choked convergent–divergent nozzle with $A_9 > A_8$

Exit Mach number, M_9	from supersonic solution of eq. 12–11 with $A/A^* = A_{98}$
Total to static T ratio at nozzle exit, μ_9	$= 1 + \frac{\gamma - 1}{2} M_9^2$
Total to static p ratio at nozzle exit, π_j	$= (\mu_9)^{\gamma/(\gamma-1)}$
Exit to ambient static pressure ratio	$x = \frac{p_9}{p_0} = \frac{\text{NPR}}{\pi_j}$
Jet velocity, $= \frac{u_9}{\sqrt{RT_{t9}}}$	$= \frac{\sqrt{\gamma} M_9}{\sqrt{1 + \frac{\gamma - 1}{2} M_9^2}}$
Mass flow rate, $\frac{\dot{m}\sqrt{RT_{t9}}}{p_{t9} A_8}$	$= \sqrt{\gamma}\left(\frac{\gamma + 1}{2}\right)^{-(\gamma+1)/2(\gamma-1)}$
Term $\frac{\dot{m}u_9}{p_0 A_8}$	$= \frac{\gamma M_9 \text{NPR}}{\sqrt{1 + \frac{\gamma - 1}{2} M_9^2}} \left(\frac{\gamma + 1}{2}\right)^{-(\gamma+1)/2(\gamma-1)}$
Term $\frac{A_9(p_9 - p_0)}{A_8 p_0}$	$= A_{98}(x - 1)$
Thrust $\frac{F}{p_0 A_8}$	= sum of two terms above or
	$= \eta_{ne}\gamma \sqrt{\frac{2}{\gamma - 1}(\mu_9 - 1)} \left(\frac{2}{\gamma + 1}\mu_9\right)^{(\gamma+1)/2(\gamma-1)}$
Expansion efficiency, η_{ne}	= given by eq. 12–26

Note: Assumed known: p_{t9}, p_0, A_8, A_9, RT_{t9}, γ. Calculate: NPR $= p_{t9}/p_0$ and A_9/A_8.

where δ^* is the displacement thickness of the boundary layer at A_8, and R is the radius of the axisymmetric nozzle. Similarly the momentum lost to skin friction on the walls at the nozzle exit is an important loss and can be estimated from

$$u_{9,\text{eff}} = u_{9,1-\text{D}}\left(1 - \frac{2\theta_9}{R_9}\right) \tag{12–31}$$

The boundary layer thickness length scale, θ, is the momentum thickness. Here, the one-dimensional mean flow velocity, $u_{9,1-\text{D}}$, must be computed from the reduced flow area at the nozzle exit. This area is related to the geometric area by an equation like eq. 12–30 with subscript 9 instead of 8.

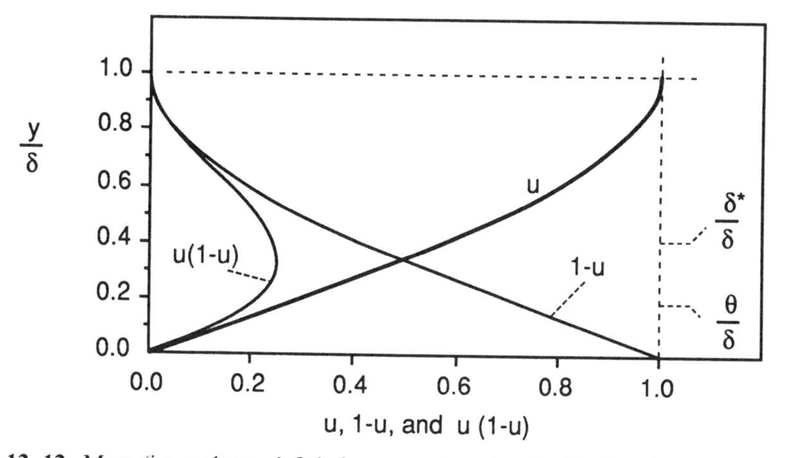

FIG. 12–12. Mass (i.e., volume deficit for a constant density flow) and momentum flux deficits in a boundary layer.

The integral length scales δ^* and θ are given by integrals given in eq. 12–32 and represent the mass flow and momentum deficits shown in Fig. 12–12 for the velocity distribution shown. The upper edge of the boundary layer is at $y = \delta$. Graphically, the integral lengths are proportional to the areas to the left of the curves labeled $(1 - u)$ and $u(1 - u)$ respectively where u stands for u/U. If the flow is incompressible, the density of the boundary layer flow (ρ) and the freestream (ρ_∞) are identical and do not need to be considered further in the interpretation of δ^* and θ given by

$$\delta^* = \int_0^\delta \frac{\rho}{\rho_\infty}\left(1 - \frac{u}{U}\right) dy \quad \text{and} \quad \theta = \int_0^\delta \frac{\rho}{\rho_\infty}\frac{u}{U}\left(1 - \frac{u}{U}\right) dy \quad (12\text{–}32)$$

Since the characteristics of the boundary layer and its integral length scales are the result of viscous effects, it is to be expected that the departure from ideal 1-D flow will be dominated by Reynolds number as a parameter.

Nonuniform Flow

Flows upstream of nozzles may be nonuniform in total pressure and in total temperature due to nonuniformity in heat input or work transfer and due to variations in irreversibility suffered by the different streamtubes that make up the total flow. Analyses that examine such flows usually approach the problem by accepting a distribution of p_t and T_t ahead of the nozzle and assuming that the flow proceeds through the nozzle as if the adjacent streamtubes cannot or do not interact to exchange energy or momentum. Thus they proceed in parallel, sharing only the local static pressure. Such descriptions are termed as quasi-one-dimensional (because the static pressure is then a function of axial position only) compound nozzle flow. Nonuniformities have an effect on the jet thrust and on the choking condition at the nozzle throat. Section 12.6 describes the nonuniform nozzle flow theory and shows the importance of mixing which may be made to occur by providing a long nozzle entry.

Rapid Flow Area Variation

For the sake of light weight in nozzle design, the convergent section of sonic or supersonic flow nozzle is usually kept relatively short with a relatively steep converging flow section. Such designs violate the one-dimensional flow assumption by allowing significant traverse flow velocities. The impact of the geometric and gas dynamic parameters on the choking condition and consequently on the nozzle performance are described in Section 12.7.

Imperfect Gas and Chemically Reacting Flow

In nozzles with large pressure and, consequently, temperature changes, the assumption that C_p remains constant may not be valid. In that case, any formulation involving the use of γ is incorrect because the basis of many flow property relations is the integrated relation between temperature and pressure for an isentropic process such as eq. 12–3. The more general relation between p and T is as given in Section 4.5.

The general description of real gases flowing through a nozzle also involves the possibility of chemical reactions. For example, the changing temperature and pressure dictate that the chemical composition might change. Whether it does, depends on the gas composition; that is, are the temperature changes significant compared to the chemical energy changes that make them possible? It may also depend on the time available for changes to actually take place. In this case the time scale for chemical reactions becomes important. For so-called "frozen" flows, the gas dynamic processes take place so rapidly that composition remains constant. For such flows, real gas considerations are limited to temperature variation of C_p. On the other hand, when the chemical processes are so fast that the possible chemical reactions can keep up with the changing conditions, the flow is described to be in "shifting equilibrium." These two limiting cases are isentropic and therefore describable by the methods developed to this point. The procedure is outlined in Section 12.8. Real flows through nozzles tend to start in equilibrium because of the high temperature and pressure in the stagnation region of the nozzle, and change to a frozen flow in the cool, low-density portion of the flow.

12.5.5. *Performance Indices Used in Practice*

In industrial practice, a number of performance indices are used to quantify the performance of nozzle configurations. For example,

$$C_d = \text{discharge coefficient} \equiv \frac{\text{actual } \dot{m}}{\text{ideal } \dot{m} \text{ for same NPR}} \tag{12–33}$$

The ideal mass flow rate is computed using the geometric area A_8 and the equations cited in Tables 12–1 or 12–2 for the appropriate NPR.

Because the momentum term in the expression for thrust dominates, taking the ratio of measured thrust to actual mass flow rate gives an effective jet velocity. Such a parameter arises if the means are available to measure the mass flow rate independently and accurately. Thus one defines

$$C_v = \text{velocity coefficient} \equiv \frac{F}{(\text{actual } \dot{m}) \times (\text{ideal } u_9 \text{ for same NPR})} \tag{12–34}$$

The ideal jet velocity is u_9 and is obtained from eq. 12–15. The velocity u_{9i} is not used so that C_v measures only those effects not easily accounted for by one-dimensional flow analysis.

The ratio of measured thrust to that calculated by the 1-D flow procedure outlined above is termed a gross thrust coefficient, C_g, defined by

$$C_g = \frac{F}{F \text{ calculated for same NPR}} \tag{12–35}$$

Evidently it follows that

$$C_g = C_d C_v \tag{12–36}$$

These coefficients are functions of NPR and of Reynolds number. They often accounts for 2-D flow effects which are important when nozzles are designed for compactness and light weight.

12.6. NONUNIFORM FLOWS

In many realistic applications, the flow through a nozzle may be such that the idea of uniform, one-dimensional flow is invalid (ref. 12–3). This may be encountered in the use of a single nozzle to expand flows from two sources such as the fan and core of a turbofan engine. Other examples include nonuniformly heated flow (as in an afterburner), flow created by a nonadiabatic process, flows with coolant injection, or flows with wakes. The ideas and relations developed for classical 1-D flow must be modified to accommodate such nonuniformities when these are significant. The following is an examination of the effect of flow uniformity on the performance of a single-throat nozzle as a thrust production device. References 12–4 and 12–5 and Section 12.6.7 gives the general calculational procedure required to describe nonuniform nozzle flow. From this it is generally difficult to draw useful conclusions except through implementation on a computer. For a discussion of fluid physics involving energy exchange between adjacent streams, the reader is referred to ref. 12–6. The objective is to determine the impact of nonuniform flow on design and performance of nozzles in a way that states conclusions clearly, using appropriate linearization if necessary. For further mathematical ease, the situation is limited to examination of a flow with two regions that differ in total pressure and temperature. Two situations are compared: one where the flow is expanded through a nozzle sufficiently rapidly so that mixing does not occur and, two, where mixing to a single uniform flow occurs ahead of the nozzle throat. The problems that arise with nonuniform flow include the following:

1. A fixed geometry nozzle processes a mass flow which is uncertain in magnitude due to its nonuniformity. This is equivalent to determining a discharge coefficient, C_d (eq. 12–33), associated with the nonuniformity.
2. A known mass flow rate is processed and the requirement is to find the correct flow areas that must be provided.
3. When the proper geometry and known mass flow rate are provided, what is the effect of the nonuniformity on thrust performance? This is equivalent to identifying a velocity coefficient, C_v (eq. 12–34), associated with the nonuniformity. It is the goal of this section to examine the two flow cases in the light of these questions.

Extension of the approach to a multiplicity of streamtubes and nonuniform properties such as γ and R is not considered, although this extension can be readily incorporated into a machine computation. The emphasis is on exposition of the variables that determine whether a flow should be considered nonuniform and under what circumstances mixing is significant in affecting design variables of the nozzle (A^*, for example) and its performance. For completeness, the performance of a thrust production system that uses two nozzles to expand two dissimilar flows is considered in Section 12.6.6.

12.6.1. *Compound Flow*

A flow may be assumed quasi-one-dimensional when it may be said that streamtubes with differing stagnation properties proceed through a duct in such a way as to share a common static pressure. This is the case when the streamtube curvature is small (see Section 12.7.1. on short nozzles). The fact that the static pressure is the link between the two flows makes it a key parameter in describing the flow, in contrast to the Mach number which is commonly used to describe 1-D flows.

Consider the flow of two streams with different total temperatures and pressures exiting a common choked nozzle (Fig. 12–13). The flow plane where the streams are in contact is the nozzle charging station and * is used for conditions at the choked throat. The nozzle is assumed properly expanded so that for an unchoked flow situation, the diverging portion shown in Fig. 12–13 would be absent. Thus the exit static pressure p_e equals p_0 (ambient).

At any station in this nozzle, the two flows are described as a primary flow (subscript 1) and a secondary (2). For convenience, the stagnation conditions of the primary flow are used as reference. Thus T_{t1}, p_{t1}, and $\dot{m}_1$ are used to form nondimensional parameters that characterize similar quantities

FIG. 12–13. Schematic of a two-stream nozzle. The primary stream serves as a reference.

in the other flow:

$$T_{t2} = \tau T_{t1}; \qquad p_{t2} = \pi p_{t1}; \qquad \dot{m}_2 = \alpha \dot{m}_1 \tag{12–37}$$

The goal is to highlight the role of nonuniformities as a contrast to the results obtained for uniform 1-D flow. This is accomplished by computing expressions for quantities of interest in forms that clearly illustrate the departure from 1-D flow.

Since the static pressure is shared by the two flows, an expression for the total mass flow rate is best written in terms of the static pressure as a measure of location along the nozzle. It turns out that a particularly convenient indicator for this pressure is an entropy-like parameter defined by

$$s \equiv \left(\frac{p}{p_{t1}}\right)^{(\gamma-1)/\gamma} = \left(1 + \frac{\gamma - 1}{2} M_1^2\right)^{-1} \tag{12–38}$$

The use of this quantity minimizes the complexities associated with γ's in the resulting expressions. The Mach number M_1 is that in the reference streamtube. The best nondimensional measure of velocity in the streamtube $i = 1$ or 2, is the ratio

$$V_i = \frac{u_i}{u_{1,\max}} = \frac{u_i}{\sqrt{\dfrac{2\gamma}{\gamma - 1} R T_{t1}}} \tag{12–39}$$

The energy equations for the two streamtubes relate the static temperatures to the velocities as

$$\frac{T_i}{T_{ti}} = 1 - \frac{V_i^2}{\tau_i} = \frac{1}{1 + \dfrac{\gamma - 1}{2} M_i^2} \tag{12–40}$$

where $\tau_1 = 1$ and $\tau_2 = \tau$ from the notation definition in Fig. 12–13. The velocities in the two flows are related to the "pressure" through

$$V_1 = \sqrt{1 - s} \qquad \text{and} \qquad V_2 = \sqrt{\tau}\sqrt{1 - \frac{s}{\sigma}}; \qquad \sigma \equiv \left(\frac{p_{t2}}{p_{t1}}\right)^{(\gamma-1)/\gamma} \tag{12–41}$$

Here, σ is used as a convenient parameter for description of the total pressure nonuniformity. The static and stagnation states for the two flows are related to the local Mach numbers required in the choking condition expression. Thus with eq. 12–40 one can write the Mach numbers at any station where the pressure is given by s as

$$M_1^2 = \frac{2}{\gamma - 1}\frac{1 - s}{s} \quad \text{and} \quad M_2^2 = \frac{2}{\gamma - 1}\frac{\sigma - s}{s} \tag{12–42}$$

The mass flow rates are given in terms of the local static pressure by the mass conservation relation

$$\dot{m}_\mathrm{i} = \left[\frac{p}{\sqrt{RT_{\mathrm{t1}}}}\sqrt{\frac{2\gamma}{\gamma - 1}}\right]\left(\frac{T_{\mathrm{t1}}}{T_{\mathrm{ti}}}\frac{T_{\mathrm{ti}}}{T_\mathrm{i}}V_\mathrm{i}A_\mathrm{i}\right) = b\frac{p}{p_{\mathrm{t1}}}A_\mathrm{i}\frac{V_\mathrm{i}}{\tau_\mathrm{i} - V_\mathrm{i}^2} \tag{12–43}$$

where b is a characteristic ρu for the primary flow:

$$b = \sqrt{\frac{2\gamma}{\gamma - 1}}\frac{p_{\mathrm{t1}}}{\sqrt{RT_{\mathrm{t1}}}} \tag{12–44}$$

The total mass flow rate (i = 1, 2) for the two flows can be written in terms of s, and the velocities as

$$\dot{m}_{\mathrm{Tot}} = bs^{\gamma/(\gamma-1)}A_\mathrm{T}\left\{\frac{A_1}{A_\mathrm{T}}\left(\frac{V_1}{1 - V_1^2}\right) + \frac{A_2}{A_\mathrm{T}}\left(\frac{V_2}{\tau - V_2^2}\right)\right\} = \dot{m}_1(1 + \alpha) \tag{12–45}$$

The mass flow rate ratio, α, introduced in eq. 12–37 is given by:

$$\alpha \equiv \frac{\dot{m}_2}{\dot{m}_1} = \frac{A_2}{A_1}\left(\frac{V_2}{\tau - V_2^2}\right)\left(\frac{1 - V_1^2}{V_1}\right) \tag{12–46}$$

and can be used to calculate the streamtube area ratio A_2/A_1. The total mass flow rate can then be written in terms of only the "pressure," s, as

$$\dot{m}_{\mathrm{Tot}} = bA_{\mathrm{Tot}}s^{1/(\gamma-1)}\left\{\frac{1 + \alpha}{\dfrac{1}{\sqrt{1 - s}} + \dfrac{\alpha\sqrt{\tau}}{\sigma}\dfrac{1}{\sqrt{1 - s/\sigma}}}\right\} \tag{12–47}$$

This expression is valid for any value of the total temperature nonuniformity, τ, and total pressure nonuniformity as embodied in σ. It reduces to the expected result when there is only one flow ($\alpha = 0$) or the flow is uniform, $\pi = \tau = 1$.

Choking

A uniform flow exhibits the phenomenon of choking, that is, maximum mass flow rate, is obtained for a value of

$$s^* \text{ (uniform)} = s_u^* = \frac{2}{\gamma + 1} \tag{12–48}$$

For nonuniform flow, the more general expression for s^* may be found by a null differentiation of $\dot{m}$ (eq. 12–27) with respect to s. The resulting expression is

$$\frac{\gamma - 1}{2} s^* = \frac{\left\{\dfrac{1}{(1 - s^*)^{1/2}} + \dfrac{\alpha\sqrt{\tau}}{\sigma} \dfrac{1}{(1 - s^*/\sigma)^{1/2}}\right\}}{\left\{\dfrac{1}{(1 - s^*)^{3/2}} + \dfrac{\alpha\sqrt{\tau}}{\sigma^2} \dfrac{1}{(1 - s^*/\sigma)^{3/2}}\right\}} \tag{12–49}$$

which is seen to be algebraically implicit and allows simplification if the pressure nonuniformity term σ is close to 1.

Two Stream Flow with a Small Degree of Nonuniformity

MASS FLOW RATE THROUGH A FIXED AREA NOZZLE. Restriction to this situation allows an overview of a number of instructive results. From the definition of σ (eq. 12–41), it follows for $\pi - 1 \ll 1$:

$$\sigma = 1 + \frac{\gamma - 1}{\gamma} (\pi - 1) \tag{12–50}$$

and eq. 12–49 gives

$$s^* = \frac{2}{\gamma + 1}\left(1 + \frac{\alpha\sqrt{\tau}}{1 + \alpha\sqrt{\tau}} \frac{\gamma - 1}{\gamma} (\pi - 1)\right) \tag{12–51}$$

Here and in the work ahead, the frequent appearance of the *secondary flow temperature corrected mass fraction* justifies a short-hand definition which is used henceforth:

$$w_R \equiv \frac{\alpha\sqrt{\tau}}{1 + \alpha\sqrt{\tau}} \tag{12–52}$$

The choked flow area for the two-stream case can be related to the uniform flow case (eq. 12–48) for the *same mass flow rate* as

$$\frac{A^*(2\,\text{stream})}{A^*\,(\text{uniform})} = \left[\frac{1 + \alpha\sqrt{\tau}}{1 + \alpha}\right](1 - w_R(\pi - 1)) \tag{12–53}$$

which shows a factor associated with only the temperature nonuniformity (first term, right side) and a term modifying $(\pi - 1)$. This relation may be used to determine effective values of total pressure and temperature for the determination of mass flow rate since the factor $p_t A^*/\sqrt{RT_t}$ is a constant for the same mass flow rate.

The Mach numbers *in the choked flow throat* may be determined using eqs. 12–42 and 12–51 which shows that the flow with larger total pressure is supersonic whereas the other is subsonic:

$$\left.\begin{aligned} (M_1^*)^2 &= 1 - \frac{\alpha\sqrt{\tau}}{1+\alpha\sqrt{\tau}}\frac{\gamma+1}{\gamma}(\pi-1) \\ &\text{and} \\ (M_2^*)^2 &= 1 + \frac{1}{1+\alpha\sqrt{\tau}}\frac{\gamma+1}{\gamma}(\pi-1) \end{aligned}\right\} \tag{12–54}$$

The (linearized) expression for the mass flow rate for the unchoked case:

$$\dot{m}_{\text{Tot}} = bA_{\text{Tot}}s^{1/(\gamma-1)}\sqrt{1 - s\frac{1+\alpha}{1+\alpha\sqrt{\tau}}\left(1+\left(1+\frac{1}{\gamma}\left(\frac{1}{M_{1,e}^2}-1\right)w_R(\pi-1)\right)\right)} \tag{12–55}$$

$$s = s_{1,e} = \left(1 + \frac{\gamma-1}{2}M_{1,e}^2\right)^{-1} \tag{12–56}$$

where A_{Tot} is the flow area and $M_{1,e}$ is the Mach number reached by the primary flow on expansion to the pressure characterized by s. In the limit of $M_{1,e} = 1$ ($s_{1,e} = s^*$), this expression yields the expected result for choked flow. These results are consistent with the work of ref. 12–5.

A useful result that may be obtained from these equations is the ratio of choked $\dot{m}$ for a given area compared to $\dot{m}$ for a uniform flow. This ratio is a discharge coefficient based on the reference (primary) flow conditions, C_d:

$$C_d = \left\{\frac{\dot{m}(2\text{ stream})}{\dot{m}(\text{uniform})}\right\}^{\text{choked}}_{\text{Fixed area}} = \frac{1+\alpha}{1+\alpha\sqrt{\dfrac{\tau}{\sigma}}\sqrt{\dfrac{1-s^*}{\sigma-s^*}}} \cong \frac{1+\alpha}{1+\alpha\sqrt{\tau}}\{1+\psi_m w_R(\pi-1)\} \tag{12–57}$$

and will be used for the problem of judging the role of nonuniformity on the mass-flow-passing capability of a nozzle with fixed area. Here the mass flow

Mach number function is defined by:

$$\psi_m = 1 + \frac{1}{\gamma}\left(\frac{1}{M_{1,e}^2} - 1\right) \quad \text{for } M_{1,e}^2 \leq 1 \qquad \text{and} \qquad \psi_m = 1 \qquad \text{for } M_{1,e}^2 \geq 1 \tag{12–58}$$

with $M_{1,e} = 1$ in the choked flow case.

JET MOMENTUM AND THRUST PER UNIT MASS FLOW RATE. With the mass flow rate known and held constant as flow properties are varied, the thrust may be investigated by calculation of the thrust per unit mass flow rate, $F/\dot{m}$. For a properly expanded flow, $F/\dot{m}$ is the mass average velocity of the exiting stream which depends on total quantities and on the static pressure. The exit pressure ($s_{1,e}$) is such that the primary flow reaches $M_{1,e}$. The velocities V_1 and V_2 are computed from eq. 12–41, from which the mass flow weighted velocity is a velocity coefficient *based on the primary flow conditions*, C_v (two-stream):

$$C_v(2\text{ stream}) \equiv \frac{\bar{V}}{V_1} = \frac{V_1 + \alpha V_2}{(1+\alpha)V_1} = \frac{1 + \alpha\sqrt{\tau}\sqrt{\dfrac{1 - s_{1,e}/\sigma}{1 - s_{1,e}}}}{1+\alpha}$$

$$\cong \frac{1+\alpha\sqrt{\tau}}{1+\alpha}\left[1 + \frac{w_R(\pi - 1)}{\gamma M_{1,e}^2}\right] \tag{12–59}$$

The last term is for the $|\sigma - 1| \ll 1$ approximation. Thus when π and τ differ from unity, changes in the jet momentum are given by eq. 12–59. Note that the temperature effect influences the pressure effect through its presence in the weighting term, w_R, (eq. 12–52).

12.6.2. *Ideal, Constant Pressure Mixer Flow*

In the following, a *constant pressure* mixer is used for simplicity. The mixer is assumed ideal in that friction losses are neglected, although losses that do occur may be charged to the entering flow before mixing takes place. The mixer is assumed to fully combine the entering flows to a single exit flow. Figure 12–14 shows the geometry and identifies the important parameters. The subscript "m" is used to describe the mixed flow.

The objective is to obtain an expression for the mixture total pressure as well as the area ratio across the mixer. The theory for the mixer is an adaptation of that developed by Oates (ref. 12–1). The conservation of energy statement gives the mixed total temperature as

$$\tau_m = \frac{1+\alpha\tau}{1+\alpha} \tag{12–60}$$

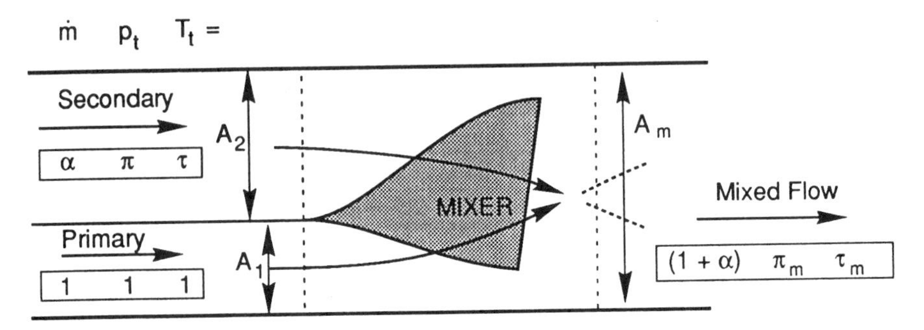

Fig. 12–14. Schematic of a full mixer. Values of the variables noted are given for each flow.

The conservation of momentum statement gives the mixed flow velocity. For any of the three flow conditions, the energy equation is used to relate the velocity to the static pressure. Thus one has (following eq. 12–41):

$$V_m^2 = \tau_m\left(1 - \frac{s}{\sigma_m}\right); \qquad \sigma_m \equiv \left(\frac{p_{tm}}{p_{t1}}\right)^{(\gamma-1)/\gamma} \tag{12–61}$$

σ_m describes the mixed-flow stagnation pressure. In the mixer, s is a measure of the local pressure which can be written in terms of primary flow Mach number there. Thus

$$s = s_{1,m} = \left(1 + \frac{\gamma - 1}{2} M_{1,m}^2\right)^{-1} \tag{12–62}$$

From the force-free conservation of momentum, the mixed flow velocity, V_m, can be written in terms of the incoming velocities. Eliminating V_m between the resulting equation and eq. 12–61 gives an expression for the desired total pressure parameter of the mixed flow, $\sigma_m = f(\sigma \text{ and } s_{1,m})$ or

$$\frac{1 - \dfrac{s_{1,m}}{\sigma_m}}{1 - s_{1,m}} = \frac{\left\{1 + \alpha\sqrt{\tau}\sqrt{\dfrac{1 - \dfrac{s_{1,m}}{\sigma}}{1 - s_{1,m}}}\right\}^2}{(1 + \alpha)(1 + \alpha\tau)} \tag{12–63}$$

where $s_{1,m}$ is given by eq. 12–62.

Linearized Mixer Analysis

In eq. 12–63, σ_m cannot be stated in terms of $(\pi - 1)$ without linearization of σ, although a machine computation is direct and the results are shown below.

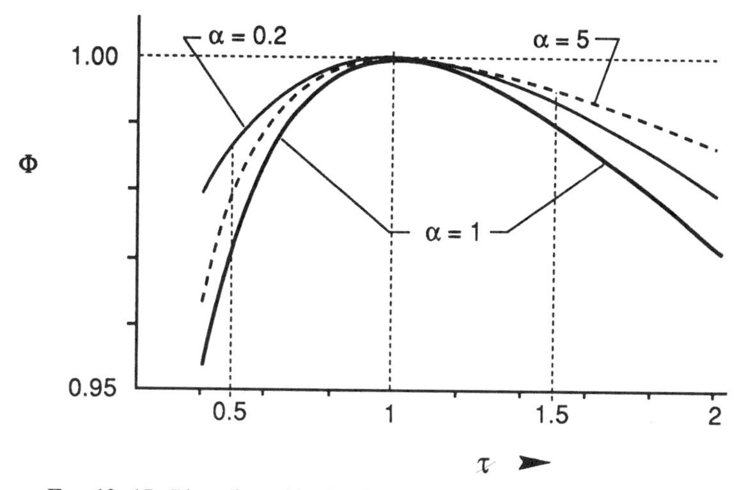

FIG. 12–15. Plot of eq. 12–65 showing Φ as a function α and τ.

The linearized result is

$$\pi_m = 1 + \frac{\gamma}{2} M_{1,m}^2(\Phi - 1) + w_R(\pi - 1)\Phi \qquad (12\text{–}64)$$

The defined function

$$\Phi \equiv \frac{(1 + \alpha\sqrt{\tau})^2}{(1 + \alpha)(1 + \alpha\tau)} \qquad (12\text{–}65)$$

is a parameter describing the total temperature nonuniformity and, in physical terms, is the ratio of the square of the momentum to the energy (per unit mass). This parameter is plotted in Fig. 12–15. Note that Φ modifies the pressure term, whereas $\Phi - 1$ modifies the mixing term. The function Φ is within 3% of unity for the parameter space ($0.5 > \tau > 2$ and $0.2 > \alpha > 5$).

From eqs. 12–64 and 12–65, the following may be concluded:

1. Two terms affect the mixed flow total pressure: one associated with heat transfer at finite Mach number, and a term due purely to the difference in the total pressures of the entering streams. The mixer Mach number term vanishes under either of two conditions: the total temperatures are equal (in which case Φ is unity) or $M_{1,m}$ vanishes.
2. The effect on the mixture total pressure is weighted in proportion to the temperature-reduced mass flow rates.

Mixer Area Ratio

The area ratio across the constant pressure mixer can be determined using the mass flow conservation equations for each of the streams, eq. 12–43 applied to the entering streams as well as the mixed stream. The resulting area ratio is, for the linearized case:

$$\frac{A_m}{A_1 + A_2} = \frac{1}{\sqrt{\Phi}}\left\{1 + (\Phi - 1)\left(\frac{\gamma - 1}{\gamma}\right)\left(\frac{\gamma}{2} M_{1,m}^2 + w_R(\pi - 1)\right)\right\} \quad (12\text{–}66)$$

The pressure function $s = s_{1,m}$ has been rewritten for clarity in terms of the flow Mach number. The small second term multiplying $(\Phi - 1)$ is a grouping that gives the importance of the total pressure nonuniformity and the role of the mixer Mach number. This term is generally small so that the first-order impact on the area ratio of the mixer is due to the total temperature distribution that affects Φ. To the extent that the factor Φ is close to unity, a constant pressure mixer is also a constant area mixer.

12.6.3. *Mixed Flow Nozzle*

A^* and Discharge Coefficient

The mass flow rate of the mixed flow through the nozzle is given by a variation of eq. 12–47 where the total pressure and temperature are π_m and τ_m respectively:

$$\dot{m}_{Tot} = bA_{Tot}\left(\frac{s}{\sigma_m}\right)^{1/(\gamma - 1)}\sqrt{1 - \frac{s}{\sigma_m}}\,\frac{\pi_m}{\sqrt{\tau_m}} \quad (12\text{–}67)$$

This nozzle is choked when $s/\sigma_m = 2/(\gamma + 1)$. The area required to pass this mixed flow is A_m^*, which can be calculated referenced to the case where π and τ (and therefore π_m and τ_m) are all unity *for the same mass flow rate.* For the linearized case, analytic expressions are obtained using eqs. 12–60 and 12–64:

$$\frac{A_m^*}{A^*(\text{uniform})} = \sqrt{\frac{1 + \alpha\tau}{1 + \alpha}}\left(1 - \frac{\gamma}{2} M_{1,m}^2(\Phi - 1) - w_R\Phi(\pi - 1)\right) \quad (12\text{–}68)$$

The mass flow rate through a nozzle of a nonuniform flow relative to a uniform one for *fixed area* is obtained from eq. 12–67 as:

$$C_d \equiv \left\{\frac{\dot{m}(\text{mixed})}{\dot{m}(\text{uniform})}\right\}^{\text{choked}}_{\text{Fixed area}} = \frac{\pi_m}{\sqrt{\tau_m}}$$

$$= \sqrt{\frac{1 + \alpha}{1 + \alpha\tau}}\left\{1 + \psi_m\left(\frac{\gamma}{2} M_{1,m}^2(\Phi - 1) + w_R(\pi - 1)\Phi\right)\right\} \quad (12\text{–}69)$$

where ψ_m is defined in eq. 12–58.

Mixed Flow Nozzle Velocity Coefficient

The energy equation for the mixed flow (eq. 12–61) gives the jet velocity when the exit pressure is specified by $s_{1,e}$ (eq. 12–56). The ratio of exit velocity referenced to the uniform case is the C_v (mixed), again based on the primary flow conditions:

$$\frac{V_m}{V_1} = \sqrt{\tau_m}\sqrt{\frac{1 - s_{1,e}/\sigma_m}{1 - s_{1,e}}}$$

$$\cong \sqrt{\frac{1+\alpha\tau}{1+\alpha}}\left(1 + \frac{1}{\gamma M_{1,e}^2}\left[\frac{\gamma}{2}M_{1,m}^2(\Phi - 1) + w_R\Phi(\pi - 1)\right]\right) \quad (12\text{–}70)$$

12.6.4. *Comparison of Mixed and Two-stream Flow Performance through a Single, Properly Expanded Nozzle*

These results allow direct comparison of the two cases, where same mass flow rate is processed by a nozzle with the difference being that the streams are coflowing or mixed. The design parameter of interest is the choked flow area ratio given by the ratio of eqs. 12–53 and 12–68:

$$\frac{A_m^*}{A^*(2\text{ stream})} = \frac{1}{\sqrt{\Phi}}\left(1 - (\Phi - 1)\left[\frac{\gamma}{2}M_{1,m}^2 + w_R(\pi - 1)\right]\right) \quad (12\text{–}71)$$

Evidently the throat areas are identical when the total temperature is uniform (i.e., when τ and therefore Φ are unity). When this is not the case, both the mixer Mach number and the total pressure nonuniformity are important, albeit the dominant effect is due to the first $\sqrt{\Phi}$ term.

The static pressures at the throat are also important so that the divergent section may be designed for complete expansion. Restating eq. 12–51 and developing the argument following eq. 12–67:

Two-stream flow: $$s^* = \frac{2}{\gamma+1}\left(1 + w_R\frac{\gamma - 1}{\gamma}(\pi - 1)\right)$$

Mixed flow: $$s^* = \frac{2}{\gamma+1}\left(1 + \frac{\gamma-1}{2}M_{1,m}^2(\Phi - 1) + w_R\frac{\gamma-1}{\gamma}(\pi - 1)\Phi\right) \quad (12\text{–}72)$$

Note the choked flow static pressure at the throat is the same for the two cases when the total temperature is uniform, $\Phi = 1$. Equations 12–58 and 12–70 allow the comparison of thrust (or thrust per unit mass flow since the mass flow is identical) for the mixed flow and coflowing cases:

$$\frac{V_m}{\bar{V}} = \frac{1}{\sqrt{\Phi}}\left(1 + \frac{(\Phi - 1)}{M_{1,e}^2}\left[\frac{\gamma}{2}M_{1,m}^2 + w_R(\pi - 1)\right]\right) \quad (12\text{–}73)$$

This thrust ratio shows that a specific relation between π and τ exists which results in equal velocities. For this linearized case, this relationship is obtained by setting the velocity ratio to unity and solving for π. One obtains,

$$\frac{\gamma}{2}\left(\left[\frac{2}{\sqrt{\Phi}+1}\right]M_{1,e}^2 - M_{1,m}^2\right) = \frac{\alpha\sqrt{\tau}}{1+\alpha\sqrt{\tau}}(\pi - 1) \qquad (12\text{–}74)$$

Typically, $M_{1,m} \ll M_{1,e}$ and Φ is close to unity so that the term in brackets is of order one. This condition fails to conform to the small $(\pi - 1)$ requirement for linearization and cannot be used except in uninteresting, very small M cases. A numerical solution of the equations is therefore required to establish the relationship between π and τ for equal thrust per unit mass flow rate.

The mass flow rate change associated with consideration of coflowing and mixed flows through a given area nozzle is given by the ratio of eqs. 12–57 and 12–69. This change is given by

$$\left\{\frac{\dot{m}(\text{mixed})}{\dot{m}(\text{2-stream})}\right\}_{\text{fixed area}} = \sqrt{\Phi}\left\{1 + \psi_m(\Phi - 1)\left[\frac{\gamma}{2}M_{1,m}^2 + w_R(\pi - 1)\right]\right\} \qquad (12\text{–}75)$$

which bears an inverse relation to eq. 12–71 and is more general because it also applies to unchoked flow. One may think of this ratio as the change in C_d experienced when the flow changes from unmixed to mixed. This change is dominated by the total temperature nonuniformity through Φ.

12.6.5. *Large π Results*

The following results are from a solution of the nonlinear equations under conditions where the linearization is invalid. For illustrative purposes, the parameter values examined are $\gamma = 1.4$, $\alpha = 1$, $\tau = 0.5, 1, 2$, and π ranging from 1 to 2.5. The mixer primary Mach number is fixed at 0.5, since this value is representative and the results can be shown to be weakly sensitive to this choice.

Two-Stream Flow

The two-stream solution is obtained by iterative solution of eq. 12–49 for the parameter s^* that leads to choking. This value of static pressure is shown in Fig. 12–16. The corresponding choked mass flow rate is then given by eq. 12–47 with $s = s^*$. The value of $\dot{m}$ thus calculated, normalized by the uniform result ($\pi = \tau = 1$), is given by eq. 12–57 and shown in Fig. 12–17. If values of $\pi < 1$ are desired, one needs only to use the secondary flow as a reference since τ both smaller and larger than unity are shown. The shifted curves result from the effective mean density changes. A result for such a plot considerably closer to unity is made possible by defining artificial averages of total pressure and temperature and using such values in the reference case. This would, however, introduce uncertainty associated with those definitions.

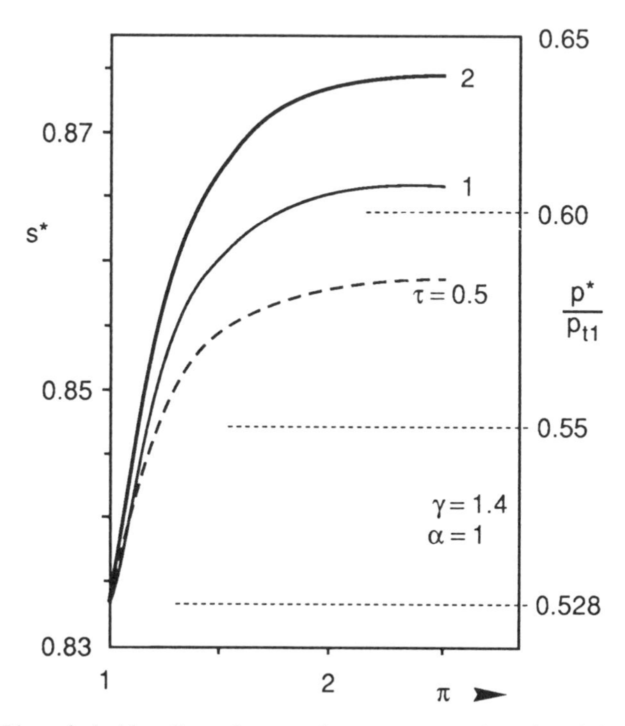

FIG. 12–16. Plot of choking-flow plane static pressure as given by s^* for two-stream flow exiting a single nozzle. Values of π and τ are given, and p^*/p_{t1} is noted on the right side scale. For $\pi < 1$, use the secondary stream as a reference.

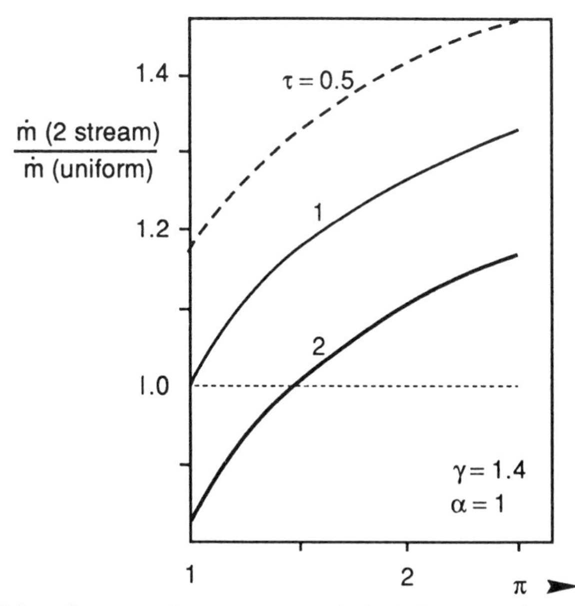

FIG. 12–17. Mass flow rate for two-stream choked flow, nondimensionalized by the uniform flow case. This ratio is given by eq. 12–57 (for the linearized case) as a discharge coefficient with the primary as a reference conditions. Fixed flow area.

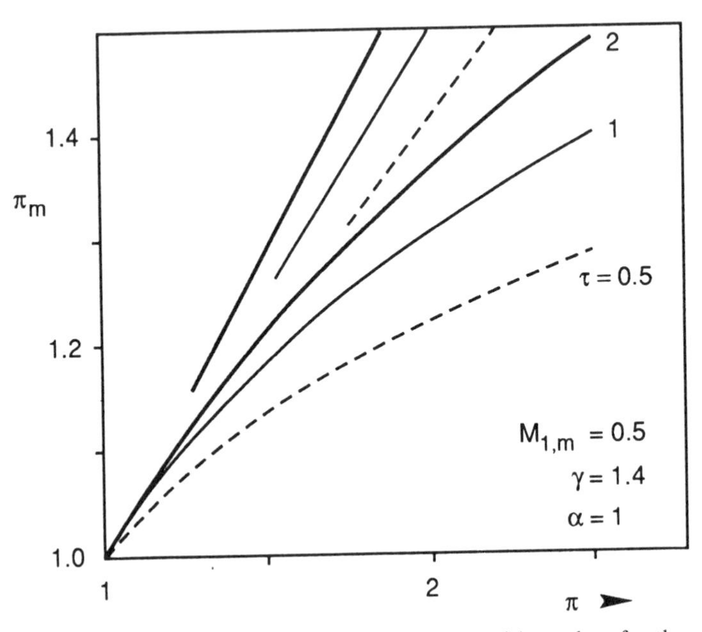

FIG. 12–18. Mixed-flow total pressure (π_m) for various π with τ values for the secondary stream ahead of the mixer as noted. Temperature-corrected mass flow weighted values of π_m are indicated as the straight lines emanating from the origin, showing error associated with use of an unjustifiable total pressure averaging method.

Mixer

The mixer requires specification of the static pressure or equivalently, the primary flow Mach number. Examination of a range of values shows a weak dependence of the effective mixed-flow total pressure on $M_{1,m}$. A value of $M_{1,m}$ of 0.5 appears representative. The mixed-flow effective total pressure for this value and π, τ as indicated is shown in Fig. 12–18. The variation of π_m is nonlinear with π and depends on the parameter τ. The figure also shows the *linear* (and therefore approximate) variations for a mean total pressure based on mass flow averaging ($\tau = 1$) and another based on temperature corrected mass flow rate. These results are consistent with eq. 12–64 near $\pi = 1$. The mean total temperature is given by eq. 12–60.

Mixed-Flow/Two-Stream Nozzle Comparison

The ratio of mass flow through a fixed area ratioed to the uniform flow case (see eq. 12–69) is very similar to the results shown in Fig. 12–17. To highlight and emphasize the difference between the two-stream and mixed flow cases, the ratio of

$$\frac{\dot{m}(\text{mixed})}{\dot{m}(2\ \text{stream})} = \frac{\pi_m}{\sqrt{\tau_m}} \frac{1 + \alpha\sqrt{\dfrac{\tau}{\sigma}}\sqrt{\dfrac{1 - s^*}{\sigma - s^*}}}{1 + \alpha} \tag{12–76}$$

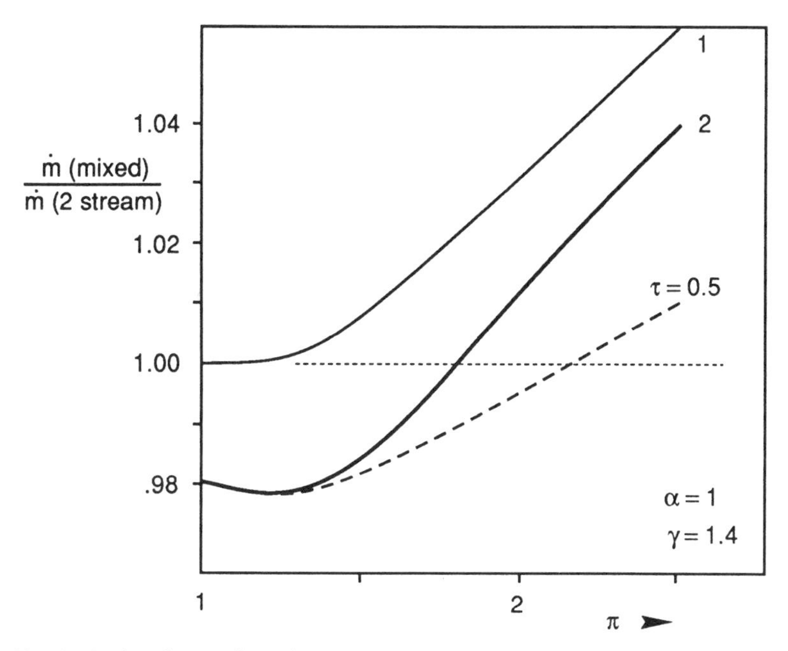

FIG. 12–19. Ratio of mass flow through a given choked nozzle with mixed flow to the mass flow rate with coflowing streams. Flow area for the two cases are equal.

is formed from eqs. 12–57 and 12–69 and plotted in Fig. 12–19. Evidently, and consistent with the linearized results, when the total temperature is uniform, the role of total pressure is small for π close to 1 and increases with increasing π. For the case of equal total pressures in primary and secondary flows, the temperature ratios $\tau = 0.5$ and 2 describe the same physical situation since choice of the reference or primary is arbitrary. The coupled effect of total pressure nonuniformity becomes evident when π is large.

C_v or Thrust Per Unit Mass Flow

The general, nonlinearized forms of eqs. 12–59 and 12–70 allow the writing of the change in velocity coefficient or thrust per unit mass flow rate. This ratio is

$$\frac{V(\text{mixed})}{V(2\,\text{stream})} = \frac{\sqrt{(1+\alpha)(1+\alpha\tau)}}{\sqrt{\dfrac{1-s_{1,\text{e}}}{1-s_{1,\text{e}}/\sigma}} + \alpha\sqrt{\tau}} \approx \frac{1}{\sqrt{\Phi}} \quad \text{as } s_{1,\text{e}} \to 0 \qquad (12\text{–}77)$$

which is plotted in Fig. 12–20 for $M_{1,\text{m}} = 0.5$ and $M_{1,\text{e}} = 3$. This ratio is greater than unity showing that a uniform flow produces greater thrust per unit mass flow than does a nonuniform one. In the limit of $s_{1,\text{e}} = 0$, that is, expansion to vacuum where M is infinite, the total pressure uniformity is irrelevant and the velocity ratio reaches the ratio noted, which is slightly larger than unity. By setting this velocity ratio (which is valid for all values of π and τ) to unity,

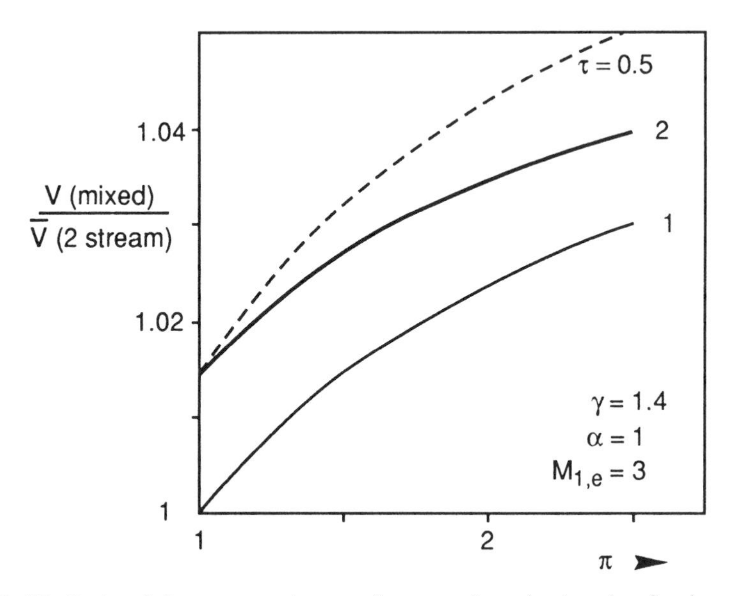

FIG. 12–20. Ratio of thrust per unit mass flow rate for mixed and coflowing streams. Mixer Mach number, $M_{1,m} = 0.5$.

the value of π (for a given τ) to give a thrust which is insensitive to mixing may be obtained. Since Φ is very close to unity, it may be used as a basis for an approximate value given by

$$\sigma\left(\frac{V(\text{mixed})}{\bar{V}(2\ \text{stream})} = 1\right) \cong 1 - \left(\sqrt{(1+\alpha)(1+\alpha\tau)}\right)\left(1 - \sqrt{\Phi}\right)(\gamma - 1)M_{1,e}^{2} \quad (12\text{–}78)$$

The correction term is positive and equals 0.09 for $\alpha = 1$, $M_{1,e} = 3$, and $\tau = 0.5$. Thus, for equal velocities, the value of σ is 0.72. For lower total pressure in the $\tau = 0.5$ stream, the mixed flow velocity is lower than the average coflowing value.

12.6.5. Summary

The calculations above were carried out for the purpose of displaying the important parameters which influence performance. To that end, total pressure nonuniformity was limited to being small so that linearization of the equations could be carried out. No such limitation is necessary in connection with the total temperature. Calculations for general π and τ are readily carried out using the results prior to linearization.

Within the limit of small total pressure variations (which are most often encountered in practice) one may conclude the following for nonuniform nozzle flows:

1. The impact on thrust per unit mass flow brought on by mixing over

coflowing unmixed through a common nozzle is dominated by the uniformity of the total temperature, eq. 12–73 and Fig. 12–20.

2. The nozzle discharge coefficient change associated with mixed or coflowing streams is affected by the same parameters as those that affect velocity coefficient: eqs. 12–73 and 12–75.
3. In nearly uniform flow, two nonuniformity parameters describe the difference in performance between mixed and unmixed flows. These are

$$\Phi = \frac{(1 + \alpha\sqrt{\tau})^2}{(1 + \alpha)(1 + \alpha\tau)} \qquad \text{and} \qquad \left\{\frac{\gamma}{2} M_{1,m}^2 + w_R(\pi - 1)\right\}$$

4. The total temperature nonuniformity plays an individual and a coupled role with the total pressure nonuniformity in determining design and performance parameters.
5. The total pressure effect is weighted in proportion to the temperature corrected mass flow fraction, w_R, eq. 12–52.
6. Constant pressure mixers are close to constant area devices as determined primarily by Φ (eq. 12–66).
7. A combination of π and τ exists such that the thrusts per unit mass flow rate for the coflowing and mixed flows are equal.

To the extent that there is an uncertainty as to whether the degree of mixing is complete, there will be an uncertainty in the design value of the thrust obtained. Additional uncertainty results if the mixing is not properly taken into account.

12.6.6. *Thrust from Two Streams through Two Nozzles*

The approach used in the development of the expressions for thrust and mass flow for the single-stream nozzle can be applied to the situation where *two* nozzles are employed to produce two jets. This is often done in practice for the case of high-bypass turbofan aircraft propulsion engines for weight reasons, although a mixed flow produces better performance.

The mass flow equations establish the relation between the throat areas for the two flows. The thrust from two choked nozzles is the sum of the (choked) mass flow rates times their respective jet velocities. Nondimensionalized by the uniform flow result, one obtains, for the linearized case:

$$\frac{F_{\text{mixed}}}{F_{\text{sep}}} = \frac{1 + \alpha}{1 + \alpha\sqrt{\tau}}\left[1 + \left(1 + \frac{1}{\gamma M_{1,e}^2}\right)\frac{\gamma}{2} M_{1,m}^2(\Phi - 1) + w_R(\pi - 1)\left(\Phi + \frac{(\Phi - 1)}{\gamma M_{1,e}^2}\right)\right] \tag{12–79}$$

The first-order effect is due to the term outside the bracket which is greater than

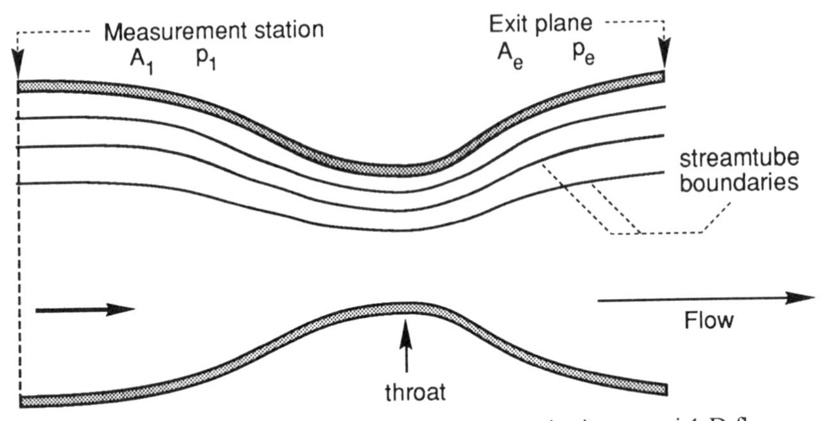

FIG. 12–21. Streamtubes with nonuniform properties in a quasi 1-D flow.

unity when the secondary flow is colder, $\tau < 1$. This is the advantage sought for exploitation in mixed turbofan engines. The small correction terms are (1) a negative term associated with the mixing process, (2) a pressure nonuniformity term, and (3) a very small cross-coupling term. The last two corrections are weighted by $\alpha\sqrt{\tau}$ as seen from eq. 12–52 defining w_{R}.

12.6.7. *General Nonuniformity Profiles in Testing of Choked Nozzles*

In practice, flows through nozzles may be nonuniform in total temperature and pressure as well as in gas properties in ways that cannot be known until an experimental determination of the flow field is made. In the discussion of the two-stream nonuniformity, it is assumed that the fractional mass associated with the distinct values of total pressure and temperature was known. In an experimental test, the inlet to a nozzle will have profiles that are known as a function of some variable such as radial location or local flow area. Before the relations such as those developed in the two-stream example can be useful, the stagnation quantities distribution must be associated with the mass flow distribution.

Figure 12–21 shows a schematic of a nozzle with representative streamtubes. One may presume for the following that at a station (7, area $= A_1$), where the flow is subsonic, a uniform static pressure p_1 is known together with profiles in p_t and T_t, as these vary with a dimensional variable such as radial distance if the geometry is axisymmetric. In order to characterize the distribution of these quantities, one may imagine that the nozzle cross-sectional area is divided into a numbered array of equal area streamtubes starting at zero where the total pressure is minimum and ending with unity where p_t is maximum. One may visualize a variable ζ which describes the streamtube number. Thus $p_t(\zeta)$ and $T_t(\zeta)$ where $0 \geq \zeta \geq 1$ may be as shown in Fig. 12–22.

The goal of an analysis of a nonuniform nozzle flow is to relate the mass flow rate and the thrust of the nozzle in terms of suitable averages of p_t and T_t measured at a station upstream from the throat. The equations

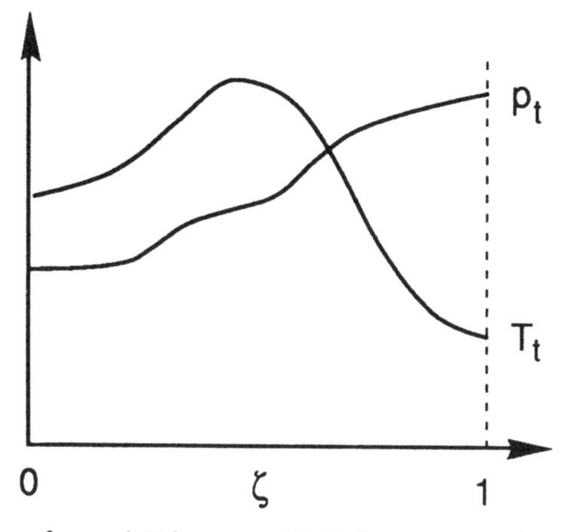

FIG. 12–22. Variation of p_t and T_t in a quasi 1-D flow measured at station 1. $\zeta = 0$ for streamtube with lowest pressure.

to be developed are equivalent to eqs. 12–23 and 12–24 or eqs. 12–45 and 12–67. A number of simplifications are used: The flow is isentropic and that expansion efficiency is unity. γ and R are assumed uniform. For questions beyond the limited investigation of the performance of a *choked, properly expanded* nozzle performance, the reader may wish to consult refs. 12–1, 12–3 and 12–4.

The mass flow rate and thrust provided by a nozzle are given by eqs. 12–23 and 12–24. These quantities involve the use of total temperature and pressure, whose appropriate averages must be determined. An *incorrect* estimate of these quantities may be made using the *area weighted averages* of p_t and T_t as given by

$$\bar{p}_t = \int_0^1 p_t \, d\zeta \qquad \text{and} \qquad \bar{T}_t = \int_0^1 T_t \, d\zeta \tag{12–80}$$

There is no reason to believe that using area weighted averages is accurate because the fundamental conservation laws must be invoked in the development of such averages: conservation of mass and of enthalpy of the flow as a whole.

The notation developed for the two stream case is used. Using $p_t(0)$ and $T_t(0)$ as reference quantities, one may define

$$\frac{p_t(\zeta)}{p_t(0)} = \pi(\zeta), \left(\frac{p_t(\zeta)}{p_t(0)}\right)^{(\gamma-1)/\gamma} = \sigma(\zeta) \qquad \text{and} \qquad \frac{T_t(\zeta)}{T_t(0)} = \tau(\zeta)$$

and

$$\left(\frac{p}{p_t(0)}\right)^{(\gamma-1)/\gamma} \equiv s \tag{12–81}$$

Thus the Mach number at any point where the "pressure" is s is given by (eq. 12–9):

$$M^2 = \frac{2}{\gamma - 1}\left[\frac{\sigma(\zeta)}{s} - 1\right] \tag{12–82}$$

The compound flow indicator for a number of streamtubes (which follows from eq. 12–13 written in terms of dp rather than du, see ref. 12–4) is

$$\sum \frac{1}{\gamma_i}\left(1 - \frac{1}{M_i^2}\right) A_i = 0$$

If the total pressure distribution were known at the throat, that is, $s^* \equiv p^*/p_t(0)$, the indicator condition can be written in terms of the $p_t(\zeta)$ with $\zeta = \zeta^*$ (i.e., at the sonic location) and the unknown static pressure (s^*) as

$$\int_0^1 \left(\frac{\sigma(\zeta^*)}{s^*} - 1\right)^{-1} d\zeta^* = \frac{\gamma - 1}{2} \qquad \text{at } A^* \tag{12–83}$$

More generally, the total pressure distribution is known at A_1 rather that at the throat A^* so that $\sigma(\zeta)$ is known as well as $s_1 = p_1/p_t(0)$. Thus the ratio σ/s_1 is available as a function of ζ. While ζ could be written as ζ_1, the subscript is omitted for clarity. The continuity equation must be invoked to relate the differential area at A_1 to that at A^*. In general,

$$d\dot{m} = \frac{p\,dA}{\sqrt{\frac{\gamma - 1}{2\gamma} RT_t}} \sqrt{\left(\frac{\sigma}{s}\right)^2 - \frac{\sigma}{s}} \tag{12–84}$$

When this relation is applied at station 1 and at A^*, the differential contribution to the mass flow rate is

$$\frac{d\dot{m}}{dA_1} = \frac{p_1}{\sqrt{\frac{\gamma - 1}{2\gamma} RT_t}} \sqrt{\left(\frac{\sigma}{s_1}\right)^2 - \frac{\sigma}{s_1}} = \frac{p^*}{\sqrt{\frac{\gamma - 1}{2\gamma} RT_t}} \sqrt{\left(\frac{\sigma}{s^*}\right)^2 - \frac{\sigma}{s^*}} \frac{dA^*}{dA_1} \tag{12–85}$$

from which T_t cancels because it remains constant along any one differential streamtube. This equation gives the differential relationship between A_1 and A^* required for other averages. Thus it follows that with $\sigma(\zeta)$ known as A_1:

$$dA^* = \frac{p_1}{p^*} \frac{\sqrt{\left(\frac{\sigma}{s_1}\right)^2 - \frac{\sigma}{s_1}}}{\sqrt{\left(\frac{\sigma}{s^*}\right)^2 - \frac{\sigma}{s^*}}} dA_1 = \left(\frac{s_1}{s^*}\right)^{(\gamma - 1)/\gamma} \frac{\sqrt{\left(\frac{\sigma}{s_1}\right)^2 - \frac{\sigma}{s_1}}}{\sqrt{\left(\frac{\sigma}{s^*}\right)^2 - \frac{\sigma}{s^*}}} dA_1$$

With the profiles determined at station 1, eq. 12–82 reads,

$$(M^*)^2 = \frac{2}{\gamma - 1}\left[\left(\frac{p_t(\zeta)}{p_1}\frac{p_1}{p^*}\right)^{(\gamma-1)/\gamma} - 1\right] = \frac{2}{\gamma - 1}\left[\left(\frac{\sigma(\zeta)}{s^*}\right) - 1\right] \quad (12\text{–}86)$$

which may be compared with eq. 12–54 for the piecewise uniform flow case.

Sonic Plane Static Pressure

The differential form of eq. 12–83 with 12–85, $dA^* = A^*\,d\zeta^*$ and $dA_1 = A_1\,d\zeta$ gives

$$\int_0^1 \frac{\left(\dfrac{\sigma}{s_1} - 1\right)^{1/2}}{\left(\dfrac{\sigma}{s_1}\left[\dfrac{s_1}{s^*}\right] - 1\right)^{3/2}}\,d\zeta = \frac{2}{\gamma - 1}\int_0^1 \frac{\left(\dfrac{\sigma}{s_1} - 1\right)^{1/2}}{\left(\dfrac{\sigma}{s_1}\left[\dfrac{s_1}{s^*}\right] - 1\right)^{1/2}}\,d\zeta \quad (12\text{–}87)$$

This equation gives the unknown pressure at the throat, s^*, in terms of the profiles in p_t, that is, $\sigma(\zeta)$ and the pressure measured as s_1 measured at station 1. In other words, eq. 12–87 gives, for a specific nonuniformity profile,

$$\frac{s^*}{s_1} = \text{fcn}\left(\frac{s_1}{\sigma}\right) \qquad \text{or} \qquad \frac{p^*}{p_1} = \text{fcn}\left(\frac{p_1}{p_t(0)}\right)$$

If this station is at the throat ($s^* = s_1$ and $\zeta = \zeta^*$), then eq. 12–87 reduces to eq. 12–83.

Sonic Flow Area

Equation 12–85 can be integrated to yield A^* or

$$\frac{A^*}{A_1} = \frac{p_1}{p^*}\int_0^1 \frac{\sqrt{\left(\dfrac{\sigma}{s_1}\right)^2 - \dfrac{\sigma}{s_1}}}{\sqrt{\left(\dfrac{\sigma}{s^*}\right)^2 - \dfrac{\sigma}{s^*}}}\,d\zeta \quad (12\text{–}88)$$

where p^* and s^* are intimately related (see equation which follows eq. 12–85) but the equation is kept as such for clarity.

Mass Flow Rate

For known $\sigma(\zeta)$ and $T_t(\zeta)$ the mass flow rate is given by

$$\dot{m} = \int_0^{A_1} \frac{d\dot{m}}{dA_1}\,dA_1 = \frac{p_1 A_1}{\sqrt{\dfrac{\gamma - 1}{2\gamma}R}}\int_0^1 \frac{\sqrt{\left(\dfrac{\sigma}{s_1}\right)^2 - \dfrac{\sigma}{s_1}}}{\sqrt{T_t}}\,d\zeta = \frac{p_{te}}{\sqrt{RT_{te}}} f(\gamma, 1)\,A^* \quad (12\text{–}89)$$

where the $f(\gamma, 1)$ term is defined in the equation for choked mass flow (eq. 12–10 with $M = 1$):

$$f(\gamma, 1) \equiv \sqrt{\gamma}\left(\frac{2}{\gamma+1}\right)^{(\gamma+1)/(2(\gamma-1))}$$

where the last portion serves to define p_{te} and T_{te}. The enthalpy balance must also be used to define these quantities.

Enthalpy Balance

The conservation of the (perfect gas) enthalpy allows a definition of mean temperature, T_{te} (with C_p uniform). Thus

$$T_{te} = \frac{1}{\dot{m}}\int_0^{A_1} T_t(\zeta)\frac{d\dot{m}}{dA_1}\,dA_1 = \frac{\displaystyle\int_0^1 \sqrt{T_t(\zeta)}\sqrt{\left(\frac{\sigma}{s_1}\right)^2 - \frac{\sigma}{s_1}}\,d\zeta}{\displaystyle\int_0^1 \frac{1}{\sqrt{T_t(\zeta)}}\sqrt{\left(\frac{\sigma}{s_1}\right)^2 - \frac{\sigma}{s_1}}\,d\zeta} \tag{12–90}$$

using eq. 12–89 for $\dot{m}$.

Effective Total Pressure

Equations 12–89 and 12–90 can then be combined to write p_{te} in terms of the known flow quantities:

$$\frac{p_{te}}{p^*} = \left(\frac{\gamma-1}{2}\right)^{(-1/2)}\left(\frac{\gamma+1}{2}\right)^{(\gamma+1)/(2(\gamma-1))}\frac{\displaystyle\int_0^1 \sqrt{\frac{T_{te}}{T_t(\zeta)}}\sqrt{\left(\frac{\sigma}{s_1}\right)^2 - \frac{\sigma}{s_1}}\,d\zeta}{\displaystyle\int_0^1 \frac{\sqrt{\left(\frac{\sigma}{s_1}\right)^2 - \frac{\sigma}{s_1}}}{\sqrt{\left(\frac{\sigma}{s^*}\right)^2 - \frac{\sigma}{s^*}}}\,d\zeta} \tag{12–91}$$

For uniform, 1-D flow the value of this expression is

$$\frac{p_{te}}{p^*} = \left(\frac{\gamma+1}{2}\right)^{\gamma/(\gamma-1)}$$

The integral in the denominator may also be expressed in terms of the ratio p^*A^*/p_1A_1 from eq. 12–88. Equation 12–91 shows that the effective total pressure is a function of not only the total pressure distribution (through σ), but also the total temperature profile. From the example to follow, however, the influence of the temperature is small for modest temperature variations. The influence of the total pressure profile on the pressure parameters (eq. 12–91) and the effective total temperatures (eq. 12–90) is significant. This has a significant impact on the mass flow handling characteristics of the nozzle and requires an accurate determination of the proper area from eq. 12–89.

Thrust

When the nozzle exit static pressure is specified and the nozzle is correctly expanded, the thrust is given by (eq. 12–19):

$$F_g = \int_0^{A_1} u_e \frac{d\dot{m}}{dA_1} dA_1 \tag{12–92}$$

where the exit velocity u_e corresponds to the exit pressure p_e. This momentum depends primarily on the pressures involved and much less on the temperatures. Recall that in the uniform flow (eq. 12–24) the thrust is independent of the total temperature. By writing u_e in terms of total temperature using the energy equation, and integrating eq. 12–92, one obtains

$$F_g = \frac{2\gamma}{\gamma - 1} p_1 A_1 \int_0^1 \sqrt{1 - \frac{s_e}{\sigma}} \sqrt{\left(\frac{\sigma}{s_1}\right)^2 - \frac{\sigma}{s_1}}\, d\zeta \tag{12–93}$$

By specializing this equation to the uniform flow case and dividing the ideal value into this thrust, one obtains a thrust coefficient for flow nonuniformity or

$$C_g = \frac{F_g}{F_{g,\text{ideal}}} = \frac{\displaystyle\int_0^1 \sqrt{1 - \frac{s_e}{\sigma}} \sqrt{\left(\frac{\sigma}{s_1}\right)^2 - \frac{\sigma}{s_1}}\, d\zeta}{\displaystyle\sqrt{1 - s_e} \int_0^1 \sqrt{\frac{T_{te}}{T_t(\zeta)}} \sqrt{\left(\frac{\sigma}{s_1}\right)^2 - \frac{\sigma}{s_1}}\, d\zeta} \tag{12–94}$$

The following example shows the very weak dependence of thrust on the temperature profile.

EXAMPLE

Consider a flow with a mean Mach number of 0.5 at A_1 and a variation which is cosine-like so that the extremes are 0.35 and 0.65, respectively. Figure 12–23 shows the variation of these quantities for $\gamma = 1.4$. The (nonmeaningful) area weighted mean total pressure is not equal to the value at $\zeta = 0.5$, but equals 1.197. The total pressure at this point is 1.186 as noted.

The static pressure at the sonic point is obtained from eq. 12–87. The square root weighting function of σ/s_1, which appears in most of the eqs. 12–88 to 12–94 varies roughly by a factor of 2, so that the large ζ portions of the integrals are emphasized. The following results follow directly ($\gamma = 1.4$, $40\Delta\zeta$ trapezoidal

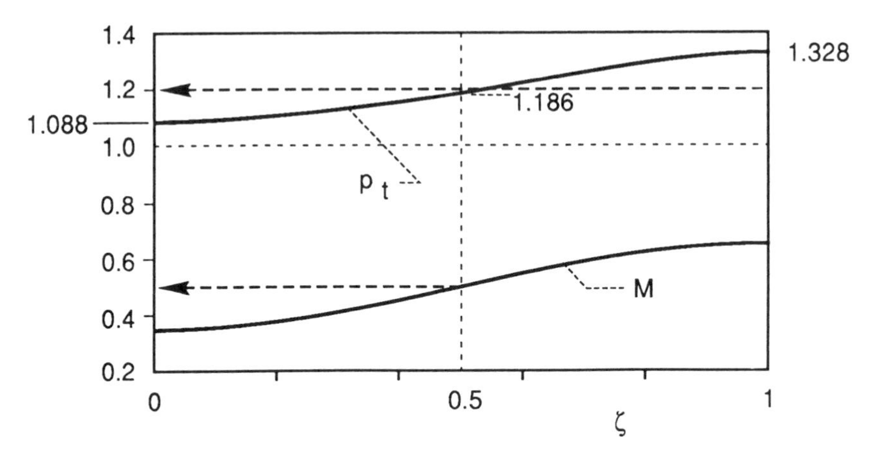

FIG. 12–23. Variation of Mach number and total pressure for the example described in Section 12.6.7.

role integration). The value of s_1/s^* is found to be 1.139 whereas the area ratio A^*A_1 is found from eq. 12–88.

	this p_t profile	uniform $M_1 = 0.5$
p_1/p^*	1.579	1.596
A^*/A_1	0.736	0.746

Thus there is a 1.3% area change in A^* associated with the nonuniformity in total pressure and a 1% change in static pressure at the choke point. One may also calculate the value of the Mach numbers at the throat. For this case they are 0.915, 0.991, and 1.086 for the three streamtubes which originally had $M_1 = 0.35$, 0.50, and 0.65, respectively. Thus a 0.15 variation in Mach number away from the mean is reduced to a still sizable 0.09 at the throat.

A number of T_t variations may be investigated. These are shown in Fig. 12–24. The constants are adjusted so that the area weighted mean T_t is uniform from case to case. The cases are

1. uniform T_t
2. T_t variation in phase with p_t,
3. T_t variation out of phase with p_t.

The variations in T_t are such that the extremes are 20% above or below the mean.

For the calculation of jet thrust, the ambient pressure is taken to be

$$\frac{p_t(0)}{p_e} = 3(s_e = 0.7306)\,\frac{p_t(0)}{p_1} = 1.0245\ (s_1 = 0.9761) \qquad \text{or} \qquad \frac{s_1}{s_e} = 1.3360$$

The temperature nonuniformity results are summarized in the following table.

Result	Profile 1 uniform T_t	Profile 2 In phase	Profile 3 Out of phase	Uniform p_t
T_{te}	1.000	1.022	0.959	
p_{te}/p^*	1.905	1.910	1.910	1.893
C_g	1.040	1.038	1.038	1.000

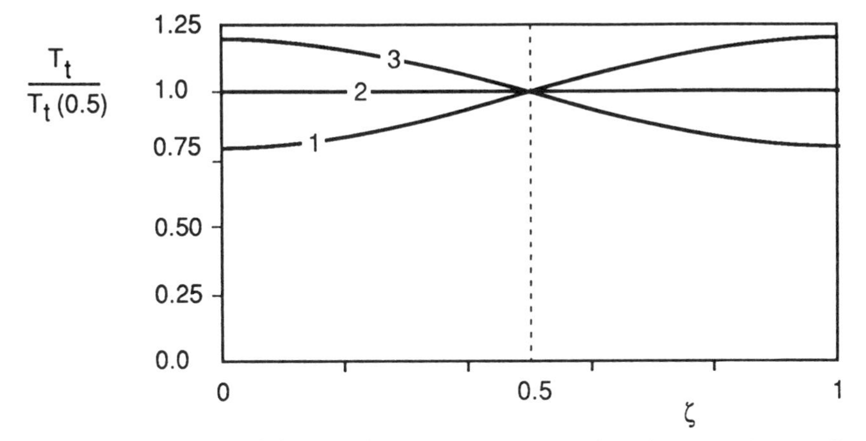

FIG. **12–24.** Variations of interest in total temperature for the example described in the text.

Thus for 20% T_t variation and 10% p_t variation away from the means, the total temperature nonuniformity by itself has a small effect on the calculated quantities, but total pressure variation can introduce measurable errors in the calculation of jet momentum (4%), flow areas required (1.3%), mean total temperature, and static pressure. This example illustrates the magnitude of the errors that might be incurred by invoking simplifications of descriptions of the flow variables of a nonuniform flow: The effect of nonuniform pressure is important in its influence on the proper nozzle area and the resultant thrust, and the influence of total temperature is weak. These results are consistent with the description of the piecewise uniform flow described in Sections 12.6.1 to 12.6.5.

12.7. NOZZLES WITH 2-D GEOMETRY AND FLOW EFFECTS

In practice, strictly one-dimensional flow through a nozzle is often impractical owing to the flow length required and the associated losses due to boundary layers and due to the added weight. Thus, nozzles may be designed to be short with relatively rapid variation in flow area. The design of aircraft converging (variable area) nozzles, for example, must take into account the need to provide acceptable nozzle performance as well as avoid separation of the external flow and minimize nozzle weight. This is very important for transonic aircraft where such drag penalties may be severe. In the following, the consequences of keeping a convergent nozzle short will be examined to illustrate the behavior of two-dimensional or axi-symmetric flow in nozzles.

12.7.1. "Soft" Choking in Short Nozzles

Consider the 2-D convergent nozzles illustrated in Fig. 12–25. If the flow is choked, then 1-D flow theory suggests that the exit plane is the sonic flow plane (left side sketch). The rapid convergence toward the centerline in the short nozzle on the right involves a necessary turning so that the pressure rises as one proceeds from the lip toward the centerline along the exit surface. The curvature of the streamtubes gives rise to a pressure gradient according to the momentum equation:

$$\left|\frac{\partial p}{\partial y}\right| = \frac{\rho U^2}{R} \tag{12–95}$$

where R is a local streamline radius of curvature and U the local velocity. The points on the centerline do not reach the sonic point pressure as early as streamlines near the edge so that the sonic line shape is not planar. A 2-D calculation or experimental measurement might show the sonic line to be as illustrated in Fig. 12–26. At the lip point Q the flow must turn further to exit

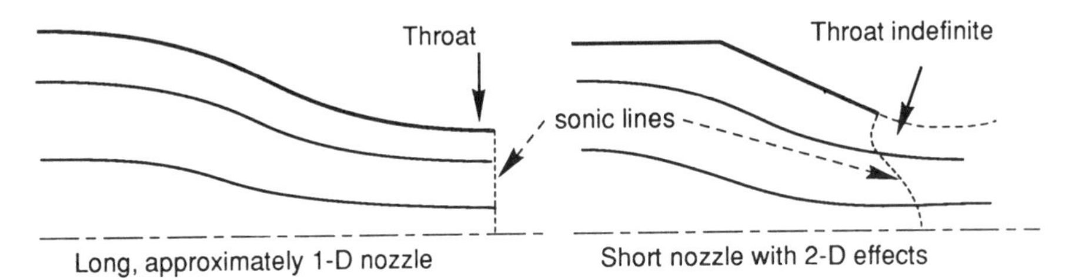

Fig. 12–25. Flow through a convergent nozzle: comparison of geometrically 1-D long nozzle and a short nozzle with 2-D effects.

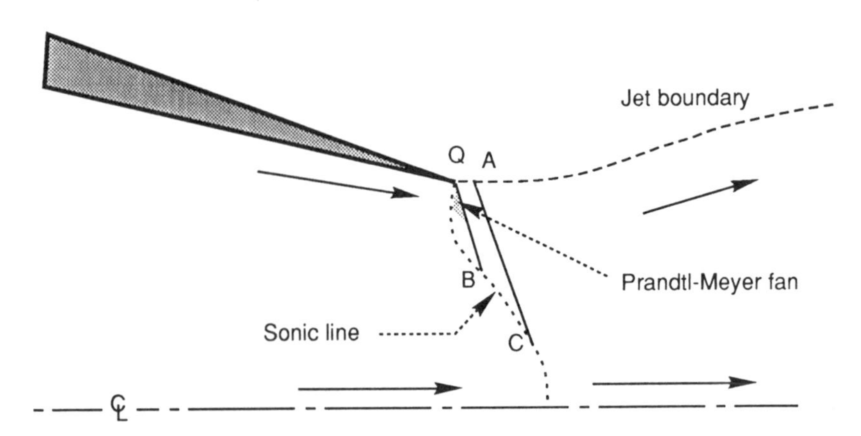

Fig. 12–26. 2-D flow through a short convergent nozzle showing sonic line and characteristics along which information can travel. Subcritical and unchoked.

the nozzle, so that a supersonic expansion fan must exist to allow this turning: A Prandtl-Meyer fan is centered on Q. In supersonic flow, information from a point can be transmitted to other points in the flow if the other points lie inside the Mach cone or characteristic surface emanating from the source point. For a flow with uniform total pressure, characteristics are constant pressure lines along which pressure information is shared with the flow field. For example, the characteristics from points Q and A intersect the sonic line (at points B and C) which means that external pressure information may be transmitted to the subsonic stream flow through a supersonic region. Such information will result in flow adjustments taking place altering, in turn, the mass flow rate because the sonic line configuration will change. This results in the interesting situation that mass flow rate is dependent on pressure ratio, even though the flow has a pressure ratio (NPR > critical NPR = 1.893 for $\gamma = 1.4$) which is great enough for the flow to be supersonic and thus ought to be choked.

As the sketch shows, the jet boundary is complex in shape suggesting that significant flow adjustments will occur outside the nozzle until the jet static pressure finally equilibrates with that of the environment. Such nozzle flows are termed "subcritical."

Suppose that the pressure ratio is raised to values that are large enough for the expansion fan to be so large in angle that the first characteristic from the lip (Q) does not contact the sonic line. The mass flow through the nozzle will then be truly choked because pressure information from outside the nozzle cannot be communicated to the mass flow determining sonic line. This "critical" flow situation is illustrated in Fig. 12–27.

In practice, the physical situation described results in a soft transition from unchoked to choked or "supercritical" flow. Figure 12–28 illustrates discharge coefficient and the sonic line location for a 25° convergent nozzle. The conical convergence angle is significant in determining the NPR where choking takes place as seen in Fig. 12–29; however, it exerts a minor impact on velocity coefficient (Fig. 12–30).

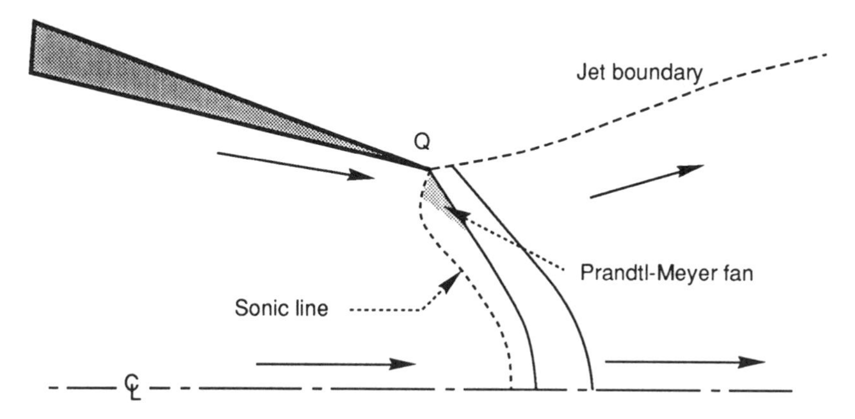

Fig. 12–27. 2-D flow through a short convergent nozzle. Supercritical and choked because characteristics in contact with ambient pressure do not reach the sonic line.

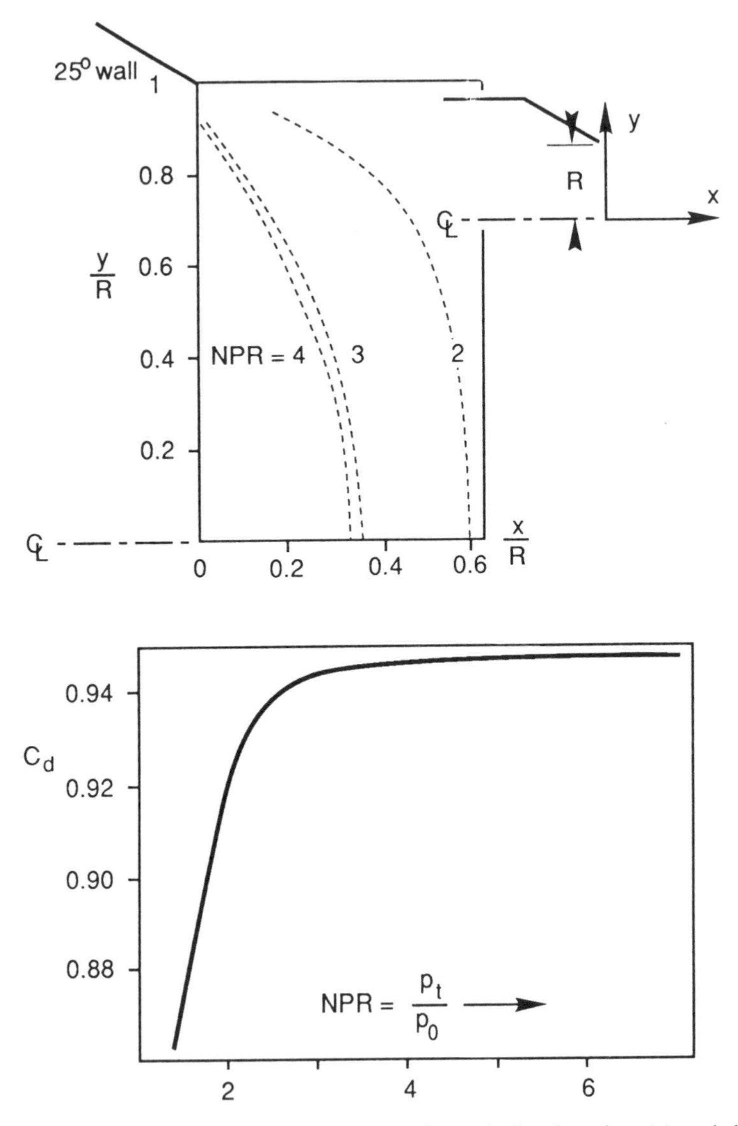

Fig. 12–28. 25° axisymmetric, convergent nozzle sonic line location (a) and discharge coefficient (b) Note nozzle pressure ratio required for choking.

12.7.2. *Plug Nozzles*

The expansion of a high pressure gas to lower pressure is best carried out with the flow exerting pressure forces on walls in such a way that all possible thrust may be realized. This argument is at the heart of proper expansion rather than underexpansion as described in Section 12.4. A practical difficulty with a convergent-divergent nozzle geometry is that the last increment of nozzle area is large, hence heavy, as the marginal benefit falls to zero near proper expansion. One way to deal with this design problem is to consider the use of a plug or

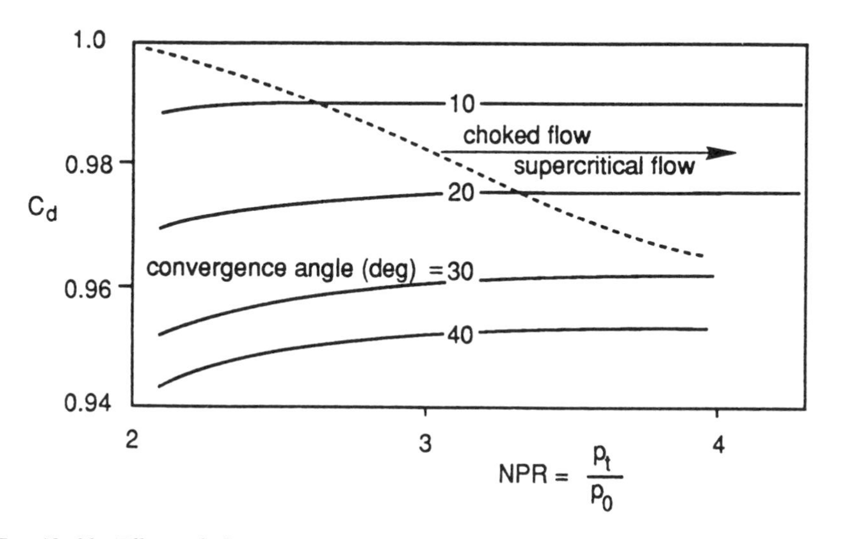

Fig. 12–29. Effect of flow convergence angle (α) on convergent nozzle discharge coefficient. Note choking condition pressure ratio. These results are for a convergence area ratio of 1.21 (radius ratio = 1.1).

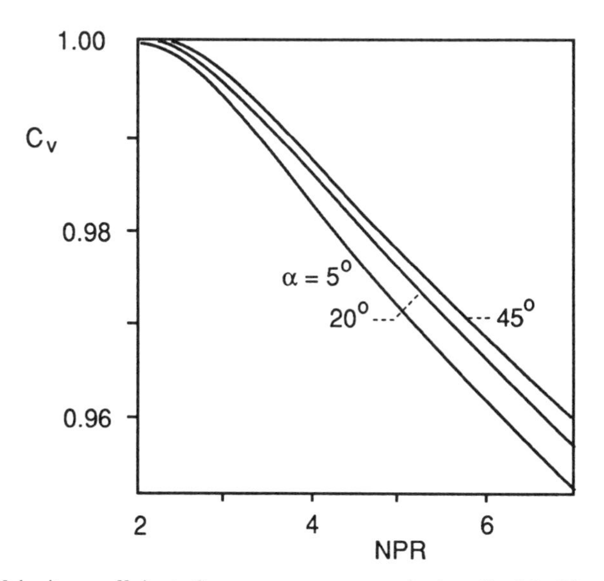

Fig. 12–30. Velocity coefficients for a convergent nozzle described in Fig. 12–29. Most of the 3% loss may be removed by contouring the nozzle to have an axial exit flow.

annular nozzle. Nozzles of this type may be axi-symmetric or two-dimensional. Such a configuration is shown in Fig. 12–31 as a symmetric device. Subsonic flow is directed around a centerbody. The outer wall forms a throat and ends with a lip. The flow at the lip is generally close to sonic but may be supersonic. The external flow has a component in the radially inward direction.

The sonic or supersonic nozzle flow expands around the lip through a

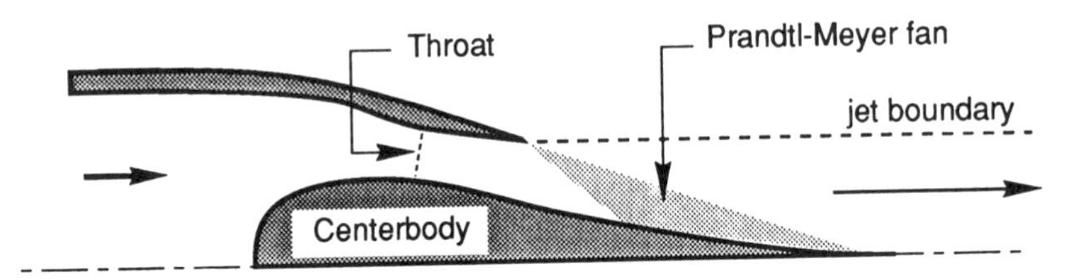

Fig. 12–31. Schematic of a plug nozzle showing the important design elements. Note that flow is initially radially inward toward the axis of symmetry.

Prandtl-Meyer fan. The centerbody is contoured to form the stream surface required to have the flow expand isentropically. The nozzle centerbody against which the decreasing pressure forces act, decreases in diameter and may be terminated early with little loss of performance.

In high-bypass turbofan engines, plug nozzles are used for the fan and/or the primary flows. For military low bypass and turbojet engines such nozzles are considered in 2-D geometries for the ability to use the centerbody to provide thrust vectoring. Such designs may be handicapped by the need to keep materials in the centerbody cool when afterburning is used. Further, 2-D designs may have serious weight concerns due in part to the necessarily flat nature of movable nozzle elements.

12.7.3. *Jet Shaping*

In commercial transport engine nozzles there is a need to meet noise emission standards. While there are a number of noise sources, an important one is the jet noise associated with the mixing between the jet and the freestream air. The noise power emitted depends on the difference in flow velocities and on the length or area of the mixing layer. In an effort to reduce the latter, particularly during the turbojet days of commercial air transports, means were used to promote the rapid mixing between jet and freestream air. The rapid mixing is promoted by lengthening the perimeter of the jet boundary so that the axial length required to complete the mixing was decreased. This was accomplished by noting that for the conditions considered, the spreading angle of the unmixed flow boundaries is rather insensitive to design or operational changes. Thus the length of the unmixed core flow depends primarily on the smallest transverse dimension of the jet. A number of designs to minimize noise using such techniques are shown in Fig. 12–32. Their performance (C_v) correlates with increased friction losses associated with the increased internal surface area on which boundary layers grow. A number of factors obviated the use of noise suppression nozzles. These include primarily the advent of the turbofan which produced a lower velocity jet for higher propulsive efficiency and takeoff thrust.

Such considerations are also relevant to the problem of infrared (IR) emissions from the hot unmixed core of the jet gas exiting a military aircraft engine. Such emissions are proportional to either the surface area of the hot core volume or the volume itself depending on the optical thickness of the radiation considered relative to the core dimensions. The emissions are

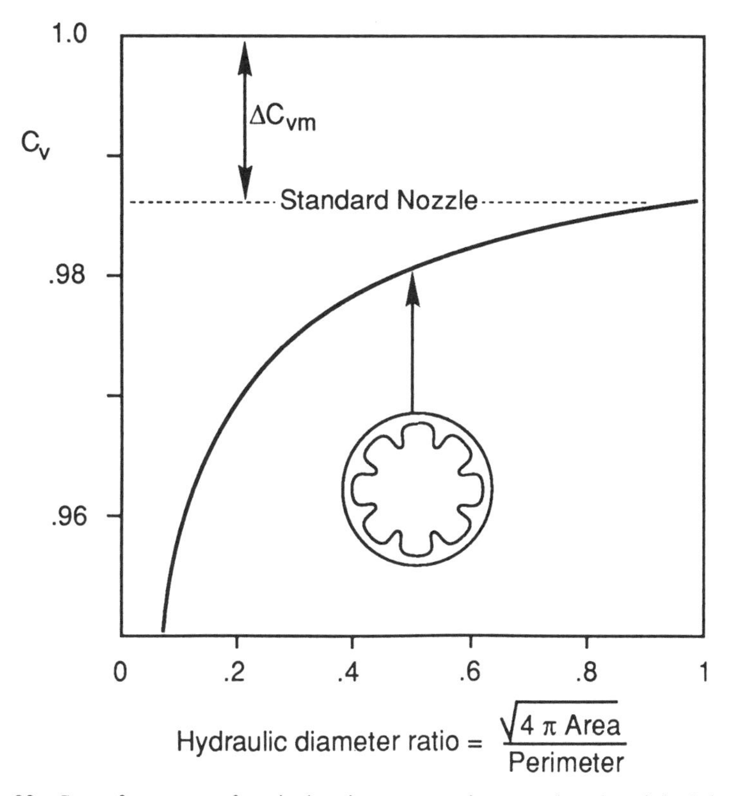

FIG. 12–32. C_v performance of typical noise suppression nozzles. An eight-lobe mixer shown is an example. The data scales approximately as $C_v = 1 - \Delta C_{vm}/\sqrt{D_h}$ ratio. The D_h ratio is the abscissa on the plot.

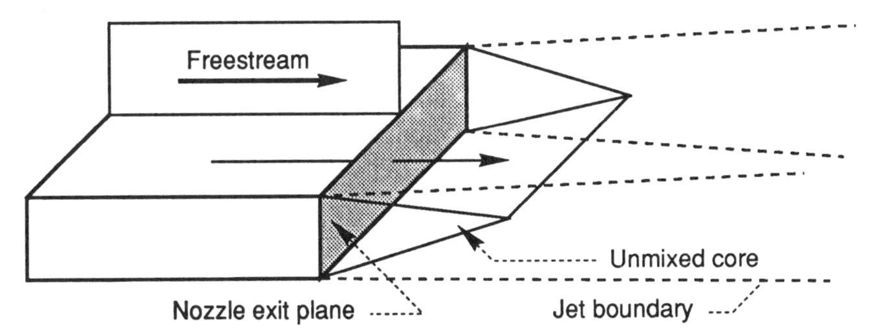

FIG. 12–33. Unmixed core of jet mixing with the freestream flow.

associated with the unavoidable radiation from the CO_2 emissions due to fuel burning and thus constitute an element of vulnerability from missiles with IR radiation-seeking capability. Nozzles for the minimization of such threats employ narrow or 2-dimensional nozzles. Reference 12–7 is a discussion of such designs as they affect the choice of thermodynamic cycle design of the basic engine. Figure 12–33 shows the geometry of the mixing process and the unmixed core, the emitter of noise, or infrared radiation. Incorporation of such design elements is an essential element of stealth in military aircraft.

12.8. NOZZLE FLOW OF IMPERFECT AND/OR CHEMICALLY REACTING GASES

Flows in propulsion devices designed to achieve high exhaust velocities necessarily operate at high (stagnation) temperatures. The work in Chapters 4 and 5 shows that, when processes over large temperature ranges are considered, the perfect gas assumption breaks down. Further, if chemical reactions take place in an expansion process, then the gas will also be nonideal because its properties are temperature dependent. An illustrative case is the discussion of a rocket engine where a high temperature and pressure gas is produced in a stagnation region that is subsequently expanded in a nozzle to produce a jet. The exit static pressure of the jet flow is assumed to be known from which all other parameters of interest are determined.

12.8.1. *Chamber or Stagnation Conditions*

Stagnation and chamber conditions in a typical (rocket) combustor will differ because the local Mach number is not zero. For purposes of simplicity, however, $M \ll 1$ is assumed in the following so that stagnation and chamber conditions are nearly identical. A typical calculation might involve knowledge of stagnation conditions, p_t, and T_t, from which composition x_{ti} (mole fractions at stagnation conditions) may be determined using the law of mass action (eq. 5–32). The composition allows determination of mixture properties, in particular, enthalpy, entropy and molar mass as follows:

Chamber Conditions Calculation Procedure given p_t *and* T_t:

law of mass action	x_{ti} (mole fraction)
h_t, s_t, and MW_t	enthalpy, entropy, and mass *per mole of mixture* are calculated using eqs. 4–65.
$\frac{h_t}{MW_t} = \tilde{h}_t;\ \frac{s_t}{MW_t} = \tilde{s}_t$	enthalpy and entropy *per unit mass*

This calculation is summarized as the *direct* process shown on the left of Fig. 12–34. Quantities in bold boxes are specified.

The determination of jet exit conditions as well as flow areas at the throat and exit are of interest for purposes of design and performance analysis. Note that the entropy *per unit mass* (not per mole) is conserved and the enthalpy *per unit mass* gives the flow kinetic energy or its velocity.

The flow from the chamber (or here the stagnation condition) is isentropic when it is adiabatic and reversible. When the flow process is very rapid, there is insufficient time for heat transfer and for the kinetics to allow the composition to change as the temperature and pressure fall. This is termed (chemically)

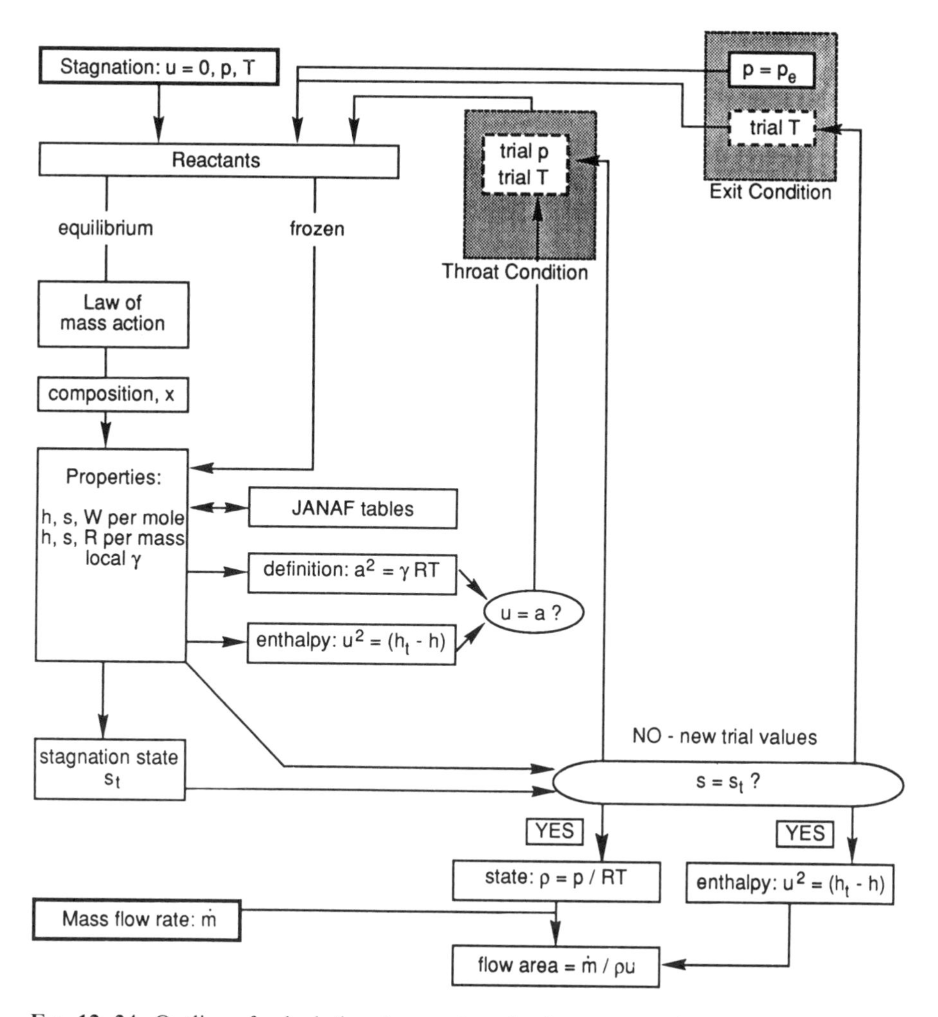

Fig. 12–34. Outline of calculational procedure for frozen or equilibrium flow through a nozzle. Bold boxes are input data, dashed bold boxes are iteration values.

frozen flow. By contrast, when the expansion process is slow (compared to kinetic time scales) the flow adjusts to the equilibrium demanded by the falling temperature and pressure. When this process is adiabatic and in chemical equilibrium, it is also isentropic. This is termed (shifting) equilibrium flow. In real nozzles, the flow generally proceeds in equilibrium during the initial stages of the expansion when intermolecular collisions are frequent enough to enforce good communication. Later as the density and temperature fall to lower value, the composition freezes. For a time between these regimes, an irreversible process governs the expansion. The calculation procedures for frozen composition, for equilibrium, and for switching from one to the other are described below. A performance model based on these processes is quite good, especially if the period of irreversible flow time fraction is brief.

12.8.2. *Frozen Flow*

The composition and molecular weight (=MW) or molar mass is invariant throughout the nozzle at the values established in the chamber. An isentropic process involving a nonreacting gas gives the following relation between T and the specified exit state pressure, p, for a single-component ideal gas or an ideal gas mixture (see Section 4.4.2):

$$s^0(T_t) - R_u \ln p_t = s^0(T) - R_u \ln p \tag{12–96}$$

where subscript "t" describes the stagnation (or chamber) condition, the unsubscripted variables might apply at any downstream position and

$$s^0(T_i) = \int_{T_{ref}}^{T_i} \frac{C_p}{T}\, dT \tag{12–97}$$

Note here that entropy is per mole as tabulated in the JANAF tables. It is acceptable to insist that entropy per mole be conserved for the *frozen flow* case because the molar mass is constant. This assumption is incorrect for the equilibrium flow case.

T may be calculated for given T_t and pressures using eq. 12–96 for the mixture. The JANAF tables then allow the calculation of enthalpy $h(T)$ (per mole) which allows determination of jet exit velocity,

$$\frac{h(T_t) - h(T)}{\mathrm{MW}} = \frac{u^2}{2} \tag{12–98}$$

The density follows from the state equation and the flow area, A_e, from the continuity equation (eq. 12–4). This procedure is summarized as the outer calculation loop in Fig. 12–34 with T(exit) being the (one) trial variable which is found by satisfaction of the conservation of entropy per unit mass statement.

Also of interest is the throat area and the conditions there. The procedure for calculating flow variables is similar except that both p and T must be trial values, and the second convergence condition is that which gives $M = 1$ (see below). This is shown as the double iterative loop in the center of Fig. 12–34. For frozen flow, the process for finding the * state is somewhat simpler than the equilibrium case because the composition is fixed which allows a simple, algebraic entropy conservation statement to be implemented directly. The following is a summary for conditions at the throat. In reality, the flow from stagnation to sonic conditions is more likely to be in equilibrium because of the high density and temperature in that portion of the flow.

Throat conditions calculation procedure:

p^* estimate (could use a constant γ value)

T^* isentropic relation $= (s^0)^{-1}\left(s^0(T_t) - R_u \ln \frac{p_t}{p}\right)$

$h(T^*)$	table data
γ^*	table data at T^*
a^*	definition of sound speed, $a^* = \sqrt{\gamma^* R T^*}$, (specific R)
ρ^*	state equation, specific R
u^*	from energy equation, $= \sqrt{2\frac{(h_t - h)}{\text{MW}}}$ (MW = molecular weight)
$u^* = a^*$?	if not, use new p^*
A^*	flow mass flow equation, $A = \frac{\dot{m}}{\rho u}$.

12.8.3. *Equilibrium Flow*

An equilibrium calculation is more complex because the composition at stagnation conditions generally includes a relatively large proportion of dissociation products, rather than the multiatomic constituents that are desired for their energy content. As the flow cools by expansion more of the desirable chemical products are formed at the cool nozzle exit. The chemical recombination makes heat available for conversion to kinetic energy so that a nozzle with equilibrium flow will have a better performance than one with frozen flow. At the exit condition: p is known and the entropy *per unit mass* is known to be the same as in the chamber. The first step in a calculation is to estimate T. A good start is a value somewhat higher than the frozen flow value.

Equilibrium flow exit condition calculation procedure:

trial T	
x_i	using law of mass action
h, s, MW	per mole, using eq. 4–65
$\frac{h}{\text{MW}} = \tilde{h}$; $\frac{s}{\text{MW}} = \tilde{s}$	per mass
is $\tilde{s}_t = \tilde{s}$?	if not, try new T
$u = \sqrt{2(\tilde{h}_t - \tilde{h})}$	energy equation (h must be per unit mass!)
ρ	state equation with specific R
A	continuity equation

This completes the computation for the exit area and exit flow conditions. The throat area may be determined by a method similar to that outlined in the frozen flow case but with the local composition determining γ and R. The calculation processes are summarized in Fig. 12–34.

When the flow freezes "suddenly" according to a criterion such as suggested by Bray (ref. 12–8), the general procedure must include the equilibrium calculation to the point in the nozzle where this transition takes place. Given such a criterion, the procedure is identical to finding the throat conditions except that the $u = a$ condition is replaced by the freezing criterion (e.g., u = characteristic nozzle dimension divided by a chemical time scale). The flow may be modeled as proceeding with fixed composition beyond this point.

12.9. BOUNDARY LAYERS

Practical internal flows cannot avoid contact with walls. There, the effect of finite fluid viscosity forces the fluid layer adjacent to the wall to be at rest relative to the wall. The layer where a gradual adjustment from zero to full freestream velocity takes place is the boundary layer. In most flows of engineering interest, particularly accelerating flows, this layer is thin. The fluid in this region is deficient in mass flux, momentum, and energy. This region of deficits responds differently to the pressures experienced by the main flow as a result of area changes and to the possible presence of waves, particularly shock waves.

In nozzles, the pressure falls in the flow direction: A *favorable* pressure gradient exists in the flow direction. This results in an acceleration of all fluid elements including those in the boundary layers. The acceleration tends to increase the shear on the wall but more importantly, it reduces the growth rate of the boundary layer over the rate that would be experienced in a flow without pressure gradient. This keeps the influence of the boundary layer small and limited to the region near the wall. In nozzles, therefore, boundary-layer effects are limited to the loss experienced by the fluid in rubbing along the wall. The energy and momentum loss can be calculated. The mass flux deficit can also be calculated and enables the designer to make the physical size adjustment necessary to account for the effective "boundary displacement."

In diffusers, the pressure rises in the flow direction. This is an *adverse* pressure gradient because the low energy and low momentum boundary layer fluid cannot adequately respond to the changing pressure. The local velocity change demanded by a pressure increase is a velocity *decrease*. Consider the flow along a wall demanding that the pressure rise, as illustrated in Fig. 12–35. The momentum equation for an incompressible flow and applied to two points along a streamline that are sufficiently close spaced so that any additional effect of viscosity can be neglected. This gives a relation between the static pressure and the velocity, known as Bernoulli's equation.

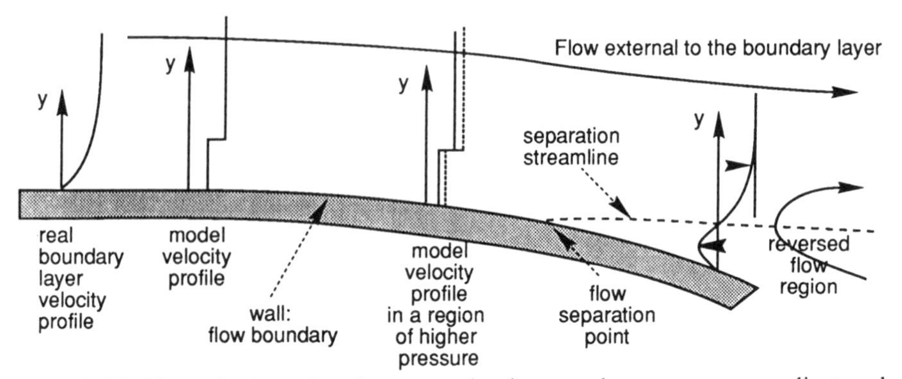

Fig. 12–35. Flow of a boundary layer experiencing an adverse pressure gradient and separating from the wall as a result. y is the local direction normal to the wall.

Bernoulli's equation in either the inviscid freestream or the nearly inviscid lower velocity flow in the boundary layer reads:

$$p + \tfrac{1}{2}\rho u^2 = \text{constant} \tag{12–99}$$

and shows how variations in p and u are related. Thus when pressure rises, velocity must decrease and this presents a difficulty for the boundary layer flow that is already very low in velocity. At some point in this flow, the fluid cannot continue the pressure rise demanded by the changing area, and the flow separates from the wall.

The use of Bernoulli's equation in this context requires a few comments. First, it is derived for incompressible flow where ρ is a constant (i.e., $M^2 \ll 1$). In nozzles, although the mean flow may certainly be far from incompressible, the boundary layer does indeed have a Mach number near zero in the vicinity of the wall. A compressible form of the momentum equation from which Bernoulli's equation is derived changes the algebraic relationship in a quantitative although not in a qualitative way, see eq. 12–9. Thus the physical conclusion from Bernoulli's expression is qualitatively valid. Second, Bernoulli's equation is derived for inviscid flow. Boundary layer flow is *created* by viscous effects, but in its progression from a low pressure state to one of higher pressure the effects of viscosity may often be neglected. When they are, Bernoulli's equation may be used.

The reader is urged to consult texts on viscous flow to obtain quantitative descriptions of boundary layer flows. To that end, integral methods of boundary layer behavior descriptions are particularly useful because they are easy to interpret physically. One can show for example, that the momentum equation for the fluid in a boundary layer can be written as an equation for the variation of momentum thickness, θ, which is influenced by shear forces due to the wall and to pressure forces transmitted from the freestream. Schlichting (ref. 12–9) shows that this equation reads

$$\frac{d\theta}{dx} = -(2\theta + \delta^*)\frac{d\ln U}{dx} + \frac{1}{\theta}\frac{\tau_{\text{wall}}}{\rho U^2} \tag{12–100}$$

θ and δ^* are the momentum and displacement thicknesses defined in eq. 12–32. Note that Bernoulli's equation in the inviscid freestream gives

$$\frac{d\ln U}{dx} = -\frac{p}{\rho U^2}\frac{d\ln p}{dx} \tag{12–101}$$

showing that increasing U, $\dfrac{d\ln U}{dx} > 0$, or decreasing p, leads to decreasing θ, but the wall shear term is always positive. For a solution, this equation requires a relation for the wall shear in terms of boundary layer profile parameters such as U, the viscosity μ, and a length scale. When the shear is written as

$$\tau_{\text{wall}} \approx \mu \frac{U}{\theta} \tag{12–102}$$

the last term in eq. 12–100 becomes a Reynolds number per unit length. Lastly, the freestream p or U variation with x must be specified. The governing equation for θ is generally not solvable in closed form but yields to numerical techniques.

The important conclusions to be drawn from an examination of eq. 12–100 are:

1. Boundary layers tend to grow as a result of wall friction experienced as the flow proceeds along a wall.
2. The pressure gradient causes the thickness to tend to decrease if the gradient is favorable and increase if adverse (i.e., increasing in the flow direction).
3. If the adverse pressure gradient is large enough, the fluid near the wall can reverse in directions leading to the situation where the flow cannot conform uniformly to the available flow area imposed by the wall geometry. More precisely stated, the flow ceases to be unidirectional and the physical manifestation is that the flow separates from the wall.

In practice, if control over flow parameters is to be maintained with a high degree of uniformity (i.e., overall efficiency), then separation must be avoided. For the designer, this limits the pressure rise that can be achieved along a wall of a specific length scale. This length scale is determined by the friction on the wall and thus the thickness of the boundary layer. In the language of the aerodynamicist, the length scale is intimately related to the flow Reynolds number.

In external flows, this tendency for the boundary layer to separate under the influence of an adverse pressure gradient leads directly to the stall of a wing. Since compressor blades are like wings, except that rotation and adjacent surfaces complicate matters, there are comparable limits to the pressure changes that can be demanded by the blading in a compressor as shown in Chapter 14.

12.10. DIFFUSERS

A diffuser is a duct where the designer is attempting to convert the kinetic energy of the flow to a greater proportion of thermal energy. Diffusers operate with severe performance requirements in jet propulsion engines, with both subsonic and supersonic freestream flow speeds. The flow passages in compressors (axial flow machines in particular) are diffusers because the

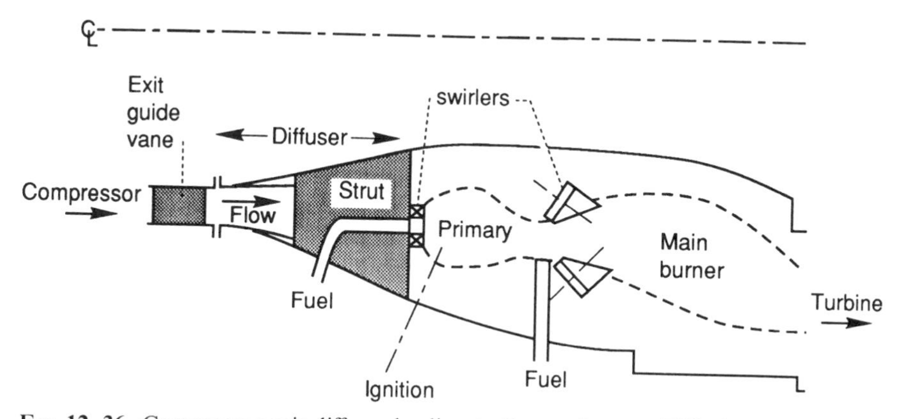

Fig. 12–36. Compressor exit diffuser leading to the combustor. Diffusion occurs in the exit guide vane and in the divergent duct leading to the swirler and mixer. (Sketch is after layout in a Pratt & Whitney JT-9D (PW-4000) turbofan engine.)

goal is to turn the flow in a manner that increases pressure at the expense of velocity. The region between compressor exit and combustor entrance is a diffuser to match the flow Mach number requirements in the compressor (~ 0.6) and the requirements in the combustor (typically less than 0.1) (Fig. 12–36).

As Fig. 12–4 illustrates, subsonic flows are slowed by providing an increasing flow area. Conversely, supersonic flow requires a decreasing flow area to diffuse the flow. Thus there are geometrical differences between subsonic and supersonic diffusers. The difference is made more significant by the fact that compression in supersonic flow takes place through waves that are absent in subsonic flow. With the possible exception of aircraft designed to fly at supersonic speeds and high-performance compressors, the most common diffusers in practice are subsonic. Since boundary layers in the adverse pressure gradient environment of the diffuser play a role in determining flow behavior, it is logical that viscous effects play a dominent role in the design and performance of diffusers. It follows that the Reynolds number is an important parameter governing diffuser performance.

12.10.1. *Subsonic Aircraft Diffusers or Inlets*

The term *inlet* is often used to describe the primary entry diffuser of an aircraft propulsion engine. Subsonic inlets, such as on commercial airliners, operate to fulfill two important roles. They match the flow Mach number desired at the entrance to the compressor or fan (of the order of 0.6) to the variable Mach number of the external flow (0 at rest to 0.85 at cruise). The rounded inlet lip allows these functions to be performed, even when crosswinds and aircraft attitude compound the flow problem. Figure 12–37 shows such an inlet with the stagnation streamlines shown at very low speed and at cruise. The rounded inlet lip allows the external flow to enter smoothly at low freestream speed, while low drag is experienced at high speed.

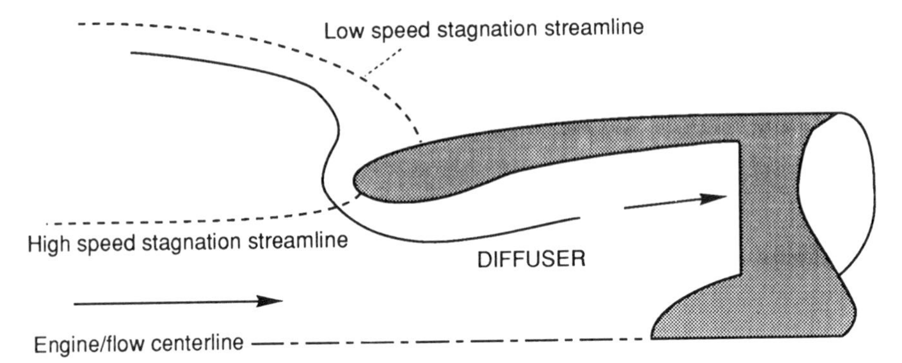

FIG. 12–37. Subsonic inlet operating under two flight conditions. Local flow Mach number is less than unity everywhere, except possibly near the lip at low speed for which case a representative streamline is shown.

PROBLEMS

1. Determine discharge, velocity, and thrust coefficients for the nonuniform flow cited in the example in Section 12.6.9 based on p_t and T_t measurements made in the "middle" of the flow, say, $\zeta = 0.5$.
2. A nozzle accelerates a flow from say $M = 0.5$ to $M = 1$. In terms of the flow parameters, find the force (appropriately nondimensionalized) on the metal structure that connects the $M = 0.5$ plane to the rest of the system.
3. One kg/sec of hydrogen is heated to 1500 K at 1 atm total pressure, $M \sim 0$, and expanded by a nozzle through a 100:1 pressure ratio.
   a. Determine the degree of dissociation of H_2 to H.
   b. Determine the power required if T is initially 298 K, assuming
      (1) No dissociation takes place, and
      (2) Composition is a mixture of H_2 and H.
   c. Find the thrust, throat area and exit areas for the following sets of circumstances: working fluid is:
      (3) Ideal and perfect H_2 only
      (4) Imperfect H_2 only
      (5) Real composition and frozen flow
      (6) Real composition and equilibrium flow
4. Find the effective total pressure and the choked flow static pressure of a flow exiting a convergent nozzle with a parabolic total pressure profile and uniform total temperature.

BIBLIOGRAPHY

Kuethe, A. M., Chow, C. Y., *Foundations of Aerodynamics*, John Wiley & Sons, New York, 1986

Liepmann, H., Roshko, A., *Elements of Gas Dynamics*, John Wiley & Sons, New York, 1957

Shapiro, A., *The Dynamics and Thermodynamics of Compressible Fluid Flow*, Ronald Press, 1953.

Fuhs, A. E., Ed., *Handbook of Fluids and Fluid Machinery*, Chapter 18.5, "Nozzle Flows." John Wiley & Sons, New York, to be published.

REFERENCES

12–1. Oates, G. C., *Aerothermodynamics of Gas Turbine and Rocket Propulsion*, AIAA, New York, 1987.

12–2. James, C. R., Kimzey, W. F., Richey, G. K., Rooney, E. C., Eds., *Thrust and Dreg: Its Prediction and Verification*, Progress in Aeronautics, Vol. 98, Am. Inst. of Aeronautics & Astronautics, New York, 1985.

12–3. Oates, G. C., Presz, W., "Effects of Nonconstant Enthalpy Addition on Fan-Nozzle Combinations," *AIAA Journal of Aircraft*, **16**(12): 891–3, 1979.

12–4. Bernstein, A., Heiser, W., Hevenor, C., "Compound-Compressible Nozzle Flow," AIAA paper 66-663, 2nd Joint Propulsion Conference, Colorado Springs, 1966. Also *J. of Applied Mech.*, **34**: 548–54, 1967.

12–5. Decher, R., "Nonuniform Flow Through Nozzles," *AIAA Journal of Aircraft*, **15**(7), July 1978.

12–6. Greitzer, E. M., Paterson, R. W., Tan C. S., "An Approximate Substitution Principle for Viscous Heat Conducting Flows," *Proc. Royal Soc., London*, A401: 163–93, 1985.

12–7. Decher, R., "Infrared Emissions from Turbofans with High Aspect Ratio Nozzles," *AIAA Journal of Aircraft*, **18**(12): 1205, December 1981.

12–8. Bray, K. N. C., "Atomic recombination in Hypersonic Wind Tunnel Nozzles," *J. of Fluid Mechanics*, **6**(1): 1–32, July 1959.

12–9. Schlichting, H., *Boundary Layer Theory*, McGraw-Hill, New York, 1980.

13

PROPELLERS AND WIND TURBINES

In contrast to nozzles where thermal energy is converted to kinetic energy in a jet, mechanical power can be added to or removed from a stream by means of a propeller. For the production of thrust, a propeller is used while a wind turbine removes flow kinetic energy as mechanical work. These devices generally have a free boundary separating the freestream flow and the flow undergoing the work interaction. Figure 13–1 shows the physical arrangement. Bounded or internal flows with work interactions are termed compressors (or turbofans) and turbines depending on whether work is added or removed. These devices are described in Chapters 14 and 15.

Mechanical power is added to a stream for the generation of propulsive thrust by rotating winglike elements mounted on a central shaft, a device termed a propeller. The desired thrust is a result of forces similar to lift on wings. Unavoidable drag forces impart tangential momentum to the flow processed by the propeller which results in a modest swirling motion. This chapter describes the key design issues underlying the satisfactory operation of propellers and wind turbines. Section 13.1 is a discussion of thrust production for vehicle applications whereas Section 13.2 centers on power extraction from moving streams. In the mathematical description of the blade aerodynamics, the reader is expected to be familiar with elementary wing aerodynamics as described in refs. 13–1 and 13–2, among others.

13.1. ANALYSIS MODELS FOR PROPELLERS

The analysis of flows with mechanical power addition may be made with models of varying sophistication. The simplest such model is the "actuator disk" where the propeller is modeled as an area of discontinuity in total pressure. A better model is to examine the propellers as wings and apply two-dimensional (2-D) airfoil theory to determine local lift and drag and from these quantities, integrated thrust and torque. Modeling the blade as the generator of a vortex system allows the determination of 3-D flow effects, induced drag, in particular. All such models may be examined with the assumptions that the flow is incompressible to highlight the effects discussed. At high flight speed and/or at high loadings, accounting for compressibility effects is required for accurate performance determination.

13.1.1. *Actuator Disk Theory*

Actuator disk theory is a one-dimensional flow theory with two streams. One is the streamtube processed by the propeller, while an external freestream is

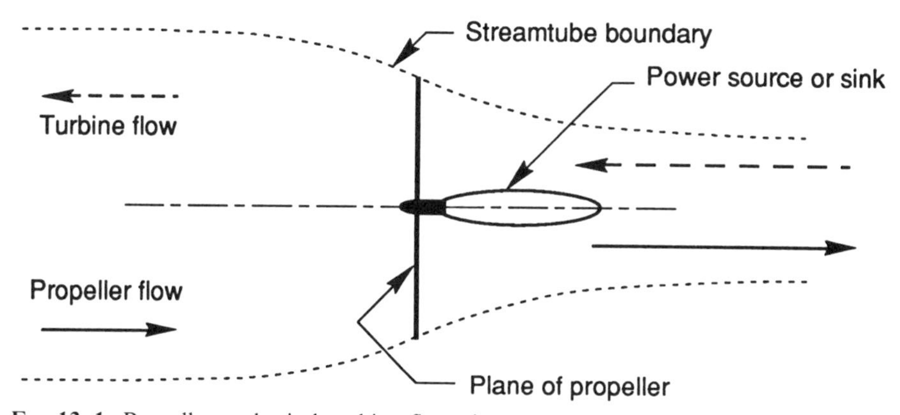

FIG. 13–1. Propeller and wind turbine flow elements, showing the flow directions for power addition and extraction and the affected streamtube.

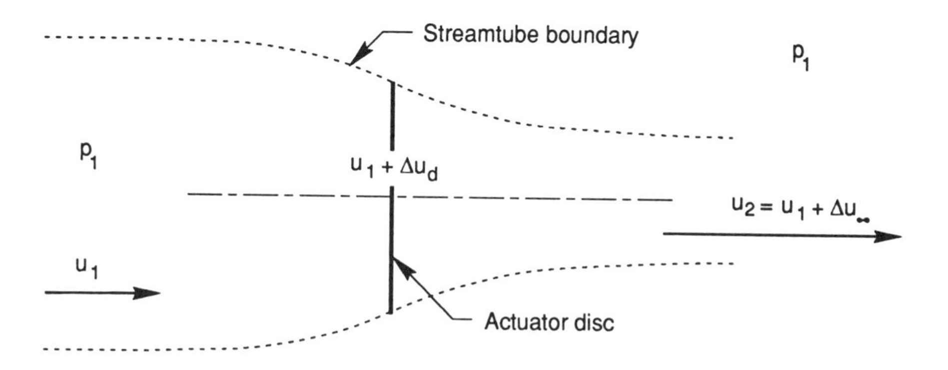

FIG. 13–2. Flow properties near and far from an actuator disk.

not, and the two flows share pressure and flow direction far up- and downstream. Figure 13–2 shows the geometry of the flow and the average state parameters used to describe the conditions at locations of interest. The result desired is a relation between the axial force on the propeller mechanism and the change in fluid velocity. From a fluid mechanical viewpoint, the boundary of the jet exiting the disk is a vortex tube created by the succession of passing blades each shedding a tip vortex into a helix. This aspect of the model is discussed in Section 13.1.7. The actuator disk theory model approximates the real hardware as having a very large number of blades so that the trailing helices form a cylindrical sheet. Further, drag associated with the airfoils is neglected so that there is no motion in the rotational direction. This is justified by the large lift/drag (L/D) ratio typical of a well-designed propeller blade.

The following summarizes the flow model assumptions:

1. The propeller is replaced by an "actuator disk" with an infinite number of blades.
2. The flow processed by the disk is purely axial. The velocity at any position

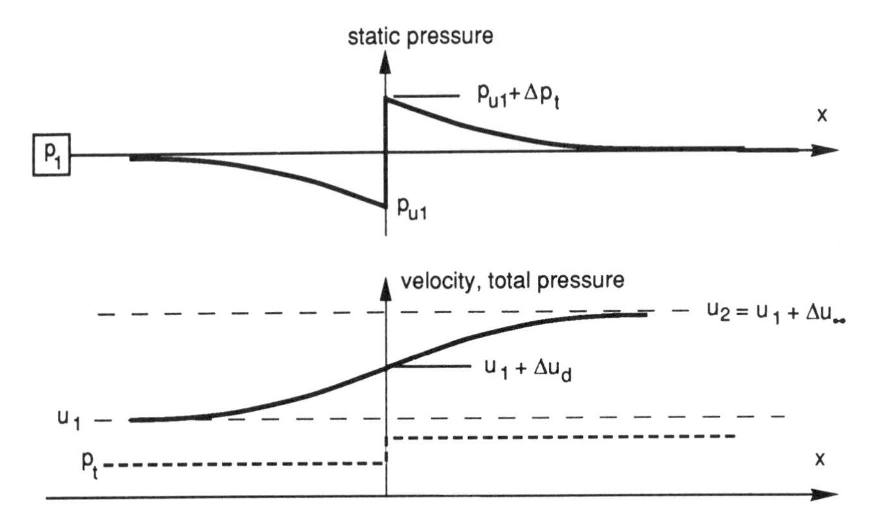

FIG. 13–3. Velocity and pressure variations through an actuator disk. $\Delta p_t = \Delta p$.

as well as the change in velocity across the disk is uniform (one dimensional).

3. The fluid is incompressible.
4. The flow is irrotational everywhere except in the vortex sheet.
5. The excess or deficit in kinetic energy in the fluid processed by the disk is the only loss.

The velocity and pressure distributions along a streamline through the disk are shown in Fig. 13–3. The pressure distribution is discontinuous across the disk whereas the velocity is continuous as mass flow is conserved in the streamtube. The enthalpy equation (eq. 2–6, adiabatic and no work interaction) relates velocity changes to changes in thermodynamic variables. The assumption of incompressible fluid allows one to state that the internal energy change and the density change can be neglected. Equation 2–8 then reduces to the Bernoulli equation (see eq. 12–99), which is an integrated momentum equation and relates static pressure and flow velocity. The definition of the total pressure for incompressible flow follows directly. Further insight into the total pressure of an incompressible flow may be found in Section 14.2.2.

Along the center line, up to the disk, the total pressure is conserved:

$$p_{t1} = p_1 + \tfrac{1}{2}\rho\, u_1^2 = p_{u1} + \tfrac{1}{2}\rho(u_1 + \Delta u_D)^2 \tag{13–1}$$

where Δu_D is the velocity increment experienced in reaching the pressure p_{u1} just ahead of the disk and the other quantities are as noted in Fig. 13–3. Similarly, downstream of the disk

$$p_{t2} = [p_{u1} + \Delta p] + \tfrac{1}{2}\rho(u_1 + \Delta u_D)^2 = p_2 + \tfrac{1}{2}\rho u_2^2 = p_1 + \tfrac{1}{2}\rho u_2^2 \tag{13–2}$$

At $x = +\infty$, the static pressure p_2 is uniform and equals p_1. The final velocity increase, Δu_∞, is given by

$$u_2 = u_1 + \Delta u_\infty \tag{13–3}$$

The change in static pressure across the disk must equal the change in total pressure since the velocity is continuous. Thus using eqs. 13–1 through 13–3, the pressure change is

$$\Delta p = p_{t2} - p_{t1} = [p_1 + \tfrac{1}{2}\rho(u_1 + \Delta u_\infty)^2] - [p_1 + \tfrac{1}{2}\rho u_1^2] = \rho\,\Delta u_\infty\left(u_1 + \frac{\Delta u_\infty}{2}\right) \tag{13–4}$$

The thrust force is therefore

$$F = \Delta p A \qquad \text{where } A = \frac{\pi D^2}{4} \qquad \text{or} \qquad F = \rho\,\Delta u_\infty A\left(u_1 + \frac{\Delta u_\infty}{2}\right) \tag{13–5}$$

From a momentum balance applied to the jet streamtube, the net thrust force must equal the rate of increase in fluid momentum:

$$F = \dot{m}(u_2 - u_1) = \dot{m}\,\Delta u_\infty = \rho(u_1 + \Delta u_{\rm D})A\,\Delta u_\infty \tag{13–6}$$

A comparison of eqs. 13–5 and 13–6 leads to the conclusion that half of the jet velocity increase occurs at the actuator disk:

$$\Delta u_{\rm D} = \frac{\Delta u_\infty}{2} \tag{13–7}$$

The remainder of Section 13.1 is concerned primarily with production of propulsive thrust. The task of power extraction from the wind shares similarities with the aerodynamics described in this section but is described separately in Section 13.2.

13.1.2. *Propulsive Thrust and Efficiency*

The useful power derived from creating the jet is the product of thrust times the velocity with which this force is moved through the medium. Thus,

$$\text{Useful power} = F u_1 = \dot{m}(u_2 - u_1)u_1 \tag{13–8}$$

The power required equals this power plus the power invested in jet kinetic energy (relative to the propulsion system!) or,

$$\text{Required power} = F u_1 + \dot{m}\tfrac{1}{2}(u_2 - u_1)^2 \tag{13–9}$$

This quantity is also the power invested in the increase in flow kinetic energy (in the absolute reference system), that is,

$$\text{Required power} = \dot{m}\tfrac{1}{2}(u_2^2 - u_1^2)$$

as can be shown using eqs. 13–8 and 13–9. The ratio of useful to required power is the propulsive efficiency:

$$\frac{\text{Useful power}}{\text{Required power}} = \eta_p = \frac{2}{1 + \dfrac{u_2}{u_1}} = \frac{1}{1 + \dfrac{\Delta u_\infty}{2u_1}} \qquad (13\text{–}10)$$

The useful and general conclusion to be drawn from the definition of propulsive efficiency is that the ideal situation where no power is wasted is realized when Δu_∞ is very small which requires a very large disk area for finite thrust (eq. 13–6). Conversely, a small (i.e., in diameter and mass flow rate) high-speed jet to produce thrust is inherently inefficient.

For a more accurate description of propulsive efficiency, other losses such as slipstream rotation, drag losses, thrust nonuniformity, blade interference, and compressibility effects must be considered in addition to the jet kinetic energy loss described by eq. 13–9.

13.1.3. *Axial Velocity Variation*

The length scale for axial variation of velocity as shown in Fig. 13–3 may be determined from an examination of the flow field induced by a semi-infinite circular cylinder of vorticity emanating from the tip of the rotor blades (see Fig. 13–14). Using ref. 13–3 for example, the axial velocity is found to be given by integrating the Biot-Savart law for this geometry. The result for the axial velocity ahead of an actuator disk is

$$u(x) = u_1 + \Delta u_D\left(1 - \frac{\xi}{\sqrt{1 + \xi^2}}\right) \qquad \text{where } \xi \equiv \frac{x}{R} \qquad (13\text{–}11)$$

Here the disk is at x or $\xi = 0$. The increment (divided by Δu_D) is shown in Fig. 13–4. Evidently, the distance over which a significant influence of the disk is felt is of the order of two radii. Compare this result with eq. 14–67 in connection with similar effects in axial flow compressors.

13.1.4. *Blade Element Analysis*

The real blades that make up the actuator disk experience forces that include drag. The objective of blade element analysis is to determine the impact of drag on efficiency. Consider a radial propeller blade rotating about an axis as shown in an isometric sketch on the right of Fig. 13–5. A view along the blade element showing the relevant velocities and forces is also shown.

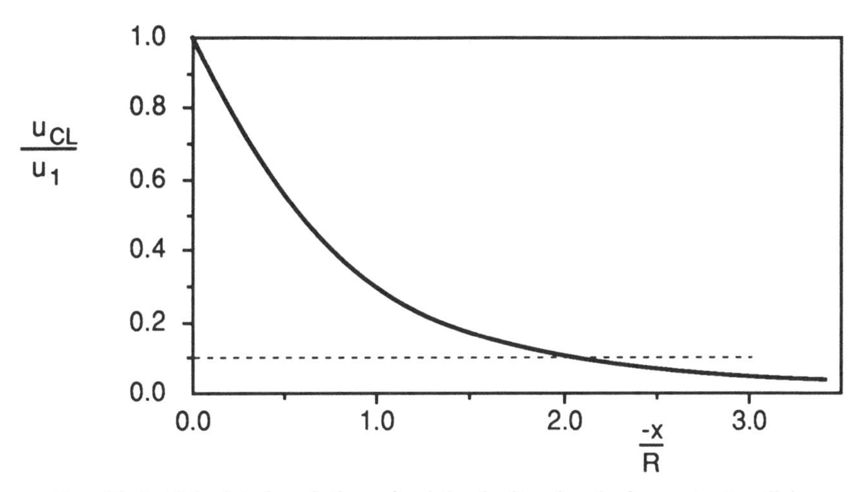

FIG. 13–4. Calculated variation of axial velocity ahead of an actuator disk.

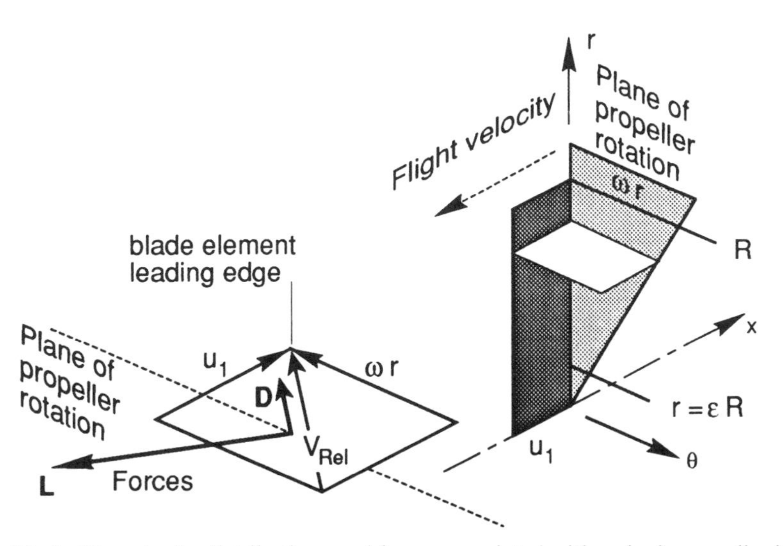

FIG. 13–5. The velocity distributions and forces associated with a single propeller blade. The sketch on the left is an expansion of the x-θ plane shown in white.

A propeller blade is similar to a wing except that the remote wind strikes the typical blade element at a radial location, r, with an angle that depends on the vector sum of aircraft forward speed, u_1, and the blade's rotational speed, ωr. The magnitude of the relative wind velocity is

$$V_{\text{REL}} = \sqrt{(\omega r)^2 + u_1^2} \tag{13–12}$$

The lift (L) and drag (D) forces on the typical element are defined as normal and parallel to the relative wind, respectively, as noted in the figure. As the

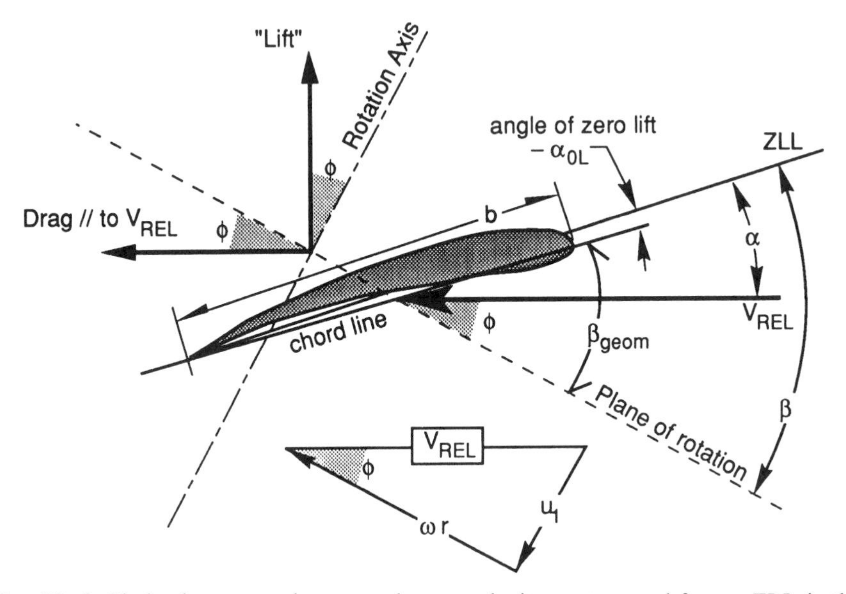

FIG. 13–6. Blade element angle nomenclature, velocity vectors and forces. ZLL is the zero lift line and α_{0L} is the angle of zero lift. Note alignment of the forces in relation to relative velocity.

propeller blade moves through the air, each point on it traces a helical path. The blade section is generally a cambered airfoil which has a chord line and a zero lift line (ZLL). The ZLL is the line orientation (in the z-θ plane) defined by the condition that the airfoil section generates zero lift when the local wind (V_{REL}) and ZLL are parallel. Figure 13–6 shows the airfoil section, its chord, the ZLL, and the plane of rotation. Several angles are also shown: α_{0L} (angle of zero lift), α (angle of attack), and the configuration angles β_{geom} (chord to rotation plane) as well as β (ZLL to rotation plane).

Figure 13–7 shows a view radially inward along a propeller blade. Two airfoil cross-sections are shown, one near the tip and another near the root. In order for the blade chord to be aligned with the local wind velocity, twist of the blade is required. This twist is described by the variation of $\beta(r)$, the angle between the plane of rotation, and the line of zero lift. This twist has a geometric component consisting of the orientation of the blade chord (β_{geom}) and an aerodynamic component due to the variation of the camber and thus α_{0L}, as shown in Fig. 13–6. Thus

$$\beta = \beta_{geom} - \alpha_{0L} \tag{13–13}$$

The lift per unit span is proportional to the angle between the relative wind and the ZLL. This force must be expressed in terms of propeller geometry parameters in order to be able to carry out an integration for thrust. The local lift can be written in terms of angles that describe the blade orientation (β, the angle between the ZLL and the plane of rotation) and the flow

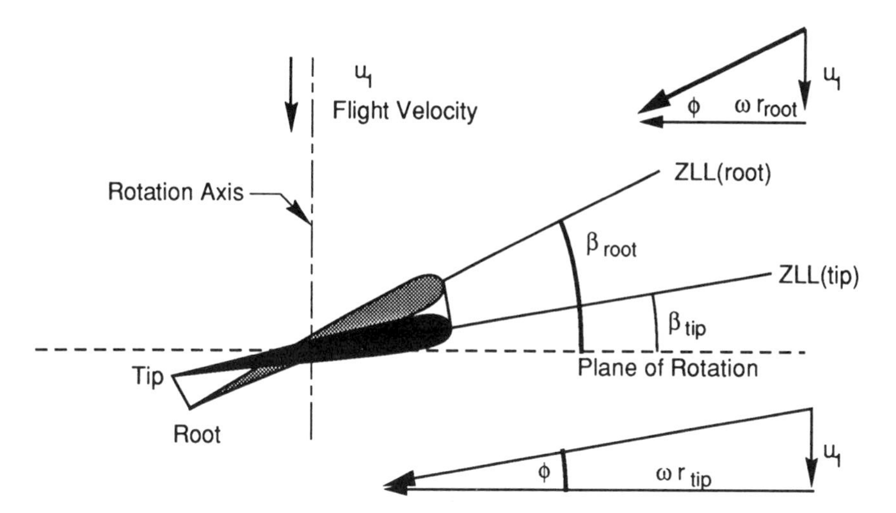

FIG. 13–7. End view of a propeller blade, showing relative wind directions at root and tip and the blade twist.

orientation (ϕ, the effective pitch angle), as

$$c_l = m\alpha = m(\beta - \phi) \tag{13–14}$$

The constant m is the airfoil section lift curve slope, which is close to its ideal value $m_0 = 2\pi$ in many cases. The angle $\phi = \phi(r)$ is given by the combination of flight velocity and angular speed:

$$\phi = \tan^{-1}\left(\frac{u_1}{\omega r}\right) = \beta - \alpha = \tan^{-1}\left(\frac{J}{\pi}\frac{R}{r}\right) \tag{13–15}$$

where the *advance ratio*, $J = u_1/ND$ describes the blade velocity field as a whole. $D(=2R)$ is the propeller disk diameter, and $N(=\omega/2\pi)$ is the angular speed expressed as rotations (rather than radians) per unit time. In eq. 13–15 all angles are functions of radial position.

Blade element theory uses the assumption that the velocity induced by the trailing vortex system is negligible. The effect is to ignore the rotation of the relative wind vector by the trailing vortex system emanating from the tips of the blade as well as others. This effect is considered in Section 13.1.7.

From Fig. 13–6, the differential force contribution parallel to the flight velocity has two vector components, one due to the "lift" and another due to the drag:

$$dF = dL\cos\phi - dD\sin\phi = q_{\mathrm{REL}}\, b(r)\, dr(c_l\cos\phi - c_d\sin\phi) \tag{13–16}$$

Here the chord width is $b(r)$, and the dynamic pressure seen by the element is:

$$q_{\mathrm{REL}} = \tfrac{1}{2}\rho V_{\mathrm{REL}}^2 \tag{13–17}$$

Similarly, the differential element of torque that must be applied is

$$dQ = r(dL \sin \phi + dD \cos \phi) = q_{\mathrm{REL}}\, rb(r)\, dr(c_l \sin \phi + c_d \cos \phi) \quad (13\text{–}18)$$

A more convenient form of these expressions is to use the dynamic pressure associated with the flight velocity, q. Thus

$$V_{\mathrm{REL}} = \frac{u_1}{\sin \phi} \quad \text{or} \quad q = \tfrac{1}{2}\rho u_1^2 = q_{\mathrm{REL}}(\sin^2 \phi) \quad (13\text{–}19)$$

Thus

$$dF = \frac{qb(r)\, dr}{\sin^2 \phi}(c_l \cos \phi - c_d \sin \phi) \quad (13\text{–}20)$$

and

$$dQ = \frac{qrb(r)\, dr}{\sin^2 \phi}(c_l \sin \phi + c_d \cos \phi) \quad (13\text{–}21)$$

For propeller with B (independent) blades, the thrust and the torque are:

$$F = qB \int_{\varepsilon R}^{R} (c_l \cos \phi - c_d \sin \phi) \frac{b(r)\, dr}{\sin^2 \phi} \quad (13\text{–}22)$$

$$Q = qB \int_{\varepsilon R}^{R} (c_l \sin \phi + c_d \cos \phi) \frac{rb(r)\, dr}{\sin^2 \phi} \quad (13\text{–}23)$$

The parameter ε is the fractional distance from the rotational axis center line to the innermost radial location of the blade. Thrust and torque are non-dimensionalized by the definitions of thrust and torque coefficients:

$$C_{\mathrm{F}} \equiv \frac{F}{\rho N^2 D^4} \quad \text{and} \quad C_{\mathrm{Q}} \equiv \frac{Q}{\rho N^2 D^5} \quad (13\text{–}24)$$

where the quantities in the denominator carry the proper units. Normally, aerodynamicists would use $q(\pi/4)D^2$ for force, $q(\pi/4)D^2 D/2 = q(\pi/8)D^3$ for torque and the angular speed, ω. Tradition has, however, evolved to use the shaft speed of the power plant (usually stated in RPM (rev/min), although used here in rev/sec), that is,

$$N = \frac{\omega}{2\pi} \quad (13\text{–}25)$$

In the industry, common practice is also found where a power loading defined as shaft power/D^2 is used in lieu of C_{Q} to which this parameter is directly related. The efficiency of the propeller as a thrust producing device is

$$\eta_{\mathrm{prop}} = \frac{\text{Thrust power}}{\text{Input power}} = \frac{Fu_1}{Q\omega} = \frac{C_{\mathrm{F}}}{C_{\mathrm{Q}}} \frac{J}{2\pi} \quad (13\text{–}26)$$

where the coefficients follow from their definitions,

$$C_{\mathrm{F}} = \frac{1}{2}\frac{BJ^2}{D^2}\int_{\varepsilon R}^{R}(c_1\cos\phi - c_d\sin\phi)\frac{b(r)\,dr}{\sin^2\phi} \qquad (13\text{–}27)$$

$$C_{\mathrm{Q}} = \frac{1}{2}\frac{BJ^2}{D^3}\int_{\varepsilon R}^{R}(c_1\sin\phi + c_d\cos\phi)\frac{rb(r)\,dr}{\sin^2\phi} \qquad (13\text{–}28)$$

The ratio $D^2/4\int b(r)\,dr$ in eq. 13–27 is a classical aspect ratio, AR, associated with wing theory. The efficiency is, in general,

$$\eta_{\text{prop}} = j\,\frac{\displaystyle\int_{\varepsilon}^{1}\left(\frac{c_1}{\tan\phi} - c_d\right)\frac{b(y)\,dy}{\sin\phi}}{\displaystyle\int_{\varepsilon}^{1}\left(c_1 + \frac{c_d}{\tan\phi}\right)\frac{b(y)y\,dy}{\sin\phi}} \qquad (13\text{–}29)$$

where y, and j are the nondimensional radius ratio and a modified advance ratio defined by

$$y \equiv \frac{r}{R} \qquad \text{and} \qquad j \equiv \frac{J}{\pi} = \frac{u_1}{\omega R} \qquad (13\text{–}30)$$

For simplicity in the resulting equations, this new definition of advance ratio (j) is more logical than J.

Insight into the parameters influencing the propeller efficiency may be gained by combining eqs. 13–26 through 13–29 with a few simplifying assumptions. With ϕ given by eq. 13–15 or Fig. 13–6, the trigonometric functions are

$$\tan\phi = \frac{j}{y} \qquad \text{and} \qquad \frac{1}{\sin\phi} = \frac{\sqrt{y^2 + j^2}}{j} \qquad (13\text{–}31)$$

First, from eq. 13–26, j is of the order of 0.5 in the limit where the drag coefficient is very small. Further, one may estimate the value of nondimensional C_{F} using eq. 13–27, the numerical magnitude is of the order of

$$C_{\mathrm{F}} \approx 2\pi^2\frac{j^2}{AR}\frac{\bar{c}_1}{\sin\phi\tan\phi} \approx \frac{2\pi^2}{AR}\bar{c}_1\bar{y}\sqrt{j^2+\bar{y}^2} \approx \frac{210}{15}(0.7)(0.75)^2 \sim 0.5$$

for $AR = 15$, a mean value of $y = 0.75$, and a small value of j.

Second, to illustrate the role of the parameter, j, in controlling performance, one may simplify the expression for efficiency by taking the blade to be nearly rectangular. $b(y)$ is then a constant and cancels out of eq. 13–29. Further, let the airfoil geometry be twisted in such a way as to have each section of the blade operate at maximum L/D at a chosen, design, value of the advance ratio. This allows one to see how the drag influences performance at conditions away from the design condition. The design value of j may be defined as j_D, so that the blade geometry is given by (eq. 13–15)

$$\beta(y) = \alpha_D + \tan^{-1}\left(\frac{j_D}{y}\right) \tag{13–32}$$

where α_D is given by a design constraint such as the maximum L/D. The value of j_D is associated with the limitations imposed by compressibility effects on the airfoil (see Section 13.1.9). At any condition characterized by value of j, the angle of attack is given by eq. 13–32 applied at the design condition and again at off-design with a constant geometric orientation $\beta(y)$:

$$\alpha = \beta(y) - \tan^{-1}\left(\frac{j}{y}\right) = \alpha_D + \tan^{-1}\left(\frac{j_D}{y}\right) - \tan^{-1}\left(\frac{j}{y}\right) \tag{13–33}$$

In the operating region below stall, the lift coefficient is

$$c_l = m_0\alpha \cong 2\pi\alpha \tag{13–34}$$

and the drag coefficient is approximately

$$c_d = c_{d0}(1 + ac_l^2) \tag{13–35}$$

where a is a characteristic of the particular airfoil section. In practice, a varies between 1 and 3. By forming the ratio of c_l/c_d and differentiating to find maximum L/D, one finds

$$c_{l,\max L/D} = \sqrt{\frac{1}{a}} \quad \text{and} \quad \left(\frac{c_l}{c_d}\right)_{\max} = \frac{1}{2c_{d0}}\sqrt{\frac{1}{a}} \tag{13–36}$$

The propeller efficiency can be evaluated using eq. 13–29 and the expression for c_l, c_d, and α as well as the trigonometric function of ϕ required. The result must be carried out numerically, and Fig. 13–8 shows such a calculation for inner radius $\varepsilon = 0.05$ and $(c_l/c_d)_{\max} = 20$. The plot shows that a relatively narrow range of j is allowed to keep the airfoil tip unstalled ($c_{l,\max} < 1.2$, say). This occurs for $j < j_D$. For j larger than design, the propeller c_l decreases and the propeller tends toward lighter loading until the tip produces negative thrust. The c_l distribution is shown in Fig. 13–9 for the conditions noted.

Note that the influence of a is relatively small. The best efficiency is seen to occur at j somewhat less than j_D. In other words, maximum section L/D does not lead to maximum thrust efficiency, although it is close. From

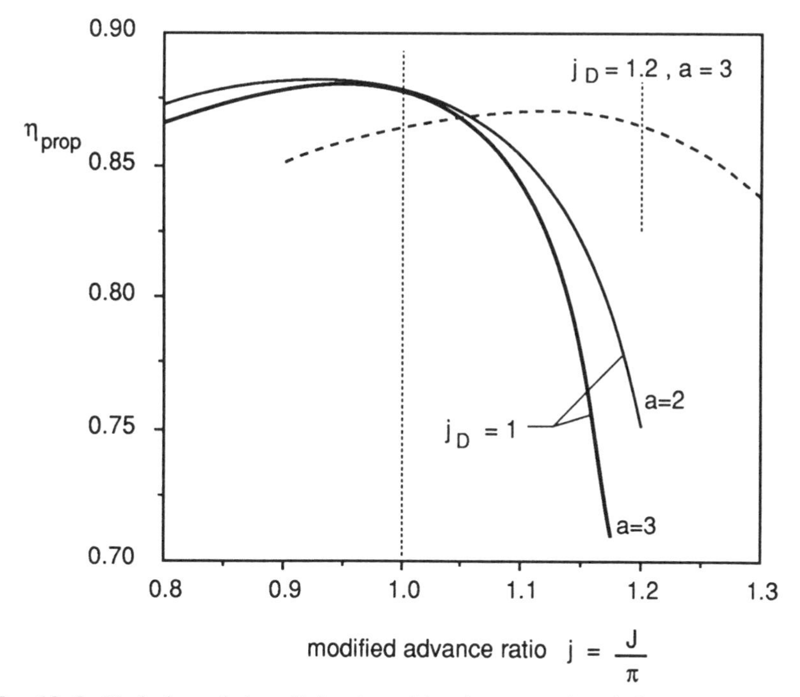

FIG. 13–8. Variation of the efficiencies with advance ratio of three propeller designs characterized by choice of j_D. Lower drag airfoil ($a = 2$) has a higher performance.

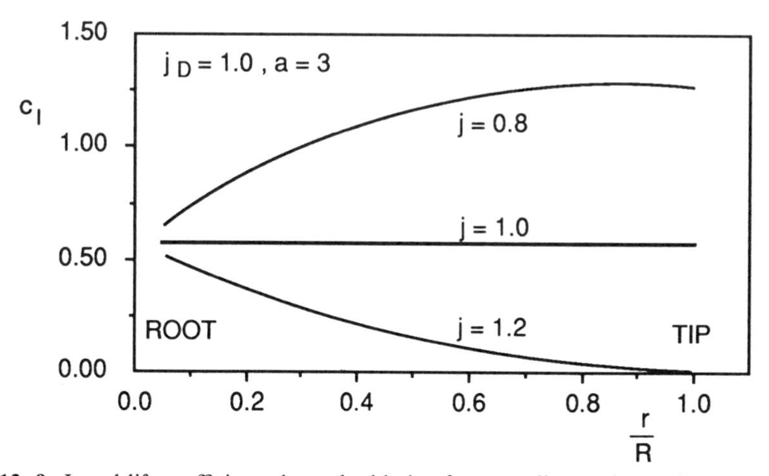

FIG. 13–9. Local lift coefficient along the blade of a propeller at three advance ratios.

additional calculations with fixed a, it may be shown that

$$\eta_{prop} \cong 1 - 2\left(\frac{c_l}{c_d}\right)_{max}^{-1} \tag{13–37}$$

This approximate result is shown in Fig. 13–10 together with more exact calculational results from the theory described here.

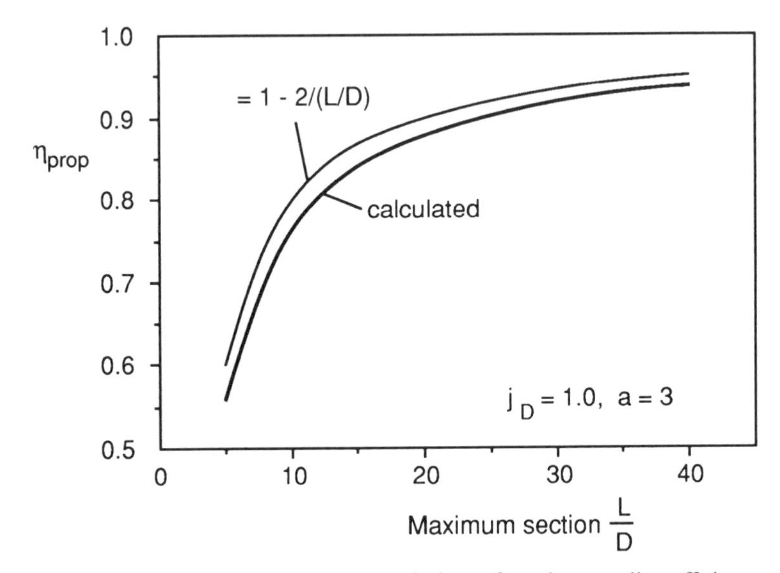

Fig. 13–10. Calculated and approximate variation of peak propeller efficiency with maximum L/D for a family of propellers characterized by the (typical) value of a noted.

13.1.5. *Computer Analysis of a Propeller of Specified Geometry*

The following procedure is used for a complete calculation of C_F, C_Q and the propeller efficiency for a given blade geometry and operating condition limited to the accuracy implied by the analysis described above.

Blade geometry, b(r), R, εR, and number of blades, B, are assumed known.

1. Fix the angle $\beta_{geom}(r)$.
2. The specified variation of zero lift angle of attack, $\alpha_{0L}(r)$

$$\text{gives } \beta(r) = \beta_{geom}(r) - \alpha_{0L}(\text{r}) \text{ [eq. 13–13]}.$$

4. Chosen u_1 and N give J from its definition and $\phi(r)$ from eq. 13–15,
5. $\alpha(r) = \beta(r) - \phi(r)$.
6. Obtain c_l and c_d from section data for $\alpha(r)$.
7. Obtain thrust and torque coefficients using eqs. 13–27 and 13–28.
8. η_{prop} follows from eq. 13–26.

For a typical fixed pitch propeller, the results of such a calculation are as shown in Fig. 13–11. Note that thrust rises to a maximum and then falls toward zero when the lift and drag terms become equal in magnitude. The torque is always non-zero due to the ever present drag. The efficiency is zero at low J (no work is done at zero forward velocity), peaks to a maximum, and falls to zero when the propeller "windmills," that is, is driven by the wind.

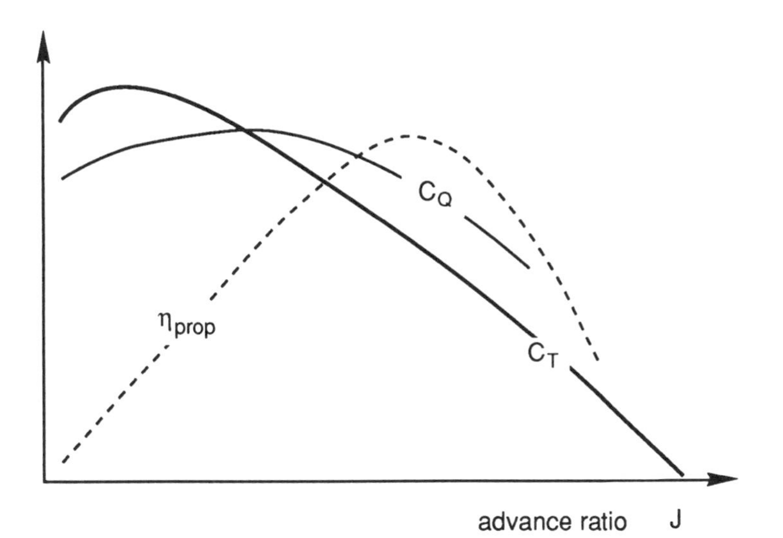

FIG. 13–11. Sketch of thrust and torque coefficients and propeller efficiency for a fixed-pitch propeller.

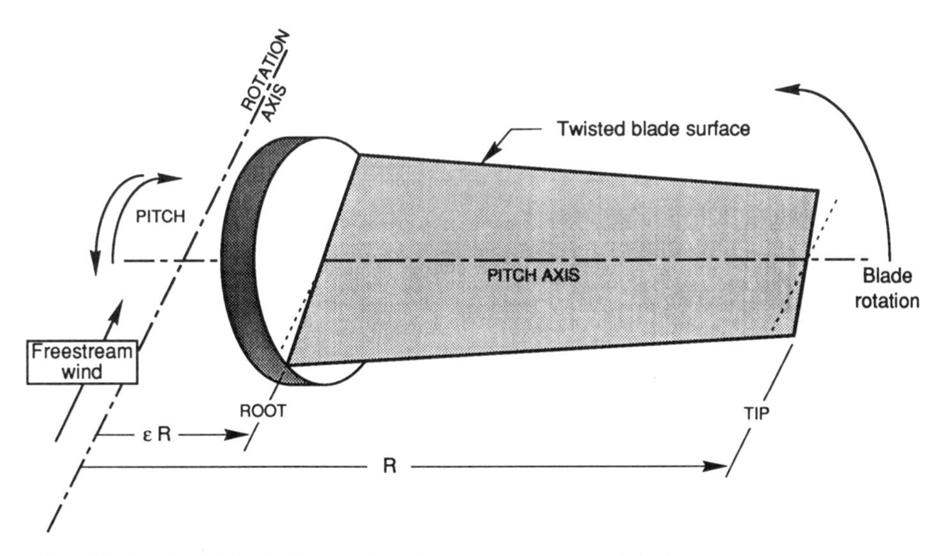

FIG. 13–12. Variable-pitch propeller showing rotation and pitch axes. Freestream wind direction is parallel to the rotation axis.

13.1.6. *Variable Pitch Propellers*

Propellers where the blade can be rotated so that β is matched to the operating condition are called variable-pitch propellers. The orientation angle at the location $r = 0.75R$ is often used as a measure of the blade orientation as a whole. Figure 13–12 shows a blade of a variable-pitch propeller and the variation of β as pitch is changed. Figure 13–13 shows the performance of such

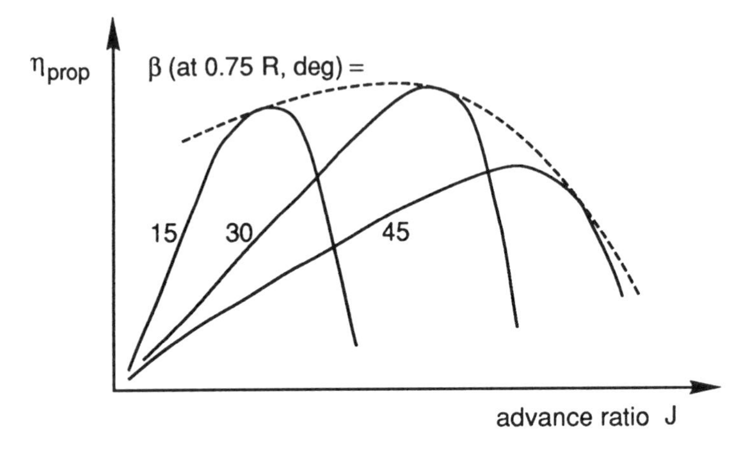

Fig. 13–13. Propeller efficiency for a variable pitch propeller at various overall incidences identified by $\beta(0.75)$. Control mechanism should seek maximum efficiency at all conditions.

a propeller as calculated by the methods above but with differing values of $\beta(0.75)$. A control mechanism should seek to maintain high efficiency at changing flight conditions as characterized by the advance ratio, J.

13.1.7. *Vortex Propeller Theory*

The analysis described in Section 13.1.4 includes the assumption that the airfoil section operates as it if were isolated in the free-stream. The actuator disk theory has shown that the overall effect of the trailing vortex sheet from the blade tip vortices is to induce a velocity increase at the actuator disk plane. For propellers which produce large thrust, it is evident that this induced flow plays a role in determining the local flow environment and thus alters the forces on the blade elements. A first order correction for this effect is to combine the results of the actuator disk theory and include them in a force calculation as described in the previous section.

The effective pitch angle ϕ must account for the induced velocity added to u_1. Thus eq. 13–15 becomes

$$\phi = \tan^{-1}\left(\frac{u_1 + \Delta u_D}{\omega r}\right) = \beta - \alpha = \tan^{-1}\left(\frac{J}{\pi}\frac{R}{r}\left(1 + \frac{\Delta u_D}{u_1}\right)\right) \qquad (13\text{–}38)$$

where the correction term $\Delta u_D/u_1$ is calculated from eq. 13–6 with $\Delta u_D = \Delta u_\infty/2$. The thrust is required to be known and the procedure outlined in Section 13.1.4 becomes iterative in the sense that thrust must be known before it can be calculated.

To carry out the calculation of the performance of a propeller with significant loading, one proceeds by solving for the velocity ratio $\Delta u_D/u_1$ using eq. 13–6, where it appears as a quadratic, and the definition of thrust coefficient (eq. 13–24). Equating the thrusts in these relations, one obtains,

$$\frac{\Delta u_D}{u_1} \equiv \psi_d = \sqrt{C + \frac{1}{4}} - \frac{1}{2} \tag{13–39}$$

where C is the combination of parameters

$$C \equiv \frac{2}{\pi}\frac{C_F}{J^2} = \frac{2}{\pi^3}\frac{C_F}{j^2}$$

Note that for small C, it equals ψ_d. Here C_F is computed from eq. 13–27 with a modification that the change from relative q to one based on far field freestream introduces the $(1 + \psi_d)^2$ term as shown below:

$$C_F = \frac{(1 + \psi_d)^2}{2}\frac{BJ^2}{D^2}\int_{\varepsilon R}^{R} (c_l \cos\phi - c_d \sin\phi)\frac{b(r)\,dr}{\sin^2\phi} \tag{13–40}$$

The solution procedure involves estimating the induced velocity ratio, ψ_d, and correcting the thrust coefficient accordingly. The calculational procedure is outlined below.

13.1.8. *Vortex Theory Implementation on a Computer*

The following quantities are assumed known or specified

$\beta(y)$ orientation of the ZLL relative to plane of rotation

$W(y) = \dfrac{b(y)}{b(0.75)}$, radial chord variation

$AR' = \dfrac{R}{b(0.75)}$, aspect ratio of the blade

This definition is more restricted than the expression associated with eq. 13–27.

$\varepsilon = r(\min)/R =$ minimum y

$m_0 =$ lift curve slope

a and $c_{d,\min} =$ constants in expression for section drag coefficient, $c_d = c_{d0}(1 + ac_l^2)$

Procedure

1. Assume ψ_d (this value is the basis for iteration convergence)

$$C = \psi_d + \psi_d^2$$

2. For an array of y from ε to 1, calculate $\phi = \tan^{-1}\left(\frac{j}{y}(1+\psi_d)\right)$

$$\alpha(y) = \beta(y) - \phi(y)$$

$$c_l = m_0\alpha$$

$$c_d = c_{d0}(1 + ac_l^2) \text{ or, from tabular data,}$$

$$C = \frac{B}{4\pi}\frac{(1+\psi_d)^2}{\text{AR}}\int_\varepsilon^1\left(\frac{c_l}{\tan\phi} - c_d\right)\frac{W(y)\,dy}{\sin\phi}$$

$$\psi_d = \sqrt{C + \frac{1}{4}} - \frac{1}{2}$$

3. Correct the value of ψ_d and, when satisfied,

$$C_F = \frac{\pi}{2}CJ^2 = \frac{\pi^3}{2}Cj^2$$

$$C_Q = \frac{BJ^2}{16}\frac{(1+\psi_d)^2}{\text{AR}}\int_\varepsilon^1\left(c_l + \frac{c_d}{\tan\phi}\right)\frac{W(y)y\,dy}{\sin\phi}$$

$$\eta_{\text{prop}} = \frac{C_F}{C_Q}\frac{\text{j}}{2}$$

The validity of results from such a calculation is limited to that range of α, where the lift and drag coefficients are described by the equations given. The choice of advance ratio, j, is limited by the compressibility effects. For example, if the blade airfoil section has a critical Mach number, M_{crit}, where drag is seen to increase as shown in Fig. 13–14, and the propeller tip is set to operate at that Mach number, then the choice of flight Mach number fixes j. The velocity triangle used to define ϕ gives the advance ratio for this condition as a function of flight Mach number, M_0:

$$j_{\text{D, critical tip}} = \left(\left(\frac{M_{\text{crit}}}{M_0}\right)^2 - 1\right)^{-1/2}$$

The value of $j_{\text{D, crit tip}}$ is 0.58 [or 1] for $M_{\text{crit}} = 0.7$ and $M_0 = 0.35$ [or 0.5]. This relation is critical for the choice of a design value of j(i.e., j_D).

As a connection between the actuator disk model and the vortex theory, it is worthy of note that the jet may be visualized to result from the induced velocity from all the trailing "wing tip" vortices (Fig. 13–15). When there are a finite number of such vortices such as those emanating from a two-, three- or more-bladed propeller, the vortex lines may be visualized as circular spirals pitched to the motion at an angle related to the advance ratio, hence the name reminiscent of machine screw terminology. On the other hand, an actuator disk model gives uniformly distributed vorticity with ringlike elements. One model

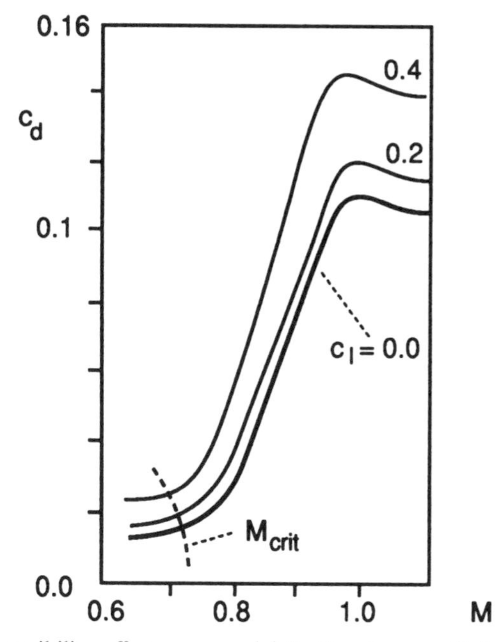

Fig. 13–14. Compressibility effects on an airfoil. The drag coefficient rises at critical local Mach number and leads to high drag and thus torque requirements.

is the derivative of the other. The semi-infinite cylindrical vortex sheet model is used to obtain the axial velocity variation given in eq. 13–11.

13.1.9. *Losses and High-Speed Effects*

Compressibility effects have a marked effect on performance when the individual blade elements, acting as wings, are exposed to a freestream velocity that approaches sonic speeds. Mach number effects become important when the "freestream" Mach number reaches the "critical" value. The critical Mach number of an airfoil is defined as that freestream M_∞ where the flow is at $M = 1$ somewhere in the field (usually on the surface) of the airfoil. The reason is that boundary-layer separation caused by compression waves leads to a drag increase (see Section 12.10.2). The magnitude of the critical Mach number depends strongly on the airfoil shape and on lift coefficient, (refs. 13–1 and 13–2). A typical variation of M_{crit} for an airfoil section is shown in Fig. 13–14. Consideration of compressibility effects on propellers leads to the following conclusions:

1. Drag results in a limitation on rotational speed. Tip speed must be less than near sonic.
2. The speed of the vehicle on which the propeller is used is also limited to speeds less than sonic.
3. Higher subsonic flight speed ($M_\infty \sim 0.7$) requires use of airfoil sections at the tip that specifically accommodate the flow compressibility. This

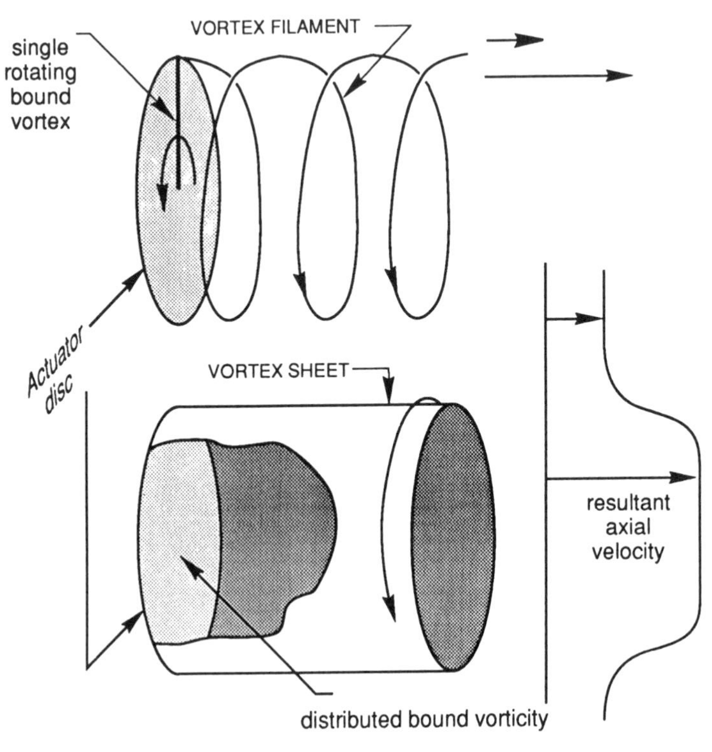

FIG. 13–15. Trailing vorticity (with vorticity in the tangential direction) as a jet boundary. The sheet is a closely spaced set of filaments. Both produce a jet as sketched on the right side of the figure.

limitation is circumvented through the use of a diffuser ahead of turbojet and turbofan engines (Fig. 12–3 and Section 12.10). Such diffusers allow commercial aircraft to fly at the critical Mach number of the wing, (typically between $0.7 > M_\infty > 0.9$) rather than a speed limited by the aerodynamic characteristics of the propulsion system.

13.2. POWER GENERATION FROM WIND ENERGY

The origin of wind power is solar power incident on the Earth (Fig. 13–16). This flux of energy drives both the global circulation patterns but also the local and diurnal winds (Fig. 13–17). From the viewpoint of energy conversion, those winds that are reliable and found where needed are of interest. The availability issue is the important element of cost: When availability is high, then the cost for conversion and storage may be made sufficiently low. Economic success of wind energy utilization is therefore site specific.

13.2.1. *Wind Characteristics*

Wind is the result of pressure gradients along the Earth's surface. The resulting air currents are variable in speed and direction. The air velocity is typically

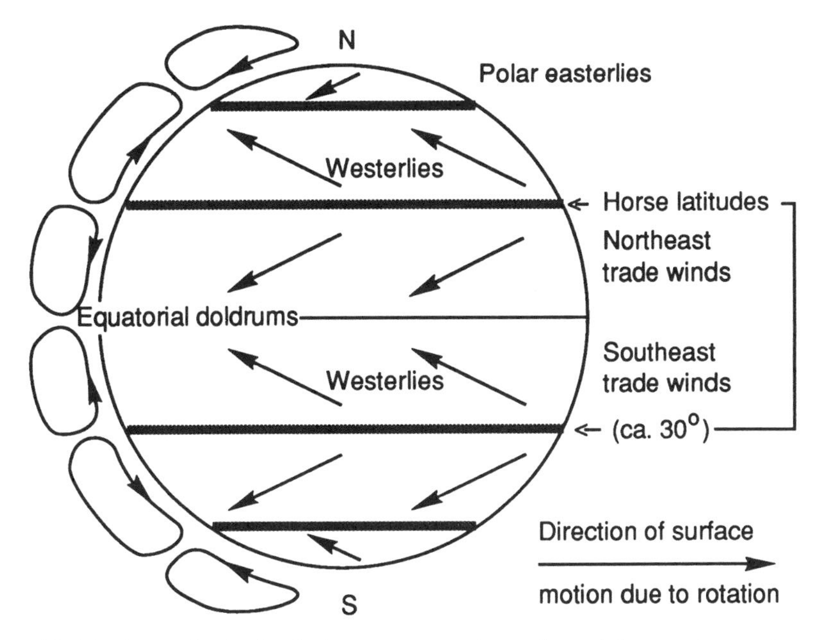

Fig. 13–16. Atmospheric circulation pattern on the surface of the rotating earth and its implications for the suitability of wind power at various locations. The contiguous United States is located largely between 30 and 50°N. The Earth's major deserts are located in the vicinity of the "horse" latitudes.

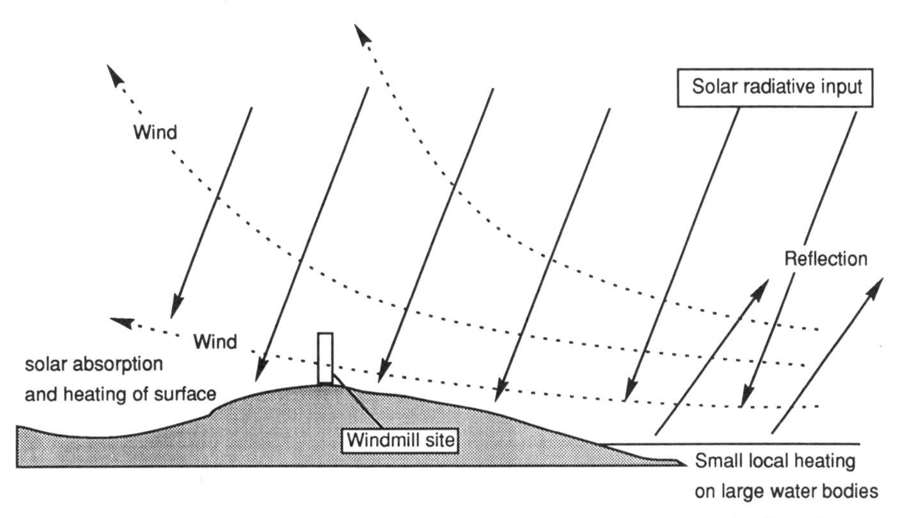

Fig. 13–17. Local preferential heating of a land mass relative to large bodies of water drives convection air currents, that is, on-shore wind during daytime and off-shore at night.

large enough for the Reynolds number to be descriptive of a turbulent flow. Over the ground surface, the air motion is describable as a boundary layer which means that the uniformity and strength of the wind depends on location, height above the ground, and size of the surface irregularities. In a typical location, the wind magnitude may be as shown in Fig. 13–18. The height of a

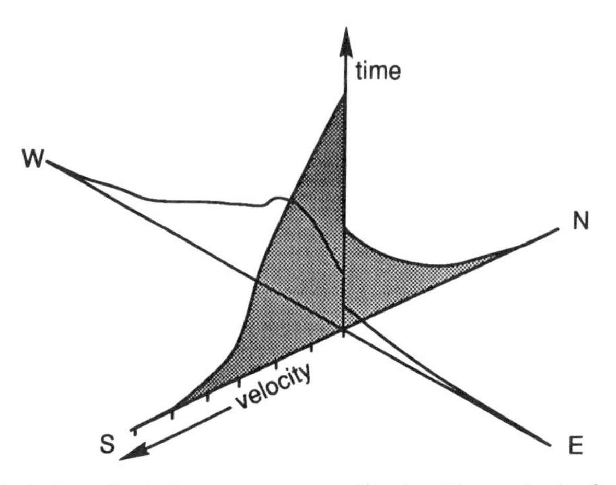

FIG. 13–18. Variation of wind power at a specific site. Shown is the fraction time (on the vertical axis) that the wind blows from the four cardinal directions and at what speed.

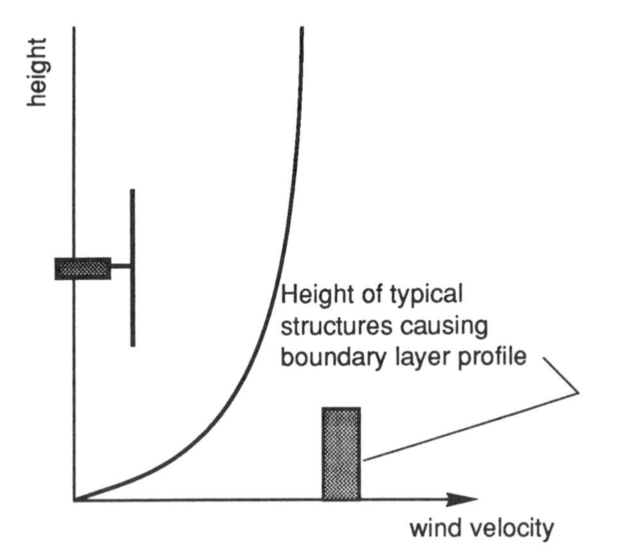

FIG. 13–19. Variation of (average) velocity in an atmospheric boundary layer.

wind turbine is important because larger velocities are experienced at larger heights, and a disk at higher elevation will experience more uniform inflow as it turns through a cycle (Fig. 13–19). The figure shows the boundary profile of the wind together with the relation between the surface roughness and the vertical extent of ground effects on the wind velocity distribution. Flow uniformity is important to help minimize cyclic structural stresses and thus fosters designs with larger average height-to-diameter ratios. On the other hand, higher elevated structures experience greater steady and unsteady loads and are subject to structural vibration which may result in increased cost for a durable design.

During times of little or no wind, there is a need by the typical user for stored

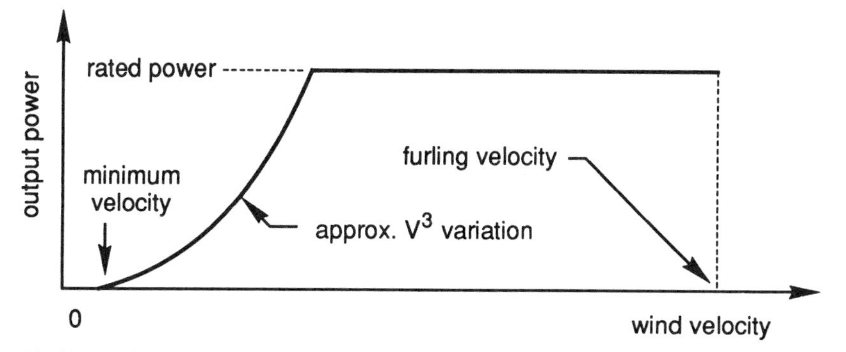

FIG. 13–20. Variation of power available to a wind turbine with increasing wind velocity.

energy, although in typical modern systems, wind-generated power may be tied into the utility grid which can serve as a backup power source. This tie-in requires not only matched voltage and phase of the AC power but also equitable arrangements for costs to be transferred between the user and the utility that serves as a power availability insurer. A way to utilize wind energy and possibly reduce the requirements for storage is to capture the energy in ocean waves. Special mechanisms must be employed to capture this form of wind energy. The reader is encouraged to consult the literature for a description of such mechanisms and estimates of the size of the resource.

All power systems are limited to that which can be accepted by the electrical generator. As wind velocity, u, increases, a minimum velocity is required to overcome frictional losses. With constant conversion efficiency, the power increases approximately with u^3 (see eq. 13–46 below) until the rated power is reached (Fig. 13–20). The turbine must then be detuned by a control system to limit the power input to the generator. Finally, above some maximum speed, the system is shut down to avoid structural or mechanical overload.

Wind characteristics may be summarized for a given location in a plot called the speed-duration curve for, say, a year's period. This curve is similar to the load-duration curve shown in Fig. 3–4 except that wind speed is plotted as a function of percent time duration. The derivative or slope of the speed duration curve with the axes inverted gives the speed-frequency curve. As an example, consider a wind speed-duration curve given by

$$\frac{t}{t_{\max}} = \exp(-u^4) \tag{13–41}$$

where u is the wind velocity. The speed-frequency curve is shown in Fig. 13–21 (labeled "f") obtained from

$$f(u) = -\frac{d\left(\dfrac{t}{t_{\max}}\right)}{du} \tag{13–42}$$

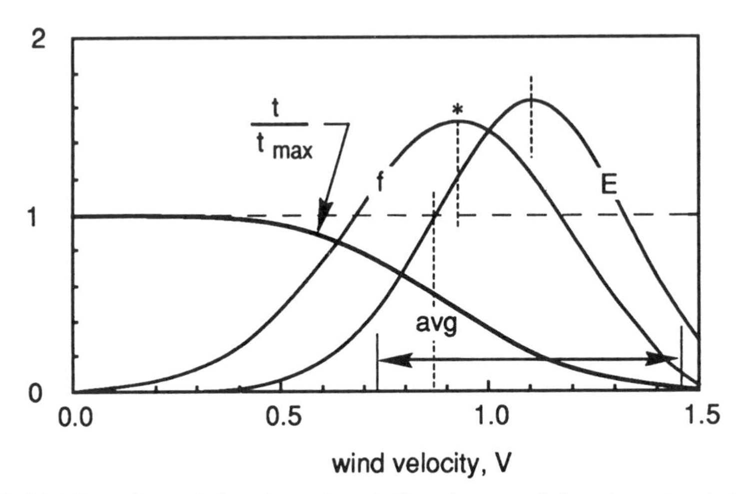

FIG. 13–21. Plot of speed functions f and E and a speed-duration plot (with axes inverted here for convenience). Average and peak velocity noted as well as range with significant power content.

The speed duration curve is also shown and labeled t/t_{max}, albeit with axes inverted. From the frequency, the most frequent speed (noted at peak f as *) and the mean speed can be determined from

$$\bar{u} = \int_0^\infty u f(u)\, du \qquad (13\text{–}43)$$

For the example, this average is close to unity on the nondimensional velocity scale and larger than the most frequent value. The energy content of the wind is obtained from the integral

$$\frac{\text{Energy during } t_{max}}{(\rho)\,(\text{disc area})} = \int_0^\infty u^3 f(u)\, du = \int_0^\infty E\, du \qquad (13\text{–}44)$$

since the atmospheric density can be assumed constant, a good approximation. For the distribution of velocities noted in the example, the integrand, E, an energy per unit speed range (eq. 13–44), is plotted in Fig. 13–21. The figure shows the velocities that have significant energy content and should form the design basis for the conversion system. The relatively low and high velocities do not contribute much to the total converted energy since this energy represented by the area under the curve near these ends (less than 0.75 or greater than 1.5 on this nondimensional velocity scale) is small.

13.2.2. *Wind Turbine Configurations*

Wind turbines are classified according to the orientation of the turbine axis. Horizontal axis machines are mounted on a tower together with the electric

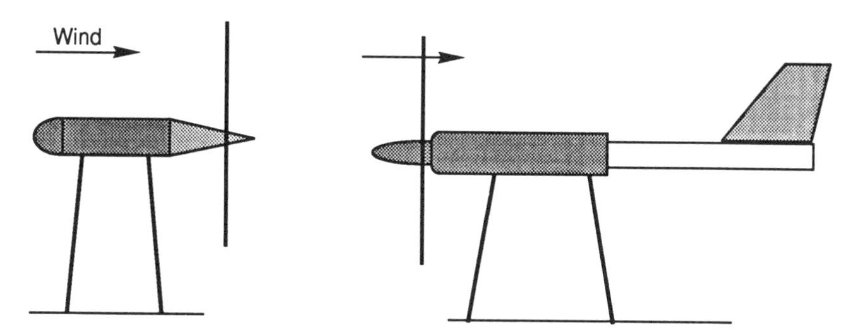

FIG. 13–22. "Pusher" (left) and "Tractor" horizontal axis turbines.

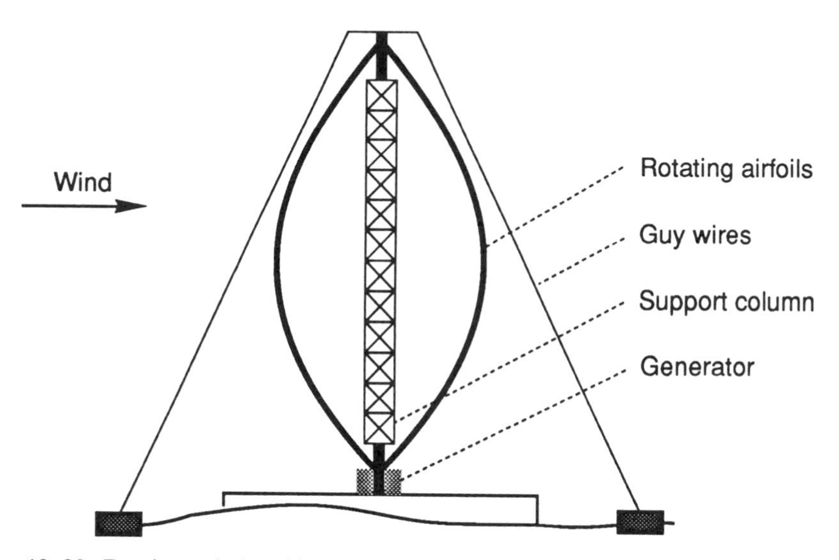

FIG. 13–23. Darrieus wind turbine showing the support mast and the two (symmetric and airfoil shaped) blades.

generator. Figure 13–22 shows two horizontal axis wind turbines (HAWT) differentiated on the basis of the relative position of the turbine and generator. The machinery is generally mounted on a rotating platform with a tail vane to point the turbine in the direction of the wind. Such machinery may or may not be responsive to wind direction changes. Inertia effects associated with turning about two axes may be important in controlling Coriolis forces on the turning blades. Most HAWTs share a common disadvantage that the mass of the electrical generator is most conveniently located on the tower. This leads to the possibility of costly support structure.

Vertical axis wind turbines (VAWT) are generally mounted with the electrical machinery on the ground. These machines are insensitive to wind direction. A form of this type is the Darrieus turbine shown in Fig. 13–23. Variations on this type are the gyromill (Fig. 13–24) and the Musgrove turbine, the latter being like the portion of the Darrieus rotor near the section midway between the top and bottom, supported by an arm. This allows a pitch control

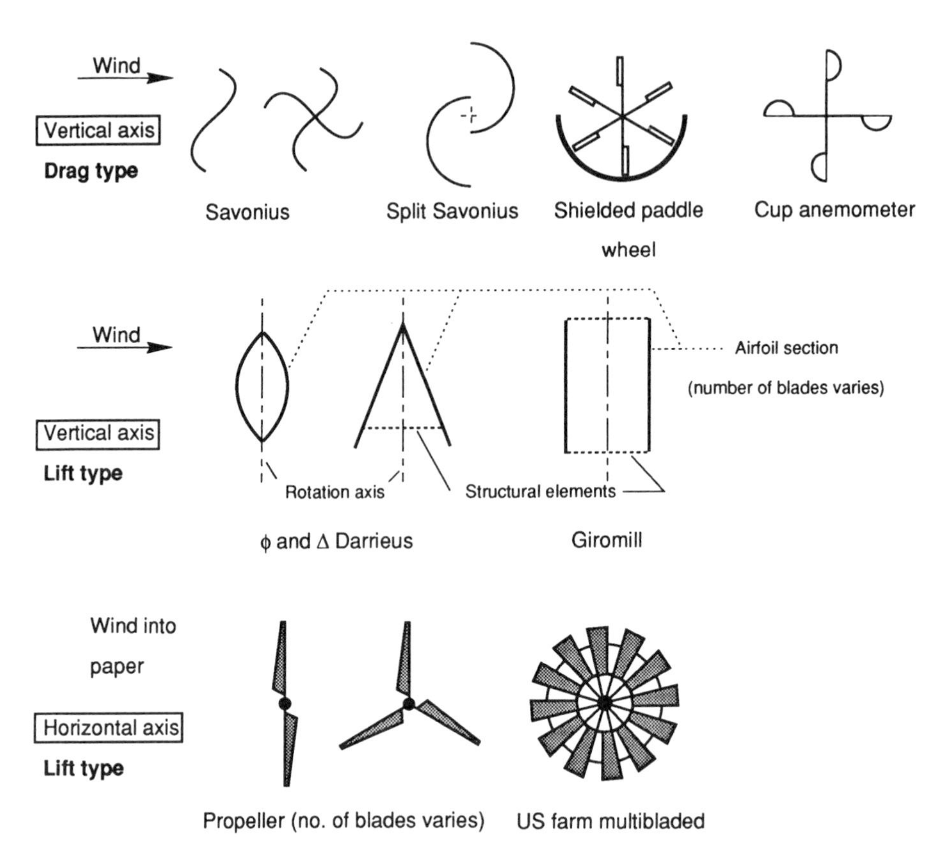

FIG. 13–24. Wind turbine configurations differentiated by axis orientation and by nature of force exerted on the active element: lift or drag. The lift machines are not self starting, hence may be used in combination with a Savonius rotor somewhere on the rotation axis.

mechanism to operate for control and optimization of performance. The authors of refs. 13–4 and 13–5 describe physical aspects of some of these machines. The detailed discussion in Section 13.2.4 is focused on the aerodynamic aspects of the Darrieus-like rotor which is descriptive of all vertical axis machines.

Performance

Wind turbines are also classified on the type of force used to interact with the wind: lift on an airfoil or drag. The simplest wind turbines are based on the drag force to produce torque. Figure 13–24 shows the Savonius rotor together with the cup anemometer and the drag plate reminiscent of a paddle wheel. A split Savonius rotor is also shown. These drag type devices, the Savonius rotor in particular, are of interest for applications where low-speed torque is desired. These machines necessarily operate with a low ratio of tip to wind speed. Other wind-power generation machines are also shown, classified according to axis orientation and to force type. Not included but worthy of mention are tracked

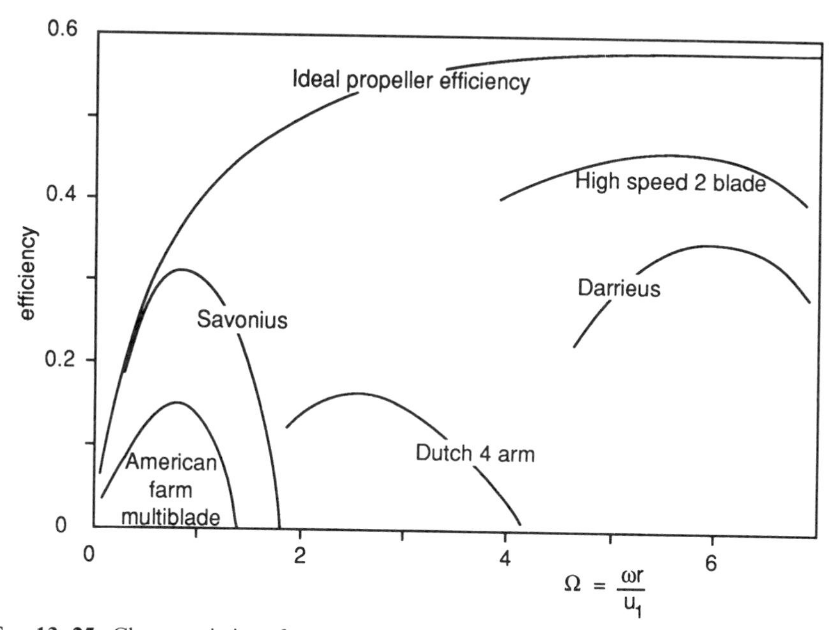

Fig. 13–25. Characteristics of a number of wind turbine types: efficiency as a function of speed ratio.

sail vehicles on the ground and Magnus effect machines, neither of which are in common use. The performance of drag type devices is considered in Section 13.3.

Lift-type turbines require higher rotational speeds to generate aerodynamic forces and generally have zero starting torque. Combinations like the Darrieus and the Savonius rotors may be built for self-starting capability. The performance of these types of devices is considered below, starting from the streamtube momentum principles developed above for the thrust producing propeller: the actuator disk.

The efficiency of wind turbines is traditionally described in terms of a power coefficient defined by

$$\text{power coefficient, } C_\text{p} \equiv \frac{\text{shaft power}}{\frac{1}{2}\rho u_1^3(\text{Area intercepted})} \tag{13–45}$$

To set the stage for the performance analysis to follow, the magnitudes of power coefficients or efficiencies of various wind turbine types found in practice are shown in Fig. 13–25 (from ref. 13–6, among others). The tip-to-wind speed ratio, like the advance ratio in the thrust-producing propeller, is the important parameter determining the efficiency of the various turbine types.

13.2.3. *Wind Turbine Efficiency: Betz Limit*

The actuator disk allows calculation of the rate of energy removal from the wind by applying the conservation equations developed in Section 13.1.2.

In contrast to the thrust-producing propeller, the velocity increments Δu_D and Δu_∞ are negative since u_1 must be greater than u_2. The momentum and energy conservation equations for the two cases are otherwise identical and hence apply here. The pressure change across the actuator disk cannot be larger than the total pressure of the incoming stream, which represents a limit to the work removal from the stream. In practice, the power removal is limited by a design balance in allowing the aerodynamic forces requiring motion to act, but not so strongly that flow is impeded.

Consider the power in the wind through a control area A, which is equal to the actuator disk area. There is, for the moment in the discussion, no propeller. The power in the wind is the product of mass flow rate times kinetic energy per unit mass, namely,

$$\dot{W}_a = (\rho u_1 A)(\tfrac{1}{2}u_1^2) = \tfrac{1}{2}\rho u_1^3 A \tag{13–46}$$

The power removed by the disk when it is in place is

$$\dot{W} = [\rho(u_1 + \Delta u_D)A]\left(\frac{u_1^2 - u_2^2}{2}\right) \tag{13–47}$$

which can be written in terms of only $(-\Delta u_\infty)$, the positive far-field velocity increment, using its definition and eq. 13–7:

$$\dot{W} = \tfrac{1}{2}\rho A\left(u_1 - \frac{-\Delta u_\infty}{2}\right)\{u_1^2 - [u_1 - (-\Delta u_\infty)]^2\}$$

$$= \tfrac{1}{2}\rho u_1^3 A\left(1 - \frac{\psi}{2}\right)(1 - (1 - \psi)^2) \tag{13–48}$$

with the parameter definition

$$\psi \equiv \frac{-\Delta u_\infty}{u_1} > 0 \tag{13–49}$$

An efficiency may be calculated based on the ratio of power extracted relative to power available using eqs. 13–46 and 13–48:

$$\eta_{WT} = \frac{\dot{W}}{\dot{W}_a} = \left(1 - \frac{\psi}{2}\right)[1 - (1 - \psi)^2] = \frac{\psi^3}{2} - 2\psi^2 + 2\psi \tag{13–50}$$

The change in total or static pressure (divided by $q_1 = (1/2)\rho u_1^2$) across the rotor disk is given by the last term in the parentheses of eq. 13–48. This ratio is plotted in Fig. 13–26 together with the wind turbine efficiency, η_{WT} (see also Fig. 13–25).

The maximum efficiency is obtained by a null derivative of eq. 13–50. The maximum efficiency (so called Betz limit) is $\eta_{WT,max} = 0.593$ at a velocity

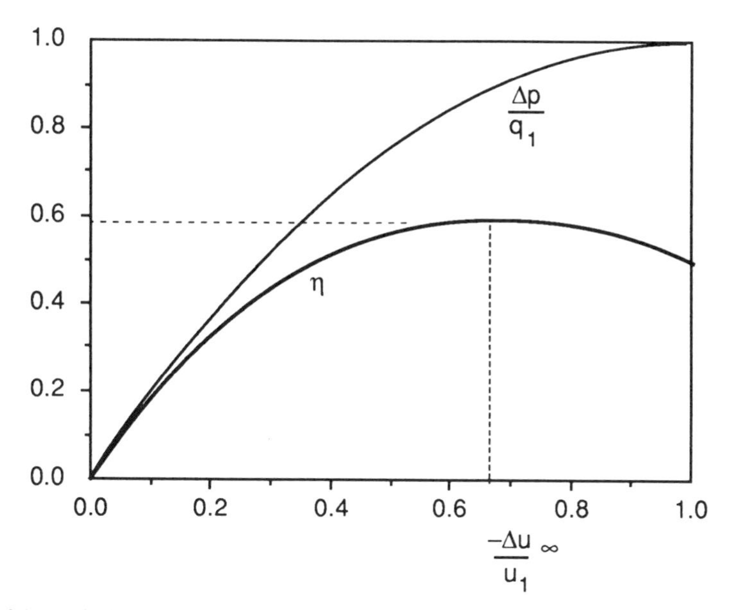

FIG. 13–26. Variation of pressure change across an actuator disk and wind turbine efficiency. The abscissa is the nondimensional velocity change experienced as a result of the turbine interaction.

parameter $\psi = \frac{2}{3}$. Irreversibilities reduce the actual maximum efficiency to lower values. The pressure change at maximum efficiency is $\frac{8}{9}$ of the freestream dynamic pressure.

13.2.4. *Darrieus Rotor Aerodynamics*

Darrieus and similar wind turbines are generally vertical axis machines. The wind flows is always normal to the rotation axis. Blackwell (ref. 13–5) gives a good review of this technology. Figure 13–27 shows a view of the plane of rotation with a single blade whose airfoil element moves in the direction shown. The airfoil section is generally symmetric because the angle of attack ranges between positive and negative stall angles.

The performance of this wind turbine is described by assuming that the wind velocity is equal to that value that exists in the absence of any interference effects associated with the turbine. Such a description is improved by correcting the wind velocity by the increment provided by an actuator disk theory or, further, by the induced velocities due to the lift of the blade(s). Such improvements are discussed subsequently.

Figure 13–27 shows the velocity vectors for a general azimuthal (θ) position. The angle of attack is seen to arise from the relation between the effective velocity vector and the rotational direction. The largest angle is experienced at $\theta = 90^\circ$ (see side sketch in Fig. 13–27), where

$$\alpha_m = \tan^{-1}\left(\frac{1}{\Omega}\right) \qquad \text{where } \Omega \equiv \frac{\omega R}{u_1} \tag{13–51}$$

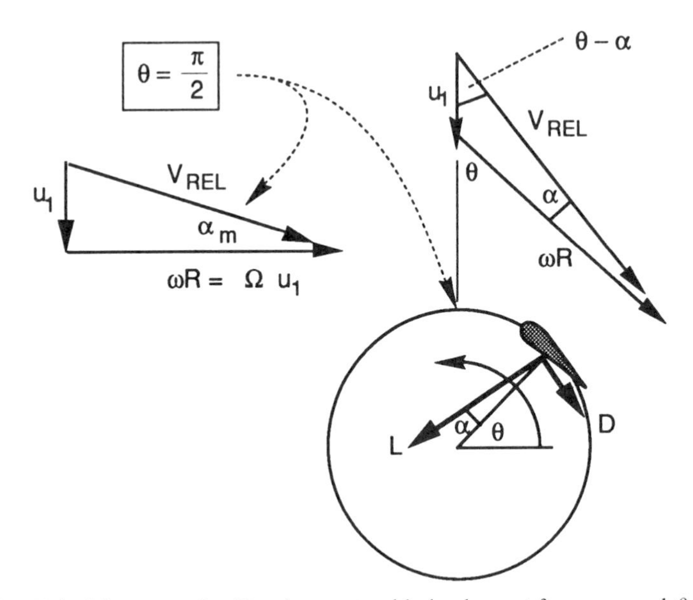

Fig. 13–27. Velocities around a Darrieus rotor blade element for a general θ. Vectors also shown at $\theta = \pi/2$ where angle of attack is largest.

Here R is the local radius of rotation and u_1 is the wind velocity. Note that *velocity ratio*, Ω, is an inverse advance ratio parameter, $\Omega = j^{-1}$ the use of which is traditional in the wind power community. Evidently, since angle of attack is limited by stall to small angles, Ω must be relatively large. For an airfoil with a lift coefficient of unity at stall

$$c_{1,m} = 1 = 2\pi\alpha_{\max} \qquad \text{gives} \qquad \Omega \approx 2\pi \tag{13–52}$$

Note the values of the speed ratio Ω in Fig. 13–25. From the velocity diagram, α is obtained from

$$\omega R \sin\theta = V_{\text{REL}} \sin(\theta - \alpha) \qquad \text{or} \qquad \Omega \sin\theta = \frac{V_{\text{REL}}}{u_1} \sin(\theta - \alpha) \tag{13–53}$$

where the relative velocity is given by

$$\frac{V_{\text{REL}}}{u_1} = \sqrt{1 + \Omega^2 + 2\Omega\cos\theta} \tag{13–54}$$

which varies between $\Omega \pm 1$ as shown in Fig. 13–28. The Darrieus rotor is not self-starting since it does not produce a torque at zero angular speed. The reason is that no lift force can be generated without sufficiently large relative motion between the wind and the blade element. If this turbine is connected to a power grid, it is generally possible to use the grid for starting. An alternative is the use of a Savonius rotor (see Fig. 13–24). Restricting the Savonius rotor to upper

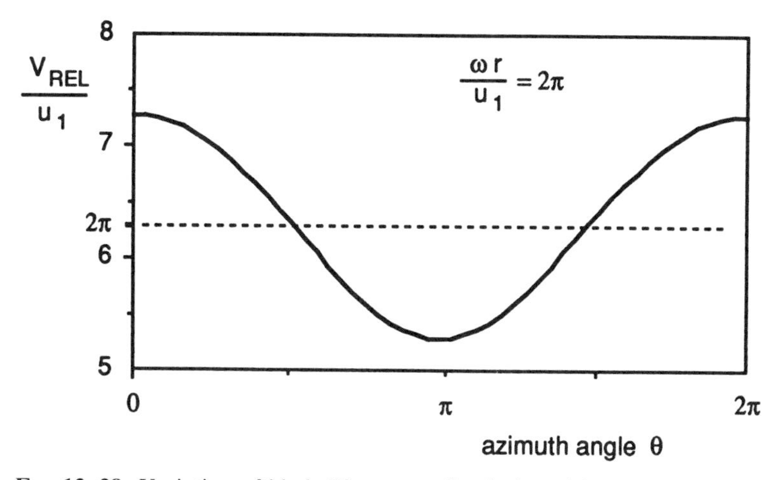

FIG. 13–28. Variation of blade "freestream" velocity with angular position.

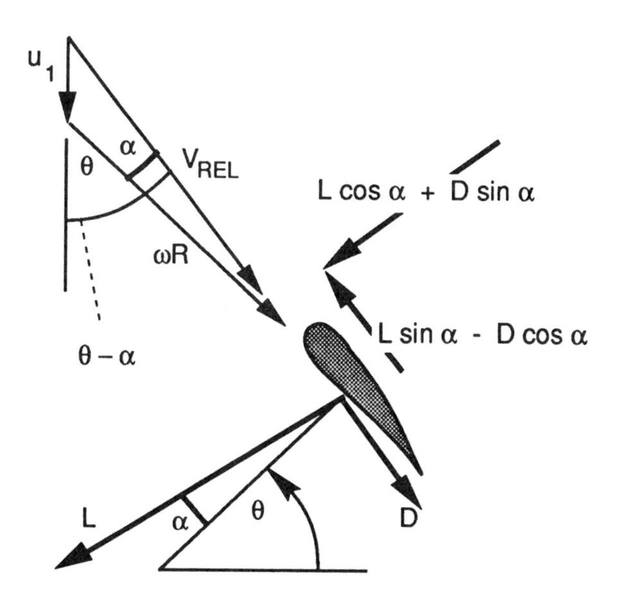

FIG. 13–29. Force vectors acting on the rotor blade element.

and lower sections (near the supports) is desirable so that wakes do not interfere with the clean flow desired by the Darrieus rotor, although a Savonius rotor on the central axis may be considered.

Combining eqs. 13–53 and 13–54 gives the variation of $\alpha(\theta)$ which, for small α, is approximately

$$\alpha = \alpha_{max} \sin \theta \tag{13–55}$$

Lift and drag forces are developed at the airfoil which are parallel and perpendicular to the relative wind. Figure 13–29 shows the forces and their components that lead to a torque about the center of rotation.

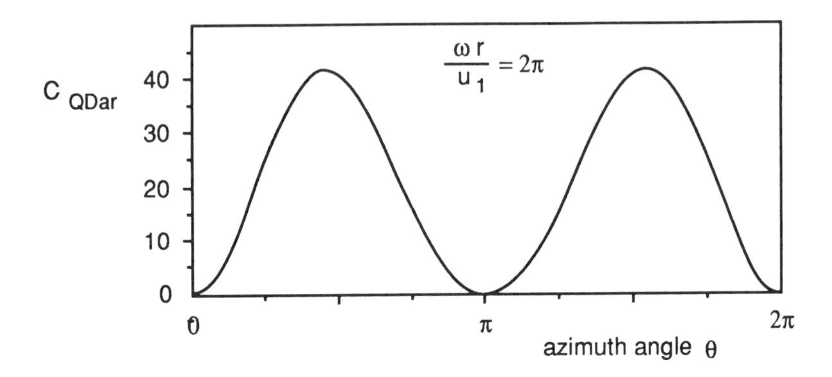

FIG. 13–30. Variation of nondimensional torque, $C_{\mathrm{Q,Dar}}$ (eq. 13–59).

The moment about the center of rotation is given by

$$Q = (L \sin \alpha - D \cos \alpha) R = q_\infty A \left(\frac{V_{\mathrm{REL}}}{u_1} \right)^2 (c_1 \sin \alpha - c_{\mathrm{d}} \cos \alpha) R \qquad (13\text{–}56)$$

where q_1 is the wind dynamic pressure and A is an element of airfoil area, chord $\times \Delta z$. For small α, and a simple algebraic description of a typical symmetric airfoil identical to eqs. 13–34 and 13–35, the torque coefficient is

$$C_{\mathrm{Q,Dar}} = \frac{Q}{q_1 A R} = (1 + \Omega^2 + 2\Omega \cos \theta)[2\pi\alpha^2 - c_{\mathrm{d0}}(1 + a(2\pi\alpha)^2)] \qquad (13\text{–}57)$$

Since α and therefore c_1 vary as $\sin \theta$, one can plot the moment as a function of θ as shown in Fig. 13–30 for $\alpha_{\max} = 1/\Omega$. $c_{\mathrm{d0}} = 0.0055$ and $a = 1$. These data correspond approximately to an NACA 0009 airfoil. $C_{\mathrm{Q,Dar}}$ is a convenient nondimensional measure of the torque for a Darrieus rotor referenced to the *wind* dynamic pressure:

$$C_{\mathrm{Q,Dar}} = (1 + \Omega^2 + 2\Omega \cos \theta)(2\pi\alpha_{\max}^2 \sin^2\theta - c_{\mathrm{d0}}(1 + a(2\pi\alpha_{\max})^2 \sin^2\theta)) \qquad (13\text{–}58)$$

For these data, the contribution due to drag is so small that the variation above is almost equal to that associated with the lift vector alone. An appreciation can be gained for the role of drag in overall performance when the moment is integrated over all 2π of the blade motion. Thus

$$\bar{C}_{\mathrm{Q,Dar}} = \frac{1}{2\pi} \int_0^{2\pi} C_{\mathrm{Q,Dar}}\, d\theta = \pi(1 - 2\pi a c_{\mathrm{d0}}) \left(1 + \frac{1}{\Omega^2} \right) - 4\pi\Omega c_{\mathrm{d0}} \qquad (13\text{–}59)$$

This value is ideally π when drag is absent and Ω is large. For the numbers cited, the small term in the first bracket is of the order of 0.03 whereas the last drag term is of the order of 0.3. This expression suggests that increasing

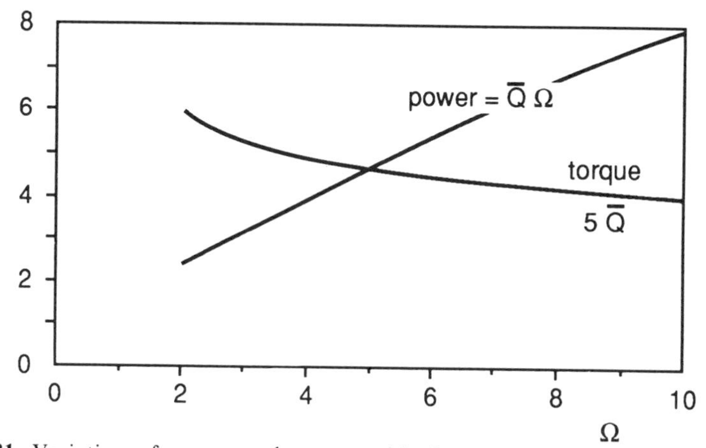

FIG. 13–31. Variation of power and torque with Ω. Torque is given by eq. 13–59, whereas power is torque times Ω.

Ω decreases the output torque. The power is obtained by multiplying Ω by the above expression. The resultant power output increases with increasing Ω as shown in Fig. 13–31.

13.2.5. *Darrieus Rotor Efficiency*

An efficiency based on the work done by the wind and the output power may be defined to account for L/D effects. The work done by the wind is the product of the force component in the wind direction times u. Thus

$$\eta_{\mathrm{L/D}} = \frac{\omega R \overline{(L \sin \alpha - D \cos \alpha)}}{\mathrm{u}\overline{(L \cos \alpha + D \sin \alpha)}} \tag{13–60}$$

where the forces are averaged over the azimuth angle, θ. The numerator involves the quantity $C_{\mathrm{Q,Dar}}$ while the denominator can be evaluated in a similar manner. The result is

$$\eta_{\mathrm{L/D}} = \frac{1 - c_{\mathrm{d0}}\left(2\pi a + \dfrac{4\Omega^3}{1 + \Omega^2}\right)}{1 + c_{\mathrm{d0}}\left(\dfrac{1}{2\pi} + \dfrac{3\pi a}{2\Omega^2}\right)} \tag{13–61}$$

Evidently, for zero drag, the efficiency is unity. The drag term in the denominator is also seen to be quite small. This efficiency modifies the equivalent actuator disk efficiency identified in eq. 13–50. Thus, to first order,

$$\eta_{\mathrm{overall}} \approx \eta_{\mathrm{act\,disk}}\, \eta_{\mathrm{L/D}} \tag{13–62}$$

For the performance of a complete Darrieus rotor, one must add the elemental contributions of blade elements that are located at various radial positions (i.e. various R). For this the shape of the blade must be known. Elements close to the top and bottom will have small R and hence small Ω. These portions of the blade arc will contribute relatively little since their airfoil sections necessarily operate at high angle of attack because the tangential velocity, and therefore Ω, are small in relation to the central section.

13.2.6. *Model Improvements*

A number of model improvements may be made to describe the forces on a Darrieus rotor more precisely. Among these are

1. Accounting for the fact that the mean flow through the rotor is not uniform in magnitude due to the deceleration imposed by the actuator disk,
2. Accounting for the increasing streamtube cross-sectional area associated with the deceleration noted above, and
3. Accounting for the shed vorticity due to
   a. The nonuniform lift, and
   b. Time or angular variation of lift.

These corrections are manifestations of the complete (bound, starting, and trailing) vortex system and as such allow calculation of adjustments to the forces on the blade elements. A complete vortex theory description typically yields results similar to those obtained above.

The following is a description of the corrections listed under 1 and 2. While this description allows calculation of the forces, the calculation is not be carried out here. Rather, a physical interpretation of the more realistic picture is given. In the light of the uncertainty of wind conditions, that is, their temporal and spatial variability, it is important to judge the relevance of more precise results: a one percentage point improvement in calculated efficiency may not be justified in view of a larger wind variability.

Figure 13–32 shows the changes in velocity vectors due to the effects described under 1.0 and 2.0 above. The velocity at the actuator disk is $u_1 + \Delta u_D$, rather than u_1, due to the slowing fluid. Here Δu_D is negative to be consistent with the disk theory presented in connection with the thrust-producing propeller. Through the region of the rotor, the velocity changes continuously. The rate of change through the disk is of the order of

$$\frac{\partial u}{\partial x} \approx \frac{\Delta u_D}{R_{max}} \tag{13–63}$$

where R_{max} is the radius of the blade element furthest from the rotational axis. The vector directions at the "plane" of the actuator disk are shown. Evidently, at $\theta = 0$ the lift is positive to contribute a positive torque. The angle of attack then goes through zero at point A. At $\theta = 90°$, the locally larger velocity leads to a larger angle of attack there compared to the value that would be calculated using local wind velocity $= u_1 + \Delta u_D$. This leads to a higher α at this point so

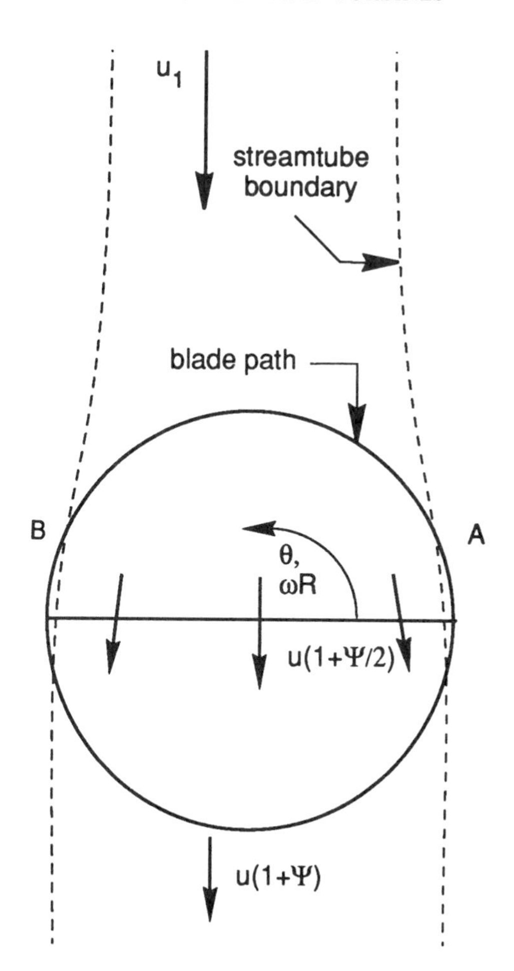

FIG. 13–32. Streamtube processed by a Darrieus wind turbine showing the one-dimensional velocity field corrections due to loading and the locally nonuniform axial velocity direction.

that more lift and therefore torque is to be expected at this point. Figure 13–33 shows the velocity vector diagram at four important angles of the cycle. Finally, lift is again zero at point B and for the remainder of the cycle, the angle of attack is smaller owing to the lower wind velocity and the local orientation. This has the effect of altering the angle of attack curve from that shown in eq. 13–51 to that in the sketch in Fig. 13–34. The correction for these effects is obviously dependent on the loading. In other words, not until the energy removed is calculated can the correction be assessed. In that sense, incorporation of the correction is iterative as it is for the propeller described in Section 13.1.7.

One way of modeling the three-dimensional effects described here is to model the rotor as two actuator disks, one in front of the other. In this way, the aerodynamic effects experienced by the rotor near $\theta = 90°$ can be separated

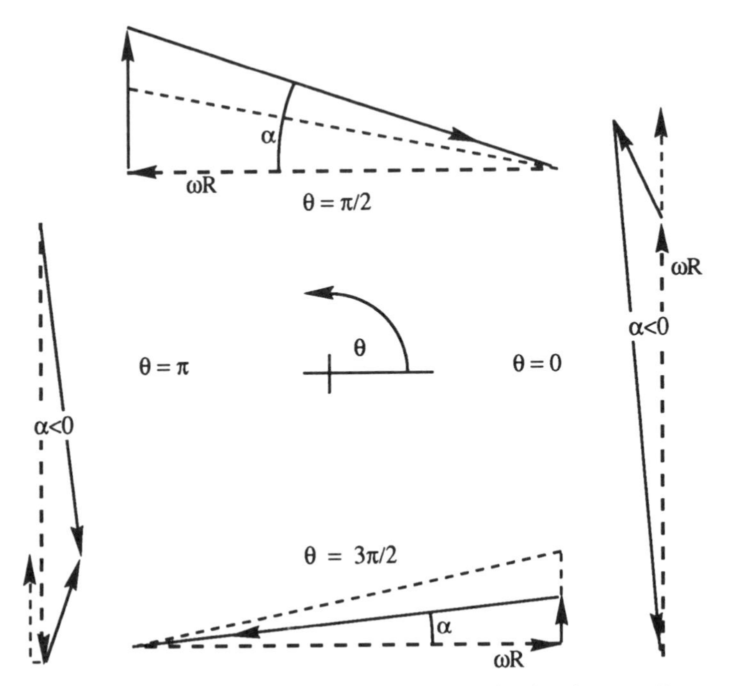

FIG. 13–33. Velocity vector relationships accounting for local nonuniform wind flow. Heavy dashed lines are the ωR vector, thinner dashed lines represent the velocity vectors with a uniform wind velocity. Solid lines are the vectors corrected for the local flow nonuniformities.

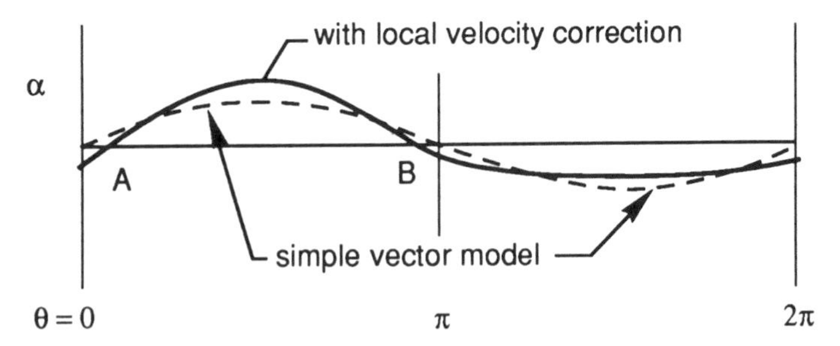

FIG. 13–34. Schematic variation of Darrieus rotor airfoil angle of attack around the cycle.

from that near $\theta = 270°$. Reference 13–7 and the section to follow are discussions of this modeling approach.

Tandem Disk Analysis

Attempts at good modeling of the Darrieus wind turbine have included greater detail in the analysis of the flow through the rotor (refs. 13–7 through 13–10). Actuator disk analyses include modeling by a single disk, by two (tandem) disks one in front of the other, and by multiple disks adjacent to each other where

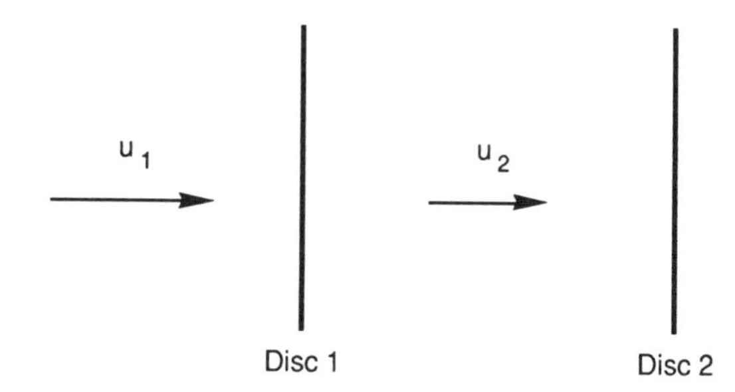

FIG. 13–35. Tandem actuator disks and their local axial velocities.

the various streamtubes are processed as required by the forces that apply locally. Lastly, combinations of tandem disks processing a multiplicity of streamtubes may be examined.

The tandem two-disk model is probably somewhat better than the single-disk model for the Darrieus turbine because it takes into account the flow field modifications imposed by the windward side of the blade-wind interaction and allows these to impact the leeward portion of the cycle. The following is a brief modeling discussion of the work that can be obtained from two disks. A simplifying assumption is to model the two disks as aerodynamically independent, that is, the inflow field to the second is taken to be the far-field downstream of the first. Consistent with this model, the velocities associated with the two disks is shown in Fig. 13–35.

At disk 1, the velocities far upstream of, at, and far downstream of, the disk are

$$u_1, u_1\left(1 - \frac{\Delta u_{\infty 1}}{2u_1}\right), \qquad \text{and} \qquad u_1\left(1 - \frac{\Delta u_{\infty 1}}{u_1}\right)$$

respectively, while for disk 2 they are

$$u_1\left(1 - \frac{\Delta u_{\infty 1}}{u_1}\right), u_1\left(1 - \frac{\Delta u_{\infty 2}}{2u_2}\right)\left(1 - \frac{\Delta u_{\infty 1}}{u_1}\right),$$

and

$$u_1\left(1 - \frac{\Delta u_{\infty 2}}{u_2}\right)\left(1 - \frac{\Delta u_{\infty 1}}{u_1}\right)$$

Here $\Delta u_{\infty i}/u_i$ is the fractional velocity change through the disk "i". If this fraction is denoted by ψ_i, the power per unit area from disk 1 is given by eq. 13–48, or

$$\dot{W}_1 = \tfrac{1}{2}\rho u_1^3\left(1 - \frac{\psi_1}{2}\right)(1 - (1 - \psi_1)^2) \tag{13–64}$$

From this, it follows that the second disk provides

$$\dot{W}_2 = \tfrac{1}{2}\rho u_1^3(1 - \psi_1)^3\left(1 - \frac{\psi_2}{2}\right)(1 - (1 - \psi_2)^2) \tag{13–65}$$

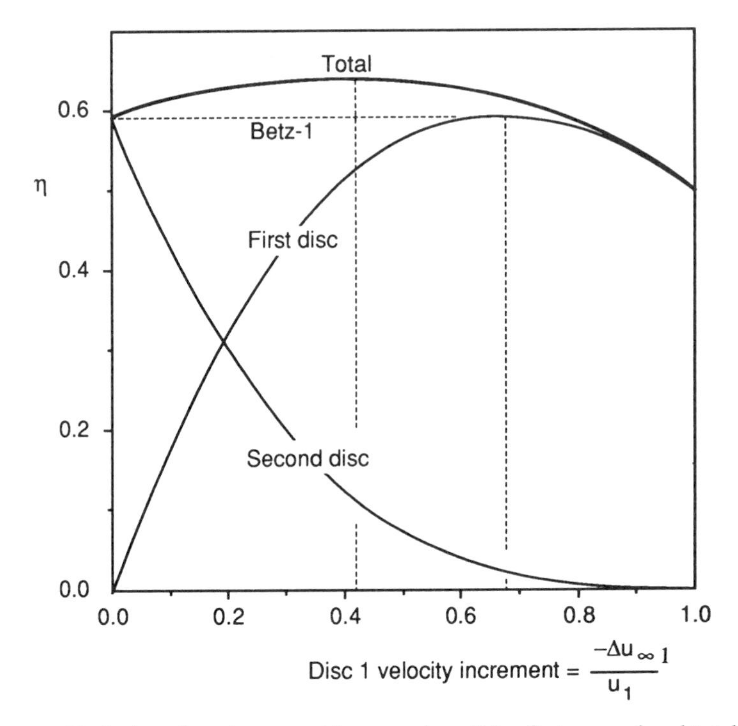

FIG. **13–36.** Variation of work removal from tandem disks, first, second and total. Betz-1 is the maximum work from a single disk.

Note that because the second disk lies in the diverged streamtube of the first, the mass flow rate processed by the second is smaller than that through the first disk. The overall efficiency of energy conversion (for equal disk areas) is

$$\eta = \frac{\dot{W}_1 + \dot{W}_2}{\frac{1}{2}\rho u_1^3} \tag{13–66}$$

The second disk allows increased overall work removal, but at conditions where the first disk removes much of the power, the second does not have much incoming air stream energy to work with. The maximum power output is obtained for the second disk operating at the Betz limit ($\psi_2 = 0.667$). The overall and individual disk's contribution to the efficiency is shown in Fig. 13–36 for varying ψ_1. The efficiency increment is of the order of 5 percentage points, or a 10% increase over the single disk. For maximum power extraction, the first disk operates at lower ψ_1 so that the second can provide a significant contribution. The overall efficiency is evidently maximized for ψ_1 of the order of 0.4 where its contribution is about 75% of the total work obtained.

13.3. DRAG-TYPE WIND TURBINES

Drag devices may be used for the extraction of power from the wind. The rotating cup anemometer is an example of a drag device. The simplest configuration has two arms whereas a self-starting one requires three. Consider the two-arm device which has cuplike plates at each end. The shape is dictated by the requirement that a net torque be experienced as a result of dissimilar forces acting on the cups. Ideally one would want large drag when the wind is "catching" the cup and zero drag on its return leg (Fig. 13–37). The shielding shown in Fig. 13–24 is intended to reduce the drag on the cup and arm advancing into the wind.

One may approach the ideal performance potential by shielding the returning cup from the wind by means of a fairing. This limit will be examined in light of a more general analysis. The cup will, in general, have a drag coefficient which varies with wind approach angle. The cup may be assumed to be immersed in a flow consisting of the vector sum of the wind plus its own rotational speed when the ratio of cup diameter to rotation arm radius is considered to be small.

The torque experienced by the arm at any azimuthal position, θ, is given by the difference of two drag forces multiplied by a moment arm which is $(R \sin \theta)$. The power is torque times angular speed, ω, so that a power coefficient can be derived as:

$$C_p = \frac{C_D \Omega}{\pi} \int_0^{\pi} [(1 - \Omega \sin \theta)^2 - c(1 + \Omega \sin \theta)^2] \sin \theta \, d\theta \qquad (13\text{–}67)$$

Here C_D is the drag coefficient of the retreating cup and the following quantities are defined:

$$C_p = \frac{\dot{W}}{\frac{1}{2}\rho u_1^3 A_{cup}}; \; C_{D(retreating)} = \frac{\text{Drag}}{\frac{1}{2}\rho u_1^2 A_{cup}}; \; c = \frac{C_{D,\,advancing}}{C_{D,\,retreating}}$$

where c is the (ideally small) drag ratio. The equation for C_p is integrated over 180 degrees of motion and the two velocity terms arise from the vector summation: the retreating cup is in a low-speed environment while the

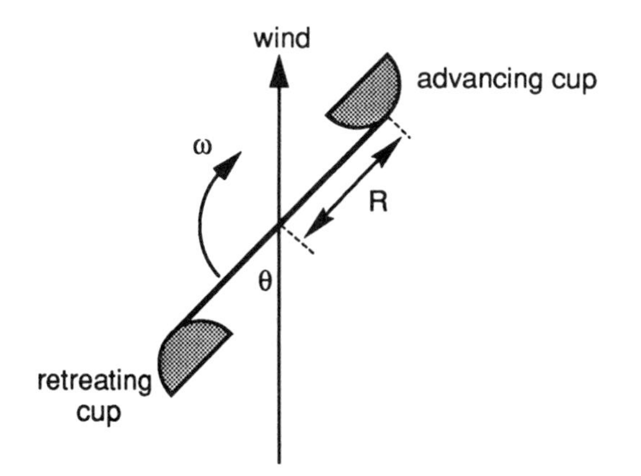

Fig. 13–37. Schematic of a drag-type wind turbine.

advancing cup is in high speed. If the drags on the two arms are equal, then the integration results in zero C_p. In this formulation the angular dependence of the two C_D's is neglected (a questionable assumption) so c is a constant.

For the special case of a fairing over the advancing cup, the term $(1 + \Omega \sin \theta)$ becomes Ω because the free wind is excluded. A special case where the advancing cup term is practically absent can also be examined. This might arise physically in a paddle wheel turbine on moving water. Here the density of the water is so large in comparison to the air that the advancing cup suffers very little drag. Results for this case arise when c is taken to be zero because the density ratio multiplies c.

The evaluation of the integral given in eq. 13–67 gives the following

$$C_p = \frac{C_D\Omega}{\pi}\left\{[2(1-c)] - [\pi(1+c)]\Omega + \left[\frac{4(1-c)}{3}\right]\Omega^2\right\} \qquad (13\text{–}68)$$

The value for maximum C_p is proportional to the retreating cup drag and is calculated for the value of Ω obtained by a mull differentiation of C_p. The results for various c ratios are shown in Fig. 13–38.

If the θ dependence in eq. 13–67 is ignored ($\sin \theta = 1$) together with the advancing cup drag ($c = 0$), the maximum C_p is readily obtained as (ref. 13–11):

$$C_p = C_D\Omega(1-\Omega)^2 \text{ for which } C_{p,\max} = \tfrac{4}{27} C_D \qquad \text{for } \Omega = \tfrac{1}{3} \quad (13\text{–}69)$$

In the case of this model with fairing, one can say that the advancing cup experiences drag according to the rotational velocity in a stagnant environment so that the performance is given by

$$C_p = C_D[\Omega(1-\Omega)^2 - \Omega^3] \qquad \text{for which } C_{p,\max} = \tfrac{1}{8} C_D \qquad \text{for } \Omega = \tfrac{1}{4} \quad (13\text{–}70)$$

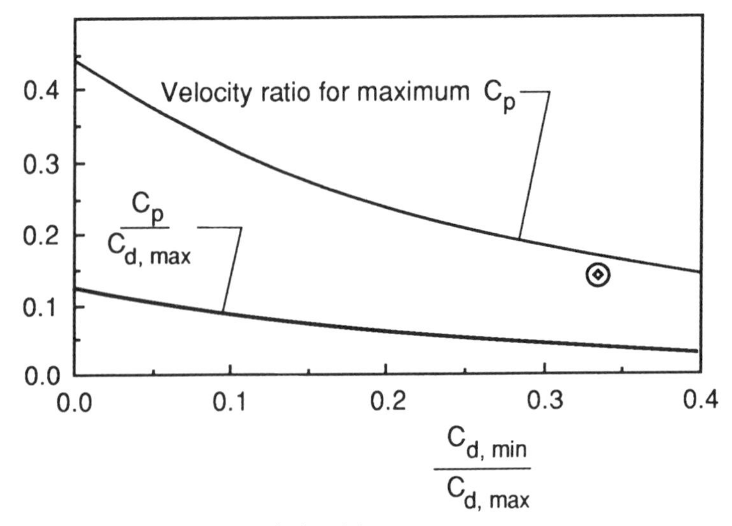

Fig. 13–38. Two-arm drag-type wind turbine performance variation with cup drag ratio.

Somewhat better performance is obtained for so-called lifting translators which are winglike devices traveling roughly normal to the wind direction along the ground. Such devices have a maximum power coefficient given by (ref. 13–11):

$$\text{Lifting translator: } C_{p,\max} = C_L \frac{2}{9}\frac{L}{D}\sqrt{1 + \frac{4}{9}\left(\frac{L}{D}\right)^2} \quad \text{at } \Omega = \frac{2}{3}\frac{L}{D} \qquad (13\text{–}71)$$

L/D for a reasonable aspect ratio wing is of the order of 10. Here Ω is really the ratio of $U(\text{vehicle})/u_1$. These devices are not described here because their performance is not particularly interesting and because the lifting translators are awkward to implement in practice.

PROBLEMS

1. An environment has a linear variation between t/t_{max} and wind speed, whose maximum is at $u = 1$. (See Fig. 13–22). Determine the distribution function $f(u)$, the most frequently observed velocity, the mean velocity, the energy content of the wind in t_{max}, and the cutoff velocities where the power is say, less than 10% of the maximum value.
2. Show that the propeller efficiency as given by eq. 13–29 is the ratio of two identical integrals when the drag is zero so that the efficiency becomes unity. When the drag is small, linearize the equation for η so that eq. 13–37 may be derived.
3. Expand the development of problem 2 to obtain an analytical expression for the maximum efficiency envelope of a variable pitch propeller, as shown in Fig. 13–16.
4. Develop an analysis procedure for examining the performance of a propeller with very poor blades (i.e. one for L/D is of order unity), and determine the efficiency of the propeller. Do not neglect the kinetic energy invested in rotation of the downstream flow.
5. Away from the central horizontal plane of largest diameter, the rotor of a Darrieus wind turbine decreases parabolically as z (the vertical coordinate) is increased or decreased. Develop a procedure for adding the contributions made at the various z's. Consider at which value the physics of the problem may no longer be valid and take advantage of any symmetry.
6. From the Biot Savart law for a vortex filament, show that eq. 13–11 follows for the geometry for which it applies.
7. Show that the two expressions for propulsive power (eq. 13–9 and the unnumbered equation following it) are equivalent.
8. Verify the results of eqs. 13–36 and 13–37.

BIBLIOGRAPHY

Cheremisinoff, N. P., *Fundamentals of Wind Energy*, Ann Arbor Science Publishers, 1979. P.O. Box 1425 Ann Arbor MI 48106.

Cobble, M. H., Lumsdaine, E., "Power and Energy in the Wind," *International J. of Energy Systems*, **7**: 1, 1987.

De Renzo, D. J., *Wind Power: Recent Developments*, Noyes Data Corp. Park Ridge, N.J., 1979

Eldridge, F. R., *Wind Machines*, NSF Research Applications Directorate, Oct. 1975.

Eggleston, D. M., Stoddard, F. S., *Wind Turbine Engineering Design*, Van Nostrand Reinhold, New York, 1987.

Farmer, P., *Wind Energy 1975–1985*, A Bibliography, Springer-Verlag, Berlin, 1986.

Freris, L. L., *Wind Energy Conversion Systems*, Prentice-Hall, Englewood Cliffs N.J., 1990.

Gustavson, M. R., "Limits to Wind Power Utilization," *Science* 204, (4388), April 6, 1979.

Kavorik, T., Pipher, C., Hurst, *J. of Wind Energy*, Domus, 1979.

Schoenmakers, R., Wrasman, B., Zwibel, H., Hinman, G., "Momentum Analysis of Tornado Wind Energy Concentrator Systems," *International J. of Energy Systems*, **5**: 2, 1985.

Wilson, R. E., Thresher, R. W., "Electrical Energy from the Wind," *Mechanical Engineering*, January 1984.

REFERENCES

13–1. Kuethe, A. M., Chow, C. Y., *Foundations of Aerodynamics*, John Wiley & Sons, New York, 1986.

13–2. Anderson, J. D., *Fundamentals of Aerodynamics*, McGraw-Hill, New York, 1991.

13–3. Batchelor, G. K., *An Introduction to Fluid Mechanics*, Cambridge University Press, 1967.

13–4. Touryan, K. J., Strickland, J. H., and Berg, D. E., "Electric Power from Vertical-Axis Wind Turbines," *AIAA J. of Propulsion and Power*, **3**(6): 481–93, 1987.

13–5. Blackwell, B. F., Sullivan, W. N., Reuter, R. C., Banas, J. F., "Engineering Development Status of the Darrieus Wind Turbine," *AIAA J. of Energy*, **1**(1): 50–65, 1977.

13–6. Savino, J. M., in Killian, H., Dugger, G. L., Grey, J., eds. *Solar Energy for Earth:* AIAA Assessment, American Institute of Aeronautics & Astronautics, New York, 1975.

13–7. Healy, J. V., "Tandem Disk Theory – with Particular Reference to Vertical Axis Wind Turbines," *AIAA J. of Energy*, **5**(4): 251–4, 1981.

13–8. Paraschivoiu, I., Delclaux, F., "Double-Multiple streamtube Model with Recent Improvements," *AIAA J. of Energy*, **7**(3): 250, 1983.

13–9. Paraschivoiu, I., Fraunie, P., Beguier, C. "Streamtube Expansion Effects on the Darrieus Wind Turbine," *AIAA J. of Propulsion and Power*, **1**(2): 150–5, 1985.

13–10. Eldridge, F. R., "Wind Machines," NSF Research Applications Directorate, Oct. 1975.

13–11. Hunt, D. V., *Wind Power*, Van Nostrand Reinhold, New York, 1981.

14

STEADY-FLOW WORK PROCESSES: GAS COMPRESSORS

Thermodynamic cycles employing gas as a working fluid require the means to raise the pressure of the gas efficiently. A number of devices are available for the task (e.g. ref. 14–1). The primary types of machines in use are aerodynamic and displacement (i.e. piston/cylinder) compressors. Because of their widespread use, aerodynamic machines are discussed first, followed by an overview of displacement and screw machine characteristics. A number of other mechanical schemes are available for the purpose of compression. Many have characteristics that make them inappropriate for use in primary energy conversion processes and hence are not detailed here. The fluid medium to which work is added may be incompressible, such as a liquid, in which case the machine is a pump; or it may be compressible in which case it is termed a compressor.

14.1. WORK INTERACTIONS WITH FLOWING MEDIA

Mechanical work may be done on a fluid element by decreasing its volume or by increasing its kinetic energy. Machines that carry out boundary displacement (to achieve volume reduction) are termed displacement machines. A piston reciprocating within a cylinder and working in conjunction with the timed action of valves carries out, in turn, induction, compression, and discharge of the working gas. By averaging over the individual motion cycles, a nearly steady supply of compressed gas can be delivered. Such machinery can compress over pressure ratios that are large, as high as 100:1. Size, weight, and efficiency of the machinery often preclude their use in applications where low weight and the ability to process very large flow rates are important. As the discussion of cycles in Chapter 9 has shown, the choice between design options for the machinery depends on the pressure ratio desired by the cycle design, on the adiabatic efficiency that the compressor can achieve, and on practical considerations.

For applications such as use in the Brayton cycle, dynamic flow machines are generally used. As a class, they are called turbo-compressors because the moving elements of the machinery turn continuously in one direction, rather than have any components suffer reciprocating motion and thus periodic stresses. Such dynamic flow machines work on steady flow by increasing the kinetic energy of the flow and then recovering the increased kinetic energy as thermal energy or, equivalently, as pressure. Because a diffusion process is necessarily involved, such machines are limited by the physics of boundary layers in adverse pressure gradient flows.

Continuous development of aerodynamic compressors over the past half century has resulted in improving performance of aircraft gas turbines and energy systems. Consequently, the field has a rich development history. Reference 14–2 is a good, more advanced, text on aerodynamic compressors to which the reader might turn for further study. Other texts are included in the aircraft propulsion bibliography listed at the end of this chapter.

The discussion of turbocompressors to follow relies on fundamental relations in the physical variables which also apply to the turbine that produces rather than absorbs mechanical power. To that end, Section 14.2 also serves as an introduction to Chapter 15 where axial flow turbines are described. In both Chapters 14 and 15, the goal is to provide insight into the design and performance characteristics of such machinery. A desired objective of this development is the understanding that allows description of the overall behavior of compressors or turbines so that part-power performance of an engine (Chapter 16) may be determined.

14.2. EULER TURBINE EQUATION

Figure 14–1 shows a generic dynamic compressor (or turbine with flow direction reversed) with a streamtube that enters and leaves a rotor. The coordinate directions and corresponding flow velocity components are as noted. The shaft power is added through a torque applied at an angular speed, ω. Newton's law states that a torque acting on a unit mass results in a change in angular momentum, $=\Delta(rv)$, where v is the tangential velocity of the streamtube. Thus, the mechanical power input (per unit mass flow) is the product of torque/mass and angular velocity, that is,

$$\frac{\dot{W}}{\dot{m}} = [\Delta(rv)]\omega \qquad (14\text{–}1)$$

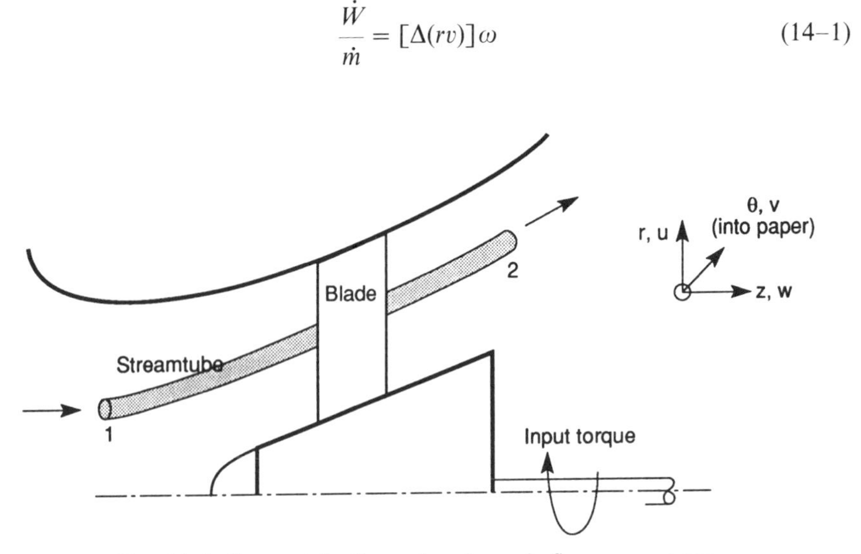

FIG. 14–1. Streamtube through a dynamic flow compressor.

The flow processed by such a device is, in general, steady, adiabatic, and irreversible. When subscripts 1 and 2 refer to a description of states ahead of and behind a blade row, the steady form of the First Law states that the power input must equal the total enthalpy increase (per unit time), or after cancellation of $\dot{m}$:

$$(r_2 v_2 - r_1 v_1)\omega = h_{t2} - h_{t1} \tag{14–2}$$

This Euler turbine equation relates the mechanical power input (primarily changes in the tangential velocity v) to the thermodynamic changes in the processed fluid medium. The subscript "t" denotes total or stagnation conditions. For a compressor, h_t increases and conversely decreases for a turbine. This equation includes no assumptions about reversibility or about fluid compressibility.

14.2.1. *Radial and Axial Flow Compressors*

Compressors and turbines are classified according to the form of the streamtube. Devices in which the streamtube turns about 90° with the exiting flow direction nearly radial, are termed *radial flow compressors*. By contrast, when the streamtube has small radial velocity components so that the principal velocities are axial (z) and tangential (θ), such devices are called *axial flow compressors*. Figure 14–2 (ref. 14–3) shows examples of the two types of machines. The radial flow compressor shown in Fig. 14–2a is a double-entry type where air enters from the left and from the right. The flows exiting the rotor are combined and turned again through a right-angle turn to enter a combustor. Figure 14–2b illustrates a single spool axial flow compressor consisting of a large number of rotating blades (rotor) and stationary blades (stator). The flow is from left to right. The connection to the shaft providing power is at the exit (right). The first set of vanes encountered by the flow are for structural support of the bearings of the rotor shaft, the second are inlet guide vanes followed by the first stage.

For reasons that will be developed below, radial flow compressors using air can be made to develop a relatively large pressure ratio (of order 3–10) with a single rotor and the associated components. On the other hand, axial flow compressors require large numbers of compression stages because the pressure ratio with one rotor (or stage) is small, of the order of 1.2. The thermodynamics of the simple Brayton cycles requires design values ranging from 5 to 40 in typical applications.

The density ratio that accompanies a small pressure change is also small, so that axial flow compressors stages may be thought of as processing an incompressible fluid. This fact makes discussion of the fluid mechanics rather direct and will be exploited in the description of axial flow compressors. Radial flow machines cannot be discussed in the same way (Section 14.5). As far as the Euler turbine equation is concerned, the primary difference between the two types of compressors is that, for the axial flow, the streamline position variation in the radial direction is small so that a single r applies to describe its radial location at both entry and exit. On the other hand, for the radial compressor, the radial location of the entry streamtube is close to the rotation axis while the exit radial location is much larger. As a result, the exit flow is close to radial in direction. The following approximate forms for eq. 14–2 may therefore be written:

$$\text{Axial flow compressor: } h_{t2} - h_{t1} \approx \omega r(v_2 - v_1)$$

$$\text{Radial flow compressor: } h_{t2} - h_{t1} \approx \omega r_2 v_2$$

14.2.2. *Pressure Variation through a Compressor with Nearly Constant Density Fluid*

In general, the total pressure ratio across a compressor is obtained from the enthalpy and the degree of reversibility (eq. 7–38). To identify the important design parameters, one assumes that the adiabatic efficiency is unity. The assumption of constant density allows use of a simple relation between pressure

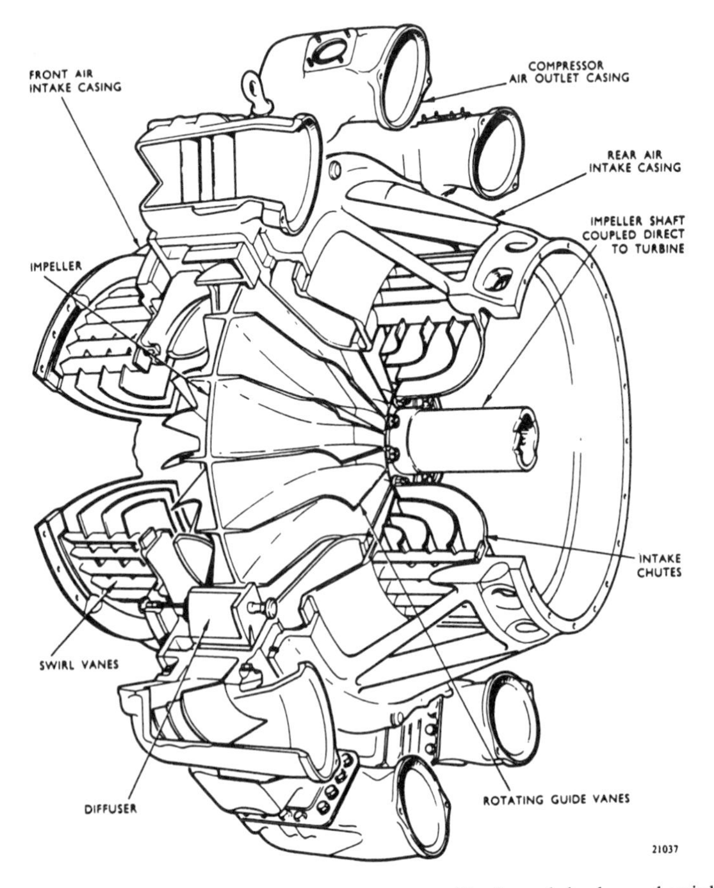

Fig. 14–2. Radial (a) and axial (b) flow compressors. Single and dual spool axial designs are shown as noted. (From ref. 14.3 with permission.)

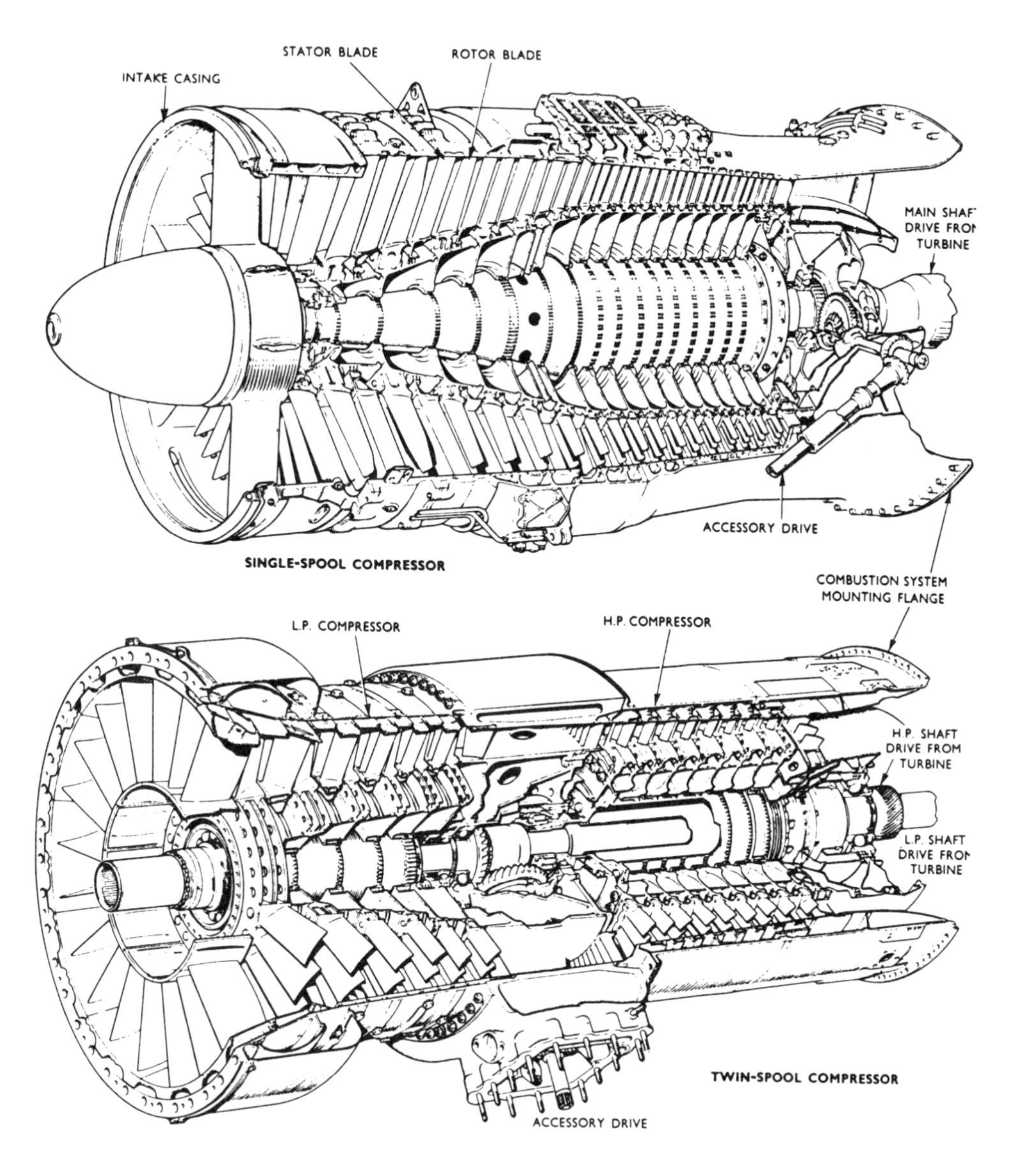

FIG. 14–2. (*continued*) (b) Axial flow compressors.

and velocity changes. While this assumption may appear limiting, it is useful for the description of pumps and of individual stages of an axial flow compressor. The liquid pump results follow directly from consideration of the axial compressor stage working with a gas. For a perfect gas, the relationship between enthalpy and temperature is:

$$h_t = h + \tfrac{1}{2}V^2 \equiv C_p T_t \tag{14–3}$$

and the total temperature is defined as noted. Changes in the static enthalpy

$$h = e + \frac{p}{\rho} \tag{14–4}$$

are reflected by changes in the pressure, p, when the density ρ and the internal energy, e, are very nearly constant. The total enthalpy per unit volume can be written as

$$\rho h_t = \rho e + p + \tfrac{1}{2}\rho V^2 \tag{14–5}$$

Thus for pumps and a single stage of an axial flow compressor, changes in h_t as stated in the Euler equation (eq. 14–1) can be written in terms of the incompressible flow total pressure defined by

$$p_t \equiv p + \tfrac{1}{2}\rho V^2 \text{ Incompressible flow} \tag{14–6}$$

as

$$p_{t2} - p_{t1} = \rho\omega(r_2 v_2 - r_1 v_1) \tag{14–7}$$

This is the Euler turbine equation for *incompressible* flow.

Incompressibility Assumption

If the function of the compressor is to change the flow density, how useful is it to say that the flow is incompressible? The answer to this question is that when pressure changes are sufficiently small, correspondingly small density changes occur. For example, the total pressure ratio through an axial compressor stage is of the order of 1.2 and the corresponding density change is about 14% ($p \sim \rho^{1/\gamma}$) when the flow is reversible. Since the pressure rise is associated with a conversion of flow kinetic energy to thermal energy, it follows that small pressure changes are the result of a small amount of kinetic energy available in the flow. The ratio of kinetic energy to thermal energy is measured by the Mach number, M, as

$$M^2 = \frac{\rho V^2}{\gamma p} \text{ since } \frac{\text{kinetic energy}}{\text{thermal energy}} = \frac{\frac{1}{2}V^2}{C_p T} = \frac{\gamma - 1}{2} M^2$$

Thus an incompressible flow is a flow with small Mach number (see also eq. 12–14). The general descriptive relation between pressure and velocity

Table 14–1. Stagnation to static pressure ratios from eqs. 14–8 and 14–9

M	exact	approximate	% error
0.5	1.186	1.175	1
0.75	1.43	1.39	3
1.0	1.89	1.70	10

(expressed as M) is given by (eq. 12–9)

$$\frac{p_t}{p} = \left(1 + \frac{\gamma - 1}{2} M^2\right)^{\gamma/(\gamma - 1)} \tag{14–8}$$

which may be simplified for small $M (M^2 \ll 1)$ to

$$\frac{p_t}{p} \approx \left(1 + \frac{\gamma}{2} M^2 + O(M^4)\right) \approx 1 + \frac{\gamma}{2} M^2 \tag{14–9}$$

For nearly incompressible flow, this equation (or its equivalent, the Bernoulli equation which states that p_t given by eq. 14–6 is constant) may be used in lieu of the more awkward eq. 14–8. The small M approximation is therefore consistent with the small density change approximation.

The error incurred through the use of the incompressible flow assumption may be gaged by comparing eqs. 14–8 and 14–9 with the M^4 term neglected. The approximate and exact values of the total to static pressure ratio are given in Table 14–1, showing that even for sonic flow the error may be acceptable.

For complete compressors consisting of a number stages, one may take the density as constant through each stage, changing value from stage to stage, consistent with the pressure changes.

14.3. AXIAL FLOW COMPRESSORS

Axial flow compressors cause a flow to undergo direction changes by means of interacting with blades that are bounded by an inner (hub) and an outer (tip) annulus, (Figs. 14–2 and 14–3). The blades are of various types:

1. Moving blades connected to the shaft providing power input (rotor)
2. Blades stationary with respect to the annulus (stator)
3. Stationary blades that direct the flow into or out of the compressor adjacent components: inlet or exit guide vanes.

A combination of a rotor and a stator blade row is called a *stage*. Typically 5–30 stages are used for an axial flow compressor. The number of stages, N,

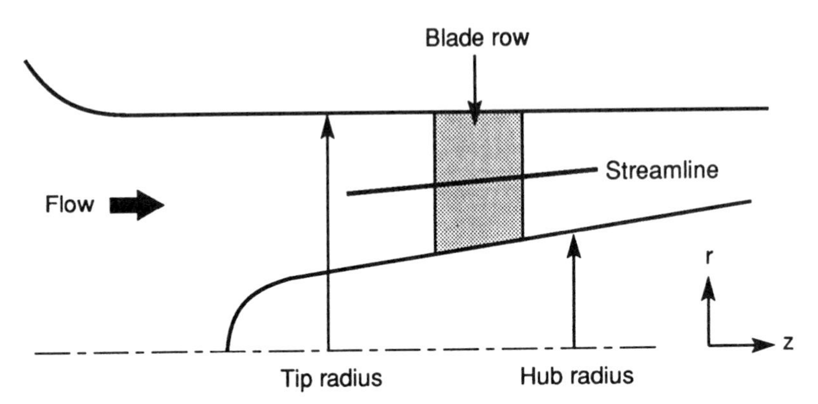

Fig. 14–3. Cross-sectional view through the axis of an axial flow compressor.

required for an overall pressure ratio (OPR) is given by

$$\pi_c = \text{OPR} = (\text{Stage PR})^N \tag{14–10}$$

The chief advantages of an axial flow compressor are large mass flow handling capability, simple flow path, high efficiency, and, for low drag installation on aircraft, small frontal area.

14.3.1. *Mean Streamtube Behavior through Blade Rows*

Consider the flow through a surface bounded by an outer or "tip" annulus and an inner or "hub" annulus (Fig. 14–4). This annular space contains blades that provide the flow direction changes. The velocity vector has two important components, one axial, which is related to the mass flow rate through the machine; and the other tangential, related to the angular momentum.

Stator blade rows are generally attached to the (outer) casing of the compressor and are thus stationary in the laboratory coordinate frame, whereas rotors are driven by the rotating central shaft. In Figs. 14–3 and 14–4, the following coordinates and velocities are defined:

radial coordinate	$= r$	velocity component $= u$
tangential	$= \theta$	velocity component $= v$
axial	$= z$	velocity component $= w$

The total velocity magnitude is: $V^2 = u^2 + v^2 + w^2$.

In an axial flow compressor, the flow is primarily along a meridional $(z - \theta)$ surface. The streamlines experience rather small changes in radial position $(r_2 = r_1 = r)$ as they pass through the stage so that the velocity component u is generally small compared to other velocity components. This is particularly true when the hub/tip radius ratio is close to unity (i.e., there is a "thin" annulus "far" from the rotational axis). Figure 14–5 shows a blade row viewed radially toward the axis of the compressor, with the important velocity vectors in the laboratory reference frame. The details of the blading are not shown in order to emphasize the overall flow changes. The blade row's function, whether rotor

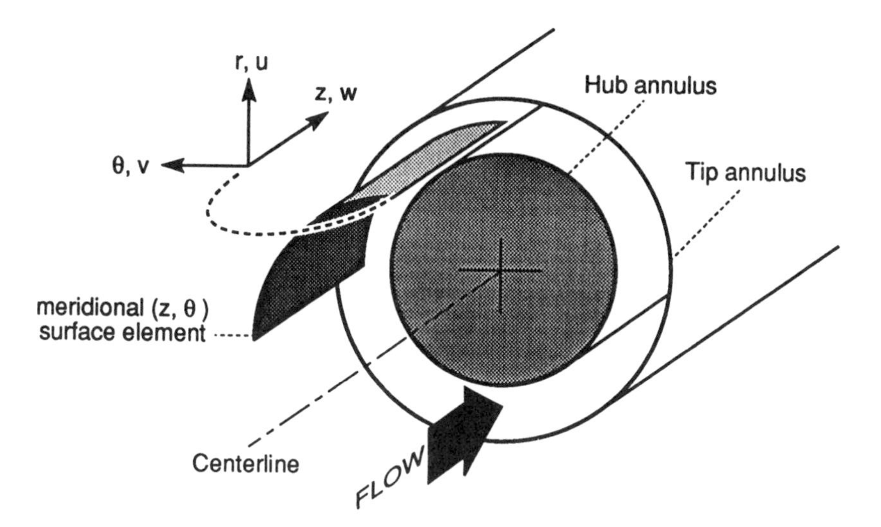

FIG. 14–4. Flow between the hub and tip annulus of an axial flow compressor.

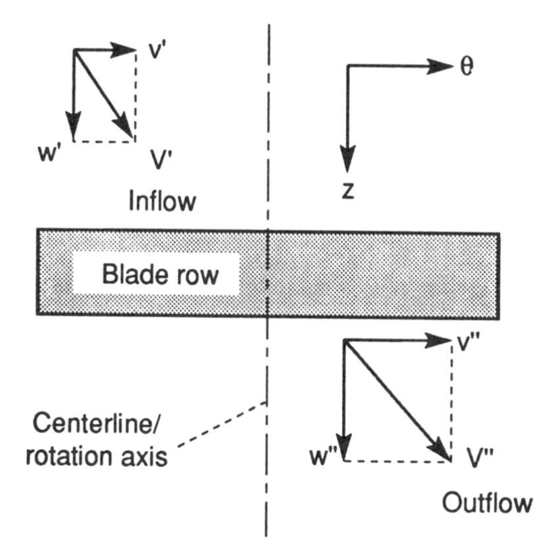

FIG. 14–5. View of z, θ (meridional) "plane" showing axial and tangential flow velocities before (′) and after (″) passage through blade row.

or stator, is to alter the direction of the mean flow and to change the magnitude of the velocity. Single and double primes denote conditions at the entry and exit of the blade row.

The continuity equation requires the mass flow rate through an annulus area (in the r-θ plane) to be constant:

$$\rho w \pi (r_{\text{tip}}^2 - r_{\text{hub}}^2) = \rho w A_{\text{i}} = \text{constant} \qquad (14\text{–}11)$$

Since r and ρ do not change significantly through the stage, the axial velocity,

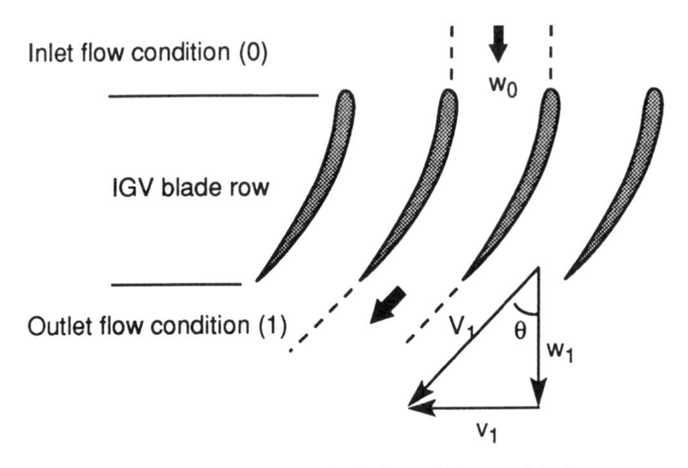

FIG. 14–6. Flow through the inlet guide vane blade row.

w, is roughly constant when the area at the inlet of a blade row equals that at the exit. In practice, an annulus area decrease is required in a compressor to accommodate density changes and thus keep the axial velocity somewhat constant. For the compressor (of many stages) as a whole, the annulus contraction therefore reflects the changing density (Fig. 14–2b).

Inlet Guide Vane

Consider first the inlet guide vane (IGV). The initially axial flow is turned to an angle consistent with good operation of the subsequent blade rows. Figure 14–6 shows the blade and the velocity vectors. Station numbers 0 and 1 are used to describe the inlet and exit of the IGV blade row. The station numbering system for the IGV and the following stage is shown in Fig. 14–7.

Since the flow area normal to the velocity vector decreases in the inlet guide vane, it follows that

1. Flow velocity increases.

$$v_1^2 + w_1^2 = V_1^2 > V_0^2 = w_0^2 \qquad \text{since } w_1 \approx w_0$$

2. Static pressure and density decrease.
3. Total pressure remains constant (in the absence of irreversibilities) since the flow experiences no work interaction.

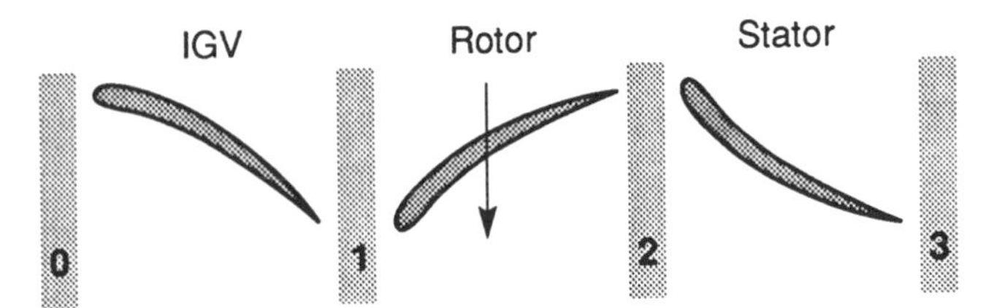

FIG. 14–7. Station numbering in axial flow compressor blading. Flow is to the right.

With $\omega = 0$, eqs. 14–1 and 14–7 show that total enthalpy and total pressure are conserved:

$$h_{t1} = h_{t0} \qquad \text{and} \qquad p_{t1} = p_{t0} \tag{14–12}$$

The lowered static pressure through the IGV requires an annulus area expansion from state 0 to 1 to keep axial velocity constant.

Rotor

A rotor follows the IGV. The flow into and out of this rotor is described by states 1 and 2 as shown in Fig. 14–8. The figure shows the blade row and the velocities relative to the *rotor*. The function of the rotor row is to *decelerate* the flow in the *laboratory* reference frame and thus realize a pressure increase. The relation between velocities relative to the blade and velocities in the laboratory are summarized in the velocity triangles shown below the blade row in the figure. The velocity exiting the IGV or a previous stator row is $V_{a,1}$ which is shown as a thin arrow vector. An observer *fixed to the blade* viewing this incoming flow sees it as the vector sum of his or her own rotating velocity (ωr)

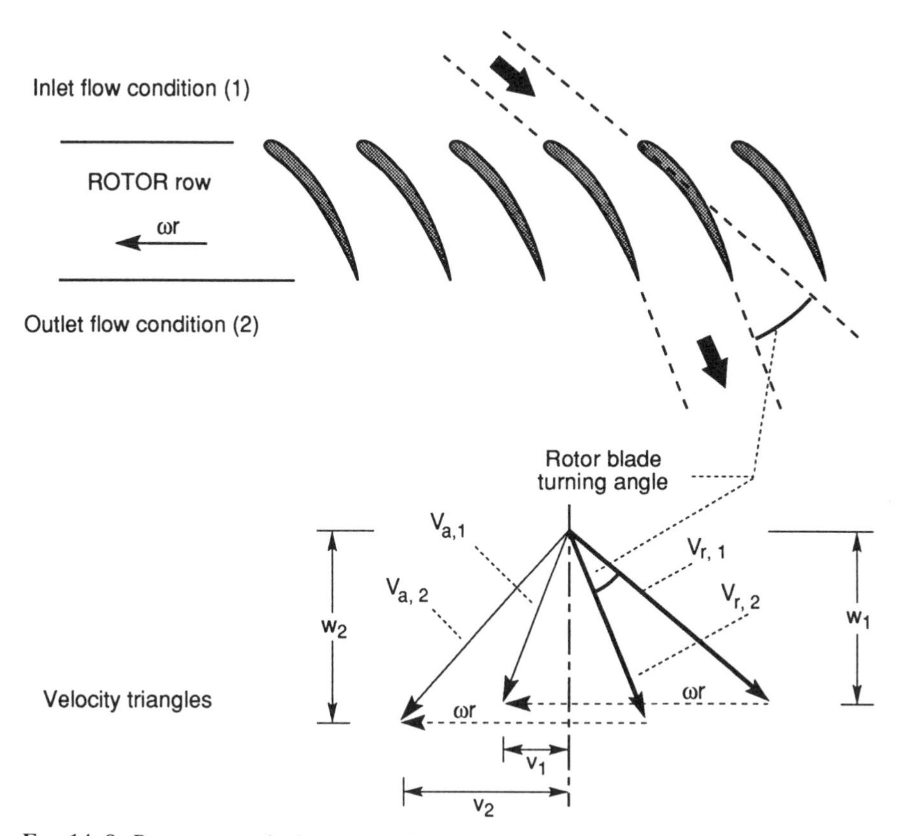

FIG. 14–8. Rotor row velocity vector diagram. Subscripts *a* and *r* describe the absolute (or laboratory) and rotor reference frames.

and the absolute velocity in the laboratory reference frame. This velocity is $V_{r,1}$ and is shown as the heavy arrow vector. The blade row forces the relative vector to be turned toward the z-axis so that its magnitude is reduced while the axial component is about constant. The reduced magnitude leads to increased pressure.

The angle θ is used to describe the flow orientation in the laboratory reference frame while β describes the flow orientation in the rotor frame. θ_1 is the turning angle of the IGV, and $\beta_1 - \beta_2 = \Delta\beta$ is that of the rotor.

Note the angles θ_1 (the exit angle of the IGV) and β_2 (rotor exit) are the flow angles imposed by the trailing edges of the blades. These so-called metal angles establish the flow direction under most circumstances involved with varying mass flow rate or rotational speed. Under conditions of heavy loading (i.e., operation at high lift), there will be departures between average flow angles and trailing edge angles. This departure is neglected in this discussion and the reader is referred to more detailed texts (refs. 14–2, 14–4, and 14–5) for a discussion of this effect.

The absolute velocity exiting the rotor blade is obtained by adding (vectorially) $+\omega r$ to the relative velocity $V_{r,2}$. Note that in the laboratory frame, the net effect of the work input is to increase the kinetic energy of the flow, $V_2 > V_1$. This velocity increase is recovered as pressure by the stator that follows the rotor.

Figure 14–8 shows the approximately constant axial velocities in and out of the stage as well as the initial tangential velocity in the laboratory reference frame, v_1 which is increased to v_2 by the work input.

Stator

The stator's function is to restore the absolute velocity exiting the rotor roughly to the orientation at the rotor inlet so that a similar following rotor row may be used to raise the pressure further. Thus the stator exit station 3 should be close to 1. Figure 14–9 shows the blading and the velocity diagram for the stator. Note that the static pressure again increases because of the decreasing

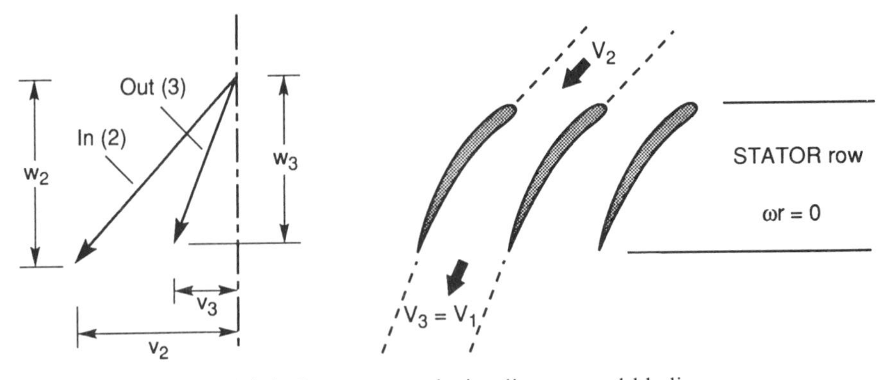

Fig. 14–9. Stator row velocity diagram and blading.

(absolute) velocity and the similarity between the velocities in the laboratory reference frame as shown in Figs. 14–8 and 14–9.

Owing to the increasing pressure environment in both sets of blade rows, an *adverse pressure gradient* is experienced by the boundary layers on the blades. This limits the pressure increase and the amount of allowable turning. Turning angles of the order of 10–30° are used.

Blade Orientation

The mean orientation angle θ is seen to play an important role in the aerodynamics of the blade row. This role will be discussed subsequently, but it is also noteworthy that θ also plays a role in determining the size of the compressor as measured by the frontal area. Consider that at any station, the mass flow rate is proportional to the axial velocity component

$$\dot{m} \sim w = V \cos\theta \approx a_t M \cos\theta \qquad (14\text{–}13)$$

which implies that mass flow rate varies as $\cos\theta$, since the Mach number associated with V is limited to values near unity. This limitation results from irreversibilities associated with shocks. As a result, angles θ of the order of 45° ensure high mass flow rate processing capability so that a reasonably compact (small tip radius) compressor can be designed for a chosen mass flow rate.

14.3.2. *Static Pressure Variation in the Blade Rows*

In the absence of losses, the Euler turbine equation for an incompressible flow through an axial blade row (eq. 14–7) is:

$$p_{t2} - p_{t1} = \rho\omega r(v_2 - v_1) \qquad (14\text{–}14)$$

which is valid for a rotor as well as a stator (for which $\omega = 0$). The total pressure increases only in the rotor row where work is added. Through the stator row p_t is constant.

By using the definition of total pressure (eq. 14–6), one may write eq. 14–14 as

$$\frac{p}{\rho} + \tfrac{1}{2}V^2 - v\omega r = \text{constant}$$

or

$$\frac{p}{\rho} + \tfrac{1}{2}(w^2 + v^2 - 2v\omega r) = \text{constant}$$

and finally since the constant $(\omega r)^2$ can be added to both sides,

$$\frac{p}{\rho} + \tfrac{1}{2}(w^2 + (v - \omega r)^2) = \text{constant} \qquad (14\text{–}15)$$

This last step involves use of the approximation that the streamline radius be constant ($r_1 = r_2$). For describing the average streamtube behavior, one may speak of an annulus design where the mean streamline proceeds at constant radius. The error associated with the difference between this model and a real device can be shown to be of the same order as the assumption of constant density. In eq. 14–15, the velocity $(v - \omega r)$ is the tangential component of the velocity in the *relative* reference frame which rotates in the case of the rotor, and is fixed in the case of the stator. It follows that Euler's equation can be written as a momentum equation for the blade passage, provided one uses the *relative* velocity,

$$\frac{p}{\rho} + \tfrac{1}{2} V_r^2 = \text{constant} \tag{14–16}$$

The conclusion is that, in the blade row, an observer will sense changes in static pressure consistent with changes in velocity *in the reference frame of the blade.*

14.3.3. *Estimate of Stage Total Pressure Ratio*

The total temperature rise through a rotor or stage is given by the Euler equation (eq. 14–1). From the velocity triangles (Fig. 14–8), the tangential velocities are given by:

$$v_i = w_i \tan \theta_i = \omega r_i - w_i \tan \beta_i \qquad i = 1 \text{ or } 2 \tag{14–17}$$

where θ is the flow orientation in the laboratory frame and β that in the rotor frame. One may simplify the calculation of the total temperature rise (and, consequently, the total pressure change) with two assumptions used earlier: (1) The axial velocity is maintained at a constant value by appropriate contraction of the flow annulus, and (2) The radial position of the average streamline does not change significantly. Thus w and r are constant through the stage. For a perfect gas, eq. 14–2 with eq. 14–17 reads:

$$\frac{T_{t2}}{T_{t1}} = 1 + \frac{w\,\omega r}{C_p T_{t1}} (\tan \beta_1 - \tan \beta_2) \tag{14–18}$$

This equation may be simplified by writing it in terms of meaningful parameters. To incorporate the possibility of irreversible flow, one may introduce a stage efficiency (eq. 7–38) to relate total pressure changes to total temperature changes.

$$\eta_s = \frac{\left(\dfrac{p_{t2}}{p_{t1}}\right)^{(\gamma-1)/\gamma} - 1}{\left(\dfrac{T_{t2}}{T_{t1}}\right) - 1} \approx \frac{\dfrac{(\gamma-1)}{\gamma}\delta}{\left(\dfrac{T_{t2}}{T_{t1}}\right) - 1} \tag{14–19}$$

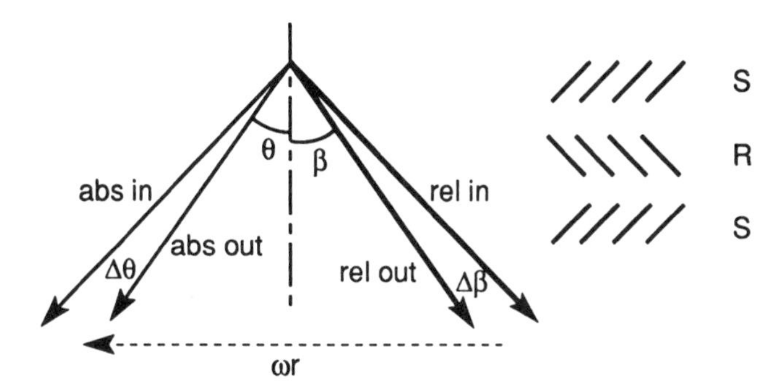

FIG. 14–10. Blading symmetry in a compressor: stator on left, rotor on right.

where $\delta(\ll 1)$ is defined as $p_{t2}/p_{t1} - 1$. The pressure ratio may be obtained by eliminating the total temperature ratio between eqs. 14–18 and 14–19:

$$\frac{p_{t2}}{p_{t1}} \approx 1 + \eta_s \gamma \left[\frac{\omega r}{\sqrt{\gamma R T_1}} \frac{w}{\sqrt{\gamma R T_1}} \frac{T_1}{T_{t1}} (\tan \beta_1 - \tan \beta_2) \right] \qquad (14\text{–}20)$$

which suggests the definitions of two Mach numbers:

$$M_\theta \equiv \frac{\omega r}{\sqrt{\gamma R T_1}} \qquad \text{and} \qquad M_z \equiv \frac{w}{\sqrt{\gamma R T_1}} \qquad (14\text{–}21)$$

The tangential Mach number, M_θ, is based on rotational blade speed and the axial Mach number, M_z, is a measure of mass flow rate. The static to stagnation temperature ratio is a function of the absolute Mach number ($M_1 \equiv V_1 \sqrt{\gamma R T_1}$). This last Mach number is typically less than about 1.4 to avoid shock wave losses.

Consider next the angle terms. For repeating stages where the pressure changes in the rotor and the stator passages are about equal, it is logical to have symmetric velocity triangles as shown in Fig. 14–10. Thus $\beta_2 = \theta_1$ and $\Delta\beta = \beta_1 - \beta_2 = \Delta\theta$.

To estimate the pressure ratio given by eq. 14–20 one needs to write the velocity components in terms of the velocity magnitude that is bounded by the absolute Mach number. Equation 14–20 may thus be written as

$$\frac{p_{t2}}{p_{t1}} \approx 1 + \eta_s \frac{\gamma M_1^2}{1 + \frac{\gamma - 1}{2} M_1^2} \times \{\cos^2 \theta_1 [\tan(\theta_1 + \Delta\theta) + \tan \theta_1][\tan(\theta_1 + \Delta\theta) - \tan \theta_1]\}$$

using

1. $$\frac{T_1}{T_{t1}} = \frac{1}{1 + \frac{\gamma - 1}{2} M_1^2}$$

2. $$\frac{\omega r}{w} = \tan(\theta_1 + \Delta\theta) + \tan\theta_1 \text{ from the velocity diagram,}$$

3. $$\frac{w_1}{V_1} = \cos\theta_1$$

4. $$\frac{V_1}{\sqrt{\gamma R T_1}} = M_1 \qquad (14\text{–}22)$$

The last square bracket term is the difference in tan θ's is associated with the turning by the blade row.

The function of θ_1 and $\Delta\theta$ in the curly bracket of eq. 14–22 is plotted in Fig. 14–11 together with the variation of w (i.e., cos θ_1 from eq. 14–13). w is of interest because it is proportional to the mass flow per unit area and thus a measure of compactness. The plot presents the performance of a family of compressors with equal blade turning angles but with different, symmetric blade orientations. The following conclusions may be drawn. First, one should note that the turning angle $\Delta\theta$ should be large for a large pressure ratio per stage. Second, a choice of θ_1 is a compromise between high stage pressure ratio and small frontal cross-sectional flow area. Angles θ_1 near 45° appear to be good choices. Therefore blading is typically as shown in Figs. 14–10 and 14–2b. Lastly, from eq. 14–22, the absolute Mach number M_1 should also be as large as possible, hence the motivation to force M_1 near sonic values.

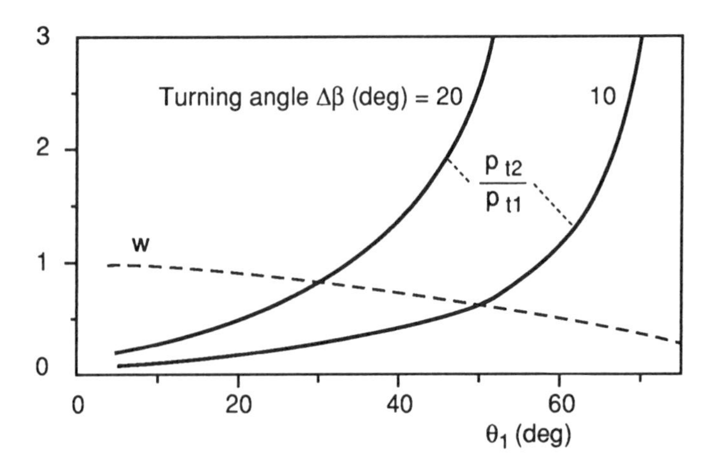

Fig. 14–11. Variation of stage performance parameters (pressure ratio and mass flow rate) for a stage with symmetric blade orientation. $\Delta\theta = \Delta\beta$. The total pressure ratio is given by eq. 14–22.

Finally, eq. 14–22 may be used to estimate the magnitude of the stage total pressure ratio. With $M_1 = 0.6$, $\eta_s = 0.9$, θ near 45°, and $\gamma = 1.4$, the equation gives the p_t ratio as 1.2 to 1.8 for $\Delta\theta = 10$ and 20° respectively.

14.3.4. *Off-design Stage Characteristics*

Having identified a particular design as appropriate, one might ask how does the stage pressure ratio of this design vary as the flow conditions are altered. A plot summarizing such behavior is termed a performance characteristic or "map." To obtain such a map one must recognize that the trailing flow tangency condition largely ensures that the trailing edge angles θ_1 and β_2 fix the flow direction out of the blade rows. It is thus appropriate to rewrite the Euler equation in its general form (compressible, and irreversible, eq. 14–3) in terms of these two fixed angles using eqs. 14–17:

$$C_p(T_{t2} - T_{t1}) = w\,\omega r(\tan\beta_1 - \tan\beta_2) = (\omega r)^2\left[1 - \frac{w}{\omega r}(\tan\theta_1 + \tan\beta_2)\right] \quad (14\text{–}23)$$

This relation shows that the total enthalpy increase due to work input depends on the wheel speed (ωr) and the mass flow rate processed (w). Shematically this relationship is shown in Fig. 14–12.

The slope of these lines is determined by the difference term $\tan\theta_1 - \tan\beta_2$. When the work input to the flow is negative, the machinery is driven by the flow and hence the term "windmilling." For a chosen ωr, increasing the axial velocity, w, decreases the angle of attack on the rotor blade. A decreased angle of attack leads to reduced lift and therefore work input to the flow. Figure 14–13 illustrates this point.

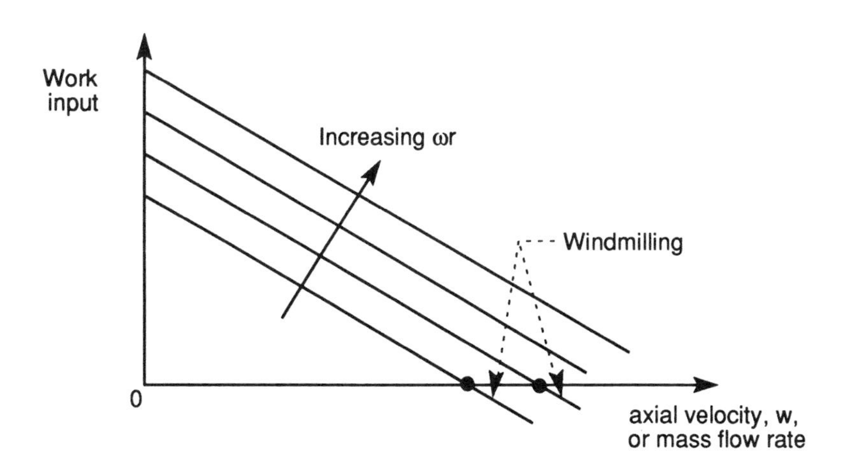

FIG. **14–12.** Ideal flow work input for a chosen stage design. Sketch of eq. 14–23. Rotor driven *by* the flow is termed "windmilling."

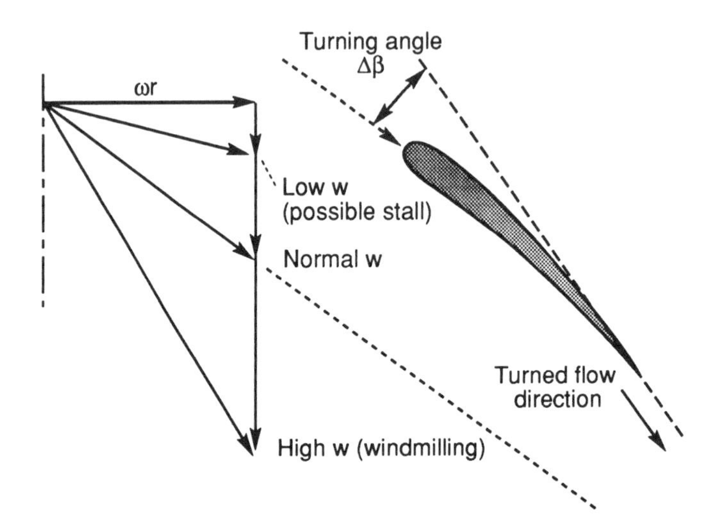

FIG. 14–13. Velocity vectors into a rotor blade for a flow with fixed ωr and three values of w.

With this physical picture in mind, it is evident that the aerodynamic performance of the blade as a force "generator" is optimal when its aerodynamic efficiency is highest. This efficiency is usually described by aerodynamicists as the lift to drag ratio (L/D), since an undesirable drag accompanies the generation of the lift force. Typically, an airfoil or blade will exhibit a maximum L/D at an angle of attack between zero and the stall value. The angle of attack is equivalent to the angle between the wind vector ahead of the blade to the trailing edge orientation (see Fig. 14–13).

It is to be expected therefore that a real compressor stage performance will differ from a design point model by the following:

1. Expected performance will not be achievable near stall (i.e., low w).
2. Stage efficiency will be highest near operation at maximum L/D, falling off to higher and lower values of w.
3. The deviation between desired flow direction and the actual direction increases as loading is increased, again at low w.

Having made the case that θ_1 and β_2 are approximately constant, eq. 14–23 can be recast as an expression for pressure ratio

$$\pi_{cs} = \frac{p_{t2}}{p_{t1}} \approx 1 + \left\{ \eta_s \frac{1}{1 + \dfrac{\gamma - 1}{2} M_1^2} \frac{(\omega r)^2}{a_1^2} \left[1 - \frac{w_1}{a_1} \frac{a_1}{\omega r} (\tan \beta_2 + \tan \theta_1) \right] \right\} \tag{14–24}$$

In summary, the results of eqs. 14–22 and 14–24 and Fig. 14–11 suggest that for a large pressure rise per stage, the following are required:

1. Large M_1 (see eq. 14–22) within the limit of shock losses which reduce efficiency
2. High efficiency
3. Large turning angle (Fig. 14–11)
4. θ, β near 45° for reasonable balance between pressure ratio and flow area

14.3.5. Compressor Parameters

Because the two Mach number parameters defined in eq. 14–21 arise naturally in the expression for total pressure ratio, one may use them to characterize the compressor stage operation on a map. By writing the speed eq. 14–24 in terms of these Mach numbers, one obtains a relation that is sketched in Fig. 14–14 where the kinship to the results in Fig. 14–12 should be evident. The line labeled (*) is meant to describe the ideal characteristic for which the real characteristic is as shown. Note that the efficiency contours lead to a maximum value along a line of constant tangential Mach number.

The values w and r in the discussion so far have been identified as mean values descriptive of the flow. For a given compressor, however, mean axial velocity is more conveniently expressed as mass flow rate, and the more meaningful radius is the tip radius of the stage which determines the tip velocity of the rotor.

Traditionally, for compressors as complete devices, the tangential (based on tip radius) and axial Mach numbers are not used. Rather, a corrected rotational speed and a corrected weight flow rate are used because these quantities are easier to measure.

Corrected Speed

The corrected speed is defined by

$$N_c = \text{Corrected speed} = \frac{N(\text{in RPM})}{\sqrt{\theta_0}} = \frac{N}{\sqrt{\dfrac{T_{t0}}{288\text{ K or }519\text{ R}}}} \qquad (14\text{–}25)$$

where θ (here θ_0) is as defined by this equation. The use of θ unfortunately conflicts with angle notation, but is traditional. The reference temperature is a nominal standard day value. The *total* temperature is used to characterize the inlet condition. The relation to the tip Mach number is not direct (but quite close) because the *static* temperature is used in the definition of Mach number. Thus

$$\frac{N}{\sqrt{\theta}} = f\, M_{\theta,\text{tip}} \qquad (14\text{–}26)$$

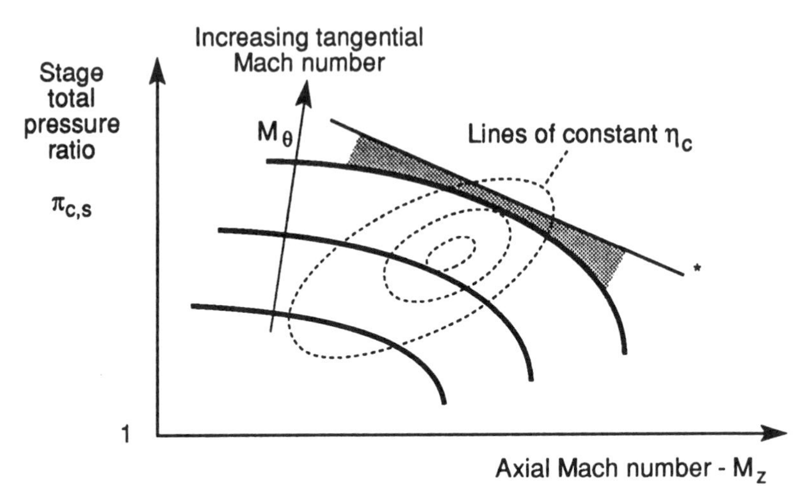

FIG. 14–14. Realistic stage characteristic similar to Fig. 14–12.

where the proportionality factor, f, is dimensioned and varies with absolute Mach number:

$$f = \frac{\sqrt{\gamma R T_{\text{ref}}}}{2\pi(60)r_t}\sqrt{\frac{T_1}{T_{t1}}} \approx \frac{0.9}{r_t}\sqrt{\frac{T_1}{T_{t1}}}$$

with r_t in meters. Here station 1 may be taken at the inlet of the first rotor.

Corrected Weight Flow Rate

The mass flow rate, $\dot{m}$, is given by the product $\rho w A$. In terms of stagnation properties of the inlet air, this gives (from eq. 12–10):

$$\left.\begin{aligned} \frac{\dot{m}}{A_1} &= \rho_1 w = \frac{p_{t0}}{\sqrt{RT_{t0}}}\sqrt{\gamma}\, M M_{z1}\left(1 + \frac{\gamma - 1}{2} M_{z1}^2\right)^{-(\gamma+1)/2(\gamma-1)} \\ &\text{or} \\ \frac{\dot{m}\sqrt{T_{t0}}}{p_{t0}} &= \left[\frac{A_1}{\sqrt{R}}\sqrt{\gamma}\right] M_{z1}\left(1 + \frac{\gamma - 1}{2} M_{z1}^2\right)^{-(\gamma+1)/2(\gamma-1)} \end{aligned}\right\} \qquad (14\text{–}27)$$

It follows that there exists a unique functional relationship between M_z and a new parameter, $\dot{m}\sqrt{T_{t0}}/p_{t0}$, for specified A, γ, and R. For given compressor and working gas A and R are constants. The relationship between $\dot{m}\sqrt{RT_{t0}}/p_{t0}A_1$ and M_{z1} is shown in Fig. 14–15. The new parameter is often normalized by standard values for the temperature and pressure as

$$\dot{m}_C \equiv \frac{\dot{m}\sqrt{\dfrac{T_{t0}}{288\text{ K or }519\text{ R}}}}{\dfrac{p_{t0}}{1.013 \times 10^5\text{ Pa or }14.696\text{ psia}}} \equiv \frac{\dot{m}\sqrt{\theta_0}}{\delta_0} \qquad (14\text{–}28)$$

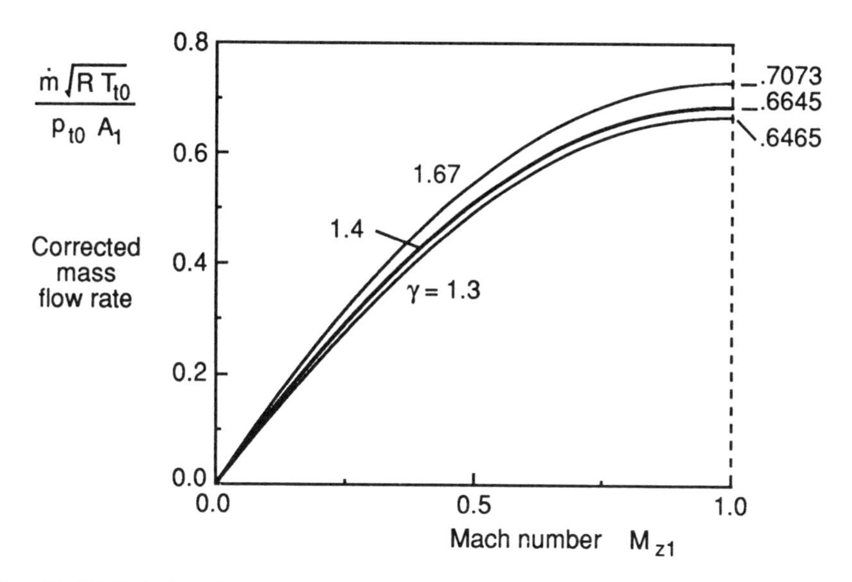

FIG. 14–15. Relation between corrected mass flow rate and axial Mach number (eq. 14–28).

This is the so-called corrected mass flow rate. δ and θ are defined here as the corrected pressure and temperature. δ and θ are unity when conditions are, in fact, standard. In common practice (in the United States, in particular), one often finds the use of corrected *weight* flow rate using the symbol, W_c, in lb/sec.

14.3.6. *Radial Variation of Turning Angle*

The above analysis describes the behavior of properties along a typical streamline. In practice, one would want to design the stage in such a manner as to realize a radially uniform total pressure profile. From eq. 14–18 or 14–20 it is evident that this requires that the product ($r\Delta \tan \beta$) be uniform in r, if w is reasonably uniform. In other words, the largest turning must be at the hub. This, in turn, results in the expression for total pressure when the tip radius is fixed being expressible as

$$\left(\frac{p_{t2}}{p_{t1}}\right) = 1 + \text{constant}\,(\tan \beta_1 - \tan \beta_2)_{\max} \frac{r_h}{r_t} \qquad (14\text{–}29)$$

Assuming that the turning angles are small and β's are of the order of 45°, the difference in $\tan \beta$ is approximately proportional to $\Delta\beta$. The pressure ratio per stage for a compressor is therefore proportional to the hub/tip radius ratio. This implies that an engine with a large frontal area (i.e., hub/tip radius ratio close to unity) can be made with a large pressure ratio while a slender engine with small radius ratio must be made with a greater number of stages for a chosen cycle pressure ratio. Figure 14–16 illustrates this idea which is made more precise (in Section 14.4) by considering the flow in two dimensions.

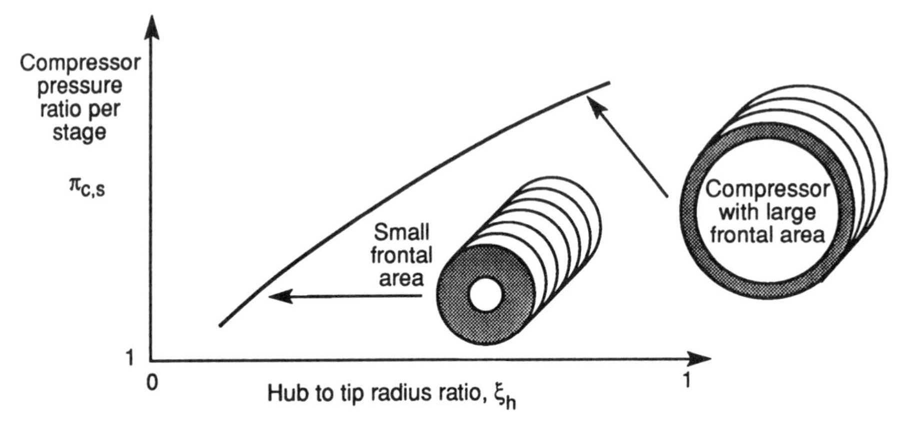

Fig. 14–16. Impact of stage total pressure ratio capability for two compressor designs with different hub/tip radius ratios. The compressors are designed for the same overall pressure ratio $\pi_c \approx \pi_{c,s}^N$ and mass flow rate. Note the differing number of stages.

Specifically it will be seen that the axial velocity, w, cannot be held to be uniform, and these conclusions must be altered, although they are qualitatively correct.

14.3.7. *Compressor Characteristics*

For a multistage compressor the characteristics are similar to those of a single stage. There are important differences, however. These differences can be attributed to the changing density brought about by changing the pressure along the compressor length. This changing density affects the local axial flow velocity because the flow rate through the machine must be constant if no flow is allowed to exit the compressor between stages.

Consider a test rig for a compressor consisting of atmospheric air entry and a tank at elevated pressure into which the compressor discharge is brought. Mechanical power is supplied as necessary. Figure 14–17 illustrates the compressor with a number of stages and this test setup.

Consider first *normal* operation of the compressor to a high exhaust pressure. The pressure will have an axial variation something like that sketched in Fig. 14–17. The density will follow with a value consistent with the stage efficiencies. At this design condition, the axial velocity is chosen to fix the flow area for the design mass flow rate.

Now imagine that the rotational speed is kept constant, and the exhaust valve on the back pressure tank is opened wider to reduce the downstream pressure somewhat. After a short adjustment time period, the flow is again steady. The lowered exit pressure now also results in a reduction in the gas density near the exit and, for the approximately constant mass flow rate, an increase in the axial velocity there. From the discussion associated with Fig. 14–13, the increased velocity leads to local unloading of the stages near the exit so that the overall pressure *change* demanded is accommodated by the

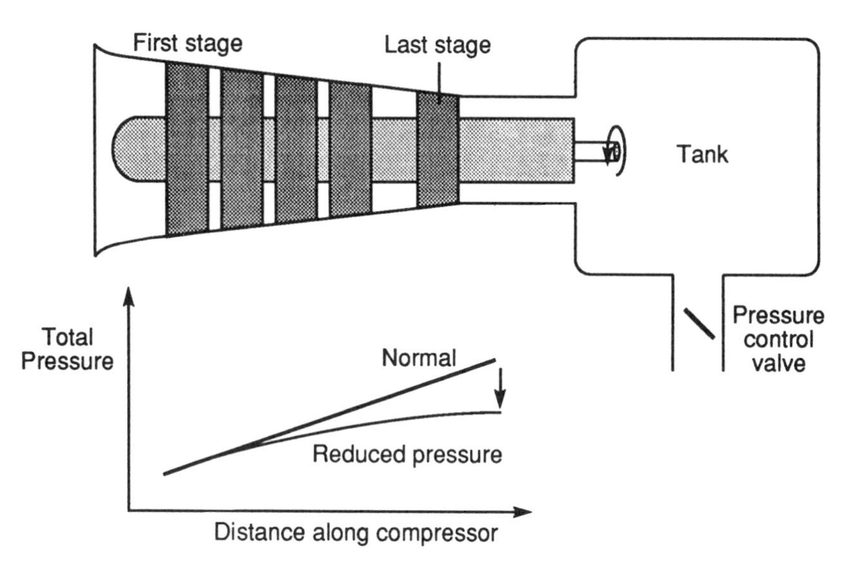

FIG. 14–17. A test rig for a multistage axial flow compressor. Pressure variation along the compressor as back pressure is changed.

last stages. The front stages do not experience a significant change in operating condition so that the mass flow rate stays constant.

This explains why constant speed lines on a multistage compressor map are similar to those of a single stage map but are much steeper. In contrast to a stage, the mass flow rate through a compressor is consequently much less sensitive to pressure ratio changes. Figure 14–18 shows a typical compressor map. The overall features of the compressor are described in the caption. It is typical of most compressors. Worthy of note are the following:

1. Maximum pressure ratio increases with increasing rotational speed, N.
2. For all constant N lines, a maximum pressure ratio is reached where the compressor stalls. The locus of these points is the stall line.
3. At a given N, the pressure ratio decreases when mass flow rate (or w) is increased. Unlike a compressor stage, the complete compressor has a steep constant N line, particularly at high N.
4. There is a point on the map where efficiency is highest, falling to lower values, as would be expected from the discussion of stage performance.

14.3.8. *Design for Reduced Power Operation*

A compressor is often integrated into a gas turbine engine and that engine must be able to operate at less than design power. In Chapter 16, the integrated operation of an engine is considered. The resulting combined performance of the compressor, turbine and load is that the compressor operating points lie along a so-called *operating line*. Satisfactory operation of less than maximum pressure ratio is important because an engine may spend a significant fraction of its operating life time at such conditions. At the very least, the operating

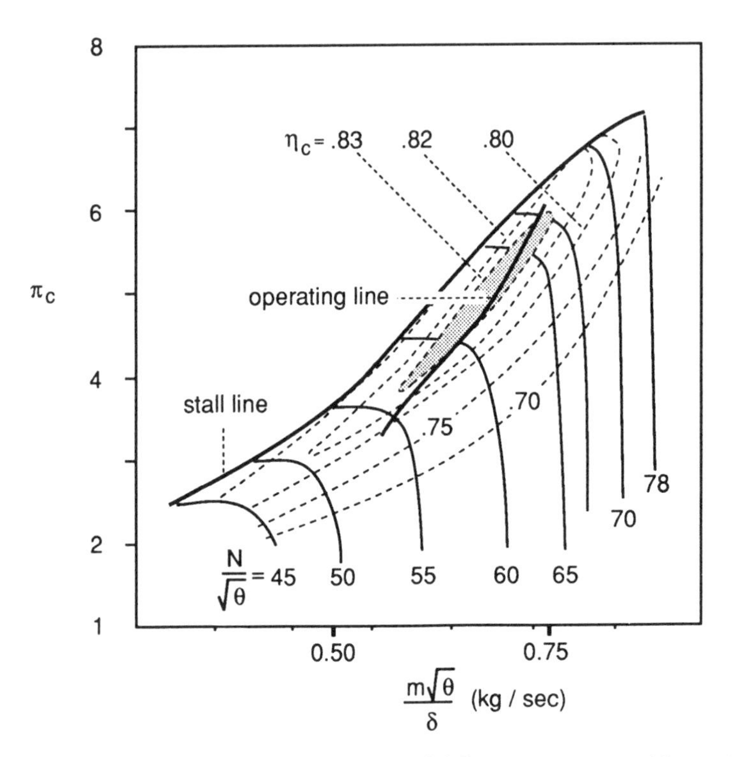

Fig. 14–18. Compressor map of a multistage axial flow compressor with a radial flow last stage. Note that size inferred by the mass flow rate and the rotational speed is small. Rotational speeds are in thousands of RPM. Operating line is for a compressor-turbine combination in an engine.

characteristics at reduced power affect the the power requirements for the starting process of such engines.

A typical operating line is as sketched in Fig. 14–18. Generally, mass flow rate, rotational speed, and pressure ratio decrease away from design point. The consequence for the front of the compressor (i.e., its first stages) is that two effects (see Fig. 14–13) compete: (1) Reduced w leads operation toward stall, whereas (2) Reduced ωr tends to unload the blades by decreasing the rotor turning angle as shown in Fig. 14–8.

Overall, the tendency is for the first to dominate so that the front of the compressor tends to stall. At the rear, the locally lower pressure leads to lower density there and therefore higher axial velocity. This means that the rear is ineffective as a compressor because the blades are "unloaded." At the higher throughflow choking is approached. The velocity diagrams are as shown in Fig. 14–19.

To minimize such effects and allow the engine to operate satisfactorily over a broad set of operating conditions, a number of means may be used:

1. The compressor can be designed to operate near stall at the rear and lightly loaded at the front at the design condition,

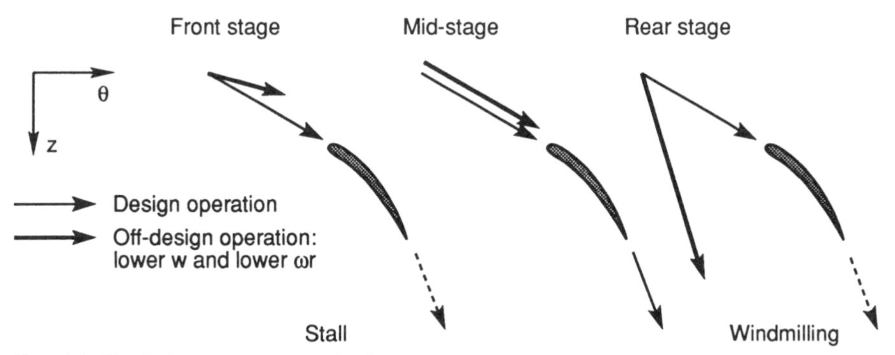

Fig. 14–19. Axial compressor velocity vectors at design and reduced power of an engine. Three locations of interest. Note relation to Fig. 14–13 where ωr is constant.

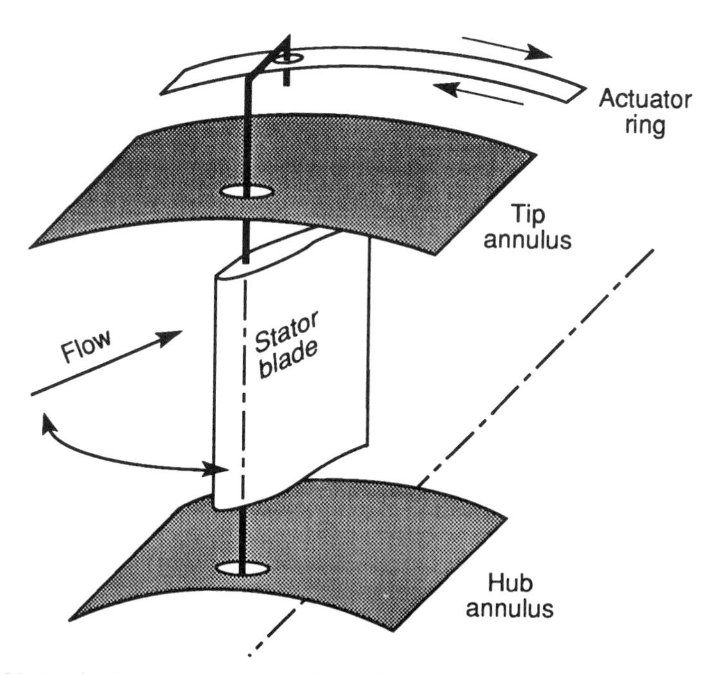

Fig. 14–20. Variable geometry stator driven by an actuator ring that is moved by a control mechanism to vary the incidence of the blades.

2. The compressor may be built with variable stator geometry which is controlled by measured engine performance parameters correlated to the gas flow angles. Figure 14–20 shows such an arrangement. A linear actuator moves the ring through a few azimuth degrees as required and thereby rotates all stator blades of the stage. Such control of *rotor* blades is generally impractical.
3. Multiple spools may be (and often are) employed to allow the front (low pressure or LP) portion of the compressor to remain at relatively high speed and simultaneously allow the high pressure (HP) portion to slow to

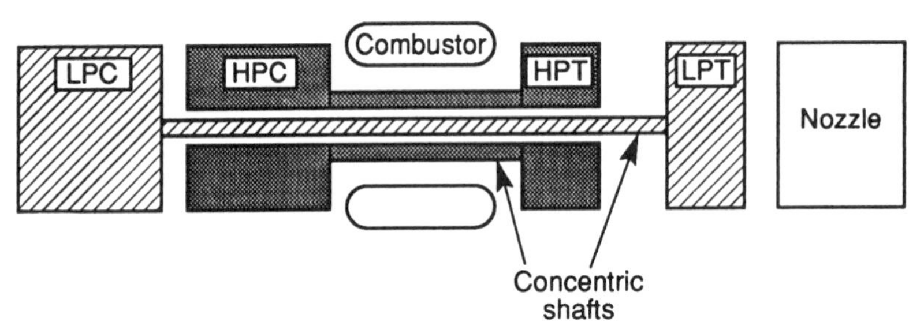

FIG. 14–21. Twin spool jet propulsion engine that operates at two shafts speeds. The letters refer to low (L) and high (H) pressure (P) compressor (C) or turbine (T), as commonly used in industrial practice.

lower speed than it would if the compressor were a single unit (Fig. 14–2b). In practice, two spools are common in aircraft engines. The Rolls Royce RB-211 uses three spools. Figure 14–21 illustrates a twin spool turbo jet engine. In a way, the combination of HPC, combustor, and HPT may be thought of as an engine (or better, as an enthalpy source) within an engine.

4. For starting, where the pressures in the compressor, at the exit in particular, are very low, the excess velocity at the rear of the compressor may be so large that reduction can only be accomplished when part of the flow is vented out of the compressor during low pressure operation. This reduces w at the rear and therefore avoids windmilling and flow choking there. Midcompressor bleed valves are in common use and are shut once the engine is in self-sustaining operation, although they may be open during idling. The opening at the top center of the compressor in Fig. 14–2b is such a bleed valve.

14.4. AXIAL COMPRESSOR THROUGHFLOW THEORY

The discussion to this point has stressed the behavior of a typical or mean streamtube. In compressors with small hub/tip radius ratio (compact machines), considerable variation in velocity vector orientation is associated with the radial pressure gradients experienced. This section is a discussion of the stage aerodynamics for a blade row operating at or near its design point. The objective is to highlight the role played by the hub/tip radius ratio in determining the geometric design of the blading.

14.4.1. *Radial Pressure Gradient*

The theory developed here is to illustrate the consequences of the radial pressure gradient that must exist because the annular flow has a component in the tangential direction and thus experiences a centripetal acceleration. The algebraic development initially assumes that the variation of axial velocity from

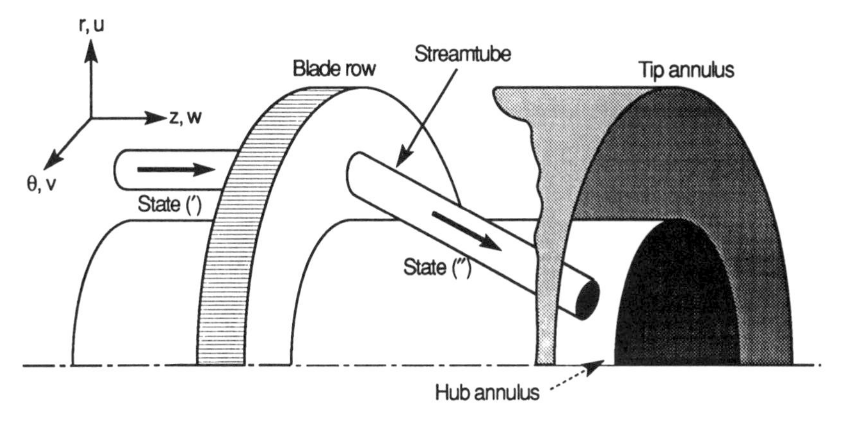

FIG. 14–22. Blade row configuration. (′) (″) denote initial and final states which are sufficiently far from the row for flow to be steady.

hub to tip is small to allow a linearization of the equations. At the conclusion of this section, this assumption is dropped and results are given for the more general case.

Consider a single blade row (rotor or stator) and assume the flow is incompressible (ρ = constant). Figure 14–22 shows the "far-field" interaction between a streamtube and a blade row. The Euler turbine equation for incompressible and loss-free flow is applied across the blade row. The more general description of compressible flow with entropy production is treated elsewhere. See, for example, refs. 14–6 and 14–7. Here, the emphasis is on the primary physical variable behavior. The notation for the description of states between blade rows is identical to that used in Section 14.2.2, i = 0, 1, 2, 3 for the entry to the IGV, to the rotor, to the stator, and again to the rotor respectively (Fig. 14–7). Thus a generalized eq. 14–15 is

$$p_i + \tfrac{1}{2}\rho[u_i^2 + w_i^2 + v_i(v_i - 2\omega r)] = p_{i+1} + \tfrac{1}{2}\rho[u_{i+1}^2 + w_{i+1}^2 + v_{i+1}(v_{i+1} - 2\omega r)] \tag{14–30}$$

This relation applies to both rotor and stator, with $\omega = 0$ in the latter case. Far upstream and far downstream of the blade row, the centrifugal forces must be balanced by radial pressure gradients, that is,

$$\frac{\partial p_i}{\partial r} = \frac{\rho v_i^2}{r_i} \tag{14–31}$$

which applies for a hub/tip radius ratio, r_H/r_T, greater than about 0.6. For the axial stage, far from the blade row, one may assume the following:

1. small u (i.e., $u \ll V$)
2. $\partial/\partial z = 0$

3. $\partial/\partial\theta = 0$ (large number of blades around the annulus)
4. $r_{i+1} = r_i$ (small streamtube displacement constrained by the geometry)

The assumptions 2 and 3 together imply that $\partial/\partial r$ becomes d/dr. Then by differentiating eq. 14–30, and substituting the expressions for the pressure gradients (eq. 14–31), one obtains

$$\frac{\rho v_i^2}{r} + \frac{\rho}{2}\left(2w_i \frac{dw_i}{dr} + 2v_i \frac{dv_i}{dr} - 2\omega \frac{d(v_i r)}{dr}\right) = \text{same terms with } (i+1) \quad (14\text{–}32)$$

With the identity

$$\frac{v^2}{r} + v\frac{dv}{dr} = \frac{v}{r}\left[v + r\frac{dv}{dr}\right] = \frac{v}{r}\frac{d}{dr}(vr) \quad (14\text{–}33)$$

the Euler equation becomes

$$w_{i+1}\frac{dw_{i+1}}{dr} - w_i\frac{dw_i}{dr} = \left(\frac{v_i}{r} - \omega\right)\frac{d}{dr}(v_i r) - \left(\frac{v_{i+1}}{r} - \omega\right)\frac{d}{dr}(v_{i+1} r) \quad (14\text{–}34)$$

Thus, if one prescribes the inlet profile (v_i and w_i), and the outlet tangential velocity, v_{i+1} (or work), as functions of r, $w_{i+1}(r)$ can be computed. The following examples of elementary velocity profiles are of interest: free vortexlike swirl and solid-body rotation. Linear combinations of these profiles as well as others may be considered using the approach described.

14.4.2. *Blade Row Velocity Increments*

Free Vortex Blading

A tangential velocity profile that varies as $1/r$ is termed free vortex-like. Consider a blade row with such an inflow and a geometry that causes a similar outflow:

$$v_i r = B' w_0 r_T, \qquad v_{i+1} r = B'' w_0 r_T \quad (14\text{–}35)$$

where w_0 and r_T are constants introduced to nondimensionalize the B's which are the nondimensional scaling factors that govern the magnitude of the tangential velocities. The tangential velocity profiles are as shown in Fig. 14–23.

It follows from eq. 14–34 that for such blading,

$$w_{i+1}\frac{dw_{i+1}}{dr} - w_i\frac{dw_i}{dr} = 0 \qquad \text{or} \qquad w_{i+1}^2 - w_i^2 = \text{constant} \quad (14\text{–}36)$$

and from the equation for the stage total pressure rise (eq. 14–30),

$$p_{t,i+1} - p_{t,i} = \rho\omega r(v_{i+1} - v_i) = \rho w_0 \omega r_T (B'' - B') = \text{constant} \quad (14\text{–}37)$$

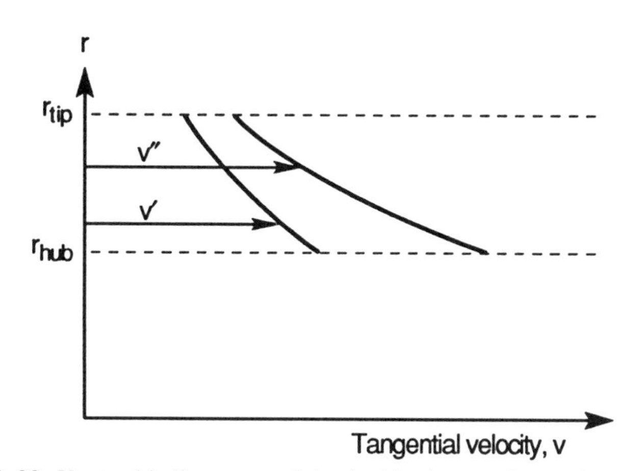

FIG. **14–23.** Vortex blading tangential velocities into and out of blade row.

Thus for free-vortex blading the total pressure increment is *uniform in r*. This uniformity is desirable to minimize subsequent mixing losses. In compressors where the flow is designed to have a uniform total pressure at the exit, this uniformity may be achieved by having each stage produce a uniform fluid.

Solid-body Rotation

For the following, it is convenient to define nondimensional geometric variables by

$$\xi \equiv \frac{r}{r_T} \quad \text{and} \quad \xi_h \equiv \frac{r_H}{r_T} \quad \text{so that } \xi_h \le \xi \le 1 \tag{14–38}$$

and scale all velocities by the uniform IGV inlet velocity, w_0.

In order to maintain approximate geometrical similarity of the velocity triangles at different radii, IGV blades should introduce a tangential velocity which increases with increasing radius. The most basic form of this type is a linear relationship between v_1 and r. Other forms can also be used. Thus, to illustrate the effect of such swirl distributions, let

$$\left.\begin{aligned} v_i = v_0 = 0; \qquad v_{i+1} = v_1 = Aw_0\frac{r}{r_T} \\ \text{or} \qquad \frac{v_1}{w_0} = A\xi \end{aligned}\right\} \tag{14–39}$$

where the subscripts 0, 1, 2, 3 correspond to the conditions shown in Figs. 14–6 through 14–9 and are summarized in Fig. 14–7. A describes the magnitude of the amount of swirl introduced.

The first task is to find $w_1(r)$ at the IGV exit. The entry conditions are given in eq. 14–39 and $\omega = 0$. The governing equation (14–34) is then

$$w_1 \frac{dw_1}{dr} = -\frac{Aw_0}{r_T}\frac{d}{dr}\left[Aw_0 \frac{r^2}{r_T}\right]; \qquad \tfrac{1}{2}w_1^2 = -(Aw_0)^2\left(\frac{r}{r_T}\right)^2 + \frac{1 - 2A^2c}{2} w_0^2$$

where the second equation is obtained after integration and gives, after rearrangement,

$$\left(\frac{w_1}{w_0}\right)^2 = 1 - 2A^2[\xi^2 - c_0] \tag{14–40}$$

The constant of integration is written in this form for convenience. Note that the axial velocity, w_1, is close to uniform across the blade row when the correction term on the right is sufficiently small. To evaluate the constant, one must apply conservation of mass, assuming that the flow areas across the IGV blade row (i.e., areas $A_1 = A_0$) are constant. Thus

$$2\pi\rho \int_{r_H}^{r_T} w_1 r \, dr = 2\pi\rho w_0 \left(\frac{r_T^2}{2} - \frac{r_H^2}{2}\right) \tag{14–41}$$

The quadratic form of the expression for w_1/w_0 (eq. 14–40) is difficult for substitution into eq. 14–41. With $w_1/w_0 \approx 1$, however, one may linearize eq. 14–40 to obtain a form of eq. 14–41 which is easy to integrate. The result is

$$\left(\frac{w_1}{w_0}\right) \cong 1 - A^2[\xi^2 + c_0] \tag{14–42}$$

Substituting this for w_1 in eq. 14–41, the constant is found to be

$$c_0 = -\frac{r_T^2 + r_H^2}{2r_T^2} = -\frac{1 + \xi_h^2}{\mathbf{2}} \tag{14–43}$$

Thus

$$\frac{w_1}{w_0} = 1 + \frac{A^2}{2}\{1 + \xi_h^2 - 2\xi^2\} \tag{14–44}$$

This velocity distribution is shown in Fig. 14–24 for $A = 1(\approx 45^\circ$ turning) and $\xi_h = r_H/r_T = 0.6$ and 0.8. Note the smaller hub/tip radius ratio makes the axial velocity more nonuniform. Note also that a smaller swirl reduces this non-uniformity. Similar results without the linearization are shown in Section 14.4.3.

In general, the retardation of the flow at the tip leads to a requirement for blade twist which may be undesirable because a twisted blade is subject to greater deformation than an untwisted blade under centrifugal loading. Essentially the twist results from relatively large axial velocities at the root of the blade, where tangential velocities are low, and vice versa. The twist may be visualized as the angle β, although the chord orientation angle may be more

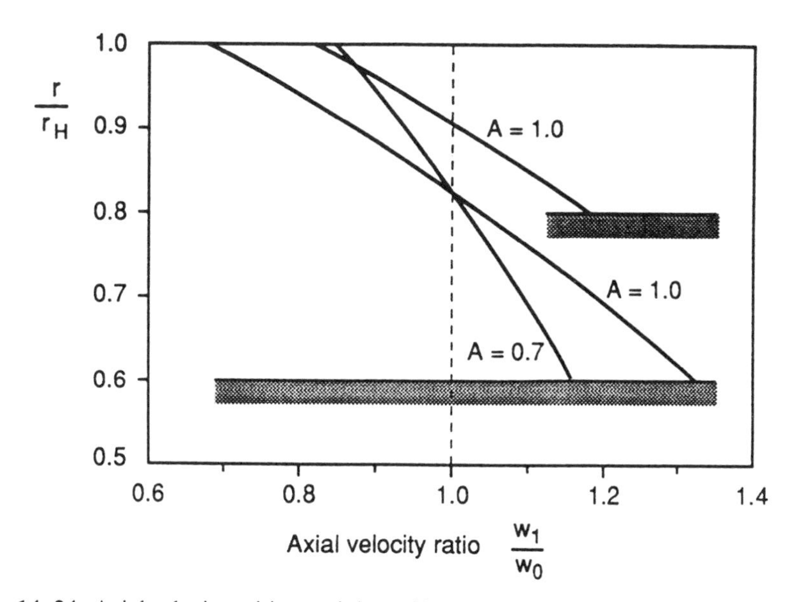

Fig. 14–24. Axial velocity exiting an inlet guide vane that initiates solid body rotational motion. Linearized theory. Note roles of A and ξ_h.

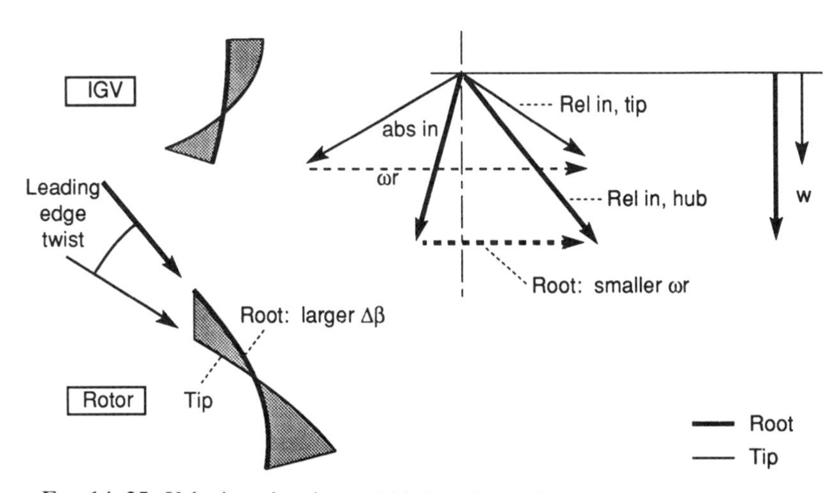

Fig. 14–25. Velocity triangles and blade schematic at root and tip of a rotor.

useful when one looks at a blade from the structural viewpoint. In any case, compressor blades tend to have sufficiently small turning angles and the two viewpoints are closely related. Thus the rotor blade geometry tends to be as shown schematically in Fig. 14–25.

The trailing edge angle θ_1 from the IGV (or a stator) and the rotor inlet angle β_1 are shown in Fig. 14–26 for the conditions noted. The figure is a plot of the velocity vectors $[v + w]$ for the stator and $[(v - \omega r) + w]$ for the rotor so that θ_1 and β_1 are as shown. The heavy lines are the locus of the vector end points: solid for $\xi_h = 0.6$ and dashed for $\xi_h = 0.8$. Note that the twist of the

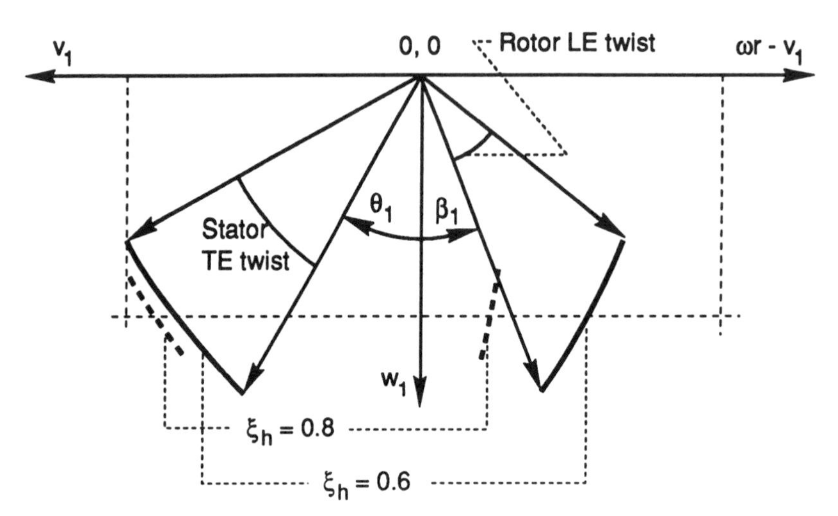

FIG. 14–26. Velocity vectors seen at stator (or IGV) exit (trailing edge = TE), left, and rotor inlet (leading edge = LE), right. Axial velocity is downward. Vectors at hub have small θ and β. Note smaller twist for hub/tip radius ratio $\zeta_h = 0.8$ contrasted to 0.6. A = 1 and $\omega r/w_1$ (at the hub) = 1.

rotor blade (as measured by the leading edge orientation and the vectors indicated) is much more severe for smaller hub/tip ratio, of the order of 30° for $\xi_h = 0.6$. This twist is of the order of 10° for $\xi_h = 0.8$ for which the vectors are not shown to keep the figure readable. Note how $A = 1$ gives approximately 35–45° orientation for the median streamline when $\xi_h = 0.6$ in both stator and rotor. For $\xi_h = 0.8$ this picture becomes somewhat more asymmetrical: The mean rotor angle is reduced to near 15°.

Rotor Vortex Increment and Solid Body Rotation through the Rotor

For the *rotor* the subscript "i" corresponds to state 1. The flow conditions into the rotor are, from eqs. 14–39 and 14–44 (linearized result),

$$\frac{v_1}{w_0} = A\frac{r}{r_T} = A\xi \qquad \text{and} \qquad \frac{w_1}{w_0} = 1 + \frac{A^2}{2}(1 + \xi_h^2 - 2\xi^2) \tag{14–45}$$

When a free-vortex-like velocity increment is added by the rotor to the solid-body rotation produced by the guide vanes, one has

$$\frac{v_2}{w_0} = A\frac{r}{r_T} + B\frac{r_T}{r} = A\xi + \frac{B}{\xi} \tag{14–46}$$

To realize a typical total pressure rise seen in practice, the scaling constant, B, can be estimated to be of the order of 0.3 from eq. 14–37. The rotor work will result in a radially uniform total pressure exiting the rotor within the assumption of incompressible and loss free flow. The axial rotor exit velocity is obtained from application of eq. 14–34:

$$w_2 \frac{dw_2}{dr} - w_1 \frac{dw_1}{dr} = \frac{Bw_0 r_T}{r^2} \frac{d}{dr}\left(Aw_0 \frac{r^2}{r_T}\right) = -2ABw_0^2 \frac{1}{r}$$

Integrating and factoring

$$\frac{(w_2 - w_1)}{w_0} \frac{(w_2 + w_1)}{2w_0} = -2AB \ln(r) + c_1 \tag{14–47}$$

Since the axial velocity would not be expected to vary by a large amount, it is reasonable to assume that the axial velocities w_0, w_1, and w_2 are of the same order. Consistent with the linearization made to calculate c_0, eq. 14–47 simplifies to read:

$$\frac{w_2 - w_1}{w_0} \cong -2AB \ln(r) + c_1 \tag{14–48}$$

This assumption is not employed in Section 14.3.3 for a numerical result to be compared with the present, more physical approach. The mass flow into the rotor equals the mass flow out:

$$\int_{r_H}^{r_T} \frac{w_2 - w_1}{w_0} r\, dr = 0 = \int_{r_H}^{r_T} [-2AB \ln(r) + c_1] r\, dr \tag{14–49}$$

Thus

$$c_1 = 2AB\left[\frac{r_T^2 \ln(r_T) - r_H^2 \ln(r_H)}{r_T^2 - r_H^2} - \frac{1}{2}\right] \tag{14–50}$$

and

$$\frac{w_2 - w_1}{w_0} = 2AB\left[\frac{r_T^2 \ln(r_T) - r_H^2 \ln(r_H)}{r_T^2 - r_H^2} - \frac{1}{2} - \ln(r)\right]$$

Simplifying the expression,

$$\frac{w_2 - w_1}{w_0} = AB\left[\frac{-\xi_h^2}{1 - \xi_h^2} \ln \xi_h - \tfrac{1}{2} - \ln \xi\right] = AB \,\mathrm{fcn}\left(\frac{r}{r_H}, \frac{r_H}{r_T}\right) \tag{14–51}$$

For $r_H/r_T = 0.6$, one obtains the results shown in Fig. 14–27. The velocity increment is divided by AB for generality. Note that for smaller ξ_h (=hub/tip radius ratio), the axial velocity variation is larger. The velocity distributions given by eq. 14–51 are shown schematically in Fig. 14–28.

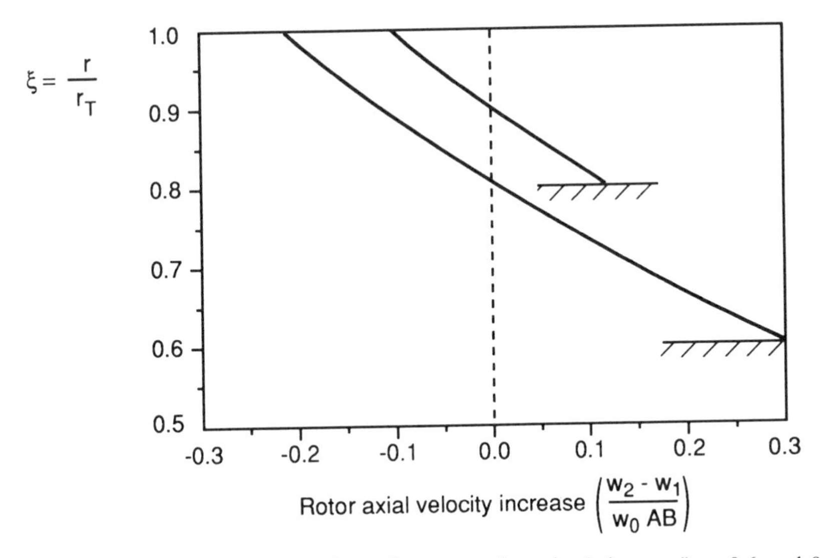

FIG. **14–27.** Axial velocity change through a rotor, linearized theory. $\xi_h = 0.6$ and 0.8.

Stator

The results for the rotor shown above are independent of ω, that is, the axial velocity change $w_2 - w_1$ is a function of radial location only:

$$\frac{w_2 - w_1}{w_0} = ABf(r) \tag{14–52}$$

Thus the stator, for which $\omega = 0$, can be used to return state 2 back to state 1 (state 3 = state 1). To do this one must add a tangential velocity change such that

$$\frac{w_3 - w_2}{w_0} = -\frac{w_2 - w_1}{w_0} = -ABf(r) \tag{14–53}$$

Since A has to do with the inlet guide vane, it is fixed and the sign of $w_2 - w_1$ may be changed by having B (stator) = $-$B (rotor). Thus v_2 and w_2 are the stator inlet conditions and v_3 and w_3 are equal to v_1 and w_1 with the sign of B changed. Since B does not appear, one has:

$$\left.\begin{aligned} &\frac{v_3}{w_0} = \frac{v_1}{w_0} = A\left(\frac{r}{r_T}\right) = A\xi \\ \text{and} \quad &\frac{w_3}{w_0} = \frac{w_1}{w_0} = 1 + \frac{A^2}{2}[1 + \xi_h^2 - 2\xi^2] \end{aligned}\right\} \tag{14–54}$$

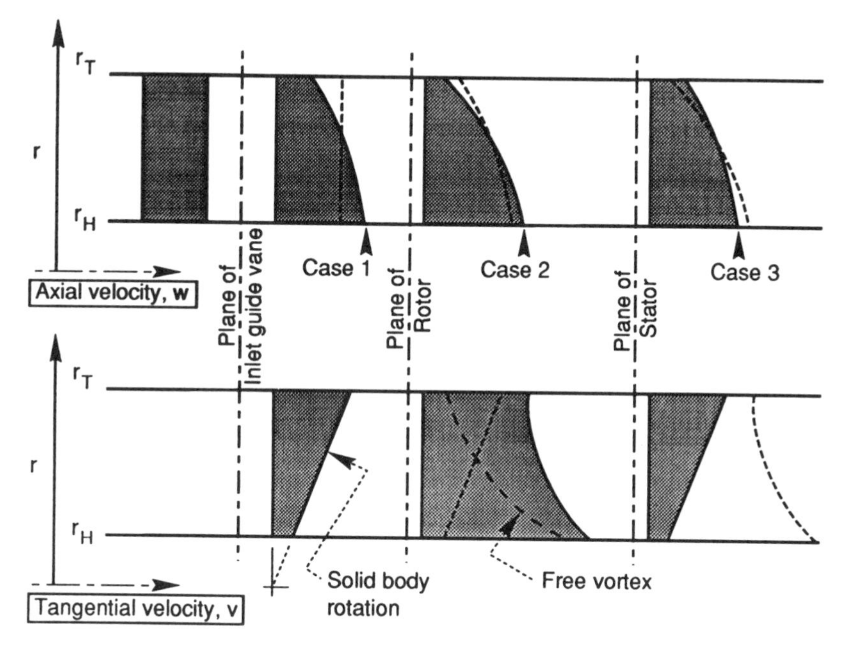

FIG. **14–28.** Summary of flow velocities through an IGV, a rotor and a stator with radial pressure equilibrium. Upper portion shows axial velocities and lower shows tangential velocities. The cases refer to the calculation of the unknown velocity distributions discussed in the text.

The result, as determined by eqs. 14–39 and 14–47 is, as shown in Fig. 14–28, that a vortical swirl is added by the rotor and then removed by the stator. The figure summarizes the velocity vectors through IGV, rotor, and stator. For this blading system, the radially uniform stagnation pressure rise is realized.

In addition to the swirl distributions described as solid-body rotation and free vortex, Horlock introduced (ref. 14.6) the so-called exponential distributions given by

$$rv = B + Cr$$

which is simply the inclusion of an element that is uniform in r (i.e. $v = C$) to the distributions considered in order to lessen the severity of the profiles.

14.4.3. *Numerical Results*

Flow through the Inlet Guide Vane

Equation 14–40 can be written without the linearization as

$$\frac{w_1}{w_0} = \sqrt{2A^2(c_2 - \xi^2)}; \qquad c_2 \equiv \frac{1}{2A^2} - c_0 \tag{14–55}$$

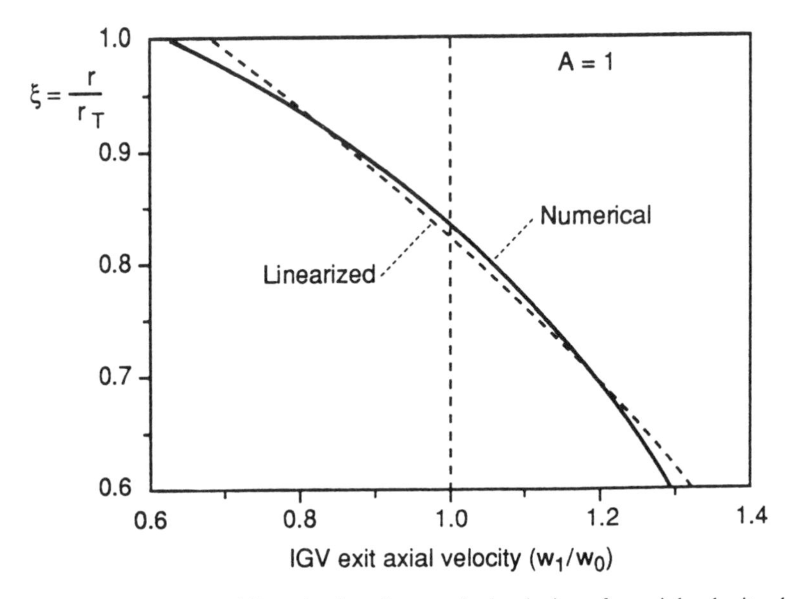

Fig. 14–29. Comparison of linearized and numerical solutions for axial velocity downstream of the IGV for $\xi_h = 0.6$.

so that the continuity equation (eq. 14–41) reads

$$\int_{\xi_h}^{1} \sqrt{2A^2(c_2 - \xi^2)}\xi \, d\xi = \tfrac{1}{2}(1 - \xi_h^2) \tag{14–56}$$

This is an equation for the constant c_2, which can be determined by evaluating the integral. The resulting nonlinear equation is

$$(c_2 - \xi_h^2)^{3/2} - (c_2 - 1)^{3/2} = \frac{3}{2\sqrt{2}} \frac{1}{A}(1 - \xi_h^2) \tag{14–57}$$

For the case examined above (i.e., $\xi_h = 0.6$ and $A = 1$), the constant $c_2 = 1.198$ which is about 1.5% larger than the value obtained from the linearized approach. A comparison of the corresponding velocity distributions is shown in Fig. 14–29. In general, the agreement is quite good, although the more exact solution gives lower values at both hub and tip radii.

Flow through the Rotor

The continuity equation applied to the flow behind the rotor states that

$$\int_{\xi_h}^{1} \frac{w_2}{w_0} \xi \, d\xi = \tfrac{1}{2}(1 - \xi_h^2) \tag{14–58}$$

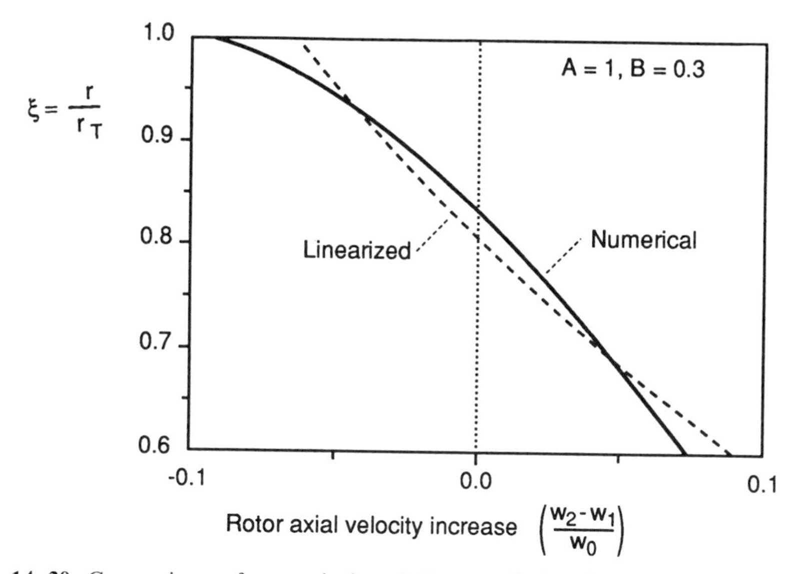

FIG. 14–30. Comparison of numerical and linear solution for rotor axial velocity change.

where the axial velocity is given by an integration of eq. 14–47:

$$\left(\frac{w_2}{w_0}\right)^2 = \left(\frac{w_1}{w_0}\right)^2 - 2AB \ln (\xi) + c_1 \tag{14–59}$$

The integration in eq. 14–58 must be evaluated numerically. For the conditions stated earlier the axial velocities in the annulus computed from the two approaches are shown in Fig. 14–30. Using a 20-step trapezoidal rule approximation to the integral, the constant $c_1 = -0.1087$. Qualitatively, the agreement is reasonable in spite of a significant difference in the curvatures of the profiles. Further, the error is somewhat large at the tip.

14.4.4. *General Analysis for Radial Equilibrium Flows*

The analysis described above is mathematically direct and leads to physical insight into the behavior of flow through the compressor and stators, limited by the approximations made to incompressible and loss-free flows. These same phenomena may be examined more generally by reducing the number of governing equations to one for an unfortunately greater mathematical abstraction. This discussion is a short review of the work of Oates as described in his text (Section 9.4) and papers (refs. 14–7 and 14–8), which is summarized here briefly to highlight the approach.

The governing equations are, as before, the momentum equation in the radial direction and the continuity equation. The latter is satisfied identically if the stream function is adopted as a replacement for the radial coordinate, r.

In one dimension, the axial velocity is related to the streamfunction, ψ, by

$$w = -\frac{1}{r}\frac{d\psi}{dr} \text{ from which } \frac{d}{d\psi} = -\frac{1}{wr}\frac{d}{dr} \tag{14–60}$$

Differential changes in total enthalpy are related by (eq. 14–3):

$$dh_t = dh + w\,dw + v\,dv \tag{14–61}$$

where dh is given by the Gibbs equation (eq. 4–14) for an isentropic process,

$$dh = \frac{dp}{\rho} \tag{14–62}$$

For radial equilibrium, the radial pressure gradient is balanced by an acceleration

$$\frac{v^2}{r} = \frac{1}{\rho}\frac{dp}{dr} \tag{14–63}$$

Combining these equations by eliminating dp/dr one obtains,

$$\frac{dh_t}{dr} = \frac{v^2}{r} + w\frac{dw}{dr} + v\frac{dv}{dr} \quad \text{or} \quad -w\frac{dw}{dr} = -\frac{1}{rw}\frac{dh_t}{dr} - \frac{v}{r}\left(\frac{-1}{rw}\frac{drv}{dr}\right) \tag{14–64}$$

In the same way that eq. 14–34 gives a solution for the variation of w, so does this, provided the variation of h_t and the angular momentum, rv, are specified. The last form of eq. 14–64 allows direct substitution of the stream function for r using eq. 14–60 so that

$$\frac{1}{r}\frac{d}{dr}\left(\frac{1}{r}\frac{d\psi}{dr}\right) = \frac{dh_t}{d\psi} - \frac{v}{r}\left(\frac{drv}{d\psi}\right) \tag{14–65}$$

The analysis problem is then reduced to specifying h_t and rv as functions of ψ, which is equivalent to specifying $v(r)$ in Section 14.4.2 where the total enthalpy was also specified to be uniform. This approach is evidently more general, allowing forms for h_t and rv to be of the more general form

$$\left.\begin{aligned} h_t &= a_1\psi + a_2 \\ \text{and} \qquad (rv)^2 &= (a_3\psi + a_4)^2 + a_5\psi \end{aligned}\right\} \tag{14–66}$$

The five constants allow significant flexibility in specifying the flow configuration changes to be carried out by the blade row. These forms, substituted into eq. 14–65, yield a differential equation which can be solved to give $\psi(r)$ in terms of Bessel functions. From this, axial velocities and thus flow angles, pressures, etc., may be determined. The reader may consult ref. 14–8, Section 9.4, for a discussion of several examples as well as a description of a general computation procedure. The reader should be cautioned that, however useful the approach may be from the viewpoint of understanding the physical implications of a situation described by complex equations, the practical fact is that numerical methods in widespread industrial use have eclipsed analytic approaches.

14.4.5. Blade Solidity

The discussion to this point has focused on the primary design variables that play a significant role in the performance of the axial flow compressor as a gas compression device. These design variables include hub/tip radius ratio, axial and tangential Mach numbers, together with the total Mach number, as well as the parameters that describe the loading on the airfoil such as the constant B which is a lift coefficient-like parameter of the blade. For the compressor to operate efficiently, the lift/drag ratio of the blade must be large. Further, for the compressor to be compact, it must have few stages (i.e., have high loading per stage). The requirements for low weight and high efficiency tend to be in conflict. In the aircraft propulsion field, there is a tendency to favor lightweight designs, whereas in the energy conversion field, high efficiency, consistent with low cost, tends to be more important.

As an example, consider that the compressor blade may be modeled as a bound vortex, much like the actuator disk described in Chapter 13. The axial influence of the disk decreases with increasing distance away from the disk in either the up- or downstream directions. If the disk imparts a change in axial velocity Δw, then the local influence on w falls as (ref. 14–9)

$$w = w_1 + \Delta w \exp\left(\frac{\pi z}{r_t - r_h}\right) \tag{14–67}$$

This implies that an aircraft compressor with a large annulus height $(r_t - r_h)$ may have to deal with local flow effects from one stage experienced at the next, if the spacing z is small. The failure to deal with such effects may result in an efficiency penalty. As long as this penalty has a smaller detrimental impact on the system than, say, allowing for greater spacing between stages and thus greater weight, this is a satisfactory compromise. In stationary terrestrial energy systems, this compromise may not have to be made in the same way.

The determination of an optimum design for a given task is difficult and draws heavily on experience. The field of compressor design is rich in the development of understanding and experience. It is particularly interesting to consider the limits that can be placed on the loading of the stage. The limit is intimately tied to the behavior of boundary layers. From a design point of view, greater adherence to the desired flow orientation may be enforced through use of blades with larger solidity, especially at high loading. The solidity is the ratio of blade chord to mean spacing between blades (see eq. 14–72 below), as shown in Fig. 14–31.

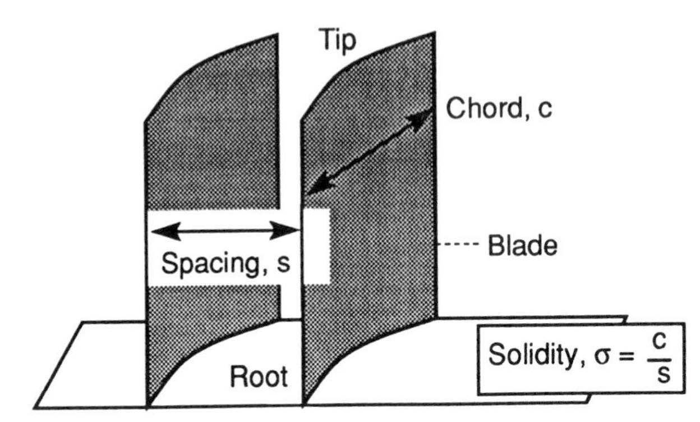

FIG. 14–31. Blade solidity definition. Spacing is measured at mid span.

The larger the solidity, however, the larger the surface to volume ratio of the duct which ultimately leads to efficiency reduction due to the larger influence of viscous effects. The solidity is the design parameter used to ensure that the average departure between blade trailing edge angle and flow angle is acceptably small (refs. 14–4, 14–5).

14.4.6. Boundary Layer Effects

For axial flow compressors, the importance of losses associated with boundary layers may be estimated very roughly by regarding the airfoils as thin flat plates on which two boundary layers grow. Such an estimate compared to experimental data shows that simple boundary layer flow friction is not the only irreversibility mechanism, nor even the dominant mechanism, and provides the motivation for further understanding of phenomena in axial flow compressors. It also exposes the role of blade design parameters as they affect viscous losses. The assumption of flat plate boundary layer on the blades is crude because 3-D flow effects, pressure gradients, and compressibility play important roles in determining boundary layer characteristics and thus blading performance.

Consider an axial flow compressor, with an inlet hub/tip radius ratio, ξ_{h0}. The compressor has a fixed tip radius, and is designed to maintain a constant axial velocity. The flow along each blade is assumed to be laminar for analytical ease and is justified by the typical value of the Reynolds number, particularly for smaller engines. This assumption is crude because the turbulence level in these machines is high, and inappropriate for larger machines wherein the flow is naturally turbulent.

The questions to be addressed are: What is the role of size (power) on the relative importance of boundary layer effects? How do geometric parameters affect this role? How important are boundary layer effects at the various locations of a compressor?

The momentum thickness of the laminar boundary layer shed from the trailing edges of a flat plate is given by (ref. 14–10)

$$\theta_c = \frac{0.664}{\sqrt{Re_c}} c \tag{14–68}$$

where the Reynolds number is based on the chord length, c. If the fluid processed is air, one can approximate the temperature dependence of the viscosity and thus write the Reynolds number as

$$Re_c = \frac{\rho}{\rho_0}\left(\frac{\mu}{\mu_0}\right)^{-1} \frac{\rho_0 w_0 r_{t0}}{\mu_0} \frac{c}{r_{t0}} \tag{14–69}$$

Here the 0 subscript is associated with entry conditions and r_{t0} is the constant tip radius of the machine. The path from stage to stage involves changes of the thermodynamic conditions that affect the ratio ρ/μ. For air, μ is close to linear in T so that this ratio may be expressed in terms of changing $p/p_0 = \pi_c$ and $T/T_0 = \tau_c$ or just either one of these when a polytropic relation between them is used. The use of τ_c and π_c here is roughly consistent with total property ratios defined earlier, when the Mach number is roughly uniform through the compressor and/or it is sufficiently small. Thus the Reynolds number is

$$Re_c = \frac{\pi_c}{\tau_c^2} Re_0 \frac{c}{r_{t0}} \tag{14–70}$$

The blade geometry is described by a blade length and the chord which are related through a (uniform) aspect ratio, AR:

$$AR = \frac{r_{t0} - r_h}{c} = \frac{r_{t0}}{c}(1 - \xi_h) = \frac{r_{t0}}{c_0}(1 - \xi_{h0}) \tag{14–71}$$

This equation relates the hub/tip radius ratio (ξ_h) to the changing chord, c. The solidity of the blading gives the number of blades in the annulus, n, which varies from front to rear in terms of the mean radius, r_m, as given in Fig. 14–31:

$$\sigma = \frac{c}{s} = \frac{nc}{2\pi r_m}; \qquad \text{where } r_m = r_{t0}\frac{(1 + \xi_h)}{2} \tag{14–72}$$

For the flow along a flat plate, the momentum thickness (θ, see eq. 12–32) is a measure of the momentum loss experienced by the flow. Hence the local importance of viscous effects is the ratio of θ (two sides per blade) to the mean circumferential spacing:

$$f = 2\theta_c \frac{n}{2\pi r_m} = \frac{1.328\sigma}{\sqrt{Re_0}} \frac{\tau_c}{\sqrt{\pi_c}} \frac{\sqrt{AR}}{\sqrt{1 - \xi_h}} \approx \left[\frac{1.33\sigma\sqrt{AR}}{\sqrt{Re_0}}\right] \frac{(\pi_c)^{-(1/\gamma - 1/2)}}{\sqrt{1 - \xi_h}} \tag{14–73}$$

The ratio f should be much less than unity. Equation 14–73 shows the ratio of f to the value at the compressor entry. The losses increase with increasing σ and $\sqrt{AR}$. The continuity equation fixes the variation of ξ_h

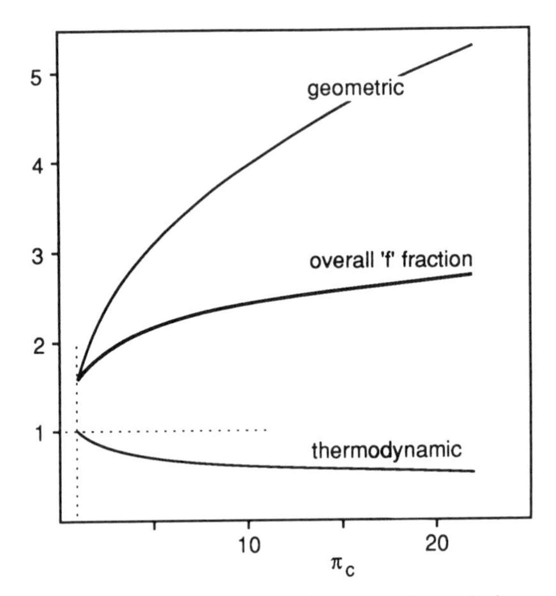

Fig. 14–32. Variation of π_c term describing the property variation and the geometric $1/\sqrt{1-\xi_h}$ terms in eq. 14–73 as well as the product of these terms showing how the relative importance of viscous effects increases slowly along a compressor.

for fixed axial velocity:

$$\left.\begin{aligned} \dot{m} &= \rho w_0 \pi r_{t0}^2(1-\xi_h^2) = \rho_0 w_0 \pi r_{t0}^2(1-\xi_{h0}^2) \\ \text{or} \quad (1-\xi_h^2) &= (\pi_c)^{-1/\gamma}(1-\xi_{h0}^2) \end{aligned}\right\} \qquad (14\text{–}74)$$

Figure 14–32 shows the variation of f as influenced by the changing pressure (numerator of the π_c fraction in eq. 14–73 and labeled *thermodynamic* in the figure) and the changing radius ratio, ξ_h (denominator of the same fraction, labeled *geometry*). These effects compete with the result that the influence of friction dominates and the friction effects are relatively more important near the compressor exit.

Overall, the magnitude of the effect is controlled by the entry Reynolds number which is related to the mass flow rate and thus to the size through r_{t0}. Thus $Re_0 = \rho_0 w_0 r_{t0}/\mu_0 \sim \sqrt{\dot{m}}$ since the mass flow rate varies as r_{t0}^2 for fixed values of ξ_{h0} and entry condition parameters. For variously sized engines, the importance of friction effects increases as $\dot{m}^{-1/4}$. If the transition Reynolds number ($Re_c = 0.5 \times 10^6$, or $Re_0 = 3 \times 10^6$) is chosen to size the engine for which this analysis is applicable, it follows that an engine working with standard atmospheric air at $M_{z0} = 0.5$ processes 15 kg/sec. The stream tube of air is of the order of 0.4 m in diameter. For such numbers, the f factor varies from 0.003 at the inlet to 0.006 at the exit of a 15:1 compressor. Since the polytropic efficiency of modern compressors is about 0.9 (i.e., deviates from unity by about 0.10), the simple view that mean flow boundary layer friction is responsible for irreversibilities is naive. Further, the disparity between these numbers suggests that improvements may be realized.

For higher Reynolds numbers with turbulent boundary layers, the dependence of θ on Reynolds number changes from a power of -0.5 to about -0.2. This change in exponent affects both sets of terms describing f in eq. 14–73 and thus controls the rate of increase. It does not, however, change the conclusion that viscous effects are generally more important in the rear of the compressor, or the way in which geometric parameters affect the result.

A more accurate modeling approach must include description of the momentum thickness θ in terms of the pressure variations on both blade surfaces.

In addition to boundary layers being an important source of irreversibility, the flow through the space between blades is also subject to secondary flows resulting from the pressure field. This is especially important in the rotor because it leads to vorticity which may cause flow separation when it is deformed into an axial orientation. The secondary flow picture is described in Section 14.6 following a discussion of the radial flow compressor where secondary flow effects are also present. It is beyond the scope of this text to examine the performance limits associated with boundary layer and secondary flows in detail. They may de described empirically, through the use of diffusion factors or they can be identified through numerical flow calculations for specific circumstances. The capability of carrying out such calculations has progressed significantly in recent years, particularly with the advent of modern computer codes which can handle the large variety of phenomena at play: flow compressibility, boundary layers with severe pressure gradients, waves, secondary flow, separated flows, heat transfer, leakages, etc. That is to say that many situations cannot be satisfactorily examined, but a good deal of understanding is in hand. The successful design of such components is still a matter of considerable development. The reader is referred to the specialist literature for an appreciation of the current state of the art.

14.5. RADIAL FLOW COMPRESSORS

The design basis of the radial flow compressor is the Euler turbine equation specialized to the case where the inlet flow with no angular momentum enters a rotor and is turned through 90° to exit in the radial direction as shown in Fig. 14–2a. The rotor angular speed is ω. Figure 14–33 shows the elements of a single entry radial flow compressor: the rotor and the diffuser. The discussion in this section is limited to an exposition of the design parameters, and how these relate to overall performance.

14.5.1. *Rotor*

The rotor shown in Fig. 14–33 provides a surface along which vanes form passages. These passages force the fluid to turn from an axial to a radial direction. A centripetal acceleration results in a body force as the flow is turned.

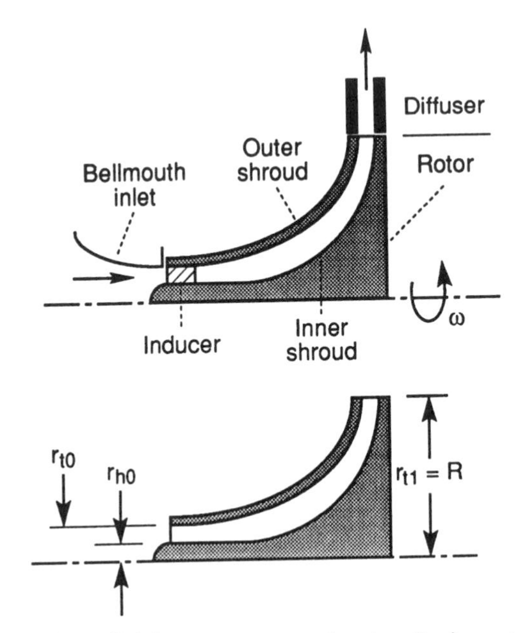

FIG. 14–33. Schematic of a radial flow compressor showing the important dimensions.

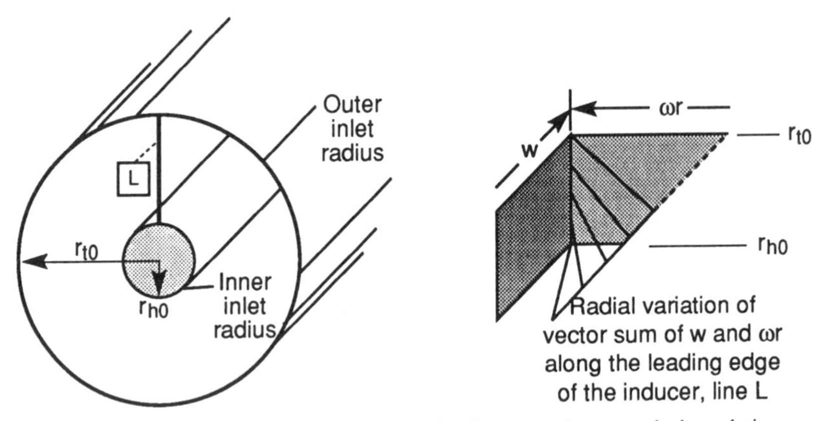

FIG. 14–34. Inducer velocity vectors in the space between hub and tip.

The vane initially takes the moving fluid onto the rotor by means of an inducer. The orientation of the inducer is determined through the vector sum of uniform incoming flow (w) and the local blade element speed, ωr, up to the impeller shroud radius (Fig. 14–34). A view of the passage looking toward the rotation axis is shown in Fig. 14–35. It should be evident that off-design operation (i.e., at reduced or increased mass flow rate) will involve a mismatch between the flow vector angle composed of w and ωr and the physically fixed inducer angle. This flow incidence misalignment can lead to irreversibility if it is sufficiently severe.

Several geometric parameters describe the rotor as shown in Fig. 14–33. The

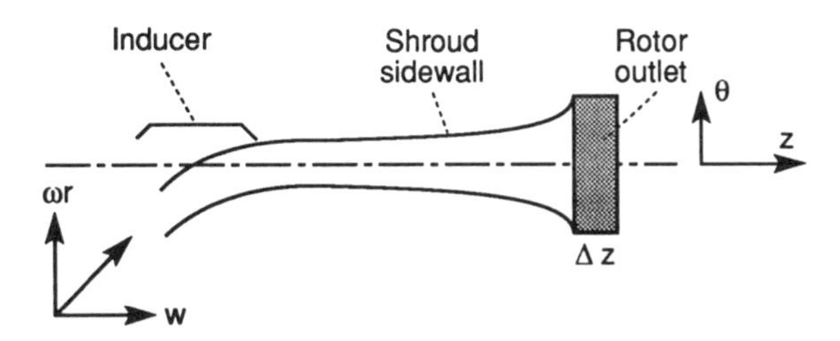

FIG. 14–35. View of passage toward axis, z-θ plane. Inducer is the curved initial section.

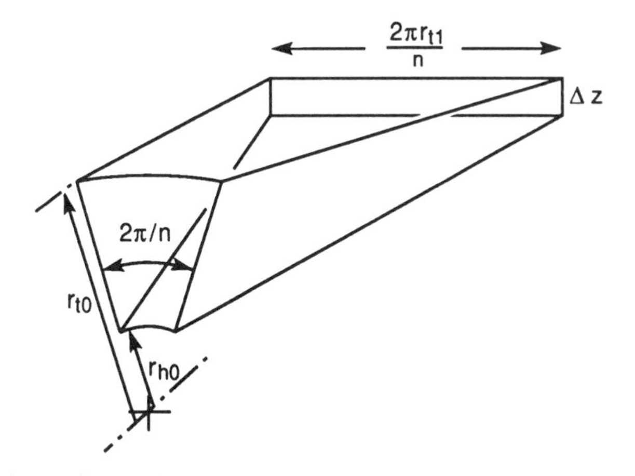

FIG. 14–36. Approximate shape of the straightened passage along a rotor. n is the number of vanes.

inlet has a minimum, a mean, and a maximum radius. The exit radius is the other important dimension. The mass flow handling capability of the compressor is given by the inlet conditions:

$$\dot{m} = \rho_0 w_0 \pi (r_{t0}^2 - r_{h0}^2) \tag{14–75}$$

where the uniform axial velocity at the inlet is w_0. At the outlet

$$\dot{m} = \rho_1 u_1 2\pi r_{t1} \Delta z \tag{14–76}$$

Here Δz is the width of the exit duct, assuming that the vanes take up a negligible volume. The passage along which the flow is processed is a tube with changing aspect ratio as illustrated in Fig. 14–36. Although it is somewhat approximate to describe the flow in the passage as one dimensional, it is a useful approach to estimate the compressor performance. The area variation is tailored to ensure a flow Mach number variation that avoids choking.

14.5.2. *Performance*

The radial flow compressor pressure ratio can be estimated using a 1-D approach which models the effect on the flow as a body force in the flow

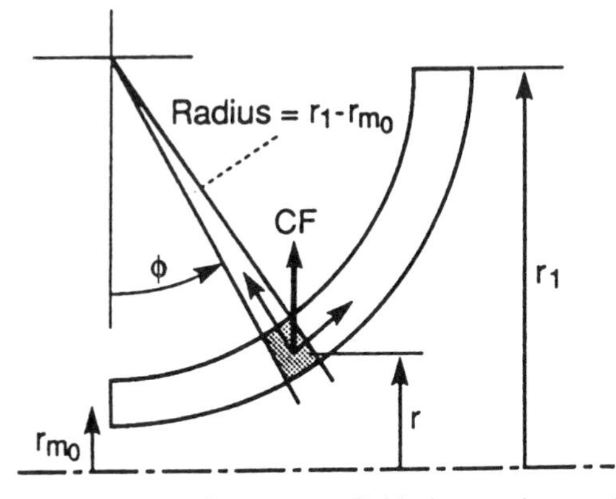

FIG. **14–37.** Geometry and body forces on a fluid element in a radial flow compressor rotor passage.

direction. Assume that the hub and shroud radii are describable as circular arcs as shown in Fig. 14–37. The radial location of the flow element shown in Fig. 14–37 is

$$r = r_{m0} + (r_1 - r_{m0}) \sin \phi \qquad (14\text{–}77)$$

and the body force per unit volume is $\omega^2 r$ (labeled CF) with components along the passage as well as normal to it. r_{m0} is the mean radius at the rotor entry. The 1-D momentum equation for the path along the duct is

$$\rho u \frac{du}{ds} + \frac{dp}{ds} = \rho \omega^2 r \sin \phi \qquad (14\text{–}78)$$

where the path length $ds = (r_1 - r_{m0})\, d\phi$. The continuity equation applied to the duct gives the variation of ρu with s, once the area variation is specified.

This momentum equation provides insight into the physical contributions to the pressure rise. The contribution by the du/ds term tends to be small. The largest contribution to an increase in the pressure is due to the changing radius, r, as given by eq. 14–77. The pressure thus rises along the flow path and so does the density. Thus ρ, r, and $\sin \phi$ increase with ϕ, and the greatest rate of increase in the pressure will be experienced near the rotor exit where $\phi \sim \pi/2$.

The flow in the passage also experiences a body force component that acts normal to the flow direction and affects the bulk of the flow by allowing the formation of circulation cells with the boundary layers in the sidewalls. Such secondary flows, which are described in Section 14.6, weaken the 1-D flow assumption.

Passage Flow Static Pressure Variation

By making a number of simplifying assumptions about the passage flow, one may estimate the pressure distribution through the rotor of the compressor.

Consider the passage to be 1-D, adiabatic and irreversible. The momentum equation (eq. 14–78) can then be written as

$$\rho u\, du + dp = \rho \omega^2 r_1^2 \left(\frac{r}{r_1} \sin \phi \frac{ds}{r_1} \right) \tag{14–79}$$

where the term in parentheses is related to the amount of turning. By letting the flow path be a part of a circular arc as described above, this last term can be shown to be an increasing function of ϕ (and termed a turning angle function, which is $d\phi' = \sin^2 \phi \, d\phi$ for very small radius ratio, $r_{m0} \ll r_1$) and rewritten as

$$\frac{r}{r_1} \sin \phi \frac{ds}{r_1} \rightarrow \left(1 - \frac{r_{m0}}{r_1}\right)^2 \left[\frac{\pi}{4} - \left(\frac{r_1}{r_{m0}} - 1\right)^{-1}\right] \tag{14–80}$$

when the integration is carried out to $\phi = \pi/2$, a maximum value. Here the role of the radius ratio is seen and a large value of r_1/r_{m0} is evidently desirable.

On the right side of eq. 14–79, the tangential Mach number based on the wheel speed is also to be found. This term is proportional to

$$M_{\theta 1}^2 = \frac{\rho_0 \omega^2 r_1^2}{\gamma p_0} \tag{14–81}$$

which is convenient when the pressure and density on the right side are normalized by the inlet values. Clearly, a large Mach number based on wheel speed is desirable for a large change in the momentum (primarily pressure) of the fluid.

Lastly, the variation of pressure may be determined when the momentum terms on the left side of eq. 14–78 are examined. The energy equation gives a relation between pressure and density when the flow is adiabatic, though not necessarily reversible. This gives a polytropic relation between the pressure and density as

$$\frac{p}{p_0} = \left(\frac{\rho}{\rho_0}\right)^n \tag{14–82}$$

where the exponent n is related to the degree of reversibility through eq. 7–46.

The specification of the flow area along the rotor relates the local velocity and density. In general, the increasing pressure results in an increase of the density and the area distribution is tailored to avoid the Mach number in the rotor reference frame from becoming sonic at which point the flow chokes. Using the continuity equation for a variable area passage ($\rho u A = \text{cst}$) and the definition of a static pressure ratio π_{cs}, defined by:

$$\pi_{cs} \equiv \frac{p}{p_0} \tag{14–83}$$

the momentum equation (eq. 14–79) can be written in the following form:

$$\frac{d\pi_{cs}}{\pi_{cs}} = \frac{M_{\theta 1}^2 \pi_{cs}^{(1/n-1)}\left[\mathrm{fcn}\left(\frac{r_{m0}}{r_1}, \phi\right) d\phi\right] + \gamma M^2\, d\ln A}{\left(1 - \frac{\gamma}{n} M^2\right)} \tag{14–84}$$

Here M is the passage flow Mach number based on the 1-D flow velocity, u. This Mach number depends on the local static pressure through the fact that the total *temperature* is conserved in the rotor reference system. If the flow were reversible, this would imply that the total pressure is constant as well and a relationship between static pressure and Mach number is established, eq. 12–9. An accurate determination of the static pressure ratio across the rotor depends on substituting $M(\pi_{cs})$ dependence and integrating this equation, usually numerically.

For some rotor designs, the Mach number may be small enough for the terms involving it to be small, especially if, in addition, the area variation is small. With the integral of the angle function in square brackets as given by eq. 14–80 above, an analytic expression for the pressure ratio is obtained by integration of eq. 14–84:

$$\pi_{cs}\left(\phi = \frac{\pi}{2}\right) = \frac{p_1}{p_0} = \left\{1 + \frac{n-1}{n}\gamma M_{\theta 1}^2\left[\frac{\pi}{4} - \left(\frac{r_1}{r_{m0}} - 1\right)^{-1}\right]\left[1 - \left(\frac{r_{m0}}{r_1}\right)^2\right]\right\}^{n/(n-1)} \tag{14–85}$$

The pressure rise is seen to be larger for increasing $M_{\theta 1}$, and hub/tip radius ratio should be small, conflicting with the ability to process a large mass flow rate through a unit of compressor frontal area. For $n = \gamma = 1.4$, radius ratio $= 0.1$, and $M_{\theta 1} = 1$, the pressure ratio is 2. Raising the wheel tip Mach number to 1.4 increases the pressure ratio to 3.5. It should be noted that for high performance rotors, the pressure ratio achievable is increased significantly by the $(1 - M^2)$ term in the denominator of eq. 14–84.

In practice, the area variation is contracted to keep up with the increasing density while avoiding choked flow in the rotor passage. The determination of the performance of real flows in radial flow compressor rotors, particularly high-performance machines, requires numerical solution of the governing equations because:

1. The r-ϕ path may not be simple and may vary significantly from streamline to streamline, specifically from rotor hub to stationary outer shroud.
2. The flow may be close to sonic, perhaps locally supersonic.
3. The wall geometry variation may be sufficiently rapid to disallow a 1-D flow assumption.

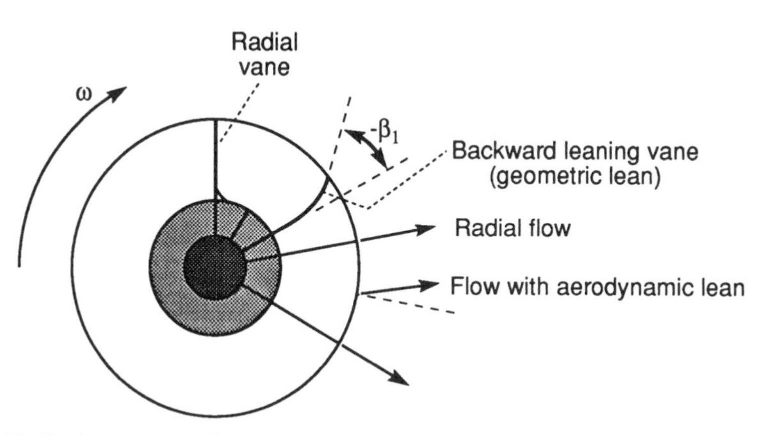

Fig. 14–38. Rotor vane blading near the outlet. All blades on a rotor are identical. Examples of zero geometric lean (12 o'clock) and backward geometric lean (2 o'clock) on a blade. Aerodynamic lean on a radial vane results from a failure of the vane to impose purely radial flow.

4. The flow may be nonuniform owing to waves and boundary layers.
5. Secondary flow may be important owing to the action of centripetal acceleration of boundary layer flow.

The presence of boundary layers in such machines of rather large surface/volume ratio is almost always important in determining the performance of radial flow compressors. As a result, efficiencies tend to be lower than for axial flow machines.

Blade Lean

Figure 14–38 shows various aspects of blade lean. In a given rotor, all blade orientations are naturally identical, although the figure shows various vane geometries for illustration purposes. The simplest blade geometry orientation on the rotor is radial as shown for a vane near the 12 o'clock position in the rotor of Fig. 14–38. In the laboratory reference frame, the exit velocity from the rotor is the sum of the radial outflow velocity, which is generally sonic or less so, and the tangential wheel velocity. This velocity combination may be manipulated by forcing the flow to follow a blade that "leans" either in the rotation direction or against it. Figure 14–38 shows a backward leaning blade (at the two o'clock position). Such lean will increase the tangential velocity when it is forward or vice versa. In forward lean with the angle β_1 negative as shown, the pressure ratio capability of the compressor is larger than for $\beta_1 > 0$ because the exit flow has a higher kinetic energy which can be diffused to higher pressure. In practice, however, forward leaning blading is not used extensively because the compression process tends to be unstable because the compressor map operating line may have a positive slope.

Total Temperature and Pressure Ratios

The role of lean on the performance of a radial flow compressor may be determined by examination of the total temperature ratio across the rotor. The Euler equation for zero inlet swirl is

$$C_p(T_{t1} - T_{t0}) = \omega r_1 v_1 \qquad \text{or} \qquad \tau \equiv \frac{T_{t1}}{T_{t0}} = 1 + (\gamma - 1)\frac{(\omega r_1)^2}{a_{t0}^2}\left[1 - \frac{u_1}{\omega r_1}\tan\beta_1\right] \tag{14–86}$$

where u and v are the radial outflow velocity and tangential velocities in the laboratory reference frame. This expression resembles the static pressure rise in the rotor. The work input or total temperature rise is of interest for the overall performance, while the static pressure performance is important when the flow details in the rotor, boundary layers in particular, are considered. Independent of aerodynamic issues, the tip speed is controlled by structural considerations of the spinning material. Values of ωr_1 of the order of 600 m/sec with a 300 m/sec inlet stagnation sound speed gives a total temperature ratio of the order of 2.5. The total pressure ratio that follows from the isentropic diffusion is of the order of 25! This is indeed a large value, showing the potential for high-pressure ratio, single-stage machines. In practice, values over 15 have been reported, although a range between 3 and 10 is more common (ref. 14–11). The performance of a complete compressor must include consideration of flow irreversibility in the rotor and, more importantly, in the diffuser.

Equation 14–86 also gives an indication of the shape of the compressor map of a radial flow compressor, expecially when irreversibilities are not severe. For example, a backward leaning bladed rotor ($\beta_1 > 0$) will have a characteristic that decreases with increasing mass flow rate (since u_1 and w_0 are closely related). A radially vaned rotor may be expected to have a reasonably flat characteristic (see Fig. 14–39 for two extreme examples). At large mass flow rates, the irreversibilities decrease the total pressure as choking is approached.

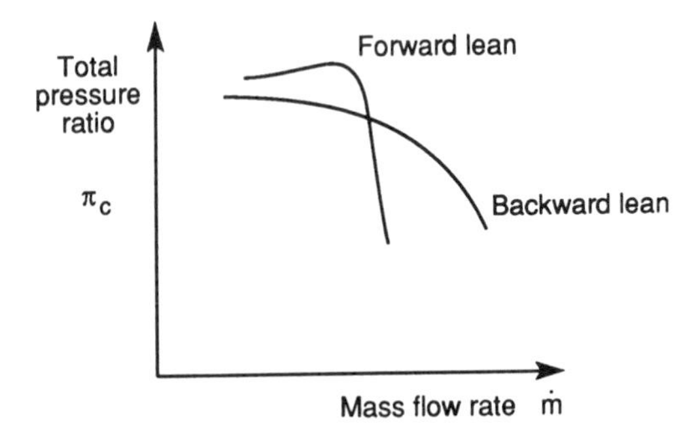

FIG. 14–39. Effect of lean on a radial flow compressor map at a given wheel speed.

14.5.3. *Rotor Exit Mach Number*

In the laboratory reference frame, the Mach number of the flow exiting the rotor is given by

$$M_1 = \frac{\sqrt{(\omega r_1 - u_1 \tan \beta_1)^2 + u_1^2}}{a_1} = \frac{(\omega r_1)}{a_{t0}} \sqrt{\left(1 - \frac{u_1}{\omega r_1} \tan \beta_1\right)^2 + \left(\frac{u_1}{\omega r_1}\right)^2} \frac{a_{t0}}{a_{t1}} \frac{a_{t1}}{a_1} \tag{14-87}$$

Here the ratio of wheel tip speed to stagnation sound speed is related to the total temperature ratio given by eq. 14–86, and the latter ratio is a function of M_1. These two equations may be combined to give an equation that may be used to calculate M_1:

$$\left.\begin{aligned} \frac{\frac{\gamma - 1}{2} M_1^2}{1 + \frac{\gamma - 1}{2} M_1^2} &= \frac{1}{2}\left(1 - \frac{1}{\tau}\right)\left(\left(1 - \frac{u_1}{\omega r_1} \tan\beta\right)^2 + \left(\frac{u_1}{\omega r_1}\right)^2\right) \equiv Z \\ \text{or} \qquad M_1 &= \sqrt{\frac{2}{\gamma - 1} \frac{Z}{1 - Z}} \end{aligned}\right\} \tag{14-88}$$

For $u_1/\omega r_1 = 1$ and $\tau = 2.5$, the exit Mach M_1 is found to be 1.5 (and may be larger). This supersonic value shows that there is a considerable challenge to the design of high-pressure ratio flow radial compressors because the diffuser must accept a very high speed flow. To make matters worse, the flow is likely to be nonuniform in velocity as it exits the rotor. The vanes force a flow tangency condition in their own vicinity which is weaker midway between the vanes and results in an oscillating exit velocity vector in the laboratory reference frame. Consequently, the diffuser inlet flow will be unsteady.

The time-averaged flow from the rotor will consist of flow near vanes and flow midway between vanes which tends to lag. Such a lag, termed an aerodynamic lean, is a typically (backward) contribution to the geometrical lean. This is sometimes referred to as "slip". This slip varies with operating condition and is heavily influenced by the boundary layer characteristics and secondary flow in the region near the exit of the rotor. Evidently, the slip and its consequence on performance are Reynolds number dependent.

14.5.4. *Efficiency*

The processes fundamental to the rotor are so different from those in the following stator or diffuser that, for the radial flow compressor, one often quantifies efficiency in terms of parameters that describe the rotor's ability to produce a high *static* pressure output. The diffuser then takes the high-speed gas motion and recovers the maximum total pressure. A total to static efficiency

(η^{ts}) may therefore be defined for the rotor by

$$\eta_T^{ts} \equiv \frac{1 - \dfrac{T_{t2}}{T_{t0}}}{1 - \left(\dfrac{p_2}{p_{t0}}\right)^{(\gamma-1)/\gamma}} \tag{14–89}$$

where the subscript "t0" represents the inlet stagnation condition and 2 the diffuser entry static condition. This is in contrast to the total-to-total efficiencies which are discussed elsewhere in this text and defined by

$$\eta_T^{tt} \equiv \frac{1 - \dfrac{T_{t2}}{T_{t0}}}{1 - \left(\dfrac{p_{t2}}{p_{t0}}\right)^{(\gamma-1)/\gamma}} \tag{14–90}$$

The numerator in both cases is the work done while the denominator is the work obtainable by expansion from p_{t0} to either p_2 or p_{t2}. Since p_2 is less than p_{t2} by an amount related to the amount of kinetic energy in the exit flow, it follows that $\eta^{ts} > \eta^{tt}$. For radial flow compressors η^{ts} ranges between 0.70 and 0.85, with the lower values associated with high pressure ratio devices.

14.5.5. *Diffuser*

The diffuser is a duct to receive the radial flow, slow it to a reasonable velocity, and finally redirect it to the following component. The exit flow direction may be to the side, as shown in Fig. 14–40, or in a direction parallel to the

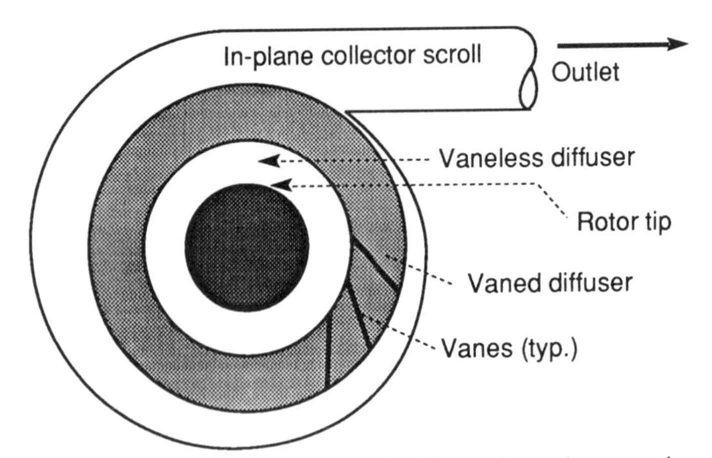

Fig. 14–40. A diffuser surrounding a rotor (shaded area has a large number of vanes, a few of which are shown). In this sketch, compressor discharge is in the plane of rotation and through a single outlet.

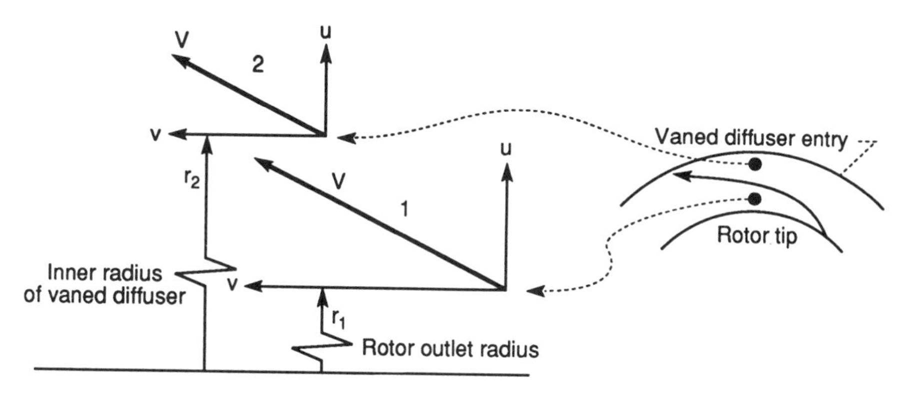

FIG. **14–41.** Vaneless diffuser velocities.

rotation axis as shown in Fig. 14–2a, which would be normal for an aircraft engine application.

The vaneless diffuser is a region where angular momentum is conserved:

$$\dot{m}rv = \text{constant, or } v \sim \frac{1}{r} \tag{14–91}$$

If the region is constant in width Δz, as shown in Fig. 14–31 (top figure), then the radial velocity is governed by the continuity equation:

$$\dot{m} = (2\pi r \Delta z)\rho u = \text{constant, or } u \sim \frac{1}{r} \tag{14–92}$$

The total velocity consisting of these components thus decreases as $1/r$ and the pressure increases correspondingly. The velocity vectors for the vaneless diffuser section are shown in Fig. 14–41. Evidently, the (inviscid) flow seeks to maintain a constant angle (relative to the radial direction) as it traverses this region. The radial (and adverse) pressure gradient acting on the wall boundary layers limits the amount of diffusion that is possible without a departure from steady and 1-D flow.

The final diffusion step is a classical diverging duct formed by the vanes shown in Fig. 14–40 and discussed briefly in Chapter 12.

14.6. SECONDARY FLOWS IN RADIAL FLOW AND IN AXIAL FLOW TURBOMACHINES

Consider a segment of a flow passage midway between the entry and the exit of a radial flow compressor rotor. At this location, the outer shroud, the inner hub and two vanes define a flow area. The body force acting on an element of the flow field has components (1) acting in the flow direction that led to the pressure rise and (2) normal to the flow (Fig. 14–37). All walls, especially the vanes, have developing boundary layers. If one imagines the boundary layer flow to be a region of very low momentum, then the pressure gradient experienced by the mean flow in the normal direction will cause the boundary layer fluid to respond by motion in the direction of lower pressure. The resultant radially inward flow along the vane wall must be returned in the free stream so that a circulation cell is established. This flow situation is present near the inlet of a radial flow compressor, although there the boundary layers are thin if the flow is well behaved (i.e., near its design condition). At the exit end, the pressure gradient is aligned with the flow direction, and the contribution to secondary flow near the exit is relatively small. However, the vorticity that has been added to the flow early in the passage tends to persist near the exit so that its consequences continue to be important.

The flow pattern just described for the radial flow machine is also present in axial flow compressors. Here the pressure gradient driving the secondary flow is as described by eq. 14–31. When the blades are viewed as airfoils generating forces, it is evident that pressure gradients also exist across the blade row in the direction of rotation, which causes the boundary layers on the hub and on the tip casings to respond by motion. For the boundary layer on the hub, there is a tendency for the low momentum air flow to respond with motion toward lower pressure (i.e. from the pressure side to the suction side of the blade) and thus alter the axial vorticity field. The casing or tip region boundary layer's response is complicated by the fact that the wall moves when seen from the rotor perspective. The combination of radial flow in the side wall and azimuthal flow at the hub and tip leads to a secondary flow pattern sketched in Fig. 14–42, showing an axial flow rotor. A similar

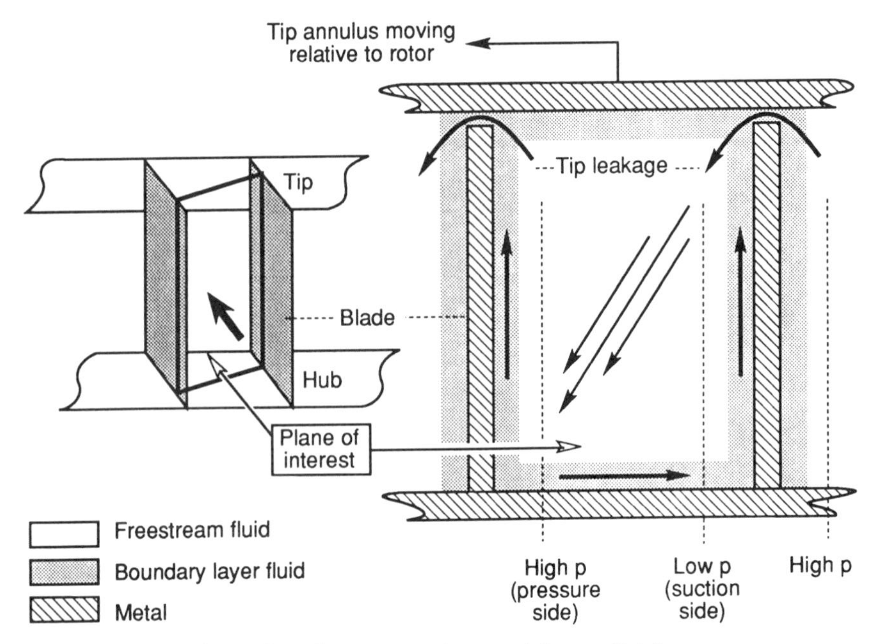

Fig. 14–42. Secondary flow pattern in an axial or radial flow compressor.

picture describes a radial compressor with differing relative importance of the various secondary flow contributions.

The radial flow compressor owes its relatively lower efficiency (compared to the axial machine) to the effects describing the shortcomings of simple 1-D modeling mentioned here. It is unfortunate for the radial flow compressor that the largest secondary flow effects are seen at the rotor inlet where they can affect the later flow. The axial flow compressor starts anew with each new blade row and contends primarily with the resulting turbulence, which can be reduced by allowing larger blade row spacing. The fact that each stage may be optimized for best performance is an important aspect of the relatively better efficiency of the axial compressor when compared to the radial one.

The moving outer annulus of both compressor types imposes a shear on the fluid streamtube which compounds the problems associated with the secondary flow. Lastly, leakages past the blade or vane tips can introduce axial vorticity and cause flow separation on the suction side of the blade and thus further influence performance.

14.7. OTHER COMPRESSOR TYPES

For compressing gases at modest flow rates, displacement compressors may be considered. Although a number of special-purpose compression devices exist (ref. 14.1), the purpose of this section is to highlight the operating characteristics of such devices as a contrast to turbomachinery, rather than discuss the special-purpose compressors that are available for various functions.

Turbomachinery performance at low mass flow rate suffers because of the increasing role of viscous friction. Further, when the size is reduced in order to reduce mass flow rate, wheel angular speed increases (for a constant near sonic flow speed or stress limitation on tip speed). This high rotational speed may make it difficult to match to a load (other than another turbocomponent) without the use of gearing.

No matter how displacement compressors are designed and built, they are functionally equivalent to the piston-cylinder arrangement. A piston is connected to a crankshaft by means of a connecting rod. The crankshaft executes a rotary motion so that the piston is forced to follow linear reciprocating motion (see Chapter 7). This motion is approximately sinusoidal when the ratio of crankshaft radius to connecting rod length is small.

Fresh gas is taken into the compression space between the piston top and the cylinder end through a valve. The flow control valves may perform their role in one of two ways. The valves can be actuated by a mechanism that performs opening and closing functions at predetermined cyclic angles of the crankshaft. An example of such a method is the use of the camshaft valve control found in the Otto or Diesel cycle engines. Alternatively, the flow processes across ports can be controlled by pressure forces. An example is the reedlike valve shown in Fig. 14–43. Here a pressure difference across a spring held valve opens the valve and permits flow when the pressure difference is large enough to overcome the spring force.

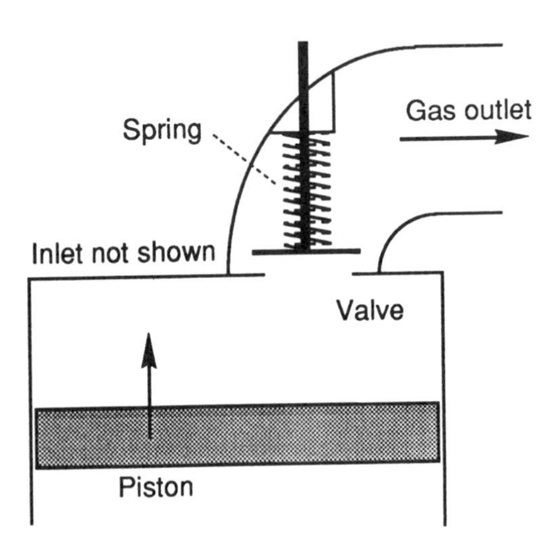

FIG. 14–43. Reedlike valved piston cyclinder compressor showing only the exhaust valve.

The springs on intake and discharge valves serve to prevent the compressed gas from returning to the low-pressure reservoir and/or backflow from the high-pressure reservoir into the cylinder. In operation, the low-pressure gas is taken into the cylinder, at maximum displacement, the valve closes, and the compression of the gas in the cylinder proceeds by means of a volume reduction. When the cylinder gas has been pressurized to an amount greater than the pressure in the outlet duct, the spring, which holds the discharge valve disk in place, allows an opening that is proportional to the pressure difference.

The difference in the operating characteristics of the compressor with the two types of valving is that the valve-actuated system is best suited to delivery of a gas at a predetermined and invariant pressure level because the irreversibility associated with overcompression for delivery to a lower-pressure reservoir or reprocessing of partially compressed gas may be unacceptable. The possibility of mechanically changing the angle where compression is to be terminated has been explored (e.g., ref. 14–12), but the mechanisms involved are complex and present potential reliability problems. In any case, none have yet found their way into widespread commercial use.

14.7.1. *Flow Pressure Losses in Valved Engines*

Whether the valve is reedlike or involves mechanical actuation, the flow into and out of the cylinder volume is controlled by the pressure difference across the port through which the flow is to take place. From a fluid flow viewpoint, one may model the process as a steady, reversible flow from the high-pressure reservoir to a static pressure at the area minimum equal to the downstream pressure followed by mixing with the gas in the cylinder volume. This situation is sketched in Fig. 14–44. The figure applies to either valve

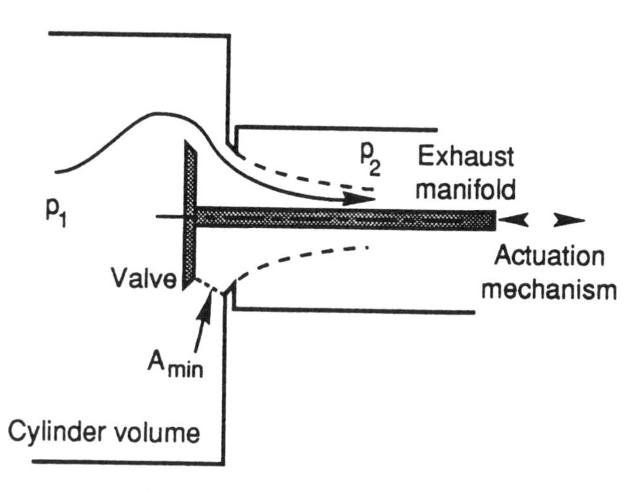

FIG. 14–44. Flow past an open valve.

situations described above. The flow separates at the area minimum and thereafter mixes irreversibly with the downstream fluid. At the separation point, the static pressure in the downstream space and the upstream total pressure determine the velocity, and thus the flow rate. For incompressible flow, as this situation may be, the mass flow rate is given by the steady flow continuity equation with the velocity given by Bernoulli's equation:

$$\dot{m} = A_{\text{min}} \rho_1 \sqrt{\frac{2(p_1 - p_2)}{\rho_1}} \tag{14–93}$$

Real flows, especially at high rotational speed, will necessarily involve compressibility effects, and these may be incorporated readily into an analysis. Here the focus is on simpler modeling to identify the parameters at play.

The difference between the reed valve control case and the mechanical control is that A_{min} is separately determined in the latter, whereas it is a function of the pressure difference $(p_1 - p_2)$ in the former. The spring constant of the retaining spring enters the analysis as a parameter:

$$F = kx \qquad \text{or} \qquad (p_1 - p_2)A_{\text{valve}} = kA_{\text{flow}} = kA_{\text{min}} \tag{14–94}$$

Reference 14–13 is an examination of this special case. It is concluded there that the rotational and therefore flow speeds should be as small because the pressure differences generated are the source of irreversibility.

The history of the cylinder pressure is given by the mass and energy equations for the open cylinder control volume. In a time interval, $dt = d\theta/\omega$, the mass of the cylinder gas is augmented by the entering fluid (during intake). θ is the angle describing the crankshaft rotation. Similarly, the energy is altered by the work done by the piston motion and enthalpy brought in.

The forms of the mass and energy conservation equations for intake and exhaust differ as noted

1. *Intake:*

$$\left.\begin{aligned} &\text{mass:} && \frac{d(\rho V)}{d\theta} = -\frac{\dot{m}_0}{\omega} \\ \text{and} \\ &\text{energy:} && \frac{d(\rho V C_v T)}{d\theta} = -\frac{\dot{m}_0 C_p T}{\omega} - p\frac{dV}{d\theta} \end{aligned}\right\} \quad (14\text{–}95)$$

2. *Exhaust:*

$$\left.\begin{aligned} &\text{mass:} && \frac{d(\rho V)}{d\theta} = +\frac{\dot{m}_i}{\omega} \\ \text{and} \\ &\text{energy:} && \frac{d(\rho V C_v T)}{d\theta} = +\frac{\dot{m}_i C_p T_4}{\omega} - p\frac{dV}{d\theta} \end{aligned}\right\} \quad (14\text{–}96)$$

Here the unsubscripted variables describe the state in the cylinder, 4 the steady flow combustor outlet state, i and 0 identify inlet and outlet mass flow rates, respectively. These are given by appropriate forms of eq. 14–93. The largest pressure difference between the cylinder gas and the environment will be experienced when the cylinder volume change rate demands a large mass transfer and the limited valve or port area cannot provide it without a significant pressure loss. In the Diesel and Otto cycle engines, this is avoided by opening and closing valves when the cylinder is nearly stationary. In this way the valve is as open as it can be during the portion of the cylinder motion when its velocity is high. Engines where a full sweep of the piston is dedicated to the intake and another to the exhaust processes are termed *four-stroke* engines. Two additional strokes are for compression and expansion with the valves closed. The four-stroke engine minimizes the total pressure loss associated with flow entry and exhaust by maximizing the available flow area. This issue is also involved with the use of multiple intake and exhaust valves (four valves per cylinder, for example) in aircraft engines and, more recently, in automotive engines. For example, the circumference of two circles (two valves) inscribed inside the cylinder circle is about 60% of the circumference of four circular valves inside the same cylinder circle. This circumference is directly proportional to the available flow area for a given valve travel, and increasing it reduces pressure losses.

The influence on cycle performance is governed by the fractional pressure loss. By equating mass flows for the intake of a four-stroke engine to the flow rate past the valve, one arrives at

$$\frac{p_1 - p_2}{p_1} = \frac{\Delta p}{p_1} \approx \frac{A_{\text{cyl}}}{A_{\text{min}}} \frac{\rho_1 V_{\text{piston}}^2}{p_1} \quad (14\text{–}97)$$

where the velocity is a mean velocity of the piston and A_{min} is the flow-restricting area associated with the open valve. The right side of this expression is a Mach number-like term which must be small for the entry pressure loss to be small. This Mach number parameter based on piston speed is directly related to the Mach number past the small valve area which, in turn, is limited to choked performance. This phenomenon limits engine air flow rate by limiting allowable piston speed and thus the power of the engine of a given displacement, see Section 3.3.1. In practice, the mean piston speed used is the order of 13 m/sec or 2500 ft/min. This gives a value for the second fraction of the order of 0.002, allowing a value of 10–100 for the area ratio to obtain a modest pressure loss.

An important distinction between the compression process in the Otto or Diesel engine and that in a steady flow displacement compressor is that the compressor's exhaust valve must open when the piston motion has reached the point where the desired pressure is reached. At this point, the piston velocity in that compressor is not small, and the situation arises where the initially small valve flow area leads to pressure losses due to continuing motion of the piston. This results in temporary overcompression of the cylinder gas. Subsequently, the area will be larger and the flow pressure loss will be smaller. The description of this physical situation also leads to a Mach number-like parameter based on piston speed which characterizes the irreversibility (refs. 14–14, 14–15 and 14–16). The net result is that the mean pressure and therefore density of the gas charge trapped in the cylinder are less than they could be. This reduction results in a smaller than ideal mass being processed, and consequently, the output power is lower than ideal. Furthermore, the work done during the intake process is a function of the pressure difference between cylinder pressure and the pressure on the back side of the piston, which is nominally atmospheric. Thus a lowered pressure in the cylinder results in additional work having to be supplied, as is evident in the p-V diagram Fig. 8–4.

An analogous picture also applies to the expander (ref. 14–14), where the loss takes place during the closing of the valve after the gas to be expanded has been admitted to the cylinder and is to be expanded further. The intake process for the steady flow displacement compressor and the discharge process of the expander is limited in the same way as those processes in the internal combustion engines.

In summary, piston speed, as characterized by a Mach number-lie parameter, limits the mass flow processing ability of all devices which use displacement processes. Steady-flow, displacement-work components have additional losses that are associated with the high-pressure mass transfer processes. These are also limited by piston speed.

The conventional friction losses associated with moving seals sliding on the inner surfaces of the cylinder walls are a direct mechanical work penalty. This loss is also a function of mean piston speeed. In the internal combustion engine community, this loss is generally described in terms of a work loss per cycle, termed the "friction mean effective pressure," or friction m.e.p., (refs. 14–15 and 14–16). Such a quantity is calculated by integrating the work around a cycle, accounting for the varying pressure on the seal rings which generally affect the friction force.

14.7.2. *Screw Compressors*

A discussion of the subject of compressors for energy systems is not complete without a short description of a displacement compressor which is in wide service in the refrigeration and chemical process industries. This is the so-called (helical or spiral) screw (or Lysholm) compressor. This compressor consists of a shell and two lobed rotors whose axes are parallel. The lobes are sculptured in a helical pattern with male and female elements that interact to enclose a volume of gas against the end wall of the shell. As the rotors turn, this volume is reduced. At the same time, the end wall section of the stationary shell seen by the gas moves and compression ends when an opening in the end wall permits flow from the compression space to the delivery duct. Fig. 14–45 shows the male and female rotors and a cross-section of the case. A top view of the two rotors with the two gear-driven shafts is shown in Fig. 14–46. The evolution of the space between the rotors is shown in Fig. 14–47 (ref. 14–17),

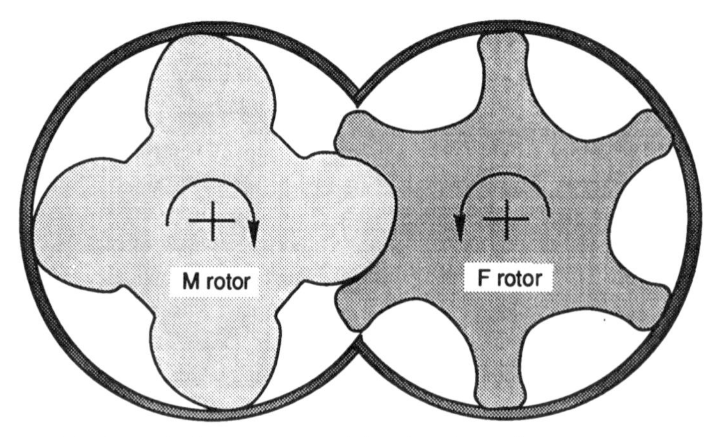

Fig. 14–45. End view of male and female rotors of a screw compressor.

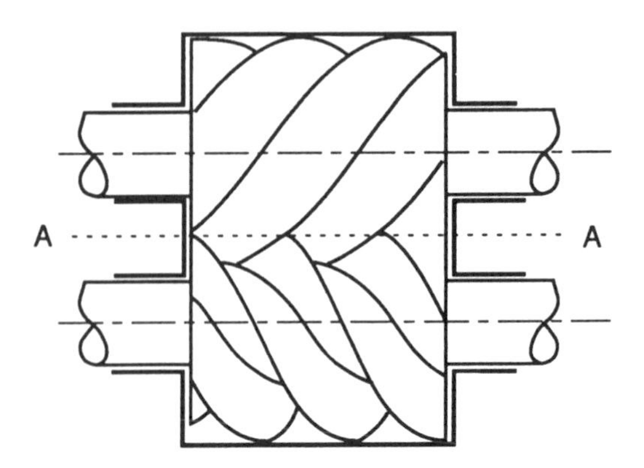

Fig. 14–46. Top view of male and female rotors of a screw compressor. The two geared and driven shafts are shown. Intake and discharge ports are typically on top and bottom, respectively, of the compressor case, astride line *A–A*.

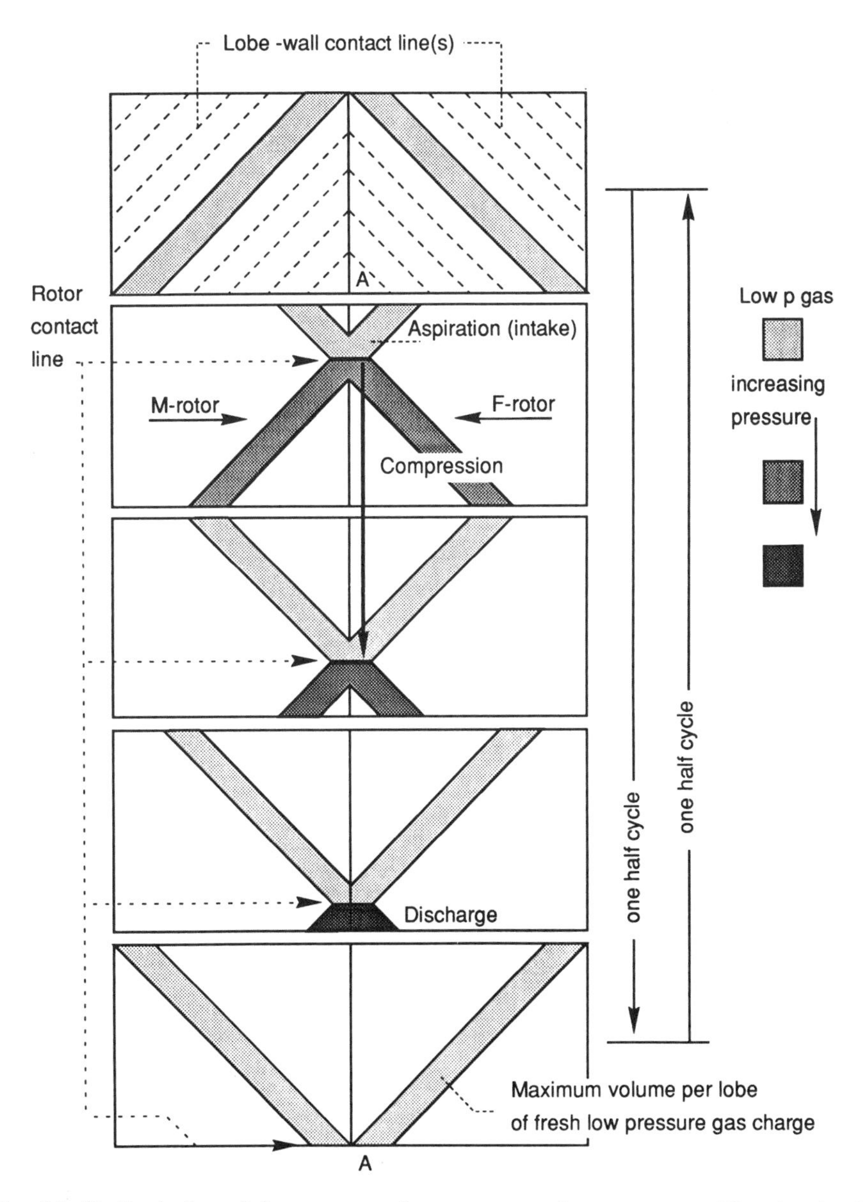

FIG. 14–47. Evolution of the compressed gas space as the rotor turns. The view is of the developed cylindrical surface around the rotors. *A–A* refers to the line shown in Fig. 14–46 on the top, and the left and right edges are the same line on the bottom. Note that inlet and exit ports are located to allow the desired function. During the return part of the cycle shown as upward, the fresh charge is transported to the initial (i.e., top) configuration.

which shows where the entry and the exit openings would have to be to execute the intake and discharge functions. The contact line forms the equivalent of a piston and the remaining exposed lobe surfaces form the equivalent of a cylinder.

In order to form an effective seal, the lobed rotors are in relatively close contact. The clearance between them must support a pressure difference which is a fraction of the outlet to inlet difference for the compressor. As a result, the rotors must have close tolerances, which are maintained by high structural stiffness. Such machines are therefore heavy in comparison to turbomachinery of the same capacity. The seal between rotors is improved considerably by an oil film in the near-contact region. The large amounts of oil involved in so-called oil-flooded compressors generally requires recovery of the oil if the compressor is used for gas (air) compression as in a heat engine cycle. In refrigeration systems, the liquid phase of the working fluid may be present to help with the sealing task.

The adiabatic efficiency of these devices is thermodynamically interesting, with percentages ranging to the low 80s for modest pressure ratio devices, lower for higher pressure ratio where heat losses are more important owing to the elevated exit temperature and the relatively high surface-to-volume ratio of the compression space. This value also depends on size (i.e. flow capacity). The pressure ratio is designed into the compressor by orientation of the ports and the helix angle, which is equivalent to the stroke of a cylinder with rectilinear motion. The fixed geometric relationship between the rotor and the inlet and exit ports allows the compressor to work efficiently at design conditions. Performance loss at off-design is associated primarily with dissipation associated with a lack of pressure equilibrium between compressed

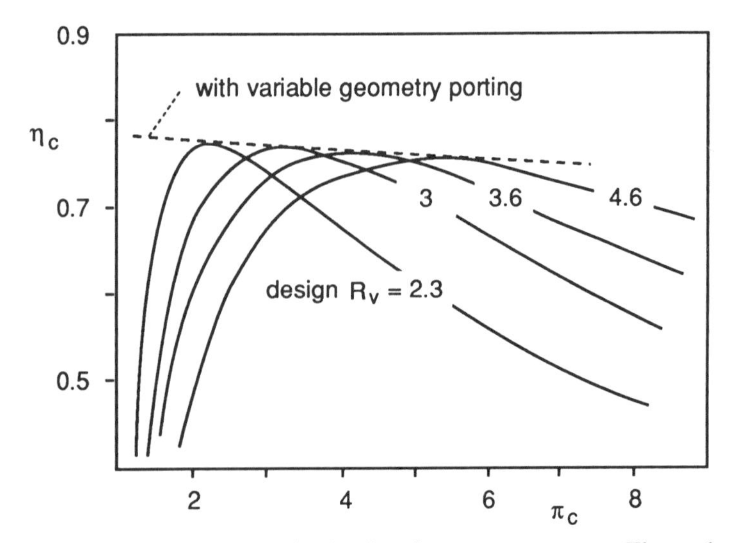

Fig. 14–48. Adiabatic efficiency of a family of screw compressors. The peak efficiency envelope of the curves may be obtained if the ports have a variable geometry. R_v is the design volume ratio.

gas space and the delivery space, and on whether the compressor is operated on- or off-design: the fixed pressure in the final volume just prior to delivery may be larger or smaller than the pressure in the space to which gas is supplied. Figure 14–48 shows a typical performance plot, from ref. 14–18.

Screw compressors tend to have acceptable design point efficiency at relatively low compressor ratio. In spite of the advantage of unidirectional motion and the attendant benefit of long service life (ref. 14–19), screw compressors tend not to be used in power cycle machines because higher pressure ratios are generally required. For Ericsson cycle engines, the screw machine may be an interesting choice. Also worthy of note is that the relatively good efficiency suggests operation as an expander is appropriate, for example, for organic Rankine cycles (ref. 14–20). Screw machines have found extensive application in the compressed air supply, food processing and chemical industries as well as in refrigeration, heat pump cycle machines.

PROBLEMS

1. Derive the relationship between the constant B in eq. 14–35 or 14–37 and the stage total pressure ratio. Also, relate B to the blade lift coefficient.
2. The static pressure is uniform ahead of the inlet guide vane of the compressor. Calculate and plot the variation of static pressure between the IGV and rotor and between rotor and stator. Use $A = 1$ and $B = 0.3$. For this blading, plot the radial distribution of flow angles.
3. By examining the laboratory frame flow angles through a multistage compressor, trace the flow path of a fluid element through such a device. Note the role of the IGV.
4. Consider the flow through an axial flow compression stage with IGV. The IGV introduces a swirl which is gentler than a linear profile described in the text, say

$$\frac{v_1}{w_0} = A\sqrt{\frac{r}{r_T}}$$

Obtain expressions for nondimensionalized w_1, w_2, and v_2 using linearized theory, and compare to numerical results.
5. In Section 14.4.6 a laminar boundary layer was used to used to estimate the role of viscous effects through a multistage compressor. Alter the analysis to use entirely turbulent boundary layer results, and obtain a figure corresponding to Fig. 14–32.
6. Provide a relation between static p and M as required in eq. 14–86 when the flow is irreversible. Assume T_t is constant.
7. For a piston constrained to move in a cylinder with a crank obtain an expression for the sine-like motion of the piston cylinder in the cylinder.

8. To estimate the heat transfer losses from the fluid in the piston-cylinder, it is necessary to have a model for the velocity distribution in the control volume. For the case of an intake stroke though a small central port, obtain an expression for the flow field by simply adding the appropriate velocity contributions. The result is not realistic because viscous forces play an important role in setting up vortical motion, but the model provides an idea where heat transfer and friction might be most important.

BIBLIOGRAPHY

Cumpsty, N. A., *Compressor Aerodynamics*, Longman Scientific and Technical, Essex, U.K., 1989.

Hawthorne, W. R., Ed., *High Speed Aerodynamics and Jet Propulsion*, Vol. 10: Aerodynamics of Turbines and Compressors, Princeton University Press, Princeton, N.J., 1964.

REFERENCES

14–1. El-Wakil, M. M., *Power Plant Technology*, McGraw-Hill, New York, 1984.

14–2. Cumpsty, N. A., *Compressor Aerodynamics*, Longman Group, U. K., 1989.

14–3. *The Jet Engine*, Rolls Royce, PLC, Derby U.K., 1973.

14–4. *Aerodynamic Design of Axial Flow Compressors*, NASA SP-36, US Government Printing Office, Washington, D.C., 1965.

14–5. Kerrebrock, J. L., *Aircraft Engines and Gas Turbines*, 2nd edition, MIT Press, Cambridge, Mass., 1992.

14–6. Horlock, J. H., *Axial Flow Compressors*, Reprint edition with Suppl., Krieger, Huntington, N.Y., 1973.

14–7. Oates, G. C., "Actuator Disc Theory for Incompressible Highly Rotating Flows," *Journal of Basic Engineering*, 94 (ser. D): 613–21, Sept. 1972.

14–8. Oates, G. C., *Aerothermodynamics of Gas Turbines and Rockets*, AIAA, New York, 1984.

14–9. Marble, F. E., "Three Dimensional Flow in Turbomachines," Section C.7 in *High Speed Aerodynamics and Jet Propulsion*, Vol. 10: Aerodynamics of Turbines and Compressors, Princeton University Press, Princeton, N.J., 1964.

14–10. Schlichting, H., *Boundary Layer Theory*, McGraw-Hill, New York, 1979.

14–11. Dean, R. C., "The Centrifugal Compressor," Gas Turbine International, March–April May–June 1973.

14–12. "Clemson Camshaft," *Mechanical Engineering*, December 1991, p. 16.

14–13. Decher, R., "Displacement Compressors with High Performance Flow Control Valves," *International Journal of Energy Research*, **13**: 327–38, 1989.

14–14. Decher, R., "Port Passage Losses in Britalus Compressor and Expanders," *International Journal of Turbo and Jet Engines*, **2**(2): 149–56, 1985.

14–15. Taylor, C. F., *The Internal Combustion Engine in Theory and Practice*, Vols. I and II, MIT Press, 1966.

14–16. Heywood, J. B., *Internal Combustion Engine Fundamentals*, McGraw-Hill, New York, 1988.

14–17. Favrat, D., Ecole Polytechnique Fédérale de Lausanne, LTT, Lausanne Switzerland, personal communication.
14–18. Stoecker, W. F., *Industrial Refrigeration*, Business News Publishing, Troy, Mich., 1988.
14–19. Bloch, H. P., Noack, P. W., *Screw Compressors*, Chemical Engineering, 2/92, pp. 100–8.
14–20. Verein Deutscher Ingenieure, *Schraubenmaschinen '87*, Bericht No. 640, Conference in Dortmund, Germany, 1987, VDI Verlag Düsseldorf. In German.

15

AXIAL FLOW TURBINES

A turbine is a rotating device for extracting power from a continuous supply of high-pressure fluid. This chapter is a description of the aerodynamics of the axial flow gas turbine as a shaft power producer. The goal is to develop an understanding of design considerations and the performance characteristics so that a complete engine using it may be analyzed.

15.1. TURBINES

The fluid processed by a turbine may be a liquid or a gas. For a liquid such a turbine is a hydraulic turbine whereas for the latter it is termed a gas turbine. A steam turbine differs from a gas turbine because the properties of the vapor demand special considerations. For example, steam at high temperature and pressure is an oxidant, droplets may form in flow passages, and for analytic purposes, the proximity of states to the saturation line makes the description of the fluid properties a bit more troublesome. Hydraulic turbines are used for converting the power from water stored in a reservoir, such as that behind a hydroelectric dam on a river. The design of hydroturbines is not considered here. Rather, the emphasis is on turbines used in thermodynamic cycles. Further attention is restricted to axial flow machines, even though for small devices of less than about 100 kw, the use of radial flow turbines may be more appropriate. The radial flow turbine is in common use in ICE turbochargers (e.g., Section 8.7). The description of the gas dynamics of radial flow turbines involves ideas developed in connection with the discussion of the Euler turbine equation as it was applied to the radial flow compressor and the axial flow turbine discussed here.

The available power in gas turbines that use a compressible fluid is proportional to the enthalpy or temperature of the incoming stream (eqs. 2–6, 7–29 and 9–2). Thus, an important design element of high-performance turbines is protection of the blades from the transfer of heat which may raise the material temperature to levels where long life and structural integrity are at risk. The practical approach to this challenge is from two fronts: development of high temperature materials and use of sophisticated cooling techniques. High temperature metals have been improved over the decades that turbines have been in use. The increase in metal temperature capability has had to be obtained with acceptable fracture toughness. An important development goal in the industry has been the development of blades with ceramic materials rather than metals to allow higher temperature capability.

The cooling problem is principally that of maintaining material temperatures at sufficiently low levels while minimizing the performance penalty associated with the use of a coolant. In Section 9–5 (or ref. 15–1) the thermodynamics of the use of coolant from the cycle is developed. Further, the temperature history (particularly on the very short RPM time scale) of material element must be sufficiently smooth to avoid thermal stress fatigue and failure. To date, cooling is performed only with gases, usually the cycle working fluid. At some point in the future, the use of liquid coolants may become practical. Other important design considerations in turbines are reliability and the economic lifetime of the turbine. The design of turbines necessarily incorporates numerous features, which are consequences of the large temperature variations experienced during startup and operation, resulting in thermal expansion and stressing of the various material elements.

As devices that interact with the flow by means of aerodynamic surfaces, turbines differ from compressors in several ways. The aerodynamics of turbines is dominated by large pressure drops through the blade rows, resulting in thin boundary layers that have little tendency to separate. Thus the pressure ratio per axial turbine stage may be considerably higher than a corresponding ratio through an axial compressor stage. The large change in pressure implies that there will be significant density changes, and the fluid must therefore be treated as compressible. The following discussion is of the two-dimensional, steady-flow gas dynamics of the axial flow turbine. The typically small cooling flows and their effects are neglected.

15.2. TURBINE NOZZLE

The Euler turbine equation (eq. 14–2) states that mechanical shaft power is removable from a flow when it has angular momentum. Thus the first step in the power extraction process by the turbine is the generation of a large angular velocity in the fluid. This is accomplished by a nozzle which turns the (generally axial) flow entering at low Mach number from the heat source in the tangential direction, where it has a high Mach number with a significant tangential velocity component. Figure 15–1 shows a sketch of a nozzle and the associated physical quantities. The following discussion is limited to description of a mean streamtube as representative of the flow as a whole. This is particularly appropriate when the hub/tip radius ratio is close to unity. The coordinate direction definitions follow compressor nomenclature.

The nozzle blade with its thin trailing edge acts like a compressor inlet guide vane and establishes the direction of the exiting flow. The blade contour may be such that the flow effectively experiences a convergent–divergent passage and reaches a supersonic velocity at the exit. Under such circumstances, the flow is choked at a throat area which is generally identified as A_4^*, consistent with overall engine state point notation. The flow seen by the following rotor may or may not be supersonic because the rotor moves away from the flow with a velocity which reduces the velocity relative to the blade (Fig. 15–1). The flow

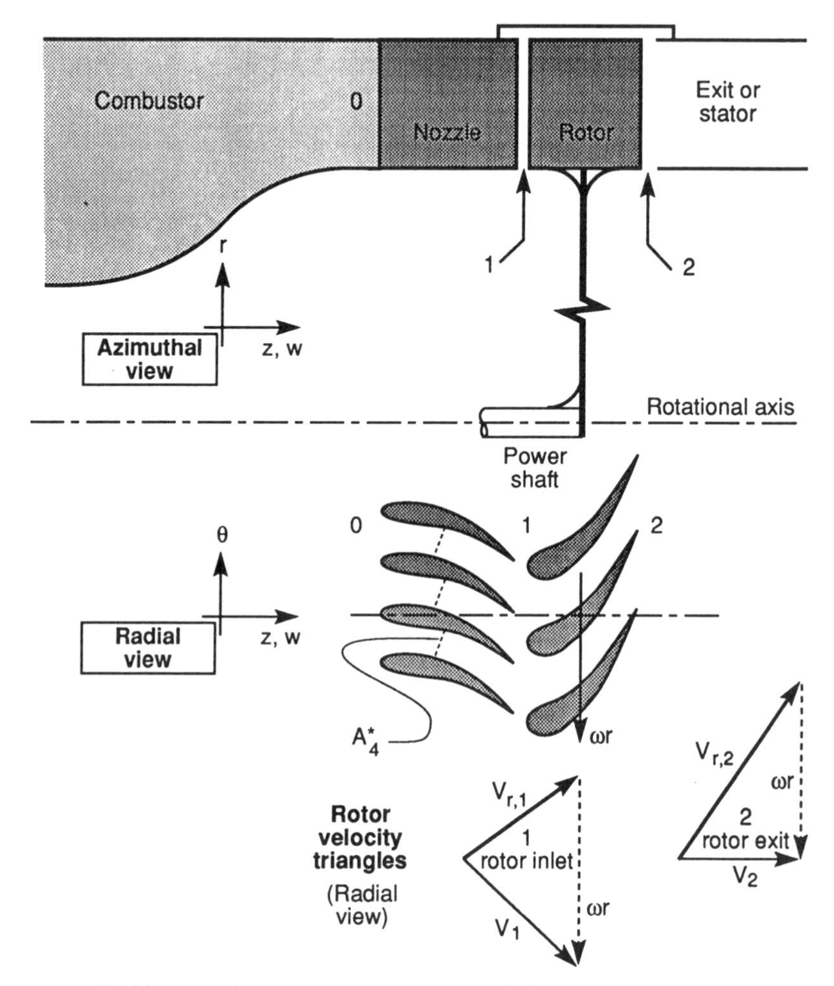

FIG. 15–1. Turbine nozzle and rotor. Top part of figure is a cross-sectional view whereas lower portion is viewed radially toward the rotation axis. Note the area A_4^* in the nozzle. In the velocity triangles, the subscript r refers to "relative" in the reference frame of the rotor.

states associated with the blading are described by a local station numbering system also shown in the figure.

Unlike the compressor where the boundary layers limit aerodynamic forces, the work from the turbine is not limited by blade loadings. The swirl, $v_1 = V_1 \sin \alpha$, should therefore be as large as possible in order to permit extraction of the largest work per stage.

The combustor is normally designed for low Mach number and is generally terminated by an area contraction prior to entry into the turbine nozzle. This nozzle entry point in the flow is designated here as station 0 where the Mach number is M_0, a design variable. Station 1 is the nozzle exit condition.

The flow from station 0 to 1 is adiabatic so that the stagnation temperature is constant (eq. 12–8):

$$T_{t1} = T_{t0} \quad \text{or} \quad T_1\left(1 + \frac{\gamma - 1}{2} M_1^2\right) = T_0\left(1 + \frac{\gamma - 1}{2} M_0^2\right) \tag{15–1}$$

The continuity equation (mass conservation) allows determination of the relation between the flow parameters and the turning angle:

$$\rho_0 w_0 A_0 = \rho_1 w_1 A_1 = \rho_1 V_1 (\cos\alpha) A_1 \tag{15–2}$$

where conditions ahead of the nozzle are assumed known. A is the flow area normal to w. An annulus area ratio is defined by

$$b = \frac{A_1}{A_0} \tag{15–3}$$

which may be as high as 1.5 in some designs, although 1.0 is closer to typical. The swirl is given by

$$v_1 = V_1 \sin\alpha = a_1 M_1 \sin\alpha = \frac{a_{t1} M_1 \sin\alpha}{\sqrt{1 + \frac{\gamma - 1}{2} M_1^2}} \tag{15–4}$$

The last form is expressed in terms of total sound speed (temperature) because that quantity is conserved in the flow through the stationary nozzle ($a_{t1} = a_{t0}$ from eq. 15–1). From the continuity equation (eq. 15–2), it should be evident that as α is increased from small values, V_1, and consequently, the Mach number, M_1, must also increase. With the flow isentropic (a realistic assumption), density changes follow temperature changes:

$$\frac{\rho_1}{\rho_0} = \left(\frac{T_1}{T_0}\right)^{1/(\gamma - 1)} \tag{15–5}$$

Equations 15–1 and 15–5 combine to give the density ratio required in eq. 15–2 as

$$\frac{\rho_1}{\rho_0} = \left(\frac{1 + \frac{\gamma - 1}{2} M_0^2}{1 + \frac{\gamma - 1}{2} M_1^2}\right)^{1/(\gamma - 1)} \tag{15–6}$$

The nozzle exit velocity is obtained from

$$\frac{V_1}{w_0} = \frac{M_1 a_1}{M_0 a_0} = \frac{M_1}{M_0}\left(\frac{1 + \frac{\gamma - 1}{2} M_1^2}{1 + \frac{\gamma - 1}{2} M_0^2}\right)^{-1/2} \tag{15–7}$$

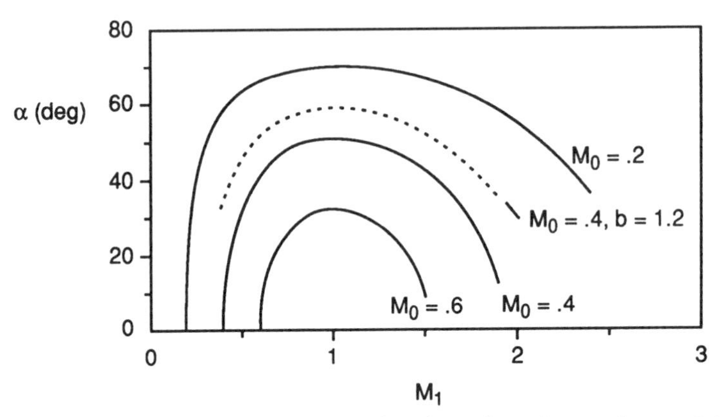

Fig. 15–2. Nozzle exit Mach number as a function of turning angle. $\gamma = 1.4$ $b = 1.0$ except as noted. Plot of eq. 15–8.

A combination of eqs. 15–6 and 15–7 into the continuity equation (eq. 15–2) yields the exit Mach number in terms of the turning angle, although not explicitly:

$$\cos\alpha = \frac{1}{\mathrm{b}}\frac{M_0}{M_1}\left(\frac{\mu_1}{\mu_0}\right)^{(\gamma+1)/2(\gamma-1)} \qquad \text{where } \mu_i \equiv 1 + \frac{\gamma-1}{2}M_i^2 \qquad (15\text{–}8)$$

The function μ is defined for convenience. When M_1 is plotted for a chosen M_0 the variation of α is as shown in Fig. 15–2.

The figure shows that for zero swirl angle (and constant annulus area, $b = 1$), the exit Mach number is M_0 ($\alpha = 0$), and increases to sonic value at some maximum angle where the flow is choked. Once choked, the flow must be returned toward the axial direction if supersonic flow is desired. Evidently, the turning angle in a swirling flow plays the same role as the flow area in a one-dimensional flow with area changes.

The desired swirl (v_1, eq. 15–4) is seen to be proportional to

1. $M_1 \sin\alpha$ (which increases with M_1), and
2. $a_1 = a_{t1}\left(1 + \frac{\gamma-1}{2}M_1^2\right)^{-1/2}$ which decreases with M_1

Hence v_1 must have a maximum with varying α or M_1 as shown in Fig. 15–3. In this figure, M_1 is chosen over α as the independent variable because flow losses in the nozzle become important when M_1 exceeds about 1.4. The reason for the limitation on M_1 is that shock waves from the trailing edge intersect the boundary layer on the adjacent blade and may cause flow separation. As an illustration, Fig. 15–4 shows a sketch from a turbine cascade experiment which shows (1) lines of constant Mach number, (2) the trailing edge shock wave, (3) the reflected weaker wave from the suction surface of the adjacent blade, and (4) the wake. The turning angle of the flow is about 35°.

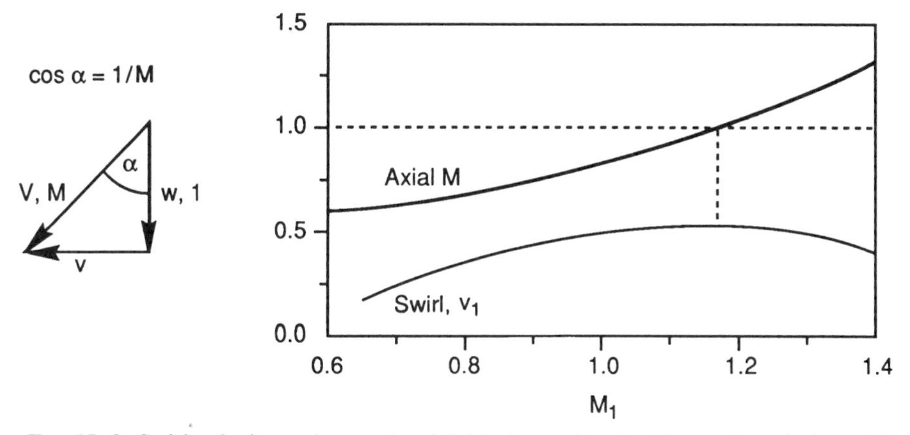

Fig. 15–3. Swirl velocity, v_1/a_{t0}, and axial Mach number ($=M_1 \cos \alpha$) variation with nozzle exit Mach number. Sonic axial Mach number noted gives maximum swirl. ($\gamma = 1.4$, $M_0 = 0.6$, $b = 1.0$). Velocity triangle is for the maximum swirl condition.

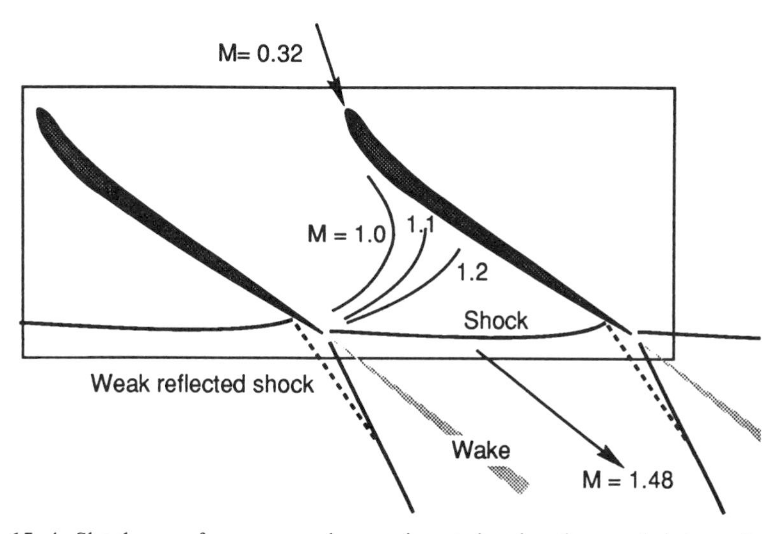

Fig. 15–4. Shock wave from a cascade experiment showing the wave's intersection with the adjacent blade. Conditions are noted. (From ref. 15–2.)

Figure 15–3 is a plot of v_1/a_{t0} and the axial Mach number versus M_1 for the specific case noted in the caption. Note that maximum swirl is obtained for a flow Mach number (M_1) in excess of unity. The condition for maximum swirl is attained when the axial Mach number is unity as shown below. To find the maximum swirl condition, one must evaluate

$$\frac{d}{dM_1}\left\{\frac{a_t M_1}{\sqrt{\left(1+\dfrac{\gamma-1}{2}M_1^2\right)}}\sqrt{(1-\cos^2\alpha)}\right\}=0$$

or with the elimination of α with eq. 15–8, this is equivalent to:

$$\frac{d}{dM_1^2}\left[\frac{\mu_0}{\mu_1} M_1^2 - M_0^2 \left(\frac{\mu_1}{\mu_0}\right)^{2/(\gamma-1)} \frac{1}{b^2}\right] = 0 \tag{15–9}$$

from which the value of M_1 for maximum swirl ($M_{1,v}$) follows:

$$M_{1,v}^2 = \frac{2}{(\gamma-1)}\left[\frac{\mu_0}{m_b^2} - 1\right] \tag{15–10}$$

Here μ is as defined in eq. 15–8 and m_b is a convenient inlet Mach number-contraction parameter:

$$m_b \equiv \left(\frac{M_0}{b}\right)^{(\gamma-1)/(\gamma+1)} \tag{15–11}$$

The maximum value of the swirl is

$$\left[\frac{v_1}{a_{t0}}\right]_{max} = \left[\frac{V_1 \sin\alpha}{a_{t0}}\right]_{max} = \sqrt{\frac{2}{\gamma-1}\left(1 - \frac{\gamma+1}{2}\frac{m_b^2}{\mu_0}\right)} \tag{15–12}$$

From the condition for maximum swirl (eq. 15–10) and the continuity equation (eq. 15–8), it follows that

$$\frac{\mu_{1,v}}{\mu_0} = m_b^{-2} \qquad \text{so that} \qquad \cos\alpha_v = \frac{1}{M_{1,v}} \tag{15–13}$$

This last result implies that for maximum swirl the *axial* velocity should be sonic, as shown in the velocity triangle and plot in Fig. 15–3.

A plot of the maximum swirl (Fig. 15–5) shows that for low values of M_0, the swirl may be made quite large. The disadvantage is that flow area A_0 required to pass the flow becomes large for small M_0 since the continuity equation can be expressed (for $M_0 \ll 1$) as:

$$\dot{m} = \rho_t a_t M_0 A_0 \left(1 + \frac{\gamma-1}{2} M_0^2\right)^{-(\gamma+1)/2(\gamma-1)} \approx (\text{cst})\ M_0 A_0 \tag{15–14}$$

Hence a choice is available between designs that have large cross sectional areas (small M_0) with few turbine stages or smaller flow area turbines with more stages (large M_0). The turbine flow area is often critical in determining the engine diameter.

As mentioned above, a limitation on the performance of the nozzle is its exit Mach number. In addition to the possible boundary layer separation, when

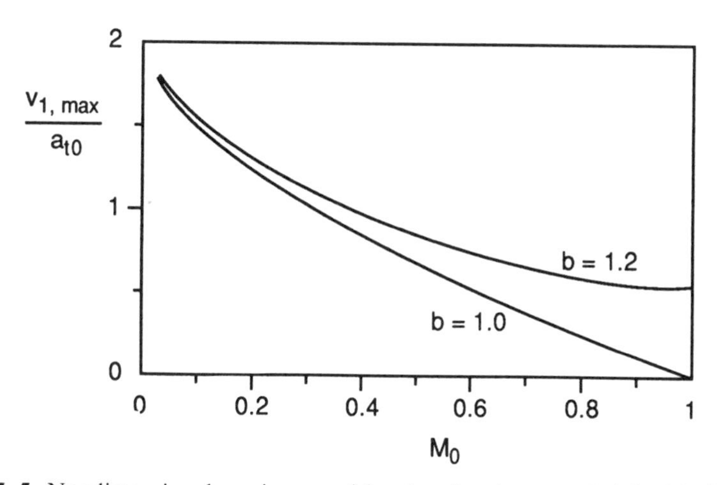

FIG. 15–5. Nondimensional maximum swirl, v_1/a_{t0}, for given nozzle inlet Mach number and area ratio. Plot of eq. 15–12 for $\gamma = 1.4$.

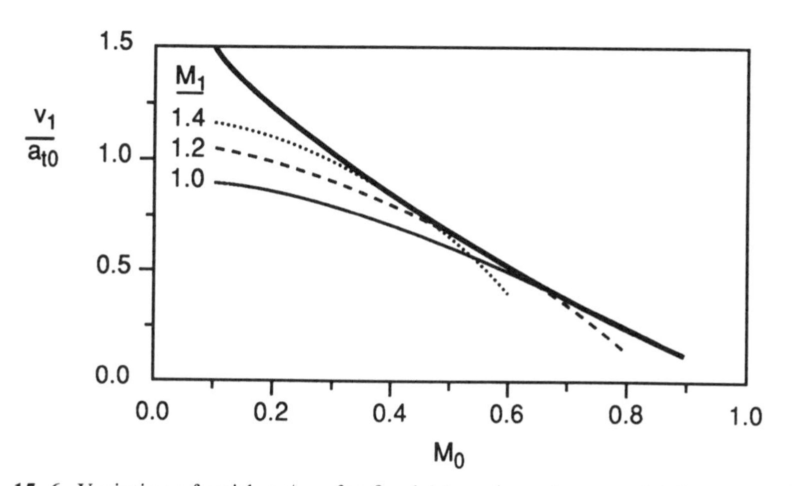

FIG. 15–6. Variation of swirl, v_1/a_{t0}, for fixed M_1 and maximum swirl (heavy envelope curve). $\gamma = 1.4$, b = 1.0.

the flow is turned through a large angle, the wakes from the trailing edges constitute a large fraction of the total flow area, also leading to a loss of average total pressure. The swirl that can be obtained with a limited M_1 is shown in Fig. 15–6. Also shown is a plot of the maximum swirl envelope associated with M_0 and the amount of swirl which can be achieved when M_1 has the values noted. The role of the annulus contraction in affecting v_1 is shown in Figs. 15–7 and 15–8.

In Fig. 15–7, the heavy line envelope curves are the same as shown in Fig. 15–5. Designs for fixed $M_1(=1.2$ here) yield the same swirls when M_0 is small but allow greater swirl when the nozzle area increases.

The variation of M_0 required for limited M_1 is thus a function of the nozzle

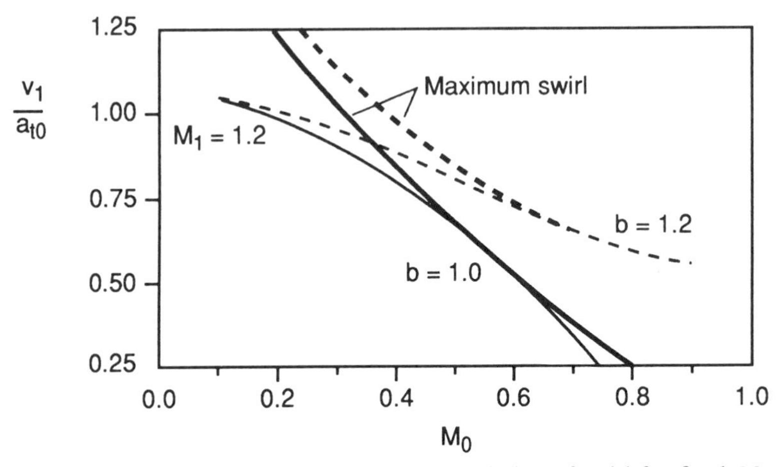

FIG. 15–7. Influence of nozzle area ratio on the variation of swirl for fixed $M_1 = 1.2$ (thin line) and maximum swirl (heavy line). $\gamma = 1.4$.

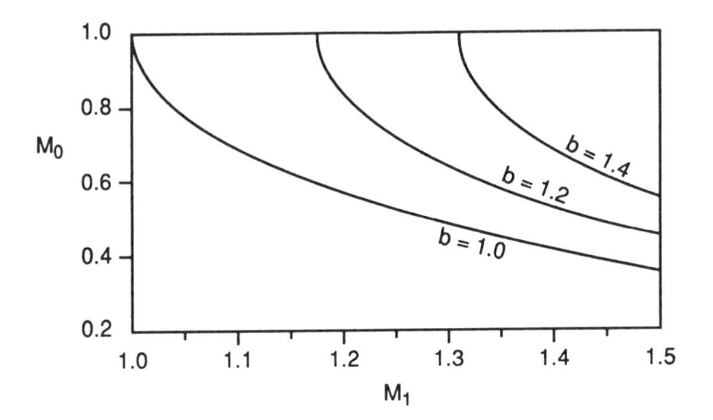

FIG. 15–8. Variation of M_0 required for maximum swirl and limited M_1. $\gamma = 1.4$.

annulus area ratio and γ. The influence of γ is small. For the results shown here, $\gamma = 1.4$. Figure 15–8 shows that the annulus area ratio must be used to simultaneously meet constraints on M_0 (turbine flow area) and on M_1 when maximum swirl is desired. This design value of M_0 shifts to larger values when b is increased. For example, with an increase in b from 1 to 1.2, M_0 for maximum swirl increases from ~ 0.6 to ~ 0.8 at $M_1 = 1.2$.

15.3. TURBINE ROTOR

The blading in the turbine rotor may be designed to change the flow direction and/or its magnitude. Rotors are classified according to the degree to which they affect the fluid static pressure. Thus, one may consider designs that decrease the pressure as well as the kinetic energy, or only decrease the kinetic energy, of the fluid. In this connection one may define:

$$\text{degree of reaction, d.r.} = \frac{(p_2 - p_1)}{(p_2 - p_0)}, \text{ often expressed in \%} \quad (15\text{–}15)$$

A 0% reaction turbine, where only the kinetic energy is reduced, is called an *impulse* turbine, whereas a *reaction* turbine has a specified degree of reaction. Thus one speaks of, say, a 30% or 50% degree of reaction. Although the pressure is awkward to manipulate in the flow equations (p varies as T^n), it is physically easy to measure and is critical to understanding boundary layer behavior. Its use to describe blading geometries is therefore appropriate.

As a measure of the fractional work output from the stage, one defines the recovery factor

$$C_R = \frac{\text{work out of stage}}{\text{KE into rotor} + \Delta\text{KE across rotor}} \quad (15\text{–}16)$$

as a measure of the work output of the stage for a given static pressure ratio across the stage, consisting of rotor and stator. For an axial flow turbine where the streamtube proceeds at a constant radius, r, the terms in the definition of C_R are:

$$\text{work output} = C_p(T_{t2} - T_{t1}) = \omega r(v_2 - v_1) \quad (15\text{–}17)$$

$$\text{KE into rotor} = \tfrac{1}{2}(v_1^2 + w_1^2) \quad (15\text{–}18)$$

$$\Delta\text{KE across rotor} = \tfrac{1}{2}\{w_2^2 + (\omega r - v_2)^2 - [w_1^2 + (\omega r - v_1)^2]\} \quad (15\text{–}19)$$

The ΔKE term is the difference of *relative* velocities squared. When $w_1 = w_2$, as is often the case, this term simplifies:

$$\Delta\text{KE across rotor} = \tfrac{1}{2}[(\omega r - v_2)^2 - (\omega r - v_1)^2] \text{ in general, and}$$
$$= 0 \text{ for an impulse turbine}$$

For the discussion to follow, the axial velocity is assumed constant. This assumption is easily relaxed, but it introduces an additional descriptive parameter that is omitted for clarity.

15.3.1. *Impulse Turbine*

Since the impulse turbine suffers no static pressure drop through the rotor, the magnitude of the velocity relative to the rotor must be invariant. The argument for this statement is embodied in eq. 14–15 or 14–16. The velocity triangle for impulse blading is consequently as shown in Fig. 15–9: $V_{r,1} = V_{r,2}$. The case shown is a nonoptimal case, with swirl remaining in the rotor exit flow ($v_2 \neq 0$).

With the lengths of the velocity vectors $V_{1,\text{Rel}}$ and $V_{2,\text{Rel}}$ equal and $w_1 \approx w_2$, one has the requirement $-\beta_1 = \beta_2$ which implies $v_1 - \omega r = \omega r - v_2$. Thus the

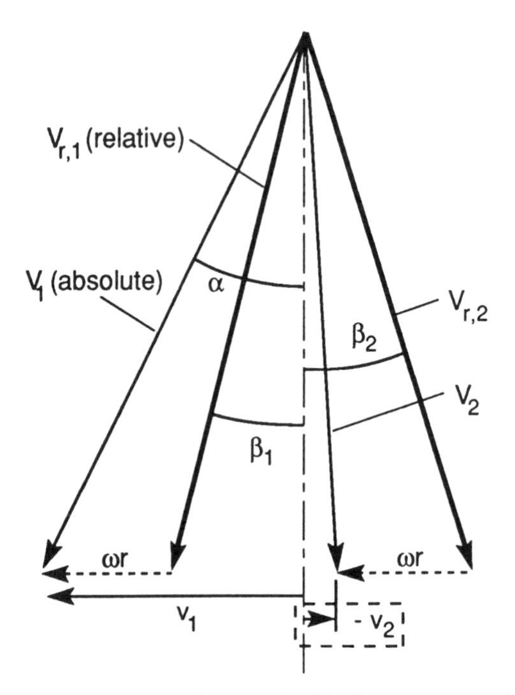

FIG. 15–9. Velocity vector diagram for impulse blading. Note equal magnitudes of $V_{r,1}$ and $V_{r,2}$.

exit swirl for impulse blading is

$$v_2 = 2\omega r - v_1 \qquad \text{[impulse blading]} \qquad (15\text{–}20)$$

15.3.2. *Reaction Turbine*

In a full (100%) reaction turbine, the entire pressure drop occurs in the rotor (i.e., there is no nozzle). It has a velocity diagram such as shown in Fig. 15–10. A wind turbine is a full reaction device.

Such a turbine cannot produce torque without leaving a swirl in the gas. The swirl represents an important loss which can be removed by a stator, in which case, the degree of reaction is no longer 100%. Thus full reaction turbines are not in common usage. A more balanced design is one with pressure drops in both nozzles and rotor. If the static pressure drops across stator and rotor are equal, the stage is termed a 50% reaction turbine. The velocity diagram must reflect the requirement that the velocity changes in the rotor and the stator are approximately equal. Figure 15–11 shows the velocity diagram for 50% reaction blading. The qualification with "approximate" has to do with the fact that, in *compressible* flow, velocity (or V^2) changes are not linearly related to pressure changes.

For multiple stages, it is desirable to have the diagram symmetrical ($\theta = \beta_1$

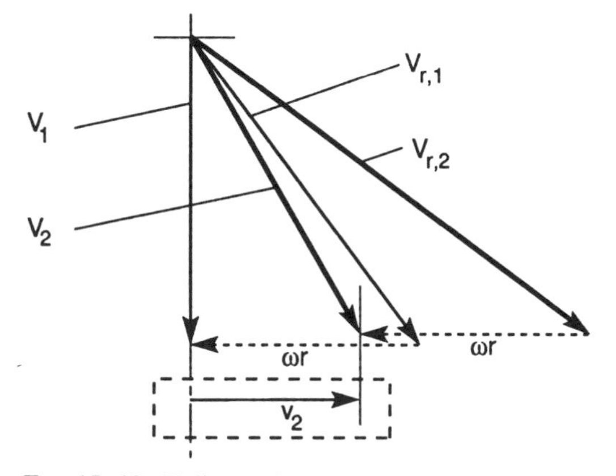

FIG. 15–10. Full reaction turbine velocity diagram.

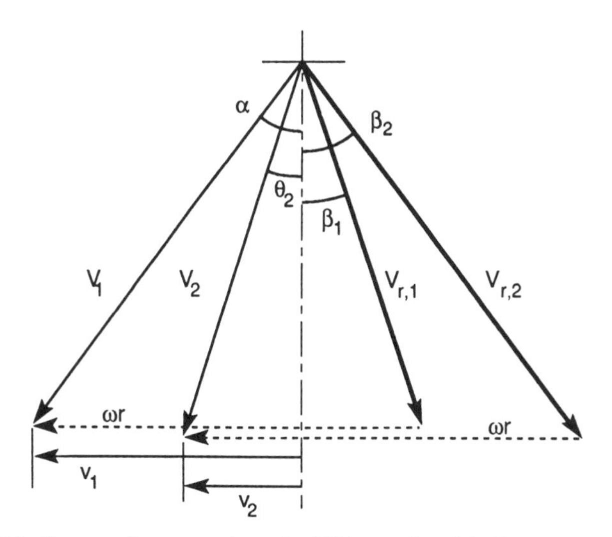

FIG. 15–11. Symmetric, approximately 50% reaction, blading velocity diagram.

and $\alpha = \beta_2$). Thus the exit swirl for symmetric reaction blading is

$$v_2 = -(v_1 - \omega r) \qquad \text{[symmetric reaction blading]} \tag{15–21}$$

This situation will be shown to approximate the 50% d.r. condition.

15.3.3. *Parametric Description of Degree of Reaction*

The stage exit swirl velocity, v_2, can be written for these two cases as

$$\frac{v_2}{\omega r} = 2(1 - \delta) - \frac{v_1}{\omega r} \tag{15–22}$$

Here δ is 0 for an impulse turbine and 0.5 for the symmetric blading reaction turbine. The optimum value of v_1 may found by examining the recovery factor, C_R, given by eq. 15–16.

The work output is given by eq. 15–17 with v_2 from eq. 15–22:

$$C_p(T_{t2} - T_{t1}) = 2(\omega r)^2\left[(1-\delta) - \frac{v_1}{\omega r}\right] \quad (15\text{–}23)$$

The kinetic energy from the nozzle (eq. 15–18) is:

$$KE_1 = \tfrac{1}{2}(\omega r)^2\left(\frac{v_1}{\omega r}\right)^2 \frac{1}{\sin^2\alpha} \quad (15\text{–}24)$$

Here the ratio of velocities is written in terms of the fixed nozzle angle:

$$\frac{v_1^2}{v_1^2 + w_1^2} = \sin^2\alpha \quad (15\text{–}25)$$

The change in kinetic energy through the rotor (from eq. 15–19) is

$$\Delta KE = \tfrac{1}{2}(\omega r)^2\left[\left(\frac{v_1}{\omega r} - (1-2\delta)\right)^2 - \left(\frac{v_1}{\omega r} - 1\right)^2\right] \quad (15\text{–}26)$$

The ΔKE term is evidently zero for the impulse turbine. The terms above can be collected in the recovery factor as:

$$C_R = \frac{\left[\dfrac{v_1}{\omega r} - (1-\delta)\right]}{\left(\dfrac{v_1}{\omega r}\right)^2 \dfrac{1}{4\sin^2\alpha} + \delta\left[\dfrac{v_1}{\omega r} - (1-\delta)\right]} \quad (15\text{–}27)$$

For a chosen blading type (δ fixed), the v_1 ratio for maximum work output can be obtained by differentiating this expression with respect to v_1 and setting the result to zero. The resulting value of v_1 is:

$$\left[\frac{v_1}{\omega r}\right]_{\max CR} = 2(1-\delta) \quad \text{and} \quad \left[\frac{v_2}{\omega r}\right]_{\max CR} = 0 \quad (15\text{–}28)$$

The rotor exit swirl velocity, v_2, is found from eq. 15–22. Thus maximum work extraction for a given pressure change through the stage is obtained for *zero exit swirl*, independent of δ. The corresponding maximum recovery

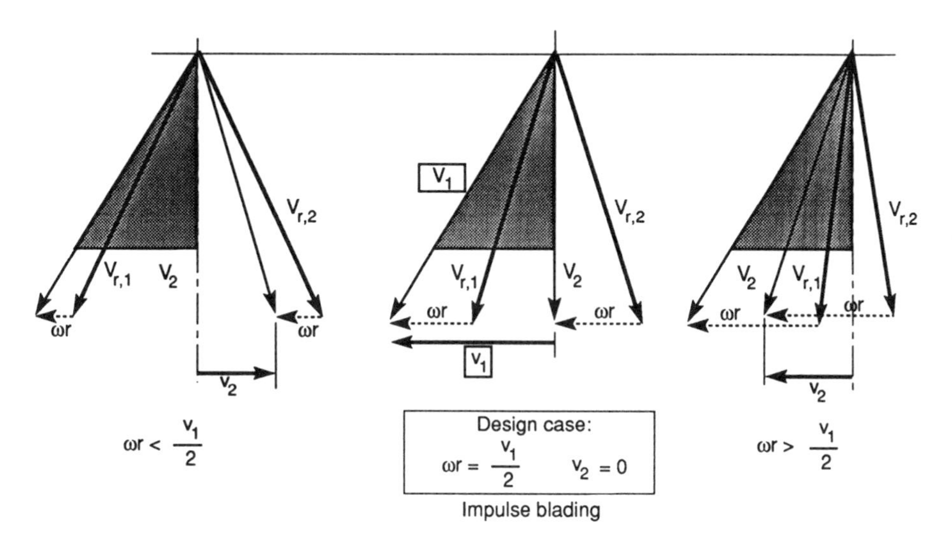

Fig. 15–12. Impulse blading with optimal (center) and nonoptimal $\omega r/v_1$. Nozzle geometry is fixed (shading) so that V_1 and v_1 are the same for all three cases.

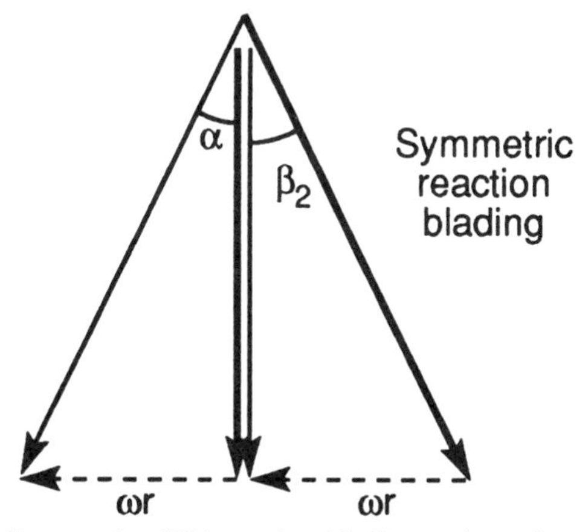

Fig. 15–13. Velocity diagram for 50% reaction blading and maximum work removal. Heavy line vectors are relative velocities, the thin line, laboratory reference frame.

factor, C_R, is:

$$\left.\begin{aligned} C_{R,\max} &= \frac{\sin^2\alpha}{(1-\delta)+\delta\sin^2\alpha} \\ &\text{or} \\ \frac{C_{R,\max}}{\sin^2\alpha} &= 1,\ \text{impulse};\ = \frac{2}{1+\sin^2\alpha} \geq 1,\ 50\%\ \text{d.r.} \end{aligned}\right\} \qquad (15\text{–}29)$$

The velocity diagrams for impulse and symmetric reaction blading are shown in Figs. 15–12 and 15–13. Figure 15–12 for the impulse blade also shows the situation for less than maximum C_R. In all cases, the relative velocity vector lengths are equal. On the left of the figure ($\omega r < v_1/2$), v_2 is to the right whereas on the right, ($\omega r > v_1/2$), v_2 is to the left.

Note that for the symmetric blading, the rotor exit may be viewed as a nozzle inlet when there is more than one stage, and thus rotor and stator (i.e., nozzle) cause the velocity vector to undergo similar orientation changes.

15.3.4. *Stage Total Temperature Ratio*

The stage total temperature ratio, in terms of wheel speed, is given by the Euler equation (eq. 15–17), with eq. 15–28 for v_1 as:

$$C_p T_{t1}\left(\frac{T_{t2}}{T_{t1}} - 1\right) = -(\omega r)^2 \frac{v_1}{\omega r} = -2(1-\delta)(\omega r)^2$$

or

$$\frac{T_{t2}}{T_{t1}} = 1 - \frac{(\omega r)^2}{C_p T_{t1}} 2(1-\delta) \qquad (15\text{–}30)$$

For a given wheel speed and T_{t1}, therefore, the impulse turbine extracts the greater work, twice as much as a symmetric blading ($\sim 50\%$ reaction) turbine.

The total temperature ratio can also be written in terms of M_0, using eq. 15–12 for the maximum nozzle exit swirl condition:

$$\frac{T_{t2}}{T_{t1}} = 1 - \frac{v_1^2}{2C_p T_{t1}} = 1 - \frac{(V_1 \sin\alpha)^2}{a_{t0}^2}\frac{a_{t0}^2}{2C_p T_{t1}}$$

$$= \frac{1}{1-\delta}\left(\left(\frac{M_0^2}{b^2}\right)^{-2/(\gamma+1)} \frac{\frac{\gamma+1}{2}M_0^2}{1+\frac{\gamma-1}{2}M_0^2} - \delta\right) \qquad (15\text{–}31)$$

This equation is plotted for $b = 1$ and 1.2 in the Fig. 15–14. Note that a low value of M_0 corresponding to large swirl is required to remove 50% of the total enthalpy as work. It is important to note that the wheel speed varies between the two δ cases, and the comparison must be interpreted with care.

For the same M_0, greater work removal is obtained from the symmetric ($\sim 50\%$ d.r.) blading owing to its larger wheel speed. For the same ωr, the impulse turbine has the disadvantage of larger velocities through the blade passages and hence greater losses. The boundary layer on the suction (low pressure) side of the impulse blade is generally closer to separation.

15.3.5. *Efficiency*

An examination of the pressure distribution along the streamlines in a rotor allows a qualitative statement about the tendency for flow separation and therefore efficiency. The streamtube processed by the impulse turbine tends to have a constant area because the pressure remains constant. By

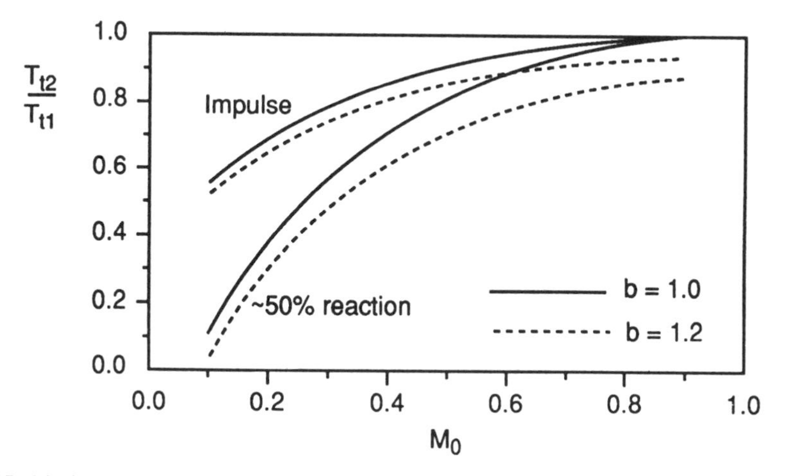

Fig. 15–14. Stage total temperature ratio for maximum nozzle swirl, zero rotor exit swirl, and $\delta = 0.0, 0.5$. Wheel speeds ωr vary, and the effect of annulus area variation is noted.

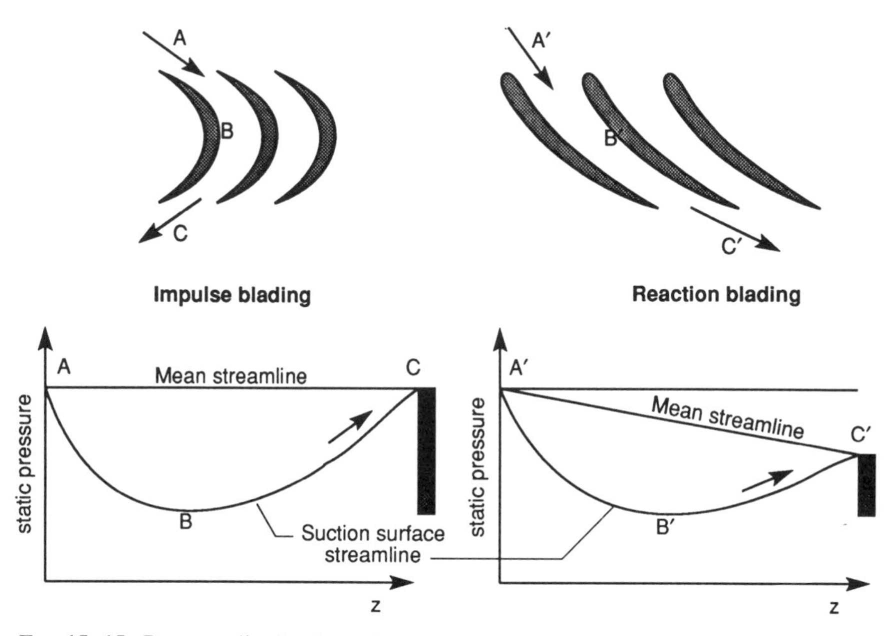

Fig. 15–15. Pressure distributions along surfaces of impulse and ~50% reaction blades. Note the relatively larger pressure rise B–C compared to B′–C'.

contrast, the streamtube through the reaction turbine tends to expand owing to the falling pressure.

Figure 15–15 shows the surface pressure distributions on the "airfoil" sections of the blades. The pressures A, and A' are the upstream static pressures on the mean flow, whereas the downstream pressures are C and C'. The mean streamline pressure distribution is given by the line A–C or A'–C', whereas that

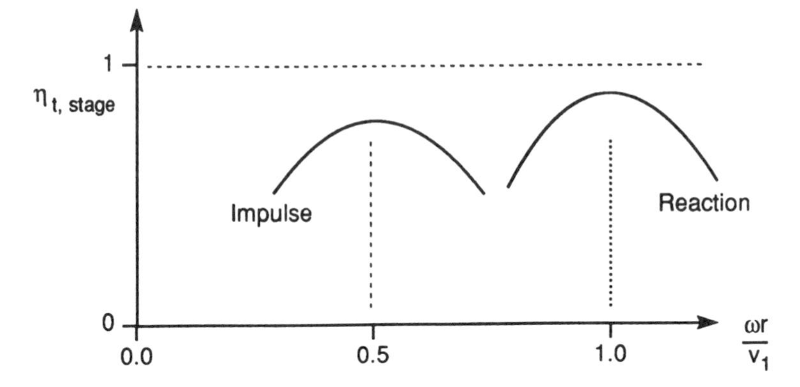

FIG. **15–16.** Typical efficiency characteristics for impulse and ~50% reaction symmetric turbine states.

near the suction side of the blade may experience something closer to A–B–C. From the large pressure *rise* demanded of the boundary layer between B and C, it may be expected that impulse turbines are closer to separation, have thicker boundary layers, and hence, have lower efficiencies. 50% reaction blades require a more modest pressure rise on the rear of the blade suction surface (B'–C').

A typical comparison of turbine stage efficiency might be as shown in Fig. 15–16. The decrease in performance away from the optimum design operation is due to the swirling kinetic energy remaining in the flow. In summary, maximum efficiency tends to increase with increasing degree of reaction.

15.3.6. *Rotor Stagnation Temperature*

An important factor is the stagnation temperature seen by the rotor, since this determines the heat transfer rate and thus peak metal temperature. One defines T_{tR} as the stagnation temperature measured by an observer on the rotor. Thus

$$C_P T_{tR} = C_P T_1 + \tfrac{1}{2} V_{r,1}^2 \tag{15–32}$$

The relative velocity for the rotor can be written for the case of maximum C_R, ($\beta_1 = 0$), with eq. 15–28, as:

$$V_{r,1}^2 = w^2 + v_1^2\left(\frac{\omega r}{v_1} - 1\right)^2 = w^2 + \frac{v_1^2}{4}\left(\frac{1-2\delta}{1-\delta}\right)^2 \tag{15–33}$$

The laboratory reference frame total temperature at the nozzle exit is

$$C_p T_{t0} = C_p T_{t1} = C_p T_1 + \tfrac{1}{2} V_1^2 = C_p T_1 + \tfrac{1}{2}(v_1^2 + w^2) \tag{15–34}$$

The two total temperatures described may related by subtracting eq. 15–34 from the combination of eqs. 15–32 and 15–33 and substituting $V_1 \sin \alpha$ in place of v_1:

$$\frac{T_{tR}}{T_{t1}} = 1 - \frac{\frac{3}{4} - \delta}{(1-\delta)^2} \frac{\frac{\gamma - 1}{2} M_1^2 \sin^2 \alpha}{1 + \frac{\gamma - 1}{2} M_1^2} \tag{15-35}$$

This ratio is less than unity because the blade is retreating from the flow exiting the nozzle. The factor associated with the degree of reaction has the following limits:

$$\frac{\frac{3}{4} - \delta}{(1-\delta)^2} = \tfrac{3}{4}, \text{ impulse}; = 1, \text{ symmetric } (\sim 50\% \text{ d.r.})$$

This factor reduces the rotor stagnation temperature to a lower value for the $\sim 50\%$ reaction blade when both have similar nozzles (which is not a realistic comparison as shown below). The magnitude of the Mach number grouping in eq. 15–35 when $M_1 \approx \sqrt{2}$ and $\alpha \approx 45°$ is about 0.14, showing that the rotor stagnation temperature is about 10–15% lower than the initial gas stagnation temperature in the laboratory reference frame.

15.3.7. *Degree of Reaction*

To this point, the degree of reaction has been dealt with parametrically through the specification of δ. The actual values of the pressures may be calculated from the *temperature* changes which take place through the blade passages. For the stator (from eq. 15–1),

$$\frac{T_1}{T_0} = \frac{\mu_0}{\mu_1} = f(M_0 \text{ and } M_1) \tag{15-36}$$

whereas for the rotor, the conserved total enthalpy in the rotor frame of reference gives

$$C_p T_1 + V_{r,1}^2 = C_p T_2 + V_{r,2}^2 \tag{15-37}$$

where (for $v_2 = 0$)

$$V_{r,1}^2 = w^2 + (v_1 - \omega r)^2 \qquad \text{and} \qquad V_{r,2}^2 = w^2 + (\omega r)^2 \tag{15-38}$$

In terms of δ, the rotor inlet relative velocity is given by eq. 15–33, and the outlet velocity is

$$V_{r,2}^2 = w^2 + v_1^2\left(\frac{\omega r}{v_1}\right)^2 = w^2 + \frac{v_1^2}{4}\left(\frac{1}{1-\delta}\right)^2 \tag{15-39}$$

The static temperature ratio may be calculated for the rotor by specification of the wheel speed. This is best done nondimensionally through specification of a wheel speed parameter,

$$\Omega \equiv \frac{(\omega r)^2}{C_p T_{tR}} \left[\approx (\gamma - 1) \left(\frac{500 \text{ m/sec}}{a_{air}(\text{at } 1250 \text{ K})} \right)^2 = 0.21 \right] \tag{15–40}$$

Thus eqs. 15–37 through 15–40 can be arranged to read

$$\frac{T_2}{T_1} - 1 = \frac{(\omega r)^2}{C_p T_{tR}} \frac{T_{tR}}{T_{t1}} \frac{T_{t1}}{T_1} \left[\left(\frac{v_1}{\omega r} \right)^2 - 2 \frac{v_1}{\omega r} \right] = -\Omega \frac{T_{tR}}{T_{t1}} \mu_1 [4(\delta - \delta^2)] \tag{15–41}$$

Note that the nozzle aerodynamics enter the rotor static pressure change through μ_1–a weak effect since this quantity is always close to unity. The rotor total temperature is given by eq. 15–35.

With the isentropic relations applied to the state changes, the degree of reaction (eq. 15–15) may be determined as

$$\text{d.r.} = \frac{\dfrac{p_1}{p_0}\left[\dfrac{p_2}{p_1} - 1\right]}{\dfrac{p_2}{p_1}\dfrac{p_1}{p_0} - 1} = \frac{\left(\dfrac{T_1}{T_0}\right)^{\gamma/(\gamma-1)} \left[\left(\dfrac{T_2}{T_1}\right)^{\gamma/(\gamma-1)} - 1\right]}{\left(\dfrac{T_2}{T_1}\dfrac{T_1}{T_0}\right)^{\gamma/(\gamma-1)} - 1} \tag{15–42}$$

Figures 15–17 and 15–18 are plots of d.r. as a function of δ. Note d.r. = 0 for $\delta = 0$, whereas for $\delta = 0.5$, the calculated d.r. is somewhat lower. The relationship is also not linear, owing to the isentropic relation between p and T and the quadratic relation between T and velocity. This nonlinearity makes δ such

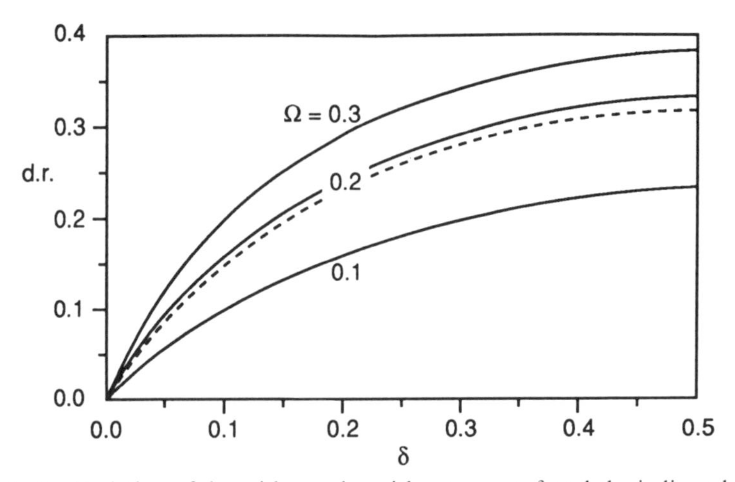

Fig. 15–17. Variation of d.r. with nozzle swirl parameter, δ and the indicated value of the wheel speed parameter, Ω. $M_0 = 0.5$, $\gamma = 1.4$, $b = 1.0$ and maximum swirl. Middle curves are for $\Omega = 0.2$, and the slightly lower (dashed) curve is for M_1 at 0.85 times the value for maximum swirl.

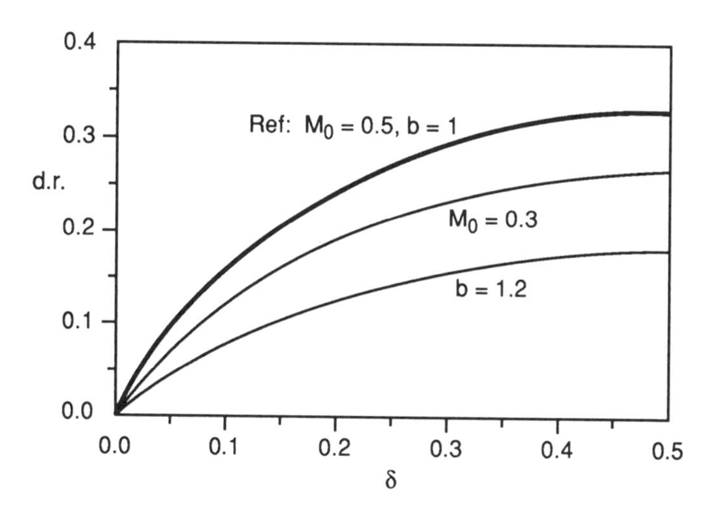

FIG. 15–18. Variation of d.r. as in Fig. 15–17, except that the reference (heavy line) parameters noted are sequentially changed as noted. Maximum swirl case. $\Omega = 0.2$, $\gamma = 1.4$.

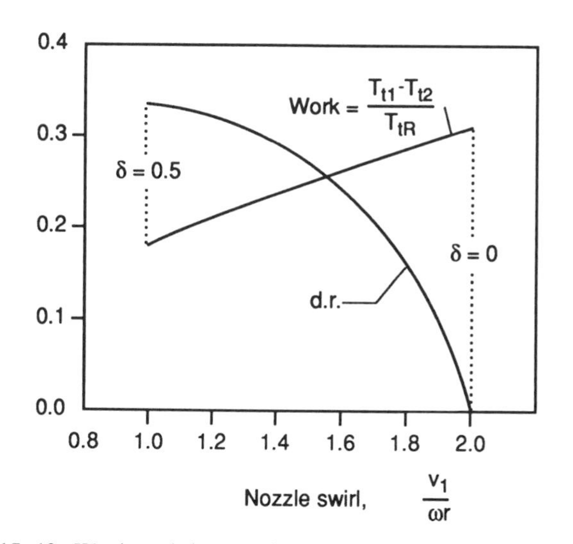

FIG. 15–19. Work and degree of reaction for various blading designs.

that it can be interpreted only as an indicator of the d.r. It appears that d.r. is roughly proportional to the cube root [or $\sim\gamma/(\gamma-1)$ power] of δ. For the display of parametric results, numerical values of $\Omega = 0.1$, 0.2, and 0.3 can be interpreted as modest, state of the art and advanced measures of wheel speed.

Evidently from Figs. 15–17 and 15–18, a high d.r. is obtained for large M_0, relatively small b, and large wheel speed, Ω. Figure 15–19 summarizes the design balance that must be struck between large work output and high efficiency for a family of designs with maximum nozzle swirl. The plot shows the ratio of the

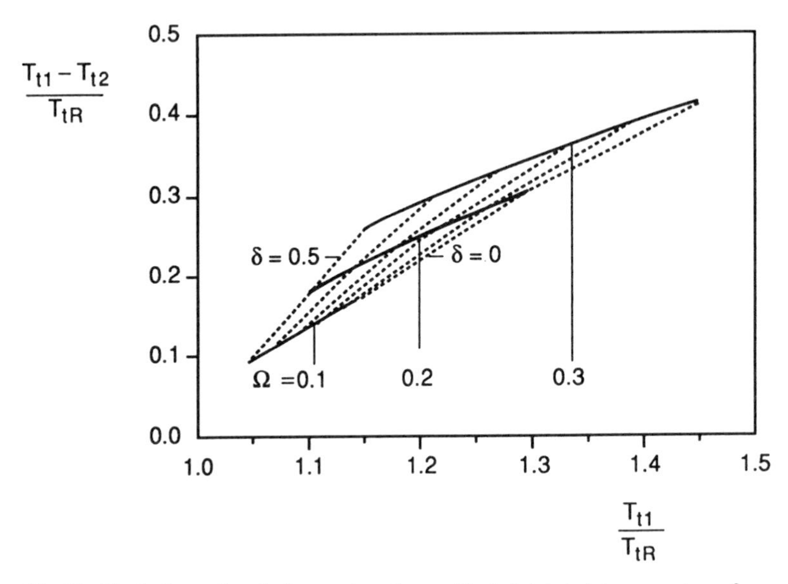

FIG. **15–20.** Variation of enthalpy extraction with inlet total temperature for various blading designs characterized by δ. The solid lines are lines of constant Ω, whereas the dashed lines are for constant δ which varies for 0.0 to 0.5 in increments of 0.1.

enthalpy extraction to rotor stagnation temperature, defined in eq. 15–43 and labeled here as "work." In Section 15.3.5 the efficiency is shown to vary monotonically with the degree of reaction. The ordinate is the ratio $v_1/\omega r$, with corresponding δ's indicated.

The Euler turbine equation (eq. 15–17) can be written to determine the work output as follows

$$\frac{T_{t1} - T_{t2}}{T_{tR}} = \frac{(\omega r)^2}{C_p T_{tR}} \frac{v_1}{\omega r} = 2\Omega(1 - \delta) \tag{15–43}$$

Thus wheel speed and blade reaction parameter determine the enthalpy extracted from the stream. These parameters also determine the turbine inlet temperature to which the entry gas must be raised. For any situation where $v_2 = 0$, the rotor stagnation temperature (eq. 15–32) can be written in terms of Ω and δ as

$$\frac{T_{t1}}{T_{tR}} = 1 + 2\Omega(\tfrac{3}{4} - \delta) \tag{15–44}$$

These parameters are plotted in Fig. 15–20 for parametric assignments of Ω and of δ. The dotted lines are for constant $\delta (0 \leq \delta \leq 0.5$ in increments of 0.1), whereas the solid ones are for constant Ω. Evidently large wheel speed and low degree of reaction are desirable for large enthalpy extraction. The turbine inlet stagnation temperature is permitted to be larger than the rotor stagnation temperature by a significant fraction, especially for impulse turbines.

15.4. COMPARISON OF IMPULSE AND REACTION TURBINES

An interesting means of displaying the differences between works produced by the two turbines is to form

$$\text{Rotor Work Parameter} = \left(\frac{T_{t2}}{T_{t1}} - 1\right)\frac{C_p T_{tR}}{(\omega r)^2} \tag{15–45}$$

so that a comparison can be made of work output for similar wheel speed and T_{tR}. This quantity is (using eqs. 15–30 and 15–35):

$$\left(\frac{T_{t2}}{T_{t1}} - 1\right)\frac{C_p T_{tR}}{(\omega r)^2} = 2(1-\delta)\frac{\left[1 + \frac{\gamma-1}{2}M_1^2\left(1 - \frac{\frac{3}{4}-\delta}{(1-\delta)^2}\sin^2\alpha\right)\right]}{1 + \frac{\gamma-1}{2}M_1^2} \tag{15–46}$$

With $\cos\alpha = 1/M_1$ for maximum swirl, the right side of eq. 15–46 can be written in terms of only M_0, b, and γ. The rotor work parameter defined by eq. 15–45 is plotted in Fig. 15–21 for impulse and $\delta = 0.5$ reaction turbines, for $b = 1.0$ and 1.2, as a function of M_0. The parameter allows one to see how wheel speed, rotor stagnation temperature, degree of reactions, and work output can be traded against one another. This maximum swirl formulation is, however, somewhat less interesting than an M_1-limited case because the losses are important and not uniformly included.

Thus with M_1 fixed, eq. 15–46 shows that the influence of degree of reaction on the rotor work parameter is dominated by the $(1-\delta)$ term. The large fraction in eq. 15–46 is close to unity. The $\sin^2\alpha$ term is of the order of 0.5 and in the limit of $\delta = 0.0$ and 0.5, the δ grouping is 0.75 and 1, respectively.

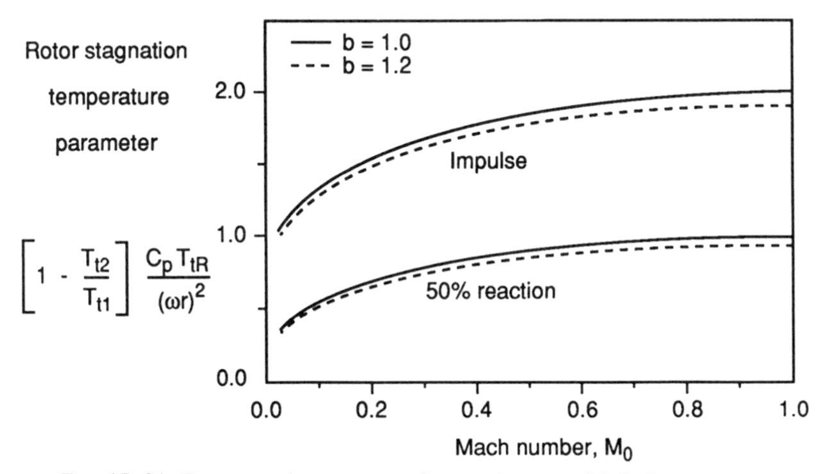

FIG. 15–21. Rotor work parameter for maximum swirl designs. $\gamma = 1.4$.

Table 15–1. Overall comparison of symmetric reaction and impulse blading

	Symmetric ~50% d.r.		Impulse
$C_{R\,max}$	$\dfrac{2\sin^2\alpha}{1+\sin^2\alpha}$	>	$\sin^2\alpha$
$\left(1-\dfrac{T_{t2}}{T_{t1}}\right)_{max\,swirl}$	$\dfrac{(\omega r)^2}{C_p T_{t1}}$	<	$2\dfrac{(\omega r)^2}{C_p T_{t1}}$
$\dfrac{T_{tR}}{T_{t1}}$	$1-\dfrac{\dfrac{\gamma-1}{2}M_1^2\sin^2\alpha}{1+\dfrac{\gamma-1}{2}M_1^2}$	<	$1-\dfrac{3}{4}\dfrac{\dfrac{\gamma-1}{2}M_1^2\sin^2\alpha}{1+\dfrac{\gamma-1}{2}M_1^2}$
$-\left[\dfrac{T_{t2}}{T_{t1}}-1\right]\dfrac{C_p T_{tR}}{(\omega r)^2}\left[1+\dfrac{\gamma-1}{2}M_1^2\right]_{max\,swirl}$	$=\dfrac{\gamma+1}{2}$	<	$2\left[1+\dfrac{\gamma-1}{2}\dfrac{M_1^2+3}{4}\right]$
Efficiency (relative)	High	>	Low
$\dfrac{\omega r}{v_1}$	1		0.5
Velocity diagram			

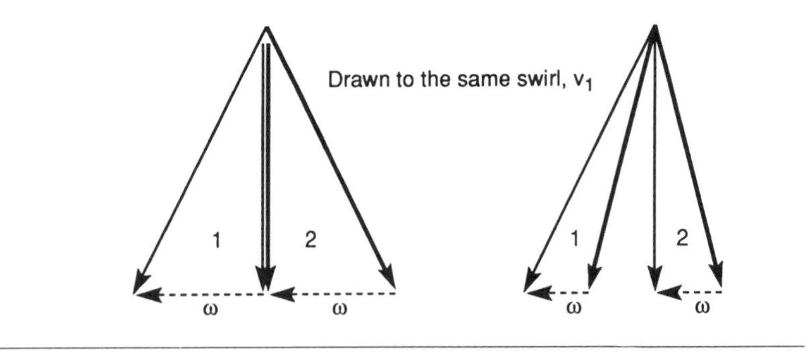

In summary, it may be said that the rotor work parameter varies between 2 and 1, depending on δ.

Table 15–1 summarizes the performance characteristics in light of design aspects of these two blading types. Because of the better efficiency of reaction blading, most modern turbines are designed for 30% to 50% reaction. Where very light weight is desired, an impulse design may be used.

15.5. BLADING ANALYSIS PROCEDURE

This section describes a calculational procedure that can be readily implemented on a spreadsheet program or in a standard computer program language to calculate all relevant performance parameters identified in the discussion above. The procedure is limited only by the following assumptions:

1. Total pressure losses are neglected.
2. Axial velocity through the stage is constant.
3. The swirl velocity is zero at the rotor exit.

The parameter, δ, is used as descriptive parameter of the turbine blading. In order for the user to judge the relative merits of various stage designs, information about a relation between the degree of reaction and the efficiency must be available.

15.5.1. *Axial Flow Turbines—Analysis Calculation Summary*

Given gas properties, allowable wheel speed, ωr, and rotor stagnation temperature, T_{tR}.

1. Calculate $\Omega \equiv \dfrac{(\omega r)^2}{C_p T_{tR}}$, eq. 15–40
2. Choose M_0 and b
3. Calculate $M_{1,v}$ M_1 for max v_1, eq. 15–10

   $\alpha_{\max v}$ $\cos \alpha_{\max v} = \dfrac{1}{M_{1,v}}$
4. Choose $M_1 < M_{1,v}$(or $\alpha < \alpha_{\max v}$)

   $v_{1,\max}$ $= (V_1 \sin \alpha)_{\max}$ eq. 15–12

   $\dfrac{p_1}{p_0}$ from $\dfrac{p_1}{p_0} = \left(\dfrac{T_1}{T_0}\right)^{\gamma/(\gamma-1)}$, eq. 15–1

   $\alpha(M_1)$ eq. 15–8
5. Choose blading parameter $\delta \equiv 1 - \dfrac{v_1}{2\omega r}$, $=0$: impulse, ≈ 0.5: reaction
6. Calculate T_{t1} $\dfrac{T_{t1}}{T_{tR}} = 1 + 2\Omega(\tfrac{3}{4} - \delta)$

   Δh_t $\dfrac{T_{t2} - T_{t1}}{T_{tR}} = -2\Omega(1 - \delta)\dfrac{T_{t1}}{T_{tR}}$

   $\dfrac{T_2}{T_1}$ $\dfrac{T_2}{T_1} = 1 - 4\Omega\delta(1 - \delta)\mu_1 \dfrac{T_{tR}}{T_{t1}}$

   $\dfrac{p_2}{p_1}$ from $\dfrac{p_2}{p_1} = \left(\dfrac{T_2}{T_1}\right)^{\gamma/(\gamma-1)}$
7. d.r. eq. 15–15

If d.r. is not satisfactory, the value of δ must be altered.

15.6. TURBINE SIMILARITY

Just as for the compressor, the similarity parameters for the turbine can be deduced by noting that α and β_2 are the (nearly) constant flow angles at the trailing edges, so that

$$C_p(T_{t2} - T_{t1}) = \omega r(v_2 - v_1) = \omega r(\omega r - w_1 \tan \beta_2 - w_1 \tan \alpha)$$

$$= (\omega r)^2 \left[1 - \frac{w_1}{\omega r} (\tan \beta_2 + \tan \alpha) \right]$$

$$\frac{T_{t2}}{T_{t1}} = 1 - (\gamma - 1) \frac{(\omega r)^2}{\gamma R T_{t1}} \left[-1 + \frac{w_1}{\omega r} (\tan \beta_2 + \tan \alpha) \right] \quad (15\text{–}47)$$

Thus, for fixed α and β_2, and since T_{t2}/T_{t1} is related to p_{t2}/p_{t1}, the stage total pressure ratio can be written in the form

$$\pi_t = f\left(\frac{(\omega r)}{\sqrt{\gamma R T_{t0}}}, \frac{w_0}{\sqrt{\gamma R T_{t0}}} \right) \quad (15\text{–}48)$$

Analogously to the compressor discussion (see Section 14.3.6), $\omega r/\sqrt{\gamma R T_{t0}}$ is related to a corrected speed $N/\sqrt{\theta_4}$. Here $\theta_4 = T_{t4}/T_{ref}$ and the station numbering system for the *engine* is used to describe the entry conditions. State point 4 describes the conditions at combustor exit and turbine entry. Similarly, $w_0/\sqrt{\gamma R T_{t0}}$ is related to the corrected weight flow rate given by $W_4\sqrt{\theta_4}/\delta_4$. Here, as for the compressor, $\delta_4 = p_{t4}/p_{ref}$ is introduced, and the traditional use of the weight flow rate W_4 is retained. Note also π_t is the total temperature ratio for the turbine as a whole. Thus the overall turbine characteristic may be summarized in terms of engine parameters as

$$\pi_t = \text{fcn}\left(\frac{W_4 \sqrt{\theta_4}}{\delta_4}, \frac{N}{\sqrt{\theta_4}} \right) \quad (15\text{–}49)$$

The functional relationship of the form of eq. 15–49 is termed a turbine map (Fig. 15–22). A plot is usually made with $1/\pi_t$ as the abscissa to parallel the way information is presented on a compressor map.

From the velocity diagram, it is evident that either or both the flows in the nozzle and in the rotor may choke, depending on the static pressure changes taking place in the blade rows. The exit velocities in both rows are larger than the entry velocities so that choking occurs near the exit. Thus as $1/\pi_t$ becomes greater than about 2.5 for a single stage turbine, locally sonic velocities may be expected, and consequently, the corrected weight flow $W_{4c} = W_4 \sqrt{\theta_4}/\delta_4$ will approach a maximum (see eq. 14–27). From eq. 15–47 it may be concluded that at a given speed ($\omega r = \text{cst}$), an increasing amount of work is removed by

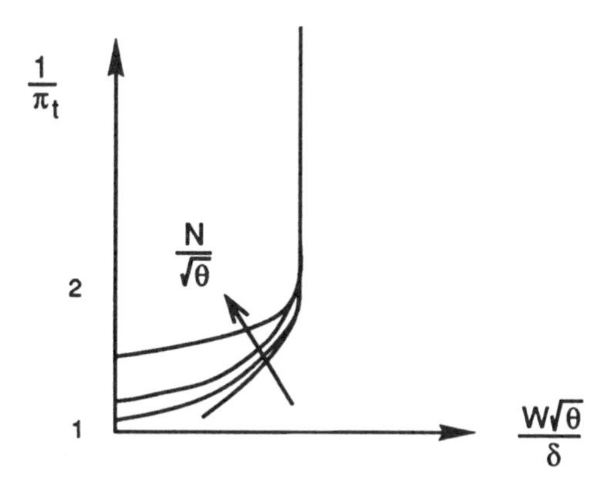

FIG. 15–22. Turbine map: lines of constant corrected speed on a plot of pressure ratio and corrected mass flow rate.

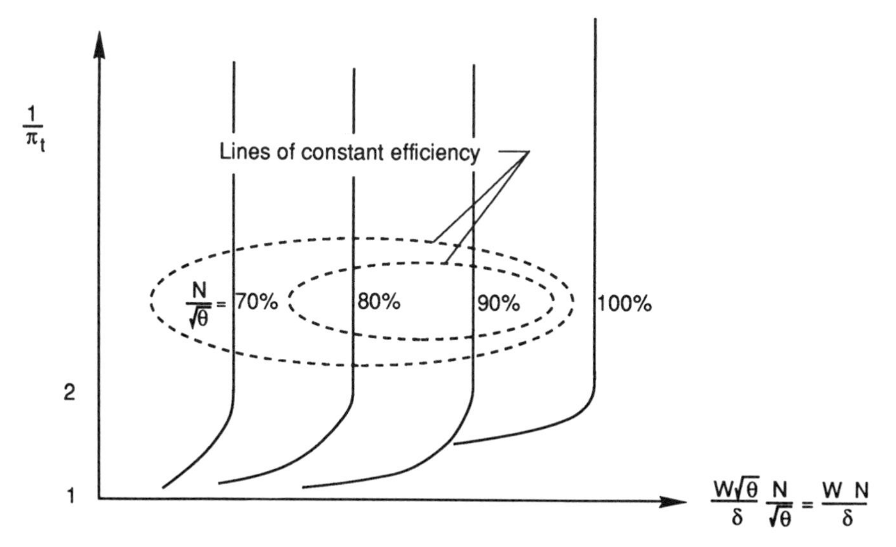

FIG. 15–23. Expanded scale turbine map to reveal constant efficiency lines.

the turbine for larger corrected weight flow since W_{4c} is proportional to w_1 and $v_1 = w_1 \tan \alpha$. Thus at constant $N/\sqrt{\theta_4}$ one has the characteristic shown in Fig. 15–22. At higher $N/\sqrt{\theta_4}$ more work is removed.

This plot is often recast as in Fig. 15–23 to show the constant speed and efficiency lines more clearly. For preliminary design purposes, it often is adequate to take $W_4\sqrt{\theta_4}/\delta_4$ and η_t as constants at the maximum power condition.

PROBLEMS

1. By allowing the axial velocity to vary through the rotor of an axial blade rotor row, write the recovery factor (eq. 15–16) in terms of a parameter b_r defined as w_2/w_1. Note the influence of the sign of b_r on the various terms in C_R.
2. Derive the relation between v_2 and v_1 (following the idea of eq. 15–22) when b is not unity. The idea of velocity similarity may be defined in terms of angles, rather than swirl velocity components.
3. Show that the maximum swirl is given by eq. 15–12 by carrying out the differentiation indicated in eq. 15–9.
4. Calculate and plot the maximum swirl as a function of A_0, *not* making the approximation indicated in eq. 15–14.

BIBLIOGRAPHY

Hill, P. G., Peterson, C. R., *Mechanics and Thermodynamics of Propulsion*, Addison-Wesley, Reading, Mass., 1965 and 1991.

Glassman, I., Ed., *Turbine Design and Application*, Vols. 1–3, NASA SP-290, Scientific and Technical Information Office, NASA, Washington D.C. 1973.

REFERENCES

15–1. Oates, G. C., Ed., *Aerothermodynamics of Aircraft Engine Components*, AIAA, New York, 1985.

15–2. Bölcs, A., Ecole Polytechnique Fédérale de Lausanne, LTT, Lausanne, Switzerland, private communication.

16

PART POWER PERFORMANCE: BRAYTON CYCLE

Section 3.3 examined the role played by the efficiency in the economics of an energy conversion system operating at various power settings. Engines of all kinds commonly deliver power that varies in time. The efficiency of most heat engines depends on the fraction of maximum power being produced because the thermodynamic cycle configuration and the component operating conditions differ from those at full power. The question addressed in this chapter is: How is the efficiency of a steady-flow cycle calculated for a part power condition? This question is discussed in Chapter 8 for the Otto and Diesel cycle engines. For an analysis of steady-flow cycles, either the Rankine or the Brayton cycle engines may be considered. The Brayton is more interesting because the recirculating work fraction is relatively large. The compressor in the Brayton cycle therefore plays a much more important role than the pump in the Rankine cycle in determining part power efficiency. In fact, for the Rankine cycle, the efficiency is dominated by the aerodynamics of the turbine and the heat losses in the other components. By contrast, the part power performance characteristics of the Brayton cycle depend critically on the characteristics of the work components developed in Chapters 14 and 15 and their interaction.

A general procedure for calculating part power performance for the cycle described in Chapter 9 is complex since the components of the cycle may be assembled in a wide variety of ways to include heat exchangers and a variety of load coupling mechanisms. Furthermore, the type of components themselves may also be of varying designs, taking advantage of either "displacing" the boundaries of a fluid element to affect volume changes or of aerodynamic forces acting on blades to result in quasi-volumetric work interactions. As a result, a number of examples are detailed using various engine component configurations to illustrate the essential features of a part load (or part power) cycle analysis. The approach taken is to examine the thermodynamic issues and the component "maps" to show how both of these elements play a role in determining the part load performance of a simple cycle consisting of compressor, heat source and expander. The analysis is restricted by the assumption of quasi-steady operation where time-dependent storage of working fluid mass in the components and inertia effects are neglected.

16.1. DESIGN AND PART LOAD OPERATION

For a system designed to deliver a range of power output, it is an arbitrary choice to designate one value of the output as being the "design" value and all else as being "off-design." It is most convenient to select the maximum power condition as the "design" point. With temperature-limited heat engines, the notion of "maximum" is intimately connected with the utilization lifetime of the engine because there is an inverse relationship between lifetime and maximum usable temperature. In an aircraft gas turbine, for example, the full-power cruise power setting is typically most interesting because it determines the performance of the aircraft in the mode for which it is designed to spend the greatest fraction of its operating life. Greater levels of performance such as "emergency" or "military" levels might be made available for appropriately short time periods, but these do not play a large role in the system design because the need is short and may never arise. Thus the view is adopted here that 100% power is the "design" condition and reduced power is "off-design."

The objective of a power system design (i.e., the process of system configuration specification) is to identify the values of the free design variable(s), given specific constraints such as performance goals at the "design" operating condition. For the Brayton cycle, the free design variables are the cycle pressure ratio and/or the cycle configuration, such as the number of intercoolers, reheat, regeneration, etc. The constants are usually the temperature limits of the cycle, the quality of the components as measured by efficiencies, etc. The performance goal that the engine is to achieve as an energy converter might be the attainment of a level of thermal efficiency, compactness, cost, or combination of these. Usually the overall requirement is for a given level of power output or thrust, in the case of a propulsion engine. The cycle analysis of the "design" condition gives values for specific power or specific thrust (i.e., output per unit mass flow rate). The size of the machinery (in terms of air or gas mass flow per unit time which must be processed through the components) is scaled by the requirements of the load. For a chosen level of flow losses, then, the design condition is important because it sizes all components of the machinery. At "off-design" or "part load," this same machinery must operate satisfactorily to provide the reduced power demanded.

The temperature T_4 for part load conditions (less than maximum power) will generally be smaller than the design value, T_{4D}. The nondimensional specific work output of a cycle configuration operating with a specified level may be written in terms of T_{4D} and chosen pressure ratio, π_{CD}, chosen to meet design point performance requirements. Thus,

$$w_D = \left[\frac{\dot{W}}{\dot{m}C_pT_0}\right]_D = \text{fcn}\left(\frac{T_{4D}}{T_0} = \theta_{4D}, \pi_{CD}, \text{work component efficiencies, etc.}\right) \tag{16–1}$$

It is assumed that the minimum cycle temperature T_0 is fixed (usually standard day: 288 K = 519 R = 59 F).

The maximum power level, $\dot{W}_D$, fixes the cycle fluid mass flow rate ($\dot{m}_D$), eq. 16–1. This mass flow rate can be interpreted as a measurement of size at various points in the cycle where density and fluid flow velocity are known (Section 3.3). In algebraic terms, this relation may be expressed (eq. 12–4) as

$$\text{flow area} = \left[\frac{\dot{m}}{\rho u}\right]_{\mathrm{D}} \tag{16–2}$$

The density and fluid velocity are equivalent one-dimensional values. The density is dictated by the cycle parameters, and the flow velocity is usually limited by loss considerations. The design task is completed with a check that, for the sizes calculated, the work component efficiencies used are realistic and accurate.

16.2. BRAYTON CYCLE PART LOAD EFFICIENCY

In order to determine the efficiency of a Brayton cycle at part load, one must ask: what changes occur when a control parameter, such as the fuel input rate, is altered to reduce the power output? The power of the Brayton cycle, $\dot{W}$, is a function of the parameters given in eq. 16–1. Evidently the power output from an engine may be reduced by (1) reducing the mass flow rate processed and/or (2) altering the thermodynamic cycle.

These two kinds of changes usually occur simultaneously because there is a relationship between the power delivered to the compressor or from the expander and the mass flow processed. The information relevant to such relationships is given in *component maps*. A map normally relates any three of the following four variables:

1. mass flow rate (usually nondimensionalized by entry conditions)
2. rotational speed
3. pressure ratio
4. enthalpy required or provided

The fourth of these variables is related to the others by means of the First Law of Thermodynamics (eq. 2–6 or 7–29). The values of adiabatic efficiency are usually displayed on such a map. The physical design of the work component plays an important role in determining the form of the map. Here, examples are described as needed.

Before considering the role of mass flow variation on the part load performance, it is instructive to examine the contribution made by the changing thermodynamics, assuming that components may be visualized or invented that can perform the function required and still process a unit mass flow. The need to consider the component characteristics is addressed subsequently. This follows from the requirement that work components must work together in a constrained manner.

16.3. PART LOAD PERFORMANCE OF A UNIT MASS BRAYTON CYCLE ENGINE

Consider the performance of a simple cycle engine, which processes a unit mass of working fluid in a unit time. For such an engine, the cycle parameters (θ_4 and π_C) determine the performance. In reality, however, the mass flow rate changes as required by the component characteristics; this compounds the influence (neglected in this section) of the cycle parameters on part load performance.

Suppose, for the present, that one is contemplating a configuration with a specified "design" value of the compressor pressure ratio, π_{cD}, and peak cycle temperature ratio, θ_{4D}. To reduce work output, one would want to reduce the area enclosed in the *T-s* plane. Two ways are available to do this:

1. Variation of the cycle temperature ratio
2. Variation of compressor pressure ratio
3. A combination of these options

These cases are discussed in the subsections to follow.

16.3.1. *Fixed Cycle Pressure Ratio*

For this case, the compressor pressure ratio, π_{cD}, remains a constant at its design value and fuel input to the cycle is reduced to reduce work output. Thus $\theta_4 < \theta_{4D}$ and an efficiency reduction is incurred because of the reduced temperature ratio (Carnot efficiency) of the cycle. Figure 16–1 shows the *T-s* diagram for this process. The work output per unit mass at design (subscript D) is

$$w_D = \eta_E \theta_{4D}\left[1 - \frac{1}{\tau_{csD}}\right] - \frac{(\tau_{csD} - 1)}{\eta_c} \tag{16–3}$$

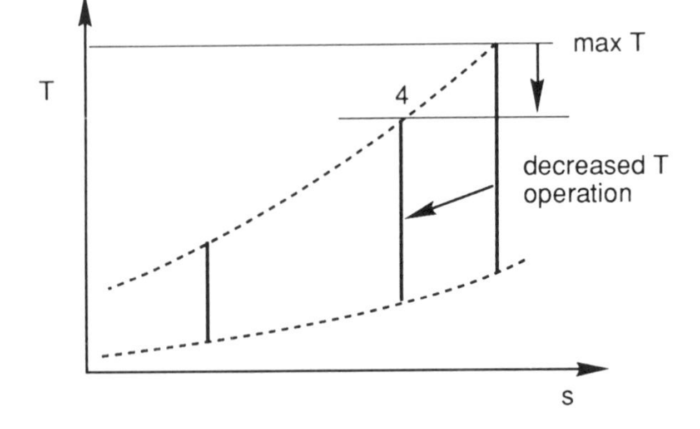

Fig. 16–1. *T-s* diagram for a Brayton cycle variation of peak cycle temperature with fixed pressure ratio.

Table 16–1. Values of "design" parameters indicated: at maximum specific work and at maximum thermal efficiency

Component efficiencies	maximum w		maximum η_{th}	
η_c, η_E	τ_{csD} (π_{cD})	w	τ_{csD} (π_{cD})	w
0.99, 0.99	2.42 (22)	2.05	4.55 (200)	1.05
0.85, 0.95	2.14 (14)	1.54	2.68 (32)	1.41

Note: $\theta_{4D} = 6$ and for the pressure ratio calculation $\gamma = 1.4$.

where constant adiabatic efficiencies are used to characterize the irreversibilities. For simplicity, the pressure ratio is written as an isentropic temperature ratio defined by:

$$\tau_{cs} = (\pi_c)^{(\gamma - 1)/\gamma} \tag{16–4}$$

The part load specific work output for the invariant pressure ratio is

$$w = \eta_E \theta_4 \left[1 - \frac{1}{\tau_{csD}} \right] - \frac{(\tau_{csD} - 1)}{\eta_c} \tag{16–5}$$

from which it follows that $w < w_D$ for $\theta_4 < \theta_{4D}$. The design values of specific work are given in Table 16–1 for $\theta_{4D} = 6$ and two sets of efficiencies, one nearly ideal, whereas the other is more realistic. The two values of the compressor pressure ratio are chosen (see Section 9.3), to give maximum specific work or maximum efficiency. For the case of maximum efficiency, the value of τ_c would be limited to θ_4 if the components were ideal. The values for w_D and τ_{csD} are given in Table 16–1. The ratio of w/w_D is shown in Fig. 16–2. The reduction in w as θ_4 is reduced is linear and steep.

The heat input to the cycle, per unit mass, is given by

$$q = \theta_4 - \tau_{cs} = \theta_4 - \tau_{csD} \text{ (constant } \pi_c \text{ case)} \tag{16–6}$$

so that a thermal efficiency ($\eta = w/q$) may be calculated. Figure 16–3 shows the thermal efficiency for the case of nearly ideal work components. This is an excellent part load characteristic because the efficiency stays high as w is reduced. The higher thermal efficiency is associated with the case where τ_{csD} is large, for which the maximum specific work is correspondingly low (see Section 9.2 and Table 9–1). For a more realistic set of component efficiencies, the same result is also shown in Fig. 16–3. Noteworthy is the fact that the level of efficiency is lower and it decreases more rapidly as the work output is reduced. The conclusion is that reduced θ_4 in a unit mass flow engine leads to reduced efficiency at part load to a degree determined by the irreversibilities in the work components.

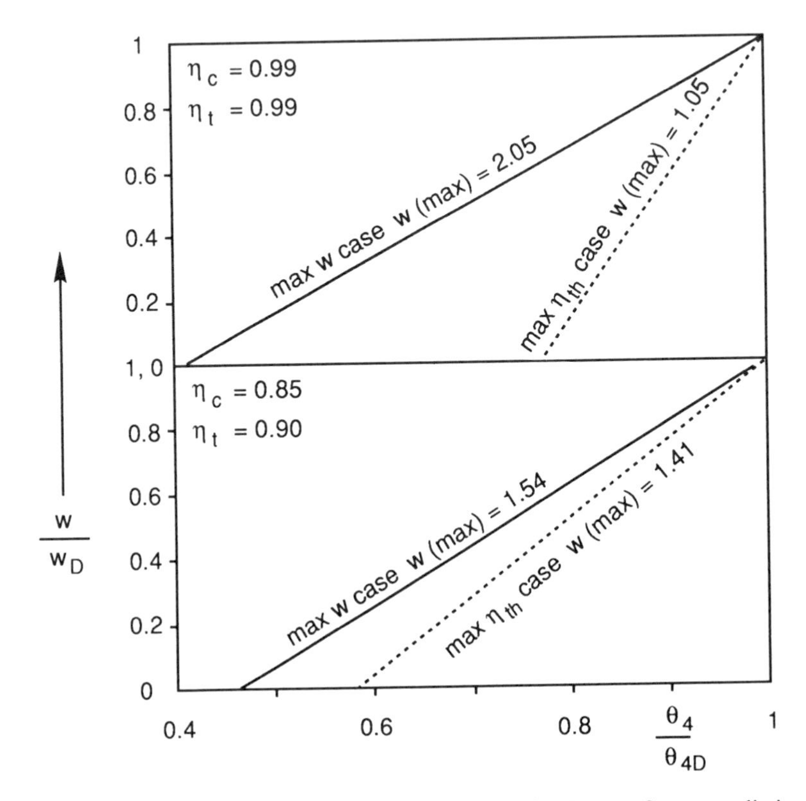

FIG. 16–2. Variation of w with θ_4 for a fixed π_c cycle. Lower figure: realistic work components. Upper figure: nearly ideal work components. Pressure ratio is chosen for maximum w or maximum efficiency at design point. Values of w for these cases are noted. $\theta_{4D} = 6$.

16.3.2. *Fixed Cycle Temperature Ratio*

An alternative way of reducing power (i.e., work per unit mass) is to keep $\theta_4 = \text{constant} = \theta_{4D}$ and increase the pressure ratio of the compressor. One might picture such an implementation by imagining a displacement compressor with a reed valve or a poppet valve whose opening timing can be varied as required. The *T-s* diagram for this cycle is shown in Fig. 16–4. The increased τ_{cs} decreases the fuel input permitted for a maximum $\theta_4 = \theta_{4D}$, and thus work output is reduced.

The part load work output (eq. 16–7) for this case is given by

$$w = \eta_E \theta_{4D}\left[1 - \frac{1}{\tau_{cs}}\right] - \frac{(\tau_{cs} - 1)}{\eta_c} \tag{16–7}$$

and the requirement is for τ_{cs} to increase above τ_{csD}. For the realistic component efficiency case, the work (referenced to the design value) given by eqs. 16–7 and 16–3 is shown in Fig. 16–5, where the control variable is $1/\tau_{cs}$ for comparative similarity to the results shown for the varying θ_4 case. Note that raising the

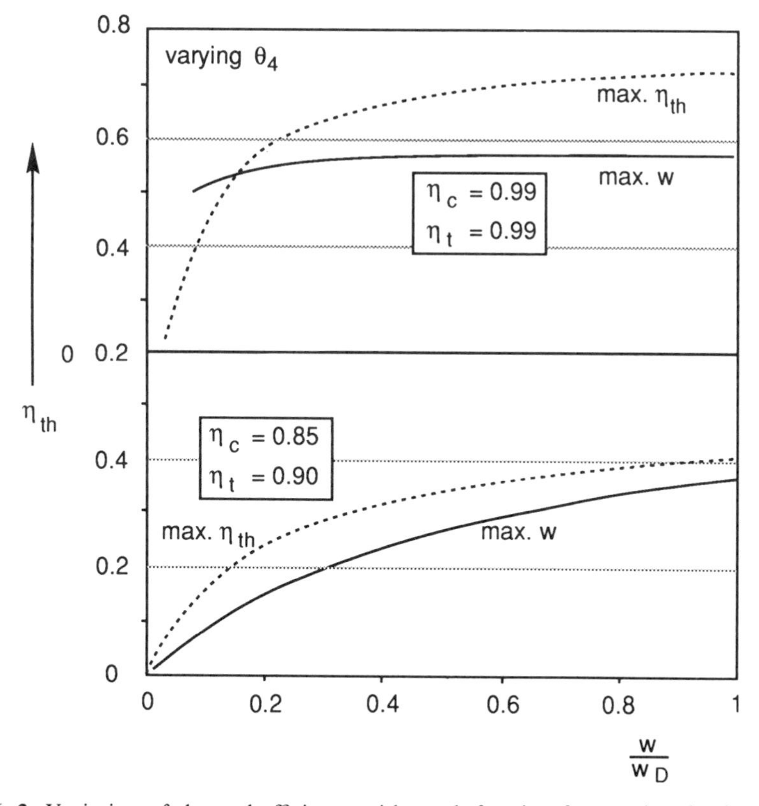

FIG. 16–3. Variation of thermal efficiency with work fraction for varying θ_4. $\theta_{4D} = 6$. Nearly ideal work components: $\eta_c = \eta_E = 0.99$ and for realistic work components: $\eta_c = 0.85$, $\eta_E = 0.90$.

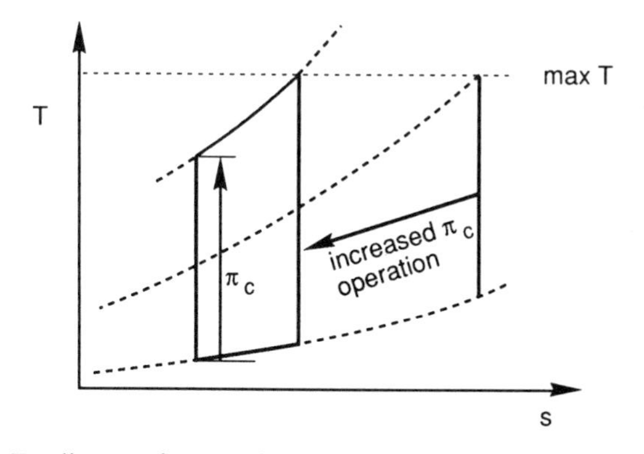

FIG. 16–4. T-s diagram for a cycle controlled by pressure ratio. θ_4 is fixed.

temperature ratio τ_{cs} by a factor of 1.5 reduces the work output of the maximum efficiency design to zero, whereas a doubling of this parameter is required for the maximum work case.

The efficiency can be calculated (as described for the constant π_c case) and is shown for varying power fraction in Fig. 16–6. Note that the efficiency for

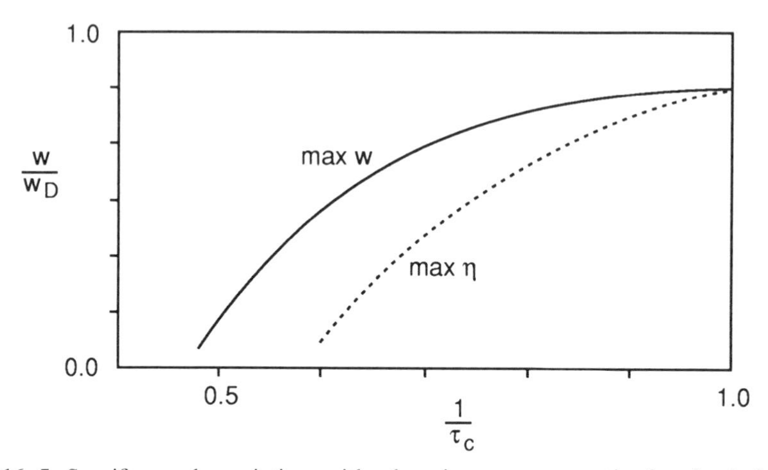

FIG. 16–5. Specific work variation with changing pressure ratio for fixed $\theta_4 = 6$. Abscissa is the reciprocal of compressor temperature ratio. $\eta_c = 0.85$, $\eta_E = 0.90$.

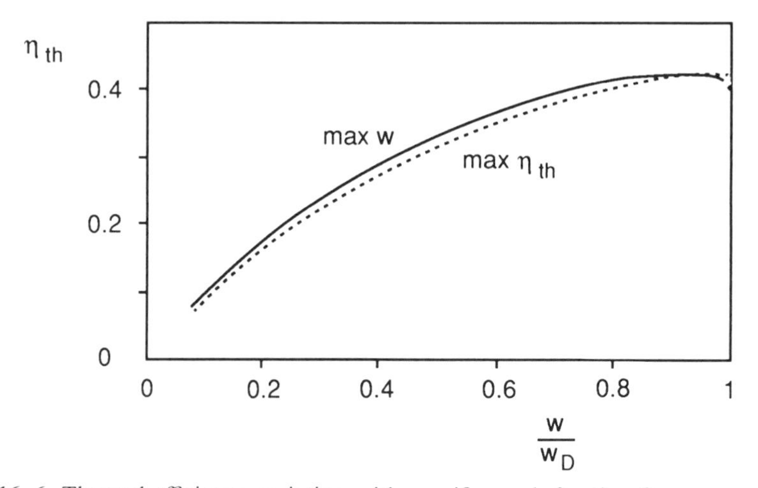

FIG. 16–6. Thermal efficiency variation with specific work fraction for pressure controlled cycles with fixed $\theta_4 = 6$. $\eta_c = 0.85$, $\eta_E = 0.90$.

the maximum thermal efficiency design case drops uniformly as output/mass is reduced, whereas the maximum specific work (w) case leads to an increase in efficiency for a slightly reduced power. A choice for τ_{csD} between these values could give a characteristic variation that is relatively "flat" near maximum work.

16.3.3. *Related Pressure and Temperature*

Attainment of a high thermal efficiency, particularly at part load, demands high component efficiencies. In the limit of ideal components, the cycle approaches a Carnot cycle when a reduced power output is obtained by increasing the pressure ratio. Between the cases of constant pressure ratio and constant peak cycle temperature (ratio), one may imagine a family of engine designs that have both θ_4 and τ_{cs} varying in a related way. For example, an engine design could be equipped with a control system or with components where τ_{cs} to θ_4 are related. Such a relationship might be

$$\frac{\tau_{cs}}{\tau_{csD}} = \left(\frac{\theta_4}{\theta_{4D}}\right)^{\beta/(\beta-1)} \tag{16–8}$$

where the parameter $0 > \beta > 1$ describes a member of a family of control system designs. $\beta = 0$ corresponds to the constant pressure ratio case, whereas $\beta = 1$ corresponds to constant θ_4. The performance of intermediate β engines would be expected to fall between that of the two cases described above.

These considerations show that thermodynamic parameters alone dictate that an efficiency variation results from changing the cycle parameters. Considerable latitude is available to alter the way in which thermal efficiency varies with percent load through design of a control system and choice of components. The changes in operating parameters also impact the functional performance of the work components in ways that are particular to their design. The role of the components is discussed in Sections 16.7.1 and 16.7.2. Before discussing the components, it is appropriate to examine the load characteristics and component arrangements.

16.4. LOAD CHARACTERISTICS

To this point the load has been assumed to be a sink for electromechanical power. For purposes of classification, one distinguishes between loads which accept the fluid material exhausting from the engine (jet engines) (Fig. 16–7a)

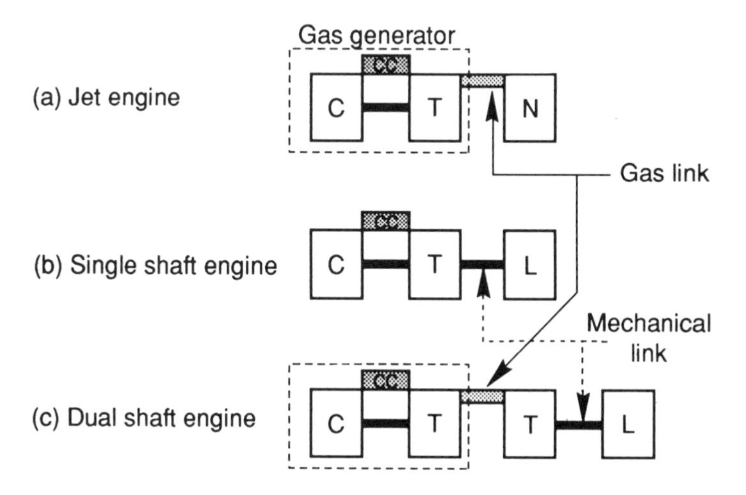

FIG. 16–7. Examples of loads for Brayton cycle engines. C and T are compressors and expanders, N a nozzle, and L is the mechanical load.

and those which absorb shaft power. In this last category, one makes a distinction between single and dual shaft engines (Figs. 16–7b, c).

16.4.1. *Jet Engines*

The most important example of this class of engines is the turbojet propulsion engine which uses the pressure and thermal energy of the fluid to create a high-momentum jet suitable for producing a thrust force on an aircraft. The mechanical assembly of components making up the engine of this type is called a "gas generator." As far as the engine is concerned, the load is the nozzle that creates this jet (Fig. 16–7a).

The nozzle's characteristic feature is that the mass flow rate is controlled by the minimum area through which the fluid travels (Section 12.4). In one-dimensional flow, this mass flow rate may be written in terms of properties at the throat area (denoted by *, see Section 12.2.1) so that

$$\dot{m} = \rho^* u^* A_{\min} \tag{16–9}$$

This may be written in terms of stagnation properties ahead of the nozzle (eq. 12–10) as

$$\dot{m} = \frac{p_t A_{\min}}{\sqrt{RT_t}} f\left(\frac{p_t}{p_0}, \gamma\right) = \frac{p_t A^* f(\gamma)}{\sqrt{RT_t}} \text{(choked flow: } M^* = 1) \tag{16–10}$$

The function $f(\gamma)$ is given in eq. 12–23 and its form is not germane to the discussion here. Choking occurs when the total to static pressure ratio (NPR) for the nozzle is sufficiently large (greater than ~ 2, see eq. 12–28), which is the case for many instances of nozzle operation. The important result from eq. 16–10 is that the load imposes a relationship between mass flow rate and two parameters that are determined by the cycle: the total temperature and pressure exiting the heat engine. Thus, while one might alter the cycle parameters to reduce the thrust, one invariably also changes (reduces actually) the mass flow rate, which further changes the thrust. A similar conclusion can be drawn for work producing shaft engines, as shown in the following section.

16.4.2. *Shaft Engines*

Mechanical shaft output power may be written as the product of torque (Q) and shaft speed. In practice, this speed (N) is measured in revolutions per minute (RPM) or radians per second in the SI system of units (ω). Thus

$$\dot{W} = Q\omega = QN\left(\frac{2\pi}{60}\right) \tag{16–11}$$

and the load speed will influence the output torque, Q, for a particular power level delivered.

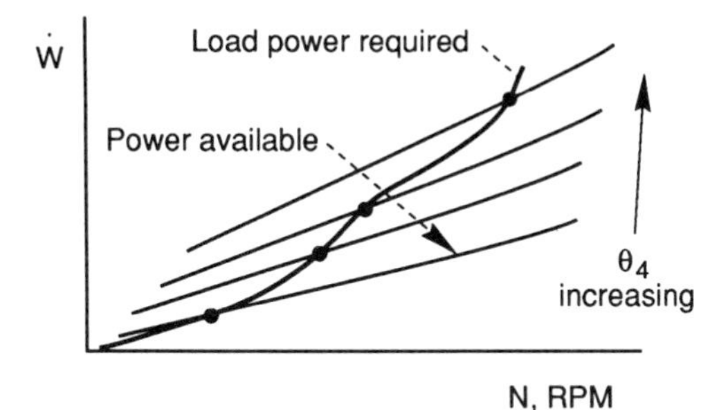

FIG. 16–8. Power plant output (available) characteristics matched to a load (required) characteristic.

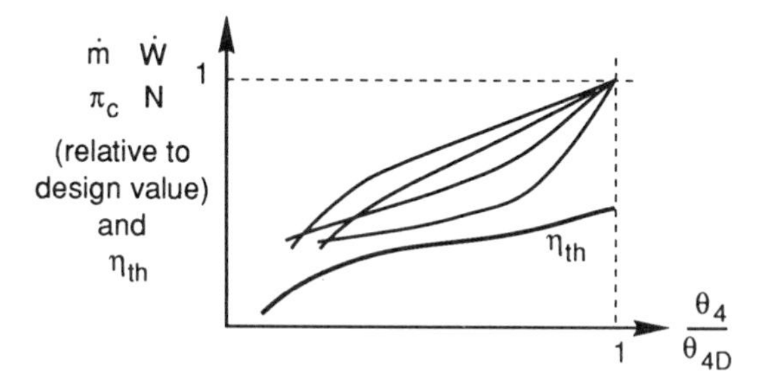

FIG. 16–9. Qualitative representation of performance of power plant and load combination.

A mechanical load will typically have a characteristic required power as a function of shaft speed which is as illustrated as the heavy line in Fig. 16–8. Such a curve may also be drawn for torque versus speed, depending on whether that representation is more convenient.

The goal of a part load performance analysis of shaft engines is to obtain power output characteristics as a function of shaft speed (N) and the relevant control variable, such as T_4 or, nondimensionally, θ_4. Figure 16–8 shows the produced power as a function of N and T_4. Such characteristics may be combined with those of the load to arrive at a relationship between N and T_4, as well as the other variables of interest (fuel consumption rate, efficiency, pressure ratio, etc.) as illustrated qualitatively in Fig. 16–9. These characteristics are specifically developed for dual-shaft and single-shaft engines in Sections 16.6 and 16.7.

16.4.3. *Shafting Configurations*

The connection between the load and the heat engine may be made in such a way as to force the components of the engine to rotate at the same speed as the load (Fig. 16–7b) or the load may be made independent of the compressor and expander by providing a second, separate expander for the load (Fig. 16–7c). Such engines are described as single-shaft and dual-shaft engines, respectively. Proposals and development efforts for engine designs with more than two shafts have also been made (ref. 16–1). When the work-producing expander is a turbine, reduction gearing may have to be provided to match the speed requirements of the load. If the expander is of the displacement type, such gear reduction may not be necessary.

Dual Shaft Engines

This engine configuration (Fig. 16–7c) resembles that of the jet propulsion engine in the sense that a coupled combination of compressor and expander supplies the gas to drive the second expander which is connected to the load. This second expander may be a turbine whose mass flow rate is also controlled by a nozzle directing the flow onto the rotor blades. Thus nozzles constrain the mass flow rate through both engine types: gas generators or dual-shaft engines. If the second expander is a displacement machine, similar, though not identical, relations apply to relate mass flow rate to upstream flow properties. The performance of dual shaft engines is examined in Section 16.6.

Single Shaft Engines

Engines of this type are simpler than the dual-shaft engines because they have fewer rotating components. The chief disadvantage is that the part load efficiency is generally inferior because the shaft speed is controlled by the load. An analysis of this class of engines is described in Section 16.7.

16.5. WORK COMPONENT MAPS

Before one is able to describe the operation of an engine as the interaction of a number of components, one must describe the components' overall characteristics. Compressors and expanders may be of the aerodynamic type or of the displacement type. In either case, the characteristic maps share a number of similarities. Although displacement work components are not used in power generation practice, they have characteristics that may make them useful in the future. The parallel between the quantities that govern their performance as a contrast to similar quantities in aerodynamic machines is of interest because the mechanics of the displacement component is easy to visualize. Brayton cycle engines with displacement work components have been studied for both open (refs. 16–2 through 16–5) and closed cycles (refs. 16–6 and 16–7), although no engine has seen complete development toward successful economic competition with existing engines.

16.5.1. *Compressors*

A compressor accepts shaft power and delivers a steady flow of fluid at a higher pressure. Good compressors are also adiabatic, so that the fluid leaves the device at an elevated temperature. Most compressors deliver greater amounts of fluid with increased shaft speed. In a displacement machine, the pressure ratio across the device may be made independent of shaft speed. It may vary with both shaft speed and mass flow rate: In an aerodynamic machine, flow angles determine the work input to the fluid. In either case, the performance of a compressor may be summarized in a "map" as shown in Fig. 14–17. Normally, lines of constant adiabatic efficiency are also shown. The figure is for a relatively small aerodynamic compressor. Details regarding such maps for displacement compressors are discussed in Chapter 14 and in Sections 16.6.2 and 16.6.3. Simple algebraic expressions to represent the data for such maps are generally not available. One can say, however, that for typical conditions, the mass flow rate is proportional to the shaft speed, N.

16.5.2. *Expanders*

The expander accepts high-pressure fluid (stagnation state: t4) which expands to perform work on its boundaries. As with the compressor, either displacement or aerodynamic devices may be used. The aerodynamic turbine converts the high-pressure fluid to a high-velocity jet, which then exerts aerodynamic forces on the moving blade surfaces (Chapter 15). The process of creating the jet usually involves accelerating the flow to supersonic speeds. At the point where sonic speeds are reached, there must be a flow area minimum, and the flow is said to be choked at this point in the turbine nozzle. In global terms, this means that the mass flow rate is related only to the total pressure and temperature of the entering flow (and not to downstream conditions) and, of course, the flow area (A_4, see Fig. 15–1). An equation like 16–10 gives the choked mass flow rate through the turbine. Chapter 15 is a detailed discussion of turbine aerodynamics, the work extraction process and, in particular, the limitations to the above discussion. Thus, one can say that a choked turbine has a "map" characteristic (Fig. 15–22) describable by

$$\dot{m} = \text{constant}\,\frac{A_4 p_{t4}}{\sqrt{T_{t4}}}, \text{ choked turbine} \qquad (16\text{–}12)$$

which is independent of N (at least under conditions close to the design point).

Displacement expanders may be visualized as processing a mass of fluid into a piston/cylinder volume and expanding that mass to the low pressure of the environment (Fig. 16–10). Clearly, the mass flow rate processed by this expander is proportional to the rotational shaft speed, the volume of the piston/cylinder at the point where the gas to be expanded is cut off from its supply (V_4), and the density of the incoming fluid ($\rho = p/RT$).

By contrast to eq. 16–12, a displacement expander map would read approximately,

$$\dot{m} = \text{constant}\,(N)\,\frac{V_4 p_{t4}}{T_{t4}}, \text{ displacement} \qquad (16\text{–}13)$$

Note the similarity of the roles played by A_4 and V_4. The total pressure role

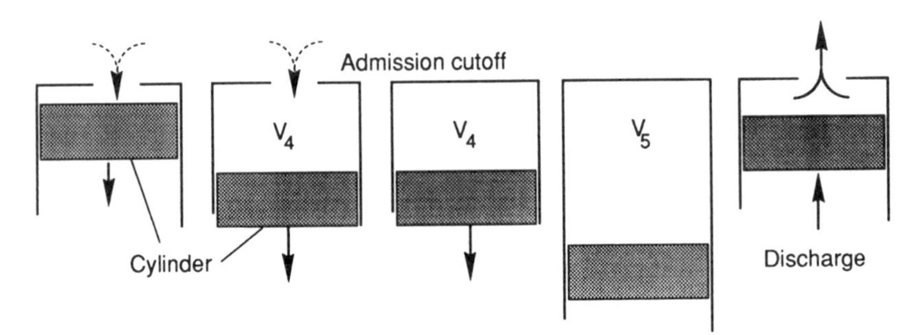

Fig. 16–10. Expansion of a gas in a displacement expander. V_4 is the volume at which communication between high-pressure supply and cylinder volume ceases.

is identical for both expander types, whereas the total temperature role is roughly similar. The presence of a rotational speed influence also distinguishes this characteristic from that of a choked turbine.

The characteristics of the expander are improved somewhat through the introduction of a mechanism that allows the volume V_4 to vary with cycle parameters (ref. 16–8). This possibility is not considered here.

The characteristics of compressors and turbines are such that the simple shafting connections described in the sections that follow are less than optimal from a performance viewpoint. To circumvent some of the limitations, unusual system modifications may be considered. These include the variable area turbine (ref. 16–9) and the variable speed drive among the work components (ref. 16–10) of the displacement type in particular. These modifications are not discussed here, and the reader should consult the original references.

16.6. PART POWER PRODUCTION OF GAS GENERATOR OR DUAL-SHAFT ENGINES

Figure 16–7c shows the component layout of a dual-shaft engine. Let the state point number description be as in Fig. 9–2 with the addition that states 5 and 6 apply to inlet and exit of the second turbine. The part load performance analysis uses the fact that the shaft speed of the load (N_2) is independent of the gas generator operating conditions and a figure like 16–9 may be generated. The power produced is that of the second expander. It is possible to derive the equations for the power output with either displacement or aerodynamic components, using the expander mass flow condition expressed as:

$$\dot{m} = \text{constant} \times p_{t4} N^{d} T_{t4}^{-(d+1)/2} \tag{16–14}$$

Here $d = 1$ corresponds to a displacement expander, whereas $d = 0$ to an aerodynamic one. Here, choked flow in the turbine expander is assumed for simplicity and the extension to consider unchoked flow is readily made, although it involves additional algebraic effort.

The power output from the engine is given by:

$$\dot{W} = \dot{m} C_p T_0 \left[\frac{T_{t5}}{T_0}\right]\left[1 - \frac{T_{t6}}{T_{t5}}\right] \tag{16–15}$$

Denoting by τ_{t1} the total temperature ratio across the first expander and τ_{t2} that for the second, one may write eq. 16–15 as

$$\dot{W} = \dot{m} C_p T_0 [\tau_{t1}\theta_4][1 - \tau_{t2}] \tag{16–16}$$

Using the assumptions that gas properties are the same through both components and that the addition of fuel mass may be neglected, the work balance between compressor and expander of the gas generator may be written (after suitable nondimensionalization) as:

$$\tau_c - 1 = \theta_4(1 - \tau_{t1}) \tag{16–17}$$

The first expander imposes a constraint on the mass flow rate entering it. Thus the mass flow rate is given by eq. 16–14 as:

$$\dot{m}_4 = C_4 \pi_c N_1^d (\theta_4)^{-(d+1)/2} \tag{16–18}$$

and for the second expander as:

$$\dot{m}_5 = C_5(\pi_c \pi_{t1}) N_2^d (\theta_4 \tau_{t1})^{-(d+1)/2} \tag{16–19}$$

In eqs. 16–18 and 16–19 the two constants are different, hence subscripted. The numerical values of the constants for a choked aerodynamic turbine and for a displacement expander with N (in RPM) are, respectively:

$$C_i = \frac{p_0 A_i}{\sqrt{RT_0}} \sqrt{\gamma} \left\{\frac{2}{\gamma + 1}\right\}^{(\gamma+1)/2(\gamma-1)} \quad \text{and} \quad C_i = \frac{p_0 V_i}{RT_0} \frac{2\pi}{60} \tag{16–20}$$

The mass flow rates, $\dot{m}_4$ and $\dot{m}_5$, must be equal in any case.

One additional condition is required before the part load performance can be calculated: The pressures at the engine compressor inlet and second turbine exit are equal, and, in the case of open cycle engines, these equal atmospheric pressure. Total pressure losses in the combustor and ducting between components are neglected, assuming that they are designed to be negligible. Hence it follows:

$$\pi_c \pi_{t1} \pi_{t2} = 1 \tag{16–21}$$

The total pressure ratios are related to the total temperature ratios: For adiabatic work components, the efficiencies are given in Section 7.6.3 as:

$$\eta_c = \frac{\pi_c^{(\gamma-1)/\gamma} - 1}{\tau_c - 1} \qquad \text{and} \qquad \eta_t = \frac{1 - \tau_t}{1 - \pi_t^{(\gamma-1)/\gamma}} \tag{16–22}$$

for the compressor and expander, respectively.

The efficiency of the heat engine is obtained from the fuel heat input (H in kJ/kg is the fuel heating value) which raises the combustor gas temperature:

$$\dot{m}_{\text{fuel}} H = \dot{m} C_p T_0(\theta_4 - \tau_c)$$

Thus

$$\eta_{\text{th}} = \frac{\dot{W}}{\dot{m}_{\text{fuel}} H} \qquad \text{or} \qquad \text{SFC} = \frac{\dot{m}_{\text{fuel}}}{\dot{W}} \tag{16–23}$$

The shaft speed N_2 may be viewed as imposed by the load, and, for the power delivered, this speed determines the load torque from eq. 16–11:

$$Q_{\text{LOAD}} = \frac{\dot{W}}{N_2} \frac{60}{2\pi} \tag{16–24}$$

The equations listed above allow calculation of the part-power output when the peak cycle temperature $\theta_4 < \theta_{4D}$ is specified. It is relatively easy to derive relatively simple equations for the power as a function of θ_4 when the components are isentropic. A more general approach is to include nonideal adiabatic component efficiencies in the procedure. It should be realized that a constant value of such efficiencies is generally simplistic because both mass flow rate and pressure ratio do influence the component efficiencies as shown in Fig. 14–17, for example. An examination of the procedures that follow shows that a varying efficiency makes the algebraic equations iterative. This is easy to handle with a computer, which is required even for the constant adiabatic efficiency case. In lieu of an algebraic development, the part-power calculation procedure is described in terms of the steps necessary for a computer calculation of performance parameters. Typical results from such calculations are then presented as examples to draw conclusions regarding the relative merits of single- and dual-shaft engines as well as between engines with aerodynamic and displacement components.

16.6.1. *Part Power Performance Calculation Procedure* (*dual-shaft*)

Design Point (maximum power) Parameters

The calculation of the design condition parameters is independent of the type of components and of the number of shafts. The difference in the characteristics arises from the mass flow eqs. 16–18 and 16–19 and the expression for the compressor mass flow rate in terms of shaft RPM and thus must be considered only at part power.

Thus, given the following:

Maximum power output, i.e., $\dot{W}_D/C_p T_0$
Known component efficiencies: η_c, η_{t1}, η_{t2}
Specified θ_{4D}, π_{cD} (design choice)
[For the displacement engine N_{1D} and N_{2D} are also required],

one calculates, using eqs.

16–22 (compressor efficiency): τ_{cD}
16–17 (N_1 shaft work balance): τ_{t1D}
16–22 (expander efficiency): π_{t1D}
16–21 (pressure balance): π_{t2D}
16–22 (expander efficiency): τ_{t2D}
16–16 (power requirement): $\dot{m}_D$

At part power, the calculation depends on the type of components in the system. Two examples are illustrated below, one with aerodynamic components and the other with displacement components. The description of these two cases should enable the reader to visualize an analysis approach to the case of mixed components.

16.6.2. *Aerodynamic Work Components*

The first case considered is that for aerodynamic components (i.e., $d = 0$). The design point information allows one to calculate the constants required in eqs. 16–18 and 16–19:

$$C_4 = \frac{\dot{m}_D\sqrt{\theta_{4D}}}{\pi_{cD}} \quad \text{and} \quad C_5 = \frac{\dot{m}_D\sqrt{\theta_{4D}\tau_{t1D}}}{\pi_{cD}\pi_{t1D}} \tag{16–25}$$

Equating the mass flows given by eqs. 16–18 and 16–19 (with $d = 0$), $\dot{m}_4 = \dot{m}_5$, gives

$$C_4 = \frac{\pi_c}{\sqrt{\theta_4}} = C_5 \frac{\pi_c\pi_{t1}}{\sqrt{\theta_4\tau_{t1}}}$$

from which π_c and θ_4 cancel out. This leads to the conclusion that π_{t1} and τ_{t1} (which are related through the turbine efficiency, eq. 16–22) are fixed hence equal to π_{t1D} and τ_{t1D}, respectively, under all load conditions.

The direct calculation of thermal efficiency (and mass flow rate) at reduced power output is not possible, and an *indirect* method must be employed. To do this one specifies the value of a key variable which can be controlled and calculates both power and efficiency.

Part Power Calculation Procedure (dual-shaft, aerodynamic components)

Specify

$$\theta_4 < \theta_{4D} \qquad \text{and} \qquad N_2$$

Calculate using eqs. [mass balance gives $\tau_{t1} = \tau_{t1D}$]

16–17 (work balance): τ_c

16–22 (compressor efficiency, iteration if η_c is not constant): π_c

16–21 (pressure balance with $\pi_{t1} = \pi_{t1D}$): π_{t2}

16–22 (turbine efficiency): τ_{t2}

16–18 (or 16–19): $\dot{m}$

16–16 (power output): $\dot{W}$

16–23 (fuel consumption rate): SFC

The procedure is summarized in the lower portion of Fig. 16–11 (labeled: Aero), where the transformations of total temperature ratios to total pressure ratios through the component efficiencies are omitted. The results relevant to the discussion of displacement components (Section 16.6.3) are also included in the figure.

Of interest is the locus of points (operating line) on the compressor map (π_c *vs* $\dot{m}\sqrt{\theta_0}/\delta_0$) so that N_1 may be determined. The compressor map also permits closure on the adiabatic efficiency of the compressor, if the variation of its magnitude is significant. Finally, the load characteristic determines either the torque or the shaft speed N_2 when the other is specified.

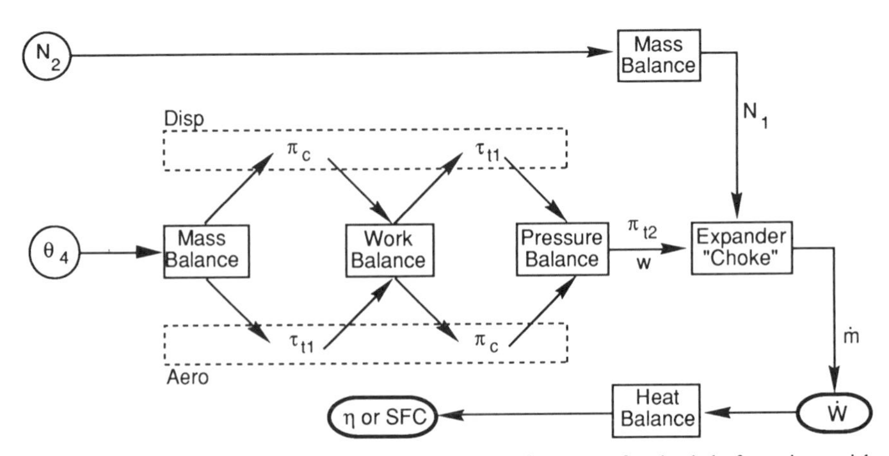

Fig. 16–11. Calculational summary of off-design performance for dual shaft engines with aerodynamic (lower arrow sequence) and displacement components (upper arrow sequence).

16.6.3. *Displacement Work Components*

The displacement, V_0, of the compressor, its rotational speed N_1, and the incoming air density determine the mass flow rate:

$$\dot{m} = \rho_0 V_0 N_1 = C_0 N_1 \qquad \text{and} \qquad \dot{m}_D = C_0 N_{1D} \tag{16–26}$$

The design mass flow rate also fixes the constants involved in eqs. 16–18 and 16–19 (with $d = 1$):

$$C_4 = \frac{\dot{m}_D}{N_{1D}} \frac{\theta_{4D}}{\pi_{cD}} \qquad \text{and} \qquad C_5 = \frac{\dot{m}_D}{N_{2D}} \frac{\theta_{4D}\tau_{t1D}}{\pi_{cD}\pi_{t1D}} \tag{16–27}$$

The shaft speed N_{2D} is fixed by the specified level of torque at the design condition.

At reduced power, equality of mass flows through the compressor, and first expander (eqs. 16–18 and 16–26) yields:

$$\pi_c = \left(\frac{C_0}{C_4}\right)\theta_4 \qquad \text{or} \qquad \frac{\pi_c}{\pi_{cD}} = \frac{\theta_4}{\theta_{4D}} \tag{16–28}$$

In words, as expander inlet temperature is reduced so is the peak cycle pressure in a linear fashion. This follows physically from the fact that the volumes processed per revolution are invariant. For $\dot{m}_4 = \dot{m}_5$, the combination of eqs. 16–18 and 16–19 applied at off-design gives

$$N_1 = N_2 \frac{C_5}{C_4} \frac{\pi_{t1}}{\tau_{t1}} \qquad \text{or} \qquad \frac{N_1}{N_{1D}} = \frac{N_2}{N_{2D}} \left(\frac{\tau_{t1D}}{\tau_{t1}} = 1\right)\left(\frac{\pi_{t1}}{\pi_{t1D}} = 1\right) \tag{16–29}$$

This equation shows that the two shafts' fractional speeds are equal.

Part Power Calculation Procedure (dual-shaft, displacement components)

Given the same design point data required in Section 16.6.1 and following determination of C_0, C_4, and C_5:

Specify

$$\theta_4 < \theta_{4D} \qquad \text{and} \qquad N_2 (= N_1 \text{ from eq. 16–29})$$

Calculate using eqs.

16–28 (gas generator mass balance): π_c
16–22 (compressor efficiency, iteration if η_c is not constant): τ_c
16–17 (work balance): τ_{t1}
16–22 (turbine efficiency): π_{t1}

16–21 (pressure balance): π_{t2}
16–22 (turbine efficiency): τ_{t2}
16–18: $\dot{m}$
16–16 (power output): $\dot{W}$
16–24 (torque): Q
16–23 (fuel consumption rate): SFC

This procedure is summarized in the upper section of Fig. 16–11, labeled "Disp." Evidently, from the discussion of these two engine types which differ in the kind of work components employed, it follows that the relationship between cycle pressure and temperature is established by the characteristics of the components. In the displacement engine, the thermodynamic relation between π_c and θ_4 is a linear one because of the assumption that the displacement expander always accepts gas volume at a rate proportional to its rotational speed. Such a scheme is associated with valving or porting control which is fixed in phase much like the valve train on an Otto cycle engine. It is possible to imagine other relationships that might permit better performance. By contrast, the aerodynamic engine's relation between the cycle parameters π_c and θ_4 depends on the compressor characteristics, which preclude stating the relation between these parameters in a simple (algebraic) way.

The variation of efficiency with the fractional power output for a gas generator of dual-shaft engine with aerodynamic and with displacement components using the method outlined above is shown in Fig. 16–12. In the latter case, the output power shaft speed is a mass flow-controlling variable, and hence for a number of shaft speeds, the various curves can be obtained. The output power is proportional to this shaft speed, N_2, so that the appropriate quantity to plot is the fractional power per unit N_2. On this basis, the

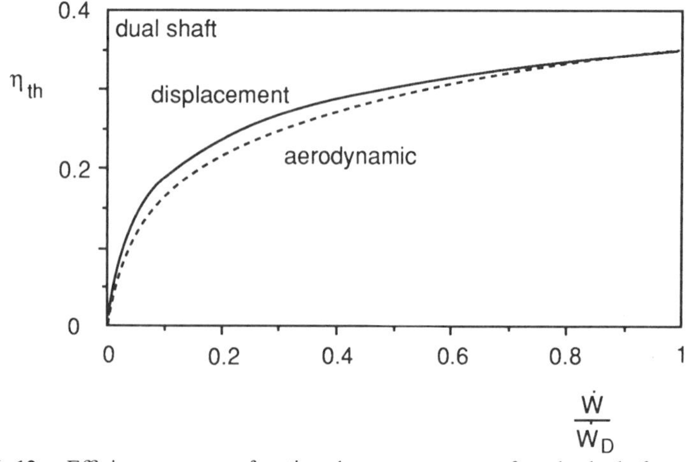

Fig. 16–12. Efficiency versus fractional power output for dual shaft engines with aerodynamic (a) and displacement (b) work components. η_c, η_{t1}, $\eta_{t2} = 0.85, 0.95, 0.95$.

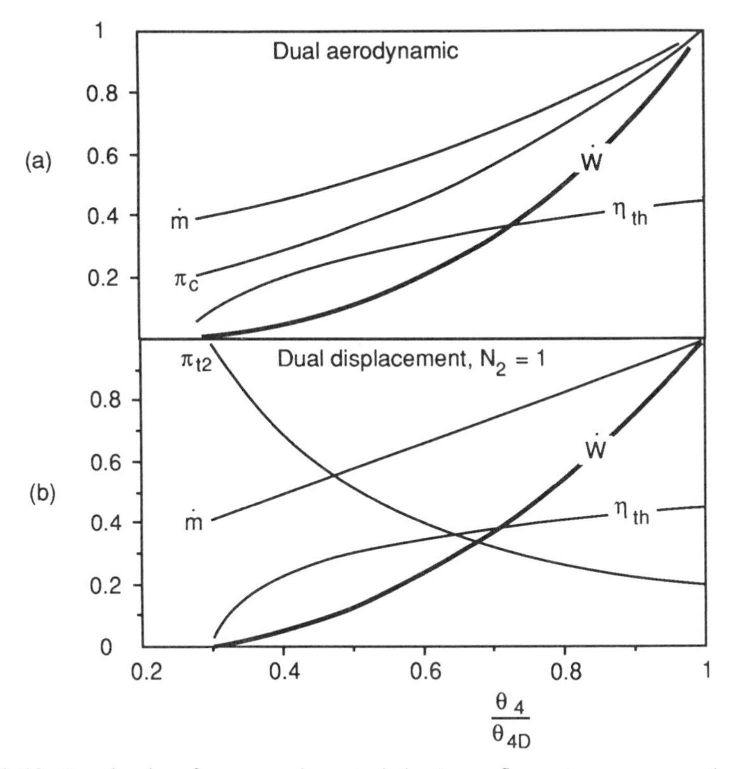

Fig. 16–13. Part load performance characteristics (mass flow rate, pressure ratio, power and efficiency) for dual shaft engines: upper figure: aerodynamic, and lower figure: displacement work components for which abscissa is also π_c/π_{cD}. η_c, η_{t1}, $\eta_{t2} = 0.85$, 0.95, 0.95. $\theta_{4D} = 6$.

displacement component engine appears to have a small advantage. A realistic comparison of the two engine kinds is possible only when the load characteristics are specified. In either case, an efficiency reduction is obtained at part load, which becomes quite severe at low load.

The various operating parameters of interest for the two kinds of engines, both at design speed, are shown in Figs. 16–13a (aero) and 16–13b (displ). For the aerodynamic engine, the pressure ratio is of interest (how much must it be made variable?), hence shown, whereas the first expander pressure ratio is noted for the displacement engine. For the displacement engine, θ_4 and π_c are proportional to each other. At reduced output speed ($N_2 = 0.6$ for example) the displacement engine performance is very close to that at $N_2 = 1$, except that the mass flow rate is proportionately reduced.

16.7. SINGLE-SHAFT ENGINE PART LOAD PERFORMANCE COMPARISON

The single-shaft engine shown in Fig. 16–8c is coupled directly to the load. The design performance characteristics for this engine type are the same as those for the dual shaft engine. The principal difference between the engine types is in the fact that the single-shaft engine has only one mass flow-constraining condition imposed by the single expander. The mass flow of the engine is then directly determined by the load shaft speed, and thus the compressor map consequently plays an important role in determining part load performance.

The expander pressure ratio is related to that of the compressor by

$$\pi_c \pi_t = 1 \tag{16–30}$$

A turbine pressure ratio greater than the value given by this expression ($1/\pi_t > \pi_c$) may involve an undesirable irreversibility. However, both aerodynamic and displacement expanders can be built or operated to approach satisfaction of eq. 16–30 (see, for example, ref. 16–11).

16.7.1. *Aerodynamic Work Components*

Aerodynamic Compressor Map Model

In order to obtain a sense of the behavior of single-shaft engines, it is necessary to develop approximate descriptive models of the compressor map.

The aero-compressor has constant speed lines that are nearly vertical (see Fig. 14–17), especially near the design point. To first order, this implies that the pressure ratio is independent of mass flow rate. The maximum value of the pressure is proportional to mass flow rate or shaft speed since these latter quantities are roughly proportional to each other. It is desirable to operate the engine at as high a pressure ratio as possible to have the greatest thermodynamic advantage. The line of state points on the compressor map along which an engine operates is termed an "operating line." The line connecting the maximum pressure ratio points on the various constant speed lines is termed a "stall line" because critical airfoils making up the compressor are close to stalling (see Section 14.2.6). The separation between an operating line and a stall line is termed the "stall margin." For elementary modeling purposes, one may say that it is best that the operating line and the stall line are coincident (i.e., operate with near-zero stall margin). The question is, then, How good is the part power performance of this engine operated in this way?

The aero-compressor map may be modeled in equation form as:

$$\frac{\pi_c}{\pi_{cD}} = \left[\frac{\dot{m}}{\dot{m}_D}\right] \quad \text{and} \quad \frac{\dot{m}}{\dot{m}_D} = \frac{N}{N_D} \tag{16–31}$$

where the exponent e is typically of order 1–2, that is, the relation is between linear and quadratic. See, for example, eq. 14–24, which shows the relation between π_c and $\dot{m}$ as quadratic since $\dot{m}$ is proportional to w, which is, in turn, proportional to ωr for operation with good blading aerodynamics. A sketch of the map is shown in Fig. 16–14. It will be shown that $e = 2$ describes a displacement compressor, whereas 1.5 is close to that of an aerodynamic compressor. Other functional relationships between π_c and mass flow rate may

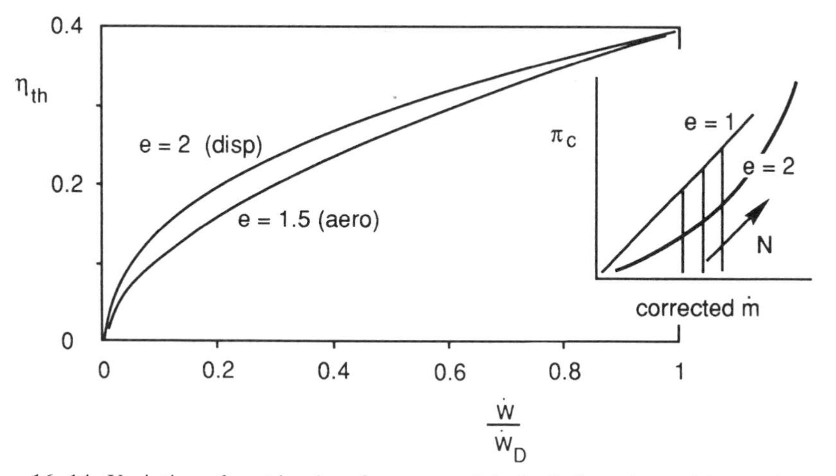

FIG. 16–14. Variation of part load performance of single shaft engines with aerodynamic (e = 1.5) compressor and a displacement compressor with e = 2. $\eta_c = 0.85$, $\eta_t = 0.90$, $\theta_{4D} = 6$.

be more accurate, but the form of eq. 16–31 is chosen to make the analysis as clear as possible. The reader may use the process developed here to include more realistic map descriptions, including tabular or curve fit forms, to arrive at more accurate numerical performance predictions.

Single-Shaft Aero-Compressor and Turbine Engine: Part Load Performance

The flow choking condition (eq. 16–10) combined with eq. 16–31 gives an expression from which the pressure ratio may be computed, provided $e > 1$:

$$\frac{\pi_c}{\pi_{cD}} = \left(\frac{\theta_4}{\theta_{4D}}\right)^{e/2(e-1)} \tag{16–32}$$

The turbine total pressure ratio may be calculated using eq. 16–30 as well as the total temperature ratios for both components using the definitions of the adiabatic efficiencies. The power output is then obtained from:

$$\frac{\dot{W}}{\dot{W}_D} = \frac{\dot{m}}{\dot{m}_D}\frac{\theta_4(1-\tau_t)-(\tau_c-1)}{\theta_{4D}(1-\tau_{tD})-(\tau_{cD}-1)} \tag{16–33}$$

from eqs. 16–12 and 16–32, the mass flow rate is

$$\frac{\dot{m}}{\dot{m}_D} = \left(\frac{\theta_4}{\theta_{4D}}\right)^{1/2(e-1)} \tag{16–34}$$

The fuel consumption rate may be calculated using the combustor energy balance (eq. 16–23) from which efficiency follows directly (see Fig. 16–14).

16.7.2. *Displacement Work Components*

Displacement Compressor Model

The displacement compressor is not constrained to a maximum pressure in the same way that an aerodynamic one is because there is no equivalent to the stall mechanism. Thus it is possible to imagine mechanical arrangements that either maintain the same pressure ratio as the design value or whatever value is demanded by the system. The first instance might be obtained if the compressor operates with conventional valves which always open at a chosen point in the rotation cycle, whereas the second would result if the valve is actuated by pressure forces. The latter design is thermodynamically superior because the compression process may be made to avoid unnecessary overcompression. An example is a design where the valve opens in response to pressure differences across it (see ref. 16–11). Thus in an examination of the best configuration of single shaft displacement engines, the compressor may be modeled as a variable pressure ratio device with mass flow rate proportional to shaft RPM.

Consistent with the dual-shaft analysis described earlier, the appropriate description of the expander is that the expander processes a mass of gas that is proportional to the gas density at the combustor exit as described by eq. 16–13.

Single-Shaft Displacement Engine: Part Load Performance

The performance of the displacement engine is calculated from eq. 16–33 so that knowledge of the cycle pressure ratio and the mass flow rate is required. The compressor map may be stated in equation form as

$$\frac{\dot{m}}{\dot{m}_{\mathrm{D}}} = \frac{N}{N_{\mathrm{D}}} \tag{16–35}$$

whereas the mass flow relation is obtained from eq. 16–13:

$$\frac{\dot{m}}{\dot{m}_{\mathrm{D}}} = \frac{N}{N_{\mathrm{D}}} \frac{\pi_{\mathrm{c}}}{\pi_{\mathrm{cD}}} \left[\frac{\theta_4}{\theta_{4\mathrm{D}}} \right]^{-1} \tag{16–36}$$

Combining these last two equations leads to the conclusion that, as in the dual shaft engine, the pressure ratio must vary as the temperature, θ_4:

$$\frac{\pi_{\mathrm{c}}}{\pi_{\mathrm{cD}}} = \frac{\theta_4}{\theta_{4\mathrm{D}}} \tag{16–37}$$

This statement is equivalent to eq. 16–32, which was derived for the aero-compressor. In fact, if the exponent e equals 2, then the relations between pressure ratio and temperature for the two kinds of components are identical. This makes the procedure for calculating the performance for the two types of machinery identical as well. Further, $e = 2$ may therefore be used to describe the displacement compressor. The extent to which an aero-compressor's e differs from 2 will determine the difference between the two compressor types and the effect on part-power efficiency.

How does one best describe the performance? Thermal efficiency is clearly important. This quantity is derived only from cycle consideration and is therefore independent of mass flow rate. The power output does depend on mass flow rate and therefore on the speed of the machinery. The speed, however, is determined by the load and thus may be considered to be an independent variable. Thus an appropriate measure of power is the power per unit shaft speed. In both kinds of devices, the shaft speed and mass flow rate are proportional to each other so that the specific work is an effective measure of power or the fraction of maximum power. A performance calculation can be summarized by a plot of thermal efficiency versus specific work (normalized by the maximum, or design, value). Such a plot allows comparison of the two machinery types.

The calculation procedure is as follows:

1. Calculation of design point quantities
2. Specification of $\theta_4 < \theta_{4D}$
3. Calculation of the pressure ratio by either of eq. 16–32 or 16–37
4. Calculation of the component temperature ratios, using the pressure balance equation (16–30) and the definitions of component efficiencies
5. Calculation of specific work
6. Calculation of thermal efficiency

16.7.3. *Single-shaft Engine Performance Comparison*

The component types are differentiated in the use of appropriate equations (eq. 16–32 or 16–37). The resulting plot for an example calculation is shown in Fig. 16–14. For the aero-compressor the chosen value of e is 1.5 to show the influence of the compressor map. The conditions, as well as the (constant) values of the adiabatic component efficiencies, are noted. The design condition is that which gives maximum specific power. The relations between pressure ratio and θ_4 show specifically how the choice of components impacts performance. In general, one may conclude that the thermal efficiency decreases uniformly over a large fractional power range. The displacement engine appears to have a modest efficiency advantage. At very low power, the decrease is severe for both component types.

The assumption that adiabatic component efficiencies are constant is approximate, particularly for the aero-compressor where the design point is often chosen so that compressor efficiency is a maximum. As pressure ratio, speed, and/or mass flow rate are reduced, the efficiency falls and thereby reduces the thermal efficiency at lower power settings to values even lower than those shown. Thus one might conclude that if the adiabatic efficiencies of the displacement components decrease more slowly as pressure ratio is reduced, then the efficiency performance of that engine is superior to the gas turbine. Realistic variations of component efficiencies can be included in a performance

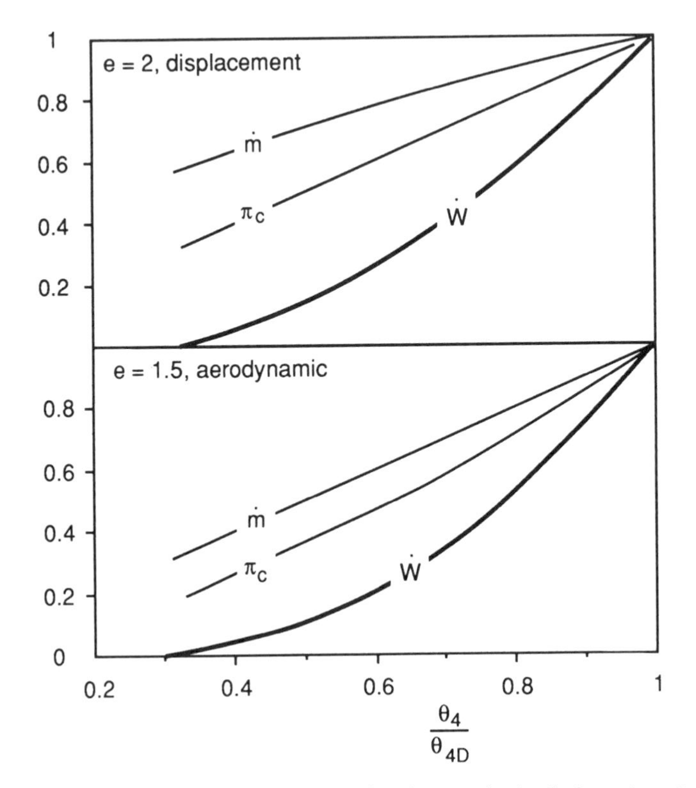

FIG. 16–15. Part load performance parameters for the two single shaft engines described in Fig. 16–14.

calculation when they are given in an appropriate functional form for use in a computer program and where iterative approaches may have to be used to solve the system of equations described here.

A more realistic analysis will also include consideration of the mechanical friction which is speed dependent and probably much more important in engines with displacement work components. Reference 16–5 is an examination of scaling considerations with this aspect as an important issue. The net output power from an engine is the calculated thermodynamic power minus friction power.

The last item of interest in an analysis of Brayton cycle engines at part power is the variation of various parameters as the control variable (fuel input rate or equivalently θ_4) is reduced. In addition to pressure ratio, the mass flow rate and the actual power, which is the product of specific work and mass flow rate, can be computed. Figure 16–15 shows the important parameters for two values of e. For $e = 2$ (disp), the θ_4 variation of pressure is linear whereas for $e = 1.5$ (aero), that of the mass flow is linear. This power output plotted as a function of speed can then be combined with the load characteristic to arrive at combined system characteristics.

The variations of thermal efficiency for one- and two-shaft engines ($e = 2$) are compared in Fig. 16–16 using the results of Figs. 16–12 and 16–14. The

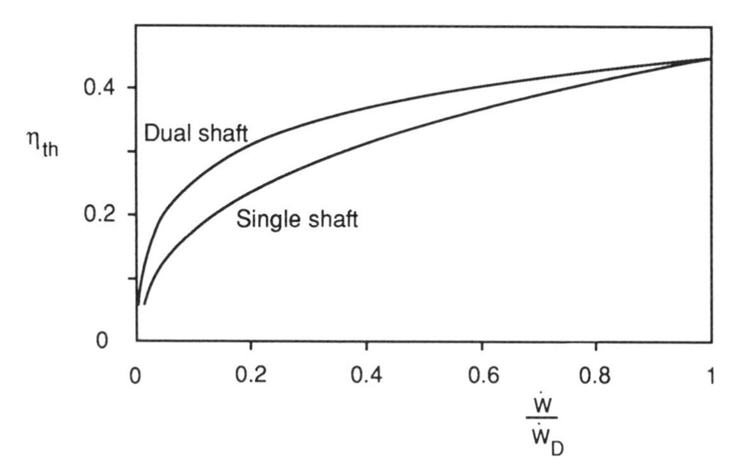

Fig. 16–16. Performance comparison of single and dual shaft engines at part power. The compressor is characterized by the e = 2. $\eta_c = 0.85$. For dual shaft engine $\eta_{t1}, = \eta_{t2} = 0.95$, while for the single shaft, $\eta_t = 0.959$, which is chosen to match the maximum w case calculated for the dual shaft engine at full power.

figure shows the efficiency at part load for the single- and dual-shaft engines with turbine (or expander) efficiency of the single-shaft engine chosen to match the maximum specific work case calculated for the dual-shaft case. Note the steepness of the decrease of efficiency with power for two cases. While it is difficult to draw conclusions regarding the relative merit of the two engine configurations, it does appear that two shaft machines are superior. Load characteristics and the variations of component efficiencies with speed that have been assumed constant here are generally important.

16.8. POWER CONTROL OF CLOSED CYCLE ENGINES

Closed cycles present unique issues and opportunities for modulating power output from power systems. Two methods are described and their features are discussed. These systems have been of interest for use in the nuclear power industry.

16.8.1. *Heat Source Bypass*

Figure 16–17 shows a schematic for a regenerated Brayton cycle. Proposed commercial power systems using this cycle offer the advantage of enabling power production with a time variability to match that of the load. To that end, a number of power modulation schemes are available: bypass control, temperature modulation, and inventory control. It is important to be able to produce the part load power at high efficiency and in such a way as to minimize the thermal stress impact on the heat source, especially if the source is a gas-cooled nuclear reactor. For a given machine, the cycle pressure ratio is nominally fixed by the compressor.

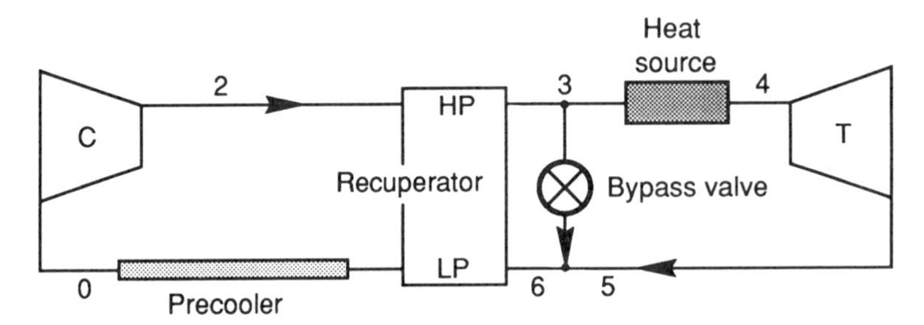

FIG. 16–17. Bypass power level control of a closed Brayton cycle.

The bypass control is exercised through the bleed of high-pressure gas to short-circuit the heat source and the turbine, as shown in Fig. 16–17. The throttling process is obviously a source of irreversibility so that use of such a scheme results in reduced part power efficiency. The cycle temperatures can be held constant with just the thermal power input matching that required to maintain the cycle temperatures at the reduced mass flow through the reactor. This has the advantage in that stresses associated with temperature gradients in the metals may be held close to constant.

The impact on performance is readily calculated since the cycle temperature may be taken to remain fixed. The regenerator will process equal masses on both sides at all times, which implies that the ideal design situation of $T_3 = T_6$ is maintained at part power. This is due to the fact that with constant T_4 and constant compressor pressure ratio, $T_3 = T_5$. The enthalpy balance on the mixer gives the temperature at state 6 which is, for an ideal and perfect gas, trivial. The cycle analysis is merely a work accounting with the full mass flow processed by the compressor and less in the heater and turbine. An ideal cycle analysis yields

$$\eta_{th} = \left[1 + \frac{w}{w_{max}}\left(\frac{\theta_4}{\tau_c} - 1\right)\right]^{-1}$$

where

$$w_{max} = \left(\frac{\theta_4}{\tau_c} - 1\right)(\tau_c - 1) \tag{16–38}$$

The symbols θ_4 and τ_c are as defined in Chapter 9. For values of these parameters of 4 and 1.5, respectively, the efficiency is as shown in Fig. 16–18.

The modulation of T_4 in an ideal cycle gives efficiency results that are identical to those of bypass control. The implication is that bypass and peak temperature reduction have the same thermodynamic merit. Temperature modulation and bypass may therefore be used together if the resulting performance is acceptable. In practice, accurate part power performance must be evaluated with significantly greater consideration of the irreversibilities than has been exercised here.

Lastly, the significantly (thermodynamically) superior method of controlling power through control of the gas inventory in the cycle is considered.

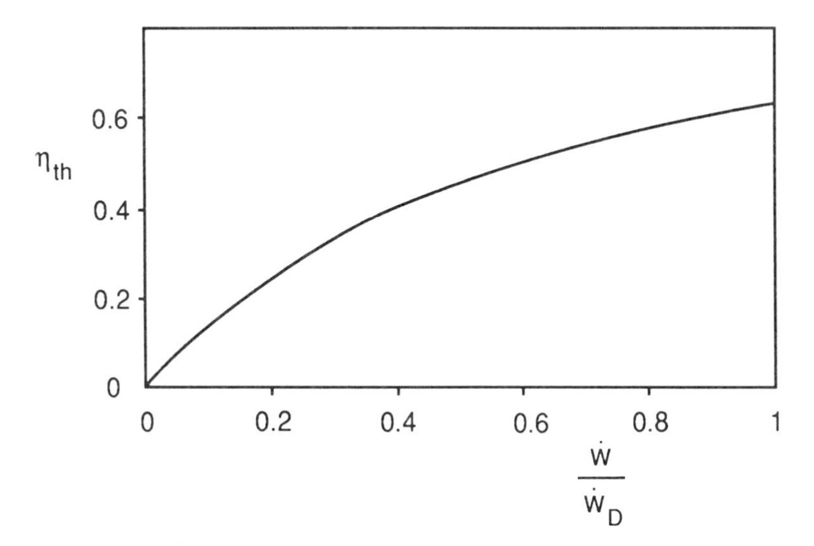

FIG. 16–18. Plot of η_{th} of a regenerated Brayton cycle (with ideal components) as a function of power fraction with a bypass power control, $\tau_c = 1.5$, $\theta_4 = 4$.

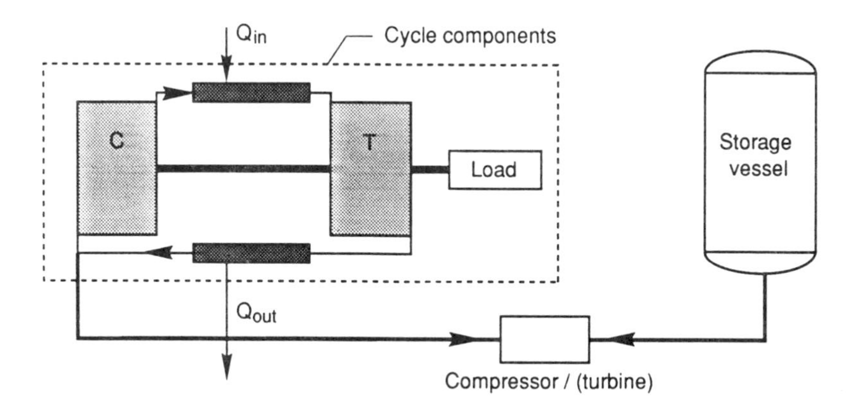

FIG. 16–19. Closed cycle with inventory control of the working fluid. Note the reversible compressor/turbine and the storage vessel.

16.8.2. *Inventory Control*

A good method of producing part-load power is available to closed cycle engines where the pressure and thus the density of the working fluid may be controlled by connecting the cycle fluid to a storage vessel, as shown in Fig. 16–19. The compressor shown in the figure is used to pump the working fluid out of the system of working components. The reduced mass of the circulating fluid results in a smaller mass flow rate which, in turn, reduces power output from the system. Means are also provided to allow the return of the fluid to the cycle when power is to be increased. In order to minimize heat transfer in the storage component, the fluid is

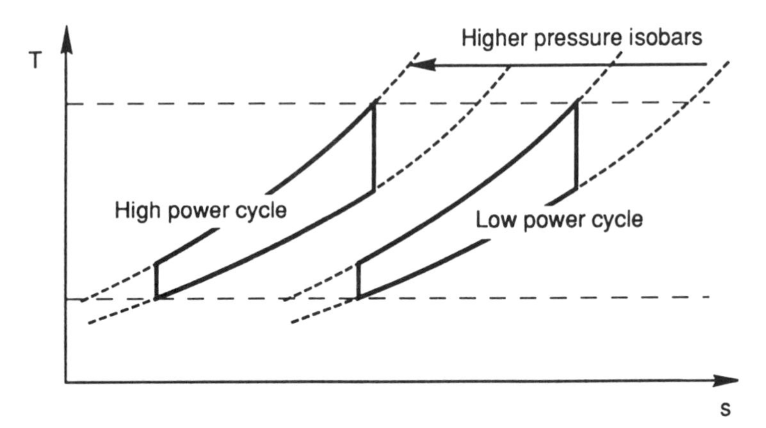

FIG. 16–20. *T-s* diagrams for part and full power for the inventory controlled closed Brayton cycle.

removed from the lowest temperature point in the cycle with appropriate means for cooling.

The operation of the cycle at reduced mass flow rate allows operation with the same temperatures and pressure ratio. This means that the heat engine operates with the same thermodynamic cycle, resulting in approximately constant efficiency and specific work.

The fact that the temperatures remain invariant as the mass flow rate is reduced implies that the local sound speeds are constant. Blading and flow-passage geometric design fix the local Mach numbers so that local flow velocities are everywhere constant to first order. With velocities constant, the mass flow rate is proportional to the density which, for constant temperature, is also proportional to the absolute pressure. The thermodynamic cycle operating at various pressure levels can be shown as in Fig. 16–20. The variation of cycle performance is a function of the working fluid properties. These properties are insensitive to absolute pressure when the gas is monatomic. For other gases, the effect of changing pressure may be significant in affecting specific work and efficiency.

It is expected, therefore, that the relation between efficiency and fractional power is relatively flat (Fig. 16–21), and a small fractional power output can be obtained by operating at low absolute pressure. In practice, the fluid frictional losses are slightly altered because the decreased density also decreases the flow Reynolds numbers. This increases the importance of viscous losses. The effect is to reduce efficiency slightly as the power output is reduced because component efficiencies are reduced.

The peak and part-power cycle efficiency noted for bypass and temperature modulation control on a realistic analysis (Fig. 16–21) are about half the value noted for an idealized analysis for the same temperature extremes, showing the importance of the irreversibilities. Further, the serious degradation of efficiency with bypass control is noted as a disadvantage in relation to inventory control. It should be noted, however, that the severity is important only if the fraction of time spent at less than full power is significant.

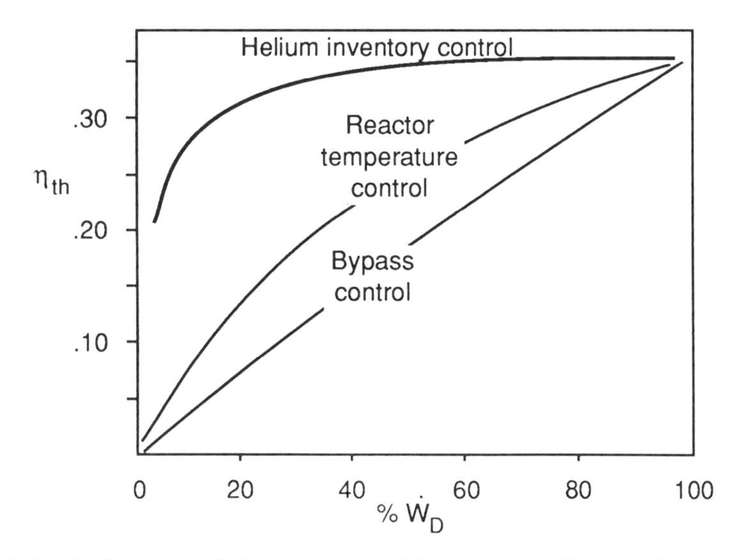

Fig. 16–21. Performance of the inventory and bypass controlled closed Brayton cycle.

PROBLEMS

1. Determine the derivatives or influence coefficients for the variation of w and thermal efficiency for the Brayton cycle at part power with a reduction of θ_4. Assume that the relation between π_c and θ_4 are as given by eq. 16–8 so that constant θ_4 and constant π_c cases are obtainable with appropriate values of β. Find $dw/d\theta_4$ and $d\eta_{th}/d\theta_4$ or $d \ln \theta_4$ and $d \ln \eta_{th}/d \ln \theta_4$ evaluated at $\theta_4 = \theta_{4D}$, for a given β and specified polytropic efficiencies. Assume a maximum specific work design.
2. Determine the derivatives of efficiency for a Carnot engine as in the problem above.
3. Establish the validity of eq. 16–38 by deriving it from first principles.
4. In the choked flow through the turbine as described by eq. 16–18, the flow area A^* is assumed constant. Describe qualitatively and quantitatively the performance improvement that might be realized at part power if the area could be tied to the value of a variable such as pressure ratio. Obtain the part power efficiency variation for such a "variable area turbine."
5. A Brayton cycle engine is designed with a turbocompressor and turbine driving a power expander of the displacement type, using a two-shaft design. Assume the component efficiencies are 0.85 for the compressor and 0.95 for each of the expanders. θ_4 at maximum power is 5.0. If the pressure is to maximize specific work, determine the pressure ratio across each of the expanders at full power. For a 100 kw power engine what are the displacement of the second expander and the volume expansion ratio? What are the thermal efficiency and the specific work? Determine

the values of the same parameters for pressure ratio chosen for maximum efficiency. Compare the size of the expander in this engine configuration to that of a Diesel to produce the same output power; show that with both at maximum power, the ratio of Diesel power to Brayton power for the same displacement (of the low pressure turbine in the case of the Brayton) is given by

$$\frac{w_{\text{Diesel}}}{w_{\text{Brayton}}}\frac{\rho_0}{\rho_6}$$

where ρ_6 is the density of the expanded Brayton cycle working fluid. w is the specific work.

6. Assume that a second burner is now added behind the first turbine of the engine in the previous problem. In that burner the temperature is again raised to $\theta_5 = 5$. How much additional (maximum) power is obtained for the two designs? How is the thermal efficiency affected, and by now much?
7. In the quasi-steady formulation of the part-load performance, it was assumed that changes in condition take place over long times. By writing the appropriate conservation equations, identify the physical variables that form the time scales describing the finite volume in the combustor and in the kinetic energy of the rotating mass of the turbomachinery.
8. Consider single- and dual-shaft engines whose performance is given by eqs. 16–16 and 16–33 respectively. By assuming that all turbines operate choked, all compressors are ideal, gases are ideal and perfect, find numerical values of all performance parameters [τ_c, τ_{t1}, τ_{t2} (or τ_t for the single shaft), and π_c] at the design point which is for maximum specific work at $\theta_{4D} = 6$. Now assume that $\theta_4 = \theta_{4D}(1 + \varepsilon)$ where ε is small compared to unity so that linearization of the equations is justified. The ratio of power to its design value is the product of a number of factors and can ultimately be written as $1 + C\varepsilon$, where C is made up of terms arising from various parameter changes (mass flow rate or specific work, for example) and gives the performance decrease with decreasing peak cycle temperature. Compare C for the single- and dual-shaft engines. Determine the C's for the efficiency decrease with power and compare to Fig. 16–16.

BIBLIOGRAPHY

Stoecker, W. F., *Design of Thermal Systems*, 3rd Edition, McGraw-Hill, New York, 1989.

REFERENCES

16–1. Kronogard, S. O., "Automotive Turbine—Advantages of Three Shaft Configurations," *Gas Turbine International*, November–December 1975.

16–2. Fryer, B. C., Smith, J. L., "Design, Construction and Testing of a New Valved, Hot Gas Engine," IECEC 739074, 1973.
16–3. Warran, G. B., Bjerklie, J. W., "Proposed Reciprocating Internal Combustion Engine with Constant Pressure Combustion," SAE 690045, 1969.
16–4. Decher, R., "The Britalus Brayton Cycle Engine," *International Journal of Turbo and Jet Engines*, **2**(2): 135–40, 1985.
16–5. Decher, R., "Power Scaling Characteristics of a Displacement Brayton Cycle Engine," *International Journal of Turbo and Jet Engines*, **2**(2): 141–8, 1985.
16–6. Fraas, A. P., "Nuclear Gas Engine," Oak Ridge National Laboratory Report 58-9-12, 1958.
16–7. Rosa, R. J., "Characteristics of a Closed Brayton Cycle Piston Engine," *AIAA Journal of Energy*, **7**(2), March–April 1983.
16–8. Decher, R., "Part Load Operation of a Displacement Brayton Cycle Engine with Variable Volume Ratio Expander." *International Journal of Energy Systems*, **9**(1): 55–9, 1989.
16–9. Oates, G. C., Ramsay, J. W., "Potential Operating Advantages of a Variable Area Turbine Turbojet," ASME paper 72-WA-Aero-4, 1972.
16–10. Decher, R., "Displacement Brayton Cycle Component Coupling with a Variable Speed Drive," *International Journal of Turbo and Jet Engines*, **5**, 265–70, 1988.
16–11. Decher, R., "Displacement Compressors with High Performance Flow Control Valves," *International Journal of Energy Research*, **13**, 327–38, 1989.

17

ENERGY STORAGE

The utilization of energy to produce power requires the ability to produce the power at the time desired. This is desirable for many work production processes and essential for transport systems where the stored energy itself needs to be transported. The variety of methods employed in storing energy is a reflection of the wide array of energy need situations. The discussion of this chapter is limited to those situations that have an impact on the way energy is used as a commodity in the world's economy. This specifically excludes situations with a practical capability of less than, say, 10 MJ. Because a kilogram of hydrocarbon fuel has a thermal energy content of approximately 40 MJ, this limitation restricts discussion to relatively large systems. In particular, discussions of springs (which store mechanical work) and of small electronic components such as capacitors and inductors (which play an important role in storing energy long enough to accomplish phase shifting and similar operations in AC circuits) are omitted here.

This chapter is a description of the parameters that characterize various energy storage devices, their design characteristics as well as their applications. The economics of power supply requires that the consideration of storage must include its cost.

17.1. WHY ENERGY STORAGE?

Energy is required in stored form whenever there is a temporal mismatch between supply and demand. Figure 17–1 shows the elements of the situation requiring storage. In a discussion of storage, the period of interest is that over which the demand supply cycle is approximately periodic. For example, the data of Figs. 3–1 to 3–3 show, in part, that periodicity is not exact and therefore one generally deals with average or typical cycles. The time-averaged values of power supplied to a conversion system and that used by it will generally be equal. A true, or ideal, storage device is neither a producer of energy nor a user. In addition to accepting and delivering power, a real storage device interacts with its external environment through losses (Fig. 17–1). At any time, the stored energy in the system is the difference between supply and the sum of demand and losses. Figure 17–2 shows the variation of energy in a storage system over a cycle time of length t_{max}. This curve is the integral of the net (input less output) power integrated over time and includes an amount of energy always present in the system. This quantity of energy may or may not be significantly large compared to the processed amount and serves to maintain its energy density

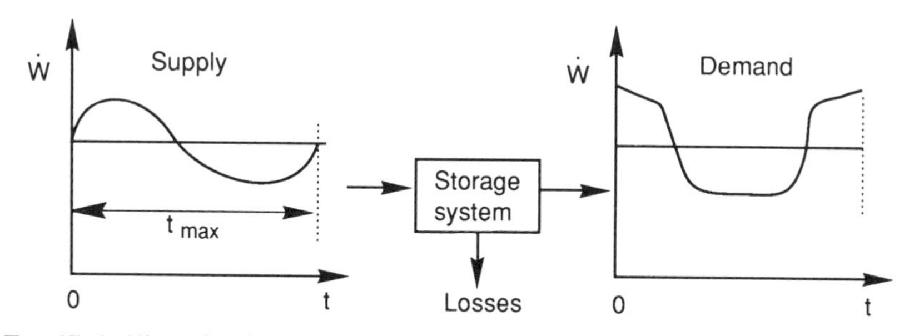

FIG. 17–1. The role of an energy storage device between energy supply and user.

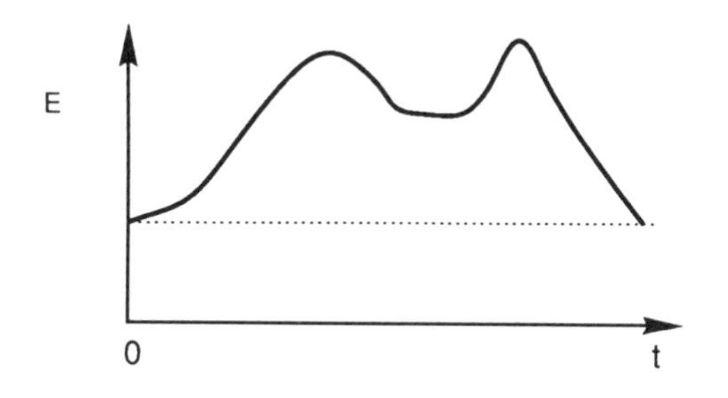

FIG. 17–2. Variation of energy in storage during one cycle period.

at an acceptable level so that the fractional variation of state parameters is appropriate.

The most commonly stored energy form is the hydrocarbon fuel. In the automobile, for example, to accommodate the extreme variation in the power demanded owing to traffic, the heat engine simply processes the amount of fuel at a rate corresponding to the instantaneous power needs. The stored energy must also be light in weight and high in density, as well as possess other practical features associated with cost, safety, environmental impact, etc.

A utility grid supplies power output variation that fluctuates on a time scale of hours, consistent with the activities of the customers (see Figs. 3–1 and 3–2). A large-scale thermal power plant, either coal or nuclear heated, has a limited ability to operate at lower than design power levels for reasons having to do with cycle thermodynamics (and therefore cost) and a much more limited ability to alter the physical state of the system to provide a lowered output on a time scale that matches changes in user need. In practice, the steady base load is generally provided by thermal power plants which are operated continuously at close to full power. The means with which power utilities deal with the problem of matching time-dependent supply to a variable load include:

1. Selling low-cost power to interruptible users such as industrial plants (aluminum reduction for example),
2. Calling on rapid startup of standby power sources that can meet load requirements in minutes (gas turbine or Diesel engines for example),

3. Calling on power supplied from interruptible energy supply (such as a hydroelectric reservoir),
4. Selling excess capacity to users with different demand characteristics (users to the east or west to take advantage of the time zone differences in the load cycle or users from a northern supplier to the south where there are climatic differences which, in turn, have an impact on power needs),
5. Wasting excess power, or
6. Employing combinations of the above.

At the other extreme of the energy storage situations is the problem of variable supply from sources such as solar or wind power. Demand may be largest at times when supply is small or unavailable or when weather interferes with the power capture process while the supply to a steady load is attempted.

17.1.1. *Description of a Power System with Storage*

In an ideal storage system, the energy in the storage system, E_s, is governed by

$$\frac{dE_s}{dt} = \dot{W}_i(t) - \dot{W}_0(t) - \dot{Q}(t) \tag{17–1}$$

Here $\dot{W}_0$, is the output power demanded by the user whereas $\dot{W}_i$ is the time-dependent input power. The input must be such as to make up for losses, $\dot{Q}(t)$ to ensure that over a cycle time t_{max}, the change in energy E_s is zero. Generally, the losses are a function of system state properties (such as temperature, or speed) and of time. A storage system thus satisfies

$$\int_0^{t_{max}} \frac{dE_s}{dt}\,dt = 0 \qquad \text{or} \qquad \int_0^{t_{max}} \dot{W}_0(t)\,dt = \int_0^{t_{max}} (\dot{W}_i(t) - \dot{Q}(t))\,dt \tag{17–2}$$

A storage efficiency may be defined on the basis of the energies processed as

$$\eta_s = \frac{E_{out}}{E_{in}} = \frac{\int_0^{t_{max}} \dot{W}_0(t)\,dt}{\int_0^{t_{max}} \dot{W}_i(t)\,dt} \tag{17–3}$$

to quantify the importance of the loss.

From a performance viewpoint, storage systems are characterized primarily by the efficiency, which measures the energy penalty that must be paid to have the ability to store energy and by the process capability on either the input or the output side or both. This latter feature controls the power that can be accepted or delivered by the system and as such controls the size and cost of the hardware. This capacity design issue is illustrated in Fig. 17–3.

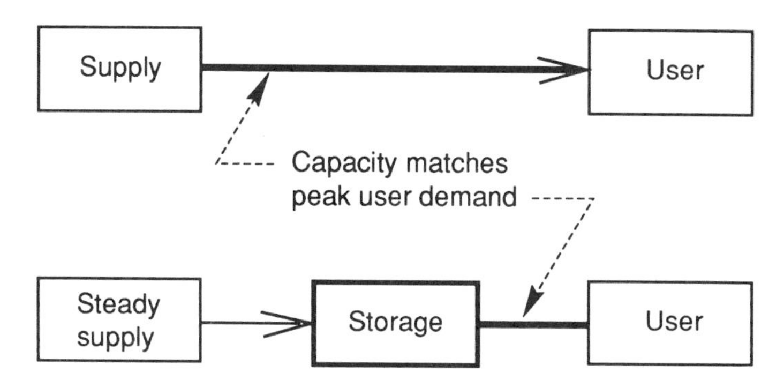

FIG. 17–3. Storage permits reduction of energy transport capacity requirements.

The cycle time, t_{max}, is clearly important in determining the total amount of energy stored.

The availability of fossil fuels for transport, for heating, for electric power, and for industrial processes has minimized the need for alternative storage means. The capability of storing energy is costly, and the need for it has been avoided through the adaptation and development of heat engines (Chapters 7 and 8) and heaters to operate as needed and when needed. In a future time, when fossil resources are less available, alternative and cost-effective means of matching source to user needs will have to be developed. This is true because the energy supply may become more time-variable through solar and wind power contributions. Further, the ability to store energy reduces capacity requirements for the energy transport system, and this reduces cost (Fig. 17–3). As an example, consider that the size of a wire to deliver electricity can be small if a small current flows continuously with storage at the user site. The alternative where the wire must handle the peak power requires a costlier supply network.

Energy storage in the form of fuels makes transportation systems possible. Without an attractive means of carrying fuel and using oxygen from the atmosphere, the modern automobile, aircraft and ship could not function as they do today. The rocket vehicle stores not only the fuel but the oxidizer as well to enable the production of very large amounts of power. Special-purpose processes such as the creation of very high temperatures (as in a fusion experiment) require rapid delivery of stored energy, from capacitors, for example. In transportation, practical means of using electricity produced by a stationary power plant are in use only for the train, and that only because travel is limited to its right of way, which allows the use of a third rail or overhead wire. Because the electric train carries no fuel, it has a transport efficiency (as defined in eq. 3–37 since $W_F = 0$) of unity. Limited-performance vehicles propelled by means other than fuels (i.e., stored or beamed energy) may yet come to be in the future, perhaps driven by the environmental impact of fuel usage or the necessity to utilize thermal resources other than fossil fuels.

Probably the most economically interesting situations requiring the storage of energy are the production of electric power by a utility system and, in the future, the transport vehicle. The utilization of solar and wind power requires the means of storing energy because of the usually serious mismatch between availability and need. The application of energy storage to these and other examples is discussed in this chapter with an eye toward the practical means for doing so.

17.2. DESIGN AND PERFORMANCE PARAMETERS

Reversibility

The process of storing energy for use at a later time should be as reversible as possible to avoid unnecessary expenditure of primary thermal energy resources. The irreversibility is measured by eq. 17–3 as work, heat, or mass carrying energy that is lost and not recoverable in the discharge part of the cycle. Such losses are most commonly friction (fluid or mechanical) of moving systems, heat transfer due to high temperature, material losses such as leakage, joule (I^2R) losses in electrical systems, etc.

Cost

The system wherein the energy is stored should be low in cost. Its cost is always a penalty that increases the cost of a unit of energy to the user over direct use of the energy. In part this is the underlying reason for energy storage to be avoided whenever possible and motivates the use of fuel using heat engines. The low cost must include consideration of the useful life of the storage system since all practical systems deteriorate in some way as they are utilized.

Energy density

For transportation systems, the desirable characteristics of stored energy should be low mass per unit energy stored or, equivalently, high energy density, see Section 3.4. Fuels have energies of the order of 40 MJ/kg (or 20,000 BTU/lb), and any scheme that is to compete must have numbers that are interesting by comparison, at least as long as fossil fuels are widely and cheaply available. Similarly, the volume associated with a typical storage cycle should be as small as possible. This may have serious consequences for the system, particularly when it is used in transport. An example is the use of hydrogen as fuel in an automobile or in an aircraft. In both cases, the cargo volume may be seriously reduced, and the required volume increases drag.

Scalability

A good storage system should be buildable in various sizes (i.e., scalable). As an example, a storage system which has a finite minimum mass for safety reasons, say, may not be useful in a small system because of the weight penalty associated with the minimum requirement.

Table 17–1. Safety of stored energy

Storage mode	Critical issue in failure
dam	hydraulics of the break and stream
fuel tank	mixing of air and fuel
flywheel	container deformation and containment
pressure vessel	material properties, containment

Demand responsiveness

The energy must be recoverable at the *rate* demanded. For example, if energy is stored in the form of liquid hydrogen, utilization may require conversion to a gaseous state. Sufficient heat exchanger area and processing pumps must be provided. This problem can be severe in the storage of heat in phase change or hydration processes where the solid storage medium cannot easily maintain thermal contact with heat exchange surfaces.

Safety

Energy is the capacity to do work. Stored energy must be held with maximum *safety*, tailored to the risk. A system must be designed to prevent a rapid and unsafe release that may well be destructive to life and property. The problem of safety in storage is usually an important concern in the design of such systems. Examples of disasters associated with uncontrolled energy release are common, and it is important to understand the process that limits the rate. For example, consider the failure of the modes listed in Table 17–1 and some of the processes that limit the impact of the failure.

17.3. ENERGY STORAGE METHODS

Energy can be stored in any of the forms listed in Section 1.1, that is, kinetic, potential, strain, thermal, chemical, electromagnetic, and nuclear. In the following, a few comments regarding the practical applications of these methods are offered.

In practice, the power flow from resource to user can be intercepted and the energy stored anywhere along the flow path. The energy cascade described in Section 2.5 describes the possibilities. This chapter emphasizes storage of electromechanical work as energy. The common storage of energy as fuel serves as a convenient reference to judge the means of electromechanical energy storage.

17.3.1. *Fuel (or chemical) Energy Storage*

Fossil fuels are utilized and stored in the many forms in which they are found in nature: gaseous, liquid, and solid; see Chapter 1. Further, they are processed and graded to have specific physical properties for use in various applications that require them. For example, the Otto (O) and Diesel (D) cycle engines as well as the gas turbine (GT) require liquid fuels with properties that relate to ignition (O and D), lubrication (D), and chemical purity (GT), in addition to a large number of secondary and important requirements. Economic considerations generally prescribe that the market deal with standard quantities, and thus fossil energy resources are classified into standard fuel forms.

The generation of alcohol and hydrogen as fuels is a means of converting primary energy resources that are readily available to the manufacturing of these useful fuel forms. For example in an environment where the land area is available and the location is suitable for the rapid growth of plants, alcohol may be an attractive stored form of solar energy. The economics of this process depend critically on the price of petroleum and on the ability of the automobile to utilize such a fuel. Because of the solubility of water in alcohol, engine and tankage material compatibility problems are present, as are problems of quality assurance in the marketplace.

Hydrogen is considered as a medium for distribution of energy in a future economy less dominated by fossil fuel. This gas can be produced by thermochemical and electrolysis reactions, distributed in the natural gas pipeline network, and transformed to hydrocarbon fuels by the methanation reaction which reads overall:

$$C + 2H_2 \rightarrow CH_4 \tag{17–4}$$

The carbon is derived from the relatively abundant coal resources. The material properties of methane and higher molecular weight products make them suitable, in fact necessary, as fuels for aircraft and automotive vehicles.

Batteries constitute a means of obtaining electrical power from stored chemical energy. As such, the materials used are the storage medium. The performance of these devices is noted in Table 17–5. The goal of having a satisfactory chemical storage battery with energy density and cost comparable to that of fuel and engine is the goal of many serious development efforts. In fuel cells, a wide range of materials may be considered as fuels, including metals such as aluminum. Rockets are possible only because of the relatively high chemical energy density of the fuels used.

17.3.2. *Mechanical Energy Storage*

Stress

Mechanical energy may be stored as strain energy in devices like springs or in any deformable structure. The amount of energy per unit mass (as is the cost) of spring material is small and this method is therefore uninteresting for most purposes, save the windup toy. At the material level, the energy in deformation is

$$\frac{\text{Energy}}{\text{volume}} = \frac{1}{2} E\varepsilon^2 = \frac{1}{2}\frac{\sigma^2}{E} \tag{17–5}$$

where E is the elastic modulus and σ, ε the stress and strain respectively. For a strong steel, this parameter is of the order of 10^6 J/m^3 or about 20,000 ft-lb/ft^3, if all of the material could be stressed to the yield point. When one considers the material density of steel ($\sim 10^4$ kg/m^3), the energy density of a stressed steel is of the order of 100 J/kg, which is very small compared to the energy density of a typical hydrocarbon fuel ($\sim 10^7$ J/kg).

Flywheels

Of greater interest is the storage of energy as kinetic energy in flywheels. Such means have been developed for vehicular applications where periodic "refills" from an electric power source are used to reenergize the flywheel which then allows travel to the next refill by transferring the flywheel kinetic energy to the tractive wheels of a car or bus. Although not strictly necessary, the means to increase the flywheel kinetic energy is generally electrical: A motor to accelerate the flywheel and a generator connected to a motor that generates a propulsive force. The elements of the flywheel energy storage system, shown in Fig. 17–4, consist of the rotating mass, the motor/generator to convey (electrical) power to or from the wheel, bearings, and an enclosure.

In addition to serving as a safety vessel, the enclosure allows operation of the turning wheel in an environment of reduced pressure so that the aerodynamic drag may be made small. The wheel has a kinetic energy in rotation and an energy per unit mass that is:

$$\mathrm{KE} = \tfrac{1}{2} I\omega^2 \qquad \text{and} \qquad \frac{KE}{m} = \tfrac{1}{2} R_G^2 \omega^2 \tag{17–6}$$

The radius of gyration of the flywheel, R_G, is defined by the combination of these expressions:

$$R_G^2 = \frac{I}{m} = \frac{\int_0^{R_{max}} r^2\, dm}{\int_0^{R_{max}} dm} \tag{17–7}$$

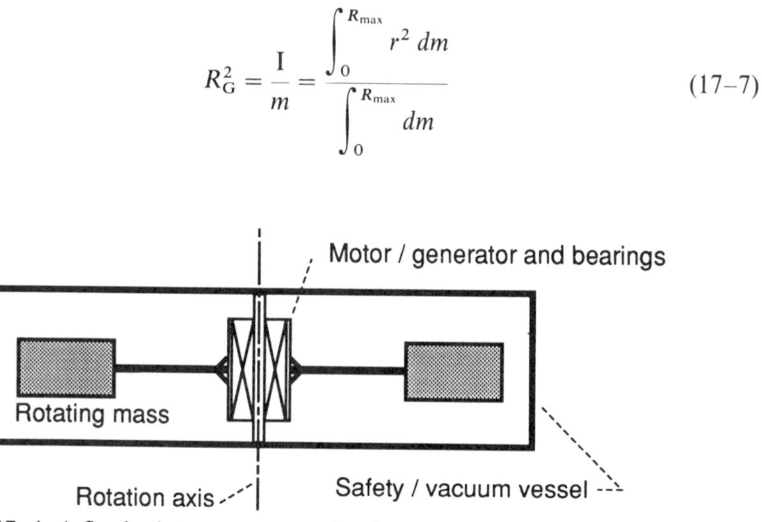

FIG. 17–4. A flywheel storage system showing enclosure and rotating mass.

The product of ωR_{max} is the "tip" speed, V_T of the outermost element. Thus

$$\frac{KE}{m} = \frac{1}{2}\left(\frac{R_G}{R_{max}} V_T\right)^2 \tag{17-8}$$

In the design of a high-performance flywheel, two features are therefore necessary: The ratio of R_G/R_{max} should be as large as possible, and a high tip speed is required. The radius ratio can be made large by using a thin annular ring which, unfortunately, has the disadvantage of a low energy stored per unit volume.

The speed is ultimately limited by the force balance on an element of rotating matter which experiences a centrifugal force which is resisted by the stress of the material. Without examining the details of the element's geometry, one can say that the force balance reads

$$\frac{[\rho_m R^3] V_T^2}{R} = (\text{geometric factor})\sigma R^2 \tag{17-9}$$

The term in brackets is the element mass of density ρ_m. The left side of the equation is the force that leads to the generation of a stress in the material. The length scale cancels out of this balance, and the allowable tip speed is thus a function of material properties and the geometry, or

$$V_T \sim \sqrt{\frac{\sigma}{\rho_m}} \equiv a_m$$

Here a_m defines a speed associated with the material properties. The indicated proportionality factor depends on the rotor geometry. a_m is similar to a sound speed, hence the use of a. A rearrangement of eqs. 17–8 and 17–9 allows the energy per unit mass to be written as

$$\frac{KE}{m} = K_s \frac{\sigma}{\rho_m} \equiv K_s a_m^2 \tag{17-10}$$

The search for good flywheel designs consists of the selection of good materials, that is, large a_m and good geometric design, large K_s. Table 17–2 and Fig. 17–5 give a summary of both of these aspects of flywheel design.

In practical applications of flywheel systems into vehicular applications, the reaction forces arising from the possible reorientation of the angular momentum vector through turning of the vehicle may have to be considered as an important design constraint. The importance of this issue may be reduced by use of two counter-rotating masses.

The design of the wheel has an important influence on the maximum tolerable acceleration which is felt by linear elements (such as bars or brushes as shown in Fig. 17–5) as bending stresses. This aspect is important when rapid energy input or output is involved.

Table 17–2. Representative values for properties of flywheel materials

Material σ/ρ	a_m watt-hr/kg*	m/sec
Aluminum	10	190
Steel	22	280
E-glass fiber	90	570
Carbon fiber	100	600
S-glass fiber	120	660
fused silica	400	1200

* Multiply by 3.6 to obtain kJ/kg..

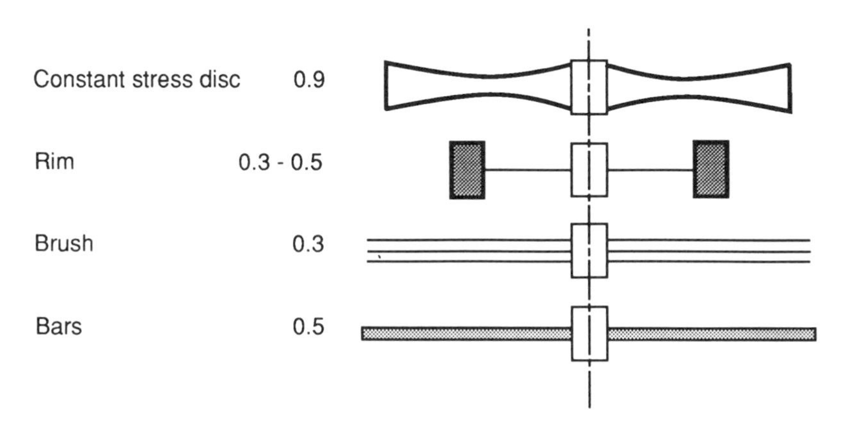

Fig. 17–5. Geometric parameters in flywheel design. Values of K_s as defined in eq. 17–10 are noted.

Pumped Hydrostorage

A mass of water at an elevated height may be, and is, used as an energy reservoir. The interruption (and storage) of a river stream may be used as a primary source of power with the advantage that the power output can be modulated as required by the user. It is often used as the source of the time varying component of power (peak load), with the base power supplied by a thermal plant. In some instances where rivers with damming possibility do not exist, it is advantageous to use excess capacity to pump water to an elevated reservoir and recover that energy investment at a later time (Fig. 17–6). Such a system is termed *pumped hydro* storage, and a number of such storage systems exist around the world. The pumps used for the uphill water movement are often reversible and thus function as turbines to reduce the capital investment required. The recovery efficiency of such systems is nominally of the order of 0.75 or better. Expended underground mines may be used as low elevation reservoirs in pumped hydro systems.

The energy density of the water stored in a reservoir is $g\Delta z$ per unit mass. Considering an elevation change Δz of 500 m to be practical, the energy density

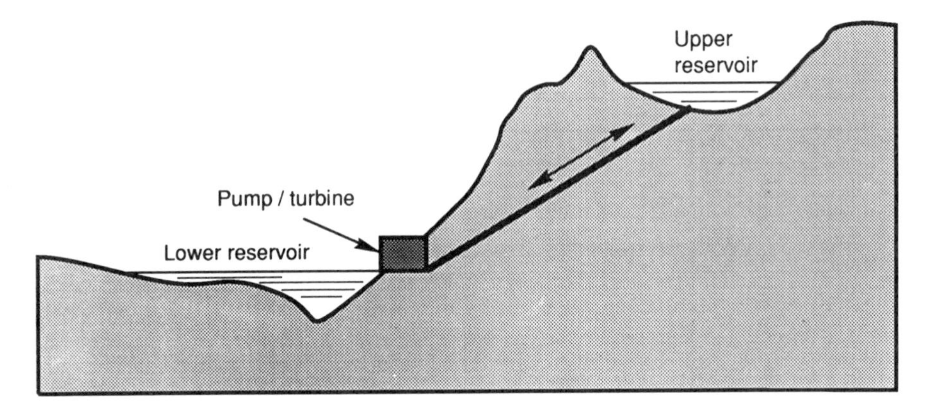

Fig. 17–6. Two level reservoir system for pumpted hydro energy storage.

is 5000 J/kg or about 1.5 watt-hr/kg, which is evidently not a large quantity compared to other possibilities of storing energy. The cost of the water is obviously small and the system cost above that of the machinery is dominated by the land area or the caves that are used.

17.3.3. *Compressed Air*

A vessel of compressed gas, most typically air, represents stored energy. The energy of a vessel is determined by the internal energy of stored gas or, for a perfect gas,

$$E = mC_v T = \frac{1}{\gamma - 1} pV \tag{17–11}$$

where the latter is for an ideal gas as well. Note that the energy content is larger for a gas whose value of γ is smaller (i.e., consisting of a more complex molecule). Evidently the pressure is proportional to the energy per unit volume, and a high pressure is desirable. If a thin-shelled vessel is used to contain the high-pressure gas, then the mass of the vessel must be included in a fair assessment of energy density. The stress level in the spherical shell determines the shell thickness. From a force balance like that carried out in connection with eq. 6–57 or Fig. 9–50, the wall thickness is given by

$$t = \frac{p}{\sigma}\frac{R}{2} \tag{17–12}$$

The mass of the shell is therefore

$$m_s = 4\pi R^2 \left(\frac{p}{\sigma}\frac{R}{2}\right) \rho_m \tag{17–13}$$

where ρ_m is the shell material density. The total mass of pressurized gas and shell is

$$m_{\text{tot}} = m_{\text{g}} + m_{\text{s}} = p\,\frac{4\pi}{3}\,R^3\left(\frac{\gamma}{a^2} + \frac{3}{2}\,\frac{\rho_{\text{m}}}{\sigma}\right) \tag{17–14}$$

Here the gas temperature is written as a sound speed to avoid the introduction of the gas constant (R) and because the ratio $\sigma/\rho_{\text{m}} = (a_{\text{m}})^2$, defined above in eq. 17–10, is also a characteristic speed squared for the vessel material. For a steel-like material, the value of a_{m} is of the order of 350 m/sec. It can be significantly larger for wound vessels made from carbon, glass, or other fibers (see Table 17–1). The energy density of a compressed gas vessel can thus be written as

$$\frac{\text{Energy}}{m_{\text{tot}}} = C_{\text{v}}T\left(1 + \frac{3}{2\gamma}\,\frac{a^2}{a_{\text{m}}^2}\right)^{-1} \qquad \text{where } a_{\text{m}}^2 \equiv \frac{\sigma}{\rho_{\text{m}}} \tag{17–15}$$

For a compressed gas similar to air, the speed of sound ratio is of order unity so that the energy density of the gas alone is reduced by about a factor of 2 when the vessel is included. The penalty is reduced if better materials are chosen, and made worse if the gas temperature is elevated because the allowable value of σ/ρ_{m} generally decreases with increasing temperature.

Large-Scale Underground Air Energy Storage

The idea of using large, natural underground cavities (salt caves, for example) has been developed into a practical energy storage method. Excess eletrical power from a grid drives compressors that pump air into the cavern. This air is then emptied and the grid is supplied via turbine driven generators. Figure 17–7 shows a typical compressed air energy storage (CAES) system. Its performance necessarily involves the irreversibilities of the compression and expansion processes as well as any heat or mass leakage to the environment from the stored compressed gas. The expansion process allows the addition of heat to the gas just prior to expansion so that net power (contrary to the notion of an ideal storage system discussed above) can be generated by this system. In effect, the system is a gas turbine engine with a temporal separation of the compression and expansion processes. The compression process is effectively nonadiabatic if heat loss is experienced during storage. Reference 17–1 describes the performance of such systems, showing the performance potential of such a gas turbine as an integral element of a utility system.

Noteworthy are the governing equations of the storage system. Since mass and energy are processed, conservation equations for these two quantities for a vessel of volume V, and state variables p, ρ, and T, are, respectively:

$$\frac{d(\rho V)}{dt} = \dot{m}_{\text{i}} - \dot{m}_{\text{e}} \tag{17–16}$$

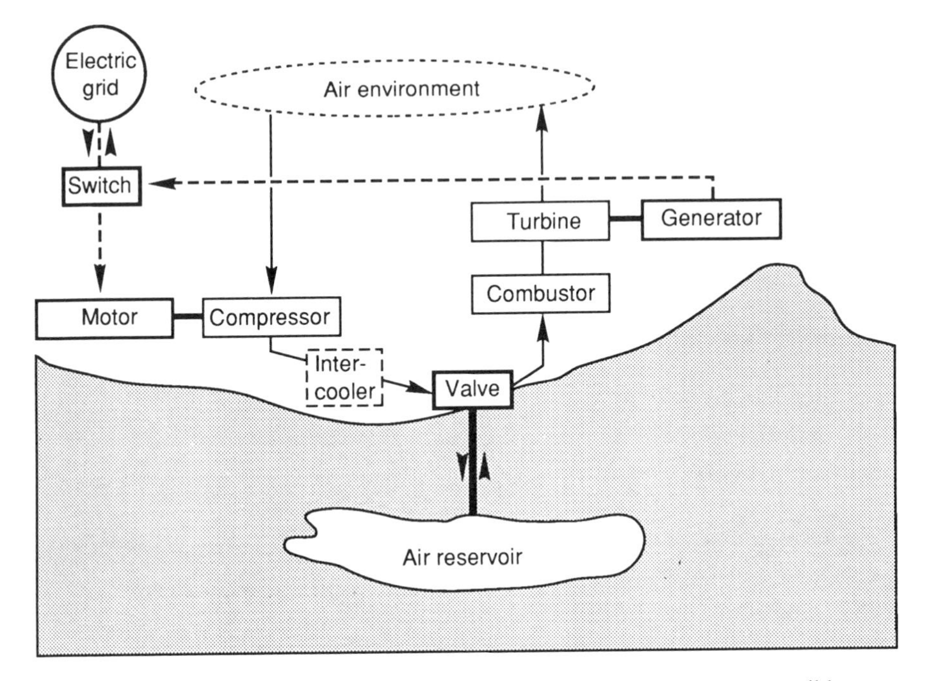

FIG. 17–7. Compressed air energy storage system. Motor, compressor, possible combustor, turbine, and generator are noted. The switches operate in different ways during input and output portions of the cycle.

and

$$\frac{d[\rho V(C_v T)]}{dt} = \frac{1}{\gamma - 1}\frac{d(pV)}{dt} = \dot{m}_i h_i - \dot{m}_e h_e + \dot{Q} \tag{17–17}$$

In effect, the mass conservation equation describes the density whereas the energy equation describes the pressure in the vessel. Here the $\dot{m}_j$ terms are time-varying mass flows with one or the other generally equal to zero, depending on the operating mode—input or output. V is nominally constant, although it may vary in systems where underground water masses are displaced. Mass leakage is neglected. Heat that may be added to the vessel, or that which may leak from it is accounted for with the last term. The power being directed to storage is related to the mass flow rate by (perfect gas):

$$\dot{W}_i = \dot{m}_i C_p (T_2 - T_0) = \frac{\dot{m}_i C_p T_0 [\pi_c^{(\gamma - 1)/\gamma} - 1]}{\eta_c} \tag{17–18}$$

The nomenclature used here follows that introduced in Chapters 7 and 9.

The energy withdrawn from the vessel is simply the enthalpy flux leaving during the withdrawal portion of the cycle. Since the power available from a turbine is proportional to the turbine inlet temperature, it generally pays to burn fuel in the exiting stream or heat it by convection and therewith produce additional power. Production of more energy than is absorbed during input means that the system is a net producer of power in addition to serving in a storage function. Because of this shift a longer time period is devoted to storage imput compared to the case where the system uses no fuel heat. The fuel heat added is:

$$\dot{Q}_f = \dot{m}_e C_p (T_4 - T) \tag{17–19}$$

and the power produced:

$$\dot{W}_e = \eta_T \dot{m}_e C_p T_4 (1 - \pi^{-(\gamma-1)/\gamma}) \tag{17–20}$$

Here π is the nondimensional pressure ($=p/p_0$) and T is that in the vessel. T_4 is the limited turbine inlet temperature. These relationships can be manipulated to yield a description of the storage vessel thermodynamic variables as a function of time during startup when the heat loss (or gain through resistive heating from excess grid power) is known and the size of the vessel is specified in relation to the mass input per cycle. The ratio of mass input to stored mass determines the pressure variation in the vessel. This variation, in turn, determines possible changes in compressor efficiency because the downstream pressure varies. Reference 17–1 is a discussion of such calculations. The parameter that controls pressure variability is

$$(\gamma - 1)\frac{\text{Energy Input per cycle}}{p_{max} V}$$

which should ideally be small ($\sim$0.20 by using a large vessel so that the pressure variation over the cycle time is small (10% variation in temperature or 20% variation in vessel pressure).

As an energy storage device, the CAES is a net producer of energy, and its storage efficiency must be evaluated in an unusual way. If no additional fuel is used, the conventional energy recovery efficiency describes the quality of the system, eq. 17–3. With the addition of fuel energy, however, one must recognize that the value of the electric power being stored is determined by the price of the fuel (most likely coal or nuclear) used to generate it while more costly distillate fuel must be used for the turbine. An appropriate measure of recovery efficiency for a fuel-using storage system is therefore

$$\eta_R = \frac{W_e}{W_i + \eta_b Q_f \dfrac{C_f(\text{distillate})}{C_f(\text{coal})}} \tag{17–21}$$

where η_b is the base plant thermal efficiency and the C_f is thermal cost of fuels. This recovery efficiency depends on the configuration of the cycle involved. Detailed examination of a number of possibilities yields the following observations. First, intercooling is desirable if heat losses from the vessel are experienced, since these are driven by the temperature of the compressor outlet or stored gas. Intercooling is also useful when there are temperature limits on the stored gas. Second, vessel pressures (π) of the order of 5–20 appear to give good efficiency with modest turbine performance capability. As one might expect, the lower pressure ratios favor a regenerated cycle configuration and result in a lower peak stored gas temperature. Typically, the recovery efficiency as defined by eq. 17–21 ranges from 0.8 to greater than unity, certainly in a range of interest to commercial users so that the cost of the system becomes the singular factor determining the practicality of implementation.

Compressed air energy storage costs and performance can be enhanced in a number of interesting ways by integrating direct power production and thermal storage. The integration of air storage with the gas turbine cycle is described above. Other possibilities considered include steam injection, integration with combined cycles, with high-pressure thermal storage, with coal gasification, and with circulating fluidized bed combustors. Reference 17–2 is a discussion of such possibilities.

17.3.4. *Thermal Energy Storage*

Thermal energy storage (for later reuse) is used primarily for solar power systems where heat is the energy form captured, and for systems where storage of the primary or incoming form of energy is not possible or attractive. The chief characteristic of thermal storage is that insulation must be provided to prevent heat loss. The temperature of the stored medium determines its economic practicality.

Latent and Sensible Heat

The storage of thermal energy can be classified according to the temperature of the medium and/or according to whether the medium undergoes phase changes (latent heat) or temperature changes (sensible heat) or a combination of the two. In general, heat stored is used at a desired temperature. Thus latent heat storage systems accept and give up the phase change energy at a temperature that is characteristic of the material.

Sensible heat systems have to be large or massive in order not to suffer significant temperature changes during an input or an output cycle. The most interesting applications for these storage methods have to do with provision of comfort heat for residents in a dwelling. Under such circumstances, it may be possible to use low-cost materials such as water or dry rocks in gravel form to moderate high daytime thermal input for nighttime use. These methods of bringing warmth for the cool evening and night may require active management of solar input and of internal air circulation.

The latent heat systems have the advantage of requiring smaller mass. They suffer from the fact that the low-temperature phase-change reactions used are typically hydration reactions, which alternate from solution to solid state. Once the material is in the solid state, transferring the heat becomes a problem as the contact area tends to diminish in time because the liquid flows away from its generation site on the surface of the heat exchanger. Table 17–3 gives some

Table 17–3. Properties of low temperature phase change reaction materials

Compound, n	Transition Temperature		Heat of Fusion	
	°F	°C	BTU/lb	kJ/kg
$CaCl_2$, 6	84–102	29–39	75.0	174.0
$Ca(NO_3)_2$, 4	104–108	40–42	90.0	209.0
$Na_2S_2O_3$, 5	120–124	49–51	90.0	209.0
Na_2SO_4, 10	90	32	104.0	242.0
Na_2HPO_4, 12	97	36	114.0	265.0
Na_2CO_3, 10	90–97	32–36	115.0	267.0

of the phase-change reactions of interest to low-temperature systems and the corresponding phase change temperature. The reactions are typically as follows:

$$\text{compound}\cdot n\text{H}_2\text{O} \Leftrightarrow \text{compound} + n\text{H}_2\text{O} + \text{C} \tag{17–22}$$

High-temperature storage of heat is primarily of interest in solar power systems, where the solar intensity is amplified through lenses or mirrors. This means of collection is motivated by the higher Carnot efficiency enabled by the conversion of such heat, but suffers also from the possibility of loss during storage unless costly means of insulation are employed. The optimization of this design is a challenge that must consider economic aspects of the installation, including the collection system, the temperature of stored energy and the heat engine that will convert it to work. The experience with such systems to date is limited to development efforts of the components and modest-scale installations. A number of related aspects of solar power are discussed in Chapter 1.

Cryogenic Fluids

Since heat engines operate between two reservoirs of different temperatures, storage of a cold fluid such as liquid nitrogen may be thought of as energy storage. An engine may be made to operate between the cold stored material and the warmer environment. Both the change phase from liquid to gas and the warming of the gas to waste heat temperature involve heat, part of which is convertible to work by the heat engine. Such a heat engine may be operated with no significant environmental impact except the release of cool nitrogen. Hertzberg (ref. 17–3) has suggested this approach for a nonpolluting automotive engine suitable for urban use. The properties of common gases near the saturation line at 1 atm pressure are given in Table 17–4.

Using a heat sink as the energy resource implies that the ratio of work done per unit heat rejected is the performance parameter of interest. This ratio is related to the classical thermal (i.e., using a heat *source*) efficiency,

Table 17–4. Physical properties of common cryogenic gases and a comparison with water and solid CO_2 which sublimes

Gas	MW	Latent heat of vaporization (kJ/kg)	Sensible heat to 298 K (kJ/kg)	Temperature (at 1 atm) (K)	Specific gravity
N_2	28.0	199	234	77	0.81
O_2	32.0	213	193	90	1.20
H_2	2.0	446	~4000	20	0.071
H_2O	18.0	2257	—	373	1.0
CO_2	44.0	571	~80	195	1.56

Note: The critical point of CO_2 is 304 K, and the pressure at 233 K (= −40 °C = −40 °F) is about 10 atm.

η_{th}, through

$$\frac{w}{q_{out}} = \frac{\eta_{th}}{1 - \eta_{th}} = \eta_s \tag{17–23}$$

η_s is the "sink efficiency" which should be as large as possible. If the fraction ϕ of the Carnot efficiency is realized, then

$$\eta_{th} = \phi\left(1 - \frac{T_{min}}{T_{max}}\right) = \phi\left(1 - \frac{1}{\theta}\right) \tag{17–24}$$

where θ is about 4 for liquid nitrogen. The sink efficiency can therefore be written

$$\eta_s = \frac{\phi\left(1 - \frac{1}{\theta}\right)}{1 - \phi\left(1 - \frac{1}{\theta}\right)} = \frac{3\phi}{4 - 3\phi} \quad \text{for } \theta = 4 \tag{17–25}$$

which ranges from 3 to 0.6 as ϕ decreases from unity to a more realistic 0.5. The ability of liquid nitrogen to absorb heat is of the order of 430 kJ/kg so that the mechanical energy density is of the order of 300–1200 kJ/kg depending on ϕ. The energy density of batteries suitable for present and future use in electric vehicles is given in Table 17–5 for comparison. Evidently, cryogenic energy storage is competitive from the viewpoint of energy density with present-day battery technology. The cost and weight of the complete energy conversion system for a specific application such as the automobile may be a significant issue that is not addressed here. Whether future battery development is successful remains to be seen.

The strong dependence of η_s on ϕ ($d \ln \eta_s / d \ln \phi \sim 2$ at $\phi = 0.5$) makes the choice and design of the engine cycle critical to the performance of such an energy conversion method.

Table 17–5. Practical primary cell energy densities of cells available and under development

Reactants	Energy density	
	watt-hours/kg	kJ/kg
$Pb\text{-}H_2SO_4$	30–60	100–200
Ag-Zn	200	600
Ni-Fe (KOH)	45	160
Li-F	6000	22,000
$Al\text{-}O_2$	3000	11,000
Mg-S	1700	6000
Na-S	800–1500	3000–6000

Nuclear Decay

Nuclear decay is the spontaneous transformation of a"radioactive" material to a more stable product. The process of conversion may involve the emission of alpha particles (helium nuclei), beta particles (electrons or positions), and/or gamma radiation (high-energy photons). These energy packets interact with the surrounding matter and share their energy with it so that heat is generated. Thus one may consider a radioactive substance as a source of heat that can be used by a heat engine to produce electric power. Such decay energy represents stored energy when one views the process of manufacturing the radioactive material in a reactor as storage of reactor power.

Nuclear decay heat sources have the characteristic that the decay reactions ultimately stop when the unstable substance is depleted. The decay of a radioactive substance is characterized by the time it takes for an initial amount of the substance to be reduced to half the initial amount. This time is the half life ($t_{1/2}$) of the reaction. This is directly related to an equivalent exponential time constant. One feature is that once the decaying material is made in a nuclear reactor, it decays and cannot be held up or "stored." The thermal power produced for a single decay reaction (i.e., there is no subsequent decay of the product produced) is

$$\dot{Q} = \dot{Q}_\text{i} \exp\left[(-\ln 2)\frac{t}{t_{1/2}}\right] \tag{17–26}$$

The decay heat from a radioactive process has been exploited for the generation of power in space. Systems consist of the packaged radioactive material and a heat engine or a thermoelectric generator to produce electric power. The use of this source is limited to environments where the radiation is of no concern or where shielding can be effective. The decay process itself makes this form of thermal energy storage useful only while the heat generation rate is sufficient for the mission.

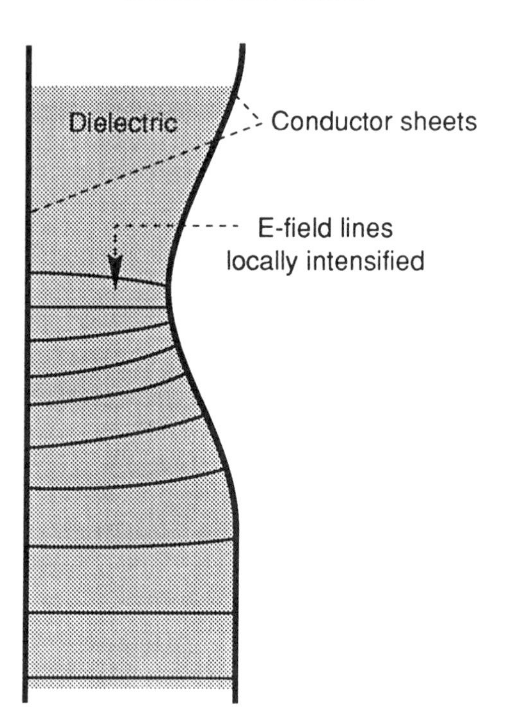

Fig. 17–8. Parallel place capacitor deformed by locally weak dielectric. A potential difference of V is applied to the two sheets.

The most notable decay power source is the SNAP (Systems for Nuclear Auxiliary Power) series. This array of power sources was and is continuing to be developed for the U.S. space program. Criteria are typically high power density, long half-life, as well as a number of other parameters including material density, shielding requirements, etc. Candidate materials for systems of various power outputs have included Cm^{242}, Sr^{90}, Po^{210}, Pm^{147}, and Pu^{238}. Plutonium is of significant interest because its power density is high (0.56 kW/kg), and its half-life is 89 years.

17.3.5. *Electromagnetic Fields*

The energy in electric and in magnetic fields is given by the expressions:

$$\text{Energy} = \tfrac{1}{2}CV^2 \qquad \text{or} \qquad = \tfrac{1}{2}LI^2 \tag{17–27}$$

respectively. Here C is the capacitance of a capacitor and L is the inductance of a coil. V is the voltage and I is the current. The devices themselves are illustrated schematicaly in Fig. 1–6. Evidently, large energies are stored when V or I are large. For the capacitor, one would design the capacitor to operate at large voltage (V), which makes it difficult to match to a load of modest resistance because the resulting current is large and the discharge time is correspondingly short. As a result, capacitors are not used for storing energy except under circumstances that can accept the peculiar features of the capacitor. An example is the use of capacitor arrays for the delivery of large amounts of power in a short time for the creation of extremely high-temperature, short-duration plasmas.

Ultimately, capacitors are limited either by the breakdown of the insulating material making up the space where the field is created and/or by the compression stress limit of this material, since the two "plates" with opposing potential are strongly attracted to each other. The breakdown and the strength issues may also be coupled, in that deformation leads to locally increased fields and thus a premature attainment of breakdown. Figure 17–8 illustrates the field lines and their influence on the capacitor material—in particular, the stress on the dielectric.

The inductor, or coil, has been proposed for storing large amounts of energy. This is made possible only because of the possibility of cooling the conductors in the coil to temperatures where resistance vanishes and the material becomes superconducting. The joule losses (I^2R) associated with ordinary conduction are unacceptable. Such coils are limited by the strength of the structure of the coil since each coil element experiences a $\mathbf{j} \times \mathbf{B}$ force that is outward and taken by the coil in tension. The possibility of coil failure generally forces the placement of the coil underground or into some secure environment where the accidental energy release is readily absorbed. Further, the superconductivity of the current carrier is affected by the B-field strength limiting its value.

The energy recovery efficiency of SMES (Superconducting Magnetic Energy Storage) is expected to be high ($>90\%$) if and when such systems are developed for commercial application (refs. 17–4, 17–5, and 17–6). However, the principal drawback is the cost that is primarily associated with the maintainance of the coolant (liquid helium near 2 K), the associated refrigeration (to counter heat leaks and joule heating of input and output connections), and thermal protection required.

17.4. AUTOMOTIVE ENGINE INTEGRATION WITH STORAGE

Of all major energy users, the automobile is probably in the greatest need of a cost-effective system for storing energy. The reason is that it has a duty cycle that involves severe and repetitive demands for power, and it is these demands that account from the level of fuel utilization performance as well as for the greatest environmental impact. Much work has been done to improve the automobile in these regards, and much remains to be done. For example, there is the thought that if automobiles could be driven by battery, then the pollution impact by the car will be transferred to the electric utility where the emissions control can presumably be easier or at least located in a region where the impact may be less severe. The automobile as a system is well developed and the switch to an automobile with a different energy source must, at the very least, have

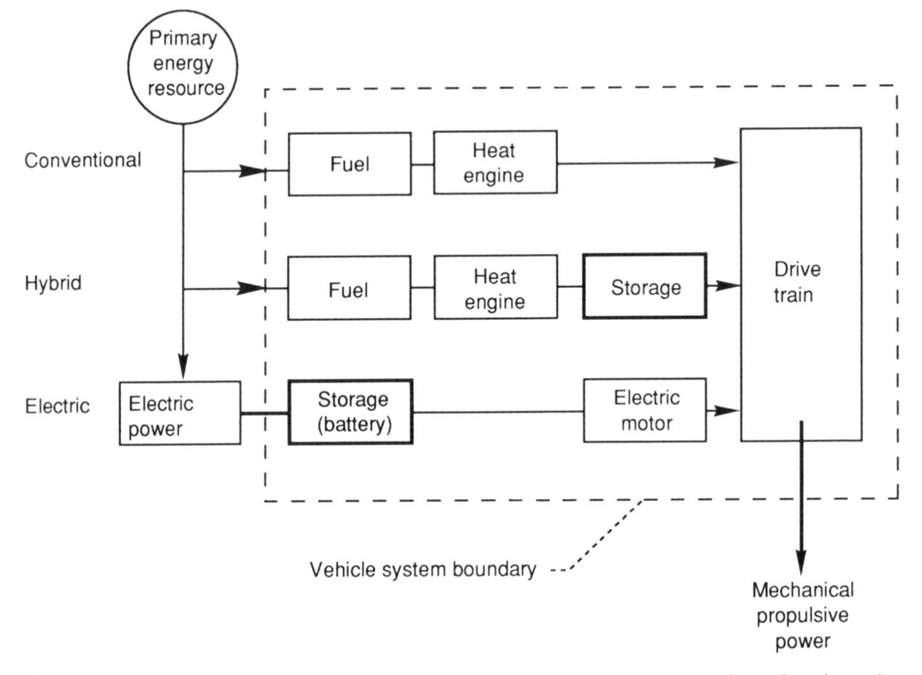

FIG. 17–9. Integration of energy storage with an automotive engine showing the conventional, hybrid, and electric vehicle propulsion methods.

performance characteristics consistent with existing vehicles (ref. 17–7): Such vehicles will necessarily share the road with gasoline powered vehicles. Batteries have shortcomings in energy density which limits range (Section 3.4.2) and require a relatively time-demanding or awkward recharge procedure. Their use in practice will be driven by the positive environmental impact and because performance is adequate for specific driving missions. Improvement in battery performance will accelerate the utilization of battery-powered vehicles.

There is another approach to the design of vehicles that is being considered. That is to recognize that the engine emissions can be reduced significantly if the primary power engine produces the *average* power needed. A storage system may be made to provide the peak need capacity with replenishment provided by a constant power heat engine and by regenerative braking when possible. Such hybrid engines may take advantage of some of the means of storing energy discussed in this chapter. The amount of stored energy is then limited to a small fraction of the mission energy. This approach allows contemplation of cycles such as Brayton and Stirling, with their ability to provide steady power output with lower-cost fuels than gasoline. Figure 17–9 illustrates the ways in which energy is stored in present and future automotive systems. Whether new approaches are adopted in the future remains to be seen. Certainly with limited fossil resources and a need for reduced environmental impact, the possibility that such engine-storage systems will find their way into the automobile cannot be excluded from consideration.

PROBLEMS

1. A coil carries a current J. Show that the stress in the coil material is proportional to $\mu_0 j^2$, where j is the conductor current density. Use the definition of the magnetic field

$$\mathbf{B}_2 = \frac{\mu_0}{4\pi} J_1 \int_1 \frac{(\mathbf{dl}_1 \times \mathbf{r}_{12})}{r_{12}^3} = -\frac{\mu_0}{4\pi} J_1 \int_1 \mathbf{dl}_1 \times \nabla \frac{1}{r}$$

and the force it causes a current-carrying region to experience

$$\mathbf{F}_2 = \int_S \mathbf{j}_2 \cdot \mathbf{dS}_2 \int_2 \mathbf{dl}_2 \times \mathbf{B}_2 = \int_2 1 \int_S (\mathbf{dl}_2 \cdot \mathbf{dS}_2)\mathbf{j}_2 \times \mathbf{B}_2 = \int_{\mathrm{Vol}} (\mathbf{j} \times \mathbf{B})\, dV$$

to arrive at an expression for the force and distribute that force over the conductor area. Note that a simple integration to find the magnetic field strength, **B**, is singular, so that care is evidently necessary. Consult a text on classical electricity and magnetism to clarify the notation inherent in the equations given.

2. In the pressurized air tank, determine the ratio of energy stored in the material walls of the vessel to energy stored in the compressed gas.
3. Consider the compressed air energy storage system that is used by filling from atmospheric pressure to 20 atm, and then complete emptying in a storage cycle. This problem is to examine the roles of various component characteristics on the overall storage performance. This problem requires computer implementation for solution of some parts of this problem.

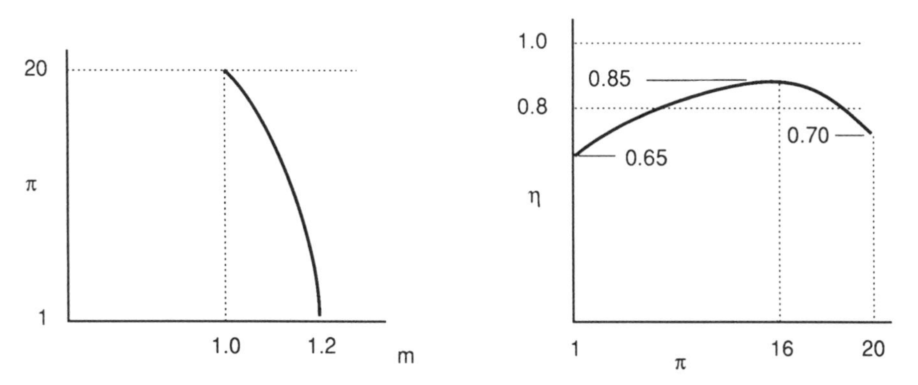

a. *Compression*

By examining the governing equation for the system, find the combination of physical parameters which governs the filling time scale. Find the number of such time scales, that is, the nondimensional time required to fill the system, the energy absorbed, and the temperature of the system at the end under the condition of constant speed compressor with:

1. Constant speed line is vertical on compressor map; efficiency is unity
2. Constant speed line varies as shown (parabolic shape in map on the left, with vertical asymptote at $\pi = 1$, $m = 1.2$), efficiency is unity
3. Constant speed line is as in part 2, except that efficiency is uniformly 0.80
4. Same as part 3, except that efficiency is as shown on the right figure

b. *Expansion*
Write the equation for the expansion process, and determine the time required in terms of turbine and vessel characteristics. For the following sets of circumstances determine the time required to empty the vessel and the energy recovered:

1. Choked, reversible turbine all the way to atmospheric pressure
2. Reversible turbine choked only to critical pressure, then unchoked
3. Uniform turbine efficiency = 0.90, choked then unchoked
4. Heat addition to $T/T_0 = 4$, with other conditions as in part 3

c. *Comparison*
Determine the energy recovery efficiencies for comparable cases in *a* and *b*.

4. Equation 17–15 gives the energy per unit mass of a system consisting of a gas and its container. If the container material has a stress capability of σ_0 at $T = T_0$ which decreases as T increases, find T for the maximum energy density. Modify this problem by noting that not all the energy is recoverable because the gas can be expanded to a pressure no lower than, say, 1 atm.
5. Set up the equations from which one could determine the optimum storage temperature of a solar collection and storage system to operate a conversion engine that realizes a fraction ϕ of the Carnot efficiency.

BIBLIOGRAPHY

Mehta, B. R., "Compressed Energy Storage," *EPRI Journal*, **11**(1), April/May 1986.

Chiu, H. H., "Mechanical Energy Storage: Compressed Air and Underground Pumped Hydro," *AIAA J of Energy*, **7**(6): 724, Nov/Dec 1983.

Vosbergh, K. G., "Compressed Air Energy Storage," *AIAA J of Energy*, **2**(2): 106, March/April 1978.

REFERENCES

17–1. Decher, R., Davis, R. N., "Performance Characteristics of Compressed Air Energy Storage Systems," *AIAA Journal of Energy*, **2**(3), May-June, 1978.

17–2. Nahkamkin, M., Swenson, E., Schainker, R., Pollak, R., "Compressed Air Energy Storage: Survey of Advanced CAES Developments," ASME 91-JPGC-NE-26, 1991.
17–3. Hertzberg, A., private communication.
17–4. "Superconducting Energy Storage," Energy Engineering, **83**(1): 57, 1986.
17–5. Kalhammer, F., "Superconducting Magnetic Energy Storage," *EPRI Journal*, **9**(4): 59, May 1984.
17–6. Boom, R. W., "Superconducting Energy Storage," *Advances in Cryogenic Engineering*, **19**: 117, 1974.
17–7. Ayres, R. U., McKenna, R. P., *Alternatives to the Internal Combustion Engine*, Johns Hopkins University Press, 1972.

18

ENVIRONMENTAL ASPECTS OF ENERGY CONVERSION

Exploitation of natural resources is a characteristic of living creatures. Man's natural hunger for power must leave its mark on the environment. The work delivered from energy resources can be useful for demanded tasks, or it can be destructive through unintended release. The question is Will the effect of harnessing energy for its desired purpose be detrimental to the environment and, if so, what must be done to mitigate the impact?

This chapter discusses man's present energy conversion activities and their unintended consequences: chemical emissions, thermal "waste" and radioactive waste. The concerns can be classified in terms of their impact on the short-term health of humans and their impact on the long-range health of the planet as a base for supporting life for humans and a diverse ecological structure. Immediate health effects are readily identifiable, but long-term health effects are more difficult to recognize, in part because these affect living creatures in varying ways. Corrective action to address environmental problems is therefore political and difficult. Forecasting the ecological heath of the planet is imprecise. The Earth as a living system is sufficiently complex as to defy modeling, now and in the near future. This is to say that predictions concerning the outcome of actions based on a model are always uncertain to some degree. Any uncertainty reduces the model's usefulness as a decision making instrument.

The subjects treated in this chapter are described more qualitatively that the matter in earlier chapters. The reasons are that complete coverage lies beyond the scope of this text. Nevertheless, the engineer concerned with energy conversion must be aware of the consequences of design and operation of the machines he deals with. Further, he should appreciate the efforts of specialists concerned with understanding the global climate and the impact of environmental change on Earth's living creatures.

An important aspect of environmental impact is that of the aesthetics of man and his machines on the plant. The low cost of power from fossil fuels has made man increasingly dependent on them, allowed significant population increases, and thus engendered a reduction of the planet's ecological diversity. The topic of the planet's evolution is a subject for serious students in other fields, and the engineer can play a role in guiding industrial society toward alternative ways to use energy and new ways to generate power.

18.1. NATURAL CYCLES

Earth's materials with their chemical characteristics are transported continually through the environment by various mechanisms. Water, for example, follows the well known hydrological cycle involving surface evaporation, rain, rivers and the attendant erosion, solution, etc. processes. To date and on a global scale, the changes made to this cycle by man's activities are relatively small. Locally, however, man many cause significant changes in the local climate, desertification, or soil salination, to mention some of the serious consequences of man's activities. Cycles similar to the hydrological cycle exist and are important from the viewpoint of energy conversion for carbon and for gases such as oxygen, nitrogen, hydrogen, chlorine, etc. Yet other cycles exist for the minerals on the planetary globe which are associated with very long (geological) time scales. The study of ecology and of the environment is the study of sources, sinks and reservoirs of the chemical constituents to which man is adding a new burden by his activities. The most interesting dimensions of these studies are the determination of the mechanisms involved in the release and removal of substances from their present location in nature and the time scale on which the mechanisms have a role to play in the evolution of the material. One hopes that the natural recycling processes for molecules relevant to fossil fuel use (water, carbon dioxide, ozone, the man-made effluents, etc.) are so dominant that man's activities will not alter these processes significantly in the relatively short time that fossil resources are available.

Use of a criterion based on changing or not changing the environment to determine whether a certain energy conversion process should or should not be exercised is very severe. First, natural changes take place even in the absence of man's activities, volcanism is one example. Second, environmental impact is unavoidable, even people involved in agricultural make an impact on the environment by the choice of the plants or animals they use.

18.2. SECURING OF RESOURCES

18.2.1. *Mining and farming*

The process of acquiring fossil energy resources from the Earth changes the environment. The impact includes the disruption of the natural equilibrium between chemical substances present in soils where mining takes place. Normally, the water passage rate through rock and soil is so slow that soluble chemicals are leached out very slowly. Structural soil disruption such as excavation exposes very large surface areas to water which increases the leaching rate to the point where comparatively large amounts of soluble matter are carried to places where it can be destructive to life, as in wilderness and where it can be useful to man, as in agriculture. Increasingly in many countries, mining permits require maintenance of the soil to a state where runoff water quality is acceptable, especially in surface or "strip" mining. Most mines are not located near the point of use, and thus coal needs to be transported to power or conversion plants. There is an environmental impact associated with the use of fuels to make that transport possible. Most fuel conversion processes, such as refining of petroleum, are not 100% efficient so that the conversion process leads to material loss and the generation of waste heat. Lastly, the end user of the energy contributes to the total heat load on the environment by the ultimate degradation of work done to heat.

To minimize environmental impact, the notion of using resources that are continuously produced from solar energy is appealing. One such modest effort is in the use of energy in plants (biomass) which grew in the recent, rather than fossil, past. The utilization of biomass energy resembles agriculture, and the impact on the environment is similar: plant species specialization at the expense of diversity, fertilizer in the atmosphere and surface water, and emissions from agricultural and transportation engines. An important advantage is that a biomass farm is a sink for CO_2. Collection efficiency for the biomass fuel relative to the solar input over a growing cycle is, unfortunately, only a few percent.

18.2.2. *Hydrostorage*

The storage of water—river water, in particular—brings with it an environmental impact. The impact is, first of all, ecological in that river stream beds are flooded and the riparian ecosystems displaced, if not destroyed. In arid climates, the storage of water leads to an increased evaporation rate, and the downstream users must deal with water that is more heavily salt laden. This affects irrigation agriculture adversely, particularly in the long term when salts have built up in the irrigated soil over time. The evaporated water loss is, itself, a penalty. Lastly, the sand normally carried downstream by rivers will settle behind storage dams. All the while the dam interrupts the sand flow, it deprives the ocean shores of sand for beach renewal. Ultimately, the fill decreases the storage capacity of the hydrodam reservoir to reduced utility unless dredging is carried out.

A biological impact of dams is the interruption of migratory fish movement. A mitigation of the blockage is the use of fish ladders. The hydroturbines should be designed to exclude fish, and the tail water should be designed to avoid the formation of deep underwater jets where the pressure is large enough to dissolve atmospheric nitrogen in the water. Water with a high nitrogen content can result in a phenomenon for fish that is similar to the bends for divers: potentially fatal nitrogen bubble formation in the bloodstream.

18.3. WASTE HEAT

Heat engines must, according to the Second Law of Thermodynamics, produce waste heat. This is low-temperature heat that is absorbed by the air or water of the environment. For a heat engine with thermal efficiency, η_{th}, the amount of waste heat developed per unit work is

$$\frac{Q_{rej}}{W} = \frac{1}{\eta_{th}} - 1 \qquad (18\text{–}1)$$

and the influence coefficient for heat rejected per unit work produced with efficiency is

$$\frac{d \ln\left(\dfrac{Q_{\text{rej}}}{W}\right)}{d \ln \eta} = -\frac{1}{1 - \eta_{\text{th}}} = -1.6 \qquad \text{for } \eta \sim 0.4 \qquad (18\text{–}2)$$

which means that every percentage point improvement in efficiency (when η_{th} is near 40%) brings a thermal waste reduction of 1.6%. This influence coefficient changes to -2.2 when η_{th} is 55%.

Fossil fuel burning transport engines deposit this waste heat in the air. The heat is almost always of minor consequence because the mixing rate with ambient air is rapid which reduces the thermal trace to negligible levels in a very short time. Compared to the natural heat budget, the contribution through the combustion of fossil fuels is small, with an impact limited primarily to local "microclimatic" changes.

Large stationary fossil fuel or nuclear power plants can also use the air as a sink for waste heat through cooling towers. These coolers may use the evaporation of water to transfer the heat in *wet* cooling towers, or they may be *dry*, in which case they are a form of counterflow or crossflow heat exchanger. The wet tower is less costly because the heat transfer area consists of the surface of all the droplets. Wet cooling towers need access to "make-up" water to replace that which is lost by evaporation. The vapor generated may alter the local microclimate by creating fog, but on the whole, the amount of water involved is such a small fraction of the natural flux that its damage impact is very small. Dry cooling towers must be used where make-up water is not available. Such a cooler consists of a water-to-air heat exchanger similar to an automotive "radiator". The heat transfer coefficient between a gas (air) and water is relatively small so that the ΔT driving the heat transfer must be relatively large for the dry process. Dry cooling generally involves a thermal efficiency penalty for the thermodynamic cycle because it requires the cycle to operate at an elevated minimum temperature. Further, the system with a dry cooling process is costlier. In the days of Rankine cycle railway locomotives, operation in arid climates required condensation of the steam for reuse. This necessitated large condensers on the locomotive structure, usually on the coal tender.

Dry and wet cooling processes may also be employed in cooling towers that differ in the methods used to bring in the ambient cooling air. A natural draft cooling tower utilizes the density difference between ambient air and the warm air that has accepted the waste heat to drive an upward convection current. Towers employing this natural draft circulation are typically rather large, structurally elegant, and characteristically shaped concrete structures (Fig. 18–1). The large capital investment and space required by this structure may be reduced when use is made of electrically driven fans to draw the flow past the heat transfer surface (Fig. 18–2). This alternative requires power and thereby reduces net efficiency, but the visual impact is smaller than that of the natural

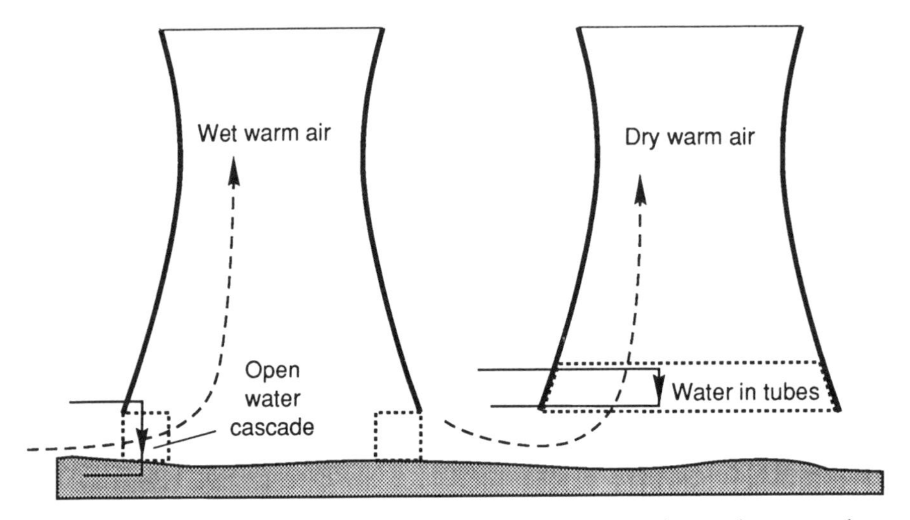

Fig. 18–1. Natural draft cooling tower for waster heat transfer to the atmosphere. Crossflow wet system on left, dry counterflow on right.

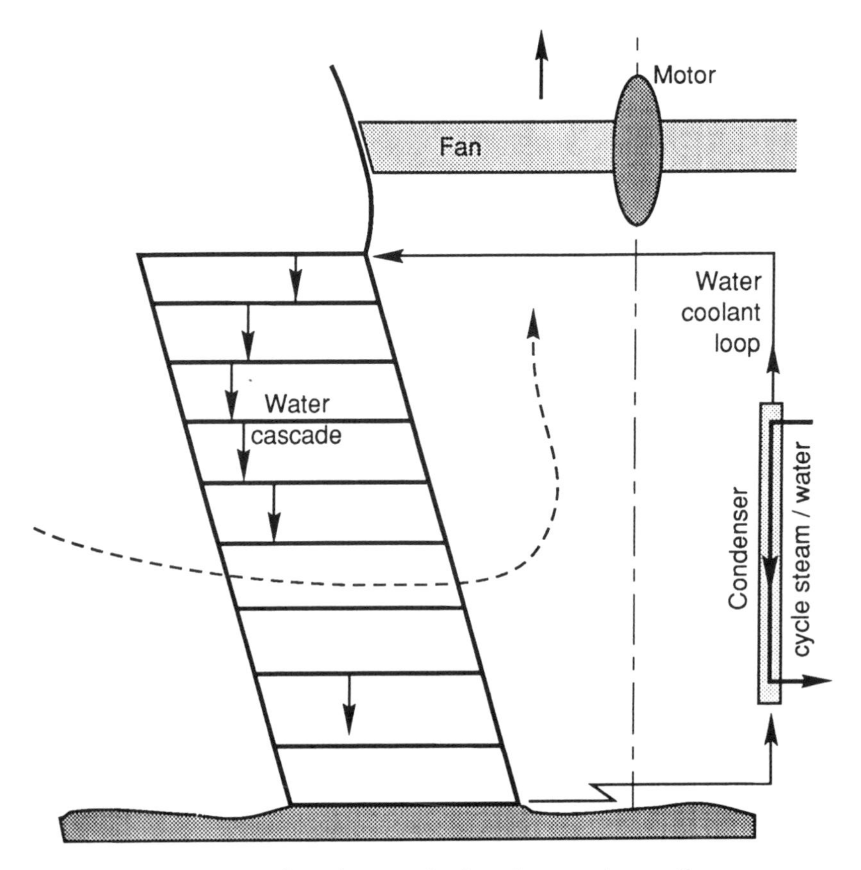

Fig. 18–2. Cross flow element of a forced convection cooling tower.

draft system. The economics of the plant and the availability of water dictate the optimum choice.

The effectiveness of heat transfer between liquids through a material wall makes it attractive to use river, lake, or ocean water as a coolant. This water is taken in by a pump, circulated through a heat exchanger, and returned to the source at a different location. Aquatic animals and the machinery must be protected by appropriate means of exclusion. Further, care must be exercised to ensure that aquatic life is not exposed to the relatively hot cooling water. Hence steps must be taken to exclude the animals from the region where mixing between the heated water and the stream is incomplete and the final temperature of the water body used must be lower than an ecologically acceptable level.

It should be noted that no central station power plant uses the coolant water used in the cycle. The high cycle pressure requires the water to be very pure as well as free of oxygen. A water-to-water heat exchanger is always employed.

18.4. CHEMICAL EMISSIONS

This section is divided into subsections that describe, first, the emission of stable molecules from combustion sources and the role these molecules may play in the environment, and second, the minor or trace emissions that may be important locally.

18.4.1. *Water and Carbon Dioxide*

The energy-releasing products of fossil fuel combustion are CO_2 and water. The water contributed to natural flows is generally very small and chemically benign so that its emissions are rarely an issue insofar as the environment is concerned. In some open cycle engines such as the steam locomotive using the Rankine cycle, the visual impact can be quite large, and would probably not be tolerated today in a newly introduced machines.

Carbon dioxide (CO_2) emissions make a contribution to a natural cycle of carbon in the Earth and its atmosphere. In the early nineteenth century, the CO_2 content of the atmosphere was measured at a level of about 285 ppm (parts per million: mole fraction $\times$ 10^6). This value has grown to a significantly larger 343 ppm in 1984 (refs. 18–1, 18–2). In the middle of the twentieth century, the increase in the rate of atmospheric CO_2 was of the order of 1 ppm per year. To date, the mean increase in atmospheric CO_2 has been of the order of 25% over the natural (pre-late nineteenth century) level, and the rate of change is increasing with time. The recorded observations (Fig. 18–3 from ref. 18–3) reflect not only the year-to-year variation in atmospheric CO_2 content but also the seasonal variation associated with plant growth in the hemisphere where the measurements are taken. The annual cyclic variation superimposed on the mean trend (noted schematically in Fig. 18–3) is about 6 ppm change during the typical year.

A very serious question currently under study by atmospheric scientists

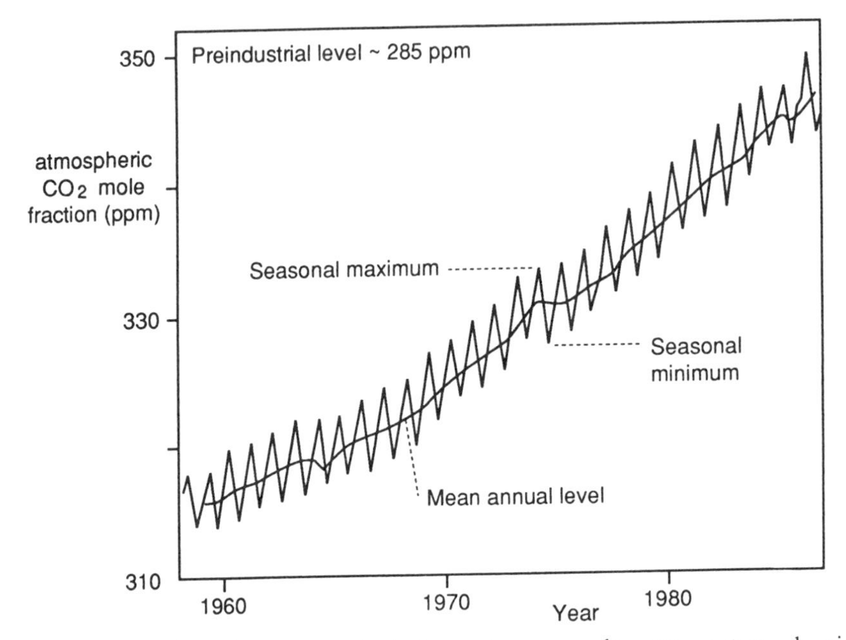

FIG. **18–3.** Variation of atmospheric CO_2 content over the past century showing seasonal variations in one (northern) hemisphere. High altitude observations on Mauna Loa, Hawaii.

deals with the impact of the long-term increase in atmospheric CO_2 on the thermal balance between Earth and sun (refs. 18–1 and 18–2, among many others). The investigation centers on determination of the changes in the mean surface temperature. This question deserves discussion and study here because it leads to a follow-on question: Given that CO_2 does have an effect on temperature, what is the related effect on the climate, wind, and weather patterns, ocean level, and ultimately its impact on various regions' ability to support agriculture? This question will continue to occupy atmospheric scientists for decades to come, unless other direction-changing events take place or can be predicted. The consequences present mankind with extraordinary questions regarding the survival of humankind on Earth, and the means humanity will employ to ensure just consideration of all living organisms. An adjunct to the CO_2 input problem is the removal rate by plants. The disruption of a production-uptake balance lies at the heart of the concern for the tropical forests.

18.4.2. *The Role of the Atmosphere in the Sun–Earth Thermal Balance*

The atmosphere of the Earth plays a number of important roles in supporting life. These include the maintenance of oxygen and water for direct life support, softening of day/night temperature fluctuations, control of the mean temperature on the Earth's surface, and the shielding of life from ultraviolet (UV) solar radiation. The role played by emissions associated with energy conversion

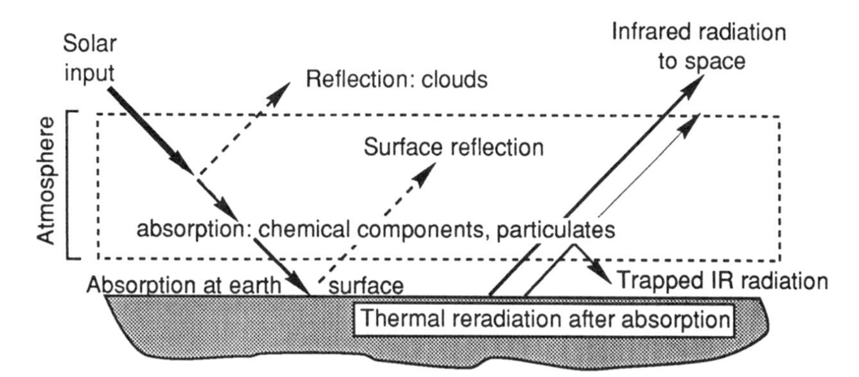

FIG. 18–4. Role of the atmosphere in controlling the thermal balance between Earth and sun.

are discussed below. Carbon dioxide is one of the critical constituents in the Earth's atmosphere regulating its ability to transmit radiative power.

The thermal budget of the Earth includes energy inputs from the following sources: radiant energy from the sun, nuclear decay heat from the Earth's interior, and energy dissipation from the Earth–moon dynamical system. The last contribution is manifest in the ocean tides and similar motion. Overall, the last two contributions are very small compared to the solar flux.

The solar input depends on materials injected into the Earth's upper atmosphere (dust and specific chemicals, e.g., sulfur and chlorine) through volcanoes etc. which can have a profound effect on the global climate. It is against this background of natural events that the importance of man's contribution must be judged. The incident solar flux is partially reflected and partially absorbed. The relative amounts of these two quantities is largely controlled by the reflectivity of the Earth as a whole (i.e. by its "albedo"), which is strongly influenced by the cloud cover. Figure 18–4 shows a simplified view of the effects of the atmosphere on the radiative fluxes to and from the Earth's surface.

Heat is rejected by the Earth by the mechanisms of thermal radiation. No other mechanism is available. This radiation leaves the Earth into cold space primarily from the Earth's *surface*. Consequently, the atmosphere serves as a window that allows solar radiation power at visible wavelengths to enter and simultaneously lets thermal radiation at lower temperatures (250–300 K) leave. The solar radiation spectrum is equivalent to blackbody radiation (see Section 18.4.3) at a temperature of 5800 K. The actual incident spectra in space and at the Earth's surface are shown in Fig. 18–5. Not coincidentally, the peak of visible radiation is near 500–600 nm (nanometers = 10^{-9} m) near the wavelength where eyes are most sensitive.

A similar plot is shown in Fig. 18–6 (from ref. 18–4) for the radiation escaping from the Earth. Shown are the blackbody spectra at 200 and 300 K as well as the emitted Earth radiation spectrum measured by a satellite in space. The Earth's thermal peak is in the infrared spectral range.

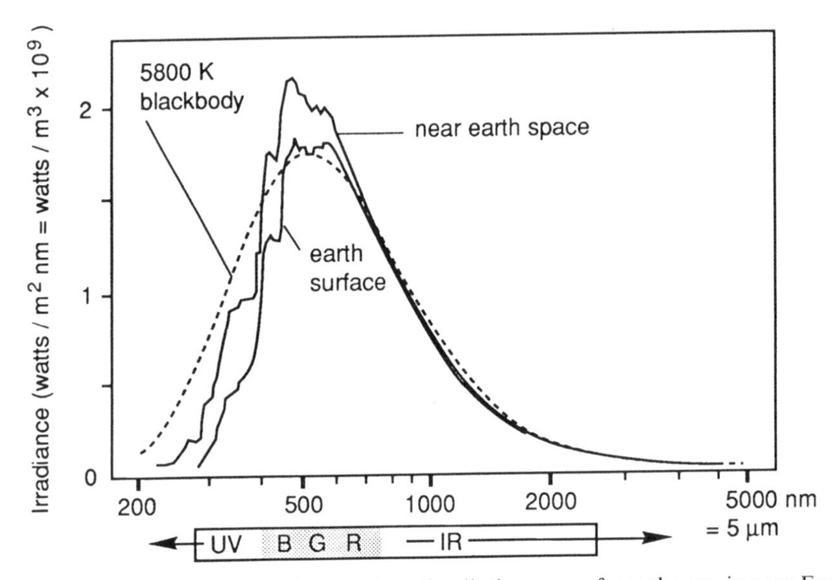

FIG. 18–5. Spectral (wavelength) variation of radiative power from the sun in near Earth space, on the ground, and from a blackbody at 5800 K. Abscissa is the wavelength in nanometers ($=10^{-9}$ m), although the use of micrometers (microns) is common in the infrared region. B, G, and R refer to blue, green, and red colors of the visible spectrum.

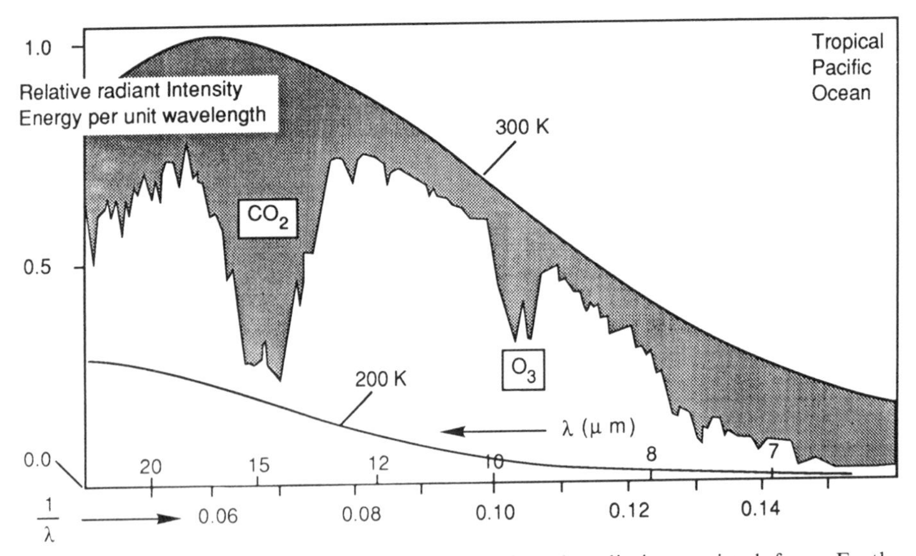

FIG. 18–6. Earth satellite measurements of infrared radiation emitted from Earth. (After ref. 18–4.)

The atmosphere's gaseous components absorb radiation at specific wavelengths. The absorption is by the processes of photo-excitation or photodissociation as exemplified by UV photon-oxygen reactions such as

$$O_2 + h\nu \rightarrow O_2^* \qquad \text{and} \qquad O_3 + h\nu \rightarrow O_2 + O \tag{18–3}$$

Although these particular reactions do not play a large role in the energy budget of the planet, they are important for the safety of living organisms. Of greater interest are the photon reactions that involve triatomic and heavier molecules that have rotational and vibrational degrees of freedom. Molecules with a number of atoms have inherently a greater number of ways in which they can absorb energy compared to diatomic molecules. The response of a molecule to an incident photon is to change its internal configuration. The rotational and vibrational configurations of a molecule differ in energy levels by amounts typically associated with infrared photons. Figure 18–6 shows the spectral regions of the atmosphere where ozone and CO_2 play important roles in the transmission of infrared radiation.

The effectiveness of the atmosphere in absorbing radiation is given by an absorption coefficient, k_ν, defined by eq. 18–9 (below). The absorption coefficient is wavelength dependent because radiation at various frequencies is absorbed according to specific reactions. In the spectral region where the absorption coefficient is zero, the atmosphere is transparent whereas a unit value of $k_\nu L$ (where L is the length scale associated with the atmosphere) describes it as opaque. Some of the reactions are reversible and serve to scatter (i.e., randomize) radiation such as the incoming solar flux. This scattering accounts for the Earth's blue sky. The atmospheric absorption variation is shown in Fig. 18–7. Clearly seen are "windows" allowing the visible light in and the Earth thermal radiation out. The model of the atmosphere serving as a window is somewhat simplistic but assists in visualizing the overall picture. In actuality, radiation is absorbed at various levels and reradiated isotropically (i.e. in all directions), some of it back out to space and some onto the surface.

The low-temperature, infrared spectral region or "window" is affected by the presence of air components CO_2, CH_4, CFCs (chlorofluoro carbons), nitrogen oxides (primarily N_2O), and O_3 (ref. 18–5). All but the CFCs are naturally present in the atmosphere. The amounts of these quantities are

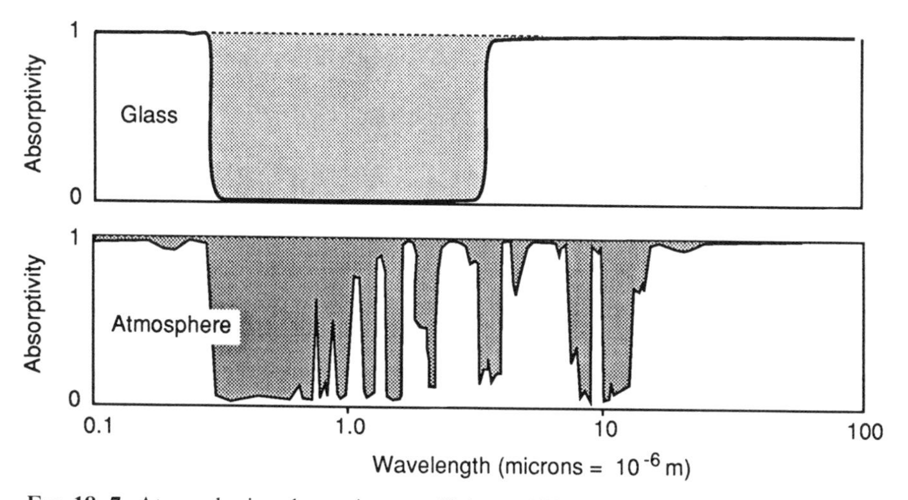

Fig. 18–7. Atmospheric absorption coefficient. Silica glass transmission (between approximately 0.2 and 2 microns) is also shown.

Table 18–1. Infrared radiation blocking atmosphere components and the amount by which they increased in the period 1975–85

Component	Percentage increase
Natural	
CO_2	4
CH_4	3
N_2O	8
CFCs	
$CFCl_3$ (R-11)	100
CF_2Cl_2 (R-12)	100
CH_3CCl_3 (methyl chloroform)	150

Source: ref. 18–5.

increasing with time as a result of man's activities. In the ten-year period ending in 1985 the CFCs and the natural components have increased by the amounts given in Table 18–1. To varying degrees these components of the atmosphere affect the transmission of radiant energy in the infrared spectral region (from the ground to space) by absorbing it and reradiating that energy both upward and downward, thereby trapping some of the energy that would otherwise escape to space. Figure 18–6 shows that CO_2 and O_3 block a significant fraction of infrared spectrum. The transmissivity of the visible window is not significantly affected by chemical components. Rather, dust particles and water droplets have an influence on the solar input by changing the reflectivity of the atmosphere and its ability to scatter light.

The low-temperature transmission block must necessarily result in a warming of the region near the Earth's surface, much like a greenhouse uses glass to admit visible light while blocking partially thermal radiation outward. The characteristics of glass are also noted in Fig. 18–7. Greater detail regarding the characteristics of various glass types may be found in ref. 4–1. The possible rise in global temperature resulting from the alteration of the atmosphere is popularly referred to as the "greenhouse" effect, and the gases involved are greenhouse gases.

A rise in global temperature can be determined as the result of applying the First Law of Thermodynamics to the global surface as a system. *If* the net input remains constant, then decreased atmospheric transparency in the infrared special region must result in an increased temperature because the amount of radiant power in a smaller frequency interval can be constant only when the temperature increases. This conclusion is valid if the stipulation of constant input is true. In reality, the atmosphere is a dynamic system whose properties respond to changes in the thermal balance: It may well be that the change in composition and transmission characteristics of the Earth modify the albedo through an alteration of the climate which may result in a greater cloud cover. If this is the case, the projected mean temperature for the Earth is lowered. Thus one has the interesting situation of understanding that activities are

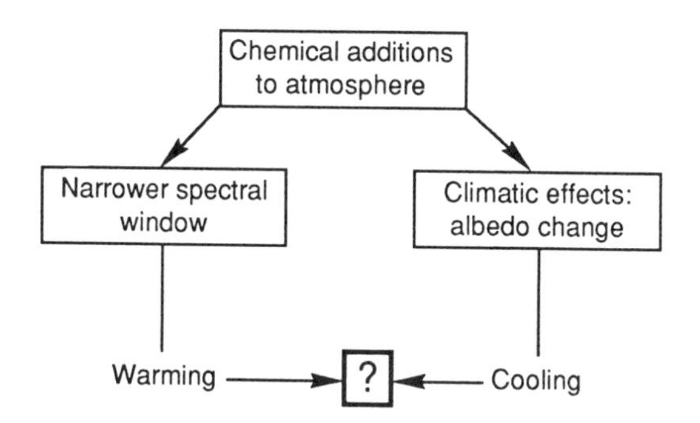

FIG. **18–8.** The global warming question.

causing changes that have competitive consequences, among themselves, and with natural phenomena. One is left with not knowing what corrective course of action, if any is available, to take (Fig. 18–8).

The consequences of inaction are that climatic alterations, for which mankind will be unprepared, will occur. These will have a profound impact on the planet, at the very least, on its agriculture. A predictive model of the global environment may well be of assistance in planning for future action. While such a model is far from available, a result of better understanding of the atmosphere has improved short-term weather forecasting. The question does remain, however, Should man, in addition to improving efficiency, switch to renewable resources or nonfossil fuel resources before dislocations become serious? If so, implementing these will have profound economic and political consequences for the management of man's affairs.

18.4.3. *Elementary Thermal Radiation Physics*

The objective of this section is to strengthen the above qualitative discussion of heat radiation with quantitative arguments. The Earth orbits the sun at a distance of one astronomical unit (A.U., see Appendix B). The sun's radiation is approximately equivalent to that from a blackbody at 5800 K. The blackbody is an ideal radiator of energy where the radiated power depends on the radiation wavelength and emitter temperature. The (wavelength) spectrum of the ideal blackbody may be described in terms of packets or quanta of energy being emitted from a cavity (the blackbody) wherein a radiation field is established by the equilibrium between the walls and space. The energy density of the space filled with packets of energy (photons) moving about randomly at the speed of light depends on the wavelength of the wave packet or photon. For the purposes of this description it is convenient to use wave frequency (ν) rather than wavelength (λ). These quantities are related by

$$\nu = \frac{c}{\lambda} \tag{18–4}$$

where c is the speed of light in vacuum. The energy distribution in "frequency space" of the photons in the cavity, B_ν, as well as those that are emitted is given by:

$$B_\nu(T) = \frac{8\pi h\nu^3}{c^3} \frac{1}{\exp\left(\frac{h\nu}{kT}\right) - 1} \tag{18–5}$$

The units of B_ν are energy/volume/frequency so that an integration over all frequencies gives energy density. This equation follows from statistical thermodynamics applied to the photon "gas." Its origin is not developed here, and the reader may wish to consult an appropriate text (e.g., ref. 18–6), to review the origin of this relation. The photon "gas" flux leaves the cavity and carries with it power which equals the energy density times $\frac{1}{4}$ of the random speed, which is the speed of light for photons. Thus the radiant flux per unit frequency per unit area (watt sec/m^2) is

$$\frac{d\dot{w}}{d\nu} = \left(\frac{c}{4}\right) B_\nu(T) = \frac{2\pi}{(hc)^2} \frac{(h\nu)^3}{\exp\left(\frac{h\nu}{kT}\right) - 1} = \frac{2\pi(kT)^3}{(hc)^2} \frac{y^3}{\exp(y) - 1} \tag{18–6}$$

The variation of the blackbody function is plotted for various temperatures in Fig. 18–9 (on a logarithmic scale). The units of the ordinate are scaled by the group of fundamental constants in eq. 18–6. The temperature is also nondimensionalized and shows that the influence of a change in temperature by half an order of magnitude, that is, a factor of $\sqrt{10} = 3.16$. The abscissa is the reduced frequency, y, shorthand for $(h\nu/kT)$ which gives for a unit value of y, $\nu = 2 \times 10^{14}$ sec^{-1} at 1000 K. The last form of eq. 18–6 shows the dependence

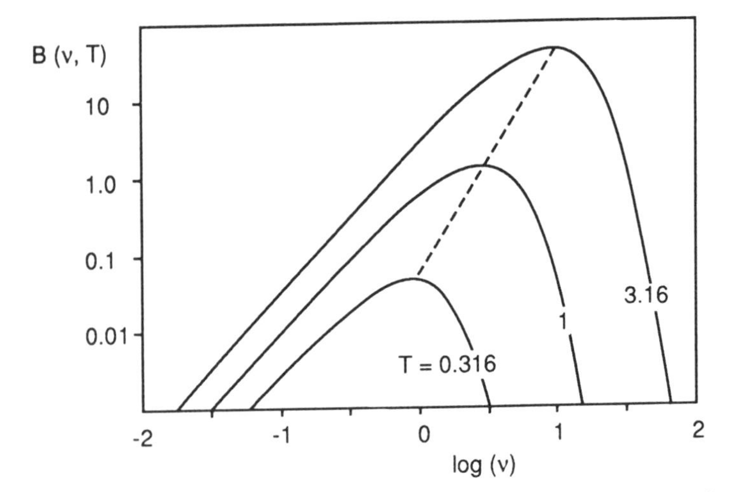

Fig. 18–9. Blackbody spectra at varying temperatures. Plot of the nondimensional form of eq. 18–6.

on reduced frequency and temperature more clearly. First, at higher temperature, more power is radiated, and this is confirmed by evaluation of an integral to be examined below. Second, at any frequency interval, higher temperature implies higher radiant flux. Third, there is a particular frequency of wavelength, where the peak flux is observed, and this peak depends on the temperature. By differentiating the blackbody function, written in terms of wavelength [i.e., $B_\lambda(T)$ which is shown in Figs. 18–5 or 18–6], one can show that the peak flux wavelength is given by (the Wien displacement law):

$$\frac{\partial B_\lambda(T)}{\partial \lambda} = 0 \rightarrow \lambda_{\max \mathrm{B}} T = 2898\ \mu\mathrm{m\ K} \tag{18–7}$$

The linear relation between (log) frequency for maximum blackbody radiative flux per unit frequency and log B_ν in Fig. 18–9 is shown in the line connecting the peaks. From the figure, it follows that at higher temperature, the peak shifts toward higher frequency, (i.e., shorter wavelengths). For example, a 5800 K blackbody has a peak wavelength of 0.5 μm (or 500 nm) which is in the visible region, whereas at 280 K the blackbody peak is near 10 μm (or 10,000 nm) which lies in the infrared part of the radiation spectrum. The frequencies that correspond to these values are 6×10^{14} and $3 \times 10^{13}\ \mathrm{sec}^{-1}$ (or Hz). The visible spectrum spans 4.6×10^{14} (red) to $7.3 \times 10^{14}\ \mathrm{sec}^{-1}$ (violet).

The total radiated power into 2π steradians of space (i.e., into a half sphere), per unit of emitter area in watts/m^2 is the summation of all frequency contributions, or the integral

$$\dot{w} = \int_0^\infty (\tfrac{1}{4}c) B_\nu(T)\, d\nu = \frac{2\pi(kT)^4}{h^3c^2} \int_0^\infty \frac{y^3}{\exp(y) - 1}\, dy = \sigma T^4 \tag{18–8}$$

With the value of the integral, the collection of fundamental constants is the Stefan-Boltzmann constant, σ (see App. B). The blackbody distribution function for a body at 5800 K is shown in Fig. 18–5 together with that of the sun which, in reality, is not a blackbody because absorption by its own atmosphere removes some of the spectral energy content. The effective blackbody temperature is determined by equating the thermal powers shown in the figure; that is, the area under the curves are matched.

The Earth's atmosphere plays an important role in controlling the energy incident on the surface by absorption. Roughly speaking, the absorption of sunlight by the atmosphere is proportional to the air mass traversed by the rays, although some frequencies are absorbed at rates that have different absorption coefficients. Such a coefficient, k_ν, is defined by

$$\frac{d\dot{w}_\nu}{\dot{w}_\nu} = -k_\nu\, dl \tag{18–9}$$

where $\dot{w}_\nu$ is the power of a beam with frequency ν and dl is the path length traversed by the beam. The reader should consult the specialist literature for the integrated and time-varying effect of absorption by the sun as it traverses the sky from sunrise to sunset in an Earth-based reference frame. This is most directly manifest, but only in part, in the daily variation of local temperature.

Earth Surface Temperature

The sun's power output is the same through every spherical shell of radius r around the sun. At $r = 1$ A.U., the Earth's disk receives about 1.4 kW/m^2. For other planets at a distance r_{planet}, the solar power varies as $(r_{\text{planet}})^{-2}$. The Earth's surface area is approximately 4 disks in area and therefore radiates on average 350 W/m^2 if there is no reflection of incident power. The blackbody temperature that corresponds to this flux is given by eq. 18–8 or,

$$T_{\text{BB}} = \sqrt[4]{\frac{\dot{w}}{\sigma}} \approx 280 \text{ K} \tag{18–10}$$

which is surprisingly close to ambient temperature in the mid-latitudes (see problem 18.3). Because the angle between the Earth's surface and the direction of the solar rays varies with latitude, the surface temperature will vary significantly from the viewpoint of life, but from a global point of view, the atmosphere is effective in distributing the energy to all points on the globe. This may be seen by noting that a polar temperature of -30°C and a tropical temperature of $+30$°C imply a variation of "merely" $\pm 10\%$ around the mean temperature. With an actual estimate of the reflected solar power of 0.35, the average thermal flux to space is about 227 W/m^2. This should be compared to the contribution of other fluxes: thermal flux from the Earth's interior (0.06), man's contribution (0.01–0.02), and tidal dissipation (0.005, all in W/m^2).

Incorporating the notion that the Earth's albedo reduces the net solar input to the Earth, one may conclude that the atmosphere plays a significant role in elevating the Earth's surface temperature to levels conducive to life. One need only look at conditions on the nearby planets Mars and Venus (refs. 18–7 and 18–8), which have similar-order solar inputs (0.45 and 2, respectively, since their mean orbital radii are about 1.5 and 0.72 A.U.), but vastly different atmospheric conditions on their surfaces.

18.4.4. Other Air Components

Global Ozone

The ultraviolet (UV) shielding by ozone (O_3, see eq. 18–3) is another aspect of the atmosphere that is affected by the addition of chemicals. Introduction of such chemicals is related to energy consumption. The list of pollutants includes chlorofluoro carbons (CFCs) which are methane family gases where Cl and F atoms are bound to carbon in place of hydrogen atoms. The concern is introduction of Cl into the atmosphere where it forms ClO, one of numerous O_3 reducing compounds. The mechanism whereby ozone is reduced is as follows. Consider the chlorofluoromethane that is photodissociated in the atmosphere [reaction (1) in eq. 18–11]. The Cl atom is very reactive and combines with an ozone molecule to produce chlorine monoxide. The ClO molecule produced reacts with atomic oxygen which is also present in the low-pressure stratosphere to reproduce the Cl atom.

$$\left.\begin{aligned} CCl_2F_2 + h\nu &\rightarrow Cl + CClF_2 \\ Cl + O_3 &\rightarrow ClO + O_2 \\ ClO + O &\rightarrow Cl + O_2 \end{aligned}\right\} \qquad (18\text{–}11)$$

The Cl atom is therefore a catalyst for the breakdown of O_3 because it is able to repeat the steps above many times before a removal reaction takes it out of circulation.

The CFCs are used as working fluids for heat pumps, air conditioning, and refrigeration cycles, and these become a problem when the machinery leaks to the environment during or after its useful life. The introduction of Cl through CFCs competes with other introductions of chlorine, salt (NaCl) from the ocean, volcanoes, industrial processes, etc. The use of CFCs as working fluids and as propellants is currently being reduced by legislation. This forces substitution by other gases that are much less effective in ozone reduction catalysis.

The consequences of ozone depletion are serious for humans in terms of the damage the UV radiation can cause. The immediate effect is an increase in skin cancer, requiring greater protection of skin from sunlight. Whether ozone depletion is an inportant issue has to be judged in the light of man's willingness to avoid Earth locales where the UV threat is greatest, in the tropics and at high altitude.

Nitrogen Oxides

To a somewhat lesser degree, the commercial electric power and transport industries are concerned about the production of nitrogen oxides (NO and NO_2, collectively referred to as NO_x) in the combustion process. The mechanism for oxidizing atmospheric nitrogen is described in Chapter 5. At high altitudes these oxides may have a role to play in the depletion of zone (ref. 18–9) whereas at ground level they are important as an element of photochemical smog (ref. 18–10 and Section 18.4.5). Section 5.13 is a description of mechanics of NO production in heated air or through the combustion process. In practice, a number of means may be used to reduce NO_x emissions from combustion sources. These include fuel-rich combustion followed by fuel-lean oxidation of CO and unburned hydrocarbons (UHC), which may themselves play a role in the creation of photochemical smog. This is the primary means of controlling NO_x from automotive engines. In gas turbines used for stationary power, water may be used to reduce flame temperature. Typically water/fuel mass flow rate ratios vary between 0.2 and 1, although for the higher values, the CO emissions become important and efficiency is reduced significantly (ref. 18–11). Other methods include ammonia injection and steam injection as well

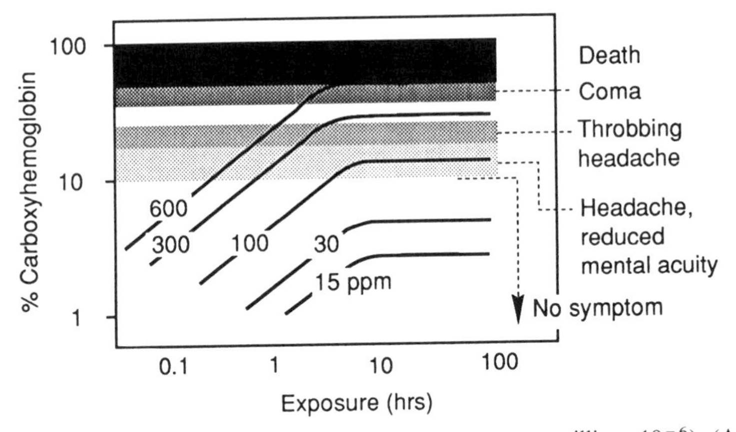

Fig. 18–10. Toxicity of carbon monoxide (ppm = parts per million, 10^{-6}). (Adapted from ref. 18–14, Seinfeld, J. H., *Air Pollution: Physical and Chemical Fundamentals*, 1975, reproduced by permission of McGraw-Hill Publishers.

as premixed gas combustion with a careful control of the temperature-time history of the resultant combustion gas (Section 5.13).

Carbon Monoxide and Sulfur Dioxide

Carbon monoxide is a component of combustion, particularly when the process is fuel rich (Section 5.1). Fuel-lean combustion in practical combustion devices also leads to the production of CO, although in very small amounts when the process is well designed and operates as designed.

CO is toxic to air respiring animals and man. The physiological effect of CO, summarized in Fig. 18–10, shows the consequence of exposure to this gas as a component of air. This plot is a typical representation of such data and shows that total dose and dose rate are important in their effect. The toxicity of CO is due to its much greater (factor of over 10^2) affinity for blood hemoglobin as compared to O_2. CO absorbed in hemoglobin is called carboxyhemoglobin (COHb). With exposure to CO, oxygen cannot be carried to the system cells, and the result is O_2 cell starvation. Levels of COHb associated with saturation at 100 ppm leads to significant physiological effects. Long-term exposure to levels greater than 600 ppm, (=0.06%!) is fatal.

Sulfur oxides (SO_2 and SO_3) are the result of combustion of fuels containing sulfur. Refined fuels for automobiles (Otto and Diesel) and for gas turbines are essentially cleaned of all sulfur content in the refining process. For the gas turbine this is necessary for metallurgical reasons associated with turbine blade life. The coal combustion process involves emission of the sulfur oxides from the sulfur that is originally present as sulfides or sulfates. As a result of the environmental protection laws limiting sulfur emissions, low-sulfur coals have a much higher economic value than does high-sulfur coal. Sulfur dioxide is readily catalyzed to the trioxide as a gas or with dissolved oxygen in water, via

$$SO_2 + \tfrac{1}{2}O_2 \rightarrow SO_3 \qquad (18\text{–}12)$$

Sulfur oxides may be captured during and after combustion by the reaction with lime (CaO from limestone, reaction 1 below) to form an insoluble, solid, and collectible calcium sulfate:

$$\left.\begin{aligned} CaCO_3 + \text{heat} &\rightarrow CaO + CO_2 \\ CaO + SO_3 &\rightarrow CaSO_4 + \text{heat} \end{aligned}\right\} \tag{18–13}$$

The process demands thermal power in the first reaction (which may be obtained, in part, from the second), liberates CO_2, and presents design and operational difficulties because handling a solid waste is required. As a product, it has commercial value as gypsum ($CaSO_4 \cdot 2H_2O$), cement aggregrates, etc. The effectiveness of sulfur removal from combustion gas is controlled by the Ca/S ratio which may vary between 1 and 4 to meet goals of 90% sulfur removal in fluidized bed reactors (ref. 18–12).

The toxicity of sulfur oxides is serious, as shown in Fig. 18–11 (adapted from ref. 18–13). The damaging effects of SO_x emissions can be noted in descriptions of old and uncontrolled smelting operations and of serious air pollution incidents of the last century. Airborne sulfur oxides react with water (and dissolved O_2) to form sulfuric acid which is a component of acid rain. The reaction is

$$H_2O + SO_2 + \tfrac{1}{2}O_2 \rightarrow H_2SO_4 \tag{18–14}$$

This acidic rainwater is a problem for locations downwind of generation sites. The affinity of the acid droplets for solid particles and their form as droplets allow the acid to penetrate deeply into lungs during respiration. At

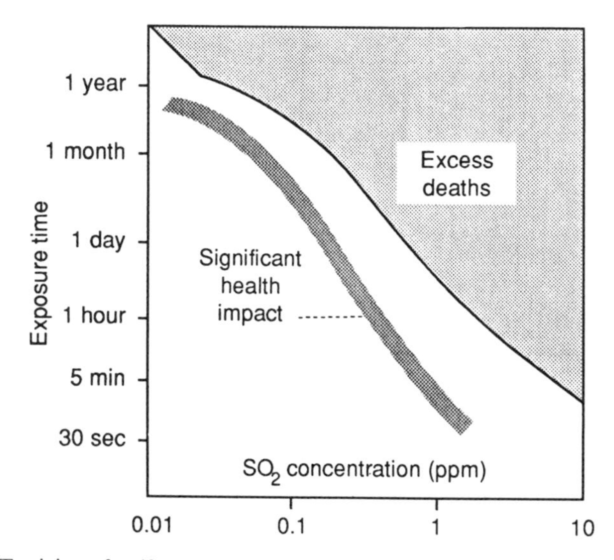

Fig. 18–11. Toxicity of sulfur oxides (ppm = parts per million, 10^{-6}). (Adapted from ref. 18–13, with permission.)

levels of 1 ppm in the atmosphere, sulfur oxides can lead to airway constriction (ref. 18–14).

Dust

A light-scattering haze may also be formed in the atmosphere by the introduction of solids in the combustion process, of coal in particular. Particles of a broad size range interfere with the transmission of sunlight and thus may be an element that changes the albedo of the Earth (Section 18.4.1). Viscous drag on these particles prevents these particles from settling out of the atmosphere rapidly. Their small size (typically microns, 10^{-6} m) gives rise to small flow velocities (U) during a steady fall in a gravitational field. The drag of an object approximately spherical in shape for this situation is given by

$$D = \tfrac{1}{2}\rho U^2 C_d(\pi R^2) \qquad (18\text{–}15)$$

Low speed and small size result in a low value for the Reynolds number describing the flow around the falling particle. The flow is therefore laminar. The drag coefficient (C_d) is given by the formula of Stokes

$$C_d = \frac{24}{Re} \qquad \text{where } Re = \frac{\rho U(2R)}{\mu} \qquad (18\text{–}16)$$

The gas viscosity, μ, is given in App. B. The drag force is balanced by the weight of the particle ($D = mg$) so that a settling velocity U may be computed if the density of the particle (ρ_m) is assumed:

$$U = \frac{\rho_m g R^2}{9\mu} \qquad (18\text{–}17)$$

Thus U is of the order of 10^{-5} m/sec for a 1-micron-diameter particle (0.3 m in 1 hr!). The natural turbulence of the atmospheric air is important in determining the actual settling time because turbulence velocity components may be significantly larger.

There is recorded evidence of the effect of dust injection by volcanoes on the climate of the Earth. Two eruptions in the nineteenth century, both in Indonesia (Tambora in 1815 and Krakatoa in 1883), have had serious impacts on global climate and consequently on agriculture. Similar effects, though smaller, have been correlated with eruptions in the twentieth century. It is against this sporadic background that man's particulate emissions on a global scale must be gauged. Locally, dust depositions can lead to consequences ranging between visual impact and serious health concerns.

18.4.5. *Photochemical smog*

Photochemical smog (smoke + fog) is a complex product that results from the action of sunlight on air containing nitrogen oxides (NO_x) and hydrocarbons.

This smog is created in cities during times of strong sunshine and was first identified with conditions in Los Angles during the 1950s. The severity of the problem is increased in cities located in geological basins that trap air, allowing the concentrations to build more than if the air were mixed with a larger air mass.

The damage caused by the smog is through its production of ozone and strong oxidants which are irritating to serous lining of tissues (eyes, nasal passages, etc.) in man and animals. These substances may be toxic to plants when present in very small concentrations. Concentrations of the order of $200\ \mu g/m^3$ or 0.1 ppm result in irritation, and values less than half of this can cause plant damage. These oxidants are complex hydrocarbon groups containing a double-bonded oxygen, which plays an important role in making these molecules reactive. Examples of such oxidants are formaldehyde, ketones, and peroxyacyl nitrates (PAN).

As far as energy conversion processes are concerned, one important issue is that NO, and NO_2 in particular, play a critical role in the low-altitude ozone generation process. The reactions of interest are

$$\left.\begin{array}{l} \text{NO production in combustion} \\ \text{NO oxidized to } NO_2 \\ NO_2 + h\nu \rightarrow NO + O \\ O_2 + O \rightarrow O_3 \end{array}\right\} \qquad (18\text{–}18)$$

Evidently, NO is a catalyst for the production of ozone because it survives the processes in the atmosphere. Naturally, there are processes that remove NO, but as shown in Table 5–2 of Section 5.13.2, the direct recombination rate of heated air is extremely slow. Note that in combustion, the atomic form of oxygen (O) plays a catalytic role in the production of NO, whereas NO is a catalyst for the production of O (and O_3) in the atmosphere.

The details of the entire smog generation process are not completely understood and beyond the scope of this text. The serious student should consult the literature on air pollution chemistry for reviews of the descriptive models. The role of hydrocarbons in the air is in the oxidation of NO to NO_2 and as building blocks in the formation of the oxidants described above. To break this cycle, environmental laws in the United States and California (earlier on and with stronger means) have focused on the reduction of NO_x and of unburned hydrocarbons emissions.

18.4.6. *Who Pays for Pollution Controls?*

As a system analysis problem, abatement of the effects of pollutants is often examined as a question of minimizing the combined costs of damage and of controls in much the same way that a system as a fuel user is optimized. As a function of plant complexity, the cost of capital equipment is high when the environment is to be kept unaltered. If the cost is to be borne by those

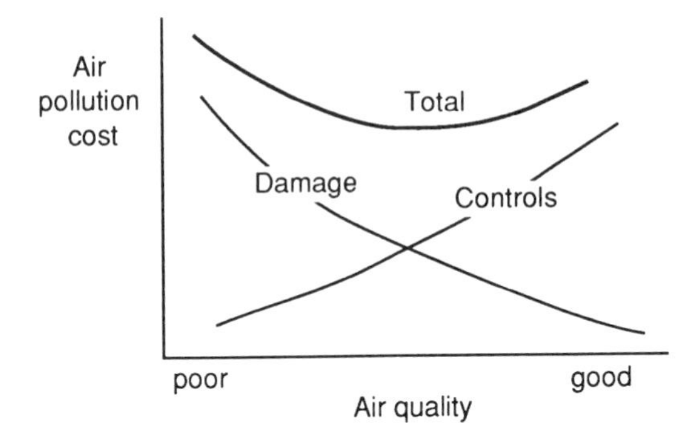

Fig. 18–12. Balance between costs for environmental controls and damage mitigation.

damaged, then the plant cost is low. Figure 18–12 illustrates the minimization problem. In reality, however, the cost of balancing these two effects is not borne by the same individual. The plant designer and operator pays for changes to the plant (from profits), whereas the environmental damage cost is borne by those affected. When this problem is severe enough to warrant attention, then the interaction between polluter and those affected requires effective legislation and enforcement so that costs are borne equitably.

18.5. NUCLEAR WASTE

Although nuclear power production is not covered in detail in this book, the effluents that are byproducts of nuclear fission are discussed here in the context of the environmental impact of other energy sources. Nuclear fission results from the reaction of neutrons with fissile isotopes of uranium or plutonium. If the fissile isotope of mass number N is indicated by the symbol ${}^{N}X$, its absorption of a neutron, ${}^{1}n$, results in the unstable isotope, ${}^{N+1}X$, which normally splits into two fission fragments and also releases two or three new product neutrons. The reaction is generally written:

$$ {}^{N}X + {}^{1}n \dashrightarrow \{{}^{N+1}X\}^{*} \dashrightarrow \mathrm{A} + \mathrm{B} + (\sim 2.5)^{1}n + Q \qquad (18\text{–}19)$$

where the asterisk (*) indicates an excited, unstable nucleus, and A and B are fission fragments. Additional aspects of the fundamentals of heat production in the fission reaction are described in Chapter 1.

The fission energy, Q, is released in the form of kinetic energy of the fission fragments and neutrons, and as radiant energy of gamma ray photons and neutrinos. The amount of energy released is approximately 1 megawatt-day (Mwd $\sim 10^{11}$ J) of thermal energy per gram of material fissioned, roughly two million times the energy released when a gram of fossil fuel burns. In SI units, the thermal energy density of fissionable nuclear material is about 10^{14} J/kg.

Uranium-233, uranium-235, and plutonium-239 are fissile isotopes. Only ^{235}U is found in nature, but the others can be produced by neutron absorption in fertile ^{232}Th or ^{238}U, and their subsequent transformation via radioactive decay to ^{233}U and ^{239}Pu.

Radioactive waste from fission reactor operation consists of various components:

1. activation products, the result of neutron activation of coolant impurities, instruments, and material located within the reactor room that may absorb neutrons and become radioactive
2. high level radioactive waste in the form of fission products, produced within the metal fuel tubes while they are in the reactor and, except for a small amount of gaseous material that may diffuse through the cladding walls, contained within the tubes until they are physically or chemically breached
3. transuranic (TRU) isotopes produced by neutron capture in uranium-238 and other actinide isotopes.

The wastes are also classified as low-level or high-level. Low-level wastes result from normal power plant operation and consist of such things as gloves, paper towels, lab coats, ion-exchange resins, filters, etc., that may contain small amounts of radioactive material. These wastes are routinely collected, monitored, reduced in volume, and packaged in 55-gallon drums that are shipped to federally licensed low-level burial ground for disposal.

By far the largest amount of radioactivity, however, is contained in high-level radioactive waste. This consists primarily of radioactive fission products and transuranic isotopes that are produced in the fission process. In the United States, where used or spent nuclear fuel is not being reprocessed but is being stored indefinitely, the spent fuel elements themselves are regarded as high-level radioactive waste. In nations that do reprocess fuel, high-level waste is in the form of glass or ceramic logs that contain the separated fission products and transuranic isotopes that resulted from fission and neutron capture reactions and have no further use.

There are key differences between nuclear waste and fossil-fueled power plant waste. One is that nuclear fission results in none of the usual combustion products that are created in fossil-fueled plants: carbon dioxide, nitrogen oxides, sulfur oxides, and, in the case of coal, substantial quantities of ash. Therefore, fission reactors do not add to atmosphere pollution or contribute to global warming. Nuclear fission does, however, result in large amounts of radioactivity. While the mass and volume of spent nuclear fuel are very small compared with the mass and volume of combustion products from equivalent size plants, the radioactivity and potential hazard of the spent fuel are enormous and long-lived, and must therefore be isolated from the biosphere for long periods of time.

Spent fuel is typically stored for several years at the reactor site where it was used so that short-lived isotopes can decay and radioactive heat generation will no longer pose a serious problem. While 100 to 200 fission products and intense levels of radioactivity are present when spent fuel is discharged, the contents change quite rapidly. Ten years after discharge from a typical

Table 18–2. Significant fission and transuranic products, 10 years after fuel discharge

Fission product	Half-life (years)	Percent of total radioactivity
Cesium-137	30.0	37.03
Plutonium-241	13.2	34.39
Strontium-90	28.1	26.45
Plutonium-238	86.0	0.98
Americium-241	458	0.77
Plutonium-240	6 580	0.23
Plutonium-239	24 400	0.14
Total		99.98

Table 18–3. Significant fission and transuranic products, 100 years after fuel discharge

Fission product	Half-life (years)	Percent of total radioactivity
Cesium-137	30.0	44.48
Strontium-90	28.1	27.19
Americium-241	458	15.82
Plutonium-238	86	4.45
Plutonium-241	13.2	4.45
Plutonium-240	6 580	2.15
Plutonium-239	24 400	1.31
Americium-243	7 950	0.07
Total		99.91

pressurized water reactor (PWR), for example, only seven isotopes, shown in Table 18–2, account for more than 99.9% of the radioactivity in the spent fuel.

It is useful to consider a single pressurized water reactor (PWR) fuel assembly as a reference. It contains about 450 kg of uranium when it is loaded into a reactor as fresh fuel and produces, during the course of its use in the reactor, about 14 000 Mwd of thermal energy. This is equivalent to the energy produced by burning 40 000 *tons* of coal. At discharge from the reactor the total radioactivity in the assembly is about 5×10^7 curies. Ten years after discharge, the activity has been reduced to approximately 10^5 curies. A curie is 3.7×10^{10} disintegrations per second.

One hundred years after discharge, the isotopes contributing most significantly to the total activity are nearly the same and are shown in Table 18–3. At this time the total radioactivity in a spent fuel assembly has decreased to approximately 1.1×10^4 curies.

In the United States, high-level radioactive wastes are currently stored by utilities either at reactor sites or at other temporary storage locations. A permanent repository has not yet been developed and licensed. A Department of Energy site at Yucca Mountain in the State of Nevada is being studied as a potential deep geological repository for the spent fuel, but will not be able to accommodate spent fuel until the twenty-first century.

The environmental impact of nuclear power reactors is due primarily to waste heat and the potential hazard of radioactivity. Radioactive emissions are hazardous because cell damage results from interactions of alpha, beta, neutron, or gamma radiations with biological material. High energy gamma radiations and fast neutrons require shielding because they penetrate biological material easily and may be very damaging. Alpha and beta particles are much less penetrating and pose less of an external radiation hazard, but may cause significant biological damage if their precursors are ingested or inhaled. They have short path lengths and exhibit high specific energy loss, causing significant local cell damage. Many transuranic isotopes such as those of plutonium decay by alpha emission. They not only have long radiological half-lives but also have long biological half-lives; in particular, they pose an inhalation hazard, since they are massive nuclei that are not readily cleared from bronchial and lung passages.

The serious student is strongly encouraged to consult the extensive literature on nuclear waste and its characteristics, on potential pathways for radioisotopes to reach the biosphere, and on the political challenge of developing a suitable repository. An extensive body of literature on biological radiation hazards is also available.

18.6. REMARKS

Harnessing of energy entails risk: There is always a possibility that it may do unintended physical harm. There is always an environmental cost of thermal, chemical, and/or nuclear emissions that are a necessary consequence of the process of doing things that benefit mankind. The environmental impact of energy conversion is reduced by increasing the efficiency of resource utilization. However, increasing the efficiency of conversion generally entails greater use of materials, labor, and more complex machines. The cost may be justified by the security of a decreased dependence on energy resources and by the social peace obtained through productive employment.

Energy independence is probably not possible for the individual members of humankind on this planet, at least not if a high standard of living is desired. It is not possible now with the present availability of low-cost fossil fuels, and it will be even less possible when either solar and/or nuclear resources carry the energy supply burden in the future. This prognosis is exacerbated by the increasing population on this planet.

PROBLEMS

1. Using the solar power incident on the Earth, calculate the power output of the sun in watts.

2. Consider that the atmosphere transmits thermal blackbody radiation at 280 K between frequencies that are 0.1 and 10 times the value for peak B_ν. What fraction of the power is radiated if the window is reduced to 0.11 and 9 times this value. To what temperature would the radiator have to rise to radiate the same power as the wider window?
3. Consider that the Earth reflects about 35% of the incident solar power and assume that it emits the remainder uniformly from all areas of its surface. Show that the average thermal flux from the Earth into space is about 227 w/m^2. With an average temperature found to be 13°C (286 K), determine the emissivity of the Earth surface.
4. Derive an expression for the blackbody distribution function as plotted as a function of wavelength in Fig. 18–5. Note the -5 power dependence on wavelength. Show that the Stefan-Boltzmann constant is given by

$$\sigma \equiv \frac{2\pi^5 k^4}{15\, h^3 c^2}$$

BIBLIOGRAPHY

Krenz, J. H. *Energy Conversion and Utilization*, Allyn and Bacon, Boston, 1976.

REFERENCES

18–1. Hammerle, R. H., Shiller, J. W., Schwarz, M. J., "Global Climate Change," *ASME J. of Engineering for Gas Turbines and Power*, **113**(3), July 1991.

18–2. Warrick, R. A., Bolin, B., Döös, B. R., Jäger, J., Eds., The greenhouse effect, climatic change and ecosystems, Scope 29, Wiley, Chichester, UK, 1986.

18–3. Woodwell, G. M., "The Effects of Rapid Global Warming on Terrestrial Ecosystems—a Positive Feedback . . . and a Solution," in *Proc. Int. Conf. on Global Warming and Climate Change*," Gupta, S., Pachavri, R. K., Eds., Tata Energy Institute, New Delhi, 1989.

18–4. Hamel, R. A., Conrath, B. J., Kunde, V. G., Probahakara, C., Revak, I., Solomonson, V. V., Wolford, G., "The Nimbus 4 Spectroscopy Experiment, 1, Calibrated Thermal Emission Specta," *J. Geophysical Res.*, **77**: 2629–41, 1972.

18–5. Ramanathan, V., Callis, L., Cess, R., Isaksen, I., Kuhn, W., Lacis, A., Mahlman, J., Reck, R., Schlesinger M., "Climate Chemical Interactions and Effects of Changing Atmosphere Trace Gases," *Rev. Geophysics*, **25**: 1441–82, 1987.

18–6. Fay, J. A., *Molecular Thermodynamics*, Addison-Wesley, Reading, Mass., 1965.

18–7. Pollack, J. B., Yung, Y., "Origin and Evolution of Planetary Atmospheres," *Annual Review of Earth and Planetary Science*, **8**, p. 825, 1980.

18–8. Kasting, J., "How Climate Evolved on the Terrestial Planets," *Scientific American*, **258**(2), February 1980.

18–9. *Atmospheric Effect of Stratospheric Aircraft*, NASA Publication 1272, First program report, Washington D.C., January 1992.

18–10. Haagen-Smit, A. J., "Chemistry and Physiology of Los Angeles Smog," *Ind. Eng. Chem.*, **44**: 1352, 1952.

18–11. Bahr, D. W., Lyon, T. F., "NO_x Abatement via Water Injection in Aircraft Derivative Gas Turbine Engines," Paper ASME 84-GT 103, American Society of Mechanical Engineers, New York, 1984.
18–12. Fennelly, P. F., "Fluidized Bed Combustion," *American Scientist*, **72**: 254–61, May-June 1984.
18–13. Williamson, S., *Fundamentals of Air Pollution*, Addison-Wesley, Reading, Mass., 1973.
18–14. Seinfeld, J. H., *Air Pollution: Physical and Chemical Fundamentals*, McGraw-Hill, New York, 1975.

APPENDIX A

CONVERSION FACTORS

Table A–1. The numerical relation between energy and power quantities

Energy

Unit	Origin	Unit Conversion
BTU	1 lb_m water, 1°F	1054 J/BTU
cal	1 g water 1°C	4.184 J/cal
ev		1.60×10^{-19} J/ev
ft-lb	mechanics	778 ft-lb/BTU 1.354 J/ft-lb
Joule	1 Nm, SI standard	—
kcal	1 kg water 1°C	0.251 kcal/BTU or 4184 J/kcal
kw-hr		3414 BTU/kw-hr
quad	defined as 10^{16} BTU	$\sim 10^{19}$ J/quad
amu		931×10^6 ev/amu
barrel of oil		$\sim 5.8 \times 10^6$ BTU $\sim 5.8 \times 10^9$ J
ft^3 of natural gas at STP		~ 1000 BTU ~ 1 MJ
metric ton of coal		$\sim 2.6 \times 10^7$ BTU $\sim 2.6 \times 10^9$ J

Power

Unit	Origin	Unit Conversion
horsepower (HP)	550 ft-lb/sec	0.746 kw/HP 1.41 HP sec/BTU
watt	1 J/sec	

APPENDIX B

FUNDAMENTAL PHYSICAL CONSTANTS

Table B–1. Values of some fundamental physical constants* (SI or MKS units)

Symbol	Value	Units	Name
m_e	9.109×10^{-31}	kg	electron mass
m_g(1 amu)	1.661×10^{-27}	kg	mass of atom of unit molecular "weight"
m_p	1.672×10^{-27}	kg	proton rest mass
m_n	1.675×10^{-27}	kg	neutron rest mass
h	6.626×10^{-34}	J-sec	Planck's constant
k	1.381×10^{-23}	J/K	Boltzmann's constant
e	1.602×10^{-19}	Coulomb	electronic charge
ε_0	8.8543×10^{-12}	$\mathrm{F\ m^{-1}}$	permittivity of free space
μ_0	$4\pi \times 10^{-7}$	$\mathrm{H\ m^{-1}}$	permeability of free space
c	2.998×10^{8}	$\mathrm{m\ sec^{-1}}$	speed of light
N_0	6.022×10^{26}	$\mathrm{(kg\text{-}mole)^{-1}}$	Avogadro's number
F	9.649×10^{7}	$\mathrm{C(kg\text{-}mole)^{-1}}$	Faraday constant
R_u	8.314×10^{3}	$\mathrm{J(kg\text{-}mole)^{-1}\ K^{-1}}$	universal gas constant
	1.987	$\mathrm{cal(mole)^{-1}\ K^{-1}}$	
σ	5.669×10^{-8}	$\mathrm{W\ m^{-2}\ K^{-4}}$	Stefan-Boltzmann constant
m_p/m_e	1836.1		proton/electron mass ratio
G	6.673×10^{-11}	$\mathrm{N\ m^2/kg^2}$	gravitational constant

Source: Rev. Mod. Phys., **41**, 375, 1969. American Institute of Physics.

Table B–2. Useful constants in energy conversion practice

Quantity	Value	Units	Name
R_E	6.38×10^{6}	m	Radius of the earth
R_S	6.96×10^{9}	m	Radius of the sun
g_0	9.806	$\mathrm{m/sec^2}$	Gravitational acceleration at earth surface
A.U. (astronomical unit)	1.496×10^{11}	m	Radius of earth orbit around sun
R	0.72	A.U.	Orbital radius of Venus
	0.98–1.02	A.U.	Orbital radius of Earth
	1.38–1.66	A.U.	Orbital radius of Mars
I_0	1.36 (Jan 21)	$\mathrm{kw/m^2}$	Solar power constant
	1.44 (Jun 21)	$\mathrm{kw/m^2}$	(no atmospheric attenuation)
ε_i	3.89	ev	Ionization potentials of Cesium
	4.54	ev	Ionization potentials of Potassium

Table B–3. Physical properties of air and water

	air	water
Composition	0.7808 N_2 0.2095 O_2 0.0093 Ar 0.0003 CO_2	H_2O
Mol. Weight	28.96	18.016
R (J/kg K)	287.1	—
γ	1.40	—
a (m/sec at 15C)	341	1500
C_p (J/kg K)	1005	4184 (triple point)
p ($N\ m^{-2}$) at S.L.	1.013×10^5	—
ρ ($kg\ m^{-3}$)	1.225 (atm, 15C)	1000 (secondary SI standard)
μ ($N\ sec\ m^{-2}$)	1.8×10^{-5}	8.9×10^{-4}
ν ($m^2\ sec^{-2}$)	1.5×10^{-5}	0.89×10^{-6}
k ($W\ m^{-1}\ K^{-1}$)	6.0×10^{-6}	0.61 [at 25C]

The viscosity and thermal conductivity of *air* are given approximately by (*)

$$\mu\left[\frac{\text{N sec}}{\text{m}^2}\right] = 1.458 \times 10^{-6}\,\frac{T^{3/2}}{T + 110.4} \qquad \text{with } T \text{ in [K]}$$

$$k\left[\frac{\text{W}}{\text{m}^2\,\text{K}}\right] = 6.325 \times 10^{-7}\,\frac{T^{3/2}}{T + 245.4 \times 10^{-12/T}} \qquad \text{with } T \text{ in [K]}$$

(*) *U.S. Standard Atmosphere*, 1962, U.S. Government Printing Office, 1962.

Table B–4. Critical constants of common gases

Gas	MW	T_{crit} K	p_{crit} atm	v_{crit} m^3/kg-mole	heat of vaporization kJ/kg (T, K at 1 atm.)
N_2	28.0	126	33.5	0.90	199 (77)
H_2	2.016	33	12.8	0.065	446 (20)
O_2	32	155	50.1	0.078	213 (90)
H_2O	18.016	647.2	218.3	0.056	2257 (373)

APPENDIX C

JANAF THERMOCHEMICAL TABLES

In the JANAF tables, the value for the reference temperature is commonly 298.15 K (with 298 used as a subscript). The reference pressure is one atmosphere ($= 1.013 \times 10^5$ N/m^2). These data are published in thermal units (cal) and in SI units. All data in this book are given in cal or kcal per mole. Reference C–1 gives the data in SI units. The value of the universal gas constant, R_u, that corresponds to the thermal system of units is $R_u = 1.987$ cal/mol K. The conversion factor between Joules and calories is exactly 4184 J/kcal. The tabular data given in the reference are described below. The following tables include only data necessary for the solution of gas combustion chemistry problems as described in Chapter 5: C_p, s^0, $H(T) - H(298)$, and K_p. The original tabular data are a listing of the following quantities:

1. Specific heat at constant pressure, $C_p = \left(\frac{\partial h}{\partial T}\right)_p$: cal/mol K

2. Entropy: Since the entropy for an ideal gas may be written as

$$s - s_{\text{ref}} = \int_{T_{\text{ref}}}^{T} C_p \frac{dT}{T} - R \ln\left(\frac{p}{p_{\text{ref}}}\right)$$

   it is the temperature integral portion of this expression that is readily tabulated as s^0.

$$s^0 \equiv \int_0^T C_P \frac{dT}{T}$$

   The units are cal/mol K. The reference state temperature for this integral is 0 K.

3. Gibbs free energy: The symbol used in these tables is F, in contrast to the use of G in this book (or g per mole or per unit mass basis).

$$F = H - Ts \text{ kcal/mol}$$

   The function tabulated is $-(F^0 - H^0_{298})/T$ which is the free energy function in the standard state at T and which is defined as

$$s^0 - \frac{H^0 - H_{298}}{T}$$

4. enthalpy $H(T) - H(298) = \int_{298}^{T} C_p\,dT$ kcal/mole (use of h in lieu of H would be consistent with the nomenclature used in this text).
5. and 6. Standard heat of formation, ΔH_f, and the free energy formation, ΔF_f, are the increments associated with the reaction of forming 1 mol of the given compound from its elements, each in its thermodynamic standard state at the given temperature. The value given is zero if the substance is already in the standard state. Values for ΔH_f are given in Table 5–1 for $T = T_{ref} = 298$ K.
7. Equilibrium constant. The quantity given is $\log_{10} K_p$ because of the extreme variation of K_p with temperature. See Section 5.9 for the background on the data given in the JANAF tables.

Table C–1 (from thermal units version of data in the source cited, with permission) is a partial listing of the JANAF data tables for these substances: C, CH_4, CO, CO_2, H, H_2, H_2O, N, N_2, NO, NO_2, O, O_2, OH.

The reader is cautioned that linear interpolation across a 500 K temperature gap may be inaccurate; the reader should therefore consult the original reference if there is concern with interpolation error. When no value for $\log K_p$ is given, it is zero, $K_p = 1$.

For computational purposes, the following curve fits to K_p data are useful and quite accurate:

$$\log(K_p)_j = A_j + B_j \frac{1000}{T}$$

where the constants are given in table C–2.

The variation of C_p with temperature for the substances listed is plotted in Fig. C–1.

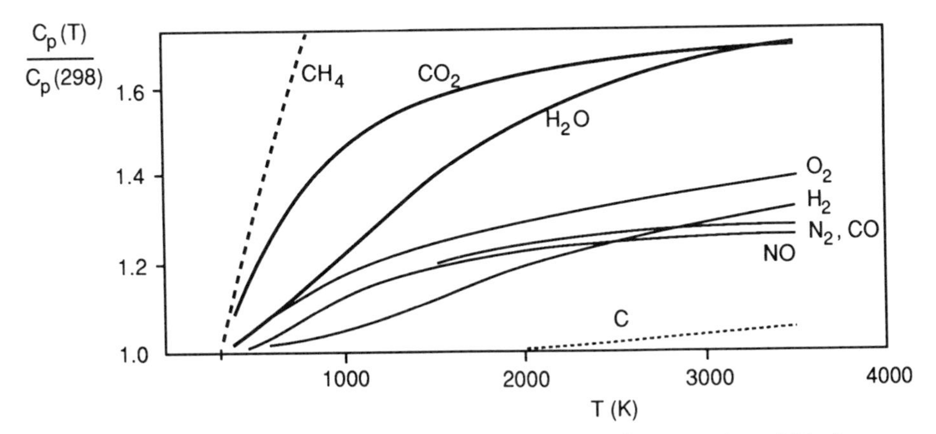

Fig. C–1. Variation of specific heats at constant pressure for a number of ideal gases. Data for CO, NO, and N_2 are quite close.

Table C–1. Abbreviated JANAF tables

	$T(K)$	C_p	$H(T) - H(298)$	s^0	$\log K_p$
C	298	4.981	0.0	37.761	−117.805
	500	4.973	1.004	40.334	−66.677
	1000	4.969	3.490	43.779	−29.202
	1500	4.975	5.975	45.794	−16.711
	2000	5.008	8.469	47.229	−10.477
	2500	5.077	10.989	48.354	−6.747
	3000	5.168	13.550	49.287	−4.266
	3500	5.261	16.157	50.091	−2.499
CH_4	298	8.518	0.0	44.490	8.902
	500	11.076	1.960	49.453	3.429
	1000	17.160	9.125	59.141	−1.011
	1500	20.688	18.679	66.840	−2.602
	2000	22.562	29.540	73.076	−3.408
	2500	23.608	41.106	78.233	−3.889
	3000	24.233	53.079	82.597	−4.206
	3500	24.633	65.302	86.365	−4.430
CO	298	6.965	0.0	47.214	24.029
	500	7.121	1.417	50.841	16.235
	1000	7.931	5.183	56.028	10.450
	1500	8.417	9.285	59.384	8.485
	2000	8.664	13.561	61.807	7.469
	2500	8.804	17.931	63.756	6.840
	3000	8.895	22.357	65.370	6.407
	3500	8.961	26.822	66.746	6.088
CO_2	298	8.874	0.0	51.072	69.095
	500	10.666	1.987	56.122	41.260
	1000	12.980	7.984	64.344	20.680
	1500	13.953	14.750	69.817	13.801
	2000	14.424	21.857	73.903	10.353
	2500	14.692	29.141	77.153	8.280
	3000	14.873	36.535	79.848	6.892
	3500	15.006	44.006	82.151	5.898
H	298	4.968	0.0	27.392	−35.614
	500	4.968	1.003	29.961	−20.159
	1000	4.968	3.487	33.404	−8.646
	1500	4.968	5.971	35.419	−4.757
	2000	4.968	8.455	36.848	−2.791
	2500	4.968	10.939	37.957	−1.601
	3000	4.968	13.423	38.862	−0.803
	3500	4.968	15.907	39.628	−0.231
H_2	298	6.892	0.0	31.208	0.0
	500	6.993	1.406	34.806	
	1000	7.219	4.944	39.702	
	1500	7.620	8.668	72.716	
	2000	8.195	12.651	45.004	
	2500	8.575	16.848	46.875	
	3000	8.858	21.210	48.465	
	3500	9.110	25.703	49.850	

Table C–1. *(continued)*

	$T(K)$	C_p	$H(T) - H(298)$	s^0	$\log K_p$
H_2O	298	8.025	0.0	45.106	40.048
	500	8.415	1.654	49.334	22.866
	1000	9.851	6.209	55.592	10.062
	1500	11.233	11.495	59.859	5.725
	2000	12.214	17.373	63.234	3.540
	2500	12.863	23.653	66.034	2.224
	3000	13.304	30.201	68.421	1.343
	3500	13.617	36.936	70.496	0.712
N	298	4.968	0.0	36.614	−79.806
	500	4.968	1.003	39.183	−46.336
	1000	4.968	3.487	42.627	−21.328
	1500	4.968	5.971	44.641	−13.217
	2000	4.969	8.455	46.070	−9.044
	2500	4.978	10.941	47.180	−6.535
	3000	5.011	13.437	48.090	−4.858
	3500	5.086	15.959	48.867	−3.656
N_2	298	6.961	0.0	45.770	0.0
	500	7.069	1.413	49.386	
	1000	7.815	5.129	54.507	
	1500	8.330	9.179	57.784	
	2000	8.601	13.418	60.222	
	2500	8.756	17.761	62.159	
	3000	8.855	22.165	63.765	
	3500	8.927	26.611	65.135	
NO	298	7.133	0.0	50.347	−15.171
	500	7.287	1.448	54.053	−8.783
	1000	8.123	5.313	59.377	−4.062
	1500	8.552	9.496	62.763	−2.487
	2000	8.759	13.829	65.255	−1.699
	2500	8.877	18.241	67.223	−1.227
	3000	8.955	22.700	68.849	−0.913
	3500	9.012	27.192	70.234	−0.690
NO_2	298	8.837	0.0	57.343	−8.977
	500	10.327	1.936	62.268	−6.669
	1000	12.468	7.730	70.215	−5.000
	1500	13.193	14.176	75.432	−4.438
	2000	13.490	20.856	79.274	−4.152
	2500	13.636	27.641	82.301	−3.979
	3000	13.718	34.482	84.795	−3.864
	3500	13.768	41.354	86.914	−3.783
O	298	5.237	0.0	38.468	−40.604
	500	5.081	1.038	41.131	−22.940
	1000	4.999	3.552	44.619	−9.807
	1500	4.982	6.046	46.642	−5.395
	2000	4.978	8.536	48.074	−3.178
	2500	4.984	11.026	49.185	−1.842
	3000	5.004	13.522	50.096	−0.949
	3500	5.041	16.033	50.870	−0.310

(continued)

Table C–1. *(continued)*

	$T(K)$	C_p	$H(T) - H(298)$	s^0	$\log K_p$
O_2	298	7.020	0.0	49.004	0.0
	500	7.431	1.455	52.722	
	1000	8.336	5.427	58.192	
	1500	8.738	9.706	61.656	
	2000	9.029	14.149	64.210	
	2500	9.301	18.732	66.254	
	3000	9.551	23.446	67.975	
	3500	9.762	28.276	69.461	
OH	298	7.136	0.0	43.918	−6.006
	500	7.049	1.430	47.582	−3.244
	1000	7.329	5.000	52.520	−1.218
	1500	7.866	8.800	55.594	−0.559
	2000	8.285	12.844	57.918	−0.236
	2500	8.573	17.063	59.800	−0.046
	3000	8.778	21.404	61.382	0.078
	3500	8.929	25.832	62.747	0.165

Note: Tables in ref. C–1 are to 6000 K in increments of 100 K.

Source: Chase, M. W., Davies, C. A., Downey, J. R., Frurip, D. J., McDonald, R. A., Syverud, A. N., *JANAF Thermochemical Tables*, 3rd edition, American Chemical Society and American Institute of Physics for National Bureau of Standards, 1985.

Table C–2. Curve fit constants for $\log(K_p)$

j	A_j	B_j
C	8.293	−37.548
CH_4	−5.313	4.267
CO	4.591	5.811
CO_2	0.057	20.582
H	2.987	−11.523
H_2O	−2.867	12.814
N	3.387	−24.805
NO	0.658	−4.718
NO_2	−3.305	−1.689
O	3.387	−13.126
OH	0.766	−2.013

Note: Temperature in degrees K.

APPENDIX D

ENERGY TRANSFER IN UNSTEADY FLOW

In steady flow, the stagnation enthalpy of a stream may be increased or decreased through heat and/or work interactions (eq. 2–6). In unsteady flows, this quantity is also altered through an unsteady variation of pressure. This may be shown by means of the 1–D flow equations. For example, for an adiabatic flow without work interaction, the energy equation reads

$$\rho \frac{Dh_t}{Dt} = \frac{\partial p}{\partial t} \tag{D–1}$$

The substantial derivative used in eq. D–1,

$$\frac{D}{Dt} \equiv \frac{\partial}{\partial t} + U \cdot \nabla = \frac{\partial}{\partial t} + u \frac{\partial}{\partial x} \text{ in 1–D flow}$$

describes changes following a fluid element as it experiences temporal changes and changes associated with convection with velocity, U. Thus the energy equation shows that the unsteady passage of a compression wave increases and an expansion wave decreases the stagnation enthalpy of a flow element.

The following is a description of unsteady flow with an eye toward application to wave machines that can serve as ICE air chargers (Section 8.10) and to steady flow engines that can extract stagnation enthalpy from gases at higher temperature than allowed for steady flow and thereby promise better performance. The model for understanding unsteady energy transfer is developed for the situation where two sections in a tube containing gases at different pressures are separated by a barrier. The unsteady flow following the removal of the barrier is of interest. A compression or shock wave is generated in the low-pressure flow, whereas an expansion wave travels into the high-pressure gas. The tube wherein these events are studied is termed a *shock tube*. Such a physical arrangement is useful for the short-term generation of flow conditions that would otherwise be difficult to obtain. The literature on shock tubes is replete with applications.

D.1. SHOCK TUBE THEORY

This discussion is limited to a simple description of shock tube performance. This analysis applies to the compression process occurring in air-charging wave machines for ICEs where the high-pressure and high-temperature exhaust

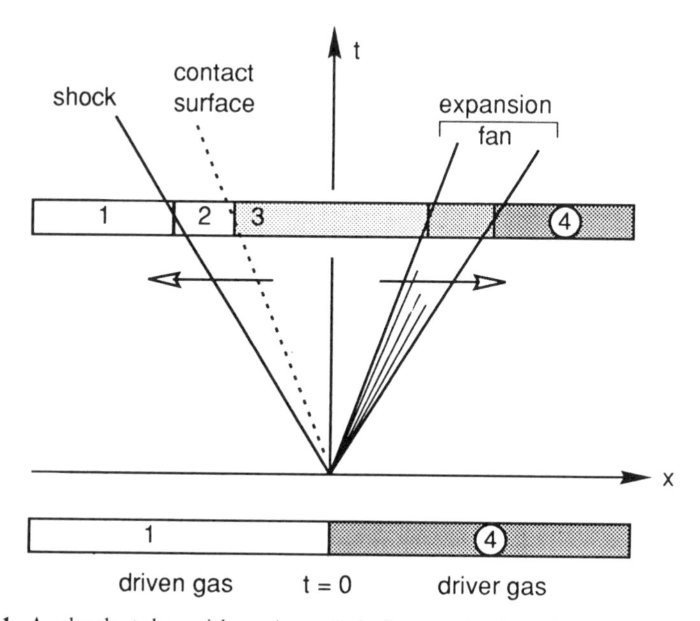

FIG. D–1. A shock tube with $p_4/p_1 > 1$ before and after ($t > 0$) removal of the diaphragm or barrier. Shown is the x-t diagram showing the contact surface, the incident shock, and the expansion wave (fan).

stream is the energy source to be transferred to a cold air charge by its compression. The model is of a shock-tube-like device where the delivered compressed gas is that created behind the reflected wave at the end of a shock tube wall.

Consider the shock tube illustrated in Fig. D-1 before and after the barrier is removed. A high-pressure (driver) gas resides at the right end of the tube, whereas the left end contains air at relatively low pressure. For simplicity, the *driver* and the *driven* gases are assumed to be air. More general derivations for differing values of γ are straightforward and may be found in the literature. Assume, further, that the unsteady wave mechanics of a constant area driver adequately models the expansion process of the high-pressure gas. In reality, the geometry of the supply manifold will affect the results to a degree that is consistent with the assumption of steady and uniform properties of the driver gas when an ICE exhaust gas is used.

At time $t = 0$, the barrier is removed, and the following processes occur. See Figs. 8–19 through 8–21 to visualize the situation being modeled. Compression waves tavel to the left and quickly form a shock wave. The shock raises the pressure and temperature from p_1 to p_2 and T_1 to T_2, respectively. An unsteady expansion wave train transmits the need for pressure to decrease at the right end at the changing local speed of sound. The gases on either side of the former barrier move to the left. This *contact surface* is shown as a dashed line in the x-t plane (Fig. D–1). States 1 and 2 describe conditions in the compression process, and 3 and 4 those in the expansion. The problem is specified by values of driver and driven gas conditions 1 and 4. To make the problem tractable, one may assume that the tube is very long compared to the tube diameter so that the events occurring on the compression side are not affected by reflections from the expansion waves interacting with geometry or condition changes in the high-pressure reservoir.

Expansion and compression waves differ in a fundamental way. The difference results in the compression wave (moving to the left in Fig. D–1) having the character of a sharp pressure discontinuity that passes an observer fixed to the tube wall. By contrast, the expansion waves (to the right) spread out over time. The time spread is the "fan" in the x-t diagram which shows that an observer located to the right of another closer to the barrier will sense a longer adjustment time interval between steady pressure levels. The lines bounding the fan, termed *characteristics*, have an inclination related to the speed at which information can travel to the right. The first characteristic travels at the sound speed of the driven fluid. The last travels at the new and lower sound speed, less the flow velocity to the left, a-u.

Similar characteristics may be thought to be associated with the compression process. The important difference is that later waves always catch up with earlier ones and thus coalesce into a single discontinuous wave, the shock. The shock wave moves steadily into the gas 1 so that in the reference frame of the wave, an observer can treat the flow as a classical shock with steady inflow and outflow and deduce therefrom the state changes that occur. These are developed in Section D.1.1.

Unsteady flow analysis, often termed the *method of characteristics*, allows one to relate conditions on either side of the expansion wave. Assuming perfect gas, one obtains

$$\frac{p_4}{p_3} = \left(1 - \frac{\gamma - 1}{2}\frac{u_3}{a_4}\right)^{-2\gamma/(\gamma - 1)} \tag{D–2}$$

This relation allows determination of the required driver/driven gas pressure ratio p_4/p_1 in terms of the pressure ratio across the shock wave p_2/p_1 since

$$\frac{p_4}{p_1} = \frac{p_4}{p_3}\frac{p_3}{p_2}\frac{p_2}{p_1} \tag{D–3}$$

Here $p_3 = p_2$ across the contact surface. The expression for p_4/p_1 is developed in Section D.1.3.

D.1.1. *Normal Shock Relations*

The jump conditions across a thin shock wave are obtained from the flow conservation equations applied to two regions of adiabatic, uniform, and constant area flow. This flow problem is also steady in the wave reference frame where the fluid enters the wave at V_1 and leaves at V_2. In the laboratory reference frame, the velocities are the shock velocity, $u_s = V_1$, and the fluid velocity after the wave passage $u_2 = V_1 - V_2 \equiv \Delta u$ since the initial velocity u_1 is zero. The mass, momentum, and energy conservation statements expressed in terms of wave fixed viewpoint and in terms of Δu are, respectively:

$$\rho_1 V_1 = \rho_2 V_2 \quad \text{or} \quad \rho_1 V_1 = \rho_2(V_1 - \Delta u) \tag{D–4}$$

$$p_2 + \rho_2 V_2^2 = p_1 + \rho_1 V_1^2 \quad \text{or} \quad p_2 - p_1 = \Delta p = \rho_1 V_1 \Delta u \tag{D–5}$$

and

$$h_2 + \tfrac{1}{2}V_2^2 = h_1 + \tfrac{1}{2}V_1^2 \quad \text{or} \quad h_2 - h_1 = \Delta h = \Delta u\left(V_1 - \frac{\Delta u}{2}\right) \tag{D–6}$$

Since the gas is assumed ideal and perfect everywhere, it follows from $p = \rho RT$ that

$$\frac{p}{\rho} = \frac{a^2}{\gamma} \quad \text{and} \quad h = C_p T = \frac{a^2}{\gamma - 1} \tag{D–7}$$

Manipulation of the conservation equations in the wave-fixed coordinate system (eqs. D–4 to D–6) as well as eq. D–7 yields, after some tedious algebra, the following (very simple) relation between the velocities:

$$V_1 V_2 = (\mathrm{a}^*)^2 \tag{D–8}$$

Here a^* is the speed of sound for the flow at sonic conditions defined by:

$$\left(\frac{V_1}{a^*}\right)^2 = \frac{M_s^2}{1 + \dfrac{\gamma - 1}{\gamma + 1}(M_s^2 - 1)} \quad \text{where } M_s \equiv \frac{V_1}{a_1} \tag{D–9}$$

Equation D–8 is the Prandtl relation, from which it may be deduced that the flow ahead of the wave is supersonic and subsonic behind it. That is,

$$\frac{V_1}{a_1} = M_s = M_1 > 1 \quad \text{and} \quad \frac{V_2}{a_2} = M_2 < 1 \tag{D–10}$$

The definitions of the Mach numbers, M_i, are in the reference frame of the wave. The shock Mach number, M_s, is a convenient parameter to quantify the strength of the shock. The mass, momentum, and energy conservation equations for this steady flow give the pressure ratio, the total temperature, and pressure ratios as well as other property changes for the shock. These are given by

$$\frac{\rho_2}{\rho_1} = \frac{V_1}{V_2} = \left(\frac{V_1}{a^*}\right)^2 = \frac{M_s^2}{1 + \dfrac{\gamma - 1}{\gamma + 1}(M_s^2 - 1)} \tag{D–11}$$

$$\frac{\Delta u}{a_1} = \frac{V_1}{a_1}\left(1 - \frac{V_2}{V_1}\right) = -\frac{2}{\gamma + 1}\left(M_s - \frac{1}{M_s}\right) \tag{D–12}$$

The negative sign is to reflect that the flow is to the left, $x < 0$.

$$\frac{p_2}{p_1} = 1 + \frac{2\gamma}{\gamma + 1}(M_s^2 - 1) \tag{D–13}$$

$$\frac{h_2}{h_1} = \frac{T_2}{T_1} = 1 + \frac{2(\gamma - 1)}{\gamma + 1}\frac{(M_s^2 - 1)(1 + \gamma M_s^2)}{M_s^2} \tag{D–14}$$

The last relation includes assumption of a perfect gas. These property ratios reduce to unity when $M_s = 1$ (i.e., when $\Delta u = 0$).

D.1.2. *Reflected Shock Wave*

The shock wave is reflected by the wall closing the left tube end, which imposes the condition that the fluid flow velocity is zero behind the wave. Figure D–2 shows the relevant features after reflection. The physics of the flow through the wave is described by its fluid inflow velocity, V_R. In the laboratory frame, the wave velocity is

$$V_R^{\text{Lab}} = V_R - u_2 = V_R - \Delta u \tag{D–15}$$

It is convenient to describe the conditions across the reflected wave in terms of the density ratios defined by

$$r \equiv \frac{\rho_2}{\rho_1} \quad \text{and} \quad r_0 \equiv \frac{\rho_5}{\rho_1} \tag{D–16}$$

The ratio r (> 1) is known from the incident shock conditions (eq. D–11), and the overall density ratio, r_0 (also > 1), is determined by the condition

$$r_0 \equiv \frac{\rho_5}{\rho_1} = \frac{p_5}{p_1}\frac{T_1}{T_5} \tag{D–17}$$

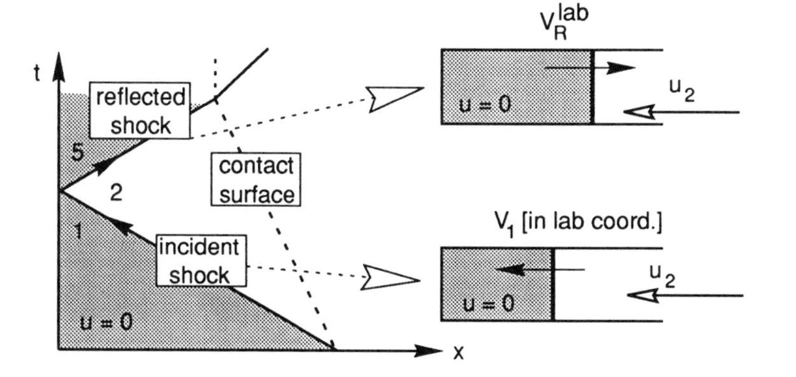

Fig. D–2. The incident and reflected shock waves near a closed tube end wall: x-t diagram near the end wall and the fluid velocities.

where the pressure and temperature ratios are themselves of interest and determined in the following. The conservation equations (eqs. D–4 to D–6) across the incident shock written in terms of the density ratio r are:

$$\frac{\Delta u}{V_1} = 1 - \frac{1}{r} \tag{D–18}$$

$$\frac{\Delta p}{p_1} = \frac{\rho_1 V_1^2}{p_1}\left(1 - \frac{1}{r}\right) \tag{D–19}$$

$$\frac{\Delta h}{h_1} = \frac{V_1^2}{2h_1}\left(1 - \frac{1}{r^2}\right) \tag{D–20}$$

For the reflected wave, the corresponding velocity ratio is

$$\frac{\Delta u}{V_R} = 1 - \frac{\rho_2}{\rho_5} = 1 - \frac{r}{r_0} \tag{D–21}$$

since Δu is the velocity change brought about by the reflected wave with an inflow of V_R. This relation allows writing, with eq. D–18:

$$\frac{V_R}{V_1} = \frac{r-1}{r}\frac{r_0}{r_0 - r} \tag{D–22}$$

The momentum equation for the reflected wave can be written by analogy to the incident wave. Thus from eq. D–4:

$$p_5 - p_2 = \rho_2 V_R \Delta u \qquad \text{or} \qquad \frac{p_5}{p_1} - 1 = \frac{p_2}{p_1} - 1 + \left(\frac{\rho_1 V_1^2}{p_1}\right)\frac{V_R}{V_1}\frac{\Delta u}{V_1}\frac{\rho_2}{\rho_1}$$

since the velocity change for both waves is Δu. From this momentum equation and eqs. D–18, D–19 and D–22, it follows that

$$\frac{p_5}{p_1} = 1 + \left[\frac{\Delta p}{p_1}\right]\left(\frac{r(r_0 - 1)}{r_0 - r}\right) \tag{D–23}$$

The energy equation for the incident wave flow (eq. D–6) and its analog for the reflected wave give the temperature behind the reflected wave. From

$$h_5 - h_1 = h_5 - h_2 + h_2 - h_1 = \Delta u\left\{\left(V_R - \frac{\Delta u}{2}\right) + \left(V_1 - \frac{\Delta u}{2}\right)\right\} \tag{D–24}$$

it follows

$$\frac{h_5}{h_1} - 1 = \left[\frac{V_1^2}{h_1}\right]\frac{\Delta u}{V_1}\left\{1 + \left(\frac{V_R - \Delta u}{V_1}\right)\right\} \tag{D–25}$$

The ratio in square brackets is obtained from eq. D–20, while the others follow from D–18 and D–22, respectively, to give:

$$\frac{h_5}{h_1} = \frac{T_5}{T_1} = 1 + \frac{\Delta h}{h_1} \frac{2r}{r+1} \frac{r_0 - 1}{r_0 - r} \quad \text{(D–26)}$$

In eqs. D–23 and D–26 the $\Delta p/p_1$ or $\Delta h/h_1$ terms are given for the incident shock by eqs. D–13 and D-14. The density ratio r is also given by its definition (eq. D–16) and by eq. D–11 in terms of M_s so that the pressure and temperature behind the reflected wave can be computed. The ratio of temperatures and pressures given by eqs. D–23 and D–26 are related to the density ratio through the state equation (eq. D–17). This equation gives r_0 explicitly in terms of the incident shock Mach number as

$$r_0 = \frac{\left[1 + \frac{2\gamma}{\gamma+1}(M_s^2 - 1)\right][1 + \gamma M_s^2]}{2\left[1 + \frac{2(\gamma-1)}{(\gamma+1)^2}\frac{(M_s^2 - 1)(1 + \gamma M_s^2)}{M_s^2}\right]\left[1 + \frac{\gamma-1}{2}M_s^2\right]} \quad \text{(D–28)}$$

which finally allows writing the following

$$\frac{p_5}{p_1} - 1 = 2\left(\frac{p_2}{p_1} - 1\right)\frac{\left[1 + \left(\frac{1}{2} + \frac{\gamma-1}{\gamma+1}\right)(M_s^2 - 1)\right]}{\left[1 + \frac{\gamma-1}{\gamma+1}(M_s^2 - 1)\right]} \quad \text{(D–29)}$$

and

$$\frac{T_5}{T_1} - 1 = 4\frac{\gamma-1}{\gamma+1}\left(1 - \frac{1}{M_s^2}\right)\left[1 + \left(\frac{1}{2} + \frac{\gamma-1}{\gamma+1}\right)(M_s^2 - 1)\right] \quad \text{(D–30)}$$

An observation may be made from these last two equations. The expressions may be linearized for M_s near 1, in which case the M_s term in the pressure equation approaches unity, whereas that in the temperature equation approaches zero. The implication is that a weak wave is simply reflected doubling the pressure rise of the incident wave. A weak wave is isentropic so that the temperature change vanishes in that limit. Comparison of a linearized form of eqs. D–29 and D–30 with the exact relations shows that linearization involves acceptable accuracy for M_s less than about 1.3. For example, $M_s = 1.3$ gives $p_2/p_1 = 2.4$, $p_5/p_1 = 5.2$ and $T_5/T_1 = 1.6$. Air-charging wave machines operate in this shock Mach number regime.

D.1.3. *Incident Shock Mach Number*

Equations D–2 and D–3 with $u_3 = u_2 = \Delta u$ allow writing the tube pressure ratio in terms of the shock Mach number. If eq. D–12 is used with the proper

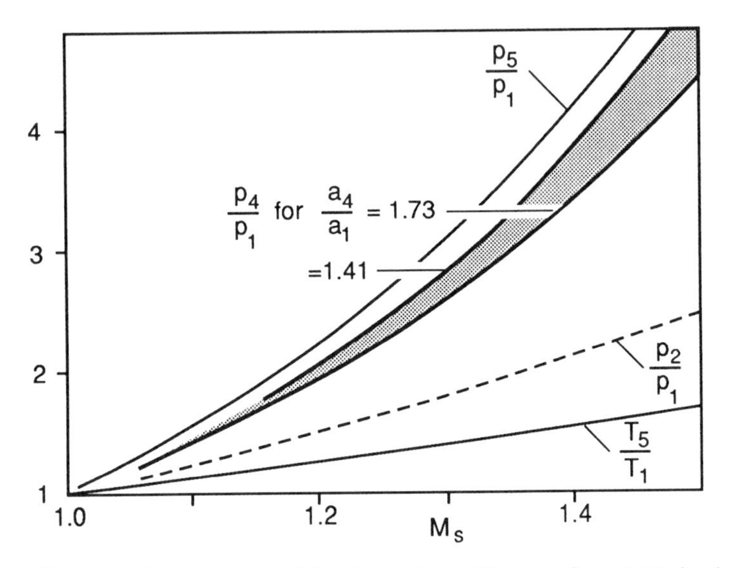

FIG. D–3. Variation of parameters behind the incident (2) and reflected (5) shock waves in terms of the shock Mach number. Also shown (for the sound speed ratio noted) is the driver/driven pressure ratio (p_4/p_1) required to establish the shock at M_s.

algebraic sense implicit in the sign, the following results

$$\frac{p_4}{p_1} = \frac{\left[1 + \left(\dfrac{2\gamma}{\gamma + 1}\right)(M_s^2 - 1)\right]}{\left[1 - \dfrac{\gamma - 1}{\gamma + 1}\dfrac{a_1}{a_4}\dfrac{(M_s^2 - 1)}{M_s}\right]^{2\gamma/(\gamma - 1)}} \tag{D–31}$$

For $M_s = 1.3$ and $a_4/a_1 = 2$, the pressure ratio p_4/p_1 required is 2.5 (=2.0 for a linearized result). Note that larger speed of sound ratio decreases the pressure ratio required for a given M_s. A plot of nondimensionalized p_2, p_5, and T_5, as well as the required p_4, is shown in Fig. D–3 for various M_s.

D.2. ENERGY EXCHANGER PERFORMANCE

The energy exchanger is supplied by a gas whose conditions are known and may be expressed in terms of the ratios p_4/p_1 and T_4/T_1. These parameters determine the shock Mach number and the subsequent fluid properties, particularly those behind the reflected shock wave.

The relation between p_5 and T_5 and the ratios above assume that the tube behaves as a closed tube at the driven end, and an open tube at the driver end. This modeling assumption is equivalent to stating that the result is limited to the case where little throughflow occurs. It should be expected that performance will decrease as the mass flow processed is increased. In other words, a plot of pressure ratio versus mass flow rate will exhibit a decreasing characteristic.

The efficiency of the process is associated primarily with the generation of entropy in the two shock waves. The estimation of that irreversibility is certainly possible, but in light of other irreversibility mechanisms present (wall friction and heat transfer), such an estimate would lack conservatism.

PROBLEMS

1. Derive eq. D–8.
2. Find the total pressure ratio p_{t2}/p_{t1} across the shock following the development leading to eq. D–14 and show that it is less than unity.
3. Using p_5 and T_5 given by eqs. D–29 and D–30, determine the efficiency of the compression process.

BIBLIOGRAPHY

Rudinger, G., *Nonsteady Duct Flow*, Dover Publications, New York, 1969.

INDEX